Agricultural Nanobiotechnology

Woodhead Publishing Series in Food Science, Technology, and Nutrition

Agricultural Nanobiotechnology

Biogenic Nanoparticles, Nanofertilizers, and Nanoscale Biocontrol Agents

Edited by

Sougata Ghosh
Department of Physics, Faculty of Science, Kasetsart University, Bangkok, Thailand
Department of Microbiology, School of Science, RK University, Rajkot, Gujarat, India

Sirikanjana Thongmee
Department of Physics, Faculty of Science, Kasetsart University, Bangkok, Thailand

Ajay Kumar
Agriculture Research Organization, Volcani Center, Ministry of Agriculture and Rural Development, Rishon LeZion, Israel

WP
WOODHEAD PUBLISHING
ELSEVIER
An imprint of Elsevier

Woodhead Publishing is an imprint of Elsevier
50 Hampshire Street, 5th Floor, Cambridge, MA 02139, United States
The Boulevard, Langford Lane, Kidlington, OX5 1GB, United Kingdom

ISBN: 978-0-323-91908-1 (print)
ISBN: 978-0-323-99936-6 (online)

For information on all Woodhead publications
visit our website at https://www.elsevier.com/books-and-journals

Publisher: Nikki P. Levy
Acquisitions Editor: Nancy J. Maragioglio
Editorial Project Manager: Lena Sparks
Production Project Manager: Sojan P. Pazhayattil
Cover Designer: Mark Rogers

Typeset by STRAIVE, India

Contents

Contributors

Kafeel Ahmad Civil Engineering Department, Jamia Millia Islamia, New Delhi, India

Sagar Bag Laboratory of Microbial Interaction, School of Biotechnology, Presidency University, Kolkata, West Bengal, India

Kaushik Banerjee National Reference Laboratory, ICAR-National Research Centre for Grapes, Pune, Maharashtra, India

Avishek Banik Laboratory of Microbial Interaction, School of Biotechnology, Presidency University, Kolkata, West Bengal, India

Sahila Beegum Nebraska Water Center, Part of the Robert B. Daugherty Water for Food Global Institute, University of Nebraska, Lincoln, NE, United States

Joorie Bhattacharya Genetic Gains, International Crops Research Institute for the Semi-Arid Tropics; Department of Genetics, Osmania University, Hyderabad, Telangana, India

Ronita Chandra Department of Microbiology, Vivekananda Mahavidyalaya, Burdwan, West Bengal, India

Anamika Chavan Amity Institute of Biotechnology, Amity University Maharashtra, Mumbai, Maharashtra, India

Ashok Choudhury Directorate of Research, Uttar Banga Krishi Viswavidyalaya, Cooch Behar, West Bengal, India

Yashodhara M. Dalal Shobhaben Pratapbhai Patel School of Pharmacy and Technology Management, SVKM'S NMIMS, Vile Parle (West), Mumbai, India

Saurav Das Department of Agronomy and Horticulture, University of Nebraska–Lincoln, Lincoln, NE, United States

Snehal Desai Amity Institute of Biotechnology, Amity University Maharashtra, Mumbai, Maharashtra, India

Ankita Dey Department of Biotechnology, Maulana Abul Kalam Azad University of Technology, Haringhata, West Bengal, India

Sougata Ghosh Department of Microbiology, School of Science, RK University, Rajkot, Gujarat, India; Department of Physics, Faculty of Science, Kasetsart University, Bangkok, Thailand

Sreejita Ghosh Department of Biotechnology, Maulana Abul Kalam Azad University of Technology, Haringhata, West Bengal, India

Sujay Ghosh AMH Energy Pvt. Ltd., Kolkata, West Bengal, India

Archita Gupta Department of Bioengineering and Biotechnology, Birla Institute of Technology, Mesra, Ranchi, Jharkhand, India

Dipak Kumar Hazra Department of Agricultural Chemicals, Bidhan Chandra Krishi Viswavidyalaya, Mohanpur, West Bengal, India

Tofazzal Islam Institute of Biotechnology and Genetics Engineering (IBGE), Bangabandhu Sheikh Mujibur Rahman Agricultural University, Gazipur, Bangladesh

Riti Thapar Kapoor Amity Institute of Biotechnology, Amity University, Noida, India

Rajib Karmakar Department of Agricultural Chemicals, Bidhan Chandra Krishi Viswavidyalaya, Mohanpur, West Bengal, India

Bhavtosh Kikani Department of Biological Sciences, P.D. Patel Institute of Applied Sciences, Charotar University of Science and Technology, Changa, Gujarat, India

Ajay Kumar Department of Postharvest Science, Agricultural Research Organization (ARO), Volcani Center, Rishon LeZion, Israel

Maushmi S. Kumar Shobhaben Pratapbhai Patel School of Pharmacy and Technology Management, SVKM'S NMIMS, Vile Parle (West), Mumbai, India

Rajesh Kumar Division of Agricultural Chemicals, ICAR-Indian Agricultural Research Institute, New Delhi, India

Kanishka Kunal Department of Bioengineering and Biotechnology, Birla Institute of Technology, Mesra, Ranchi, Jharkhand, India

Dibyajit Lahiri Department of Biotechnology, University of Engineering & Management, Kolkata, West Bengal, India

Rosalin Laishram Division of Biochemistry, ICAR-Indian Agricultural Research Institute, New Delhi; Department of Biochemistry, Faculty of Agriculture, Uttar Banga Krishi Viswavidyalaya, Cooch Behar, West Bengal, India

Jaya Lakkakula Amity Institute of Biotechnology, Amity University Maharashtra, Mumbai, Maharashtra, India

Nur Uddin Mahmud Institute of Biotechnology and Genetics Engineering (IBGE), Bangabandhu Sheikh Mujibur Rahman Agricultural University, Gazipur, Bangladesh

Sanjay Kumar Mehta Department of Bioengineering and Biotechnology, Birla Institute of Technology, Mesra, Ranchi, Jharkhand, India

Momina Civil Engineering Department, Jamia Millia Islamia, New Delhi, India

Anupam Mondal Laboratory of Microbial Interaction, School of Biotechnology, Presidency University, Kolkata, West Bengal, India

Prithusayak Mondal Regional Research Station (Terai Zone), Uttar Banga Krishi Viswavidyalaya, Cooch Behar, West Bengal, India

Kunal Mukhopadhyay Department of Bioengineering and Biotechnology, Birla Institute of Technology, Mesra, Ranchi, Jharkhand, India

Moupriya Nag Department of Biotechnology, University of Engineering & Management, Kolkata, West Bengal, India

Rahul Nitnavare Division of Plant and Crop Sciences, School of Biosciences, University of Nottingham, Leicestershire; Department of Plant Sciences, Rothamsted Research, Hertfordshire, United Kingdom

Kumaresh Pal Department of Biochemistry, Faculty of Agriculture, Uttar Banga Krishi Viswavidyalaya, Cooch Behar, West Bengal, India

Azania T. Panicker Shobhaben Pratapbhai Patel School of Pharmacy and Technology Management, SVKM'S NMIMS, Vile Parle (West), Mumbai, India

Sanjoy Kumar Paul Institute of Biotechnology and Genetics Engineering (IBGE), Bangabandhu Sheikh Mujibur Rahman Agricultural University, Gazipur, Bangladesh

Shri Prakash Department of Zoology, Kulbhaskar Ashram P.G. College Prayagraj, Prayagraj, Uttar Pradesh, India

Mohd Rafatullah School of Industrial Technology, Universiti Sains Malaysia, Penang, Malaysia

Rina Rani Ray Department of Biotechnology, Maulana Abul Kalam Azad University of Technology, Haringhata, West Bengal, India

Paritosh Chandra Roy Institute of Biotechnology and Genetics Engineering (IBGE), Bangabandhu Sheikh Mujibur Rahman Agricultural University, Gazipur, Bangladesh

Bishwarup Sarkar College of Science, Northeastern University, Boston, MA, United States

Prahlad Sarkar All India Network Project on Jute and Allied Fibres, Uttar Banga Krishi Viswavidyalaya, Cooch Behar, West Bengal, India

Manish Singh Amity Institute of Biotechnology, Amity University Maharashtra, Mumbai, Maharashtra, India

Sneha Singh Department of Bioengineering and Biotechnology, Birla Institute of Technology, Mesra, Ranchi, Jharkhand, India

Hossain Sohrawardy Institute of Biotechnology and Genetics Engineering (IBGE), Bangabandhu Sheikh Mujibur Rahman Agricultural University, Gazipur, Bangladesh

Sirikanjana Thongmee Department of Physics, Faculty of Science, Kasetsart University, Bangkok, Thailand

Anurag Tripathi Department of Zoology, Kulbhaskar Ashram P.G. College Prayagraj, Prayagraj, Uttar Pradesh, India

Nilesh S. Wagh Amity Institute of Biotechnology, Amity University Maharashtra, Mumbai, Maharashtra, India

Nanobiotechnology: Emerging trends, prospects, and challenges

1

Anurag Tripathi and Shri Prakash
Department of Zoology, Kulbhaskar Ashram P.G. College Prayagraj, Prayagraj, Uttar Pradesh, India

1.1 Introduction

Agriculture is the main source of food and nutrition for animals and humans. But the conventional technologies of agriculture are not sufficient to meet the needs of growing population. In addition, the present-day tools and techniques used in agriculture pose a serious damage to environment and ecosystem. Indiscriminate use of chemical fertilizers, traditional irrigation techniques, pesticides, and herbicides has been questioned at the scientific and policy levels that need to be replaced by sustainable techniques. At this alarming juncture, nanobiotechnology may be a promising solution along with other sustainable agricultural techniques, especially for those nations where agriculture is the main occupation of majority of the population. Nanobiotechnology is a newly explored interdisciplinary area of research by the integration of nanotechnology and biotechnology in particular and biological sciences, surface science, organic chemistry, nucleic acid chemistry (DNA-based nanodevices for precise and targeted drug delivery), proteomics, pharmaceutics, molecular science, and semiconductor physics in general for its application in various sectors including electronics, robotics, biomedical research, and agriculture. Nanotechnology may be described as the science of designing and building machines at the dimensions of 1–100 nm where the unique physical properties of nanomaterials make it novel and extraordinary (Mukhopadhyay, 2014). At the nanolevel, every atom and chemical bond is specified with its unique properties; hence, at this level, nanomachines, inorganic and organic compound-based nanomaterials, and nanodevices can be designed to manipulate matter at the atomic level. The judicious and precise application of nanotechnology in agricultural sector has far-sighting prospects at national and global levels. Specifically engineered nanotools may have a wide scope for the nations like India having their applications in enhancing the productivity and nutritional value of produces, sustainable and efficient and eco-friendly nanofertilizers, nanopesticides, preservation of food and food additives, understanding host–parasite relationship at molecular level, enhancing the quality of soil and water management to name a few. Thus, it is likely to overwhelm all spheres of agricultural activities. Various terminologies that are in vogue in agrinanobiotechnology are summarized in Table 1.1.

Agricultural Nanobiotechnology. https://doi.org/10.1016/B978-0-323-91908-1.00006-7

Table 1.1 Agrinanotechnology-related definitions.

Nanodevices	Definitions
Nanomaterial	A natural, incidental or manufactured material containing particles, in an unbound state or as an aggregate or as an agglomerate and where, for 50 % or more of the particles in the number size distribution, one or more external dimensions is in the size of nanoscale (European Commission, 2011)
Nanofertilizer	Nanomaterials which can supply one or more nutrients to the plants and enhance their growth and yields
Nanomaterial-enhanced fertilizer	Nanomaterials, which, when augmented with conventional fertilizers, can improve their performance, without directly providing crops with nutrients
Nanopesticide	Any pesticide formulation that involves either very small particles of a pesticide active ingredient or other small engineered structures with useful pesticidal properties
Nanocoating	Nanocoatings or surface coatings of nanomaterials, on fertilizer or pesticide active ingredients, can hold the material more strongly due to higher surface tension than the conventional surfaces and thus help in controlled release
Nano-enabled formulation	Nano-enabled formulation encompasses those emulsions made of smaller micelles formed with smaller amount of surfactants, or microcapsules with a well-defined nanopore network
Nanoemulsion	This is a solid sphere of nanoscale range, which acts as carrier desired elements and chemicals including agrochemicals. Its surface is amorphous and lipophilic, and it is a substitute for liposome and vesicle (Jaiswal et al., 2015)
Nanosphere	Aggregate in which the active compound is homogeneously distributed into the polymeric matrix
Nanocapsule	Aggregate in which the active compound is concentrated near the center core, lined by the polymer matrix
Nanogel	Hydrophilic, generally cross-linked, polymers, which can absorb high volumes of water
Nanosensor	Nanoscale devices capable of detecting and responding to physicochemical and biological stimuli. Nanomaterials may provide larger surface area for the immobilization of the target recognition elements or they may confer the sensor part to express their own optical-physical and electrochemical properties

Modified and adapted from Lavicoli, I., Leso, V., Beezhold, D.H., Shvedova, A.A., 2017. Nanotechnology in agriculture: opportunities, toxicological implications, and occupational risks. Toxicol. Appl. Pharmacol. 329, 96–111.

1.2 Nanobiotechnology and agriculture

Nanobiotechnology is a revolutionary technology of 21st century. The efficient, precise, and need-based application of nanotechnology has immense results in fostering sustainable agriculture with augmented food and nutritional security along with mitigating environmental challenges with precise use of water and agrochemicals.

Nanobiotechnology has the potential to protect plants, monitor plant growth, diagnose plant disease, enhance nutritional efficacy, and reduce waste (Prasad et al., 2017). Specifically designed nanotools have enabled the enhanced crop productivity with more nutritional value and less cost input. Nanotechnology is expanding in contemporary fields of agriculture like food processing, food preservation, food packing, and dairy industry. It is also being supported by Indian government by making plans to extend support for commercial application of nanobiotechnology. Some of the precise and emerging applications of nanobiotechnology in agriculture are depicted in Fig. 1.1.

1.2.1 Nanotools in agriculture

In the past two decades, many researches are in progress about the precise application of tools and techniques of nanobiotechnology in agricultural sector with many positive outcomes in the form of nanotools like inorganic compound-based nanoparticles, carbon nanotubes, quantum dots, organic dendrimers, polymer nanocapsules, nanoshells,

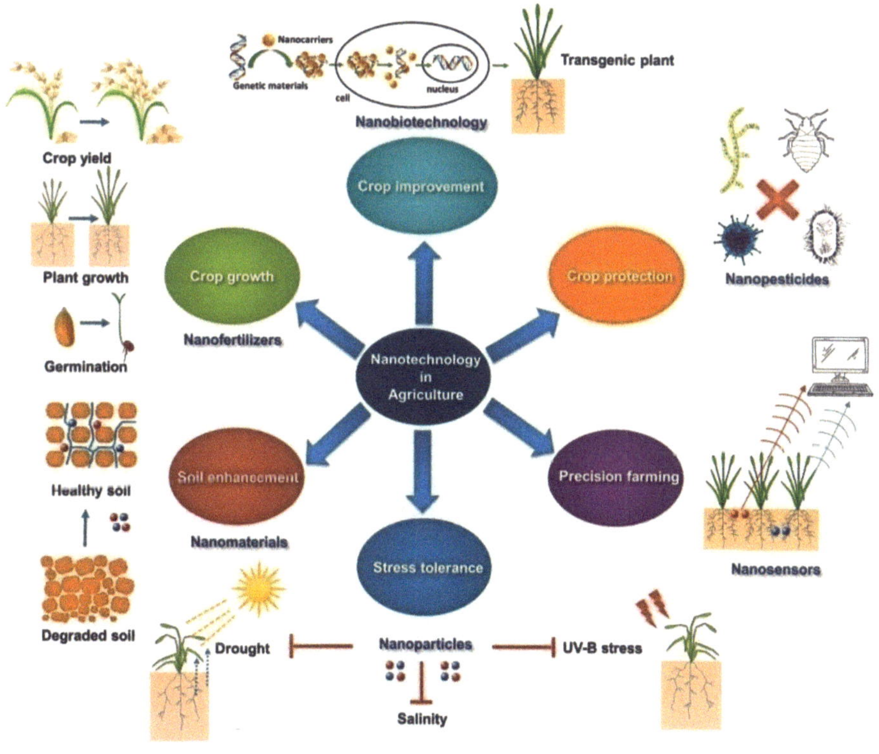

Fig. 1.1 Possible application of nanobiotechnology in agriculture.
Adapted from Shang, Y., Hasan, M.K., Ahammed, G.J., et al., 2019. Applications of nanotechnology in plant growth and crop protection: a review. Molecules 24(2558), 1–23. https://doi.org/10.3390/molecules24142558, published in Molecules; through open access policy.

and nanospheres, which are being tested for various purposes in agricultural research. Nanoparticles (NPs) can be defined as natural or engineered tiny particle of nanoscale range (1–100 nm). These are endowed with unique properties including shape, size, large surface area, surface functionalization, porosity, zeta potential, hydrophobicity, and hydrophilicity. Newly designed carbon nanomaterials (CNMs) are a class of engineered nanomaterials, which are extensively being used in the field of electronics, nanomedicine, optics, biosensors, and agriculture in minuscule, due to their specific mechanical, electrical, optical, and thermal properties. The CNM family includes carbon nanotubes (CNTs), fullerenes, nano-onions, nanobeads, nanofibers, grapheme, nanodiamonds, nanohorns, carbon dots to name a few (Mukherjee et al., 2016). These materials assume quantum properties and specific physicochemical properties at nanoscale level. Carbon nanotubes (CNTs) are cylindrical structures with open and closed ends. These can be divided into single-walled nanotubes (SWCNTs) and multiwalled nanotubes (MWCNTs) on the basis of number of concentric layers of rolled graphene sheets. The application of these composite nanomaterials in the agricultural sector is still in infant stage; nevertheless, it has given promising results. The internalization of CNTs within plant cells and cell organelles is well proven; thus, it can be used as nanotransporter. Many nanotools including buckyballs, dendrimers, and nanocapsules are being tested for precise and efficient drug delivery in form of nanoveterinary medicines (Mukherjee et al., 2016). Adhesion-specific nanoparticles are used for the removal of bacteria *Campylobacter jejuni* from poultry. In addition, iron nanoparticles are used for feeding of livestock and fisheries (Mukhopadhyay, 2014).

Many hydroponic studies have proven that accumulation of CNMs in plants increases the root length. A study made on six plant species, carrot (*Daucus carota*), tomato (*Lycopersicon esculentum*), lettuce (*Lactuca sativa*), cabbage (*Brassica oleracea*), onion (*Allium cepa*), and cucumber (*Cucumis sativus*), by using coated and uncoated CNTs at different exposures showed that these have indirect effect on plant root systems by hampering with microbial-root interactions or by altering crucial biochemical processes (Mukherjee et al., 2016). Another significant study on gram plant (*Cicer arietinum*) on exposure to citrate-coated water-soluble CNTs exhibited that they enhance water uptake efficiency, thereby increasing the plant growth. It is evident from the study on corn (*Zea mays*) and barley (*Hordeum vulgare*) that application of CNMs enhances the rate of germination in comparison to that of untreated with CNMs. Fullerol, another CNM, has been shown to enhance biomass and phytomedicinal content of a medicinal plant bitter melon (*Momordica charantia*) (Dwivedi et al., 2016). Its extracts are used for the treatment of diseases like cancer, diabetes, and acquired immunodeficiency syndrome (AIDS) (Mukherjee et al., 2016). In addition, CNMs have been shown to boost the growth of terrestrial plants also. A plenty of NPs are under experimental stage, which could efficiently deliver the nutrients administered through plant roots and rarely through foliar way. A brief summary of NPs (Chugh et al., 2021) and their efficacy in agriculture has been presented in Table 1.2. Thus, it is worth inferring that CNMs have the potential to enhance plant growth, seed germination, nutrient uptake, and fruit quality. However, more extensive studies are needed to authenticate various implications of CNMs on plants.

Table 1.2 Application of nanoparticles in delivering significant nutrients and their effects on target plants.

Nanoparticles	Nutrients	Plant species	Effect on plants
Titanium dioxide	Titanium	*Oryza sativa* (rice) *Solanum lycopersicum* (tomato) *Hordeum vulgare* (barley) *Mentha piperita* (peppermint)	Increased amino acids, fatty acids, and P in root, shoot, and grains Increased amino acids, phenolics, antioxidant capacity Increased, P, Ca, Mg, Zn, Mn, amino acids Increased N, chlorophyll, menthol, menthone
Cerium dioxide	Cerium	*Oryza sativa* (rice) *Haordieum vulgare* (barley) *Triticum aestivum* (wheat) *Cucumis sativus* (cucumber) *Coriandrum sativum* (cilantro)	Increased K, Ca, Na, protein albumin, total sugars Increased P, K, Ca, Mg, S, Cu, Fe, Zn, Mn,amino acids, fatty acids Increased P, Fe, K, amino acids, fatty acids, total sugars Increased K, Ca, Mg, S, P, Fe, Mn, total sugars, starch, proteins Increased Ce, catalase, ascorbate peroxidase activity
Zinc oxide Zn-amino acid nanocomplex	Zinc	*Pisum sativum* (pea) *Cucumis sativus* (cucumber) *Zea mays* (maize) *Solanum lycopersicum* (tomato) *Arachis hypogaea* (peanut) *Ocimum basilicum* (sweet basil)	Increased P, Fe, Zn, Mn, total sugars Increased K, Mg, Fe, Mn, Zn, S, prolamin, globulin, gluten Increased K, germination, growth, yield Increased lycopene Increased Zn, chlorophyll, root biomass, yield Increased catechin, hesperetin
Copper	Copper	*Solanum lycopersicum* (tomato) *Cucumis sativus* (cucumber)	Increased K, total proteins, vitamin C, total phenols, flavonoids, lycopene, antioxidant capacity Increased Cu, Fe, sugar, organic acids, amino acids, fatty acids

Continued

Table 1.2 Continued

Nanoparticles	Nutrients	Plant species	Effect on plants
Iron-trioxide	Iron	*Arachis hypogaea* (peanut) *Glycine* max (soybean)	Increased Zn, growth, biomass
Calcium oxide Calcium carbon trioxide	Calcium	*Arachis hypogaea* (peanut) *Vigna mungo* (black gram)	Increased Ca, root development, increased root and shoot growth, biomass
Magnesium	Magnesium	*Vigna unguiculata* (cow pea)	Increased photosynthesis, growth, yield
Silver	Silver	*Cucumis sativus* (cucumber) *Solanum lycopersicum* (tomato) *Lactuca sativa* (lettuce)	Increased growth, yield, biomass, total soluble solids in fruit Increased superoxide dismutase activity Increased Ag content

Modified and adapted from Chugh, G., Siddique, K.H.M., Solaiman, Z.M. 2021. Nanobiotechnology for agriculture: smart technology for combating nutrient deficiencies with nanotoxicity challenges. Sustainability 13(1781): 1–20, through open access policy.

1.2.2 Biomass waste-based nanomaterials and their efficacy

The fluorescent carbon dot (C dot) is one of the novel types of nanomaterials, which are prepared from biomass waste, and in comparison with quantum dots, it has significant advantage due to its excellent photostability and biocompatibility, low cytotoxicity, and easy surface functionalization. It has very wide application in the fields of bioimaging, chemical sensing, environmental monitoring, disease diagnosis, and photocatalysis as well (Kang et al., 2020; Xue et al., 2016; Zuo et al., 2016). Biomass waste is a natural organic carbon source, which mainly consists of cellulose, hemicellulose, lignin, ash, proteins, and few other compounds. Perennial grass, organic domestic waste, agricultural residues, fishery, poultry, forestry, and animal husbandry are the main sources of biomass, which often get discarded leading to environmental problems. Recently, biomass wastes are used as raw materials in the production of C dots. Various approaches like pyrolysis, solvothermal method, microwave-assisted method, and ultrasonic-assisted method are efficiently employed for preparing fluorescent C dots (Kang et al., 2020; Tripathi et al., 2016), which is depicted in Fig. 1.2. Thus, these novel approaches provide vibrant nanomaterials as well as mitigate the environmental threats.

1.2.3 Nanofertilizers (NFs)

Traditionally employed fertilizers have several limitations. Fertilizers are often used to enhance micro- and macronutrients in soil for enhancing crop productivity. Most macronutrients are insoluble in soil; hence, only a small fraction is available to plants and a large quantity of fertilizers run off, contributing to soil, air, and water pollution. Thus, it creates a long-term threat to agroecosystem despite its short-term gain. Owing to these disadvantages, chemical fertilizers are being replaced by specially formulated nanofertilizers (NFs) like hydroxyapatite NPs, zeolite, mesoporous silica NPs, nitrogen, zinc, carbon, copper, silica nanoparticles (Seleiman et al., 2021; Usman et al., 2020), in which specific micro- and macronutrients are encapsulated within nanoparticles. Meticulously designed nanofertilizers with the precise use of nanoparticles have great advantage over conventional fertilizers, and these may be an effective tool for sustainable agriculture. The possible efficacy of NFs is illustrated in Fig. 1.3. Nanofertilizers may be a plant nutrient itself, or they encapsulate the agrochemicals and micro-/macronutrients coated with a thin film, and these are delivered as nanoemulsions, which ensure a stronghold of nutrients on the plant surface due to high surface tension of the nanocoating. Additionally, they release the nutrients at slow and steady pace and can hold it up to 40–50 days (Seleiman et al., 2021). Nanozeolites and nanoclays are used for soil improvement due to their potential in release and retention of nutrients and water (Chugh et al., 2021; Seleiman et al., 2021).

Nanofertilizers are highly targeted, site-specific, and endowed with high adsorption efficiency contributing in enhanced photosynthesis and expanded leaves of plants along with reduced toxicity of the soil and prevention in water pollution. Nanofertilizers can be synthesized by two approaches: In the top-down approach, a bulk of nutrient is broken down into nanosized particles, and in bottom-up approach, NFs are synthesized

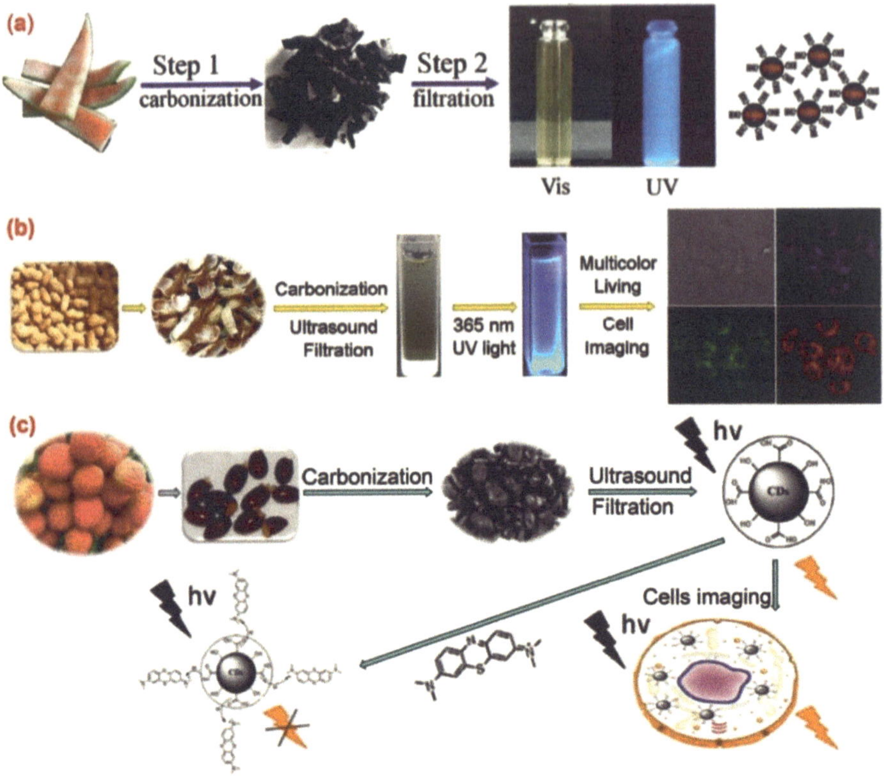

Fig. 1.2 Synthesis and application of fluorescent C dots from agricultural waste by pyrolysis: (A) Synthesis of water-soluble C dots from watermelon peel; (B) Preparation and application of C dot from lychee seeds; (C) Preparation and application of C dot from peanut shells. Adapted from Kang, C., Huang, Y., Yan, X.F., Chen, Z.P., 2020. A review of carbon dots produced from biomass wastes. Nanomaterials 10(2316), 3–24, published in Nanomaterials; through open access policy.

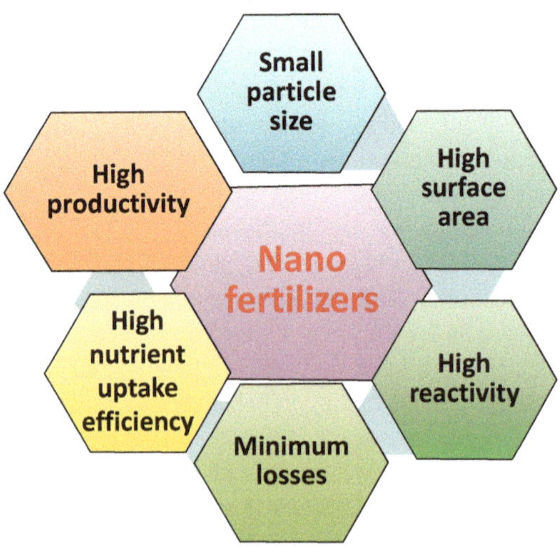

Fig. 1.3 Advantages of nanofertilizers.
Adapted from Seleiman, M.F., Almutairy, K.F., Alotaibi, M., et al., 2021. Nano-fertilization as an emerging fertilization technique: why can modern agriculture benefit from its use? Plan. Theory, 10, 1–27, published in Plants; through open access system.

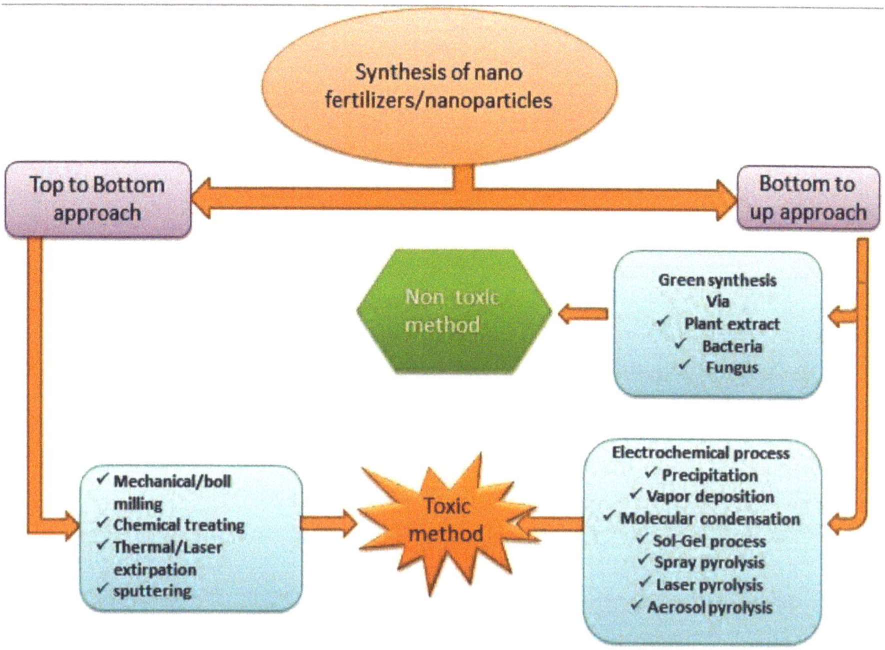

Fig. 1.4 Main approaches for the synthesis of nanofertilizers and NPs.
Adapted from Seleiman, M.F., Almutairy, K.F., Alotaibi, M., et al., 2021. Nano-fertilization as an emerging fertilization technique: why can modern agriculture benefit from its use? Plan. Theory 10, 1–27, published in Plants; through open access policy.

from atoms, molecules, and monomers (Seleiman et al., 2021). Various toxic and non-toxic approaches to synthesize NPs are depicted in Fig. 1.4. It has been shown that the treatment of TiO_2 nanoparticles on maize enhances the growth and yield, and the application of SiO_2 and TiO_2 nanoparticles increased the activity of nitrate-reductase in soybeans and intensified plant absorption capacity (Sekhon, 2014). Researches in rice plants have indicated that urea nanohybrids can deliver urea many times over a period of time leading to enhanced yield of grain. Likewise, the foliar application of potassium nanofertilizer on *Cucurbita pepo* enhanced the biomass, growth, and yield (Chugh et al., 2021). Long-term application of nanofertilizers is considered to be a "nano-bio-revolution" in the field of nanobiotechnology (Chugh et al., 2021). However, more researches are needed in this interdisciplinary area of study.

1.2.4 Nanobiosensors

Nanosensors are electronic devices, which possess a sensing part and electronic data processing part. Sensing part is highly valuable for agricultural sector because it can sense heat, light, pathogens, humidity, and chemicals, thereby managing the need-based system. Nanosensors and nanobased delivery systems are expected to improve agricultural, food, and environmental sectors. These nanodevices could help in efficient use of agricultural natural resources like nutrients, chemicals, and water. The integrated application of nanosensors and remote sensing and geographic information system could detect crop pests, stress like drought, level of nutrients and pathogens in soil to name a few. Nanofibres and nanowires are used in synthesis of nanosensors, which are used for sensing fertilizers, pesticides, herbicides, pathogens, moisture, and soil pH with high efficacy, thereby resulting in enhanced crop productivity with better agrochemical and paste management (Lv et al., 2018). In addition, the level of soil micronutrients and presence of microbes and viruses can be assessed by nanosensors. Nanosmart dust and gas nanosensors can measure the level of environmental pollution (Sekhon, 2014). Many fluorescent dye biosensors and magnetic nanoparticle-based nanosensors have been used for the detection of pathogenic bacteria. Recently, a bioluminescence oxygen biosensor has been developed, which is cheap and compatible for food packaging, and immuno-biosensors have also been developed for the detection of microbial toxins (Lugani et al., 2021). Hence, bio-nanosensors allow the quantification and rapid detection of bacteria, pathogens, and toxic materials present in crops, thereby increasing the biosafety.

1.2.5 Nanopesticides and nanoherbicides

Traditional pesticides are used to improve crop production and efficiency, but it has limited efficacy. Recently, nanopesticides are being tested to replace general pesticides, which are plant protection products at nanoscale. These are prepared through nanoformulations, which combine several surfactants, organic polymers, and inorganic metal nanoparticles including polymeric nanospheres and nanocapsules together with nanogels and nanofibres (Chugh et al., 2021; Kang et al., 2020). These nanoparticles provide microencapsulation to pesticides through nanoformulations of existing pesticides and fungicides and are efficient tool for hydrophobic pesticides with greater solubility, and these are effective in precise and targeted delivery of pesticides with high efficacy and less consumption. Silver is well-known for its antimicrobial activity, and the application of silver nanoparticles (AgNPs) has been reported to inhibit the growth of plant pathogen in dose-dependent manner (Chugh et al., 2021). The dose-dependent use of titanium-alumina-copper nanoparticles (TiO_2-Al-Cu NPs) has been reported to be effective against a range of pests enhancing the plant growth and tolerance against various abiotic and biotic stresses. The use of silica nanoparticles and copper nanoparticles has been found to be highly effective in the formulation of nanopesticides (Sekhon, 2014). Thus, research and development in the formulation of nanopesticides could be a highly efficient tool in integrated pest management, but the environmental risk assessment and biosafety issues need to be addressed in researches before its commercialization.

In the same way, the application of nanoherbicides can be a viable alternative to remove weeds, thereby increasing crop yield. The nanosilicon carrier comprising diatom frustules has been efficiently used for the delivery of pesticides and herbicides in plants. Zinc layered hydroxide and zinc-aluminum layered double hydroxide have also been demonstrated for the preparation of nanohybrid compound containing two herbicides simultaneously (Prasad et al., 2017).

1.2.6 Nanophoto-semiconductors

Titanium dioxide (TiO_2) has been proven to be the most effective photocatalyst due to its high efficacy, photochemical stability, and nontoxic nature. These unique structural, chemical, and optical properties of TiO_2 make it suitable material for designing photo-semiconductors at nanolevel, which have very wide scope in agricultural sector including degradation of pesticides, plant protection and seed germination, crop disease control, water purification, and pesticide residue detection (Wang et al., 2016). These nanomaterial-based photo-semiconductors degrade the nonbiodegradable pesticides into water, carbon dioxide, and other biodegradable and less toxic materials without secondary pollution. Many pesticides like chlortoluron, cyproconazole, and paraquat have been tested for such kind of photocatalysis (Wang et al., 2016). In addition, TiO_2 semiconductors produce superoxide and hydroxide ions in photocatalysis process, which increase the seed stress resistance and water and oxygen intake, thereby enhancing the germination of seeds. Moreover, these superoxides and hydroxides are effective antimicrobial agents. These properties of TiO_2 nanomaterials have made it an intense area of research that needs more explorations through research.

1.2.7 Nanomaterials in drought- and salinity-resistant crop production

Drought and salinity are the two important environmental stresses that lead to reduction in crop growth and yield across the world. Many nanoparticles including NFs enhancing the water-holding capacity of soil have been tested and proven to counter these stresses with high efficacy and least toxicity. During stress condition in plants, reactive oxygen species (ROS) are accumulated within plant cell contribution in lipid peroxidation, cell membrane damage, and leakage of solutes causing apoptosis. Results have shown that NPs enhance the antioxidant activity and decrease the production of superoxides (Seleiman et al., 2021). Zinc and copper NPs have been tested to enhance antioxidant activity of enzymes superoxide dismutase (SAD) and catalase (CAT), thereby keeping the physiological balance during drought condition upon plants of steppe ecotype (Taran et al., 2017). Foliar spray of iron nanoparticle (Fe-NP) was found to reduce water stress effect and to enhance yield and oil percentage in safflower (*Carthamus tinctorius*) (Seleiman et al., 2021). Selenium- and silica-based NPs have also been tested to be efficient in increasing the antioxidant enzyme and vitamin C activity and reducing the peroxide degradation and level of hydrogen peroxide in strawberry plants subjected to drought stress (Zahedi et al., 2020). In nanoparticle-treated strawberry plants subjected to drought stress, photosynthetic pigments were found to be conserved in comparison with untreated plant. Thus, these nanoparticles enhance the nutritional value and fruit quality of plants.

Another significant stress factor is salinity, and more than 20% of the globally cultivable lands are affected by salt stress. The problem of salinity stress is gradually expanding due to indiscriminate use of fertilizers, industrial pollution, and poor irrigation practices, resulting in excess salt concentration. Salt-affected soil has a low osmotic potential, resulting in nutritional and ionic imbalance. Application of silica-based NPs has been proven to be effective in reducing stress by maintaining ionic balance, regulating the sodium ion absorption, transport and distribution within tissues, regulating the polyamine levels, and enhancing the antioxidant efficacy within plants (Zhu et al., 2019; Zhao et al., 2011). Nanocalcium application on *Solanum lycopersicum* grown in salt stress condition exhibited the higher yield and more fruits (Seleiman et al., 2021). Thus, precise use of nanomaterials may be helpful in coping the stress responses in plants, which needs more exploration.

1.2.8 Nanofabrication in agriculture

With the vibrant application of nanotechnology in fabrication technology, new classes of materials are expected to be manufactured in the near future. Current engineered nanomaterials are grouped into four classes, i.e., metal-based materials, carbon-based nanotools, dendrimers, and nanocomposite materials. In nanofabrication, specific and need-based nanomaterials are fabricated and laser ablation, X-ray, arc discharge, focused ion beams, and scanning probes are being applied by industries for nanomatching of atoms (Mukherjee et al., 2016). In top-down approach of nanofabrication just by the removal of one atom at nanoscale, desired structure can be designed. In bottom-up approach, a desired structure can be fabricated by sophisticated assembly of atom by atom. In farming sector, both these methods could be employed to fabricate the nanostructures. For example, application of clay minerals in nanofabrication is in experimental stage in which covalent and ionic bonds could be manipulated for manufacturing nutrient supply device.

1.3 Nanomaterials for food and nutrition management

Application of nanoparticles in food sector has given tremendous results. It has revolutionized the whole food industry. Advanced techniques like microfluidics, microelectromechanical systems, and DNA microarrays have advanced the food sector by separation of pathogens, contaminants, smart delivery of nutrients, and nanocapsulation of nutraceuticals (Lugani et al., 2021). Nanoparticles improve the nutritional value and texture of food products, and they improve consistency and physical performance of food, prevent lump formation, remove food contaminants, enhance product shelf life, and help in lighter, stronger, and active packaging (Lugani et al., 2021; Rossi et al., 2014).

1.4 Delivery of DNA in plants through encapsulated nanomaterials

With the application of carbon nanofibres and nanocapsules, it has been possible to encapsulate the foreign DNA and chemicals and its precise and targeted delivery into specific cell and tissue for genetic engineering programs. Through the encapsulation with carbon nanofiber, precise DNA delivery has been done to genetically modify golden rice. In addition, with the integrated application of genetic engineering, nanotechnology, and bioinformatics, novel and desired plant varieties can be created and it has led to the emergence of a new branch of science, synthetic biology.

1.5 Nanomaterials and environmental hazards

Environmental hazards and pollutants are a major global challenge. Release of various contaminants in environment through agricultural practices and other commercial activities has posed a serious threat. Nanomaterials may be effective in active detection and remediation of these contaminants. Effective nanosensors are being tested to be used as analytical devices to monitor a wide variety of agrochemicals, heavy metals, organic pollutants, and pathogens (Chugh et al., 2021). Nanosensors are being fabricated for the monitoring of physicochemical properties of agroecosystems for effective management. Furthermore, climate change that refers to change in the baseline of climate over time span including temperature, water safety, cloud, salinity, alkalinity, and pollution with metal toxicity needs to be detected and monitored (Shang et al., 2019). Nanomaterials have many applications in the field of environment such as production of highly efficient renewable energy, solar cells, remediation, and nanobiosensors for pollutants with greater sensitivity, but they are also suspected to have negative impact on many species and to disrupt the ecological dynamics. According to a report of 2014, the global production of CNTs ranges between 55 and 3300 tons (Mukhopadhyay, 2014). Metal nanoparticles exert cytotoxicity depending on the surface charge of cell (Prasad et al., 2017). These materials can enter the environment accidentally or as waste discharge, which may contaminate the water, and due to their hydrophobicity, they are soluble in organic solvents, thus impacting the uptake of soil contents by plants along with co-contaminants. However, the fate of CNMs in soil depends on its physicochemical properties as well as edaphic factors also. The possible merits and demerits of CNMs are illustrated in Fig. 1.5. Thus, more studies are required to extrapolate and infer about the fate of CNMs in environment.

1.5.1 Nanotoxicity

Despite the promising results in the application of nanotools in agriculture, there are many apprehensions to be resolved. Toxicity of nanomaterials in agroecosystem is of key concern; therefore, the toxicity and impact of released NPs on environment and plants need to be addressed. The interaction of NPs and soil results in altered physicochemical properties of the soil. Few reports are there about the alteration in the pH of soil, organic soil content, and cation exchange capacity on interaction with silver nanoparticles (AgNPs). Likewise, ZnONPs application in soil caused toxicity and reduced the plant biomass. Further, it is interesting to note that the use of TiO_2 and ZnONPs altered the bacterial community structure with distinct impact on environment (Mishra et al., 2017). In addition, plants directly interact with NPs due to large surface area of leaf and root system, resulting in phytotoxicity. The minuscule-sized NPs enter the plant tissue through adsorption and pose negative impact on plant system. The phytotoxicity of NPs depends on their size and concentration. It has been

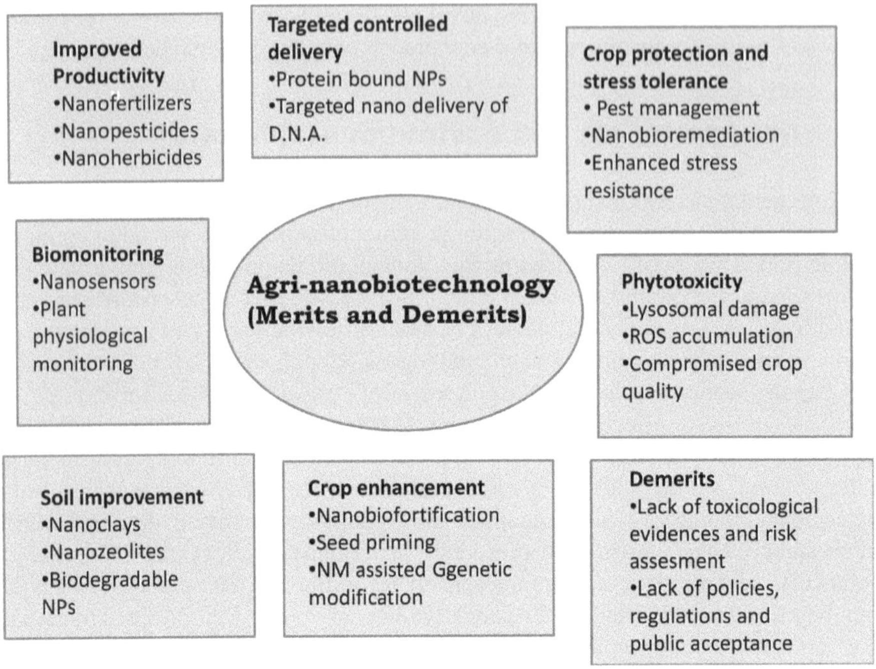

Fig. 1.5 Possible advantages and disadvantages of nanobiotechnology in agriculture. Modified and redrawn from Chugh, G., Siddique, K.H.M., Solaiman, Z.M., 2021. Nanobiotechnology for agriculture: smart technology for combating nutrient deficiencies with nanotoxicity challenges. Sustainability 13(1781), 1–20, published in Sustainability; through open access policy.

observed that NPs in the range of 5–10 nm possess greater toxicity (Mishra et al., 2017). Hence, the tripartite interaction of NPs with plants, soil, and soil microbiota must be analyzed before the commercial use of agriculture-related NPs and more emphasis should be given on fabrication and application of biologically synthesized NPs for agricultural processes (Chen and Yada, 2011).

1.5.2 Nanoremediation

Many nanoscale materials such as metal, metal oxide, CNTs, and bimetallic NPs have been investigated with greater potential in cleaning up contaminated soil and water sites and detoxification of pollutants. Iron oxides have been used for the removal of heavy metals and organic compounds. Nanoscale zerovalent iron (ZVI-NPs) has been tested for remediation due to its low toxicity and low cost. Nanoscale zerovalent iron nanoparticles (ZVI-NPs) are capable of removing heavy metals such as cadmium from aqueous solution, and chromium from polluted soil, and wastewater (Lavicoli et al., 2017; Tosco et al., 2014). Nanostructured bimetallic systems like palladium-iron (Pd—Fe), silver-iron (Ag—Fe), and nickel-iron(Ni—Fe) stabilized with polymers and surfactants have been studied to remove heavy metals dyes and to kill bacteria. In addition, porous titanium silicate and aluminum nanocomposites have been used for the removal of heavy metals like cadmium and lead (Lavicoli et al., 2017; Tosco et al., 2014). But reports are there that these bimetallic NPs generate intermediate products and toxins. The regeneration capacity, reusability, and possible environmental hazards of these NPs are less known that need to be sorted out before their large-scale application.

1.5.3 Nanomaterials and biosafety issues

Keeping in view biosafety issues, a multipronged strategy like, activation of plant enzymatic system, hormonal regulation, stress gene activation, avoiding uptake and accumulation of toxic metals within plant cells should be adopted to maintain and foster sustainable agriculture (Shang et al., 2019). Application of genetically modified crops is among the fastest adopted technology in crop improvement programs; however, there is need for a precise and sensitive detection of this unique technology before its commercial release. However, sophisticated tracer techniques are there for the analysis of NPs within the plant tissues using positron emission tomography (PET) or single-photon emission-computed tomography (SPECT) (Lavicoli et al., 2017). Moreover, an appropriate regulatory framework should be set up to assess the environmental and health-related risks (Prasad et al., 2017; Zhang and Fang, 2010).

1.5.4 Biosynthesis of nanoparticles for green economy

Though the precise application of CNMs and NPs has the scope to revolutionize the whole agricultural sector, it has many disadvantages including nanotoxicity, high cost input, and many environmental concerns. It is now possible to biologically synthesize nanomaterials with a wide range of size, shape, compositions, and physicochemical properties via eco-friendly, green chemistry-based techniques. It employs biological entities such as actinomycetes, bacteria, fungi, plants, viruses, and yeasts, which act as biological factories for the synthesis of nanomaterials (Shah et al., 2015). The size, shape, and composition of the biological entities are suitable for nanofabrication; for example, the nanocomposition of bacteriophage is suitable for the nanoencapsulation of gold and iron oxide to be used as NP. Viruses have been used in the synthesis of NPs for the delivery of silicon dioxide (SiO_2), cadmium sulfide (CdS), iron oxide (Fe_2O_3), and zinc sulfide (ZnS) (Lavicoli et al., 2017). In addition, unicellular alga *Chlorella vulgaris* was used to synthesize tetrachloroaurate ions to form algal bound gold, which was subsequently reduced into gold nanoparticles (AuNPs). Reports are there about the synthesis of silver and gold nanoparticles from algae *Chondrus crispus* and *Spirogyra insignis* (Shah et al., 2015). Plant *Aloe vera* has been used to synthesize AgNPs and AuNPs. Likewise, *Mangifera indica* has been used for the synthesis of AgNPs and *Eucalyptus macrocarpa* has been used for the synthesis of AgNPs and AuNPs (Shah et al., 2015; Mittal et al., 2013). However, the size and shape of these NPs are highly variable, which needs intense research to synthesize the NPs of desirable size and shape because these are inexpensive, nontoxic, and environmental-friendly.

1.6 Challenges in application of nanobiotechnology in agriculture

One of the biggest challenges before nanobiotechnology to be applied in the field of agriculture is the integration of farming with all spheres including geosphere, biosphere, hydrosphere, and atmosphere to manage the food and nutritional security at global level along with mitigating climate change effects and managing the natural resource. The use of nanomaterials may have inadvertent and devastating effects on soil microbes, water, and related ecosystem, which need to be taken care of. The second important challenge before agrinanobiotechnology is the high cost input. For example, according to an estimate, 50 mg nanowire is adequate for 50 million cell phones, while the requirement of nitrogen fertilizer for one hectare could be 100 kg approximately (Mukhopadhyay, 2014). In addition, the fate and behavior of nanomaterials like zinc oxide and titanium oxide as nanofertilizer are unpredictable because it will spread in the wide area of field and may create toxicity. Carbon dot, which is one of the key nanomaterials with diverse applications in multiple sectors including agriculture, is prepared from the raw biomass, thereby harvesting the renewable energy. The photoluminescence mechanism of C dot is less understood; therefore, a wide mechanism of photoluminescence of C dot is needed to expand its application from renewable biomass (Kang et al., 2020). Despite the high potential application of nanotechnology in agriculture, there are probable toxicological hazards due to the release of these products into the environment (Anjum et al., 2013). The implications of these nanomaterials on human health, environment, and biosafety issues are less known, which require a wide and intense study. There should be a precise balance between sustainable agriculture, enhanced crop yield, and sustainable environment, which is summarized in Fig. 1.6.

The cytotoxic and genotoxic effects of NPs and the accumulation of reactive oxygen species (ROS) as a by-product in plants need to be explored. Additionally, the mechanism of uptake of engineered nanomaterials and their distribution in plants and the bioavailability of micronutrients within NPs are yet to be explored (Lowry et al., 2019; Zhang, 2018). However, the explored and unexplored areas of agrinanobiotechnology that need to be addressed in the future are depicted in Fig. 1.7. Hence, the application of NPs and its control is not easy in agriculture unlike that of electronic or optical sector and it will require a mammoth and time consuming with more sophisticated research to cope with these challenges.

1.7 Conclusion and future perspectives

Being the new entrant in agriculture, there are enormous challenges before nanobiotechnology that need a highly sophisticated research and trial. Nanomaterials can be used to enrich the nutritional potential of the soil to enhance the productivity and environmental safety. Nanomaterials are to be designed, which may facilitate the uptake of nutrients by the plants rapidly and specifically. Designation and application

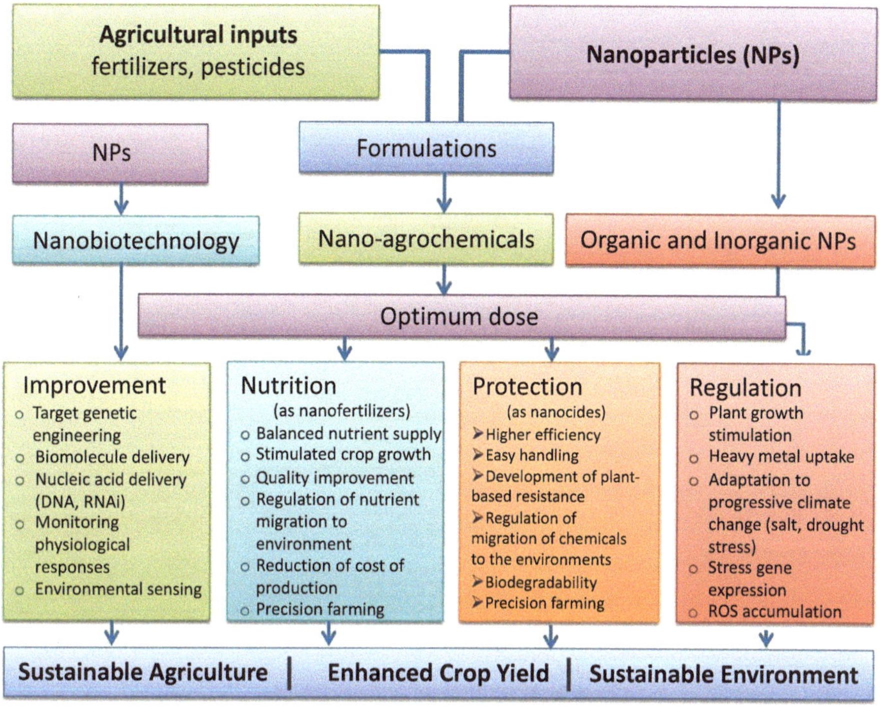

Fig. 1.6 Application of nanomaterials in sustainable agriculture.
Adapted from Shang, Y., Hasan, M.K., Ahammed, G.J., et al., 2019. Applications of nanotechnology in plant growth and crop protection: a review. Molecules 24(2558), 1—23. doi:10.3390/molecules24142558, published in Molecules; through open access policy.

of nanobiopolymers in agricultural field for coating of seeds as a soil stabilizer and protector, saving of nutrients and water as well as enhanced yield have wide prospects. Nanofabricated materials can be designed in hydrogel and suspension forms for their easy storage and convenient delivery system. By enhancing the adsorption capacity of nanoparticles such as iron and calcium carbonate nanoparticles that have great soil-binding capacity, they may be used for nanoremediation. For example, zerovalent iron nanoparticles can be used for the remediation of soil contaminated with pesticides, heavy metals, and radionuclides.

Nanomaterials may also be used for genetic improvement of plants. Specially formulated nanomaterials could be used for delivery of genes and drug molecules to target destinations within cell. Nanoarray-based technologies can be used for gene expression in plants in order to develop stress- and salinity-resistant varieties. Nanotools for the management of natural resources, smart delivery system for agrochemicals, and smart system for food processing and packaging are needed. Nanoremediation could be effective in purifying contaminated water and cleaning up large contaminated sites. Nanozeolites may be used to eradicate the acidity of soil, thereby enhancing its quality. Thus, the application of nanobiotechnology in

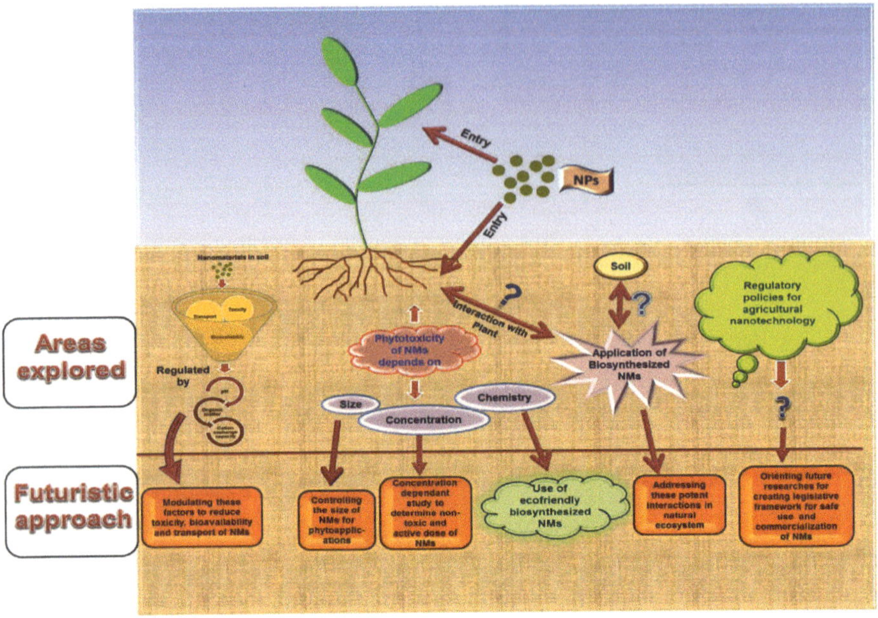

Fig. 1.7 Schematic representation of the points to be addressed in future researches on agrinanotechnology.
Adapted from Mishra, S., Keswani, C., Abhilash, P.C., Fraceto, L.F., Singh, H.B., 2017. Integrated approach of Agri-nanotechnology: challenges and future trends. Front. Plant Sci., https://doi.org/10.3389/fpls.2017.00471. published in Frontiers in Plant Sciences through open access policy.

agriculture is astonishing though it is still in infant stage. By the formulation of nanofertilizers, the nutritional value and productivity of the crops can be enhanced. The development of nanopesticides and nanoherbicides, preservation and packaging of food, removal of contaminants from soil and water, improving the durability of fruits and vegetables, strengthening the natural fibers, efficient delivery and expression of genes for genetic improvement of crops, regeneration of soil fertility and reclamation of salt-affected lands, prevention of acidification of irrigated areas, precision water management, and energy conservation are some promising tasks. Thus, in a wide perspective, enhancing the crop yield to meet the food and nutritional security, enhancing the energy efficiency with sustainable agriculture and water management along with mitigating the environmental challenges, and promotion of green economy are the other prime tasks to be accomplished through this innovative technology that are hoped to be accomplished in the near future if this technology is explored, fostered, and applied efficiently and judiciously. Moreover, the properties of nanomaterials such as size, exposure, dose, surface chemistry, immune responses, accumulation, retention, and other effects should be scrutinized properly before their commercial use (Prasad et al., 2017).

Though this innovative technology has given the tremendous benefits to the agricultural sector, a long gap has to be covered from theoretical approach to practical approach. Future research must be given due importance to resolve the risk-related issues and making the more efficient, cost-effective and eco-friendly, nontoxic nanomaterials. Hence, scientific community must work together to improve it with realistic approach. The safety limits of the use of nanoparticles must be explored and clarified for maintaining the sustenance of agriculture and to avoid toxicity. Physicochemical properties of soil and its interaction with nanomaterials must be a key area of research in order to maintain biosafety (Parisi et al., 2015; Usman et al., 2020). Emphasis on biologically synthesized nanomaterials including viruses, bacteria, fungi, algae, and higher plants may give promising results to meet the challenges regarding NPs. Hence, future researches must be targeted to develop smart nanomaterials for increasing agrochemical efficiency through targeted delivery, smart plants as sensors, improving the health and function of soil and plants for sustainability through microbiome enhancement, thereby increasing the yield with environmental safety (Lowry et al., 2019).

Thus, following points should be taken into consideration in future researches: (a) safe application of NPs at permissible level for agricultural benefits by modulating the behavior, bioavailability, and toxicity determining factors should be optimized, (b) long-term and more realistic experimental design is required for exploring the safety limits and reducing the nanotoxicity, and (c) administration of biologically synthesized nanomaterials and assessment of their advantages over nonbiologically synthesized nanomaterials (Mishra et al., 2017).

In concluding remarks, nanotechnology is the revolutionary technology, which has many expectations in agricultural sector including enhanced sustainable productivity and security of agriculturally produced food. The application of nanotools and NPs in agriculture may be helpful in effective and targeted delivery of agrochemicals (nanopesticides, nanoherbicides, nanofertilizers) in plants with enhanced quality and quantity of yields. Though still in infancy, need-based CNMs and NPs may be formulated through nanofabrication to cope with modern agriculture-related challenges. New nano-bio-industrial and agricultural products may be helpful in mitigating environmental problems with better and sophisticated pollution detection system through nanosensors and nanoremediation, thereby aiding in better environment risk assessment and agricultural management system with low cost input. However, there is still a very wide gap between laboratory results and realistic field results related to nanobiotechnology. The interaction of NPs with plant cells, tissues, and soil depends on many factors and physicobiochemical properties and is obscure. Hence, more realistic and extensive researches are needed for producing nontoxic, cost-effective, and eco-friendly nanoproducts with greater efficacy for sustainable agricultural practices.

Conflict of interest

The authors declare no conflict of interest.

References

Anjum, N.A., Gill, S.S., Duarte, A.C., Pereira, E., Ahmad, I., 2013. Silver nanoparticles in soil plant systems. J. Nanopart. Res. 15, 1896. https://doi.org/10.1007/s11051-013-1896-7.

Chen, H., Yada, R., 2011. Nanotechnologies in agricultural development. Trends Food Sci. Technol. 22, 585–594.

Chugh, G., Siddique, K.H.M., Solaiman, Z.M., 2021. Nanobiotechnology for agriculture: smart technology for combating nutrient deficiencies with nanotoxicity challenges. Sustainability 13 (1781), 1–20.

Dwivedi, S., Saquib, Q., Al-Khedhairy, A.A., Musarrat, J., 2016. Understanding the role of nanomaterials in agriculture. In: Singh, D.P., Singh, H.B., Prabha, R. (Eds.), Microbial Inoculants in Sustainable Agricultural Productivity. Springer, New Delhi, India, pp. 271–288.

European Commission, 2011. Compilation of Information Concerning Experience With the Definition (EUR 26567 EN). JRC.

Jaiswal, M., Dudhe, R., Sharma, P.K., 2015. Nanoemulsion: an advanced mode of drug delivery system. 3 Biotech 5 (2), 123–127.

Kang, C., Huang, Y., Yan, X.F., Chen, Z.P., 2020. A review of carbon dots produced from biomass wastes. Nanomaterials 10 (2316), 3–24.

Lavicoli, I., Leso, V., Beezhold, D.H., Shvedova, A.A., 2017. Nanotechnology in agriculture: opportunities, toxicological implications, and occupational risks. Toxicol. Appl. Pharmacol. 329, 96–111.

Lowry, G.V., Avellan, A., Gilbertson, L.M., 2019. Opportunities and challenges for nanotechnology in the agri-tech revolution. Nat. Nanotechnol. 14, 517–521.

Lugani, Y., Sooch, B.S., Singh, P., et al., 2021. Nanobiotechnology applications in food sector and future innovations. Microbial Technol. Food Health. https://doi.org/10.1016/B978-0-12-819813-1.00008-6.

Lv, M., Liu, Y., Geng, J.H., Kou, X.H., Xin, Z.H., Yang, D.Y., 2018. Engineering nanomaterials-based biosensors for food safety detection. Biosens. Bioelectron. 106, 122–128.

Mishra, S., Keswani, C., Abhilash, P.C., Fraceto, L.F., Singh, H.B., 2017. Integrated approach of Agri-nanotechnology: challenges and future trends. Front. Plant Sci. https://doi.org/10.3389/fpls.2017.00471.

Mittal, A.K., Chisti, Y., Banerjee, U.C., 2013. Synthesis of metallic nanoparticles using plant extracts. Biotechnol. Adv. 31, 346–356. https://doi.org/10.1016/j.biotechadv.2013.01.003.

Mukherjee, A., Majumdar, A.S., Servin, A.D., et al., 2016. Carbon nanomaterials in agriculture: a critical review. Front. Plant Sci. 7, 1–13.

Mukhopadhyay, S.S., 2014. Nanotechnology in agriculture: prospects and constraints. Nanotechnol. Sci. Appl. 7, 63–71.

Parisi, C., Vigani, M., Rodriguez-Cerezo, E., 2015. Agricultural nanotechnologies: what are the current possibilities? Nano Today 10, 124–127.

Prasad, R., Bhattacharya, A., Nguyen, Q.D., 2017. Nanotechnology in sustainable agriculture: recent development, challenges and perspectives. Front. Microbiol. 8. https://doi.org/10.3389/fmicb.2017.01014.

Rossi, M., Cubadda, F., Dini, L., Terranova, M.L., Aureli, F., Sorbo, A., et al., 2014. Scientific basis of nanotechnology, implications for the food sector and future trends. Trends Food Sci. Technol. 40, 127–148. https://doi.org/10.1016/j.tifs.2014.09.004.

Sekhon, B.S., 2014. Nanotechnology in agri-food production: an overview. Nanotechnol. Sci. Appl. 7, 31–53.

Seleiman, M.F., Almutairy, K.F., Alotaibi, M., et al., 2021. Nano-fertilization as an emerging fertilization technique: why can modern agriculture benefit from its use? Plan. Theory 10, 1–27.

Shah, M., Fawcett, D., Sharma, S., et al., 2015. Green synthesis of metallic nanoparticles via biological entities. Materials 8, 7278–7308.

Shang, Y., Hasan, M.K., Ahammed, G.J., et al., 2019. Applications of nanotechnology in plant growth and crop protection: a review. Molecules 24 (2558), 1–23. https://doi.org/10.3390/molecules24142558.

Taran, N., Storozhenko, V., Svietlova, N., et al., 2017. Effect of zink and coppor nanoparticles on drought resistance of wheat seedlings. Nanoscale Res. Lett. 12, 60. https://doi.org/10.1186/s11671-017-1839-9.

Tosco, T., Papini, M.P., Viggi, C.C., Sethi, R., 2014. Nanoscale zerovalent iron particles for groundwater remediation: a review. J. Clean. Prod. 77, 10–21. https://doi.org/10.1016/j.jclepro.2013.12.026.

Tripathi, M., Sahu, J.N., Ganesan, P., 2016. Effect of process parameters on production of biochar from biomass waste through pyrolysis: a review. Renew. Sustain. Energy Rev. 55, 467–481.

Usman, M., Farooq, M., Wakeel, A., Nawaz, A., Alam Cheema, S., Rehman, H.U., Ashraf, I., Sanaullah, M., 2020. Nanotechnology in agriculture: current status, challenges and future opportunities. Sci. Total Environ. 721, 137778.

Wang, Y., Sun, C., Zhao, X., et al., 2016. The application of nano-TiO2 photo semiconductors in agriculture. Nanoscale Res. Lett. 11, 529. https://doi.org/10.1186/s11671-016-1721-1.

Xue, M.Y., Zhan, Z.H., Zou, M.B., Zhang, L.L., Zhao, S.L., 2016. Green synthesis of stable and biocompatible fluorescent carbon dots from peanut shells for multicolor living cell imaging. New J. Chem. 40, 1698–1703.

Zahedi, S.M., Moharrami, F., Sarikhani, S., Padervand, M., 2020. Selanium and silica nanostructure based recovery of strawberry plants subjected to drought stress. Sci. Rep. 10, 17672. https://doi.org/10.1038/s41598-020-74273-9.

Zhang, W., 2018. Global pesticide use: profile, trend, cost/benefit and more. Proc. Int. Acad. Ecol. Environ. Sci. 8, 1–27.

Zhang, L., Fang, M., 2010. Nanomaterials in pollution trace detection and environmental improvement. Nano Today 5, 128–142. https://doi.org/10.1016/j.nantod.2010.03.002.

Zhao, X., Lv, L., Pan, B., Zhang, W., Zhang, S., Zhang, Q., 2011. Polymer-supported nanocomposites for environmental application: a review. Chem. Eng. J. 170, 381–394. https://doi.org/10.1016/j.cej.2011.02.071.

Zhu, Y.X., Gong, H.J., Yin, J.L., 2019. Role of silicon in mediating salt tolerance in plants: a review. Plan. Theory 8, 147. https://doi.org/10.3390/plants8060147.

Zuo, P.L., Lu, X.H., Sun, Z.G., Guo, Y.H., He, H., 2016. A review on syntheses, properties, characterization and bioanalytical applications of fluorescent carbon dots. Microchim. Acta 183, 519–542.

Crop-mediated synthesis of nanoparticles and their applications

2

Sougata Ghosh[a,c] *and Bishwarup Sarkar*[b]
[a]Department of Microbiology, School of Science, RK University, Rajkot, Gujarat, India, [b]College of Science, Northeastern University, Boston, MA, United States, [c]Department of Physics, Faculty of Science, Kasetsart University, Bangkok, Thailand

2.1 Introduction

Nanotechnology has received substantial response for its applications in diverse fields that include optics, electronics, mechanics, energy, space, catalysis, textiles, agriculture, food, and medicine (Jadoun et al., 2021; Ghosh and Webster, 2021a,b; Ghosh et al., 2021). The large surface-area-to-volume ratio and extremely small size of nanoparticles facilitate such wide applications of these nanomaterials. However, several nanoparticles show toxic properties because of which nanotechnology was integrated with green chemistry in order to synthesize nanoparticles using bacteria, fungi, algae, and plants (Bloch et al., 2021; Ranpariya et al., 2021; Ghosh and Turner, 2021; Ghosh et al., 2016a,b,c,d). These biosynthetic methods of nanoparticle synthesis highlighted a notable benefit to nature as well as environment (Duan et al., 2015).

Plants are known to be cost-effective chemical factories that require little maintenance. Plant-assisted fabrication of nanoparticles has thus higher kinetics than other biosynthetic methods such as utilizing microbial cultures (Jadoun et al., 2021). Extracts made out of various parts of plants such as leaves, fruits, roots, stem, and seeds have been reported to contain phytochemicals that can act as stabilization and reducing agent during the synthesis of nanoparticles (Narayanan and Sakthivel, 2011). Significant amount of free radical scavenging biomolecules such as polyols, phenolic compounds, nitrogen compounds, vitamins, reducing sugar, terpenoids, and some other metabolites that are rich in antioxidant activity are known to be present in these plant extracts (Rokade et al., 2018, 2017; Jamdade et al., 2019; Shende et al., 2017, 2018; Bhagwat et al., 2018; Shinde et al., 2018; Mohamad et al., 2014). Hence, natural plant extracts are generally used for reduction, capping, and stabilization during metallic nanoparticle formation, which can be observed as visible color change.

Agriculturally important cereals (*Oryza sativa, Triticum aestivum, Zea mays* L., *Hordeum vulgare, Pennisetum glaucum*), pulses (*Cajanus cajan* L., *Cicer arietinum* L.), vegetables (*Solanum tuberosum, Coriandrum sativum* L., *Zingiber officinale, Solanum melongena* L., *Spinacia oleracea, Allium cepa, Brassica oleracea, Raphanus sativus, Brassica rapa* L., *Pisum sativum*), spices (*Curcuma longa* L., *Piper nigrum Cuminum cyminum* L., *Ferula asafoetida, Elettaria cardamomum*), fruits (*Diospyros malabarica, Ficus carica,*

Agricultural Nanobiotechnology. https://doi.org/10.1016/B978-0-323-91908-1.00017-1

Cynometra ramiflora, Ziziphus spina-christi (L.) Willd., *Rubus illecebrosus*), and cash crops (*Camellia sinensis, Coffea arabica, Saccharum officinarum, Theobroma cacao, Gossypium hirsutum*) have been extensively studied for its nanoparticle synthesis ability, which are explained in this chapter and summarized in Table 2.1. These eco-friendly, bio-fabricated nanoparticles were further demonstrated to have antimicrobial, anticancer, anti-candidal, antifungal, and larvicidal activities as well as exhibit efficient photocatalytic abilities for various dye degradation.

Hence, such crop-assisted preparation of nanomaterials could be potentially useful in large-scale production of environmentally benign nanoparticles that can have a multitude of applications in biomedical, chemical, optical, and several other industries. The following section gives a comprehensive detail about various crops that are exploited for the synthesis of nanoparticles.

2.2 Cereals

Synthesis of nanoparticles from cereals has been extensively reported. Akhayere et al. (2020) in one such study synthesized magnetic nanosilica (MNS) using *Hordeum vulgare* (barley) husk waste. The average diameter of spherical MNS was 162 nm with a large surface area of approximately 120 m^2/g. This hydrophobic MNS at a concentration of 0.6 g was then observed to effectively remove 85% of 40 mg/L of petroleum and hence proposed to act as efficient sorbent for petroleum and related products from wastewater samples.

In another study, silicon nanoparticles (SiNPs) were fabricated by Bose et al. (2018) using husks of *Oryza sativa* (rice) without utilizing any external reducing agent as illustrated in Fig. 2.1. Rice husk powder was added with 1 M NaOH, followed by heating at 170°C for 1 h and removal of anthocyanin dyes by centrifugal ultrafiltration to obtain around 5% estimated yield of SiNPs. The carbon backbone materials such as cellulose, hemicellulose, and lignin present in the sample were proposed to be responsible for the reduction of silica to silicon. Ultraviolet–visible (UV–Vis) spectroscopy results demonstrated two absorption peaks at 270 nm and 350 nm corresponding to L-L transition and intrinsic direct $\Gamma - \Gamma$ band gap of silicon, respectively. Furthermore, the solution exhibited the green luminescence under UV light, while it appeared brown under white light illumination. High-resolution transmission electron microscopy (HRTEM) image of SiNPs revealed its spherical morphology. Hydrophilic nature of SiNPs was observed when organic solvents such as dichloromethane (DCM) or hexane were introduced into the mixture. Organic solvents were observed to form a separate layer, while SiNPs were shown to disperse in water. Transmission electron microscopy (TEM) results showed average size of SiNPs to be 4.9 ± 0.2 nm that were highly monodispersed. Moreover, zeta potential of SiNPs was −27.1 mV. X-ray photoelectron spectroscopy (XPS) results of SiNPs revealed the deconvoluted spectra with peaks at around 99.6, 100.7, and 103.7 eV, indicating different oxidation states of silicon, which are Si(0), Si(I), and Si(IV), respectively. Fourier transform infrared (FTIR) spectral analysis further revealed the additional peaks at 833 cm^{-1} and 857 cm^{-1} corresponding to the Si—N stretching vibrations of SiNPs that were

Table 2.1 Nanoparticle synthesis using crop extracts for various applications.

Name of crop	Type of nanoparticles	Size (nm)	Shape	Application	Reference
Cereals					
Hordeum vulgare	MNS	162	Spherical	Removal of petroleum and its by-products	Akhayere et al. (2020)
Oryza sativa	SiNPs	4.9 ± 0.2	Spherical	Used as material in WLEDs	Bose et al. (2018)
Oryza sativa	bSNPs	10–30	Irregular	Biocompatibility to human mesenchymal stem cells	Athinarayanan et al. (2014)
Oryza sativa	ZnONPs	250–700	Star-like	–	Ramimoghadam et al. (2013)
Pennisetum glaucum	AgNPs	17	Spherical	Removal of algae, catalytic oxidation of benzyl alcohol	Musere et al. (2021)
Triticum aestivum	AgNPs	17.2	Ellipsoidal	Antibacterial	Ma et al. (2016)
Triticum aestivum	CuONPs	22 ± 1.5	Spherical	Catalysis and reusability in 4-nitrophenol reduction	Buazar et al. (2019)
Triticum aestivum	AgNPs	5.06	Spherical	Peroxide reduction, nontoxic	Waghmode et al. (2013)
Zea mays L.	AgNPs	–	Spherical	Cytotoxic effect, DPPH scavenging	Rajkumar et al. (2019)
Zea mays L.	NiONPs	40	Spherical and quasi-spherical	Used as material for positive electrode in asymmetric supercabattery devices	Nwanya et al. (2020)
Zea mays L.	Cu_2ONPs and CuONPs	30–90	Spherical, pseudo-spherical, and conical	Dye degradation, antibacterial	Nwanya et al. (2019)
Pulses					
Cajanus cajan L.	AgNPs	5–60	Spherical	Antibacterial	Nagati et al. (2012)
Cajanus cajan L.	ZnONPs	40–80	Spherical	Dye degradation, antibacterial	Manjunath et al. (2014)

Continued

Table 2.1 Continued

Name of crop	Type of nanoparticles	Size (nm)	Shape	Application	Reference
Cajanus cajan L.	TiO_2NPs	10–25	–	Enhancement of antifouling ability of PVDF membrane	Arif et al. (2019)
Cicer arietinum L.	AuNPs	30–80	Spherical, pentagonal, and triangular	–	Singh et al. (2013)
Cicer arietinum L.	Fe_3O_4NPs	13	–	Biocompatibility, antibacterial	Koli et al. (2019)
Cicer arietinum L.	CuONPs	–	Hexagonal	–	Verma et al. (2019)
Vegetables					
Allium cepa	AuNPs	45.42	Spherical and triangular	Antibacterial and proteasome inhibition	Patra et al. (2016)
Brassica oleracea	AgNPs	20	Spherical	Antibacterial, cytotoxicity, antioxidant	Ansar et al. (2020)
Brassica rapa L.	AuNPs	31.4 ± 14.9	Spherical	Dye reduction	Narayanan and Park (2015)
Coriandrum sativum L.	AgNPs	11.9	Spherical	Antibacterial	Ashraf et al. (2019)
Coriandrum sativum L.	AgNPs	37	Spherical	Antiacne, antidandruff, and antibreast cancer	Sathishkumar et al. (2016)
Coriandrum sativum L.	AuNPs	32.96 ± 5.25	Spherical, triangular, and hexagonal	Antioxidant, analgesic, biocompatibility	Jiao et al. (2021)
Pisum sativum	MNPs	20–30	Spherical	Dye degradation	Prasad et al. (2017)
Raphanus sativus	ZnONPs	30	Spherical	Antibacterial	Liu et al. (2020)
Solanum melongena L.	AuNPs and AgNPs	29 (AuNPs) and 92.4 (AgNPs)	Spherical	Antibacterial	Das and Bhuyan (2019)
Solanum tuberosum	AgNPs	10–12	Spherical	Photocatalytic dye degradation	Roy et al. (2015)

Solanum tuberosum	MNPs	–	Spherical	Dye degradation	Sharma et al. (2019)
Solanum tuberosum	AgNPs	39.5–80.3	–	Antifungal, biocompatibility	Almadiy and Nenaah (2018)
Spinacia oleracea	ZnONPs	40.59	Spherical	Plant growth promotion	Lakshmi et al. (2017)
Zingiber officinale	AgNPs	10	Spherical	Antibacterial	Vijaya et al. (2017)
Zingiber officinale	AuNPs	13	Spherical	Antibacterial	Chitra et al. (2015)
Zingiber officinale	AgNPs	10	Spherical	Antimicrobial	Yang et al. (2017)
Spices					
Cuminum cyminum L.	AgNPs	100	Spherical	Anticancer	Dinparvar et al. (2020)
Curcuma longa L.	AgNPs	15–40	Spherical	Antimicrobial, nontoxicity, wound attenuation	Maghimaa and Alharbi (2020)
Curcuma longa L.	CuNPs	5–20	Spherical	Antibacterial.	Jayarambabu et al. (2020)
Elettaria cardamomum	AgNPs	18.72	Spherical	Antibiofilm, mosquito larvicidal	Vinotha et al. (2020)
Ferula asafoetida	AgNPs	39	Spherical, rod-shaped, and cuboid	Mosquito larvicidal	Saranyadevi et al. (2021)
Piper nigrum	SnO_2NPs	8.85 ± 3.5	Spherical	Anticancer	Tammina et al. (2017)
Piper nigrum	AgNPs	9–30	Spherical and hexagonal	Antibacterial	Paulkumar et al. (2014)

Continued

Table 2.1 Continued

Name of crop	Type of nanoparticles	Size (nm)	Shape	Application	Reference
Other crops					
Camellia sinensis	NiNPs	43.87–48.76	Spherical	Photocatalytic dye degradation	Bibi et al. (2017)
Coffea arabica	ZnONPs	40	Cubic	–	Abel et al. (2021)
Cynometra ramiflora	MNPs	58.5	Spherical	Photocatalytic dye degradation	Bishnoi et al. (2018)
Diospyros malabarica	AgNPs	17.4	Spherical	Antibacterial, catalysis	Bharadwaj et al. (2021)
Ficus carica	Fe_2O_3NPs	9 ± 4	Spherical	–	Demirezen et al. (2019)
Gossypium hirsutum	AgNPs	30	Spherical	Antimicrobial, anticancer	Kanipandian and Thirumurugan (2014)
Rubus illecebrosus	AgNPs	12–50	Spherical	Antioxidant	Kumar et al. (2017)
Saccharum officinarum	CuONPs	84.40	Square, rectangular, cubic, cylindrical, and spherical	Antibacterial	Mary et al. (2019)
Theobroma cacao	AgNPs	8.96–54.22	Spherical	Antibacterial, mosquito larvicidal	Azeez et al. (2017)
Ziziphus spina-christi (L.) Willd.	CuNPs	5–20	Spherical	Antibacterial, dye removal	Khani et al. (2018)

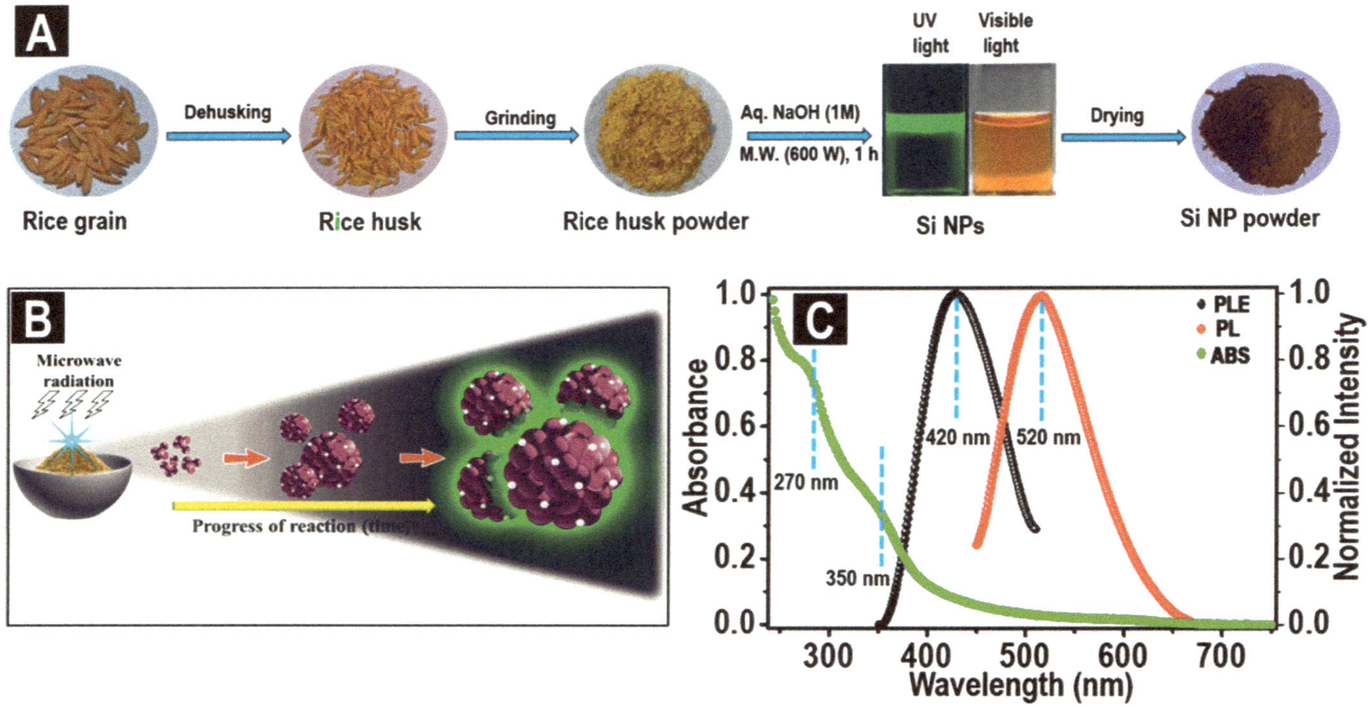

Fig. 2.1 (A) Procedures involved in the synthesis of green luminescent Si NPs. (B) Schematic illustration of nanoparticle nucleation, growth, and coalescence leading to smaller Si NPs. (C) Absorption (ABS), excitation (PLE), and emission spectra (PL) of the Si NPs.
Reprinted with permission from Bose, S., Ganayee, M.A., Mondal, B., Baidya, A., Chennu, S., Mohanty, J.S., Pradeep, T., 2018. Synthesis of silicon nanoparticles from rice husk and their use as sustainable fluorophores for white light emission. ACS Sustain. Chem. Eng. 6, 6203–6210. Copyright © 2018 American Chemical Society.

absent in the infrared (IR) spectrum of rice husk. Photoluminescence spectrum of SiNPs exhibited no change in the emission maximum for a period of 1 month, highlighting its stability. Fluorescence intensity of SiNPs varied within 9% when pH value of 12.0 was changed to 5.0, indicating effective potential function in biological media. Photostability of SiNPs was also compared with CdTe quantum dots and fluorescein isothiocyanate (FITC) dye under UV light. Photoluminescence intensity of SiNPs was observed to remain intact, while CdTe quantum dots exhibited more than 80% decrease in the first 16 h. SiNPs were also investigated in its utilization for the production of white light wherein green luminescent SiNPs that were synthesized were combined with blue-emitting SiNPs and red-emitting gold-bovine serum albumin (Au@BSA) cluster. CIE 1931 chromaticity diagram showed coordinates of produced white light to be (0.31, 0.27), which is almost similar to coordinates of pure white light. Solid-state emission of the prepared material in order to directly use it as a phosphor in white light-emitting diodes (WLEDs) was also evaluated. Cellulose was used as a supporting material, while the prepared material was incubated at 60°C for drying. Luminous efficacy of the prepared material was observed to be superior as compared to commercial WLEDs and compact fluorescent lamps (CFLs).

Likewise, Athinarayanan et al. (2014) also reported the synthesis of biogenic silica nanoparticles (bSNPs) using rice husks. TEM images revealed the irregular morphologies of obtained particles having dimensions in between 10 and 30 nm. bSNPs at a concentration of 200 µg/mL were shown to be biocompatible using 3-(4,5-dimethylthiazol-2-yl)-2,5-diphenyltetrazolium bromide (MTT) assay wherein no significant change in the viability of human mesenchymal stem cells was observed.

In another study, Ramimoghadam et al. (2013) used rice flour as biotemplate for hydrothermal synthesis of zinc oxide nanoparticles (ZnONPs). A unique star-like morphology of ZnONPs with a size range of 250–700 nm was observed in field emission scanning electron microscopy (FESEM) images when 1 g of uncooked rice was used for the synthesis of the same. Moreover, different morphologies such as flake-, flower-, rose-, star- and rod-like structures were also obtained with varying concentrations of uncooked rice.

Musere et al. (2021) also biofabricated AgNPs using *Pennisetum glaucum* (pearl millet) husk that had spherical morphology and polydispersed in nature with an average size of 17 nm. These AgNPs were then observed to exhibit toxic effects against algae with 41.31% reduction in chlorophyll concentration. Moreover, biogenic AgNPs were doped into charcoal for catalytic oxidation of 90% of benzyl alcohol in the presence of H_2O_2 into benzaldehyde.

Silver nanoparticles (AgNPs) were synthesized by Ma et al. (2016) using *Triticum aestivum* (wheat) straw biomass. Wheat straw solution having a concentration of 4 mg/mL was mixed with $AgNO_3$ to obtain a final concentration of 1 mM after which the mixture was directly exposed to the light of varying intensities at room temperature. The color of the solution was observed to change from yellow to brown that was attributed to surface plasmon resonance (SPR) of AgNPs in the visible region. Maximum yield of AgNPs was obtained after 90 min of reaction with 2 mg/mL of wheat straw biomass, 1 mM $AgNO_3$, and 60,000 lx light intensity. TEM analysis further revealed the ellipsoidal and nearly spherical morphology of AgNPs having a mean diameter of 17.2 nm and zeta potential value of −21.6 mV. X-ray diffraction (XRD) patterns of synthesized AgNPs revealed four intense peaks of diffraction having 2θ values 38.0°, 46.13°, 64.43°, and 76.61°, which highlighted the face-centered cubic (FCC) structure of AgNPs. Furthermore, FTIR spectral analysis showed peaks at 3201, 1071, 2921, 1544, 1400, and 2852 cm^{-1}, which indicated the presence of several functional groups on the surface of AgNPs such as alcohol, phenol, and amino groups. Hence, it was assumed that alcohols and proteins may be responsible for the reduction of Ag^{+} ions as well as stabilization of AgNPs. Antibacterial activity of AgNPs was also investigated against *Bacillus subtilis* and *Escherichia coli* wherein higher inhibitory effects were observed against *B. subtilis*. Therefore, it was implied that these AgNPs synthesized using wheat straw biomass could be potentially applicable in antibacterial medications and food preservation.

Copper oxide nanoparticles (CuONPs) were also synthesized using wheat seed extract by Buazar et al. (2019). The electron-rich biomolecules present in the aqueous extract of the wheat seeds were speculated to act as both reducing and capping agents for CuONPs. TEM images exhibited the spherical shape of the prepared nanoparticles with an average size of 22 ± 1.5 nm. Furthermore, biofabricated CuONPs exhibited the significant catalytic activity and reusability in the degradation of organic para-nitrophenol pollutant.

Waghmode et al. (2013) also reported the utilization of wheat plant extract in synthesis of AgNPs. Polydispersed, nanocrystalline AgNPs having a spherical shape and an average size of 5.06 nm were obtained that showed excellent activity against peroxide reduction. Moreover, preliminary toxicology investigations revealed that earthworms were stable in the presence of up to 1500 ppm AgNPs.

Zea mays L. extract (corn flour) was reported to be efficiently used for biosynthesis of AgNPs by Rajkumar et al. (2019). AgNPs were prepared using 1 mM of $AgNO_3$ solution and corn flour extract under photo-catalyzed conditions. The reaction mixture was observed to transition its color from yellow to reddish brown, which indicated the formation of AgNPs. The SPR band of AgNPs was observed at around 410–420 nm wherein maximum yield of AgNPs was observed in the presence of 1 mM of $AgNO_3$ solution. Likewise, highest AgNP production was seen at an incubation temperature of 80°C that was proposed to reduce aggregation of the particles. XRD measurements also revealed the crystalline nature of AgNPs, while scanning electron microscopy (SEM) images exhibited the spherical shape and monodispersed as well as aggregated nature of AgNPs. FTIR spectral analysis further revealed the distinct peaks at 3186, 1726, 1651, 1289, 1545, and 1068 cm^{-1} that corresponded to O—H stretching of phenolic group, carbonyl stretching vibration, C=C stretching vibration, C—O stretching vibration, aromatic and aliphatic amines, respectively. Moreover, cytotoxicity of synthesized AgNPs was evaluated against mice HepG2 cells that showed reduced viability with increased AgNP concentration. Antioxidant potential of AgNPs was also investigated using 2,2-diphenyl-1-picryl-hydrazyl-hydrate (DPPH) and 2,2′-azino-bis(3-ethylbenzothiazoline-6-sulfonic acid (ABTS) scavenging abilities. DPPH scavenging activity of AgNPs ranged from 84.50% to 97.9%, while ABTS scavenging activity only ranged from 0% to 4.94% as compared to gallic acid.

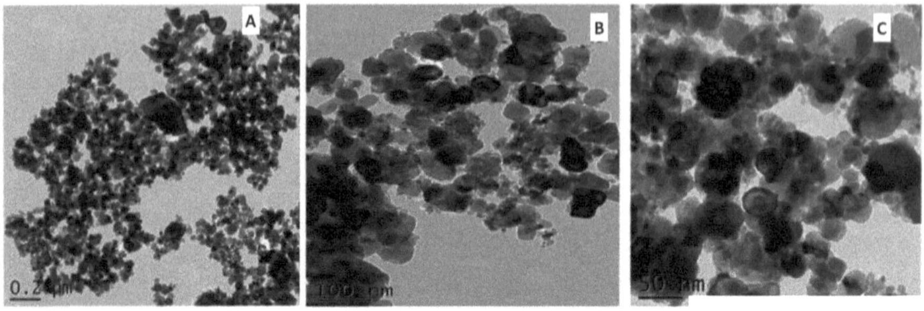

Fig. 2.2 TEM of the biosynthesized NiONPs at various magnifications. Inset represents scale bar equivalent to (A) 0.2 μm; (B) 100 nm; (C) 50 nm.
Reprinted with permission from Nwanya, A.C., Ndipingwi, M.M., Ikpo, C.O., Obodo, R.M., Nwanya, S.C., Botha, S., Ezema, F.I., Iwuoha, E.I., Maaza, M., 2020. *Zea mays* lea silk extract mediated synthesis of nickel oxide nanoparticles as positive electrode material for asymmetric supercabattery. J. Alloys Compd. 822, 153,581. Copyright © 2019 Elsevier B.V.

In another study, Nwanya et al. (2020) reported the green synthesis of nickel oxide nanoparticles (NiONPs) using maize dry silk extract. TEM images, in Fig. 2.2, revealed the spherical and quasi-spherical shapes of NiONPs having an average diameter of 40 nm. Furthermore, XRD patterns exhibited the FCC structure of the synthesized NiONPs with a lattice constant of 4.172 Å. Selective area electron diffraction (SAED) patterns showed poly-nanocrystalline nature of NiONPs as well. These biosynthesized NiONPs were also considered as positive electrodes for asymmetric supercabatteries wherein a specific capacity of $54 C g^{-1}$ was obtained along with a scan rate of $5 mV s^{-1}$ and capacity retention of 60% after 2000 galvanostatic charge–discharge (GCD) cycles. Hence, a superior electrochemical energy storage capability with a specific capacity of $29 C g^{-1}$ and a specific current of $0.25 A g^{-1}$ was observed in asymmetric supercabattery devices made up of activated carbon and NiONPs as its negative and positive electrodes, respectively.

Dry husk extract of *Zea mays* L. was also used for biosynthesis of cuprite oxide nanoparticles (Cu_2ONPs) by Nwanya et al. (2019). While the un-annealed copper oxide resulted in the formation of Cu_2ONPs, all the particles were oxidized into CuO upon annealing at 600°C for 2 h. SEM images of obtained nanoparticles revealed the spherical and conical shapes of the un-annealed Cu_2ONPs that were agglomerated, while pseudo-spherical- and spherical-shaped particles were observed on annealing at 300°C and 600°C, respectively. The particle size distribution of the nanoparticles was observed by HRTEM wherein the diameter of 600°C annealed CuONPs showed a range of 30–90 nm. CuONPs were then used for the degradation of methylene blue dye and textile effluents in the presence of visible light. The degradation rate of methylene blue dye and textile effluents in the presence of only visible light was observed to be 37% and 2%, respectively, whereas in the presence of mixture of Cu_2ONPs and CuONPs, the degradation rates increased up to 99% in the case of both methylene blue dye and textile effluents. Wastewater contaminated with methylene blue showed 80%,

87%, and 91% removal of the dye when treated with Cu_2ONPs, CuONPs (annealed at 300°C), and CuONPs (annealed at 600°C), respectively. Similarly, textile effluent wastewater exhibited 83%, 86%, and 90% degradation in the presence of Cu_2ONPs, CuONPs (annealed at 300°C), and CuONPs (annealed at 600°C), respectively. Furthermore, effective antibacterial activities of Cu_2ONPs and CuONPs (annealed at 300°C) were observed against *Escherichia coli* 518,133 and *Staphylococcus aureus* 9144 strains, respectively. Cu_2ONPs also showed effective antibacterial activity against *Pseudomonas aeruginosa* and *Bacillus licheniformis*.

2.3 Pulses

Leaf extract of *Cajanus cajan* (L.) (pigeon pea) was used for facile green synthesis of AgNPs by Nagati et al. (2012). *C. cajan* leaf extract was added with 1 mM of $AgNO_3$ that was further heated at 60°C for 15 min. The color of the extract was observed to change from light yellow to reddish brown due to SPR. FTIR analysis of prepared AgNPs revealed the peaks that corresponded with aliphatic amine and aliphatic alkenes of alkaloids and terpenoids that were suggested to be present on the surface of AgNPs. SEM images further revealed the spherical morphology of the AgNPs with a size range of 5–60 nm, while SAED pattern showed crystalline nature of the particles. Antibacterial activity of AgNPs prepared from pigeon pea leaf extract was also investigated against *E. coli* and *S. aureus*. Effective antibacterial activity of AgNPs was observed to increase in a dose-dependent manner.

ZnONPs were also fabricated by Manjunath et al. (2014) using *C. cajan* wherein solution combustion method was carried out using 1:1 ratio of zinc nitrate and pigeon pea ground powder. XRD patterns indicated hexagonal zincite structure of synthesized ZnONPs having a crystallite size of 40 nm. FTIR spectra of ZnONPs showed bands at 420, 1398, and 1735 cm^{-1} that corresponded with vibrational stretching of Zn—O, C=O, and C—O bonds, respectively. SEM images of ZnONPs revealed the agglomeration of the particles that formed spongy cave-like structures having well-defined pores as observed in Fig. 2.3. TEM images, on the other hand, showed spherical shape of the ZnONPs having a size range of around 40–80 nm. Furthermore, ZnONPs were then used as photocatalysts for degradation of methylene blue dye in the presence of sunlight as well as UV light. Maximum photocatalytic activity of ZnONPs was obtained at a pH value of 10.0 and in the presence of 5 ppm of methylene blue dye. Degradation of dye was observed to be more prominent in the presence of UV light as compared to sunlight as UV radiation has equivalent band gap with that ZnONPs and thus promotes photocatalysis. The surface area of ZnONPs was found to be 79 m^2/g that was proposed to absorb light energy, resulting in electron excitation and formation of holes in the valence band, which was suggested to play a major role in dye degradation. Moreover, coumarin was used as a probe molecule to detect the production of hydroxyl radicals during dye degradation. Photoluminescence intensity was observed to increase at around 454 nm with simultaneous increase in irradiation time, indicating increased production of hydroxyl radicals on the surface of ZnONPs.

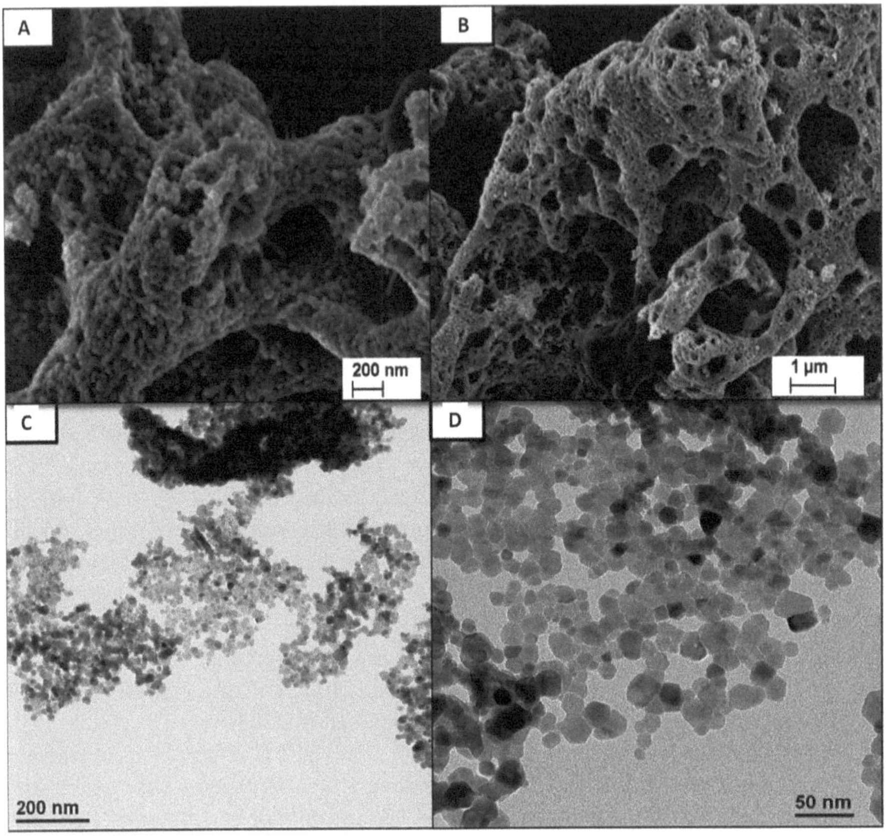

Fig. 2.3 SEM and TEM images of ZnONPs.
Reprinted with permission from Manjunath, K., Ravishankar, T.N., Kumar, D., Priyanka, K.P., Varghese, T., Naika, H.R., Nagabhushana, H., Sharma, S.C., Dupont, J., Ramakrishnappa, T., Nagaraju, G., 2014. Facile combustion synthesis of ZnO nanoparticles using *Cajanus cajan* (L.) and its multidisciplinary applications. Mater. Res. Bull. 57, 325–334. Copyright © 2014 Elsevier Ltd.

Antibacterial activity of ZnONPs was also observed against *S. aureus* and *Klebsiella aerogenes* with a minimum inhibitory concentration (MIC) value of 200 μg and 250 μg, respectively.

C. cajan extract was also used by Arif et al. (2019) for the green synthesis of titanium oxide nanoparticles (TiO_2NPs) that was used as loading materials for polyvinylidene fluoride (PVDF) ultrafiltration membranes. Aqueous extract of pigeon pea was added with 5 mM of titanium isopropoxide under optimal temperature for 7 h in order to form TiO_2NPs. The diffraction peaks obtained in the XRD pattern revealed the anatase phase of TiO_2NPs that had a crystalline size of 10.5 nm. The average particle size of TiO_2NPs was also calculated using dynamic light scattering (DLS) and was found to be in the range of 10–25 nm. Furthermore, 2D atomic force microscopy (AFM) images revealed the decrease in surface roughness in PVDF/TiO_2 composite membranes with increase in TiO_2 loading. Thus, it was implied that increase in surface smoothness would provide efficient antifouling ability to the membrane as well as reduce the penetration and adhesion of contaminants into the valleys and peaks of membrane, respectively. Moreover, permeate flux decline of composite membrane was also reduced with subsequent decrease in surface roughness. Contact angle value was also increased to 75.2° in the presence of 0.03 TiO_2NPs/g of PVDF, which suggested increased surface hydrophilicity of the membrane. Antifouling ability of TiO_2NPs incorporated membranes was confirmed using bovine serum albumin (BSA) adsorption test wherein adsorption of BSA decreased from 2.85 to 2.15 mg cm^{-2} with concomitant increase in TiO_2NP concentration on the membrane to 0.02 TiO_2NPs/g of PVDF. However, 0.03 TiO_2NPs/g of PVDF showed agglomeration of particles that resulted in increased adsorption. Water flux of the membrane was also observed to increase with enhanced hydrophilicity.

Cicer arietinum L. (chickpea) leaf extract was used for the green synthesis of gold nanoparticles (AuNPs) by Singh et al. (2013). Aqueous extract of chickpea leaves was added with 0.1 mM $HAuCl_4$, followed by incubation at room temperature for 24 h. A purple-colored solution was observed that indicated SPR of AuNPs. UV–Vis absorption spectral analysis exhibited a broadened absorption peak at 550 nm that suggested polydispersed nature of AuNPs that were formed. TEM images further revealed the spherical, pentagonal, and triangular shapes of AuNPs that had a size range of 30–80 nm.

Koli et al. (2019) also reported chickpea extract-mediated synthesis of magnetite nanoparticles (Fe_3O_4NPs). $FeSO_4{\cdot}7H_2O$ at a concentration of 0.15 M was supplemented with aqueous chickpea extract in the presence of urea and kept at 50°C for 1 h under stirring conditions. The change in color of the solution from brown to dark gray was suggested to be an indicator of Fe_3O_4NP formation. XRD patterns of the obtained Fe_3O_4NPs corresponded with pure Fe_3O_4 phase having cubic structure and crystalline nature with characteristic peaks revealing inverse spinel structure of magnetite. The presence of both urea and chickpea extract further showed a smaller crystallite size of particles of around 4.6 nm as compared to Fe_3O_4NPs prepared using either urea or chickpea extracts. Furthermore, XRD analysis also revealed the increased stability of Fe_3O_4NPs made using chickpea extracts. FTIR analysis revealed the surface coating of Fe_3O_4NPs by pectin. TEM micrographs depicted an average size of Fe_3O_4NPs to be around 13 nm when prepared using both chickpea extract and urea as evident from Fig. 2.4. Moreover, Fe_3O_4NPs showed superparamagnetic behavior having negligible coercivity and retentivity. Additionally, Fe_3O_4NPs were shown to exhibit optimum heat generation capacity near to therapeutic temperature of 43°C. Antimicrobial activity of Fe_3O_4NPs was evident against *S. aureus* and *P. aeruginosa* with zones of inhibition found to be 9 mm and 3 mm, respectively.

In another study, Verma et al. (2019) synthesized chickpea pod extract-mediated CuONPs and reported its characterization. Cupric acetate monohydrate was added with chickpea pod extract after which the reaction was carried out for 14–16 h under low-temperature conditions. UV–Vis spectroscopy results revealed the SPR of CuONPs at 270 nm, while absorption peak at 266 nm was suggested to indicate π-π^* transition of polyphenols that are present in the chickpea extract. FTIR spectra also showed the

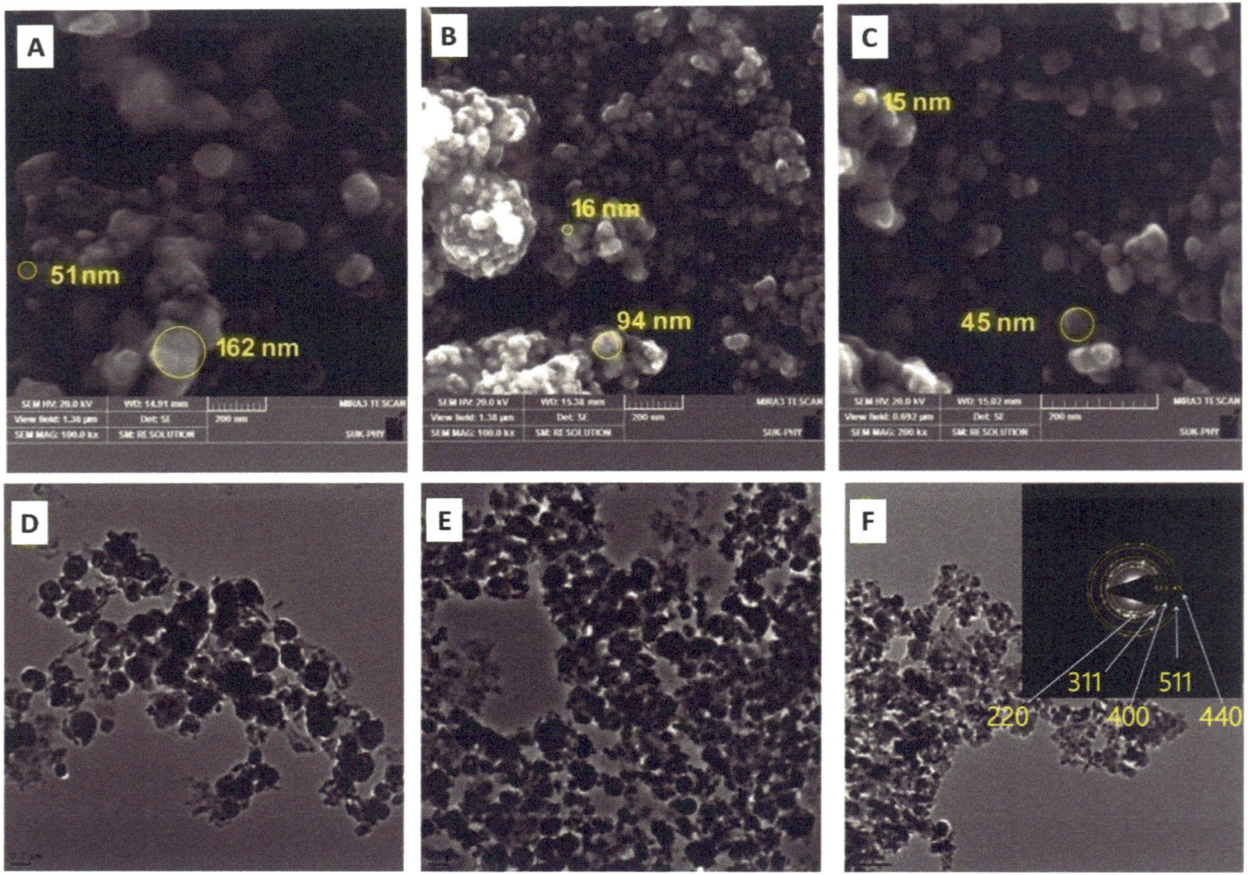

Fig. 2.4 (A–C) FESEM micrographs of as-prepared (dried powder) samples A, B, and C, respectively, (D–F) TEM micrographs of water-dispersed samples A, B, and C, respectively. Inset of (F) shows SAED pattern for sample C.
Reprinted with permission from Koli, R.R., Phadatare, M.R., Sinha, B.B., Sakate, D.M., Ghule, A.V., Ghodake, G.S., Deshpande, N.G., Fulari, V.J., 2019. Gram bean extract-mediated synthesis of Fe_3O_4 nanoparticles for tuning the magneto-structural properties that influence the hyperthermia performance. J. Taiwan Inst. Chem. Eng. 95, 357–368. Copyright © 2018 Taiwan Institute of Chemical Engineers. Published by Elsevier B.V.

presence of functional groups such as N—H, O—H, C—H, C=O, C=C, C=N, alkynes, and nitriles. XRD patterns revealed the crystalline nature of CuONPs wherein the average crystallite size of CuONPs was found to be 12.76 nm. SEM images of CuONPs showed hexagonal morphology of the particles, while energy-dispersive X-ray (EDX) spectroscopy results exhibited the presence of 68.05% and 30.95% of copper and oxygen in the nanoparticles by weight, respectively. Magnesium and silicon were also observed to be present that were proposed to act as stabilizing and capping agents for CuONPs.

2.4 Vegetables

Allium cepa (onion) peel extract was efficient in biofabrication of AuNPs as reported by Patra et al. (2016). SEM images of obtained AuNPs revealed an average size of 45.42 nm having spherical or triangular morphology. FTIR spectral analysis further showed the presence of water-soluble phenolics and cysteine derivatives that were supposed to be involved in the reduction of gold ions as well as stabilization of AuNPs. Moreover, AuNPs showed synergistic antibacterial activity against *S. aureus* and *E. coli* when combined with rifampicin resulting in 22.49 mm and 9.99 mm zones of inhibition, respectively. Likewise, synergistic anticandidal activity was observed against five pathogenic species of *Candida* when combined with amphotericin B wherein zones of inhibition ranged from 10.23 to 16.06 mm. AuNPs at a concentration of 100 μg/mL further demonstrated antioxidant ability with 14.44% DPPH scavenging activity. Similar concentration of AuNPs also displayed 30.89% nitric oxide scavenging ability. Additionally, AuNPs at a concentration of 10 and 100 μg/mL also showed 25.40% and 29.16% proteasome inhibitory activities, respectively.

Ansar et al. (2020) in their study described an eco-friendly process for AgNP synthesis using *Brassica oleracea* (cabbage) leaf extract. Highly dense AgNPs were formed that were relatively spherical in shape as seen in TEM images. The average size of the particles was 20 nm, which showed effective antibacterial activity against *Staphylococcus epidermidis* and *P. aeruginosa* showing inhibition zone values of 14.33 ± 0.57 and 12.0 ± 0.20 mm, respectively. Cytotoxicity of biogenic AgNPs was also tested against MCF-7 cells using MTT assay that demonstrated prominent effect at 100 μg/mL with an IC_{50} value of 55 μg/mL. Antioxidant ability of AgNPs was also observed wherein maximum DPPH scavenging ability of 79% was obtained at a concentration of 200 μg/mL of biogenic AgNPs, while superoxide radical and hydroxyl scavenging ability of AuNPs was recorded to be 70% and 35%–71%, respectively.

Likewise, Narayanan and Park (2015) synthesized AuNPs using leaf extract of *Brassica rapa* L. (turnip) and investigated its dye degradation ability. The average size of spherical AuNPs was found to be 31.4 ± 14.9 nm and was stable due to capping with amine groups and ortho-substituted aromatic biomolecules that were present in the turnip leaf extract. These crystalline AuNPs were then observed to efficiently reduce methylene blue dye to leucomethylene blue in the presence of sodium borohydride wherein the reaction rate constant was $0.372\,min^{-1}$ and the reaction followed a pseudo-first-order kinetics.

Coriandrum sativum L. (coriander) leaves were reported to be efficient in the synthesis of AgNPs by Ashraf et al. (2019) wherein 1 mM of $AgNO_3$ extract was added into coriander leaf extract after which the reaction was carried out at 75°C for 25 min. Visible color change of the solution from yellow to brown was observed that was indicative of SPR by AgNPs that were synthesized. The zeta potential of AgNPs was found to be −18.6 mV. FTIR analysis of AgNPs further showed peaks at 3275.64 cm^{-1} that corresponded with H-bond and O—H stretching vibrations of phenols and alcohols. Another peak at 1632.43 cm^{-1} indicated C=O and -NHCO extensions, while other peaks revealed the presence of alkanes, amines, anhydrides, ethers, and carboxylic groups. Therefore, it was suggested that metabolite terpenoids and proteins of coriander leaf extract were attached to the surface of AgNPs, which prevent the aggregation of the particles. SEM images of AgNPs displayed the spherical shape, while XRD patterns showed crystalline nature of AgNPs with a mean size of 11.9 nm. Antibacterial activity of AgNPs was also characterized against *B. subtilis*, *Pasteurella multocida*, *Enterobacter aerogenes*, and *S. aureus* wherein effective zones of inhibition of around 14, 10, 11, and 12 mm were observed in the presence of 200 μL of AgNPs.

In a similar study, Sathishkumar et al. (2016) biosynthesized AgNPs using coriander leaf extract and investigated its antiacne, antidandruff, and antibreast cancer activities. A brown color solution was obtained after incubating coriander leaf extract with 1 mM of $AgNO_3$ at room temperature, highlighting AgNP formation. UV–Vis spectrum also indicated maximum absorption at 430 nm, which confirmed SPR of AgNPs. Total flavonoid content in coriander leaf extract was observed to be 27.25 mg/g that was suggested to be primarily involved in the synthesis of AgNPs as the hydroxyl and carbonyl groups of flavonoids behave as both reducing and capping agents to reduce Ag^+ ions and prevent aggregation of particles, respectively. FESEM images displayed the spherical shape of AgNPs with an average particle size of around 37 nm. FTIR spectra exhibited the peaks at 3500 and 3400 cm^{-1} representing N—H stretching, while other peaks at 2925, 1638, and 1100 cm^{-1} denoted C—H stretching of alkyl groups, C=O and C-X vibrations, respectively. FCC nature of AgNPs was confirmed by XRD pattern as well, while the surface area of AgNPs was found to be 33.72 m^2/g. Furthermore, antibacterial activity of AgNPs against *Propionibacterium acnes* was observed wherein 11.3 ± 0.3, 13.7 ± 0.4, and 17.5 ± 0.2 mm inhibition zones were seen in the presence of 4, 8, and 12 μg of biogenic AgNPs, respectively. MIC value of AgNPs against *P. acnes* was thus found to be 3.1 μg/mL. Similarly, antifungal activity against *Malassezia furfur* also showed 25.7 ± 0.3, 29.2 ± 0.5, and 32.4 ± 0.7 mm inhibition zones in the presence of 20, 40, and 60 μg of AgNPs, respectively, with an MIC value of 25 μg/mL. Finally, antibreast cancer activity of green-synthesized AgNPs was also investigated against MCF-7 (human breast cancer) cells. The IC_{50} value of AgNPs was found to be 30.5 μg/mL, while total cell growth inhibition was observed in the presence of 100 μg/mL of AgNPs that was speculated to be a result of mitochondrial respiratory chain disruption in the cells through the formation of reactive oxygen species (ROS) and adenosine triphosphate (ATP) synthesis interruption.

Jiao et al. (2021) also reported utilization of coriander leaf extract for the synthesis of AuNPs and investigated its antioxidant, analgesic activity. Tetrachloroauric (III) acid trihydrate having a concentration of 10 mM was added with coriander leaf extract, which resulted in color change from light brownish to purple, indicating AuNP development. UV–Vis spectra revealed sharp peak at 540–550 nm range, while XRD pattern confirmed FCC structure of synthesized AuNPs. The average size of particles was 32.96 ± 5.25 nm with spherical, triangular, and hexagonal shapes as observed in TEM images in Fig. 2.5. The zeta potential of AuNPs was further found to be −21.4 mV, indicating appropriate stability of the particles. Moreover, antioxidant ability of eco-friendly, biogenic nanoparticles was revealed wherein 85.73% and 96.83% of DPPH and ABTS radical scavenging assay were, respectively, observed in the presence of 800 μM of AuNPs. Analgesic activity of AuNPs was also evaluated that revealed decrease in paw licking by Swiss albino mice in the presence of 200 mg/kg of AuNPs, suggesting potential inhibition of pain mediator's activity. Additionally, no cytotoxic effects of AuNPs were seen after treatment of MCF-7 and HeLa cells with the same, thus displaying its biocompatibility for future application in molecular imaging and drug delivery.

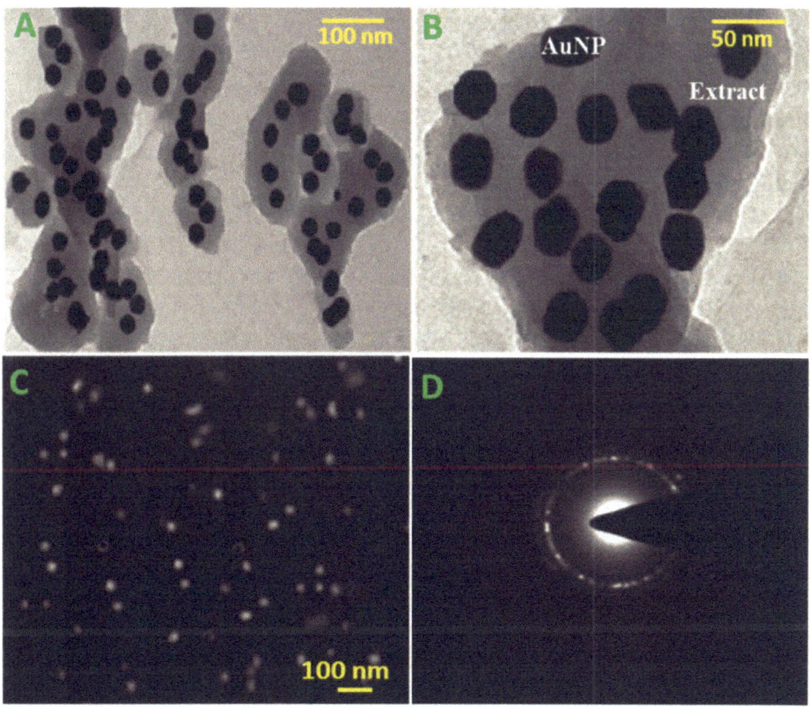

Fig. 2.5 HRTEM images (A, B), FESEM image (C), and SAED pattern (D) of green-synthesized AuNPs.

Reprinted with permission from Jiao, Y., Wang, X., Chen, J.H., 2021. Biofabrication of AuNPs using *Coriandrum sativum* leaf extract and their antioxidant, analgesic activity. Sci. Total Environ. 767, 144,914. Copyright © 2021 Elsevier B.V.

Pisum sativum (pea) peel extract has also been used as reducing and capping agents for the biosynthesis of MNPs by Prasad et al. (2017). Agglomerated MNPs that were obtained had an average size range of 20–30 nm as seen in TEM images with a high surface area of around 17.6 m^2/g. Additionally, MNPs were shown to exhibit ferromagnetic nature with a saturation magnetization value of 64.2 emu/g. These eco-friendly MNPs were then investigated for its catalytic properties in which efficient degradation of methyl orange dye was observed.

Raphanus sativus (radish) root extract was also considered as reducing agent by Liu et al. (2020) for the synthesis of ZnONPs. HRTEM images of synthesized ZnONPs showed a mean particle size of around 30 nm that were spherical in shape and had a negative zeta potential value of −33 mV. In addition, ZnONPs displayed the exceptional antibacterial activity against pathogenic strains of *P. aeruginosa*, *E. coli*, *S. aureus*, and *Acinetobacter baumannii* that were known to cause diabetic foot ulcers.

Likewise, other vegetables have also been reported to efficiently synthesize different nanoparticles that have a wide range of applications. In one such study, Das and Bhuyan (2019) developed AuNPs and AgNPs using fruit peel aqueous extract of *Solanum melongena* L. (brinjal) under microwave irradiation conditions. The average diameter, polydispersity index (PDI), and zeta potential of AuNPs were 29 nm, 0.214, and −18.5 mV, while the respective values of AgNPs were 92.4 nm, 0.159, and −20.4 mV. Furthermore, AgNPs demonstrated significant antibacterial activity against *Pseudomonas fluorescens* and *Bacillus amyloliquefaciens*.

Biogenic AgNPs were prepared by Roy et al. (2015) using *Solanum tuberosum* (potato) infusion. Potato infusion was added into 20 mM of $AgNO_3$ solution, followed by incubation at room temperature for 24 h. The color of the solution was observed to alter from colorless to dark yellow, which indicated SPR of AgNPs. Furthermore, UV–Vis spectroscopy exhibited the maximum absorbance at 430 nm, while XRD pattern highlighted the FCC structure and crystalline nature of synthesized AgNPs. TEM micrographs further displayed the spherical nature of AgNPs with an average size of 10–12 nm that was confirmed by DLS results. FTIR spectra of AgNPs revealed a less intense shift in peaks of C=O, C—H, and C—N vibrations of amide and amine groups that were speculated to be due to the capping of AgNPs by these groups. Later on, biogenic AgNPs were investigated for its photocatalytic activity in the degradation of methyl orange dye in the presence of sunlight. In the presence of AgNPs, the characteristic peak of methyl orange at 460 nm was observed to decrease significantly within 8 h of incubation.

In a similar study, Sharma et al. (2019) devised an experiment to synthesize magnetic nanoparticles (MNPs) using potato extract and evaluated its potential in dye degradation. Ferrous sulfate heptahydrate salt was mixed with starch-rich aqueous potato extract and kept at 80°C. The color of the mixture was observed to change from yellow to deep black, which was indicative of MNP formation. The synthesized MNPs were found to be superparamagnetic in nature having a saturation magnetization value of 74.4 emu/g. FTIR analysis of the MNPs also revealed a strong band that corresponded to Fe—O bond stretching along with other peaks that highlighted the coating of starch over MNPs. XRD patterns displayed the crystalline nature of MNPs, whereas SEM and TEM images highlighted the spherical morphology of the particles. Additionally,

MNPs were investigated for its application as a peroxidase mimetic catalyst for the degradation of rhodamine B dye in the presence of H_2O_2 as oxidant. The absorption maximum value of rhodamine B dye at 554 nm was observed to gradually decrease with increasing MNPs contact time as well as ultrasonic radiation, and within 30 min, the intensity of absorption was negligible.

AgNPs were also prepared by Almadiy and Nenaah (2018) using potato extracts containing steroidal alkaloids and checked for their antifungal activity. Leaves of potato plant were taken for extraction, purification, and identification of total potato alkaloids (TPAs). Yield of TPAs was found to be 70.6%, out of which 35.0% was observed to be α-chaconine, while 28.2% was α-solanine. *Z*-diameter of AgNPs was found to range within 39.5–80.3 nm. Moreover, a pronounced antifungal activity of AgNPs made from TPAs was observed against *Alternaria alternate, Rhizoctonia solani, Botrytis cinerea*, and *Fusarium oxysporum f.* sp. *lycopersici* with an effective concentration (EC_{50}) value ranging between 10.9 and 16.1 μg/mL. Furthermore, germination bioassay of *T. aestivum* L. and *Raphanus sativus* L. seeds was carried out in the presence of AgNPs. Germination percentage and radicle and shoot growth in both plants were observed to be intact, highlighting the absence of phytotoxic properties of biogenic AgNPs.

Similarly, *Spinacia oleracea* (spinach) leaf extract was utilized for the fabrication of ZnONPs by Lakshmi et al. (2017). The average particle diameter of ZnONPs was found to be 40.59 nm. Treatment of *Vigna radiata* seeds with 125 ppm of biogenic ZnONPs resulted in early seedling growth and positively affected growth characteristics of *V. radiata* plants.

Vijaya et al. (2017) used dried root extract of *Zingiber officinale* (ginger) for the fabrication of AgNPs and studied its antibacterial properties. Microwave irradiation method was carried out using 5 mM of $AgNO_3$ solution and dried ginger root extract to synthesize AgNPs. A yellow-colored solution was obtained that was indicative of AgNPs formation, which was also confirmed in UV–Vis spectrum that revealed a characteristic intense peak of AgNPs at 425 nm. FTIR peaks at 1112, 931, 859, 761, and 601 cm^{-1} corresponded with heterocyclic compounds, such as alkanoids, flavonoids, and alkaloids present in the root extract that was speculated to act as capping agent for AgNPs. HRTEM results revealed the diameter of spherical AgNPs to be 10 nm and also confirmed the role of ginger root extract as reducing, capping, and stabilizing agents for AgNPs. Efficient antibacterial activity of AgNPs was also observed against Gram-positive organisms such as *Bacillus cereus*, *S. aureus*, *Micrococcus luteus*, *B. subtilis*, and *Enterococcus* sp. with zones of inhibition found to be 18, 16, 16, 20, and 18 mm, respectively. Likewise, inhibition zones of around 21, 17, 22, and 17 mm were observed in the presence of AgNPs against Gram-negative organisms such as *P. aeruginosa*, *Salmonella typhi*, *E. coli*, and *K. pneumonia*, respectively.

Ginger extract was also reported to be utilized by Chitra et al. (2015) for AuNP synthesis using 0.5 mM of auric chloride along with poloxamer 188 that was used as stabilizing agent. Maximum absorption spectrum of prepared nanoparticles was observed at 518–533 nm, thus confirming SPR of AuNPs. FTIR spectral analysis of AuNPs exhibited a slight shift in vibrational frequencies compared to ginger extract along with additional peaks at 900–1500 cm^{-1}, indicating the presence of poloxamer

188 as well. Furthermore, DLS results exhibited the average particle size of AuNPs to be around 13 nm and 8 nm that were prepared using 0.5% and 1.0% *w/v* of poloxamer 188, respectively. XRD patterns of AuNPs exhibited the crystallite nature of the particles, which corresponded to FCC structure of pure gold. TEM images also displayed the spherical shape of AuNPs having similar size as obtained in DLS. Antibacterial activity of prepared AuNPs was then checked against common human pathogens, namely *S. aureus*, *E. coli*, and *K. pneumonia*, wherein 24%, 20%, and 27% inhibition zones were observed in the presence of 75 μL of AuNPs prepared using ginger extract and poloxamer 188, respectively.

In another similar study, Yang et al. (2017) prepared biogenic AgNPs using ginger extract and evaluated its antibacterial activity against aquatic pathogens. $AgNO_3$ solution having a concentration of 1 mM was mixed with ginger extract and heated at 85°C after which the color of the solution turned golden yellow. This color change was attributed to SPR of AgNPs, which displayed an intense peak at 423 nm. TEM micrographs also revealed the spherical shape of the AgNPs that had an average size of 10 nm and were monodispersed in nature. Moreover, FTIR spectra of AgNPs were found to be similar to those of the ginger extract that indicated the presence of bioactive molecules on the surface of AgNPs. In addition, significant inhibition zones of around 15.8 ± 0.05, 14.1 ± 0.10, 13.5 ± 0.07, 14.9 ± 0.09, 11.1 ± 0.02, and 13.6 ± 0.03 mm were observed against *Vibrio anguillarum*, *Vibrio alginolyticus*, *Aeromonas punctata*, *Vibrio parahaemolyticus*, *Vibrio splendidus*, and *Vibrio harveyi* in the presence of 20 μL of biosynthesized AgNPs, respectively. Furthermore, the MIC value of AgNPs against *V. anguillarum* was found to be 12.9 μg/mL. Hence, these highly stable and dispersed AgNPs were proposed to act as effective antimicrobial agents against such aquatic pathogens.

2.5 Spices

Apart from the crops mentioned above, species have also been reported to effectively synthesize nanoparticles. For instance, Dinparvar et al. (2020) fabricated AgNPs using *Cuminum cyminum* L. (cumin) seed extract and reported its anticancer activity as well. SEM images showed the spherical shape of biogenic AgNPs having an average size of around 100 nm. The IC_{50} value of AgNPs was 1.25 μg/mL against J774 macrophage cells. Furthermore, anticancer activity of eco-friendly AgNPs was observed against MCF-7 and AU565 (human breast adenocarcinoma) cells wherein 87.5% and 96% of growth inhibition were observed, respectively, in the presence of 12.5 μg/mL of biogenic AgNPs.

Maghimaa and Alharbi (2020) reported the green synthesis of AgNPs using *Curcuma longa* L. (turmeric) leaf extract and evaluated for its antimicrobial applications. The reaction solution composed of 0.1 M of $AgNO_3$ solution and turmeric leaf extract altered its color from brown to intense black that highlighted AgNP synthesis. UV–Vis absorption spectra of prepared AgNPs displayed the highest peak at 350 nm, and FTIR spectra revealed the presence of carboxylic, hydroxyl, alkenes, and carbonyl groups. HRTEM images then displayed the spherical morphology of the AgNPs with size range of 15–40 nm. These biogenic AgNPs were then embedded onto cotton fabrics whose SEM images showed semi granulated patterns on the surface of the fabric. EDX spectra also demonstrated the presence of AgNPs on the surface of cotton fabric as compared to uncoated fabric. AgNPs also displayed the effective antimicrobial activity of synthesized AgNPs against *P. aeruginosa* and *S. aureus* with 19 mm and 18 mm zones of inhibition, respectively. AgNP-loaded cotton fabrics also demonstrated similar antimicrobial activities against these pathogens. Cytotoxicity of formulated AgNPs was also investigated against fibroblast (L929) cells via MTT assay. The presence of 35 μg/mL of AgNPs displayed very mild cytotoxicity against L929 cells, highlighting its nontoxic nature. Wound healing potential of AgNPs was also observed, wherein 15 and 20 μg of AgNPs demonstrated appreciable cell migration toward the wounded area when incubated at 37°C for 24 h.

Aqueous extract of turmeric powder was also reported by Jayarambabu et al. (2020) using copper nanoparticles (CuNPs). Microwave method was carried out using 0.1 M of copper acetate dihydrate and turmeric powder extract, which resulted in color change from yellow to brick brown, indicating the formation of CuNPs. FESEM images exhibited the size range of CuNPs to be around 5–20 nm. XRD patterns of prepared CuNPs confirmed the purity of prepared nanoparticles with an average crystallite size of 16.77 nm. TEM image of spherical CuNPs showed size range of particles to be 5–20 nm. Antibacterial activity of CuNPs was also tested against *B. subtilis* and *E. coli* wherein 23 ± 0.89 and 22 ± 0.76 mm zones of inhibition were observed in the presence of 250 μL of CuNPs.

Elettaria cardamomum or cardamom (EC) seed extract was also reported by Vinotha et al. (2020) to be useful in the preparation of ZnONPs. Homogenous and spherical ZnONPs were fabricated having an average size of 18.72 nm, which showed SPR at 373 nm. Moreover, efficient antibiofilm activity of ZnONPs was observed against *Enterococcus faecalis*, *S. aureus*, *P. aeruginosa*, and *Proteus vulgaris*. Mosquito larvicidal activity of ZnONPs was also demonstrated against *Aedes aegypti* and *Culex tritaeniorhynchus* wherein LC_{50}, LC_{90} values were found to be 13.27, 25.36 and 15.09, 29.70 μg/mL, respectively, as seen in Fig. 2.6.

In another recent study, aqueous resin extract of *Ferula asafoetida* (asafoetida) was used by Saranyadevi et al. (2021) to mediate AgNP synthesis. SEM images displayed the spherical, rod-shaped, and cuboid morphology of the fabricated AgNPs that had an average size of 39 nm. These biogenic AgNPs were then investigated for its larvicidal activity against *Aedes aegypti* larvae. The lethal concentration (LC_{50} and LC_{90}) of AgNPs was found to be 153.23 and 303.13 μg/mL, respectively.

Piper nigrum (pepper) seed extract was demonstrated to synthesize tin oxide nanoparticles (SnO_2NPs) by Tammina et al. (2017). Fabrication of SnO_2NPs was carried out using 0.1 M $SnCl_2.2H_2O$, which gave a brown-colored precipitate after 30 min of the addition of pepper seed extract that was calcined at 300°C, 500°C, and 900°C. XRD patterns demonstrated the tetragonal rutile structure of SnO_2NPs that were pure and crystalline in nature. The mean diameter of SnO_2NPs was found to be 8.85 ± 3.5 nm. Maximum UV–Vis absorption spectra of SnO_2NPs calcined at 300°C, 500°C, and 900°C were found to be 270, 289, and 293 nm, respectively. Moreover, luminescent properties of SnO_2NPs were studied wherein 300°C calcined

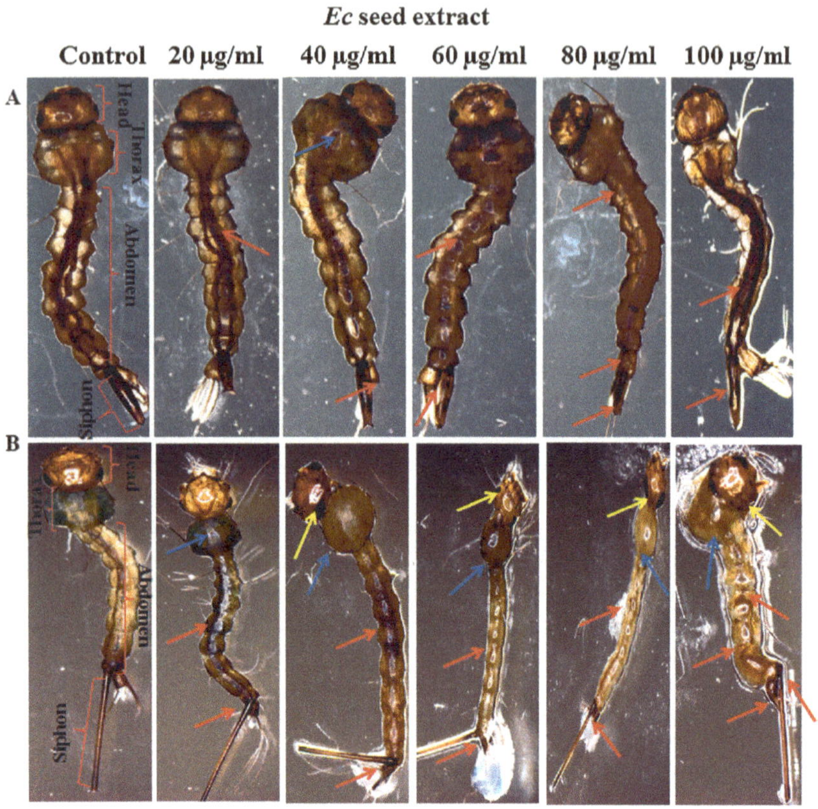

Fig. 2.6 Morphological changes of (A) *Ae. aegypti* and (B) *Cx. tritaeniorhynchus* mosquito larvae after treatment of *E. cardamomum* seed extract. Arrows indicate the morphological aberrations in head (yellow arrow), thorax (blue arrow), and abdomen (red arrow) regions. Reprinted with permission from Vinotha, V., Yazhiniprabha, M., Raj, D.S., Mahboob, S., Al-Ghanim, K.A., Al-Misned, F., Govindarajan, M., Vaseeharan, B., 2020. Biogenic synthesis of aromatic cardamom-wrapped zinc oxide nanoparticles and their potential antibacterial and mosquito larvicidal activity: An effective eco-friendly approach. J. Environ. Chem. Eng. 8, 104,466. Copyright © 2020 Elsevier Ltd.

SnO_2NPs showed significant emission at 373 nm when excitation wavelength was 280 nm. Likewise, the photoluminescence of 500°C and 900°C calcined SnO_2NPs samples showed prominent emission bands at 399 and 422 nm, respectively. Additionally, DLS results showed PDI values of 300°C, 500°C, and 900°C calcined SnO_2NPs to be 0.279, 0.313, and 0.498, respectively, highlighting monodispersed nature of all three samples. FTIR spectra of all three SnO_2NPs exhibited bands at 1732–1737 cm^{-1}, 1365 cm^{-1}, and 1213–1217 cm^{-1} that corresponded to C═O stretching vibrations of ketone, covalent ester or ether linkage, and C—N planar vibrations of aliphatic amines, respectively. Hence, these results were suggested to demonstrate the capping of SnO_2NPs by bioactive molecules that are present in pepper seed extract. In addition, the cytotoxic activity of SnO_2NPs against human colorectal carcinoma cells

(HCT116) and adenocarcinomic human alveolar basal epithelial cells (A549) was investigated. Cell viability was observed to decrease with concomitant increase in SnO_2NPs concentration. The IC_{50} values of SnO_2NPs calcined at 300°C, 500°C, and 900°C were 165, 174, and 208 μg/L against HCT116 cells and 135, 157, and 187 μg/L for A549 cells, respectively. Furthermore, dichlorodihydrofluorescein (DCF) method was carried out for the evaluation of ROS generation in the presence of SnO_2NPs. Intensity of DCF at 523 nm was clearly observed to increase with subsequent increase in SnO_2NPs concentration. Thus, it was implied that SnO_2NPs induce ROS generation in cancer cell lines in order to confer cytotoxicity.

Paulkumar et al. (2014) also used leaf and stem extracts of *P. nigrum* for green synthesis of AgNPs and investigated its antibacterial activity. The reaction mixture containing 1 mM of $AgNO_3$ and pepper extract showed a color change from pale yellow to brown, which is indicative of AgNP formation. SEM images of AgNPs revealed its predominant spherical shape along with few hexagonal-shaped particles, while TEM images showed a size range of 9–30 nm. Antibacterial activity of biosynthesized AgNPs was evaluated against two known phytopathogens, namely *Citrobacter freundii* and *Erwinia cacticida*. The presence of 50 μL of AgNPs that were synthesized only by leaf extract of pepper in combination with chloramphenicol exhibited 15.853 ± 0.153 and 16.295 ± 0.203 mm zones of inhibition against *C. freundii* and *E. cacticida*, respectively. Likewise, AgNPs synthesized only using stem extract when combined with chloramphenicol displayed 12.833 ± 0.441 and 14.533 ± 0.291 mm zones of inhibition against *C. freundii* and *E. cacticida*, respectively.

2.6 Other crops

Apart from cereals, pulses, vegetables, and spices, several fruits and cash crops have also been shown to assist in the formation of nanoparticles. For example, nickel nanoparticles (NiNPs) were synthesized using *Camellia sinensis* (tea) leaf extract by Bibi et al. (2017). Uniformly distributed spherical-shaped NiNPs were produced having a size range of 43.87–48.76 nm and a magnetization saturation value of 0.073 emu/g. Moreover, 99.5% of crystal violet dye was observed to be degraded, indicating its efficient photocatalytic activity.

Aqueous extract of *Coffea arabica* (coffee) leaf was also used as a reducing agent by Abel et al. (2021) for fabricating crystalline, cubic-shaped ZnONPs. SEM images further demonstrated the average diameter of synthesized ZnONPs to be around 40 nm.

Similarly, MNPs were prepared using inedible *Cynometra ramiflora* fruit extract by Bishnoi et al. (2018). SEM images displayed the spherical shape and size of 58.5 nm of obtained MNPs, while the surface area was 107.97 m^2/g. Photocatalytic ability of MNPs was also demonstrated wherein efficient reusability and stability in methylene blue degradation were observed in the presence of sunlight.

Bharadwaj et al. (2021) recently reported one such study wherein *Diospyros malabarica* fruit extract was used to synthesize AgNPs. TEM images of spherical

AgNPs that were obtained showed an average size of 17.4 nm that were polydispersed in nature. Efficient antibacterial activity of AgNPs was also observed against pathogenic strains of *S. aureus* and *E. coli* with an average zone of inhibition of 8.4 ± 0.3 mm and 12.1 ± 0.5 mm, respectively, in the presence of 500 μg/mL of AgNPs. Furthermore, anticancer activity of AgNPs was investigated against U87-MG (human primary glioblastoma) cells using MTT assay, which provided an IC_{50} value of 58.63 ± 5.74 μg/mL. Additionally, catalytic reduction of toxic nitrophenols was also mediated by AgNPs to produce less toxic aminophenols.

Likewise, in another study, Demirezen et al. (2019) synthesized iron oxide nanoparticles (Fe_2O_3NPs) using dried fruit extract of *Ficus carica* (common fig). Fe_2O_3NPs that were obtained had an average diameter of 9 ± 4 nm that were spherical in shape as seen in TEM images. Elemental analysis further revealed 71.35% oxygen, 8.46% iron, 19.19% chlorine, and 1.01% potassium, while the zeta potential was found to be 20.7 mV.

Gossypium hirsutum (cotton) leaf extract was utilized by Kanipandian and Thirumurugan (2014) for the biosynthesis of AgNPs. Cotton leaf extract was considered as phytoreducer and capping agent that produced spherical AgNPs having a size of 30 nm, as observed in FESEM and HRTEM images. XRD patterns also demonstrated FCC and crystalline nature of the obtained AgNPs. Antimicrobial activity of AgNPs was also observed against pathogenic strains of *E. coli, P. aeruginosa, Rhodococcus rhodochrous, S. aureus*, and *C. albicans*. Furthermore, cytotoxicity of AgNPs was seen against A549 and HBL100 (normal) cells.

AgNPs were reported to be synthesized by Kumar et al. (2017) using *Rubus illecebrosus* (Andean blackberry) fruit extract. Crystalline and spherical-shaped AgNPs were obtained with a size range of 12–50 nm as seen in Fig. 2.7. In addition, more than 78% antioxidant ability against DPPH was observed in the presence of 0.1 mM of biogenic AgNPs.

Juice of *Saccharum officinarum* (sugarcane) was also shown to act as a stabilizing agent and mediate CuONP synthesis by Mary et al. (2019). TEM micrographs of prepared CuONPs showed the presence of square-, rectangular-, cubic-, cylindrical-, and spherical-shaped particles with an average size of 84.40 nm, while XRD analysis revealed the monoclinic crystalline structure of CuONPs. Furthermore, the bactericidal activity of CuONPs against *E. coli*, *P. aeruginosa*, *S. aureus*, and *B. subtilis* wherein considerable zones of inhibition were observed is evident in Fig. 2.8.

Similarly, another cash crop, namely *Theobroma cacao* (cocoa bean) extract, was reported by Azeez et al. (2017) to be applicable in AgNP formation. Cocoa bean extract demonstrated the fabrication of polydispersed and spherical AgNPs with a size range of 8.96–54.22 nm and maximum absorption at 438.5 nm. FTIR spectral analysis further indicated the role of phenolics and proteins of cocoa bean extract in the synthesis and stabilization of AgNPs. EDX and SAED patterns also depicted crystalline face-centered cubic phase of AgNPs. Antibacterial activity of biogenic AgNPs was also observed against multidrug-resistant strains of *E. coli*, *S. aureus*, and *K. pneumoniae* producing zones of inhibition ranging from 10 mm to 14 mm. These activities of AgNPs were further enhanced by 42.9%–100% when combined with

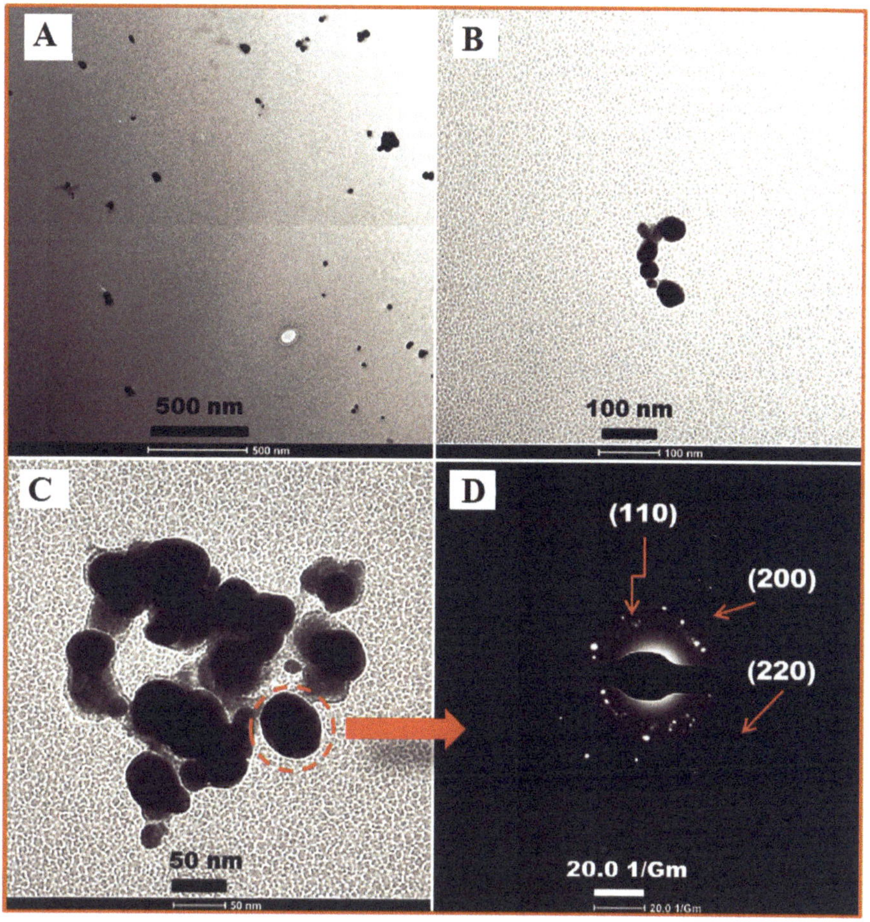

Fig. 2.7 (A–C) TEM and (D) SAED images of AgNPs.
Reprinted with permission from Kumar, B., Smita, K., Cumbal, L., Debut, A., 2017. Green synthesis of silver nanoparticles using Andean blackberry fruit extract. Saudi J. Biol. Sci. 24(1), 45–50. Copyright © 2015 The Authors. Production and hosting by Elsevier B.V. on behalf of King Saud University.

ampicillin, cefuroxime, cefixime, and erythromycin. Moreover, efficient larvicidal activity of AgNPs against *Anopheles gambiae* mosquito larvae was obtained with an LC_{50} value of 44.37 μg/mL.

Khani et al. (2018) also green-synthesized CuNPs using *Ziziphus spina-christi* (L.) Willd. (Christ's thorn jujube) fruit extract and reported its antibacterial assay and triphenylmethane dye degradation. FESEM images of biosynthesized CuNPs exhibited the spherical, well-dispersed, and homogenous particles having an average size range of 5–20 nm. CuNPs at a concentration of 80 mg further showed the efficient removal of crystal violet from aqueous solutions with 95% of dye removed within 7.5 min and

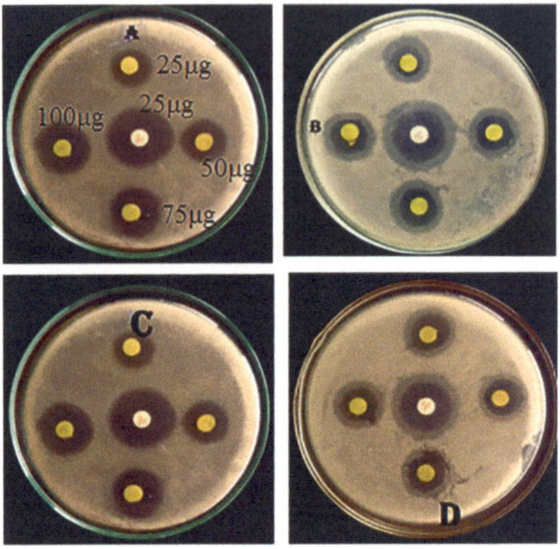

Fig. 2.8 (A–D) Antibacterial activity of CuONPs: Zone of inhibition of (A) *Escherichia coli*, (B) *Pseudomonas aeruginosa*, (C) *Staphylococcus aureus*, and (D) *Bacillus subtilis*.
Reprinted with permission from Mary, A.A., Ansari, A.T., Subramanian, R., 2019. Sugarcane juice mediated synthesis of copper oxide nanoparticles, characterization and their antibacterial activity. J. King Saud Univ. Sci. 31(4), 1103–1114. Copyright © 2019 Production and hosting by Elsevier B.V. on behalf of King Saud University.

exhibiting a high adsorption capacity of 37.5 mg/g. Additionally, the significant antibacterial activity of CuNPs was also seen against *E. coli* and *S. aureus*.

2.7 Conclusions and future perspectives

Crops are not only nutritionally rich but also storehouse of diverse bioactive principles that can have promising therapeutic potential like anticancer, antidiabetic, antimicrobial, and antioxidant activities. Hereby, more recently, the extracts from various parts of crop plants are extensively explored for the synthesis of several nanoparticles with broad-spectrum applications. It is important to note that the physicochemical and biomedical properties of nanoparticles are solely shape- and size-dependent (Ghosh, 2018). However, the major drawback of the biogenic synthesis of nanoparticles using crop plants is the uncontrolled size and shape. This anisotropy can be fine-tuned, and careful optimization of the reaction parameters like duration, temperature, pH, metal salt concentration, and plant extract concentration can result in the fabrication of monodispersed nanoparticles with more uniform size and shape (Ghosh et al., 2015a,b,c).

Further, biogenic nanoparticles can be functionalized with bioactive principles from medicinal plants for effective delivery to the site of the disease (Ghosh et al., 2015d; Kitture et al., 2015). This strategy for developing multifunctionalized nanoparticles using a biogenic carrier that is more biocompatible will be advantageous over the available chemically synthesized nanocarriers. Anticancer drugs and antibiotics can be functionalized on the crop-generated nanoparticles for site-specific targeting.

Also, the crop-synthesized nanoparticles can be encapsulated within polymeric beads for water treatment processes such as removal of hazardous dyes, heavy metal, and pesticides (Robkhob et al., 2020; Karmakar et al., 2020; Adersh et al., 2015). Such encapsulation or entrapment strategies can enhance the reusability, recyclability, and regeneration of the nanocomposite for effective wastewater treatment for a longer duration with minimal decrease in efficiency. Although biogenic nanoparticles synthesized using crops can be promising materials for so many biomedical and environmental applications, its toxicity, retention in the environment, and fate should be thoroughly studied and understood before bringing them to practice.

References

Abel, S., Tesfaye, J.L., Shanmugam, R., Dwarampudi, L.P., Lamessa, G., Nagaprasad, N., Benti, M., Krishnaraj, R., 2021. Green synthesis and characterizations of zinc oxide (ZnO) nanoparticles using aqueous leaf extracts of coffee (*Coffea arabica*) and its application in environmental toxicity reduction. J. Nanomater. 2021, 3413350.

Adersh, A., Kulkarni, A.R., Ghosh, S., More, P., Chopade, B.A., Gandhi, M.N., 2015. Surface defect rich ZnO quantum dots as antioxidant inhibiting α-amylase and α-glucosidase: a potential anti-diabetic nanomedicine. J. Mater. Chem. B 3, 4597–4606.

Akhayere, E., Vaseashta, A., Kavaz, D., 2020. Novel magnetic nano silica synthesis using barley husk waste for removing petroleum from polluted water for environmental sustainability. Sustainability 12, 10646.

Almadiy, A.A., Nenaah, G.E., 2018. Ecofriendly synthesis of silver nanoparticles using potato steroidal alkaloids and their activity against phytopathogenic fungi. Braz. Arch. Biol. Technol. 61. https://doi.org/10.1590/1678-4324-2018180013.

Ansar, S., Tabassum, H., Aladwan, N.S., Ali, M.N., Almaarik, B., AlMahrouqi, S., Abudawood, M., Banu, N., Alsubki, R., 2020. Eco friendly silver nanoparticles synthesis by *Brassica oleracea* and its antibacterial, anticancer and antioxidant properties. Sci. Rep. 10, 1–12.

Arif, Z., Sethy, N.K., Kumari, L., Mishra, P.K., Verma, B., 2019. Green synthesis of TiO_2 nanoparticles using *Cajanus cajan* extract and their use in controlling the fouling of ultrafiltration PVDF membranes. Korean J. Chem. Eng. 36 (7), 1148–1156.

Ashraf, A., Zafar, S., Zahid, K., Shah, M.S., Al-Ghanim, K.A., Al-Misned, F., Mahboob, S., 2019. Synthesis, characterization, and antibacterial potential of silver nanoparticles synthesized from *Coriandrum sativum* L. J. Infect. Public Health 12 (2), 275–281.

Athinarayanan, J., Periasamy, V.S., Alhazmi, M., Alatiah, K.A., Alshatwi, A.A., 2014. Synthesis of biogenic silica nanoparticles from rice husks for biomedical applications. Ceram. Int. 41 (1), 275–281.

Azeez, M.A., Lateef, A., Asafa, T.B., Yekeen, T.A., Akinboro, A., Oladipo, I.C., Gueguim-Kana, E.B., Beukes, L.S., 2017. Biomedical applications of cocoa bean extract-mediated silver nanoparticles as antimicrobial, larvicidal and anticoagulant agents. J. Clust. Sci. 28, 149–164.

Bhagwat, T.R., Joshi, K.A., Parihar, V.S., Asok, A., Bellare, J., Ghosh, S., 2018. Biogenic copper nanoparticles from medicinal plants as novel antidiabetic nanomedicine. World J. Pharm. Res. 7 (4), 183–196.

Bharadwaj, K.K., Rabha, B., Pati, S., Choudhury, B.K., Sarkar, T., Gogoi, S.K., Kakati, N., Baishya, D., Kari, Z.A., Edinur, H.A., 2021. Green synthesis of silver nanoparticles using *Diospyros malabarica* fruit extract and assessments of their antimicrobial, anticancer and catalytic reduction of 4-nitrophenol (4-NP). Nanomaterials 11, 1999.

Bibi, I., Kamal, S., Ahmed, A., Iqbal, M., Nouren, S., Jilani, K., Nazar, N., Amir, M., Abbas, A., Ata, S., Majid, F., 2017. Nickel nanoparticle synthesis using *Camellia sinensis* as reducing and capping agent: growth mechanism and photo-catalytic activity evaluation. Int. J. Biol. Macromol. 103, 783–790.

Bishnoi, S., Kumar, A., Selvaraj, R., 2018. Facile synthesis of magnetic iron oxide nanoparticles using inedible *Cynometra ramiflora* fruit extract waste and their photocatalytic degradation of methylene blue dye. Mater. Res. Bull. 97, 121–127.

Bloch, K., Pardesi, K., Satriano, C., Ghosh, S., 2021. Bacteriogenic platinum nanoparticles for application in nanomedicine. Front. Chem. 9, 624344.

Bose, S., Ganayee, M.A., Mondal, B., Baidya, A., Chennu, S., Mohanty, J.S., Pradeep, T., 2018. Synthesis of silicon nanoparticles from rice husk and their use as sustainable fluorophores for white light emission. ACS Sustain. Chem. Eng. 6, 6203–6210.

Buazar, F., Sweidi, S., Badri, M., Kroushawi, F., 2019. Biofabrication of highly pure copper oxide nanoparticles using wheat seed extract and their catalytic activity: a mechanistic approach. Green Process. Synth. 8, 691–702.

Chitra, K., Reena, K., Manikandan, A., Antony, S.A., 2015. Antibacterial studies and effect of poloxamer on gold nanoparticles by *Zingiber officinale* extracted green synthesis. J. Nanosci. Nanotechnol. 15, 4984–4991.

Das, R.K., Bhuyan, D., 2019. Microwave-mediated green synthesis of gold and silver nanoparticles from fruit peel aqueous extract of *Solanum melongena* L. and study of antimicrobial property of silver nanoparticles. Nanotechnol. Environ. Eng. 4, 5.

Demirezen, D.A., Yıldız, Y.Ş., Yılmaz, Ş., Yılmaz, D.D., 2019. Green synthesis and characterization of iron oxide nanoparticles using *Ficus carica* (common fig) dried fruit extract. J. Biosci. Bioeng. 127 (2), 241–245.

Dinparvar, S., Bagirova, M., Allahverdiyev, A.M., Abamor, E.S., Safarov, T., Aydogdu, M., Aktas, D., 2020. A nanotechnology-based new approach in the treatment of breast cancer: biosynthesized silver nanoparticles using *Cuminum cyminum* L. seed extract. J. Photochem. Photobiol. B Biol. 208, 111902.

Duan, H., Wang, D., Li, Y., 2015. Green chemistry for nanoparticle synthesis. Chem. Soc. Rev. 44, 5778–5792.

Ghosh, S., 2018. Copper and palladium nanostructures: a bacteriogenic approach. Appl. Microbiol. Biotechnol. 101 (18), 7693–7701.

Ghosh, S., Turner, R.J., 2021. Editorial: Nanomicrobiology: emerging trends in microbial synthesis of nanomaterials and their applications. Front. Microbiol. 12, 751693.

Ghosh, S., Webster, T.J., 2021a. Metallic nanoscaffolds as osteogenic promoters: advances, challenges and scope. Metals 11, 1356.

Ghosh, S., Webster, T.J., 2021b. Mesoporous silica based nanostructures for bone tissue regeneration. Front. Mater. 8, 692309.

Ghosh, S., More, P., Nitnavare, R., Jagtap, S., Chippalkatti, R., Derle, A., Kitture, R., Asok, A., Kale, S., Singh, S., Shaikh, M.L., Ramanamurthy, B., Bellare, J., Chopade, B.A., 2015a. Antidiabetic and antioxidant properties of copper nanoparticles synthesized by medicinal plant *Dioscorea bulbifera*. J. Nanomed. Nanotechnol. S6 (007). https://doi.org/10.4172/2157-7439.S6-007.

Ghosh, S., Nitnavare, R., Dewle, A., Tomar, G.B., Chippalkatti, R., More, P., Kitture, R., Kale, S., Bellare, J., Chopade, B.A., 2015b. Novel platinum-palladium bimetallic nanoparticles synthesized by *Dioscorea bulbifera*: anticancer and antioxidant activities. Int. J. Nanomedicine 10 (1), 7477–7490.

Ghosh, S., Jagtap, S., More, P., Shete, U.J., Maheshwari, N.O., Rao, S.J., Kitture, R., Kale, S., Bellare, J., Patil, S., Pal, J.K., Chopade, B.A., 2015c. *Dioscorea bulbifera* mediated synthesis of novel $Au_{core}Ag_{shell}$ nanoparticles with potent antibiofilm and antileishmanial activity. J. Nanomater. 2015, 562938.

Ghosh, S., More, P., Derle, A., Kitture, R., Kale, T., Gorain, M., Avasthi, A., Markad, P., Kundu, G.C., Kale, S., Dhavale, D.D., Bellare, J., Chopade, B.A., 2015d. Diosgenin functionalized iron oxide nanoparticles as novel nanomaterial against breast cancer. J. Nanosci. Nanotechnol. 15 (12), 9464–9472.

Ghosh, S., Patil, S., Chopade, N.B., Luikham, S., Kitture, R., Gurav, D.D., Patil, A.B., Phadatare, S.D., Sontakke, V., Kale, S., Shinde, V., Bellare, J., Chopade, B.A., 2016a. *Gnidia glauca* leaf and stem extract mediated synthesis of gold nanocatalysts with free radical scavenging potential. J. Nanomed. Nanotechnol. 7, 358.

Ghosh, S., Harke, A.N., Chacko, M.J., Gurav, S.P., Joshi, K.A., Dhepe, A., Dewle, A., Tomar, G.B., Kitture, R., Parihar, V.S., Banerjee, K., Kamble, N., Bellare, J., Chopade, B.A., 2016b. *Gloriosa superba* mediated synthesis of silver and gold nanoparticles for anticancer applications. J. Nanomed. Nanotechnol. 7, 4.

Ghosh, S., Chacko, M.J., Harke, A.N., Gurav, S.P., Joshi, K.A., Dhepe, A., Kulkarni, A.S., Shinde, V.S., Parihar, V.S., Asok, A., Banerjee, K., Kamble, N., Bellare, J., Chopade, B.A., 2016c. *Barleria prionitis* leaf mediated synthesis of silver and gold nanocatalysts. J. Nanomed. Nanotechnol. 7, 4.

Ghosh, S., Gurav, S.P., Harke, A.N., Chacko, M.J., Joshi, K.A., Dhepe, A., Charolkar, C., Shinde, V.S., Kitture, R., Parihar, V.S., Banerjee, K., Kamble, N., Bellare, J., Chopade, B.A., 2016d. *Dioscorea oppositifolia* mediated synthesis of gold and silver nanoparticles with catalytic activity. J. Nanomed. Nanotechnol. 7, 5.

Ghosh, S., Pardesi, K.R., Satriano, C., Basavanna, A.M., 2021. Editorial: microbial fabrication of nanomaterials and their applications. Front. Chem. 9, 739739.

Jadoun, S., Arif, R., Jangid, N.K., Meena, R.K., 2021. Green synthesis of nanoparticles using plant extracts: a review. Environ. Chem. Lett. 19, 355–374.

Jamdade, D.A., Rajpali, D., Joshi, K.A., Kitture, R., Kulkarni, A.S., Shinde, V.S., Bellare, J., Babiya, K.R., Ghosh, S., 2019. *Gnidia glauca* and *Plumbago zeylanica* mediated synthesis of novel copper nanoparticles as promising antidiabetic agents. Adv. Pharm. Sci. 2019, 9080279.

Jayarambabu, N., Akshaykranth, A., Rao, T.V., Rao, K.V., Kumar, R.R., 2020. Green synthesis of Cu nanoparticles using *Curcuma longa* extract and their application in antimicrobial activity. Mater. Lett. 259, 126813.

Jiao, Y., Wang, X., Chen, J.H., 2021. Biofabrication of AuNPs using *Coriandrum sativum* leaf extract and their antioxidant, analgesic activity. Sci. Total Environ. 767, 144914.

Kanipandian, N., Thirumurugan, R., 2014. A feasible approach to phyto-mediated synthesis of silver nanoparticles using industrial crop *Gossypium hirsutum* (cotton) extract as stabilizing agent and assessment of its in vitro biomedical potential. Ind. Crop Prod. 55, 1–10.

Karmakar, S., Ghosh, S., Kumbhakar, P., 2020. Enhanced sunlight driven photocatalytic and antibacterial activity of flower-like $ZnO@MoS_2$ nanocomposite. J. Nanopart. Res. 22, 11.

Khani, R., Roostaei, B., Bagherzade, G., Moudi, M., 2018. Green synthesis of copper nanoparticles by fruit extract of *Ziziphus spina-christi* (L.) Willd.: application for adsorption of triphenylmethane dye and antibacterial assay. J. Mol. Liq. 255, 541–549.

Kitture, R., Chordiya, K., Gaware, S., Ghosh, S., More, P.A., Kulkarni, P., Chopade, B.A., Kale, S.N., 2015. ZnO nanoparticles-red sandalwood conjugate: a promising anti-diabetic agent. J. Nanosci. Nanotechnol. 15, 4046–4051.

Koli, R.R., Phadatare, M.R., Sinha, B.B., Sakate, D.M., Ghule, A.V., Ghodake, G.S., Deshpande, N.G., Fulari, V.J., 2019. Gram bean extract-mediated synthesis of Fe_3O_4 nanoparticles for tuning the magneto-structural properties that influence the hyperthermia performance. J. Taiwan Inst. Chem. Eng. 95, 357–368.

Kumar, B., Smita, K., Cumbal, L., Debut, A., 2017. Green synthesis of silver nanoparticles using Andean blackberry fruit extract. Saudi J. Biol. Sci. 24 (1), 45–50.

Lakshmi, S.J., Bai, R.R., Sharanagouda, H., Ramachandra, C.T., Nadagouda, S., Doddagoudar, S.R., 2017. Biosynthesis and characterization of ZnO nanoparticles from spinach (*Spinacia oleracea*) leaves and its effect on seed quality parameters of green gram (*Vigna radiata*). Int. J. Curr. Microbiol. App. Sci. 6 (9), 3376–3384.

Liu, D., Liu, L., Yao, L., Peng, X., Li, Y., Jiang, T., Kuang, H., 2020. Synthesis of ZnO nanoparticles using radish root extract for effective wound dressing agents for diabetic foot ulcers in nursing care. J. Drug Deliv. Sci. Technol. 55, 101364.

Ma, Q., Li, J., Xu, L., Sun, Q., Yong, B., Wang, Y., 2016. Biosynthesis of silver nanoparticles using wheat straw biomass under light radiation and their antibacterial activity. Bioresources 11 (4), 10190–10200.

Maghimaa, M., Alharbi, S.A., 2020. Green synthesis of silver nanoparticles from *Curcuma longa* L. and coating on the cotton fabrics for antimicrobial applications and wound healing activity. J. Photochem. Photobiol. B Biol. 204, 111806.

Manjunath, K., Ravishankar, T.N., Kumar, D., Priyanka, K.P., Varghese, T., Naika, H.R., Nagabhushana, H., Sharma, S.C., Dupont, J., Ramakrishnappa, T., Nagaraju, G., 2014. Facile combustion synthesis of ZnO nanoparticles using *Cajanus cajan* (L.) and its multidisciplinary applications. Mater. Res. Bull. 57, 325–334.

Mary, A.A., Ansari, A.T., Subramanian, R., 2019. Sugarcane juice mediated synthesis of copper oxide nanoparticles, characterization and their antibacterial activity. J. King Saud Univ. Sci. 31 (4), 1103–1114.

Mohamad, N.A.N., Arham, N.A., Jai, J., Hadi, A., 2014. Plant extract as reducing agent in synthesis of metallic nanoparticles: a review. Adv. Mat. Res. 832, 350–355.

Musere, P.S., Rahman, A., Uahengo, V., Naimhwaka, J., Likius, D., Bhaskurani, S.V., Jonnalagadda, S.B., 2021. Synthesis of silver nanoparticles using pearl millet (*Pennisetum glaucum*) husk and its application for the removal of algae in the water and catalytic oxidation of benzyl alcohol. J. Clean. Prod. 312, 127581.

Nagati, V.B., Koyyati, R., Donda, M.R., Alwala, J., Kundle, K.R., Padigya, P.R.M., 2012. Green synthesis and characterization of silver nanoparticles from *Cajanus cajan* leaf extract and its antibacterial activity. Int. J. Nanomater. Biostruct. 2 (3), 39–43.

Narayanan, K.B., Park, H.H., 2015. Homogeneous catalytic activity of gold nanoparticles synthesized using turnip (*Brassica rapa* L.) leaf extract in the reductive degradation of cationic azo dye. Korean J. Chem. Eng. 32 (7), 1273–1277.

Narayanan, K.B., Sakthivel, N., 2011. Green synthesis of biogenic metal nanoparticles by terrestrial and aquatic phototrophic and heterotrophic eukaryotes and biocompatible agents. Adv. Colloid Interface Sci. 169 (2), 59–79.

Nwanya, A.C., Razanamahandry, L.C., Bashir, A.K.H., Ikpo, C.O., Nwanya, S.C., Botha, S., Ntwampe, S.K.O., Ezema, F.I., Iwuoha, E.I., Maaza, M., 2019. Industrial textile effluent treatment and antibacterial effectiveness of *Zea mays* L. dry husk mediated bio-synthesized copper oxide nanoparticles. J. Hazard. Mater. 375, 281–289.

Nwanya, A.C., Ndipingwi, M.M., Ikpo, C.O., Obodo, R.M., Nwanya, S.C., Botha, S., Ezema, F. I., Iwuoha, E.I., Maaza, M., 2020. *Zea mays* lea silk extract mediated synthesis of nickel oxide nanoparticles as positive electrode material for asymmetric supercabattery. J. Alloys Compd. 822, 153581.

Patra, J.K., Kwon, Y., Baek, K.H., 2016. Green biosynthesis of gold nanoparticles by onion peel extract: synthesis, characterization and biological activities. Adv. Powder Technol. 27 (5), 2204–2213.

Paulkumar, K., Gnanajobitha, G., Vanaja, M., Rajeshkumar, S., Malarkodi, C., Pandian, K., Annadurai, G., 2014. *Piper nigrum* leaf and stem assisted green synthesis of silver nanoparticles and evaluation of its antibacterial activity against agricultural plant pathogens. Sci. World J. 2014, 829894.

Prasad, C., Yuvaraja, G., Venkateswarlu, P., 2017. Biogenic synthesis of Fe_3O_4 magnetic nanoparticles using *Pisum sativum* peels extract and its effect on magnetic and methyl orange dye degradation studies. J. Magn. Magn. Mater. 424, 376–381.

Rajkumar, T., Sapi, A., Das, G., Debnath, T., Ansari, A., Patra, J.K., 2019. Biosynthesis of silver nanoparticle using extract of *Zea mays* (corn flour) and investigation of its cytotoxicity effect and radical scavenging potential. J. Photochem. Photobiol. B Biol. 193, 1–7.

Ramimoghadam, D., Hussein, M.Z.B., Taufiq-Yap, Y.H., 2013. Hydrothermal synthesis of zinc oxide nanoparticles using rice as soft biotemplate. Chem. Cent. J. 7, 136.

Ranpariya, B., Salunke, G., Karmakar, S., Babiya, K., Sutar, S., Kadoo, N., Kumbhakar, P., Ghosh, S., 2021. Antimicrobial synergy of silver-platinum nanohybrids with antibiotics. Front. Microbiol. 11, 610968.

Robkhob, P., Ghosh, S., Bellare, J., Jamdade, D., Tang, I.M., Thongmee, S., 2020. Effect of silver doping on antidiabetic and antioxidant potential of ZnO nanorods. J. Trace Elem. Med. Biol. 58, 126448.

Rokade, S.S., Joshi, K.A., Mahajan, K., Tomar, G., Dubal, D.S., Parihar, V.S., Kitture, R., Bellare, J., Ghosh, S., 2017. Novel anticancer platinum and palladium nanoparticles from *Barleria prionitis*. Glob. J. Nanomed. 2 (5), 555600.

Rokade, S., Joshi, K., Mahajan, K., Patil, S., Tomar, G., Dubal, D., Parihar, V.S., Kitture, R., Bellare, J.R., Ghosh, S., 2018. *Gloriosa superba* mediated synthesis of platinum and palladium nanoparticles for induction of apoptosis in breast cancer. Bioinorg. Chem. Appl. 2018, 4924186.

Roy, K., Sarkar, C.K., Ghosh, C.K., 2015. Photocatalytic activity of biogenic silver nanoparticles synthesized using potato (*Solanum tuberosum*) infusion. Spectrochim. Acta A Mol. Biomol. Spectrosc. 146, 286–291.

Saranyadevi, S., Suresh, K., Mathiyazhagan, N., Muthusamy, R., Thirumalaisamy, R., 2021. Silver nanoparticles synthesized using asafoetida resin, characterization of their broad spectrum and larvicidal activity. Ann. Rom. Soc. Cell Biol. 25 (4), 15035–15049.

Sathishkumar, P., Preethi, J., Vijayan, R., Yusoff, A.R.M., Ameen, F., Suresh, S., Balagurunathan, R., Palvannan, T., 2016. Anti-acne, anti-dandruff and anti-breast cancer efficacy of green synthesised silver nanoparticles using *Coriandrum sativum* leaf extract. J. Photochem. Photobiol. B Biol. 163, 69–76.

Sharma, R.K., Yadav, S., Gupta, R., Arora, G., 2019. Synthesis of magnetic nanoparticles using potato extract for dye degradation: a green chemistry experiment. J. Chem. Educ. 96, 3038–3044.

Shende, S., Joshi, K.A., Kulkarni, A.S., Shinde, V.S., Parihar, V.S., Kitture, R., Banerjee, K., Kamble, N., Bellare, J., Ghosh, S., 2017. *Litchi chinensis* peel: a novel source for synthesis of gold and silver nanocatalysts. Glob. J. Nanomed. 3 (1), 555603.

Shende, S., Joshi, K.A., Kulkarni, A.S., Charolkar, C., Shinde, V.S., Parihar, V.S., Kitture, R., Banerjee, K., Kamble, N., Bellare, J., Ghosh, S., 2018. *Platanus orientalis* leaf mediated rapid synthesis of catalytic gold and silver nanoparticles. J. Nanomed. Nanotechnol. 9, 2.

Shinde, S.S., Joshi, K.A., Patil, S., Singh, S., Kitture, R., Bellare, J., Ghosh, S., 2018. Green synthesis of silver nanoparticles using *Gnidia glauca* and computational evaluation of synergistic potential with antimicrobial drugs. World J. Pharm. Res. 7 (4), 156–171.

Singh, A., Sharma, M.M., Batra, A., 2013. Synthesis of gold nanoparticles using chick pea leaf extract using green chemistry. J. Optoelectron. Biomed. Mater. 5 (2), 27–32.

Tammina, S.K., Mandal, B.K., Ranjan, S., Dasgupta, N., 2017. Cytotoxicity study of *Piper nigrum* seed mediated synthesized SnO_2 nanoparticles towards colorectal (HCT116) and lung cancer (A549) cell lines. J. Photochem. Photobiol. B Biol. 166, 158–168.
Verma, P.R., Khan, F., Chandra, C., Dewangan, P.K., 2019. Biofabrication of CuO nanoparticles from *Cicer arietinum* pod extract and characterization. J. Emerg. Technol. Innov. Res. 6 (5), 213–215.
Vijaya, J.J., Jayaprakash, N., Kombaiah, K., Kaviyarasu, K., Kennedy, L.J., Ramalingam, R.J., Al-Lohedan, H.A., Mansoor-Ali, V.M., Maaza, M., 2017. Bioreduction potentials of dried root of *Zingiber officinale* for a simple green synthesis of silver nanoparticles: antibacterial studies. J. Photochem. Photobiol. B Biol. 177, 62–68.
Vinotha, V., Yazhiniprabha, M., Raj, D.S., Mahboob, S., Al-Ghanim, K.A., Al-Misned, F., Govindarajan, M., Vaseeharan, B., 2020. Biogenic synthesis of aromatic cardamom-wrapped zinc oxide nanoparticles and their potential antibacterial and mosquito larvicidal activity: an effective eco-friendly approach. J. Environ. Chem. Eng. 8, 104466.
Waghmode, S., Chavan, P., Kalyankar, V., Dagade, S., 2013. Synthesis of silver nanoparticles using *Triticum aestivum* and its effect on peroxide catalytic activity and toxicology. J. Chem. 2013, 265864.
Yang, N., Li, F., Jian, T., Liu, C., Sun, H., Wang, L., Xu, H., 2017. Biogenic synthesis of silver nanoparticles using ginger (*Zingiber officinale*) extract and their antibacterial properties against aquatic pathogens. Acta Oceanol. Sin. 36 (12), 95–100.

Algae-assisted synthesis of nanoparticles and their applications

3

Azania T. Panicker, Yashodhara M. Dalal, and Maushmi S. Kumar
Shobhaben Pratapbhai Patel School of Pharmacy and Technology Management, SVKM'S NMIMS, Vile Parle (West), Mumbai, India

3.1 Introduction

Nanotechnology is a quickly expanding interdisciplinary field of science that involves observing, quantifying, predicting, and manipulating matter at the atomic, molecular, and supramolecular scales. It has been developed comparatively recently and is already being applied to almost every field of science, including engineering, physics, computer science, materials science, chemistry, and biology (Bayda et al., 2020). The reduction in size has several advantages in terms of economic effectiveness and acceleration of mechanical, chemical, and biological components. Nanotechnology is used in the field of manufacturing at the nanolevel due to its feature of self-assembly (Emerich and Thanos, 2003). Self-assembly is the concept of spontaneous arrangement of molecules into defined three-dimensional shapes under specific conditions.

Nanoparticles (NPs) are those nanomaterials that are defined as having one structural dimension of less than 100 nm and surrounded by an interfacial layer. They are comparable in size to cell organelles and proteins. This enables their ready incorporation into biological systems and hence justifies their applications in DNA probing, drug and gene delivery, as well as tissue engineering. Green synthesis of NPs is a growing field in nanobiotechnology and green nanotechnology. Natural biodegradable components such as enzymes, vitamins, polysaccharides, plant extracts, biodegradable polymers, and microorganisms are used in the synthesis. The demand for dependable and environmentally friendly procedures to manufacture metal and metal oxide NPs is invariably present. Biological synthesis does this by reducing or even avoiding the usage of harmful and toxic substances.

Algae-assisted synthesis is one of the many biosyntheses of NPs. These are polyphyletic, unorganized, and artificial assemblages of photosynthetic organisms that produce oxygen (exceptions of colorless members, related to pigmented forms). This definition could also include plants in the algal division, yet they have very few similarities being only limited to storage compounds produced and defense strategies. Except for seaweeds, which demonstrate specialization and differentiation of their vegetative cells, most algae are not differentiated into roots, stems, leaves, or well-defined vascular tissues; they do not produce embryos and have both monogenetic

Agricultural Nanobiotechnology. https://doi.org/10.1016/B978-0-323-91908-1.00001-8

and digenetic life cycles. They also come in a variety of shapes and sizes, including single cells, multicellular loose or filmy aggregates, matted or branching colonies, and more elaborate leafy or blade forms (Barsanti and Gualtieri, 2014). They are classified into three main types, Chlorophyceae (chlorophyll-containing green algae; Chlamydomonas, Spirogyra, and Chara), Phaeophyceae (predominantly marine, chlorophyll-a,-c, carotenoids and xanthophyll pigment-containing brown algae; Dictyota, Laminaria, and Sargassum), and Rhodophyceae (red r-phycoerythrin pigment-containing algae; Porphyra, Gracilaria, and Gelidium).

Algal synthesis is preferred for various reasons. Due to their increased biological activity and availability, green algae are chosen over plants. Algae also grow at a far quicker rate than plants. The oceans, which span more than 70% of the earth's surface, represent a vast reservoir of natural resources (Snelgrove, 2016; Kim et al., 2018). Hence, it is more suited to large-scale generation of NPs of varied sizes and shapes, and it removes the need for cell culture maintenance (Singh et al., 2013; Honary et al., 2012; Gunalan et al., 2012). This method seems to be cost-effective and eco-friendly, with quicker results, and does not involve any toxic chemicals (Mohanpuria et al., 2008). The advantages of employing these biomaterials for the production of NPs include their accessibility, safety, and a wide range of metabolites that may assist in reduction (Chandran et al., 2006). When selecting an alga for synthesis, it is important to evaluate the organism's therapeutic characteristics as well as whether or not capping or reducing agents are present. The size, shape, stability, and synthesis of NPs are all affected by a variety of parameters, including the capping agents used, the pH of the reaction, the templates used, and the temperature of the reaction. The production and size of intracellular particles may be regulated by adjusting important parameters such as pH, temperature, substrate concentration, and substrate exposure duration (Gericke and Pinches, 2006).

The proposed mechanism of NP synthesis will be discussed in this chapter, along with the variations with different metals and algae. We will also elaborate on the advantages of the products formed from this method. The uses of the product obtained from the biosynthesis consist of applications as antimicrobial agents, in vitro antidiabetics, antioxidants, and even as anticancer agents (Asha et al., 2015; Singh et al., 2019). We will also discuss the use of algae in the agricultural domain and their ability to improve the activity and quality of irrigation water, which in turn would facilitate a better harvest, especially against the accumulation of heavy metals.

3.2 Biological synthesis of nanoparticles using algae

Physical and chemical nanoparticle production methods have been widely used, but they are expensive and utilize hazardous substances, limiting their use. To address these concerns, eco-friendly green techniques that are both safe and environmentally beneficial are being developed. Green synthesis has numerous advantages over chemical and physical approaches, including being ecologically sound, cost-effective, and readily scaled up for large-scale NP syntheses; also, there is no need for high temperature, pressure, energy, or harmful chemicals (Salam et al., 2012). These green techniques involve the use of biological agents such as bacteria (Bhuyar et al., 2020), fungi (Clarance et al., 2020), algae (Singh et al., 2019), and plants (Ebrahimzadeh et al., 2020). Despite the fact that several research and reviews have been published on nanoparticle biosynthesis, the exact mechanism is still not known. This may be because the diverse biological agents that react with the metal ions may have various interactions based on the biomolecules present in them. Marine algae are rich in lipids, minerals, some vitamins but also other bioactive compounds such as polysaccharides, proteins, polyphones and are well-recognized as functional food, as well as prospective drug applications for diseases such as cancer and thrombosis (Perez et al., 1998; Nishino et al., 1999; Miyashita, 2009; Namvar et al., 2012). As a result, their phytoconstituent functional groups (carboxyl, hydroxyl, and amino) operate as effective reducing and capping agents for metal NP, and the reactions occur in a single step, making the procedure robust (Mahdavi et al., 2013). Many factors influence NP synthesis, including pH, temperature, substrate concentration, stirring, and static conditions. To prevent NPs of different shapes and sizes during the synthesis reaction, the buffering action must be strong to ensure a stable pH. The SPR peak widens and deflects into a longer wavelength area at low pH. Thus, spherical NP formation in the reaction is aided by a high pH. Many large NP pearls are formed in alkaline conditions, being much more stable than the cluster of NPs formed in acidic conditions. The synthesis of NPs using green methods necessitates a temperature of nearly 100 degrees Celsius. At high temperatures, a high rate of AuNP formation was observed, according to studies. When their conversion rate rose, their average size shrank. An increase in contact time/interaction leads to a higher SPR peak intensity and quicker NP synthesis (Uzair et al., 2020). The process occurs either inside the cell or outside the cell in microorganisms. In the intracellular method, the cell wall plays a major role as the metal ion passes through it into the cell and is consequently reduced to stable metal form. The extracellular method may involve reductive enzymes being released by the microorganism. The bioreduction mechanism involves the activation phase, growth phase, and termination phase. In the activation phase, the metal ion is reduced to stable metal, and then, nucleation occurs. The growth phase comprises the spontaneous organization of the nearby small NPs into a larger particle along with thermodynamic stability (Ostwald ripening). When the NP reaches its ultimate shape, it is said to be at the termination phase (Shankar et al., 2016) (Fig. 3.1). When it comes to algae-assisted NP synthesis, the most common products that have been studied are gold (AuNPs) and silver nanoparticles (AgNPs). A change in the color of the culture media can be used to observe the synthesis, which is a deep brown final solution for AgNPs and a change from a ruby red solution to a deep purple color for AuNPs. Other products include copper oxide nanoparticles (CuONPs), zinc (Zn), iron (Fe), cadmium (Cd), palladium (Pd), chromium (Cr), and lead (Pb). These are mentioned in Table 3.1 and discussed in detail further in the chapter.

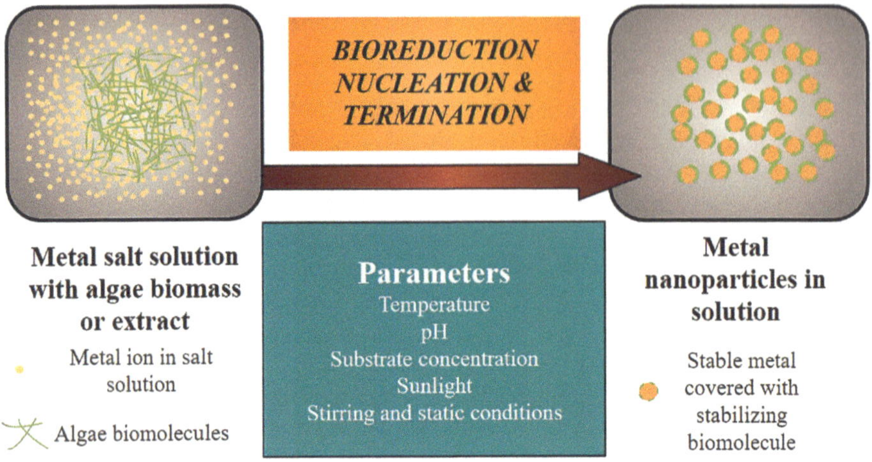

Fig. 3.1 Mechanism of algae-assisted synthesis of nanoparticles hypothesized by Shankar et al. (2016).

3.2.1 Gold nanoparticles (AuNPs)

Many of the early studies have been based on synthesizing gold nanoparticles (AuNPs). Singaravelu et al. (2007) first used *Sargassum wightii* seaweed as the algae-assisted method to synthesize highly stable AuNPs by the extracellular process. Stable gold metal Au^0 was formed from aqueous $HAuCl_4$ solution by treating with 1 g of seaweed powder. 95% Product was formed within 12 h of stirring. This was further researched, and the variability with microorganism was studied. In a comparison of the green algae *Rhizoclonium hieroglyphicum* and the blue-green algae *Lyngbya majuscula* and *Spirulina subsalsa*, quick binding of gold ions to the *Rhizoclonium* (40%–80%) cells was seen within 10 min of exposure to radioactive and stable gold solutions, whereas in *L. majuscula* (35%–60%) and *S. subsalsa* (20%–35%) cells, a moderate to very slow initial metal uptake was noted. This can be correlated with metabolism-independent passive adsorption (extracellular) and metabolism-dependent active absorption (intracellular), followed by reduction within 72 h of being exposed in the two cases, respectively. Intracellular synthesis is thought to be influenced by physicochemical factors such as substrate concentration, temperature, and pH. Through –NH2 (amine) groups, surface-bound proteins and their residual amino acids play a critical role in NP capping and stability at basic pH. As a result, changes in the optimal pH for Au sorption by different algae species might be linked to changes in cell surface chemical composition, resulting in varying chemical interactions (Mehta and Gaur, 2005; Chakraborty et al., 2009).

Brown algae have a high metal absorption compared to other microbes like fungi and other algae, which might be attributed to their mucilaginous polysaccharide-rich cell wall. Metal recovery is aided by the functional groups present, particularly carboxyl, which represent 60%–80% of the biomass dry weight. Ecklonia cava, a marine brown alga, was used in a study to produce AuNPs within 1 min at 80°C by reducing

Table 3.1 Phycogenic nanoparticles (NPs) and their properties.

NPs	Organism	Type	Size (nm)	Characteristics	Reference
Gold nanoparticles (AuNPs)					
Au	*Sargassum wightii*	Marine brown seaweed	11		Singaravelu et al. (2007)
Au	*Rhizoclonium hieroglyphicum* >*Lyngbya majuscula* >*Spirulina subsalsa*	Green algae Unicellular bacteria Blue-green multicellular filamentous	<20	–	Chakraborty et al. (2009)
Au	*Kappaphycus alvarezii*	Red algae	10–40	Spherical	Rajasulochana et al. (2010)
Au	*Tetraselmis kochinensis*	Green algae	5–35	–	Senapati et al. (2012)
Au	*Stoechospermum marginatum (kützing)*	Brown seaweed	4–80	Antibacterial	Rajathi et al. (2012)
Au	*Prasiola crispa*	Green algae	5–25	Near spherical	Sharma et al. (2014a)
Au	*Ecklonia cava*	Brown seaweed	30 ± 0.25	Biocompatibility with the human keratinocyte Antimicrobial activity	Venkatesan et al. (2014)
Au	*Lemanea fluviatilis (L.)*	Red algae	Average 5.9	Antioxidant in a biomatrix	Sharma et al. (2014b)
Au	*Rhizoclonium fontinale*	Green algae	~16	Spherical Biochemical changes	Parial and Pal (2015)
Au	*Spirulina platensis*	Blue-green multicellular filamentous	~5	Antibacterial G+	Suganya et al. (2015)

Continued

Table 3.1 Continued

NPs	Organism	Type	Size (nm)	Characteristics	Reference
Au	*Pithophora oedogonia (Montagne)*	Green algae	32.06	Electrochemical performance for determining carbendazim in soil	Li and Zhang (2016)
Au	*Calothrix* sp.	Blue-green multicellular filamentous	30–120	Truncated	Kumar et al. (2016)
Au	*Egregia* sp.	Brown algae	~8	–	Colin et al. (2018)
Au	*Dunaliella salina*	Green halophilic microalgae	Average 22.4	In vitro anticancer activity on breast cancer cell line	Singh et al. (2019)
Silver nanoparticles (AgNPs)					
Ag	*Sargassum wightii*	Marine brown seaweed	8–27	Antibacterial	Govindaraju et al. (2009)
AgCl	*Sargassum plagiophyllum*	Marine brown seaweed	~18–42	Antibacterial	Dhas et al. (2014).
Ag	*Colpomenia sinuosa*	Brown algae	54–65.	Antidiabetic	Kiran and Murugesan (2014)
Ag	*Enteromorpha flexuosa* (*Wulfen*)	Green algae	2–32	Antibacterial G+ > G-	Yousefzadi et al. (2014)
Ag	*P. oedogonia* (*Montagne*)	Green algae	34.03	Antibacterial	Sinha et al. (2015)
Ag	*Sargassum polycystum*	Marine brown seaweed	–	Antibacterial, anticancer (cytotoxic)	Asha et al. (2015)
Ag	*Caulerpa racemosa*	Marine green seaweed	~25	Catalytic degradation of methylene blue	Edison et al. (2016)

Ag	*D. salina*	Green halophilic microalgae	Average 15.26	Anticancer potential	Singh et al. (2017)
Ag	*Spyridia fusiformis (Børgesen)*	Red seaweed	5–50	Antibacterial	Murugesan et al. (2017)
Ag	*Caulerpa serrulata*	Marine green seaweed	Average 10 ± 2	Antibacterial Catalytic reduction of Congo red	Aboelfetoh et al. (2017)
Ag	*Laurencia catarinensis*	Red algae	39.41–77.71	Several shapes	Abdel-Raouf et al. (2018)
Ag	*Neochloris oleoabundans*	Green algae	16.63	Photo-induced	Bao and Lan (2018)
Ag	*Padina pavonica*	Brown algae	49.58–86.37	Bactericidal	Abdel-Raouf et al. (2019)
Ag	*Caulerpa scalpelliformis*	Marine green seaweed	16.17–47.78	Spherical Antiproliferative and wound healing	Manikandan et al. (2019)
Ag	*Padina* sp.	Brown algae	~ 25–60	Spherical Antibacterial	Bhuyar et al. (2020)
Ag	*Chlorella ellipsoidea*	Green algae	220.8 ± 31.3	Nearly spherical Photocatalytic degradation of methylene blue and methyl orange Antibacterial	Borah et al. (2020)
Ag	*Sargassum myriocystum*	Marine brown seaweed	20 ± 2.2	Hexagonal Antibacterial, larvicidal, anticancer, catalytic degradation of methylene blue	Balaraman et al. (2020)

Continued

Table 3.1 Continued

NPs	Organism	Type	Size (nm)	Characteristics	Reference
Ag	Laminaran (ScL) and fucoidan (ScF) from *Saccharina cichorioides* and fucoidan (FeF) and alginate (FeA) from *Fucus evanescens*	Brown algae	AgNPs-FeA; 57 ± 9 AgNPs-FeF: 64 ± 6 AgNPs-ScF: 53 ± 4 AgNPs-ScL: 45 ± 2	Spherical Antibacterial: <fucoidan Cytotoxic: fucoidan < alginate and laminaran Catalytic activity: alginate < fucoidan < laminaran	Yugay et al. (2020)
Ag	*Noctiluca scintillans*	Green algae	4.13	Spherical Antimicrobial, anticancer, colorimetric Sensing of hydrogen peroxide	Elgamouz et al. (2020)
Ag	*Gelidium corneum*	Marine red algae	20–50	Centric cubic Antibiofilm and antimicrobial	Öztürk et al. (2020)
Both AgNPs and AuNPs					
	Oscillatoria sp. and *Spirulina platensis*	Blue-green algae	Ag: 44.2 Au: 16.6	Ag_2O/AgONPs: Spherical AuNPs: octahedral, pentagonal, and triangular Antiviral activity	El-Sheekh et al. (2020)
	Neodesmus pupukensis (MG257914)	Green algae	Ag: 52–179 Au: 5–34	Spherical Antimicrobial (AgNPs: G+ & G-) Antioxidant	Omomowo et al. (2020)

Others					
CdS	*Phaeodactylum tricornutum, Bohlin*	Diatom; microalgae	0.9–1.1	–	Scarano and Morelli (2003)
CuO	*Bifurcaria bifurcata*	Brown algae	5–45	Antibacterial	Abboud et al. (2014)
Pd	*Sargassum bovinum*	Marine brown seaweed	5–10	Octahedral Electrocatalytic action toward H_2O_2	Momeni and Nabipour (2015)
Ag, Cr, Pb, Zn, Fe	*Spirulina platensis*	Blue-green multicellular filamentous	–	Antimicrobial	Gunasundari et al. (2017)
Fe_3O_4	*Turbinaria decurrens*	Brown seaweed	15–18	Navy blue absorption, Langmuir adsorption	Khaleelullah et al. (2017)
CuO	*Cystoseira trinodis*	Brown algae	Average 6–7.8	Monoclinic Antibacterial, antioxidant Catalytic degradation of methylene blue	Gu et al. (2018)

Au: Gold; Ag: Silver; AgCl: Silver chloride; Cr: Chromium; Pb: Lead; Zn: Zinc; Fe: Iron; CuO: Copper (II) oxide; Fe_3O_4: Iron (III) oxide; G+: Gram-positive; G−: Gram-negative chloroauric acid. The AuNPs produced were biocompatible with the human keratinocyte cell line and showed good antimicrobial activity (Venkatesan et al., 2014). Other noteworthy studies include the synthesis of photoluminescent AuNPs using the red alga, *Lemanea fluviatilis* (Sharma et al., 2014b), and the recent characterization of AuNPs synthesized photocatalytically (discussed in AgNPs) from the halophilic green microalgae, *Dunaliella salina*, showing anticancer activity on breast cancer cell line in vitro (Singh et al., 2019). Some of these AuNPs mentioned in Table 3.1 are illustrated in Fig. 3.2.

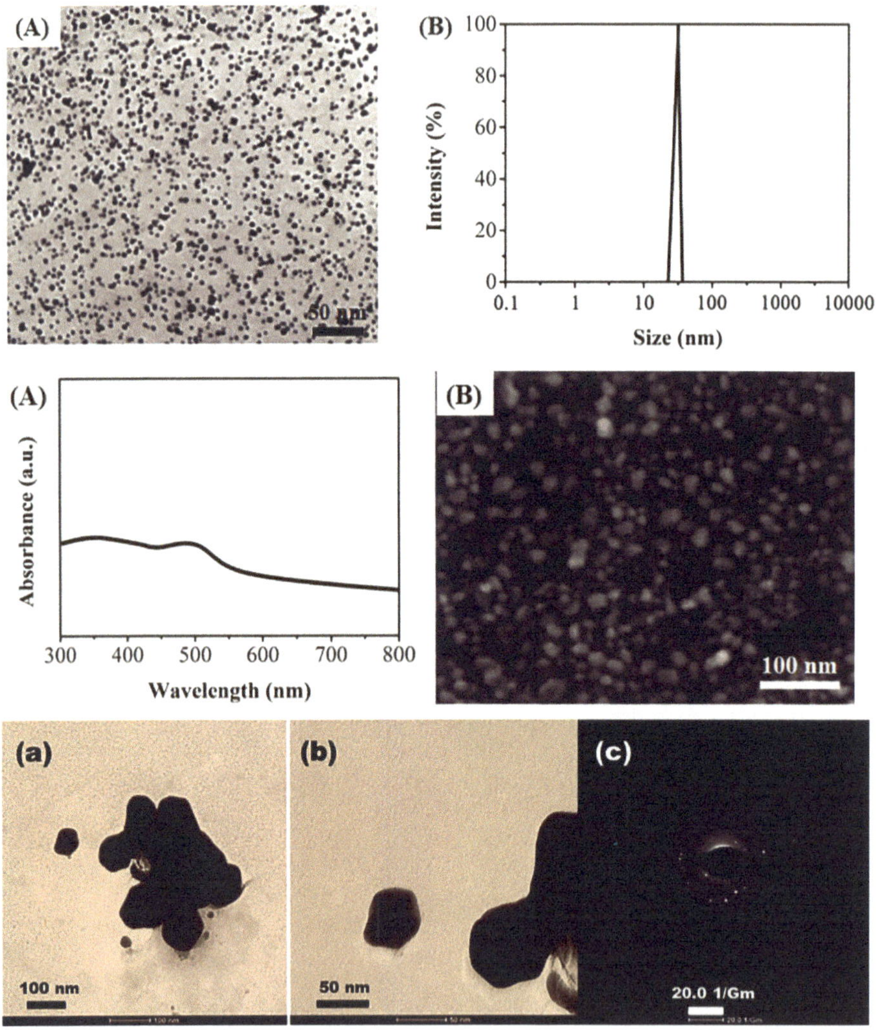

Fig. 3.2 Some of the AuNPs synthesized using (i) *Pithophora oedogonia* (Montagne) (A) TEM,(B) SEM (Li and Zhang, 2016) (ii) *Calothrix* sp. under the influence of ultrasound irradiation (A) and (B); SAED (selected area (electron) diffraction) spots in (C) clearly verify the existence of polycrystalline nature of AuNPs (Kumar et al., 2016).

3.2.2 Silver nanoparticles (AgNPs)

Silver nanoparticles (AgNPs) were studied later and seem to have similar synthesis mechanisms. Additionally, several studies seem to suggest that the synthesis of AgNPs extracellularly in microbes is nitrate reductase-mediated (Durán et al., 2005; Ingle et al., 2008). Photo-induced biosynthesis of AgNPs by *Dunaliella salina* was first discussed by Singh et al. (2017). The product was truncated in shape and showed promising anticancer potential, which was later confirmed by Singh et al. (2019). The mechanism of this photocatalytic bioreduction has been studied using *Neochloris oleoabundans* cell extract. It is a disrupted cell aqueous extract (DCAE) obtained by sonication of the green algae biomass solution. The effect of various production parameters was studied, and the mechanism was hypothesized. The presence of methanol-extracted chlorophyll accelerated the reaction. Oversonication possibly leads to the destruction of other reactive agents, which are only activated by light of specific wavelengths. The mechanism proposed involves the absorption of photons of those wavelengths by chlorophyll molecules to produce excited electrons, which consequently reduce the Ag^+ ions to stable Ag^0 (Bao and Lan, 2018). Other parameters such as pH, temperature, concentration of extract, and time of exposure and their effect on the dimensions of AgNPs synthesized using aqueous extract of *Caulerpa serrulata,* a green marine alga, have been examined. Spherical AgNPs with crystalline nature and mean particle size of 10 ± 2 nm were formed when higher concentration of extract was used for synthesis at higher temperatures. For this alga, the rate of reduction of Ag^+ was more rapid at higher temperatures (95°C) and alkaline pH ranges. Catalytic activity of these AgNPs was studied by measuring the reduction of Congo red characterized by fading or bleaching. On adding the AgNPs to the Congo red and $NaBH_4$ mixture, fast reduction was observed within 5 min with the effectiveness being affected by concentration of the extract, $NaBH_4$ concentration, and catalyst quantity. Additionally, the AgNPs showed effective antibacterial action against both Gram-negative and Gram-positive microorganisms (Aboelfetoh et al., 2017). Some of these AgNPs mentioned in Table 3.1 are illustrated in Fig. 3.3.

3.2.3 Others

Research on algae-assisted synthesis of other metal NPs (other than AuNPs and AgNPs) is limited; however, they are quite interesting and must be explored. Scarano and Morelli (2003) demonstrated that cadmium sulfide (CdS) nanoparticles can be synthesized using *Phaeodactylum tricornutum,* which contribute phytochelatins (PC) peptides, involved in the cellular detoxification mechanism of plants, algae, and yeasts, to form stable metal/PC complexes. Sulfide ions (S^{2-}) can be integrated into complexes to stabilize PC-coated CdS nanocrystallites, according to the findings.

The electrocatalytic activity of palladium (Pd) nanoparticles obtained from brown seaweed, *Sargassum bovinum,* was investigated by measuring the electrochemical reduction of hydrogen peroxide (H_2O_2). This was done using a modified carbon ionic liquid electrode (PdNPs/CILE), a nonenzymatic sensor developed for the determination. For H_2O_2 detection, the sensor provides a rapid response, high sensitivity and

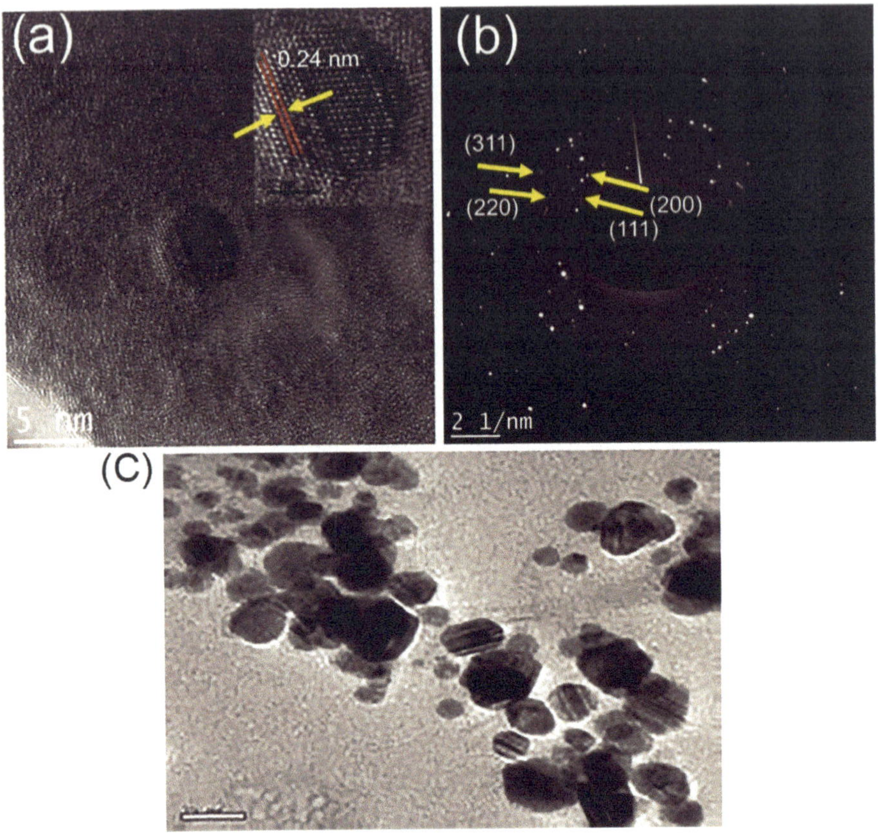

Fig. 3.3 Some of the AgNPs synthesized using (i) *Noctiluca scintillans* (A) HRTEM showing the spherical morphology and crystalline nature. The inset is showing a magnified view of the NP. (B) SAED pattern showing the polycrystalline nature of NP (Elgamouz et al., 2020); and (ii) *Spyridia fusiformis* HRTEM images (Murugesan et al., 2017).

selectivity, a wide linear range, and a low detection limit. This could be further explored as a promising electrochemical sensing platform (Momeni and Nabipour, 2015).

In an ultrasonic-assisted synthesis of metal NPs using *Spirulina platensis*, when compared to all other metal NPs (silver, palladium, and iron), chromium and zinc NPs demonstrated stronger antibacterial activity against bacterial and fungal species (Gunasundari et al., 2017). Monoclinic copper oxide nanoparticles (CuONPs) were generated within 90 min in a similar process utilizing the aqueous extract of *Cystoseira trinodis*, a brown alga, at room temperature. The product gave a brown-colored precipitate of highly crystalline, spherical CuONPs having a diameter around 6–7.8 nm. They were effective photocatalysts for the breakdown of methylene blue. The activity was highest at pH 4. The amount of degradation was inversely related to dye concentration and directly related to the amount of catalyst used. Moreover, the CuONPs showed good antibacterial and antioxidant activity as well (Gu et al., 2018). The extract of the brown seaweed *Turbinaria decurrens* was used to produce superparamagnetic iron oxide nanoparticles (SPIONs). The crystalline ferric oxide (Fe_3O_4) particles produced had an average size of 18 nm. The adsorption capacity of the bio-SPIONS was studied using navy blue dye, and it was found to be enhanced because of certain functional groups contributed by the seaweed (Khaleelullah et al., 2017).

The increasing demand for nanoparticles warrants research into efficient synthesis methods. These techniques/methods must be economical and applicable on an industrial scale. Biological methods can fulfill the demand for environment-friendly methods required for large-scale production. As discussed, there exists a multitude of biological materials in nature that can act as reducing agents for the process. The application of NPs depends mostly on their size and shape, which is directed by the existing reducing and capping agents. For example, the catalytic activity of the polysaccharides used by Yugay et al. (2020) is alginate < fucoidan < laminaran for the synthesis of AgNPs. Out of these, the fucoidan NPs showed distinct antibacterial activity and the alginate and laminaran showed much higher cytotoxicity in rat C6 glioma cells compared to the other polysaccharides. Thus, if one can manipulate these characters, these approaches have the potential to be employed as promising approaches for large-scale synthesis in the future.

3.3 Agriculture and water management

Water is one of the major elements required for the preservation and nurturing of agriculture as well as mankind (Oster et al., 2006). However, there are quite a few sources for agricultural water like the groundwater system and irrigation facilities and they are all interconnected to each other. Thus, a single source of water when contaminated shows a degrading effect on the rest of the water systems. This degrading effect is an aftermath of unrestricted dumping of industrial and residential wastes. These untreated wastes contain heavy metals like zinc (Zn), cadmium (Cd), copper (Cu), manganese (Mn), mercury (Hg), lead (Pb), arsenic (As), and chromium (Cr; hexavalent > trivalent), which cause major damage to the vegetation and in turn human beings (Nordberg, 1992; Bhattacharya et al., 2006; Parvathi et al., 2007; O'Connell et al., 2008; El-Shafey, 2010; Pellera et al., 2012; Rangabhashiyam and Selvaraju, 2015). The poor quality of agricultural and irrigation water leads to high salinity and alkaline levels of the groundwater (Oster, 1994; Datta and Dayal, 2000). Additionally, the use of the water contaminated by the untreated effluents of the industries may result in the aggregation of chemical and microbiological pollutants in the soil and water, which later is imbibed and concentrates in the harvest and finally in the livestock and humans. This may lead to an advanced degree of antimicrobial resistance. Thus, prior to the use of poor quality and disposed water, the effluvium is treated as it can prove to be a valuable source of nutrients and water. The necessity to improve the quality of water lies in the ability of water to provide security for food and improve livelihood (Food and Agriculture Organisation of the United Nations, 2017). Chemical precipitation, adsorption, ion exchange, membrane separation, and other traditional effluent treatment technologies are available (Eccles, 1999).

However, these processes have a number of drawbacks, including partial metal removal, costly equipment, high reagent or energy needs, and the formation of hazardous sludge or other waste products that must be disposed of. This warrants an alternative process that avoids these problems. The use of algae for treatment by accumulation is one of them.

As discussed above, the metal ions being positively charged interact with the negatively charged functional groups on the algae or inside the algae if accumulated intracellularly. Metal recovery is aided by the functional groups present, particularly carboxyl, which represent 60%–80% of the biomass dry weight (Volesky, 2003; Romera et al., 2007; Mata et al., 2008) as mentioned earlier. In order to improve previously contaminated groundwater as well as improve nutrient levels of irrigation water, there has been a wide study in the field of algae and their application on the enhancement of irrigation water quality. Algae have been known to be a beneficial evaluating tool for any alterations, which may take place in the ecosystem. The alterations include not only changes on land but also the quality of water. Algae act as a primary biological indicator that helps in a swift evaluation of the water quality (O'Farrell et al., 2002). Algae are known to work in water bodies like lakes and reservoirs as the nutrient level in these sources affects and limits primary productivity rates. The concentration of algae in the water bodies is proportional to the water quality, which in turn affects the biomass, productivity rates, and community structure (Alvarez-Gongora and Herrera-Silveira, 2006; Celik, 2006).

Algal growth was reported in water bodies with abnormal levels of nutrients like phosphorus and nitrogen. A 0.1 mg/L of phosphate increase in freshwater bodies and streams commences the growth of algae and other unwanted or undesirable aquatic life forms, which work to clog irrigation systems, thus making the water transported and used for vegetation and agriculture unsuitable to produce superior quality stock. Furthermore, algae as mentioned before act as a primary biological indicator toward the temperature of water (growth when temperatures are above 20°C), the calmness and clearness of the waters, and pH levels (alkaline). Algae are registered as a primary indicator due to the fact that they have been found to be present in water bodies even in the absence of the abovementioned factors (Agriculture NSW Water Unit, 2017). Algae also play an important role in agriculture through their ability to act as biofertilizers (seaweeds), which results in a decreased nitrogen and phosphorus runoff, resulting in an increase in the water quality that flows into the rivers and oceans post agricultural use.

Maerl, a calcareous red alga found off the coast of the United Kingdom and France, has been reported to reduce the acidity of the soils (Blunden et al., 1981). Microalgae are used for improving the fertility of the soil (Abdel-Raouf et al., 2012). Cyanobacteria work as a nitrogen-fixing agent in rice fields (Vargas and Novelo, 2007; Huang and Chow, 1988). Furthermore, cyanobacteria are classified into heterocystous and nonheterocystous, both of which are capable of fixing nitrogen in the soil (Abdel-Raouf et al., 2012). The nitrogen fixed by cyanobacteria can be freed and incorporated by higher plants (Stewart, 1970). Moreso, on further investigation into the paddy fields of India researchers, Gupta and Lata (1964) have observed and reported the ability of cyanobacteria to accelerate the process of seed germination as well as promote the growth of the seed. In addition to this, it was also observed that the protein content of the seeds improved and thus overall improved the quality of the seeds. Algae are also useful post death due to the decay and formation of organic matter (mucilage), which when mixed with the soil binds the soil and increases humus content (helps retain moisture) (Bolyshev and Novichkova-Ivanova, 1978). This results in a more habitable environment for the growth of plants (Marathe and Chaudhari, 1975). Algae are one of the major components, which form the arid and semiarid ecosystems. Additionally, they act as an indicator toward the health of the environment, thus reducing the pollution levels as well as improving the health of the soil and in turn the plants. Algae, although known for their primary photosynthetic activity, are ecologically not completely defined (Abdel-Raouf et al., 2012).

3.4 Other applications and future prospects

A few more applications of the above-discussed NPs are discussed in detail below.

3.4.1 Antimicrobial activity

For centuries, silver has been used to cure burns, wounds, and bacterial infections. This use declined as more and more antibiotics were developed. Now that antibiotic resistance is on the rise and a looming problem, there is a resurgence in need for the use of silver as an antimicrobial agent. AgNPs draw interest due to their relatively nontoxic nature in low concentrations and broad spectrum of activity (Omomowo et al., 2020). Kailasa et al. (2019) discuss the potential of AgNPs as an antimicrobial in detail. It states that size, capping agent, and concentration of AgNPs play essential roles in inhibiting microbial growth. The increase in demand warrants for efficient synthesis methods, which biosynthesis can provide. Several of the algae-associated syntheses produce AgNPs with good antimicrobial activity and have great potential in this sector (Murugesan et al., 2017; Aboelfetoh et al., 2017; Abdel-Raouf et al., 2019). There are even studies showing antibacterial activity in some drugs with AgNPs. Surwade et al. (2019) showed that AgNPs act synergistically with ampicillin and the antibacterial activity seemed to be more prominent with a lower concentration of ampicillin. Fatima et al. (2020) synthesized environmentally friendly AgNPs that can be potentially used as a replacement to commercially available antibiotic medication in the treatment of fish diseases showing the highest activity for pathogens *Vibrio harveyi and Vibrio parahaemolyticus*. Ag_2O/AgONPs (silver oxide NPs) and AuNPs synthesized from Oscillatoria sp. and Spirulina platensis inhibited the replication of the herpes simplex (HSV-1) virus in a dose-dependent manner to Vero cells, producing a 90% reduction in the cytopathic effect (CPE) of the virus at 31.25 μL (El-Sheekh et al., 2020).

Another utility of antimicrobial activity of metal NPs is in food packaging. CuONPs and Zn NPs have been used in food preservation because of their antibacterial action (Rajamanickam et al., 2012; Gautam and Mishra, 2017; Lomate et al., 2018). CuONPs have a potential for forming antimicrobial nanohybrids, and they seem to increase the antimicrobial activity that may be present in the nanocomposite material (Almasi et al.,

2018). The method of synthesis by Gu et al. (2018) can be used to produce the CuONPs in a green and efficient manner. The simplicity of the process suggests an easy way to scale up to meet industry needs. It was also mentioned that this approach might be utilized as a backup to traditional preservation methods (Rajamanickam et al., 2012).

3.4.2 Novel drug delivery systems

Nanotechnology is becoming extremely relevant in the development of medication delivery systems. Natural polysaccharides may be found in a variety of environments, including algae, animals, plants, and microorganisms. Seaweed polysaccharides in particular consist of alginate, carrageenan, ulvan, laminarin, etc. They have a lot of resources, and they have been used in disciplines including medicine delivery, tissue engineering, and biosensors (Venkatesan et al., 2015; Muinao et al., 2019). Various techniques, including ionic gelation, emulsion, and polyelectrolyte complexing, can be used to transform natural polymers into NPs (Venkatesan et al., 2016). Surface groups such as hydroxyl, carboxyl, and sulfate groups can be found on these polysaccharides that are hydrophilic in nature and interact with biological tissues. Thus, these properties give them an advantage to their use in drug delivery systems. Furthermore, polysaccharide-based materials have a number of benefits that might be considered when assessing their use as a drug delivery vehicle (Posocco et al., 2015). Additionally, AuNPs coated with protein biomolecules are also highly important because of their vast application potential in areas of diagnostics and drug delivery (Gole et al., 2001).

3.4.3 Cancer treatment

Cancer is a terminal condition that is constantly being researched. AuNPs and AgNPs formed from *Dunaliella salina* were studied, and their cytotoxic activity was compared with cisplatin, a known cancer drug. Both types of biosynthesized NPs showed good activity against MCF7 (cancer) cell lines and were relatively nontoxic to the MCF10A (normal, control) cell lines as compared to cisplatin (Singh et al., 2017; Singh et al., 2019). AuNPs synthesized using *Ecklonia cava* showed biocompatibility with the human keratinocyte cell line and good antimicrobial activity against pathogenic bacteria (Venkatesan et al., 2014). This indicates that AuNPs (along with the biomolecules associated with them) have anticancer activity with good selectivity for cancer cells. They are being explored for being used in nano-photo-thermolysis using lasers. AgNPs ranging from 10 to 100 nm respond to incoming radiation causing them to absorb and scatter light. This is known as plasmon resonance. This effect destroys tissue by local heating or can be used to release molecules of therapeutic importance at particular sites. Gold nanoparticles can also be used for providing possibilities for targeting specific cells (Devi and Awasthi, 2019).

Some more uses include the possible application of AgNPs in the treatment of diabetes (Kiran and Murugesan, 2014) and as a hydrogen peroxide (H_2O_2) colorimetric sensor which is used as an antiseptic (Elgamouz et al., 2020). In an experimental rat model, the green AgNPs demonstrated strong healing effects on wounds by improving wound closure (Manikandan et al., 2019). AuNPs also showed electrocatalytic activity, which was used to determine carbendazim level in the soil (Li and Zhang, 2016).

3.5 Conclusion

The above-discussed algae-assisted biosynthesized nanoparticles have an expansive scope of uses in an assortment of fields and added advantages due to the biomolecules associated with them. The concept of this method is now better understood than it was 30 years ago, but there is still more to explore. There are still some species of microorganisms that have still not been investigated for their potential in this field. We are still in the dark about exactly how the bioreduction and capping process takes place in this type of synthesis. Controlling the shape and size is still of prime importance and must be researched more. This will be required in the case of large-scale manufacturing. The downstream processing of the synthesized NPs should also be green and cost-effective. We need to consider all of these factors and further the research in this very promising field that can be a norm in the future if understood well enough.

References

Abboud, Y., Saffaj, T., Chagraoui, A., El Bouari, A., Brouzi, K., Tanane, O., Ihssane, B., 2014. Biosynthesis, characterization and antimicrobial activity of copper oxide nanoparticles (CONPs) produced using brown alga extract (*Bifurcaria bifurcata*). Appl. Nanosci. 4 (5), 571–576.

Abdel-Raouf, N., Al-Homaidan, A.A., Ibraheem, I.B.M., 2012. Agricultural importance of algae. Afr. J. Biotechnol. 11 (54), 11648–11658.

Abdel-Raouf, N., Alharbi, R.M., Al-Enazi, N.M., Alkhulaifi, M.M., Ibraheem, I.B.M., 2018. Rapid biosynthesis of silver nanoparticles using the marine red alga *Laurencia catarinensis* and their characterization. Beni Suef Univ. J. Basic Appl. Sci. 7 (1), 150–157.

Abdel-Raouf, N., Al-Enazi, N.M., Ibraheem, I.B.M., Alharbi, R.M., Alkhulaifi, M.M., 2019. Biosynthesis of silver nanoparticles by using of the marine brown alga *Padina pavonia* and their characterization. Saudi J. Biol. Sci. 26 (6), 1207–1215.

Aboelfetoh, E.F., El-Shenody, R.A., Ghobara, M.M., 2017. Eco-friendly synthesis of silver nanoparticles using green algae (*Caulerpa serrulata*): reaction optimization, catalytic and antibacterial activities. Environ. Monit. Assess. 189 (7), 349.

Agriculture NSW Water Unit, 2017. Farm Water Quality and Treatment: Algae, third ed. Department of Primary Industries. Prime Fact 1506.

Almasi, H., Jafarzadeh, P., Mehryar, L., 2018. Fabrication of novel nanohybrids by impregnation of CuO nanoparticles into bacterial cellulose and chitosan nanofibers: characterization, antimicrobial and release properties. Carbohydr. Polym. 186, 273–281.

Alvarez-Gongora, C., Herrera-Silveira, J.A., 2006. Variations of phytoplankton community structure related to water quality trends in a tropical karstic coastal zone. Mar. Pollut. Bull. 52 (1), 48–60.

Asha, K.S., Johnson, M., Chandra, P.K., Shibila, T., Revathy, I., 2015. Extracellular synthesis of silver nanoparticles from a marine alga, *Sargassum polycystum C. agardh* and their biopotentials. WJPPS 4, 1388–1400.

Balaraman, P., Balasubramanian, B., Kaliannan, D., Durai, M., Kamyab, H., Park, S., Chelliapan, S., Lee, C.T., Maluventhen, V., Maruthupandian, A., 2020. Phyco-synthesis of silver nanoparticles mediated from marine algae *Sargassum myriocystum* and its potential biological and environmental applications. Waste Biomass Valoriz. 11 (10), 5255–5271.

Bao, Z., Lan, C.Q., 2018. Mechanism of light-dependent biosynthesis of silver nanoparticles mediated by cell extract of *Neochloris oleoabundans*. Colloids Surf. B Biointerfaces 170, 251–257.

Barsanti, L., Gualtieri, P., 2014. Algae: Anatomy, Biochemistry, and Biotechnology. CRC Press.

Bayda, S., Adeel, M., Tuccinardi, T., Cordani, M., Rizzolio, F., 2020. The history of nanoscience and nanotechnology: from chemical–physical applications to nanomedicine. Molecules 25 (1), 112.

Bhattacharya, A.K., Mandal, S.N., Das, S.K., 2006. Adsorption of Zn (II) from aqueous solution by using different adsorbents. Chem. Eng. J. 123 (1–2), 43–51.

Bhuyar, P., Rahim, M.H.A., Sundararaju, S., Ramaraj, R., Maniam, G.P., Govindan, N., 2020. Synthesis of silver nanoparticles using marine macroalgae *Padina* sp. and its antibacterial activity towards pathogenic bacteria. Beni Suef Univ. J. Basic Appl. Sci. 9 (1), 1–15.

Blunden, G., Barwell, C.J., Fidgen, K.J., Jewers, K., 1981. A survey of some British marine algae for anti-influenza virus activity. Bot. Mar. 24 (5), 267–272.

Bolyshev, N.N., Novichkova-Ivanova, L.N., 1978. The algal flora of initial stages of soil formation on blanket loam under the condition of a lysimeter (Moscow & Leningrad). Bot. Zh. 63, 690–696.

Borah, D., Das, N., Das, N., Bhattacharjee, A., Sarmah, P., Ghosh, K., Chandel, M., Rout, J., Pandey, P., Ghosh, N.N., Bhattacharjee, C.R., 2020. Alga-mediated facile green synthesis of silver nanoparticles: photophysical, catalytic and antibacterial activity. Appl. Organomet. Chem. 34 (5), e5597.

Celik, K., 2006. Spatial and seasonal variations in chlorophyll-nutrient relationships in the shallow hypertrophic lake Manyas, Turkey. Environ. Monit. Assess. 117 (1–3), 261–269.

Chakraborty, N., Banerjee, A., Lahiri, S., Panda, A., Ghosh, A.N., Pal, R., 2009. Biorecovery of gold using cyanobacteria and a eukaryotic alga with special reference to nanogold formation–a novel phenomenon. J. Appl. Phycol. 21 (1), 145.

Chandran, S.P., Chaudhary, M., Pasricha, R., Ahmad, A., Sastry, M., 2006. Synthesis of gold nanotriangles and silver nanoparticles using Aloevera plant extract. Biotechnol. Prog. 22 (2), 577–583.

Clarance, P., Luvankar, B., Sales, J., Khusro, A., Agastian, P., Tack, J.C., Al Khulaifi, M.M., AL-Shwaiman, H.A., Elgorban, A.M., Syed, A., Kim, H.J., 2020. Green synthesis and characterization of gold nanoparticles using endophytic fungi *Fusarium solani* and its in-vitro anticancer and biomedical applications. Saudi J. Biol. Sci. 27 (2), 706–712.

Colin, J.A., Pech-Pech, I.E., Oviedo, M., Águila, S.A., Romo-Herrera, J.M., Contreras, O.E., 2018. Gold nanoparticles synthesis assisted by marine algae extract: biomolecules shells from a green chemistry approach. Chem. Phys. Lett. 708, 210–215.

Datta, K.K., Dayal, B., 2000. Irrigation with poor quality water: an empirical study of input use, economic loss, and coping strategies. Indian J. Agric. Econ. 55 (1), 26–37.

Devi, M., Awasthi, S., 2019. Gold nanoparticles in drug delivery systems: therapeutic applications. In: AIP Conference Proceedings. vol. 2142(1). AIP Publishing LLC, p. 150013.

Dhas, T.S., Kumar, V.G., Karthick, V., Angel, K.J., Govindaraju, K., 2014. Facile synthesis of silver chloride nanoparticles using marine alga and its antibacterial efficacy. Spectrochim. Acta A Mol. Biomol. Spectrosc. 120, 416–420.

Durán, N., Marcato, P.D., Alves, O.L., De Souza, G.I., Esposito, E., 2005. Mechanistic aspects of biosynthesis of silver nanoparticles by several *Fusarium oxysporum* strains. J. Nanobiotechnol. 3 (1), 8.

Ebrahimzadeh, M.A., Naghizadeh, A., Amiri, O., Shirzadi-Ahodashti, M., Mortazavi-Derazkola, S., 2020. Green and facile synthesis of Ag nanoparticles using *Crataegus pentagyna* fruit extract (CP-AgNPs) for organic pollution dyes degradation and antibacterial application. Bioorg. Chem. 94, 103425.

Eccles, H., 1999. Treatment of metal-contaminated wastes: why select a biological process? Trends Biotechnol. 17 (12), 462–465.

Edison, T.N.J.I., Atchudan, R., Kamal, C., Lee, Y.R., 2016. *Caulerpa racemosa*: a marine green alga for eco-friendly synthesis of silver nanoparticles and its catalytic degradation of methylene blue. Bioprocess Biosyst. Eng. 39 (9), 1401–1408.

Elgamouz, A., Idriss, H., Nassab, C., Bihi, A., Bajou, K., Hasan, K., Abu Haija, M., Patole, S.P., 2020. Green synthesis, characterization, antimicrobial, anti-Cancer, and optimization of colorimetric sensing of hydrogen peroxide of algae extract capped silver nanoparticles. Nanomaterials 10 (9), 1861.

El-Shafey, E.I., 2010. Removal of Zn (II) and hg (II) from aqueous solution on a carbonaceous sorbent chemically prepared from rice husk. J. Hazard. Mater. 175 (1–3), 319–327.

El-Sheekh, M.M., Shabaan, M.T., Hassan, L., Morsi, H.H., 2020. Antiviral activity of algae biosynthesized silver and gold nanoparticles against Herps simplex (HSV-1) virus in vitro using cell-line culture technique. Int. J. Environ. Health Res. 32 (3), 1–12.

Emerich, D.F., Thanos, C.G., 2003. Nanotechnology and medicine. Expert Opin. Biol. Ther. 3 (4), 655–663.

Fatima, R., Priya, M., Indurthi, L., Radhakrishnan, V., Sudhakaran, R., 2020. Biosynthesis of silver nanoparticles using red algae *Portieria hornemannii* and its antibacterial activity against fish pathogens. Microb. Pathog. 138, 103780.

Food and Agriculture Organisation of the United Nations, 2017. Water Pollution from and to Agriculture. Water Action Decade. Accessed 1 July 2021.

Gautam, G., Mishra, P., 2017. Development and characterization of copper nanocomposite containing bilayer film for coconut oil packaging. J. Food Process. Preserv. 41 (6), e13243.

Gericke, M., Pinches, A., 2006. Microbial production of gold nanoparticles. Gold Bull. 39 (1), 22–28.

Gole, A., Dash, C., Ramakrishnan, V., Sainkar, S.R., Mandale, A.B., Rao, M., Sastry, M., 2001. Pepsin-gold colloid conjugates: preparation, characterization, and enzymatic activity. Langmuir 17 (5), 1674–1679.

Govindaraju, K., Kiruthiga, V., Kumar, V.G., Singaravelu, G., 2009. Extracellular synthesis of silver nanoparticles by a marine alga, *Sargassum wightii Grevilli* and their antibacterial effects. J. Nanosci. Nanotechnol. 9 (9), 5497–5501.

Gu, H., Chen, X., Chen, F., Zhou, X., Parsaee, Z., 2018. Ultrasound-assisted biosynthesis of CuO NPs using brown alga *Cystoseira trinodis*. characterization, photocatalytic AOP, DPPH scavenging and antibacterial investigations. Ultrason. Sonochem. 41, 109–119.

Gunalan, S., Sivaraj, R., Venckatesh, R., 2012. *Aloe barbadensis* Miller mediated green synthesis of mono-disperse copper oxide nanoparticles: optical properties. Spectrochim. Acta A Mol. Biomol. Spectrosc. 97, 1140–1144.

Gunasundari, E., Kumar, P.S., Christopher, F.C., Arumugam, T., Saravanan, A., 2017. Green synthesis of metal nanoparticles loaded ultrasonic-assisted *Spirulina platensis* using algal extract and their antimicrobial activity. IET Nanobiotechnol. 11 (6), 754–758.

Gupta, A.B., Lata, K., 1964. Effect of algal growth hormones on the germination of paddy seeds. Hydrobiologia 24, 430–434.

Honary, S., Barabadi, H., Gharaei-Fathabad, E., Naghibi, F., 2012. Green synthesis of copper oxide nanoparticles using *Penicillium aurantiogriseum*, *Penicillium citrinum* and *Penicillium waksmanii*. Dig. J. Nanomater. Biostruct. 7 (3), 999–1005.

Huang, T.C., Chow, T.J., 1988. Comparative studies of some nitrogen-fixing unicellular cyanobacteria Isolated from rice fields. Microbiology 134, 3089–3097.

Ingle, A., Gade, A., Pierrat, S., Sonnichsen, C., Rai, M., 2008. Mycosynthesis of silver nanoparticles using the fungus *Fusarium acuminatum* and its activity against some human pathogenic bacteria. Curr. Nanosci. 4 (2), 141–144.

Kailasa, S.K., Park, T.J., Rohit, J.V., Koduru, J.R., 2019. Antimicrobial activity of silver nanoparticles. Nanoparticles in Pharmacotherapy, 461–484. https://doi.org/10.1016/B978-0-12-816504-1.00009-0.

Khaleelullah, M.M.S.I., Murugan, M., Radha, K.V., Thiyagarajan, D., Shimura, Y., Hayakawa, Y., 2017. Synthesis of super-paramagnetic iron oxide nanoparticles assisted by brown seaweed *Turbinaria decurrens* for removal of reactive navy-blue dye. Mater. Res. Exp. 4 (10), 105038.

Kim, J.H., Lee, J.E., Kim, K.H., Kang, N.J., 2018. Beneficial effects of marine algae-derived carbohydrates for skin health. Mar. Drugs 16 (11), 459.

Kiran, M.V., Murugesan, S., 2014. In vitro cytotoxic activity of silver nano particle biosynthesized from *Colpomenia sinuosa* and *Halymenia poryphyroides* using DLA and EAC cell lines. World J. Pharm. Sci. 2 (9), 926.

Kumar, B., Smita, K., Sánchez, E., Guerra, S., Cumbal, L., 2016. Ecofriendly ultrasound-assisted rapid synthesis of gold nanoparticles using *Calothrix* algae. Adv. Nat. Sci. Nanosci. Nanotechnol. 7 (2), 025013.

Li, L., Zhang, Z., 2016. Biosynthesis of gold nanoparticles using green alga *Pithophora oedogonia* with their electrochemical performance for determining Carbendazim in soil. Int. J. Electrochem. Sci. 11, 4550–4559.

Lomate, G.B., Dandi, B., Mishra, S., 2018. Development of antimicrobial LDPE/Cu nanocomposite food packaging film for extended shelf life of peda. Food Packag. Shelf Life 16, 211–219.

Mahdavi, M., Namvar, F., Ahmad, M.B., Mohamad, R., 2013. Green biosynthesis and characterization of magnetic iron oxide (Fe3O4) nanoparticles using seaweed (*Sargassum muticum*) aqueous extract. Molecules 18 (5), 5954–5964.

Manikandan, R., Anjali, R., Beulaja, M., Prabhu, N.M., Koodalingam, A., Saiprasad, G., Chitra, P., Arumugam, M., 2019. Synthesis, characterization, anti-proliferative and wound healing activities of silver nanoparticles synthesized from *Caulerpa scalpelliformis*. Process Biochem. 79, 135–141.

Marathe, K.V., Chaudhari, P.R., 1975. An example of algae as pioneers in the Lithosere and their role in rock corrosion. J. Ecol. 63, 65.

Mehta, S.K., Gaur, J.P., 2005. Use of algae for removing heavy metal ions from wastewater: progress and prospects. Crit. Rev. Biotechnol. 25 (3), 113–152.

Miyashita, K., 2009. The carotenoid fucoxanthin from brown seaweed affects obesity. Lipid Technol. 21 (8–9), 186–190.

Mohanpuria, P., Rana, N.K., Yadav, S.K., 2008. Biosynthesis of nanoparticles: technological concepts and future applications. J. Nanopart. Res. 10 (3), 507–517.

Momeni, S., Nabipour, I., 2015. A simple green synthesis of palladium nanoparticles with *Sargassum* alga and their electrocatalytic activities towards hydrogen peroxide. Appl. Biochem. Biotechnol. 176 (7), 1937–1949.

Muinao, T., Boruah, H.P.D., Pal, M., 2019. Current advances on polysaccharide-based nanoconjugates for efficient and targeted drug delivery in cancer treatment. In: Polysaccharide Carriers for Drug Delivery. Woodhead Publishing, pp. 413–441.

Murugesan, S., Bhuvaneswari, S., Sivamurugan, V., 2017. Green synthesis, characterization of silver nanoparticles of a marine red alga *Spyridia fusiformis* and their antibacterial activity. Int J Pharm Pharm Sci 9 (5), 192–197.

Namvar, F., Mohamed, S., Fard, S.G., Behravan, J., Mustapha, N.M., Alitheen, N.B.M., Othman, F., 2012. Polyphenol-rich seaweed (*Eucheuma cottonii*) extract suppresses breast tumour via hormone modulation and apoptosis induction. Food Chem. 130 (2), 376–382.

Nishino, T., Fukuda, A., Nagumo, T., Fujihara, M., Kaji, E., 1999. Inhibition of the generation of thrombin and factor Xa by a fucoidan from the brown seaweed *Ecklonia kurome*. Thromb. Res. 96 (1), 37–49.

Nordberg, G.F., 1992. Cadmium in the human environment: toxicity and carcinogenicity. IARC Sci. Publ. 118, 1–470.

O'Connell, D.W., Birkinshaw, C., O'Dwyer, T.F., 2008. Heavy metal adsorbents prepared from the modification of cellulose: a review. Bioresour. Technol. 99 (15), 6709–6724.

O'Farrell, I., Lombardo, R.J., de Pinto, P.T., Loez, C., 2002. The assessment of water quality in the lower Luján River (Buenos Aires, Argentina): phytoplankton and algal bioassays. Environ. Pollut. 120 (2), 207–218.

Omomowo, I.O., Adenigba, V.O., Ogunsona, S.B., Adeyinka, G.C., Oluyide, O.O., Adedayo, A.A., Fatukasi, B.A., 2020, March. Antimicrobial and antioxidant activities of algal-mediated silver and gold nanoparticles. In: IOP Conference Series: Materials Science and Engineering. 805(1). IOP Publishing. 012010.

Oster, J.D., 1994. Irrigation with poor water quality. Agric Water Manag 25 (3), 271–297.

Oster, J.D., Clothier, B.E., Wichelns, D., 2006. History of agricultural water management, Agric. Agric Water Manag 86 (1–2), 3–8.

Öztürk, B.Y., Gürsu, B.Y., Dağ, İ., 2020. Antibiofilm and antimicrobial activities of green synthesized silver nanoparticles using marine red algae *Gelidium corneum*. Process Biochem. 89, 208–219.

Parial, D., Pal, R., 2015. Biosynthesis of monodisperse gold nanoparticles by green alga *Rhizoclonium* and associated biochemical changes. J. Appl. Phycol. 27 (2), 975–984.

Parvathi, K., Kumar, R.N., Nagendran, R., 2007. Biosorption of manganese by *aspergillus Niger* and *Saccharomyces cerevisiae*. World J. Microbiol. Biotechnol. 23 (5), 671–676.

Pellera, F.M., Giannis, A., Kalderis, D., Anastasiadou, K., Stegmann, R., Wang, J.Y., Gidarakos, E., 2012. Adsorption of Cu (II) ions from aqueous solutions on biochars prepared from agricultural by-products. J. Environ. Manage. 96 (1), 35–42.

Perez, G.R.M., Zavala, S.M.A., Perez, G.S., Perez, G.C., 1998. Antidiabetic effect of compounds isolated from plants. Phytomedicine 5 (1), 55–75.

Posocco, B., Dreussi, E., De Santa, J., Toffoli, G., Abrami, M., Musiani, F., Grassi, M., Farra, R., Tonon, F., Grassi, G., Dapas, B., 2015. Polysaccharides for the delivery of antitumor drugs. Materials 8 (5), 2569–2615.

Rajamanickam, U., Mylsamy, P., Viswanathan, S., Muthusamy, P., 2012. Biosynthesis of zinc nanoparticles using actinomycetes for antibacterial food packaging. In: International Conference on Nutrition and Food Sciences. vol. 39. Singapore.

Rajasulochana, P., Dhamotharan, R., Murugakoothan, P., Murugesan, S., Krishnamoorthy, P., 2010. Biosynthesis and characterization of gold nanoparticles using the alga Kappaphycus alvarezii. Int. J. Nanosci. 9 (05), 511–516.

Rajathi, F.A.A., Parthiban, C., Kumar, V.G., Anantharaman, P., 2012. Biosynthesis of antibacterial gold nanoparticles using brown alga, *Stoechospermum marginatum* (kützing). Spectrochim. Acta A Mol. Biomol. Spectrosc. 99, 166–173.

Rangabhashiyam, S., Selvaraju, N., 2015. Efficacy of unmodified and chemically modified *Swietenia mahagoni* shells for the removal of hexavalent chromium from simulated wastewater. J. Mol. Liq. 209, 487–497.

Salam, H.A., Rajiv, P., Kamaraj, M., Jagadeeswaran, P., Gunalan, S., Sivaraj, R., 2012. Plants: green route for nanoparticle synthesis. Int. Res. J. Biol. Sci. 1 (5), 85–90.

Scarano, G., Morelli, E., 2003. Properties of phytochelatin-coated CdS nanocrystallites formed in a marine phytoplanktonic alga (*Phaeodactylum tricornutum*, Bohlin) in response to Cd. Plant Sci. 165 (4), 803–810.

Senapati, S., Syed, A., Moeez, S., Kumar, A., Ahmad, A., 2012. Intracellular synthesis of gold nanoparticles using alga *Tetraselmis kochinensis*. Mater. Lett. 79, 116–118.

Shankar, P.D., Shobana, S., Karuppusamy, I., Pugazhendhi, A., Ramkumar, V.S., Arvindnarayan, S., Kumar, G., 2016. A review on the biosynthesis of metallic nanoparticles (gold and silver) using bio-components of microalgae: formation mechanism and applications. Enzyme Microb. Technol. 95, 28–44.

Sharma, B., Purkayastha, D.D., Hazra, S., Gogoi, L., Bhattacharjee, C.R., Ghosh, N.N., Rout, J., 2014a. Biosynthesis of gold nanoparticles using a freshwater green alga, *Prasiola crispa*. Mater. Lett. 116, 94–97.

Sharma, B., Purkayastha, D.D., Hazra, S., Thajamanbi, M., Bhattacharjee, C.R., Ghosh, N.N., Rout, J., 2014b. Biosynthesis of fluorescent gold nanoparticles using an edible freshwater red alga, *Lemanea fluviatilis* (L.) C. Ag. and antioxidant activity of biomatrix loaded nanoparticles. Bioprocess Biosyst. Eng. 37 (12), 2559–2565.

Singaravelu, G., Arockiamary, J.S., Kumar, V.G., Govindaraju, K., 2007. A novel extracellular synthesis of monodisperse gold nanoparticles using marine alga, *Sargassum wightii* Greville. Colloids Surf. B Biointerfaces 57 (1), 97–101.

Singh, M., Kalaivani, R., Manikandan, S., Sangeetha, N., Kumaraguru, A.K., 2013. Facile green synthesis of variable metallic gold nanoparticle using *Padina gymnospora*, a brown marine macroalga. Appl. Nano 3 (2), 145–151.

Singh, A.K., Tiwari, R., Kumar, V., Singh, P., Khadim, S.R., Tiwari, A., Srivastava, V., Hasan, S.H., Asthana, R.K., 2017. Photo-induced biosynthesis of silver nanoparticles from aqueous extract of *Dunaliella salina* and their anticancer potential. J. Photochem. Photobiol. B Biol. 166, 202–211.

Singh, A.K., Singh, V.K., Singh, M., Singh, P., Khadim, S.R., Singh, U., Koch, B., Hasan, S.H., Asthana, R.K., 2019. One pot hydrothermal synthesis of fluorescent NP-carbon dots derived from *Dunaliella salina* biomass and its application in on-off sensing of Hg (II), Cr (VI) and live cell imaging. J. Photochem. Photobiol. A Chem. 376, 63–72.

Sinha, S.N., Paul, D., Halder, N., Sengupta, D., Patra, S.K., 2015. Green synthesis of silver nanoparticles using fresh water green alga *Pithophora oedogonia* (Mont.) Wittrock and evaluation of their antibacterial activity. Appl. Nano 5 (6), 703–709.

Snelgrove, P.V., 2016. An ocean of discovery: biodiversity beyond the census of marine life. Planta Med. 82 (09/10), 790–799.

Stewart, W.D.P., 1970. Algal fixation of atmospheric nitrogen. Plant Soil, JTSOR 32, 555–588.

Suganya, K.U., Govindaraju, K., Kumar, V.G., Dhas, T.S., Karthick, V., Singaravelu, G., Elanchezhiyan, M., 2015. Blue green alga mediated synthesis of gold nanoparticles and its antibacterial efficacy against gram positive organisms. Mater. Sci. Eng. C 47, 351–356.

Surwade, P., Weikel, C., Luxton, T., Peloquin, D., Fan, X., Shah, V., 2019. Augmented antibacterial activity of ampicillin with silver nanoparticles against methicillin-resistant Staphylococcus aureus (MRSA). J. Antibiot. 72 (1), 50–53.

Uzair, B., Liaqat, A., Iqbal, H., Menaa, B., Razzaq, A., Thiripuranathar, G., Fatima Rana, N., Menaa, F., 2020. Green and cost-effective synthesis of metallic nanoparticles by algae: safe methods for translational medicine. Bioengineering 7 (4), 129.

Vargas, R., Novelo, E., 2007. Seasonal changes in periphyton nitrogen fixation in a protected tropical wetland. Biol. Fertil. Soils 43, 367–372.

Venkatesan, J., Manivasagan, P., Kim, S.K., Kirthi, A.V., Marimuthu, S., Rahuman, A.A., 2014. Marine algae-mediated synthesis of gold nanoparticles using a novel *Ecklonia cava*. Bioprocess Biosyst. Eng. 37 (8), 1591–1597.

Venkatesan, J., Lowe, B., Anil, S., Manivasagan, P., Kheraif, A.A.A., Kang, K.H., Kim, S.K., 2015. Seaweed polysaccharides and their potential biomedical applications. Starch-Stärke 67 (5–6), 381–390.

Venkatesan, J., Anil, S., Kim, S.K., Shim, M.S., 2016. Seaweed polysaccharide-based nanoparticles: preparation and applications for drug delivery. Polymers 8 (2), 30.

Yousefzadi, M., Rahimi, Z., Ghafori, V., 2014. The green synthesis, characterization and antimicrobial activities of silver nanoparticles synthesized from green alga *Enteromorpha flexuosa* (wulfen). J. Agardh. Mater. Lett. 137, 1–4.

Yugay, Y.A., Usoltseva, R.V., Silant'ev, V.E., Egorova, A.E., Karabtsov, A.A., Kumeiko, V.V., Ermakova, S.P., Bulgakov, V.P., Shkryl, Y.N., 2020. Synthesis of bioactive silver nanoparticles using alginate, fucoidan and laminaran from brown algae as a reducing and stabilizing agent. Carbohydr. Polym. 245, 116547.

Biogenic nanoparticles from cyanobacteria and their applications

4

Sreejita Ghosh[a], *Moupriya Nag*[b,*], *Dibyajit Lahiri*[b,*], *Sujay Ghosh*[c], *Ankita Dey*[a], *and Rina Rani Ray*[a]
[a]Department of Biotechnology, Maulana Abul Kalam Azad University of Technology, Haringhata, West Bengal, India, [b]Department of Biotechnology, University of Engineering & Management, Kolkata, West Bengal, India, [c]AMH Energy Pvt. Ltd., Kolkata, West Bengal, India

4.1 Introduction

Over the last few years, significant advancement has been observed in nanotechnology in the area of fabrication of nanomaterials with excellent physicochemical properties. Nanotechnology has been used in almost every field of science for the creation of newer approaches to solve various research-associated tailbacks. The field of nanobiotechnology has evolved with the aim to design nanoparticles (NPs) using biological counterparts influencing the physical properties. NP synthesis with a variety of shapes and sizes stimulated a great deal of interest because of their unique characteristics in comparison with their massive counterparts. Stability in terms of physicochemical, biochemical, and chemical characteristics of the NPs may vary widely at the nanoscale mostly because of the greater aspect ratio between surface area and volume (Baker et al., 2013). This promotes substantial differences in melting point, mechanical characteristics, thermal and electrical conductivities, optical absorption, and catalytic and biological activities (Chaudhary et al., 2020). NPs act as aqueducts between the molecular or atomic structures and the bulk products. The size of NPs ranges from 1 to 100 nm, and they are classified according to their structure, dimensions, and origin as depicted in Fig. 4.1.

Because of their unique physicochemical and morphological (charge, size, and shape) features of NPs, they are widely applied in almost every field of science such as energy, space, communication, defense, agriculture, and biomedicine. Nowadays, applications of NPs have significantly increased in the field of diagnosis of diseases, imaging, drug delivery, treatment, cancer therapies, and tissue engineering. Nevertheless, using NPs for biological purposes in humans is a bit critical, and so, it is better if the NPs used for biomedical purposes need to be obtained from green sources (Cai et al., 2008). Thus, considering different factors like bioavailability,

[*] Authors contributed equally.

Agricultural Nanobiotechnology. https://doi.org/10.1016/B978-0-323-91908-1.00003-1

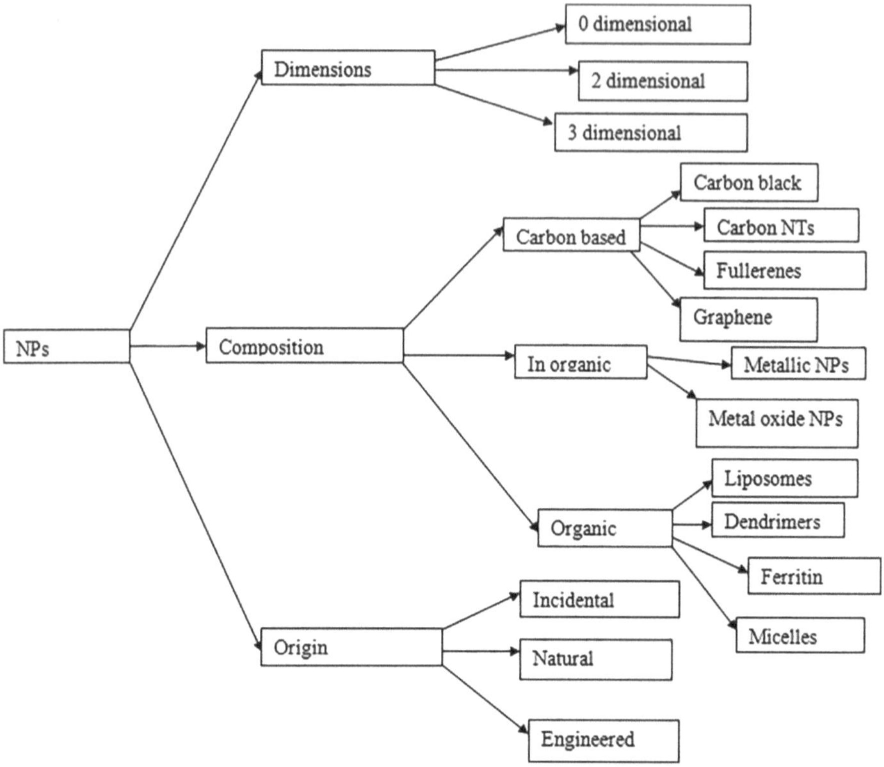

Fig. 4.1 Classification of NPs according to dimension, composition, and origin.

biocompatibility, biosafety, and biodistribution of NPs, the current NP synthesis process must shift toward the safer option. NPs are mainly produced by physical and chemical processes like mechanical milling, laser ablation, solgel, and electroexplosion, but such NPs are harmful for both human health and environment. Therefore, the urge for NP synthesis from eco-friendly, alternative, cost-effective, non-hazardous, and safer green technologies has increased. Organisms like fungi, algae, plants, and bacteria can be used as a green synthesis technology for NPs, which are free of harmful impurities. Such biological approaches consist of different biological units behaving as capping and reducing agents in either extracellular or intracellular NP synthesis (Iravani, 2011). Functional groups (thiol, carbonyl, amine, and hydroxyl), peptides, and proteins of the biological units reduce the salt precursor to NPs and are known as stabilizing or reducing agents. They stabilize and cap the NPs simultaneously, so this kind of synthesis is comparatively stable, reproducible, and simple.

NP biosynthesis is dependent on different factors such as bioextract and metal precursor concentration, pH, light source exposure, contact time in the reaction mixture, and temperature. Such factors are responsible for growth, nucleation, agglomeration, aging, and stabilization of NPs, thereby altering the crystallinity, shape, and size of the

NPs thus synthesized. Following their synthesis, NPs are subjected to different spectroscopic and microscopic characterization procedures for finding their novel physicochemical properties. To determine the actual physicochemical and morphological (distribution of particles, size, and shape) properties of these NPs is significant for using them for biomedical purposes. The mostly used microscopic technologies for finding morphological characteristics and size of NPs involve scanning electron microscopy (SEM), transmission electron microscopy (TEM), high-resolution transmission electron microscopy (HRTEM), and atomic force microscopy (AFM) (Menon et al., 2017). For analysis of crystallinity, structure and composition of the NPs, energy-dispersive spectroscopy (EDS), UV–Visible spectroscopy (UV–Vis), dynamic lights scattering (DLS), selected area electron diffraction (SAED), X-ray photoelectron spectroscopy (XPS), X-ray diffraction (XRD), and Fourier transform infrared spectroscopy (FTIR) are usually used (Shah et al., 2015).

For green methods of synthesis, the use of algae has increased since they are considered as good sources of proteins, secondary metabolites, peptides, and pigments that can act as nanobiofactories. Nevertheless, they have high rate of growth, they are easy to harvest, and their scale-up is not much expensive, thereby becoming potential entities for NP biosynthesis. Algae are the most ancient organisms that inhabit various ecosystems, and they are major photosynthetic organisms found on the Earth. Algae have the capacity of hyper-accumulating metals and then transforming them into NPs, and they are one of the best choices for green synthesis of NPs. Many of the metal oxide and metallic NPs are synthesized by different types of algae like brown algae (Phaeophyceae), blue-green algae (cyanobacteria), red algae (Rhodophyceae), and green algae (Chlorophyceae).

Over a long period of time, algae have been both commercially and industrially used as feed, food, cosmetics, additives, fertilizers, pharmaceuticals, and agents of bioremediation. Algae are used for biosynthesis of NPs because they require low input of energy, no external capping or reducing agents are required, easy reaction, and low cost. This chapter focuses on the potential of cyanobacteria in biosynthesis of NPs, their biosynthesis mechanism, types of NPs produced by cyanobacteria, characterization of those NPs, and their applications.

4.2 Screening of cyanobacteria for synthesis of NPs

Various cyanobacterial strains like *Anabaena* sp. 66-2, *Aphanizomenon* sp. 127-1, *Cylindrospermopsis* sp. 121-1, *Cylindrospermopsis* sp. USC CRB3, *Lyngbya* sp. 15-2, *Limnothrix* sp. 37-2-1, *Synechocystis* sp. 48-3, and *Synechococcus* sp. 145-6 with the capability to synthesize AgNPs were screened by the use of 2 different methods, which are as follows: suspension of the washed live cyanobacterial biomass into the solution of silver nitrate and by the addition of silver nitrate solution within a culture media devoid of cells (Patel et al., 2014). It was observed that out of these 8 strains, 7 of them (except *Limnothrix* sp. 37-2-1) could synthesize AgNPs. Mostly, AgNPs could be synthesized from both the cell-free media and the biomass itself. This proves that AgNP biosynthesis requires an extracellular component such as polysaccharides. Analysis through TEM indicated that these NPs were found to be attached within the organic matrix. The AgNPs had different sizes (13–31 nm) and shapes depending on the source organism.

Husain et al. (2015) tested 30 strains of cyanobacteria such as *Arthrospira indica* PCC7940, *Chroococcus* NCCU-207, and *Gloeocapsa gelatinosa* NCCU430 for the biosynthesis of AgNPs. Aqueous extracts of cyanobacteria were exposed for the biosynthesis of AgNPs at a temperature of 30°C. On scanning, the aqueous extracts of these cyanobacteria after synthesizing AgNPs produced a single peak in the UV–Vis range. Micrographs of SEM showed that all the cyanobacterial strains could synthesize AgNPs; however, depending on the time of reaction, their size and shape changed. Most of the NPs so produced were spherical in shape. Reaction time taken for the synthesis of NPs was between 30 and 360 h, and the size ranges from 38 to 88 nm. With respect to size, *Cylindrospermum stagnale* NCCU-104 showed best results with NPs of size 38-40 nm. However, with respect to time, *Microchaete* sp. NCCU-342 showed best results by synthesizing AgNPs within a maximum of 30 h' time.

Aqueous extracts devoid of cells from *Leptolyngbya* sp. WUC59 found from polluted water could synthesize AgNPs by the use of polyphasic method (Singh et al., 2020). The biologically synthesized NPs were characterized by different techniques such as XRD, UV–Vis, HRTEM, EDS, and FTIR. The distinct change in color and the production of a single peak at 430 nm in the UV–Vis spectra confirmed the production of AgNPs. The physical and morphological appearance showed that the AgNPs so synthesized were crystals with size ranging between 20 and 35 nm. These AgNPs possessed antibacterial activities against *E. coli* and *Bacillus subtilis* and suppressed their growth at a concentration of 10 mg/L within a period of 3 h. At a concentration of 25 mg/L, these AgNPs can remarkably increase seed germination rates and initiated premature growth of seedlings of wheat (*Triticum aestivum*) within 4–8 days. Thus, from this study, it can be concluded that these AgNPs synthesized from *Leptolyngbya* sp. WUC59 can act as an alternative to the synthetic antibacterial agents in pharmaceutical industries and also for the treatment of bacterial infections in food crops.

Selenium (Se) is an important trace element needed by all living beings, and their excess as well as deficiency can cause various diseases. Substitutes of Se in higher concentration are toxic. So, Se in form of SeNPs can be an effective way to use them, but SeNPs produced by chemical and physical methods are also toxic to human health. Thus, biogenic SeNPs are very much in need since they are less toxic than chemically synthesized SeNPs. Thus, extracts from 20 different types of cyanobacteria like *Anabaena variabilis* NCCU-441, *Gloeocapsa gelatinosa* NCCU-430, and *Arthrospira indica* SOSA-4 were screened for SeNP synthesis (Afzal et al., 2019). The SeNPs were synthesized within a period of 5 days with size ranging between 11.8 and 60 nm. These SeNPs after their characterization were further tested for their antioxidant activities.

ZnO NPs are of extreme importance as alternatives to various chemotherapeutic and pharmaceutical agents. They are also called "generally recognized as safe" (GRAS) metal oxide NPs. Hence, there is an immediate need of ZnO NP production. Extracts from 30 cyanobacteria were screened for biosynthesis of ZnO NPs (Asif et al., 2021). Biosynthesis of ZnONPs was confirmed by the change of color to pale white from blue. An absorption peak was obtained at 372 nm in UV–Vis spectrum.

Analysis through SEM in the process of screening demonstrated that *Oscillatoria* sp. could synthesize the smallest sized ZnO NPs ($\sim$ 40 nm), which can be increasingly optimized for a greater yield of NPs by altering the reaction parameters such as temperature, pH, metal precursor and extract concentration and time of reaction. The optimum conditions for ZnO NPs biosynthesis were found to be 10 mL of extract, 0.02 M zinc nitrate, pH of 8 for 3 h at 30°C. The NPs were further purified using calcination at a temperature of 350°C. Biogenic ZnO NPs showed more effective antibacterial and antioxidant activity against both Gram-negative and Gram-positive bacteria in comparison with those synthesized chemically. Synergistic activity of streptomycin in combination with ZnO NPs can be also used as a biomedical alternative.

Biofabrication of ZnO NPs can be done by harvesting the metabolites in cellular filtrate of a newly identified and isolated cyanobacteria *Arthrospira platensis* (El-Belely et al., 2021). The ZnO NPs were either crystalline or spherical in shape and had a size range of 30–55 nm at a wavelength of 370 nm. FTIR results showed different peaks of absorption in relation to formation of NPs. XPS analysis confirmed the existence of ZnO at various bending energies. Data analysis indicated that the effects of biogenic ZnO NPs were dependent on doses. These ZnO NPs had antimicrobial effects against *S. aureus*, *B. subtilis*, *E. coli*, *P. aeruginosa*, and *C. albicans*, respectively.

Copper oxide NPs (CuO NPs) were extracellularly synthesized by *Phormidium cyanobacterium* (Rahman et al., 2010). *Phormidium* can synthesize CuO NPs by extracellular hydrolysis of copper ion by various metal chelating anionic reductases/proteins produced by the cyanobacteria under normal experimental conditions such as neutral pH, aerobic ambience, and room temperature. Besides reducing copper ions into CuO NPs, the proteins also play an important role in stabilizing the NPs in the presence of room temperature.

Saran et al. (2017) used the cell-free media of *S. platensis* for green synthesis of CuONPs with size ranging from 30 to 40 nm. The presence of CuO NPs can be visually confirmed by the formation of a deep brown color of the mixture and by the presence of a surface plasmon response band obtained at 259 nm by UV–Vis spectroscopy. These CuO NPs possessed great antibacterial effect against pathogenic Gram-negative bacteria like *Proteus vulgaris* MTCC-7299, *E. coli* MTCC-9721, and *Klebsiella pneumoniae* MTCC-9751 and also against Gram-positive bacteria like *Streptococcus epidermidis* MTCC-2639, *S. aureus* MTCC-9542, and *Bacillus cereus* MTCC-9017.

4.3 Mechanism of cyanobacteria-mediated synthesis of NPs

Cyanobacteria (blue-green algae) have an exceptional condition in the biological platform, and they are of the order *Chroococcales* having 2 different families called *Entophysalidaceae* and *Chroococcaceae*. The members belonging to these 2 families can be distinguished by their colony formations in different habitats for growth. They have a dense growth pattern forming parenchymatous masses of cells over moist rock surfaces in terrestrial environment. Every cyanobacterium is microscopic and is widely found in aquatic conditions. Cyanobacteria are naturally photoautotrophic since they utilize water as electron donor and have 2 photosynthetic pigments, which are carotene and chlorophyll *a* aiding in photosynthesis. According to their morphologies, they are also considered to be the equivalent to the single-celled bacteria (Khan et al., 2019). They are the only prokaryotes producing oxygen. Cyanobacteria have been immensely exploited for the green synthesis of different types of NPs.

Spherical AgNPs with diameter of 2–8 nm are synthesized by *S. platensis*, and they are effectively used in food, health, and pharmaceutical industries. Apart from *S. platensis*, AgNPs having different shapes and sizes are also synthesized by other cyanobacteria like *Plectonema boryanum* (octahedral NPs with diameter of 200 nm), *Oscillatoria willei* (spherical NPs with diameter of 10–25 nm), *C. stagnale* (pentagonal NPs with diameter of 38–88 nm), and *Microchaete diplosiphon* (spherical NPs with diameter of 80 nm) (Husain et al., 2015).

Just like AgNPs, *S. platensis* is also capable of synthesizing gold NPs (AuNPs). Research on AuNP synthesis showed that *S. platensis* synthesizes cubic, octahedral, or spherical AuNPs extracellularly using peptides and proteins as reducing agents (Kalabegishvili et al., 2012). Another type of important cyanobacteria is *Phormidium valderianum*, which is also called as alkalotolerant *Rhodococcus* that is capable of generating monodispersive intracellular triangular-shaped AuNPs at 530 nm of wavelength and absorbance of 1.897 in UV–Vis spectrometry (Mosulishvili et al., 2002). The emergence of a single broad peak of AuNPs at 530 nm is due to the surface plasmon response (SPR) depending on the species of surface absorption, dielectric constant of the medium, and size of the particles. *P. valderianum* has also been demonstrated to extracellularly produce hexagonal-, spherical-, and face-centered cubic (FCC) AuNPs with diameter of 24 nm by the use of cytoplasmic metabolites as reducing agents (Khan et al., 2019). Apart from monometallic NPs, *S. platensis* is also involved in the green synthesis of crystalline magnetic silica NPs and core-shell Ag-AuNPs using extracellular proteins (Mahdieh et al., 2012).

4.3.1 Intracellular NP synthesis

Intracellular NP synthesis is a process, in which the bulk material is reduced to NPs, and this whole process is performed inside the host cells depending on different biological parameters. In vivo NP synthesis method in the laboratory consists of three main steps: culturing of the target microbes, interaction between the living cells and the precursor solution and isolation and purification of the NPs from the cells for their characterization by means of physicochemical processes (Lengke et al., 2007). Mixture of precipitated NPs and microbes is centrifuged at a speed of 12,000 rpm. The pellets so obtained are washed numerous times using distilled water, and then, they are directly examined and/or those cells are lysed with processes like sonication for NP extraction and their subsequent analysis.

Intracellular interaction between the microbes and the bulk material is attained by means of two different types of protocols. The first protocol includes collection of the biomass from the microbial cells in their growth phase through centrifugation, and then, this biomass is washed for the removal of trace amounts of salts from the medium. The washed biomass is then dissolved in water, followed by mixing with an appropriate concentration of the precursor solution of the bulk material (Jenaa et al., 2012). The second protocol involves individual culturing of the microbes alongside the solution of the bulk material, and they are subjected to the culture conditions for a period of 2 weeks (Merin et al., 2010).

Inside cyanobacteria, ions get reduced intracellularly through the electron transport chain (ETC) involving energy-producing reactions such as photosynthesis and to a lesser extent in respiratory electron transportation system due to the presence of enzymes like NADH-dependent reductases, and these redox reactions take place in the cell membrane, thylakoid membrane, and cytoplasm (Jeffryes et al., 2015).

In the study conducted by Lengke et al. (2007), it was found that *P. boryanum* UTEX 485 system caused the precipitation of silver nitrate, octahedral, spherical, and anhedral AgNPs both in the culture solution and inside the cells. This is because the Ag ions were either exported via a membrane transport system or get reduced through an intracellular electron donor. It can be concluded from this study that the growth phase and the metabolic state (photosynthesis or respiration) of the organism are responsible for determining its capacity of NP biosynthesis. Significantly, *P. boryanum* was also found to reduce solutions of gold (III) chloride intracellularly to AuNPs (Lengke et al., 2006). It indicates that the process of biorecovery in *P. boryanum* is due to either the acid medium or the solution of gold (III) chloride killing the cyanobacteria and liberating more amount of organic sulfur further precipitating the AuNPs. The interaction of the organic material (produced from the membrane of cyanobacteria) or the cyanobacteria with gold (III) chloride increased metallic AuNP growth.

Nitrogen-fixing enzymes such as nitrogenases are potentially used as reducing agents to fabricate Au ions into AuNPs. Some macromolecules and polysaccharides may also take part in intracellular stabilization and reduction of NPs. Some reports also demonstrated that the different pigments of cyanobacteria may act as potential bioreductants that can bring about intracellular reduction of the bulk materials to NPs (El-Naggar et al., 2017).

4.3.2 Extracellular NP synthesis

Extracellular synthesis suggests that the NP synthesis is taking place outside the cell with effluxed biomolecules like ions, pigments, enzymes, proteins, antioxidants, and hormones, which play an important part in NP reduction. In the process of biofabrication, these biological molecules serve as capping and reducing agents. There exist different methods of extracellular NP syntheses like biomolecules, extracts from cell biomass, and culture media devoid of cells.

4.3.2.1 Biomolecule-mediated synthesis of NP

4.3.2.1.1 Pigments

Microalgae, cyanobacteria, algae, and actinomycete are differentiated from each other on the basis of their cellular pigments such as R-phycoerythrin, C-phycocyanin, and carotenoid that are capable of NP biosynthesis. C-phycocyanin is defined as a blue-colored accessory photosynthetic pigment forming a portion of phycobilisomes, which are structures connected with thylakoids helping in light trapping, followed by electron transfer to photosystem II (PS II) centers of reaction. Phycocyanin is known to be involved in electron transfer and can bind with heavy metals. Some of the recent studies demonstrated that C-phycocyanin is also involved in the biosynthesis of NP. For example, Patel et al. (2014) demonstrated that phycocyanin extracts from 2 diverse species of cyanobacteria (*Limnothrix* and *Spirulina* species) led to AgNPs production having various sizes and shapes. It was concluded from this study that the variations in morphological characteristics are due to the difference in the molecular weight and purity of the 2 different C-phycocyanin extracts. It was also demonstrated in this study that there exists a correlation between the capacity of C-phycocyanin in NP synthesis and the existence of light because the pigment will not be able to reduce silver nitrate to AgNPs in the absence of light. C-phycocyanin extracts of *Spirulina* serve as a capping or protective agent in the process of reduction of silver nitrate to AgNPs (Wei et al., 2017). A protein pigment extracted from the phycocyanin of *Nostoc linckia* was shown to carry out silver nitrate reduction to AgNP in the presence of direct luminescence (2400–2600 Lux) for a period of 24 h at room temperature (El-Naggar et al., 2017). In another investigation, extracts of phycoerythrin isolated from *Nostoc carneum* acted as a reducing agent in AgNP biosynthesis (El-Naggar et al., 2018). C-phycoerythrin extracts from cyanobacterial species *Phormidium tenue* NTDM05 can successfully synthesize cadmium sulfide (CdS) NPs (MubarakAli et al., 2012). Although the mechanism through which AgNPs are synthesized from C-phycocyanin is not clearly understood, it is assumed that the mechanism is more or less similar to that through which R-phycoerythrin (red pigment) extracts produce AgNPs without the inclusion of any external reduction agents.

4.3.2.1.2 Proteins

There have been extensive studies regarding the role of proteins in reduction and stabilization of NPs. Proteins and amino acids can act as reducing and stabilizing agents in the biosynthesis process of metal NPs (MNPs). MubarakAli et al. (2011) demonstrated that the existence of sulfate, carboxyl, and amino groups in the sequence of amino acids in proteins of cyanobacteria brings about the biological reduction of very small NPs having uniform distribution and particle size. With the help of a single-cell protein (SCP) isolate from *S. platensis*, extracellular production of Ag-Au core-shell NPs along with AuNPs and AgNPs was performed (Govindraraju et al., 2008). The FTIR bands were identified at 1242, 1541, and 1653 cm^{-1} in the form of amide I-III bands of proteins or polypeptides, and therefore, it was concluded from this study that cyanobacterial proteins played a major role in NP stabilization process. In the presence of stress due to heavy metals, modified or new proteins can also perform twofold function, serve as stabilizing agents, and precursor material fabrication to CuONPs. Rahman et al. (2010) demonstrated that a protein weighing 22 kDa extracted from *Phormidium* served as the reducing agent of copper sulfate to copper oxide (CuO) NPs. AuNP production from layer of polysaccharide in heterocyst and extracellular polymeric substance present in vegetative cells of *Anabaena cylindrica* was first identified by Rosken et al. (2016).

4.3.2.1.3 Enzymes

There have been extensive studies carried out to investigate the functions of different enzymes such as NADH-dependent enzymes and nitrate reductases in NP bioreduction process. Most importantly, nitrate reductase is a metalloprotein containing ions of molybdenum as cofactors and performs catalytic reactions in sulfur, carbon, and nitrogen cycle. Ali et al. (2019) demonstrated that these enzymes are significant in potential reduction of MNPs during their biodesigning.

It was found that nitrate reductase from *Chlorella pyrenoidosa* catalyzed the extracellular NADH-dependent synthesis of AuNPs (Oza et al., 2012). Cyanobacterial nitrogenase acts as reducing agent in the biosynthesis of AgNPs, AuNPs, PdNPs, and PtNPs (Brayner et al., 2009). Three different strains were chosen: *Leptolyngbya foveolarum*, *Anabaena flos-aquae*, and *Calothrix pulvinate* for NP synthesis. *C. pulvinate* was found to synthesize Pt and Pd NPs after a period of 15 days and 30 min, respectively, while AuNPs were synthesized within a few hours' time and that for AgNP took a span of 24 h. In contrast to this, when these strains and precursor bulk materials were incubated with nitrogenases, Pt, Pd, Ag, and Au were synthesized within a period of 72 h, 2 h, and 10 min, respectively.

4.3.2.1.4 Polysaccharides

Extracellular polysaccharide (EPS) obtained from *Nostoc commune* was used for the reduction of silver nitrate to AgNPs (Morsy et al., 2014). In short, 10 g dried cyanobacteria were mixed with water, followed by suspension in 100 mL potassium phosphate buffer at pH 7, and then, it was blended at a medium velocity for a period of 10 s with subsequent overnight stirring at room temperature. After this, the suspension was again homogenized for minimum 3 times for a span of 10 s, and then, it was kept at room temperature for 10 min. The top layer of the polysaccharides was harvested by the use of a spatula, and the lower aqueous portion was centrifuged at 6000 rpm at 20°C for 10 min. The pellets were rejected, and the water-soluble supernatant EPS portion was overnight dialyzed in distilled water, followed by freeze-drying with the help of a lyophilizer. Consequently, 50 mL aqueous EPS solution at a concentration of 10 mg/mL was transferred in 30 mM solution of silver nitrate. The reaction takes place in autoclave at 15 psi pressure and 121°C temperature for a period of 5 min. In order to precipitate the biological AgNPs, the suspension was further centrifuged at 15000 rpm for a period of 30 min at a temperature of 20°C. It was reported from this study that the EPS acted as a potential agent for capping and reduction, showing that AgNPs coated with EPS restricted the particles to come in contact with oxygen, thereby enhancing their stability.

4.3.2.2 Filtrate of cell biomass

In cellular biomass filtrate synthesis procedure of NPs, the microbial cultures are centrifuged, followed by washing of the pellets for 3 times using distilled water for the elimination of remnants of excessive metals from the culture medium. After this, the washed pellets are freeze-dried with a lyophilizer or dried using an oven at approximately 40°C with subsequent crushing using a mortar and pestle. The resultant fine powder is then mixed in distilled water and filtered with Whatman filter paper. At last, the filtrate is mixed using a solution of metal precursor and then incubated in the presence of appropriate conditions. In this synthesis process, active biological moieties like enzymes and proteins present in the cellular filtrate play important roles in stabilizing and biofabricating the NPs. In another method described by Hassan et al. (2019), the cellular biomass was collected during the mid-logarithmic phase and was washed numerous times for the removal of few trace elements, followed by soaking in distilled water for a particular amount of time. Then, after centrifugation of the mixture, the supernatant is utilized in synthesis of NPs.

4.3.2.3 Media devoid of cells

NP biosynthesis can be performed by the use of microbial culture media after the biomass has been removed by centrifugation. The extracellular filtered culture media are mixed with solution of metal precursors under specific conditions for NP biosynthesis. After testing 5 different strains of cyanobacteria, it was found that cell-free media of *Synechococcus* sp. synthesized AgNPs in the presence of light (Keskin et al., 2016). It was indicated that the biological contents like antioxidants, enzymes, alkaloids, and phenolics existing within the cell-free culture media play a vital role in controlling NP bioreduction method. Microbe-derived NPs are biofabricated by using cell-free media containing biomolecules like sulfur-containing proteins and NADH-reductases (Ashmore et al., 2018).

4.4 Types of NPs synthesized by cyanobacteria

Cyanobacteria contain non-membranous organelles like unstacked thylakoids, chloroplasts, cyanophycean starch, phycobiliprotein pigments and peptidoglycan walls or matrices. In addition to this, they are also enriched with biomolecules like enzymes, proteins, vitamins, antioxidants, and polysaccharides. Many of the NPs with different shapes and sizes have been produced so far using macro- or microorganisms, algae, plants, fungi, and bacteria, but nowadays, cyanobacteria have also evolved as a potential biomachinery for the biofabrication of NPs. Cyanobacteria are capable of synthesizing various NPs like MNPs (AuNPs and AgNPs), metal oxide NPs (CuONPs and ZnONPs), and semiconductor NPs (CdS NPs).

4.4.1 AuNPs

AuNPs are good conductors of heat and possess less toxicity to be used in normal cells, which are appropriate for the utilization in photothermal therapeutic approach for cancer treatment. Besides physiochemical approaches, biological approaches can also be used to synthesize AuNPs in ecologically safe manner. *Anabaena laxa* was first detected to catalyze the reduction of hydrogen tetrachloroaurate (III) to AuNPs by Lenartowicz et al., 2017. It was reported that *A. laxa* could fabricate AuNPs intracellularly having varied sizes and morphologies by the use of 3 various auric chloride concentrations (1, 0.5, and 0.1 mM). It was found that the broadest UV spectral peak was obtained at 560 and 545 nm in case of 1 and 0.5 mM of auric chloride. Three 2θ degrees (64.6°, 44.3°, and 38.2°) were obtained via X-ray diffraction study, demonstrating crystallinity of AuNPs. Even, micrographs of TEM revealed that AuNPs that appeared at 0.5 mM were spheres, but the NPs with other different shapes like irregular, hexagonal, or triangular were obtained. The sizes of such NPs range from 0 to 30 nm, but the larger particles are of sizes between 30 and 100 nm. At high concentration of auric chloride (1 mM), irregular, hexagonal-, and triangular-shaped AuNPs were obtained within the size range between 20 and 50 nm. Additionally, living cultures of cyanobacteria could also synthesize AuNPs more effectively than the dead cells, demonstrating that the metabolic activities of these live cyanobacterial cells play a major role in NP bioreduction method.

Dahoumane et al. (2012) indicated that *A. flos-aquae* could biologically reduce AuNPs intracellularly. Rosken et al. (2016) demonstrated that *Anabaena* sp. and *A. cylindrica* biologically reduced Au ions into AuNPs without producing harmful anatoxin-a. They also showed that *A. cylindrica* intracellularly produced AuNPs after incubation for a period of 4 h in the presence of direct luminescence. XRD results of these AuNPs revealed that they have an ellipse shape having an aspect ratio of 1.15. The AuNPs obtained by green synthesis method are spherical in shape, and their average size is around 10 nm. Nevertheless, micrographs obtained from TEM indicated that the AuNPs mostly found in 50% vegetative cells in comparison with heterocyst are found within the membranes of thylakoid. It was found in this investigation that the heterocysts are not much significant for NP production because the results indicated that greater than 10 vegetative cells synthesized AuNPs, whereas only a single heterocyst produced NPs. The bioreduction effects of 3 different cyanobacterial species *Microcoleus chthonoplastes*, *P. tenue*, and *P. valderianum* were tested in auric chloride solutions in the presence of varying pH values (Parial et al., 2011). It was found that all the 3 screened strains of cyanobacteria could synthesize AuNPs, but only *P. valderianum* could biosynthesize AuNPs having various sizes and shapes like hexagonal (25 nm), triangular (24 nm), and nanosphere (15 nm) with respect to varying pH values. *Gloeocapsa* sp. was used for intracellular production of AuNPs by the use of washed whole cells (10 mL culture) mixed with 90 mL auric chloride solution, followed by incubation and stirring at room temperature (Geetha et al., 2014). Within a few hours' time, AuNPs were synthesized with UV absorbance at 547 nm. The FTIR results displayed that the cyanobacterial proteins play an important role in bioreduction process forming these NPs with broad peaks at 3424.96 cm^{-1} (with respect to N—H stretching) and 1640.16 cm^{-1} (with respect to amide I). Apart from this, TEM and SEM micrographs revealed that AuNPs possess spherical shape with diameters lesser than 100 nm.

4.4.2 AgNPs

AgNPs are the most widely used MNPs in different fields and are used for manufacturing of industrial and medical products such as surgical instruments, dressings, detergents, masks, and healthcare products. Traditionally, AgNPs have been designed using physical and chemical techniques. Because of the shortcomings of these methods such as high requirement of energy, high cost, and generation of harmful by-products, green synthesis method is used for synthesizing clean NPs, which are devoid of any toxic effects from physicochemical approaches. The temperature for reaction and injection rate are the 2 main factors for synthesizing AgNPs via polyol process and through a modification of precursor injection rate (Kim et al., 2006). Spherical AgNPs with size 17 nm were produced at 100°C. On the contrary, some investigations showed that neutral organisms can generate NPs having the smallest range of size 10, 14.9, and 15 at ambient temperature (Bin-Meferji and Hamida, 2019). AgNPs can also be synthesized via photoinduced approach (Jin et al., 2003). It was found from this study that AgNPs possess triangular shape within the size range of 30–120 nm. Poly(styrene sulfonate) and citrate were used as chemical stabilizing agents to stabilize the obtained AgNPs. However, *Arthrospira platensis* and *Arthrospira maxima SAE* could synthesize triangular-shaped AgNPs (46 and 61 nm, respectively) at a temperature of 30°C and a period of 24 h without requiring any external stabilization agents (Husain et al., 2015).

Green synthesis approach of NP fabrication has emerged increasingly to find out new microorganisms having hydrolytic capability. Among all the biological units, cyanobacteria are one of the preferred organisms for producing AgNPs. *Desertifilum* IPPAS B-1220 could biofabricate AgNPs extracellularly by the use of extracted cell biomass in the presence of direct luminescence for a period of 24 h at ambient temperature (Hamida et al., 2020). D-SNP formation was confirmed by the UV absorption peak obtained at 421 nm. XRD pattern showed 5 2θ degrees at 77.7, 64.7, 44.6, 38.4, and 29.7, indicating D-SNP nanocrystallinity. Data obtained from FTIR exhibited varied peaks at 3453.72 cm^{-1}, 2353.23 cm^{-1}, 1626.05 cm^{-1}, 1042.35 cm^{-1}, and 601.81 cm^{-1} for the D-SNPs. With respect to dominant spectra of FTIR, it was reported that the major biomolecules bringing about stabilization and bioreduction processes are polysaccharides and proteins. It was also suggested that *Desertifilum* IPPAS B-1220 sp. can synthesize spherical AgNPs with size ranging between 4.5 and 26 nm.

P. boryanum UTEX 485 was first used to perform both extracellular and intracellular AgNP syntheses (Lengke et al., 2007). It was found that AgNPs are spherical or octahedral in shape with size ranging between 1 and 200 nm. XPS and EDS revealed the conjunction of trace elements like sulfur, phosphorus, and iron with AgNPs. The process of silver nitrate biofabrication could be associated with metabolic activity by the use of nitrate and reducing nitrate to ammonium and nitrite, and the ammonium is fixed in the form of glutamine before the death of the cyanobacteria. After cyanobacterial death, organic matter was released, resulting in AgNP precipitation within the solution. Recently, the hydrolytic effect of a new cyanobacterial strain *Nostoc* sp. Bahar M for extracellular biofabrication of AgNPs was investigated (Bin-Meferji and Hamida, 2019). It was detected that the plasmon resonance of AgNPs synthesized by *Nostoc*-associated process was at 403 nm, demonstrating that the AgNPs so synthesized are extremely small. XRD results confirmed the presence of FCC crystal structure of the AgNPs. SEM and TEM studies indicated that the AgNPs synthesized by *Nostoc* sp. Bahar M had uniform particle distribution and are spherical-shaped with an average nanosize of 14.9 nm. Even, from the FTIR results, it was found that the protein molecules play a significant role in AgNP biofabrication process. In a recent study, aqueous extract from the biomass of *Oscillatoria limnetica* was investigated for their capability to synthesize AgNPs within 18 h at room temperature (Hamouda et al., 2019). Spherical AgNPs were obtained with a varied size range between 3.30 and 17.97 nm. The different abiotic parameters such as time, pH, and concentration of silver nitrate and natural reducing agents were also studied for the determination of optimum conditions for AgNP biosynthesis. The most appropriate conditions for AgNP production by *O. limnetica* were found to be time of 18 h and a pH of 6.7. However, the entire process of synthesis is dose-dependent for concentration of the extract of cyanobacteria as well as silver nitrate.

Thirty species of cyanobacteria were tested for extracellular AgNP synthesis (Husain et al., 2015). The results indicated that all 30 species of cyanobacteria could synthesize AgNPs from their precursor materials having different sizes. *C. stagnale* was found to be the best in synthesizing smallest NPs with nanosizes ranging between 38 and 40 nm. On the other hand, Sudha et al. (2013) screened various species of cyanobacteria from Muthupet mangroves including *Oscillatoria*, *Aphanothece*, *Aphanocapsa*, *Microcoleus*, *Lyngbya*, *Phormidium*, *Spirulina*, *Gloeocapsa*, and *Synechococcus* for their AgNP biofabrication capacity. Among all the screened cyanobacteria species, only *Microcoleus* was found to reduce silver nitrate forming AgNPs spherical in shape with average diameter of 55 nm.

Patel et al. (2014) carried out intracellular as well as extracellular synthesis of AgNPs with dark and light incubation utilizing various isolates of cyanobacteria like *Aphanizomenon* sp., *Anabaena* sp., *Cylindrospermopsis* sp. USC CRB3, *Cylindrospermopsis* sp. 121-1, *Limnothrix* sp., *Lyngbya* sp., *Synechococcus* sp., and *Synechocystis* sp. Except *Aphanizomenon* sp. (could not synthesize nanoscale AgNPs), all the screened cyanobacteria could reduce silver nitrate to AgNPs in the presence of direct illumination by the use of either extracts of cellular biomass or cell-free culture media. Besides this, only cellular biomass extracts from *Synechocystis* sp., *Limnothrix* sp., and *Anabaena* sp. could synthesize AgNPs in dark conditions (Fig. 4.2).

4.4.3 Metal oxide NPs

Anabaena sp. and *Nostoc* sp. could synthesize ZnONPs, while *S. platensis* and *P. cyanobacterium* could synthesize CuONPs (Rahman et al., 2010). β-iron oxyhydroxide NPs could be synthesized by *Klebsormidium* sp., *C. pulvinata*, and *A. flos-aquae* (Brayner et al., 2009). It was demonstrated that there exist 2 different mechanisms by which higher plants can acquire energy and can be denoted as strategy I and strategy II. Strategy I includes acidification of soils and lateral root formation and formation of certain transfer cells within the rhizodermis, followed by Fe^{2+} transporter and Fe^{3+} chelate reductase induction. This strategy can be found in every plant excluding grasses. Grasses follow strategy II, where iron is taken in the form of Fe^{3+} siderophore complex. According to these observations, it was concluded that the process of iron oxide biofabrication by cyanobacteria acts according to the principle of strategy II (siderophore-mediated acquisition of iron). Due to the limitation of iron availability,

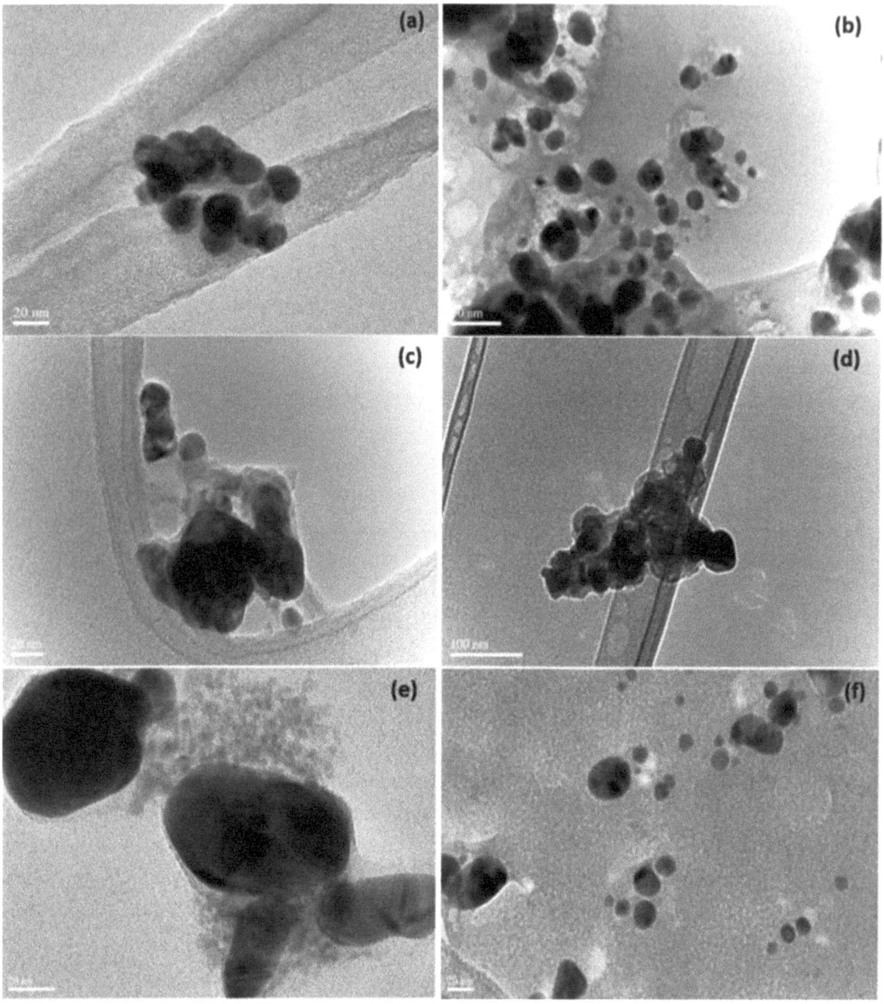

Fig. 4.2 Synthesis of Ag-NPs mediated by the biomass of (A) *Coelastrum* 143-1, (B) *Botryococcus* sp., (C) *Synechocystis* 48-3, (D) *Anabaena* 66-2, (E) *Limnothrix* sp. 37-1-2, and (F) C-phycocyanin from *Spirulina* sp. (Patel et al., 2014).

the cells produced organic Fe^{3+} metal chelating ions (siderophores), which can scavenge and solubilize iron ions from the surroundings and internalize them within the cells. Dahoumane et al. (2010) used *A. flos-aquae* for biosynthesis of akaganeite NPs. It was shown in this study that XRD analysis of the freeze-dried cyanobacteria, followed by reacting with iron salts, displayed the presence of akaganeite tetragonal NPs possessing an amorphous phase with respect to the cyanobacterial biomass. TEM of ultrathin-tested cyanobacterial sections with precursors of iron revealed the availability of dark β-FeOOH akaganeite nanorods within the cyanobacterial cells.

HRTEM gave more knowledge about the pattern of distribution of the akaganeite NPs within the cells, where the nanorods seem like a complicated arrangement of pores giving rise to a sponge-like structural formation.

4.4.4 Bimetallic NPs

Bimetallic NPs are formed due to the combination of 2 various metals, and their ratio is modified for providing unique physicochemical characteristics obtained from the constituent metals. These NPs have wide range of applications because of the excess degree of freedom they possess. In addition to AuNPs and AgNPs, Ag—Au core-shell NPs can be extracellularly synthesized with SCP *S. platensis* (Govindraraju et al., 2008). These Ag—Au bimetallic NPs exhibited absorbance maxima at 509 nm with size of 17–25 nm. *Lyngbya majuscula* was exposed to an equimolar mixture of Ag and Au (pH 4, 1 mM) for a period of 72 h for Au—Ag nanoalloy production (Roychoudhury et al., 2015). It was concluded that *Lyngbya* sp. can act as eco-friendly biofactories for biofabrication of Au—Ag nanoalloy. UV absorbance of this nanoalloy was found to be at 481 nm having a nanosize range between 5 and 25 nm.

4.5 Characterization of cyanobacteria-mediated NPs

Green synthesis methods comprise a modernized form of bottom-up nanotechnology for biofabrication of NPs. However, biological processes overcome many shortcomings to compete with the conventional physicochemical processes like synthesizing NPs with uniform sizes and shapes. The target of the green synthesis method is utilization of natural sources for conversion of bulk materials into NPs possessing novel characteristics. These natural sources used for NP biofabrication include both unicellular organisms like bacteria, diatoms, and cyanobacteria and multicellular organisms like plants and algae and the enzymes, SCPs, and pigments isolated from these organisms. The variation in effective natural sources and biomolecules allows NP production possessing different characteristics. Such NPs are made to undergo different characterization analyses for assessing their chemical and physical properties such as particle distribution, size, stability, dispersion, morphology of surfaces, and charge. These analytical methods make use of different spectroscopic methods like FTIR, UV–Vis spectroscopy, dynamic light scattering (DLS), zeta potential, and nuclear magnetic resonance (NMR) techniques for the determination of wavelength ranges and chemical constitution of biological coatings outside the biofabricated NPs and for the confirmation of NP production by characterization of their hydrodynamic diameters and their charges (Hamida et al., 2020). XPS, XRD, and energy-dispersive spectroscopy (EDS) can determine the structure, composition, and crystalline phase (Shah et al., 2015). Microscopic analyses like TEM, SEM, AFM, and HRTEM evaluate the surface morphology and size of the NPs (Khan et al., 2019). In addition to this, there exist many different techniques like vibrating sample magnetometer (VSM) and ferromagnetic resonance (FMR) for the assessment of the magnetic characteristics of NPs. These characterization techniques are further described in detail in Table 4.1.

Table 4.1 Characterization techniques of cyanobacteria-mediated NPs.

Analytical techniques	Type of analysis	Characters identified	References
X-ray analyses	XPS	Evaluation of NP structures	Dahoumane et al. (2016)
	XRD	Determination of crystal phase, structure, and composition of NPs	Shah et al. (2015)
	EDS	Evaluation of atomic structure of NPs	Shah et al. (2015)
Spectroscopic analyses	FTIR	Determination of chemical composition of biological coats outside the NPs	Hamida et al. (2020)
	UV–Vis spectroscopy	Confirmation of NP formation and identification of their wavelength range	MubarakAli et al. (2012)
	DLS	Estimation of hydrodynamic diameter	Poinern (2014)
	Zeta potential	Estimation of charge of NPs	Poinern (2014)
	NMR	Determination of purity and content of NPs	Nikolić et al. (2019)
Magnetic techniques	FMR, VSM, superconducting quantum interference device (SQUID)	Estimation of magnetization remanence, magnetic saturation, demagnetization field, and magnetic anisotropic constant of the NPs	Upadhyay et al. (2016)
Microscopic analyses	HRTEM and TEM	Determination of surface morphology and size of NPs	Quester et al. (2013)
	SEM	Determination of morphological appearance and sizes of NPs	Quester et al. (2013)
	AFM	Determination of size, geometry, roughness, and morphology of surfaces of the NPs	Khan et al. (2019)

4.6 Applications of cyanobacteria-mediated NPs

NPs synthesized biologically are generally biocompatible and devoid of harmful chemicals present on the surfaces since they do not make use of any external reducing or capping agents during NP biosynthesis, thereby possessing low toxicity in comparison with chemically obtained NPs. Cyanobacterial species also do not use any harmful chemicals for the stabilization and reduction of NPs because they have natural biomolecules imparting no or less toxicity, and therefore, these NPs can be safely used for different biomedical purposes. Various uses of these cyanobacterial NPs mostly in biomedicines are discussed in detail below.

4.6.1 Antimicrobial effects

4.6.1.1 Antibacterial effects

The overuse of antibiotics for treating bacterial diseases has led to the development of multidrug-resistant (MDR) strains of bacteria. Requirement for an efficient and safe treatment for MDR strains is a major problem faced globally. Thus, there has been a shift toward using NPs as an alternative antibacterial approach demonstrating superior and bactericidal effects. As NPs kill bacterial cells by cell membrane disruption synthesizing reactive oxygen species (ROS), they exhibit wide-spectrum antibacterial efficiency against Gram-negative and Gram-positive bacteria.

The cyanobacterial NPs have been screened for their antibacterial activities against a variety of bacteria. Colloidal-shaped stable AgNPs synthesized by the aqueous extract of *Caulerpa serrulata* showed excellent antimicrobial activity even at a low concentration against *Staphylococcus aureus*, *Shigella* sp., *Pseudomonas aeruginosa*, *Escherichia coli*, and *Salmonella typhi* (Aboelfetoh et al., 2017). The largest zone of inhibition was found to be 21 mm with 75 μL AgNPs against *E. coli*, while the smallest zone of inhibition was found to be 10 mm with 50 μL AgNPs against *S. typhi*. Similarly, AgNPs synthesized from the aqueous extract of *Pithophora oedogonia* have exhibited antibacterial effect against *Micrococcus luteus*, *E. coli*, *B. subtilis*, *S. aureus*, *P. aeruginosa*, *Vibrio cholerae*, and *Shigella flexneri* (Sinha et al., 2015). The largest zone of inhibition was found to be 17.2 mm against *P. aeruginosa* showing excellent antibacterial efficacy against MDR Gram-negative rods.

Spherical AuNPs biosynthesized from protein extracts of *S. platensis* remarkably retarded the growth of *B. subtilis* and *S. aureus* (Suganya et al., 2015). AuNPs produced by *Nitzschia* and *Ecklonia cava* showed antibacterial activity against *S. aureus*, *E. coli*, *B. subtilis*, *P. aeruginosa*, *Candida albicans*, *Aspergillus fumigates*, and *Aspergillus niger* (Borase et al., 2017). AuNPs from *Stoechospermum marginatum* exhibited excellent antibacterial effects against tetracycline resistant *Enterobacter faecalis* (Rajathi et al., 2012). *Neodesmus pupukensis* synthesized AuNPs and AgNPs that showed zone of inhibition of 24.5 mm against *E. coli*, 43 mm against *Pseudomonas*, 39 mm against *Serratia marcescens*, and 27 mm against *Klebsiella pneumoniae*, while AuNPs were found to be effective only against *S. marcescens* (28.5 mm) and *Pseudomonas* sp. (27.5 mm) (Omomowo et al., 2020). These results prove that cyanobacteria-mediated NPs can be effectively used as an alternative antibacterial agent in the future.

4.6.1.2 Antifungal effects

In a recent study, it was found that the AgNPs synthesized from *Nostoc* sp. HKAR-2 inhibited *Trichoderma harzianum* (IZ range 4–5) and *A. niger* (IZ range 3–5) (Sonker et al., 2017). AgNPs synthesized from the EPS of *N. commune* showed antifungal activity against fungal strains that infect the seeds of maize, sorghum, sesame, and soybean (Morsy et al., 2014). The grain surfaces were sterilized using AgNPs, followed by the growth of the seeds in dichloran chloramphenicol peptone agar at a temperature of 25°C using direct plate process. The growth of *Fusarium nygamai* was completely inhibited on soybean seeds and sorghum grains after sterilization with AgNPs showing an efficacy of 14.63% and 17.24% in comparison with NaOCl sterilization, which showed an efficacy of 8.7% and 7.5%, respectively.

4.6.1.3 Antialgal effects

The inhibitory effect of AgNPs synthesized from *Chlorella vulgaris*, *Scenedesmus obliquus*, and *S. platensis* was studied against *Microcystis aeruginosa* (El-Sheekh and El-Kassas, 2016). *M. aeruginosa* are responsible for causing liver cancer. From the results obtained in this study, it can be concluded that the AgNPs can be potentially used as antialgal agent for the reduction of toxicity resulting from *M. aeruginosa*. With increase in AgNP concentration, the cell viability of *M. aeruginosa* reduced to 10 mg/L because the AgNPs destroyed almost 30% of the total viable cells in 1 mg/L with an efficiency of 92%.

4.6.2 Antitumor effects

Cancer is one of the main reasons of death globally. Cancer is nowadays increasing because of different reasons such as overpopulation, aging, and lifestyles that make some individuals prone to this disease. Nanoscience advancements have led to the development of new methods for the prevention of metastasis and tumor growth.

There have been studies carried out using cyanobacteria-mediated synthesized NPs and their effects against various cell lines of cancer. Biosynthesized AgNPs by *Desertifilum* sp. drastically reduced cell viability of MCF-7 breast, HEPG2 liver, and Caco2 cell lines of colon cancer with half of the maximum inhibitory concentration (MIC) of 90, 58, and 32 μg/ mL, respectively, suggesting that they can be used as potential anticancer substances (Hamida et al., 2020). Bin-Meferji and Hamida (2019) found that AgNPs synthesized from *Nostoc* found the Caco2 viability of cells by 50% with MIC of 150 μg/ mL. This is because AgNPs caused morphological alterations like cell clustering, shrinkage of cells, and detachment of cells. The cytotoxic effects of AgNPs against Caco2 cells are due to their interactions with various cell organelles like endoplasmic reticulum (ER) and mitochondria and also biomolecules like proteins and DNA. AgNPs were also found to decrease the MCF-7 cancer cell proliferation 7 times in comparison with normal cells from human lung fibroblast (W138) and human amnion (WISH). However, the identical concentration (13.07 μg/mL) of 5-fluorouracil (standard antitumor drug) possessed increased cytotoxic activities against WISH and W138 cells against cancer cells. Hence, AgNPs synthesized biologically seem to be non-toxic to normal cells than 5-fluorouracil. Thus, it can be suggested that after injecting mice with Ehrlich ascites carcinoma with 5 mg, AgNPs lead to a remarkable decrease in volume of tumor, number of tumor cells, body weight, and white blood cells (WBCs) count. Highly stable, eco-friendly, and biological AgNPs exhibited antiproliferative actions against cells of leukemia, particularly cell lines of lymphoblastic leukemia (REH) (Roychoudhury et al., 2016). However, these AgNPs at concentrations ranging between 100 and 1000 μg/mL had no effect on peripheral blood mononuclear cells (PBMC). Cell lines of REH exhibited highest sensitivity in response to AgNPs than other leukemic cell lines lymphoblastic leukemia and myeloblastic leukemia having MIC of 1000 and 620 μg/mL, respectively. Even, the nuclear and morphological modifications of AgNP-treated REH via 4′,6-diamidino-2-phenylindole (DAPI) staining method were studied, and it was observed that AgNPs caused shrinkage, damaged nucleus, and condensed and marginalized chromatin.

In a recent study, it was shown that AgNPs synthesized from *Anabaena* sp. had strong antitumor effects against COLO205 and Dalton's lymphoma cells (Singh et al., 2014a, 2014b). It was found that these AgNPs even at a low concentration could decrease cell proliferation through reactive oxygen species (ROS) generation leading to damage of DNA and cell death. AgNPs synthesized from *Nostoc* sp. HKAR-2 possessed antitumor activities against MCF-7 cells (Sonker et al., 2017). With increased AgNPs concentration, more tumor cells were killed ensuring that they can be effectively used to treat cancerous cells.

4.6.3 Antioxidation activity

The toxicity of biological ZnONPs synthesized from cellular extracts of *Anabaena* L31 was tested against ZnONP-shinorine conjugates in vivo. ROS generation was monitored by the use of dichlorodihydrofluorescein diacetate (DCFH-DA) as probe. With contrast to control cells incubated in the presence of DCFH-DA, ROS attained a maximum effect within 12 h of incubation as observed under a fluorescence microscope. However, ROS generation was reduced in cells spiked with ZnONP-shinorine conjugate. Thus, it can be concluded that the ZnONP-shinorine conjugates possessed antioxidative activities and can be used as non-harmful sunscreen agents. MubarakAli et al. (2013) found that AuNPs synthesized from *Phormidium* sp. possessed greater antioxidant activities in comparison with citric acid. With increased AuNP concentration, their scavenging potentials also get increased.

4.6.4 Photocatalytic effects

AgNPs synthesized by *Synechococcus* sp. possessed photocatalytic actions (Keskin et al., 2016). Such AgNPs performed photodegradation of methylene blue within 4 h, and the decolorization of this organic compound reached 18% at an initial dye concentration of 20 mg/L. However, AgNPs biofabricated by *Microchaete* sp. NCCU-342 caused the degradation of methyl red by 84.60% within a period of 2 h more efficiently than the extracts from cyanobacteria (Husain et al., 2019).

4.6.5 Other uses

AuNPs synthesized from L. *majuscule* possessed antimyocardial infarction activities (Younis et al., 2019). It was found that only AuNPs or AuNPs along with extracts from cyanobacteria inhibited isoproterenol-mediated modifications in serum markers for cardiac injuries, artery pressure indices, electrocardiography (ECG), and antioxidant capacities of the heart. AgNPs obtained from phycobiliproteins of *Nostoc* sp. had antihemolytic properties (El-Naggar et al., 2018). 2,2′-Azobis (2-amidinopropane) dihydrochloride (AAPH) was used for the induction of full hemolysis of erythrocytes (100%) of the blood drawn from rat. The AgNPs were found to restrict the hemolytic effect with an efficiency of 96.4% (hemolysis of erythrocytes reduced to only 3.6%), which is identical to those induced by vitamin C (96%). It can be suggested that the potential antihemolytic activity of these AgNPs is because of the presence of *Nostoc* extracts like phenols (Ar-OH) and amino compounds ($-NH_2$) as capping agents of AgNPs.

MubarakAli et al. (2012) also found that CdS NPs synthesized from C-phycoerythrin can act as fluorescence probes for biolabeling and ultrasensitive DNA detection. In another study, it was also observed that AuNPs synthesized from *Coelastrella* sp. and *Phormidium willei* can serve as a biolabeling agent (MubarakAli et al., 2013). However, it was found that only AuNPs synthesized from *Coelastrella* sp. can actively bind DNA even at a higher concentration.

4.7 Conclusion and future perspectives

Green nanotechnology is an interesting field in advanced science. The utilization of cyanobacteria as biomachinery for green synthesis of NPs is extremely valuable because this method is non-toxic, safe, eco-friendly, cheap, and has low energy requirement and can produce NPs with unique sizes and shapes. NPs synthesized from cyanobacteria have varied chemical, physical, and biological characteristics providing versatile applications. These NPs can be potentially used as photocatalytic, anticancer, and antimicrobial agents. Cyanobacterial species including *Nostoc* sp. and *Desertifilum* sp. have been tested and used in AgNP synthesis possessing remarkable anticancer properties against different cell lines of cancer. In addition to this, AuNPs and AgNPs biofabricated by the use of *Anabaena* sp. and *Spirulina* sp. exhibited broad-spectrum inhibitory action against Gram-negative and Gram-positive bacteria. Improvement and development of cyanobacterial biosynthetic processes may provide us with new biological NPs having unique characteristics for different purposes. Their introduction in medicinal therapies will reduce side effects and increase efficacy (Hamida et al., 2020).

However, some avenues still need to be studied in a greater detail; for example, many cyanobacterial strains are still not exploited for NP synthesis. Further investigations are required regarding the biocatalytic actions of those cyanobacteria and capabilities in biotransformation of bulk precursor materials into NPs.

Although there are some studies about the cyanobacteria-mediated NP synthesis, the exact synthesis pathway is yet to be elucidated. So, it is necessary to further investigate the exact mechanism behind that biosynthesis of NPs by cyanobacteria in addition to the determination of the reason as to why different strains of the same cyanobacterial species generate morphologically different NPs. Moreover, the elucidation of the effects of the abiotic and biotic conditions like metabolic state, electrostatic interactions, biomolecules (proteins, pigments, and enzymes), and conditions of growth in this green NP synthesis process will facilitate the regulation of biological and physicochemical properties of NPs. The evolution of lethal diseases that are threatening for human health warrants the screening of these biological NPs for different applications in the medical sector for the development of new anticancer and antimicrobial agents.

References

Aboelfetoh, E.F., El-Shenody, R.A., Ghobara, M.M., 2017. Eco-friendly synthesis of silver nanoparticles using green algae (*Caulerpa serrulata*): reaction optimization, catalytic and antibacterial activities. Environ. Monit. Assess. 189 (7), 349.

Afzal, B., Yasin, D., Husain, S., Zaki, A., Srivastava, P., Kumar, R., 2019. Screening of cyanobacterial strains for the selenium nanoparticles synthesis. Biocatal. Agric. Biotechnol. 21, 101307.

Ali, J., Ali, N., Wang, L., Waseem, H., Pan, G., 2019. Revisiting the mechanistic pathways for bacterial mediated synthesis of noble metal nanoparticles. J. Microbiol. Methods 159, 18–25.

Ashmore, D., Chaudhari, A., Barlow, B., Barlow, B., Harper, T., Vig, K., 2018. Evaluation of *E. coli* inhibition by plain and polymer-coated silver nanoparticles. Rev. Inst. Med. Trop. Sao Paolo 60, e18.

Asif, N., Fatima, S., Aziz, M.N., Shehzadi, Zaki, A., Fatma, T., 2021. Biofabrication and characterization of cyanobacteria derived ZnO NPs for their bioactivity comparison with commercial chemically synthesized nanoparticles. Bioorg. Chem. 113, 104999.

Baker, S., Harini, B.P., Rakshith, D., Satish, S., 2013. Marine microbes: invisible nanofactories. J. Pharm. Res. 6 (3), 383–388.

Bin-Meferji, M.M., Hamida, R.S., 2019. Biofabrication and antitumor activity of silver nanoparticles utilizing novel Nostoc sp. Bahar M. Int. J. Nanomedicine 14, 9019–9029.

Borase, H.P., Patil, C.D., Suryawanshi, R.K., Koli, S.H., Mohite, B.V., Benelli, G., 2017. Mechanistic approach for fabrication of gold nanoparticles by *Nitzschia diatom* and their antibacterial activity. Bioprocess Biosyst. Eng. 40 (10), 1437–1446.

Brayner, R., Yéprémian, C., Djediat, C., Coradin, T., Herbst, F., Livage, J., 2009. Photosynthetic microorganism-mediated synthesis of akaganeite (β-feooh) nanorods. Langmuir 25 (17), 10062–10067.

Cai, W., Gao, T., Hong, H., Sun, J., 2008. Applications of gold nanoparticles in cancer nanotechnology. Nanotechnol. Sci. Appl. 1, 17–32.

Chaudhary, R., Nawaz, K., Khan, A.K., Hano, C., Abbasi, B.H., Anjum, S., 2020. An overview of the algae-mediated biosynthesis of nanoparticles and their biomedical applications. Biomol. Ther. 10 (11), 1498.

Dahoumane, S.A., Djediat, C., Yéprémian, C., Coute, A., Fievet, F., Brayner, R., 2010. Design of magnetic akaganeite-cyanobacteria hybrid biofilms. Thin Solid Films 518 (19), 5432–5436.

Dahoumane, S.A., Djediat, C., Yepremian, C., Couté, A., Fiévet, F., Coradin, T., 2012. Species selection for the design of gold nanobioreactor by photosynthetic organisms. J. Nanopart. Res. 14 (6), 1–17.

Dahoumane, S.A., Wujcik, E.K., Jeffryes, C., 2016. Noble metal, oxide and chalcogenide-based nanomaterials from scalable phototrophic culture systems. Enzyme Microb. Technol. 95, 13–27.

El-Belely, E.F., Farag, M.M., Said, H.A., Amin, A.S., Azab, E., Gobouri, A.A., et al., 2021. Green synthesis of zinc oxide nanoparticles (ZnO-NPs) using *Arthrospira platensis* (class: Cyanophyceae) and evaluation of their biomedical activities. Nanomaterials (Basel) 11 (1), 95.

El-Naggar, N.E., Hussein, M.H., El-Sawah, A.A., 2017. Bio-fabrication of silver nanoparticles by phycocyanin, characterization, in vitro anticancer activity against breast cancer cell line and in vivo cytotxicity. Sci. Rep. 7 (1), 10844.

El-Naggar, N.E., Hussein, M.H., El-Sawah, A.A., 2018. Phycobiliprotein-mediated synthesis of biogenic silver nanoparticles, characterization, in vitro and in vivo assessment of anticancer activities. Sci. Rep. 8 (1), 8925. https://doi.org/10.1038/s41598-018-27276-6.

El-Sheekh, M.M., El-Kassas, H.Y., 2016. Algal production of nano-silver and gold: their antimicrobial and cytotoxic activities: a review. J. Genet. Eng. Biotechnol. 14 (2), 299–310.

Geetha, S., Sathakkathul, J., Aarthi, R., Blessie, H., 2014. Green synthesis of gold nanoparticle using marine cyanobacteria *Gloeocapsa* sp and the. J. Chem. Pharm. Sci. 4, 172–174.

Govindraraju, K., Basha, S.K., Kumar, V.G., Singaravelu, G., 2008. Silver, gold and bimetallic nanoparticles production using single-cell protein (*Spirulina platensis*) Geitler. J. Mater. Sci. 43 (15), 5115–5122.

Hamida, R.S., Abdelmeguid, N.E., Ali, M.A., Bin-Meferji, M.M., Khalil, M.I., 2020. Synthesis of silver nanoparticles using a novel cyanobacteria *Desertifilum* sp. extract: their antibacterial and cytotoxicity effects. Int. J. Nanomedicine 15, 49–63.

Hamouda, R.A., Hussein, M.H., Abo-Elmagd, R.A., Bawazir, S.S., 2019. Synthesis and biological characterization of silver nanoparticles derived from the cyanobacterium *Oscillatoria limnetica*. Sci. Rep. 9 (1), 13071.

Hassan, S.E., Fouda, A., Radwan, A.A., Salem, S.S., Barghoth, M.G., Awad, M.A., 2019. Endophytic actinomycetes *Streptomyces* spp mediated biosynthesis of copper oxide nanoparticles as a promising tool for biotechnological applications. J. Biol. Inorg. Chem. 24 (3), 377–393.

Husain, S., Sardar, M., Fatma, T., 2015. Screening of cyanobacterial extracts for synthesis of silver nanoparticles. World J. Microbiol. Biotechnol. 31 (8), 1279–1283.

Husain, S., Afreen, S., Hemlata, Yasin, D., Afzal, B., Fatma, T., 2019. Cyanobacteria as a bioreactor for synthesis of silver nanoparticles-an effect of different reaction conditions on the size of nanoparticles and their dye decolorization ability. J. Microbiol. Methods 162, 77–82.

Iravani, S., 2011. Green synthesis of metal nanoparticles using plants. Green Chem. 13 (10), 2638–2650.

Jeffryes, C., Agathos, S.N., Rorrer, G., 2015. Biogenic nanomaterials from photosynthetic microorganisms. Curr. Opin. Biotechnol. 33, 23–31.

Jenaa, J., Pradhan, N., Dash, B.P., Sukla, L.B., Panda, P., 2012. Biosynthesis and characterization of silver nanoparticles using microalga *Chlorococcum humicola* and its antibacterial activity. Int. J. Nanomater. Biostruct. 1–8, 3.

Jin, R., Cao, Y.C., Hao, E., Métraux, G.S., Schatz, G.C., Mirkin, C.A., 2003. Controlling anisotropic nanoparticle growth through plasmon excitation. Nature 425 (6957), 487–490.

Kalabegishvili, T.L., Kirkesali, E.I., Rcheulishvili, A.N., Ginturi, E.N., Murusidze, I., Pataraya, D.T., Gurielidze, M., Tsertsvadze, G., Gabunia, V., Lomidze, L., Gvarjaladze, D., Frontasyeva, M., Pavlov, S.S., Zinicovscaia, I., Raven, M.J., Seaga, M.M.F., Faanhof, A., 2012. Synthesis of gold nanoparticles by some strains of Arthrobacter genera. J. Mater. Sci. Eng. A 2 (2), 164–173.

Keskin, N.O., Kilic, N.K., Donmez, G., Tekinay, T., 2016. Green synthesis of silver nanoparticles using cyanobacteria and evaluation of their photocatalytic and antimicrobial activity. J. Nanopart. Res. 40, 120–127.

Khan, A.U., Khan, M., Malik, N., Cho, M.H., Khan, M.M., 2019. Recent progress of algae and blue-green algae-assisted synthesis of gold nanoparticles for various applications. Bioprocess Biosyst. Eng. 42 (1), 1–15.

Kim, D., Jeong, S., Moon, J., 2006. Synthesis of silver nanoparticles using the polyol process and the influence of precursor injection. Nanotechnology 17 (16), 4019–4024. https://doi.org/10.1088/0957-4484/17/16/004.

Lenartowicz, M., Marek, P.H., Madura, I., Lipok, J., 2017. Formation of variously shaped gold nanoparticles by *Anabaena laxa*. J. Clust. Sci. 28 (5), 3035–3055.

Lengke, M.F., Ravel, B., Fleet, M.E., Wanger, G., Gordon, R.A., Southam, G., 2006. Mechanisms of gold bioaccumulation by filamentous cyanobacteria from gold(III)-chloride complex. Environ. Sci. Technol. 40 (20), 6304–6309. https://doi.org/10.1021/es061040r.

Lengke, M.F., Fleet, M.E., Southam, G., 2007. Biosynthesis of silver nanoparticles by filamentous cyanobacteria from a silver(I) nitrate complex. Langmuir 23 (5), 2694–2699.

Mahdieh, M., Zolanvari, A., Azimee, A.S., Mahdieh, M., 2012. Green biosynthesis of silver nanoparticles by *Spirulina platensis*. Sci. Iran. 19 (3), 926–929.

Menon, S., Rajeshkumar, S., Kumar, V.S., 2017. A review on biogenic synthesis of gold nanoparticles, characterization, and its applications. Resour. Efficient Technol. 3 (4), 516–527.

Merin, D.D., Prakash, S., Bhimba, B.V., 2010. Antibacterial screening of silver nanoparticles synthesized by marine micro algae. Asian Pac. J. Trop. Dis. 3 (10), 797–799.

Morsy, F., Nafady, N.A., Abd-Alla, M.H., El-Hady, D.A., 2014. Green synthesis of silver nanoparticles by water soluble fraction of the extracellular polysaccharides/matrix of the Cyanobacterium *Nostoc commune* and its application as a potent fungal surface sterilizing agent of seed crops. Univers. J. Microbiol. Res. 2 (2), 36–43.

Mosulishvili, L.M., Kirkesali, E.I., Belokobylsky, A.I., Khizanishvili, A.I., Frontasyeva, M.V., Pavlov, S.S., 2002. Experimental substantiation of the possibility of developing selenium- and iodine-containing pharmaceuticals based on blue-green algae Spirulina platensis. J. Pharm. Biomed. Anal. 30 (1), 87–97.

MubarakAli, D., Sasikala, M., Gunasekaran, M., Thajuddin, N., 2011. Biosynthesis and characterization of silver nanoparticles using marine cyanobacterium, *Oscillatoria willei* NTDM01. Dig. J. Nanomater. Biostruct. 6 (2), 385–390.

MubarakAli, D., Gopinath, V., Rameshbabu, N., Thajuddin, N., 2012. Synthesis and characterization of CdS nanoparticles using C-phycoerythrin from the marine cyanobacteria. Mater. Lett. 74, 8–11.

MubarakAli, D., Arunkumar, J., Nag, K.H., SheikSyedIshack, K.A., Baldev, E., Pandiaraj, D., 2013. Gold nanoparticles from pro and eukaryotic photosynthetic microorganisms–comparative studies on synthesis and its application on biolabelling. Colloids Surf. B Biointerfaces 103, 166–173.

Nikolić, V., Ilić-Stojanović, S., Petrović, S., Tačić, A., Nikolić, L., 2019. Administration routes for nano drugs and characterization of nano drug loading. In: Shyam, S.M. (Ed.), Characterization and Biology of Nanomaterials for Drug Delivery. Elsevier, pp. 587–625.

Omomowo, I.O., Adenigba, V., Ogunsona, S.B., Adeyinka, G.C., Oluyide, O.O., Adedayo, A.A., 2020. Antimicrobial and antioxidant activities of algal-mediated silver and gold nanoparticles. IOP Conf. Ser.: Mater. Sci. Eng. 805, 012010.

Oza, G., Pandey, S., Mewada, A., Kalita, G., Sharon, M., 2012. Facile biosynthesis of gold nanoparticles exploiting optimum pH and temperature of fresh water algae *Chlorella pyrenoidusa*. Adv. Appl. Sci. Res. 3 (3), 1405–1412.

Parial, D., Patra, H.K., Dasgupta, A.K., Pal, R., 2011. Screening of different algae for green synthesis of gold nanoparticles. Eur. J. Phycol. 47 (1), 22–29.

Patel, V., Berthold, D., Puranik, P., Gantar, M., 2014. Screening of cyanobacteria and microalgae for their ability to synthesize silver nanoparticles with antibacterial activity. Biotechnol. Rep. (Amst.) 5, 112–119.

Poinern, G.E.J., 2014. A Laboratory Course in Nanoscience and Nanotechnology. CRC Press.

Quester, K., Avalos-Borja, M., Castro-Longoria, E., 2013. Biosynthesis and microscopic study of metallic nanoparticles. Micron 54–55, 1–27.

Rahman, A., Ismail, A., Jumbianti, D., Magdalena, S., Sudrajat, H., 2010. Synthesis of copper oxide nanoparticles by using Phormiium cyanobacterium. Indones. J. Chem. 9 (3), 355–360.

Rajathi, F.A., Parthiban, C., Kumar, V.G., Anantharaman, P., 2012. Biosynthesis of antibacterial gold nanoparticles using brown alga, *Stoechospermum marginatum* (kützing). Spectrochim. Acta A Mol. Biomol. Spectrosc. 99, 166–173.

Rosken, L.M., Cappel, F., Korsten, S., Fischer, C.B., Schonleber, A., Smaalen, S.V., 2016. Time-dependent growth of crystalline Au(0)-nanoparticles in cyanobacteria as self-reproducing bioreactors: 2. Anabaena cylindrica. Belstein J. Nanotechnol. 7, 312–327.

Roychoudhury, P., Ghosh, S., Pal, R., 2015. Cyanobacteria mediated green synthesis of gold-silver nanoalloy. J. Plant Biochem. Biotechnol. 25 (1), 73–78.

Roychoudhury, P., Gopal, P.K., Paul, S., Pal, R., 2016. Cyanobacteria assisted biosynthesis of silver nanoparticles—a potential antileukemic agent. J. Appl. Phycol. 28 (6), 3387–3394.

Saran, S., Sharma, G., Kumar, M., Ali, M.I., 2017. Biosynthessi of copper oxide nanoparticles using cyanobacteria Spirulina platensis and its antibacterial activity. Int. J. Pharm. Sci. Res. 8 (9), 3887–3892.

Shah, M., Fawcett, D., Sharma, S., Tripathy, S.K., Poinern, G.E., 2015. Green synthesis of metallic nanoparticles via biological entities. Materials (Basel) 8 (11), 7278–7308.

Singh, G., Babele, P.K., Kumar, A., Srivastava, A., Sinha, R.P., Tyagi, M.B., 2014a. Synthesis of ZnO nanoparticles using the cell extract of the cyanobacterium, *Anabaena* strain L31 and its conjugation with UV-B absorbing compound shinorine. J. Photochem. Photobiol. B 138, 55–62. https://doi.org/10.1016/j.jphotobiol.2014.04.030.

Singh, G., Babele, P.K., Shahi, S.K., Sinha, R.P., Tyagi, M.B., Kumar, A., 2014b. Green synthesis of silver nanoparticles using cell extracts of *Anabaena doliolum* and screening of its antibacterial and antitumor activity. J. Microbiol. Biotechnol. 24 (10), 1354–1367.

Singh, Y., Kaushal, S., Sodhi, R.S., 2020. Biogenic synthesis of silver nanoparticles using cyanobacterium *Leptolyngbya* sp. WUC 59 cell-free extract and their effects on bacterial growth and seed germination. Nanoscale Adv. 2 (9), 3972–3982.

Sinha, S.N., Paul, D., Halder, N., Sengupta, D., Patra, S.K., 2015. Green synthesis of silver nanoparticles using fresh water green alga *Pithophora oedogonia* (Mont.) Wittrock and evaluation of their antibacterial activity. Appl. Nanosci. 5 (6), 703–709.

Sonker, A., Richa, Pathak, J., Rajneesh, Kannaujiya, V.K., Sinha, R.P., 2017. Characterization and in vitro antitumor, antibacterial and antifungal activities of green synthesized silver nanoparticles using cell extract of *Nostoc* sp. strain HKAR-2. Can. J. Biotech. 1 (1), 26–37.

Sudha, S.S., Rajamanickam, K., Rengaramanujam, J., 2013. Microalgae mediated synthesis of silver nanoparticles and their antibacterial activity against pathogenic bacteria. Indian J. Exp. Biol. 51 (5), 393–399.

Suganya, K.S., Govindaraju, K., Kumar, V.G., Dhas, T.S., Karthick, V., Singaravelu, G., 2015. Blue green alga mediated synthesis of gold nanoparticles and its antibacterial efficacy against gram positive organisms. Korean J. Couns. Psychother. 47, 351–356.

Upadhyay, S., Parekh, K., Pandey, B., 2016. Influence of crystallite size on the magnetic properties of Fe_3O_4 nanoparticles. J. Alloys Compd. 678, 478–485.

Wei, N., Hou, Y., Lu, Z., Yu, H., Wang, Q., 2017. Synthesis of silver nanoparticles stabilized with C-phycocyanin and for fluorimetric detection of copper ions. IOP Conf. Ser. Earth Environ. Sci. 108, 1–5.

Younis, N.S., Bakir, E.M., Mohamed, M.E., Semary, N.A., 2019. Cyanobacteria as nanogold factories II: chemical reactivity and anti-myocardial infraction properties of customized gold nanoparticles biosynthesized by *Cyanothece* sp. Mar. Drugs 17 (7), 402.

Nanobiotechnology of endophytes

5

Anupam Mondal[a], *Sagar Bag*[a], *Ronita Chandra*[b], *and Avishek Banik*[a]
[a]Laboratory of Microbial Interaction, School of Biotechnology, Presidency University, Kolkata, West Bengal, India, [b]Department of Microbiology, Vivekananda Mahavidyalaya, Burdwan, West Bengal, India

5.1 Introduction

Nanobiotechnology is a new interdisciplinary subject that combines biology with nanotechnology. The terminology implies the use of nanoscale biotechnology at atomic and molecular levels to construct new biomaterials (Porter and Youtie, 2009). This novel technology has shaped different types of nanoparticles, e.g., gold (Au), silver (Ag), platinum (Pt), zinc (Zn), nickel (Ni), titanium (Ti), palladium (Pd), and copper (Cu) at nanoscale range. Nanoparticles can be synthesized by three separate approaches, i.e., chemical, physical, and biological. Chemical and physical methods involve several environmentally hazardous chemicals and conditions. Moreover, both of these methods are usually a bit costly to perform (Iravani, 2014; Messaoudi and Bendahou, 2020). Apart from all restrictions, researchers have developed a new technique, named "green synthesis" for nanoparticle production. This method is highly significant in providing a huge application toward present humankind (Messaoudi and Bendahou, 2020). Nanoparticles synthesized through biogenic processes are the most beneficial approach that has offered many advantageous roles over chemical and physical methods. However, the biological approach is economical, environment-friendly, requires a lesser amount of energy, and can produce better-defined nanoparticles, which are exceptionally compatible and nontoxic for medical, pharmaceutical, agricultural, and environmental applications (Gour and Jain, 2019). The "biogenic" approach utilizes bacteria, fungi, yeasts, actinomycetes, and viruses for the biological synthesis of nanoparticles (Ovais et al., 2018; Rana et al., 2020). Microbes are ubiquitous all around the environment, which enables them to be conserved and recycled at a specific niche (Azmath et al., 2016; Banik et al., 2019a). In comparison with other microbial groups, "endophytes" have been significantly influenced by the secretion of various types of secondary metabolites. These biological activities provide a pathway for innovation toward the endophytes producing nanoparticles (Baker and Satish, 2015). Several endophytes such as *Aspergillus fumigatus*, *Bacillus cereus*, *Colletotrichum* sp., *Penicillium citrinum*, *Pseudomonas veronii*, *Rhodococcus* sp., and *Saccharomonospora* sp. were isolated from various origins, and all of them showed their potential for nanoparticle biosynthesis (Rana et al., 2020). This chapter aims to offer an idea about various types of endophytic bacteria producing nanoparticles, their different modes of biosynthesis along with the types of nanoparticles, their shape, size, and their application in diverse fields.

Agricultural Nanobiotechnology. https://doi.org/10.1016/B978-0-323-91908-1.00018-3

5.2 Endophytes

Endophyte is a Greek word that means "within the plant." These are nonpathogenic microbial agents (fungi and bacteria) that dwell inside the tissues of plants during their life cycle. They gain access to the plant through roots, stems, leaves, and seeds. Several recent studies demonstrate that these microbes offer a huge role in plants' health as well as in their survival. They secrete phytohormones, which help in nutrient acquisition, improve plant growth, as well as protect them from phytopathogens (Hardoim et al., 2015; Potshangbam et al., 2017; Mukherjee et al., 2020). Being facultative or obligate, their extensive definition does not decide their functional associations. Besides communalistic symbiosis, they may adapt mutualistic relationships with their host (Fesel and Zuccaro, 2016; Khare et al., 2018; Banik, 2019). During their mutualistic relationship, the endophytes are capable of colonizing both intracellularly and intracellularly to the plant tissues (Banik et al., 2016b). One of the best associations is rhizobium-plant symbiosis in which the bacterial endosymbiont regulates the nitrogen requirements of the host plant (Santoyo et al., 2016; Banik et al., 2019b). From different studies, it was considered that this association evolved about 60 million years ago. Since then, it was found that this relationship plays an incredible role in terrestrial environments. Furthermore, it also offers a huge profit to both the participants (Khare et al., 2018). Recently, several studies have demonstrated that some endophytic bacteria can accelerate defense response upon pathogens and they provide resistance against a series of harmful environmental conditions like salinity, presence of heavy metals, and drought. Endophytic microorganisms based on stable symbiosis promote the host plant for the secretion of metabolites that induce growth of the host plant as well as acclimatize them to the ecosystem. Endophytes defend plants from the biotic and abiotic stresses and help them to sustain yields and plant growth (Khare et al., 2018; Lata et al., 2018).

5.3 Ubiquity of endophytes

Presently, 300,000 plant species reside on the globe and almost all of them comprise endophytes (Smith et al., 2008). Endophytic microorganisms were found in each plant species that have been examined. Based on the distribution of rhizospheric PGPB in the environment, it was found that plants would not be able to compete with pathogenic agents in the absence of endophytes (Timmusk et al., 2011). The communities of endophytes were predominantly restricted in five different phyla named Proteobacteria, Actinobacteria, Verrucomicrobia, Planctomycetes, and Acidobacteria. Among them, Proteobacteria gather the highest numbers of endophytes, i.e., about 90%. Proteobacteria can be *Actinomycetales*, *Burkholderiales*, *Enterobacteriales*, *Flavobacteriales*, *Pseudomonadales*, *Rhizobiales*, *Sphingomonadales*, and *Xanthomonadales*. It was reported that Class α, β, and γ-Proteobacteria play a major role in the diversity analysis of endophytic microorganisms. However, some constantly found members of endophytes belong to Actinobacteria and Firmicutes class. *Bacillus*, *Micrococcus*, *Pseudomonas*,

Burkholderia, *Stenotrophomonas*, *Pantoea*, and *Microbacterium* are the endophytic bacterial genera that are reported as the most common types than other endophytic bacterial genera (Santoyo et al., 2016; Banik et al., 2016a).

5.4 Role of endophytes in plant growth promotion

5.4.1 Helps in phytostimulation

A plant requires around 16 elements for its growth and development. Among them, some of the major elements are carbon (C), hydrogen (H), oxygen (O), nitrogen (N), phosphorus (P), etc. These elements are available in the form of chemicals, which they derive from the soil, water, atmosphere, and organic substance. Endophytic microorganisms play an incredible role in uptaking these nutrients (Nair and Padmavathy, 2014). For instance, *Burkholderia vietnamiensis*, an endophytic diazotrophic bacteria isolated from *Populus trichocarpa*, produces a vital compound, named indole acetic acid (IAA) that influences plant growth promotion via cell division, differentiation, and extension (Xin et al., 2009). *Cladosporium sphaerospermum*, fungi that were first isolated from *Glycine max* (L.) Merr., contain enormous amounts of bioactive gibberellin hormones such as GA3, GA4, and GA7 that boost plant growth promotion in soybean and rice varieties (Sudha et al., 2016).

5.4.2 Good source of novel and bioactive components

Endophytes produce hundreds of natural products like flavonoids, alkaloids, polyketides, peptides, terpenoids, steroids, saponins, tannins, phenolic acids, and quinones. Most bioactive compounds from endophytic bacteria possess antiviral, antibiotics, antituberculosis, antimalarial, antiinflammatory, anticancer, biocontrol agents, and immunosuppressant properties (Joseph and Priya, 2011). The abovementioned bioactive metabolites are core sources of drug production not only in medical science but contribute promising application in food and agronomy (Godstime et al., 2014; Shukla et al., 2014).

5.4.3 Antimicrobial property

Most endophytes isolated from plant tissues play a significant role in antimicrobial activity. Endophytes that were isolated from medicinal plants possess biological efficacy for a wide range of pathogenic microorganisms (Nair and Padmavathy, 2014). Studies have revealed that *Tectona grandis* L. and *Samanea saman* Merr. possess overall 37 endophytic microorganisms; among them, only 18 produce antimicrobial agents that prevent the growth of *Bacillus subtilis*, *Escherichia coli*, and *Staphylococcus aureus* (Kumar et al., 2011).

5.5 Mechanism of plant growth promotion by endophytes

Endophytic microbes help in plant growth promotion indirectly or directly through various ways. These mechanisms are as follows (Fig. 5.1).

5.5.1 Indirect mechanisms

Plants are capable of conquering different types of biotic stresses. Along with that, they can also thrive in adverse environmental conditions such as cold, drought, pathogenesis, or hypersaline conditions. Endophytes and some root-associated microbes support plant growth through several indirect methods to manage heavy and translational metal contamination. These mechanisms are described as follows:

5.5.1.1 Bioremediation mechanism

Bioremediation is a biological process where pollutants are removed from nature. Plants are unable to neutralize toxic metabolites that are secreted from them. Endophytes and microbes thriving in the rhizosphere help in the detoxification of toxic compounds (Han et al., 2011; Chandra and Banik, 2021) and removal of heavy metallic stress from plants, reduce plant-secreted toxic metabolic compounds, and inhibit pest growth on plants (Yadav and Yadav, 2017).

5.5.1.2 Biocontrol mechanism

Biocontrol is an eco-friendly method for diminishing the number of plant pathogens via natural antagonistic microorganisms (Rybakova et al., 2016). Antagonistic rhizosphere bacteria like *Pseudomonas* and *Bacillus* are well-known biocontrol agents. Besides antibiotic production, endophytes generate numerous metabolic products that are applicable in pharmaceutical and agrochemical fields (Hayat et al., 2010).

5.5.1.3 Phytoremediation mechanism

Endophytes catabolize phytoremediation and help in plant development via solubilization of minerals, fixation of nitrogen, nutrient alteration, production of phytohormones, and siderophore by the use of 1-aminocyclopropane-1-carboxylic as nitrogen source (Stępniewska and Kuźniar, 2013). Endophytes and root-associated microbes detoxify heavy metals, promote degradation of polyaromatic hydrocarbons (PAH), and help in phytoremediation of oil-containing soils (Olson et al., 2008; Rajkumar et al., 2009; Yadav and Yadav, 2017; Banik et al., 2018).

5.5.1.4 Plant stress tolerance mechanism

The beneficial endophytic bacteria have been reported to promote plant adaptation in various environmental stresses (Yadav and Yadav, 2017). They mitigate plant stress caused by heavy metallic accumulation, drought, temperature, and solar UV-B radiation (Piccoli and Bottini, 2013). Endophytes increase cold tolerance by changing

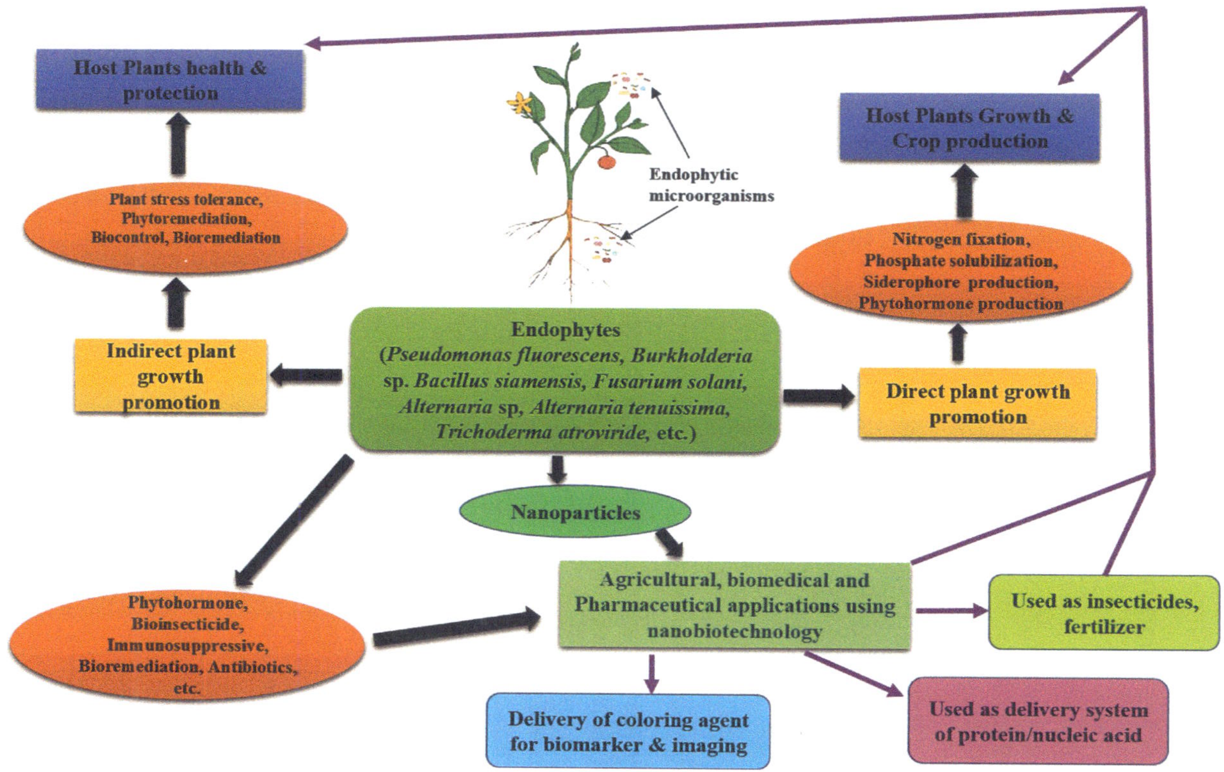

Fig. 5.1 Endophytes and their role in plant growth promotion and many other areas by nanobiotechnological approach.

plant's photosynthetic properties through deposition of cold stress-mediated metabolic compounds such as phenolics, proline, and starch (Fernandez et al., 2012; Yadav and Yadav, 2017).

5.5.2 Direct mechanisms

Endophytes directly help plants by fixing nitrogen and synthesizing many components like insecticidal by-products, antimicrobial metabolites, iron-chelating compounds, and phosphate-solubilizing components. The abovementioned significant and well-cited mechanisms are described below:

5.5.2.1 Nitrogen fixation mechanisms

Numerous symbiotic endophytes can fix atmospheric nitrogen. Till now, legume-rhizobium symbiosis has been a well-known global research topic and has focused on promoting nitrogen fixation ability through plant and bacterial genome manipulation (Yadav and Yadav, 2017).

5.5.2.2 Phosphate solubilization mechanism

Some endophytes are capable of phosphate solubilization by emitting organic acids that convert the complex phosphate form into plant usable orthophosphate form. Though phosphorus is present enormously in soils, the maximum portion of it remains inaccessible due to its insoluble nature (Miller et al., 2010). Many endophytic actinomycetes such as *Streptomyces* sp. are found to achieve phosphate solubilization to enhance plant growth (Jog et al., 2014).

5.5.2.3 Phytohormone production mechanism

Most microorganisms secrete phytohormone, which promotes plant growth and development structurally as well as morphologically (Yadav and Yadav, 2017). Endophytes boost the nonlegume plant's development by enhancing the production of plant hormones such as auxins (Dutta et al., 2014), ethylene (Kang et al., 2012), IAA (Patel and Patel, 2014), and also gibberellic acid (GA) (Khan et al., 2014).

5.5.2.4 Siderophore production mechanism

Endophytes can secrete high-affinity iron-chelating agents known as siderophores. This low-molecular-weight organic compound helps in transporting metal nutrients across the plant cell walls. These endophytes deprive pathogens of uptaking iron as their essential nutrient (Yadav and Yadav, 2017). A few endophytes secreting siderophores possess biocontrol properties like phenolate, hydroxamate, and catecholate (Rajkumar et al., 2010).

5.6 Nanobiotechnology

Norio Taniguchi first introduced "nanotechnology" in 1974 (Taniguchi, 1974). Nanotechnology deals with materials at a 1–100 nm nanoscale range that can be used to create a variety of applications in synthesizing nanomaterials (Sarkar and Sarkar, 2019). Nanomedicine is using a variety of instruments with the help of nanotechnology to progress quicker, and it is one of the most efficient solutions for increasing medical or disease management difficulties. It can defeat the challenges faced by conventional medicine as well as strengthen the understanding of several pathological and physiological procedures. An in-depth understanding of these procedures gives us novel advantages and also develops therapeutic concepts for current difficulties (Yasmeen, 2021). The development of nanomedicine-based approaches has encouraged the rational progress of targeted therapies using chronic and external therapies to improve the delivery of drugs with a rising understanding of cancer biology. Currently, nanomedicine has a variety of advantages to realize the idea that the capability to make materials and apparatus on a molecular scale can bring immediate profits in drug development and also in research (Yasmeen, 2021).

5.7 Nanobiotechnology enhancing plant resistance

Certain unfavorable environmental circumstances such as low temperatures, high temperatures, and droughts along with some agents and pathogens like viruses, bacteria, fungi, pests, insects, mycoplasmas, and nematodes are mainly responsible for plant's diseases and also lead to financial damage (Alghuthaymi et al., 2015). Disease resistance properties in plants defend them from phytopathogens. These resistant properties are involved in diminishing the growth of pathogens in or on the plant. Plants developed resistance via (i) genetic resistance, (ii) plant breeding, (iii) acquired resistance, (iv) pathogen-associated molecular pattern (PAMP)-triggered immunity (PTI), (v) signal transduction, etc. (Rana et al., 2020). Mesoporous silica nanoparticles (NPs) help in the transfer of chemicals and DNA into plants (Torney et al., 2007). Gene transfer through NPs is a relatively novel technology, and it has the strong ability to transfer genetic components directly to the plant cells. As a result, NPs behave like transgenic vehicles of nucleic acids (Rai et al., 2012). The interaction between NPs and plants should be studied well before applying it in agricultural fields for better improvement. In photosynthetic organisms, TiO_2 nanoparticles enhance chlorophyll synthesis, dry weight, and metabolisms in plants. NPs have antimicrobial characteristics, which improve the strength of the plants and also resist it from stress conditions (Rana et al., 2020).

5.8 Synthesis of nanoparticles by endophytes

Physical and chemically synthesized NPs raise many challenges for human beings as well as for the environment. Generation of NPs by biogenic path offers an alternate way to explore new biosources, which are capable of reducing metals to their nanoscales. Some microbes, such as fungi, bacteria, and plants, are engaged in the synthesis of nanoparticles. There are relatively few studies available on the production of nanoparticles from endophytic microbial agents. The synthesis of nanoparticles by endophytes has been believed to be nonhazardous, safe, and appropriate "green chemistry" methods.

It is outstanding that the saprophytic bacteria and fungi are capable of reducing the metal ions present in the environment to metallic nanoparticles (e.g., Ag, Au, Zn, and Cu), and this approach has also gained stability (Iravani et al., 2014; Joshi et al., 2017; Soliman et al., 2018; Syed et al., 2019). Under high metallic stress conditions, endophytes reduce metal ions to the extent of nanoparticle scales as their defense mechanism. AgNPs have been found to possess antibacterial, antifungal, and antibiofilm properties. Therefore, it has naturally acquired much acclimation, which in recent days has become the topic of research. Ibrahim et al. (2019, 2020) have isolated *Bacillus siamensis* C1 from *Coriandrum sativum* and *Pseudomonas poae* from *Allium sativum*. They vividly reported that it possesses the property of producing spherical AgNPs (Fig. 5.2). Au-NPs possess bactericidal activity and are reportedly found to be produced from *Pseudomonas fluorescens* 417, isolated from *Coffea arabica* (Syed et al., 2016; Messaoudi and Bendahou, 2020). Syed et al. (2019) pursued some research on AgNPs and ended up with some fortunate results. They used *Aneurinibacillus migulanus* strain isolated from the inner *Mimosa pudica* leaf segment, which synthesizes NPs of varying shapes and sizes (the average size determined was 24.27 nm). *Bacillus* sp., an endophytic bacterium, has been identified as a producer of nanoparticles. This endophyte was obtained from *Garcinia xanthochymus*, and the size of the spherical-shaped nanoparticle ranges from 20 to 40 nm (Sunkar

Fig. 5.2 Size and morphology of silver nanoparticles (AgNPs) that were synthesized by reducing 3 mM $AgNO_3$ by using culture filtrates of the endophytic bacterium *P. poae* strain CO. (A) transmission electron microscopy and (B) scanning electron microscopy.
From Ibrahim, E., Zhang, M., Zhang, Y., Hossain, A., Qiu, W., Chen, Y., Wang, Y., Wu, W., Sun, G., Li, B., 2020. Green-synthesization of silver nanoparticles using endophytic bacteria isolated from garlic and its antifungal activity against wheat Fusarium head blight pathogen *Fusarium graminearum*. Nanomaterials 10, 219. Copyright © 2020MDPI.

and Nachiyar, 2012a,b). One of the endophytic bacteria *Bordetella* sp. has been isolated from *Piper nigrum*, and they can produce AgNPs (silver nanoparticles) at room temperature by extracellular method (Thomas et al., 2012).

Along with endophytic bacteria, the beneficiary role of actinomycetes in green nanotechnology is worth mentioning. Nanoparticles synthesized by rare Actinobacteria have been reported in a few papers (El-Moslamy, 2018; El-Gamal et al., 2018). They can be used as immunosuppressives, antivirals, and insecticides (Messaoudi et al., 2015). *Streptomyces capillispiralis* and *Streptomyces zaomyceticus* Oc-5 have been used for the synthesis of Cu NPs (Hassan et al., 2018, 2019). The produced Cu-NPs hold insecticidal activity along with other properties. *Isoptericola* SYSU 333150 strain was used to synthesize AgNPs (size ranges from 11 to 40 nm) using photo-irradiation for different periods, which was effective against a panel of pathogens (Verma et al., 2016; Singh et al., 2017; Messaoudi and Bendahou, 2020). However, this field is yet to be explored to its highest potential. Green synthesis of various nanoparticles from endophytic microbes and their activity is summarized in Table 5.1.

Fungi, the other efficient candidate, on the other hand, possess metal uptake capacity. This makes them useful as a tool of biological NP production. Metal-accumulating and metal-tolerating fungi were explored by some of the concerned biologists (Moghaddam et al., 2015). Synthesis of Au-NPs from *Fusarium solani* having cytotoxic activity against cervical cancer cells (HeLa) is remarkable. The morphology of nanomaterial was striking (needle- and flower-like structures with spindle shape, and showed pink-ruby red colors) (Clarance et al., 2020). ZnO NPs, size ranging from 15 to 45 nm, was synthesized by using fungi *Alternaria tenuissima* (Abdelhakim et al., 2020). The particles were spherical and have antioxidant activity along with anticancerous properties. AgNPs range between 10 and 15 nm and have antiinflammatory activities produced using *Exserohilum rostratum* (Bagur et al., 2019) (Fig. 5.3). *Amylomyces rouxii*, an endophytic fungus, has been identified as a synthesizer of AgNPs. It was isolated from *Phoenix dactylifera*. The AgNPs showed a broad range of antimicrobial actions against both phytopathogenic fungi and Gram-positive and Gram-negative bacteria (Musarrat et al., 2010). Verma et al. (2010) in their study reported that *Aspergillus clavatus*, an endophytic fungus, synthesizes AgNPs. The spherical- or hexagonal-shaped AgNPs range in size from 10 to 25 nm. *Pestalotia* sp., an endophytic fungus that has been isolated from *Syzygium cumini* (L) leaves, can synthesize extracellular AgNPs. The effectiveness of antibiotics like sulfamethizole and gentamicin has been enhanced in conjunction with AgNPs (Raheman et al., 2011). In a study, Ramalingmam et al. (2015) reported that *Curvularia lunata*, an endophytic fungus, isolated from *Catharanthus roseus* leaves, can biosynthesize AgNPs extracellularly from the solution of silver nitrate. Furthermore, the combination of several antibiotics such as rifampicin, ampicillin, chloramphenicol, kanamycin, and erythromycin along with AgNPs has improved antimicrobial efficacy. One of the endophytic fungi, *Aspergillus niger* obtained from *Simarouba glauca*, can synthesize AgNPs of 41.9 nm size and is reported to have strong antioxidant activity (Rani and Arya, 2017). Popli et al. (2018), in their recent study, reported an endophytic fungus, *Cladosporium*, which was isolated from *Loranthus micranthus*. They can synthesize AgNPs from the aqueous extract. These

Table 5.1 Synthesis of NPs from endophytic bacteria.

Bacterial species	Host plants	Nanoparticles (NPs)	Shape	Size (nm)	Application	References
Bacillus siamensis C1	*Coriandrum sativum*	AgNPs	Spherical	~25–50	Antibacterial	Ibrahim et al. (2019)
Pseudomonas fluorescens 417	*Coffea Arabica*	AuNPs	Spherical	~5–50	Antibacterial	Syed et al. (2016)
Pseudomonas poae CO	*Allium sativum*	AgNPs	Spherical	~19.8–44.9	Antifungal	Ibrahim et al. (2020)
Streptomyces zaomyceticus Oc-5	*Oxalis corniculata*	CuONPs	Spherical	~78	Antimicrobial, antioxidant, and anticancer	Hassan et al. (2019)
Streptomyces capillispiralis Ca-1	*Convolvulus arvensis*	CuONPs	Spherical	~3.6–59	Antimicrobial and insecticides	Hassan et al. (2018)
Streptomyces sp.	*Mentha longifolia*	AgNPs	Spherical	~2.3–85	Antimicrobial	El-Gamal et al. (2018)
Streptomyces coelicolor	*Ocimum sanctum*	MgONPs	Spherical and ellipsoidal	~25	Antimicrobial	Gol et al. (2018)
Lysinibacillus xylanilyticus	*Chiliadenus montanus*	$CaCl_2$NPs	Spherical	~10–100	Transform cellulose into sugars	Yousef et al. (2019)
Bacillus cereus	*Garcinia xanthochymus*	AgNPs	Spherical	~20–40	Antibacterial	Sunkar and Nachiyar (2012a,b)
Aneurinibacillus migulanus	*Mimosa pudica*	AgNPs	Spherical	~20–60	Antimicrobial	Syed et al. (2019)
Sphingobacterium thalpophilum	*Withania somnifera*	ZnONPs	Rectangular	~11	Antimicrobial	Rajabairavi et al. (2017)
Bordetella sp.	*Piper nigrum*	AgNPs	Spherical	~100	Antibacterial	Thomas et al. (2012)

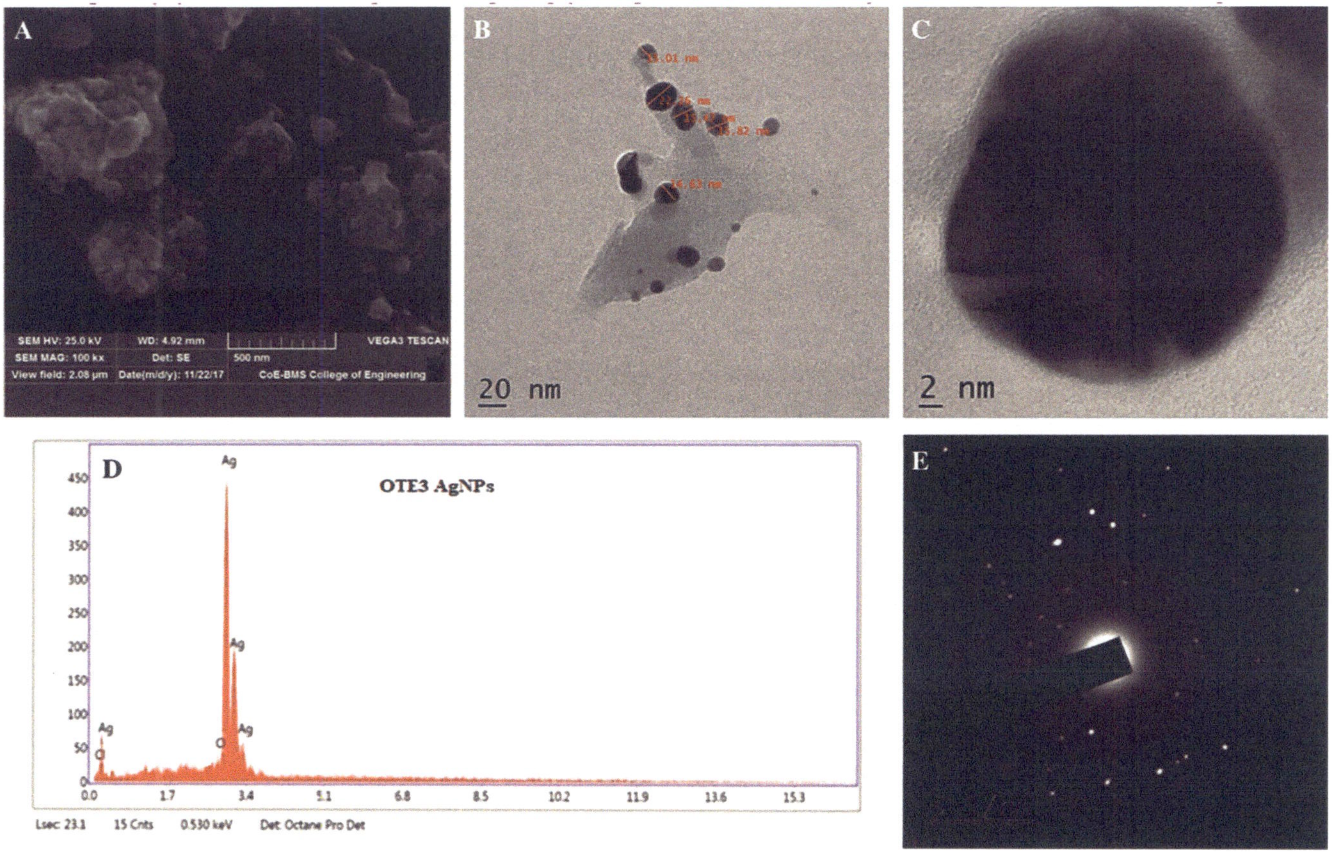

Fig. 5.3 Characterization of AgNPs (A) SEM (B, C) TEM images. (D) EDX and (E) SAED analysis of AgNPs.
Reproduced with permission from Bagur, H., Poojari, C.C., Melappa, G., Rangappa, R., Chandrasekhar, N., Somu, P., 2019. Biogenically synthesized silver nanoparticles using endophyte fungal extract of *Ocimum tenuiflorum* and evaluation of biomedical properties. J. Clust. Sci. 1–15.

AgNPs from *Cladosporium* have radical-scavenging properties as well as have shown effective antioxidant activity. Green synthesis of NPs from endophytic fungi along with their activity is summarized in Table 5.2.

5.9 Mechanisms of nanoparticle synthesis using endophytes

Nanoparticles are synthesized by numerous endophytes through the secretion of enzymes and secondary metabolites (biomolecules) that are found in extra- and intracellular regions of the bacterial cell. These secreted materials could reduce the ionic metals and convert them to their corresponding NPs. In response to the environmental stress conditions, microbes come in contact with metal ions and they secrete some secondary metabolites and enzymes. These metabolites and enzymes play a potential role in the reduction of metal salts, which lead to the detoxification of toxic metal ions to less toxic metal NPs (Singh et al., 2019). The three steps of mechanism on the biogenic synthesis of NPs using microbes are as follows (Fig. 5.4).

❖**Step 1 (Interaction of microbes and metal ions):**

Electrostatic interactions help in binding the negatively charged cell wall surface of microorganisms to the metallic ions. There is a cationic membrane transporter that transports the metal cation into the cell (Singh et al., 2019).

❖**Step 2 (Bioreduction of metallic ions):**

Here, metal ions (M^+) are converted to zerovalent metals (M°) via bioreduction procedure and this happens due to the reduction properties of either biomolecule containing functional groups such as hydroxyl group (—OH) or carboxyl (COO—) group, or by the enzyme (e.g., NADH-dependent nitrate reductase) that is synthesized itself by microorganisms. As a result, monovalent silver ions (Ag^+) are reduced and silver NPs are formed (Talekar et al., 2012; Velusamy et al., 2016). This reaction is based on the reduction of mono- or divalent metal ions states and ultimate conversion to the less toxic zerovalent states. There are several morphologically distinct shapes (cubes, spheres, triangles, hexagons, oval, etc.) of nanoparticles (NPs) that are found in nature (Chokkareddy et al., 2018).

❖**Step 3 (NPs stabilization):**

Prevention of further growth and agglomeration helps to promote stabilization of the size and shape of the nanoparticles (Singh et al., 2019). Endophytes biosynthesizing small-sized NPs have a higher surface/volume ratio, which shows the activity affirmative by its effects. And it has been proved that it is an example of good-quality NPs (Niño-Martínez et al., 2019). For biosynthesis of homogeneous (shape and size) NPs, various physicochemical parameters (pH, temperature, incubation period, agitation, metal salt concentration, nature, and concentration of carbon and nitrogen source) should be maintained strictly in the culture media (Khandel et al., 2018).

Table 5.2 Synthesis of NPs from endophytic fungus and their host plant, shape, size, application.

Fungal species	Host plants	Nanoparticles	Shape	Size (nm)	Application	References
Fusarium solani	*Chonemorpha fragrans*	AuNPs	Spindle	~40–45	Anticancer	Clarance et al. (2020)
Alternaria sp.	*Raphanus sativus*	AgNPs	Spherical	~4–30	Antibacterial	Singh et al. (2017)
Nemania sp.	*Taxus baccata*	AgNPs	Spherical or ellipsoidal	~5–70	Antibacterial	Farsi and Farokhi (2018)
Alternaria tenuissima	*Erythrophleum fordii*	ZnONPs	Spherical	~15–45	Antimicrobial, anticancer, and antioxidant	Abdelhakim et al. (2020)
Trichoderma atroviride	*Chiliadenus montanus*	AgNPs	Spherical	~10–15	Antibacterial	Abdel-Azeem et al. (2020)
Lasiodiplodia theobromae	*Cinnamomum zeylanicum*	AgNPs	Spherical to oval	~76	Antibacterial	Ranjani et al. (2020)
Talaromyces purpurogenus	*Pinus densiflora*	AgNPs	Round to triangle	~25	Antimicrobial and anticancer	Hu et al. (2019)
Pestalotia sp.	*Madhuca longifolia*	AgNPs	Angular	<40	Antibacterial	Verma et al. (2016)
Exserohilum rostrata	AgNPs	AgNPs	Spherical	~10–15	Antibacterial, anti-inflammatory, and antioxidant	Bagur et al. (2019)
Penicillium janthinellum	*Suaeda monoica*	AgNPs	Spherical	~8–14	Antibacterial	Bharathidasan and Panneerselvam (2012)
		AuNPs	Spherical	~20–35	Antibacterial	Alappat et al. (2012)

Continued

Table 5.2 Continued

Fungal species	Host plants	Nanoparticles	Shape	Size (nm)	Application	References
Colletotrichum gloeosporioides	*Bauhinia variegata*					
Cryptosporiopsis ericae	*Potentilla fulgens*	AgNPs	Spherical	~5.5	Antimicrobial	Devi and Joshi (2014)
Aspergillus terreus	*Rhizophora annamalayana*	AgNPs	Spherical	~100	Antidermatophytic	Kalaiselvam (2013)
Curvularia lunata	*Catharanthus roseus*	AgNPs	Spherical	~10–50	Antibacterial	Ramalingmam et al. (2015)

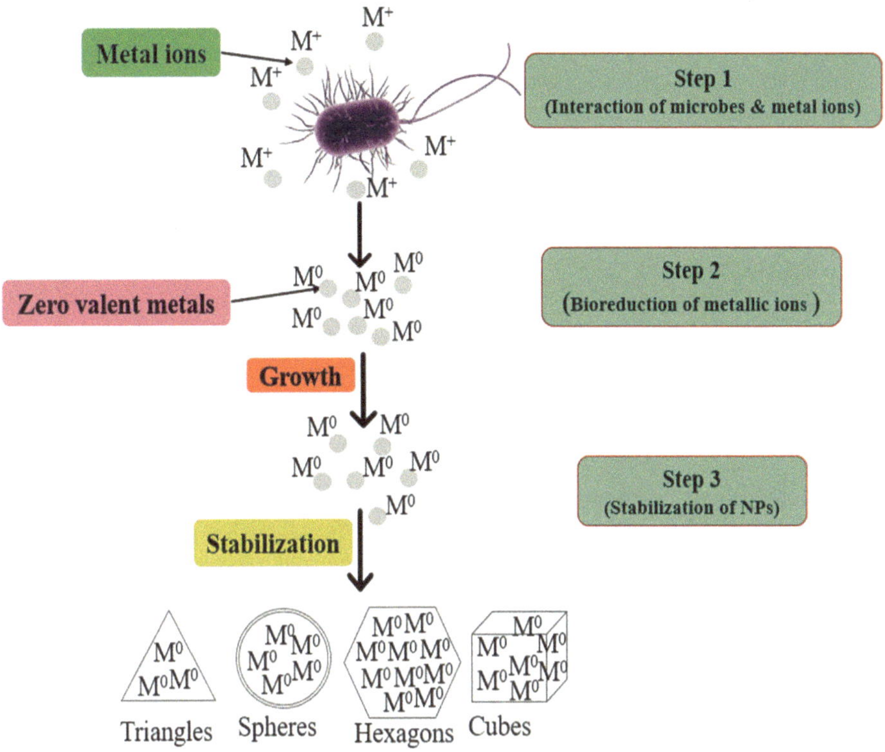

Fig. 5.4 Schematic representation of nanoparticle synthesis by biological method. Modified from Meena, M., Zehra, A., Swapnil, P., Marwal, A., Yadav, G., Sonigra, P., 2021. Endophytic nanotechnology: an approach to study scope and potential applications. Front. Chem. 9, 47 with minor modification.

5.10 Applications of nanoparticles synthesized by endophytes

Endophytes are the natural producers of several metabolites, which contribute to a vast application in numerous fields like industrial, environmental, agricultural, and pharmaceutical. Out of all the inorganic NPs, two of them (Ag and Au) are found to acquire huge importance because of their possible application in the abovementioned fields.

5.10.1 NPs antimicrobial mechanism in phytopathology

Generally, some essential metal particles are found to perform cellular functions in living organisms, but variation in metal concentration may lead to implant toxicity and can be microbicidal to most of the bacteria (Harrison et al., 2004; Nies, 2000). A number of theories have been reported against the mechanism of action of different metals. Among them, AgNPs are reported to show a number of inhibitory mechanisms against microorganisms (Le Ouay and Stellacci, 2015; Liao et al., 2019; Qais et al., 2019). These nanoparticles possess high affinity toward phosphorus and sulfur. AgNPs bind and react with the phosphorus group of DNA, which lead to the inhibition of DNA replication (Liao et al., 2019). They also inhibit protein and enzymes biosynthesis by binding to the sulfur-containing amino acids of the bacterial cell (Roy et al., 2019). All these lead to impart cell permeability, blocking of enzymes responsible for respiratory chains, and uncoupling of respiratory electron transport, which ultimately cause the death of the bacterial cell (Dakal et al., 2016). Gold nanoparticles (Au-NPs) have been also reported to possess antibacterial properties (Shamaila et al., 2016; Katas et al., 2019). But their antimicrobial efficacy is enhanced when these NPs are utilized in amalgamation with drugs, vaccines, antibiotics, and antibodies (Tao, 2018).

5.10.2 NPs antifungal mechanism in phytopathology

Some of the metallic nanoparticles are also found to acquire antifungal properties. For instance, an endophytic fungus named *Epicoccum nigrum* synthesizes AgNPs, which shows a broader antifungal spectrum than fluconazole and itraconazole. The antifungal effect of biologically synthesized AgNPs is found to be active against a large number of pathogenic fungi such as *Aspergillus flavus*, *A. fumigatus*, *Cryptococcus neoformans*, *F. solani*, *Candida tropicalis*, *Candida albicans*, *Candida krusei*, *Candida parapsilosis*, and *Sporothrix schenckii* (Qian et al., 2013).

5.10.3 NPs as nanopesticides

As NPs possess exceptional physicochemical properties, their utilization in agricultural fields as nanopesticides has happened to be more popular than the commercially synthesized pesticides. Currently, nanotechnology is rapidly gaining huge applications in food industries starting from food processing to food packaging. Also, the diseases caused in forestry and agricultural crops are controlled with the aid of nanopesticides (Bouwmeester et al., 2009). Even the NPs are used to control fungal disease by inhibiting the hyphal growth of pathogenic fungi. For instance, Park et al. (2006) analyzed the efficacy of nano-formularized silica-silver particles for controlling pathogenic fungi. The potentiality of nanopesticides was investigated for the control of phytopathogens (Jo et al., 2009).

5.10.4 NPs improving plant resistance

The economic loss and diseases in plants are caused because of various agents (viruses, bacteria, mycoplasma, fungi, nematodes, insects, pest), pathogens, and unfavorable ecological stress (drought, low temperature, and high temperature) (Alghuthaymi et al., 2015). Therefore, plants are required to be defended from phytopathogens through acquired resistance, genetic resistance, signal transduction, pathogen-associated molecular pattern (PAMP)-triggered immunity (PTI), etc.

(Punja, 2004). In order to achieve genetic modification, nanobiotechnology proposes a novel technique, which involves the integration of vertically aligned carbon nanofiber (VACNF) elements along with the cells that are biochemically manipulated using nanofiber interface. Through this modification, organization and observation of viable cells are achievable even at molecular and subcellular levels (McKnight et al., 2003). Rai et al. (2012) in their study illustrated a new method of gene transfer through NPs, which possess the ability to relocate genetic material within the plant cells, thereby serving as a transgenic vehicle for nucleic acids. Furthermore, NPs can be utilized for the improvement of agricultural crops. But it is very essential to identify as well as recognize the interaction between various NPs and plant diversity before they are applied to individual flora.

5.11 Conclusions and future perspectives

The biosynthesis of nanoparticles via endophytic microorganisms is considered a significant potential biofactory. It is one of the cost-effective and eco-friendly approaches. Previously, nanoparticles were only manufactured using chemical and physical processes, which resulted in the production of some toxic by-products that may cause a direct threat to the environment. But currently, endophytic microorganisms, both fungi and bacteria, are known for their capacity to synthesize different bioactive components as well as certain nanoparticles. The size-dependent features of nanoparticles make them appealing for use in diverse areas, including biomedicine and agriculture. AgNPs are very well-recognized for their bactericidal and antifungal properties. Nanofilms and nanoclays are often used as barriers to prevent food spoilage. The persistent use of fertilizers is causing many serious risks to humans, wildlife, and soil productivity. And when this fertilizer circulates from the soil to water, it poses a risk to marine life. The invention of nanopesticides proved that they are less toxic to the atmosphere when it is used to treat a variety of diseases. The latest breakthroughs have focused on the construction of an environmentally sustainable pathway for the production of nanoparticles utilizing endophytes. The synthesis and production of nanoparticles from endophytic microbial agents have arisen as a fascinating field of nanobiotechnology. Furthermore, a more detailed molecular procedure underlying the biosynthesis of nanoparticles by endophytes must be focused on. It is very vital to acquire more authentic information about the impact of nanoparticles or nanomaterials on endophytic microbes as well as to inhibit their detrimental effects on living beings. Future explorations and advanced technology on a larger scale are also needed.

Acknowledgment

This chapter has been financially supported by the Science and Engineering Research Board (File No. SRG/2020/000586), Department of Science and Technology, Government of India.

References

Abdel-Azeem, A., Nada, A.A., O'donovan, A., Thakur, V.K., Elkelish, A., 2020. Mycogenic silver nanoparticles from endophytic *Trichoderma atroviride* with antimicrobial activity. J. Renew. Mater. 8, 171.

Abdelhakim, H.K., El-Sayed, E.R., Rashidi, F.B., 2020. Biosynthesis of zinc oxide nanoparticles with antimicrobial, anticancer, antioxidant and photocatalytic activities by the endophytic *Alternaria tenuissima*. J. Appl. Microbiol. 128, 1634–1646.

Alappat, C.F., Kannan, K.P., Vasanthi, N.S., 2012. Biosynthesis of Au nanoparticles using the endophytic fungi isolated from *Bauhinia variegata* L. Eng. Sci. Technol. 2, 377–380.

Alghuthaymi, M.A., Almoammar, H., Rai, M., Said-Galiev, E., Abd-Elsalam, K.A., 2015. Myconanoparticles: synthesis and their role in phytopathogens management. Biotechnol. Biotechnol. Equip. 29, 221–236.

Azmath, P., Baker, S., Rakshith, D., Satish, S., 2016. Mycosynthesis of silver nanoparticles bearing antibacterial activity. Saudi Pharm. J. 24, 140–146.

Bagur, H., Poojari, C.C., Melappa, G., Rangappa, R., Chandrasekhar, N., Somu, P., 2019. Biogenically synthesized silver nanoparticles using endophyte fungal extract of *Ocimum tenuiflorum* and evaluation of biomedical properties. J. Clust. Sci., 1–15.

Baker, S., Satish, S., 2015. Biosynthesis of gold nanoparticles by *Pseudomonas veronii* AS41G inhabiting *Annona squamosa* L. Spectrochim. Acta A 150, 691–695.

Banik, A., 2019. Plant endophytes: true symbiont or opportunistic pathogens? EC Microbiol. 15 (7), 533–535.

Banik, A., Mukhopadhaya, S.K., Dangar, T.K., 2016a. Characterization of N_2-fixing plant growth promoting endophytic and epiphytic bacterial community of Indian cultivated and wild rice (*Oryza* spp.) genotypes. Planta 243, 799–812.

Banik, A., Mukhopadhaya, S.K., Sahana, A., Das, D., Dangar, T.K., 2016b. Fluorescence resonance energy transfer (FRET)-based technique for tracking of endophytic bacteria in rice roots. Biol. Fertil. Soils 52, 277–282.

Banik, A., Pandya, P., Patel, B., Rathod, C., Dangar, M., 2018. Characterization of halotolerant, pigmented, plant growth promoting bacteria of groundnut rhizosphere and its in-vitro evaluation of plant-microbe protocooperation to withstand salinity and metal stress. Sci. Total Environ. 630, 231–242.

Banik, A., Chattopadhyay, A., Ganguly, S., Mukhopadhyay, S.K., 2019a. Characterization of a tea pest specific *Bacillus thuringiensis* and identification of its toxin by MALDI-TOF mass spectrometry. Ind. Crops Prod. 137, 549–556.

Banik, A., Dash, G.K., Swain, P., Kumar, U., Mukhopadhyay, S.K., Dangar, T.K., 2019b. Application of rice (*Oryza sativa* L.) root endophytic diazotrophic *Azotobacter* sp. strain Avi2 (MCC 3432) can increase rice yield under green house and field condition. Microbiol. Res. 219, 56–65.

Bharathidasan, R., Panneerselvam, A., 2012. Biosynthesis and characterization of silver nanoparticles using endophytic fungi *Aspergillus concius*, *Penicillium janthinellum* and *Phomosis* sp. Int. J. Pharm. Sci. Res. 3, 3163.

Bouwmeester, H., Dekkers, S., Noordam, M.Y., Hagens, W.I., Bulder, A.S., De Heer, C., Ten Voorde, S.E., Wijnhoven, S.W., Marvin, H.J., Sips, A.J., 2009. Review of health safety aspects of nanotechnologies in food production. Regul. Toxicol. Pharmacol. 53 (1), 52–62.

Chandra, R., Banik, A., 2021. Detoxification and bioconversion of arsenic and chromium. In: Nanobiotechnology, Microbes and Plant Assisted Synthesis of Nanoparticles, Mechanisms and Applications, pp. 253–270, https://doi.org/10.1016/B978-0-12-822878-4.00016-X.

Chokkareddy, R., Redhi, G.G., Kanchi, S., Ahmed, S., 2018. Green synthesis of metal nanoparticles and its reaction mechanisms. In: Kanchi, S., Ahmed, S. (Eds.), Green Metal Nanoparticles: Synthesis, Characterization and Their Application. John Wiley & Sons, pp. 113–139.

Clarance, P., Luvankar, B., Sales, J., Khusro, A., Agastian, P., Tack, J.C., Al Khulaifi, M.M., Al-Shwaiman, H.A., Elgorban, A.M., Syed, A., Kim, H.J., 2020. Green synthesis and characterization of gold nanoparticles using endophytic fungi *Fusarium solani* and its in-vitro anticancer and biomedical applications. Saudi J. Biol. Sci. 27, 706–712.

Dakal, T.C., Kumar, A., Majumdar, R.S., Yadav, V., 2016. Mechanistic basis of antimicrobial actions of silver nanoparticles. Front. Microbiol. 7, 1831. https://doi.org/10.3389/fmicb.2016.01831.

Devi, L.S., Joshi, S.R., 2014. Evaluation of the antimicrobial potency of silver nanoparticles biosynthesized by using an endophytic fungus, *Cryptosporiopsis ericae* PS4. J. Microbiol. 52 (8), 667–674.

Dutta, D., Puzari, K.C., Gogoi, R., Dutta, P., 2014. Endophytes: exploitation as a tool in plant protection. Braz. Arch. Biol. Technol. 57, 621–629.

El-Gamal, M.S., Salem, S.S., Abdo, A.M., 2018. Biosynthesis, characterization, and antimicrobial activities of silver nanoparticles synthesized by endophytic *Streptomyces* sp. Egypt. J. Biotechnol. 56, 69–85.

El-Moslamy, S.H., 2018. Bioprocessing strategies for cost-effective large-scale biogenic synthesis of nano-MgO from endophytic *Streptomyces coelicolor* strain E72 as an anti-multidrug-resistant pathogens agent. Sci. Rep. 8, 1–22.

Farsi, M., Farokhi, S., 2018. Biosynthesis of antibacterial silver nanoparticles by endophytic fungus *Nemania* sp. isolated from *Taxus baccata* L. (Iranian yew). Zahedan J. Res. Med. Sci. 20.

Fernandez, O., Theocharis, A., Bordiec, S., Feil, R., Jacquens, L., Clément, C., Fontaine, F., Barka, E.A., 2012. *Burkholderia phytofirmans* PsJN acclimates grapevine to cold by modulating carbohydrate metabolism. Mol. Plant Microbe Interact. 25, 496–504.

Fesel, P.H., Zuccaro, A., 2016. Dissecting endophytic lifestyle along the parasitism/mutualism continuum in *Arabidopsis*. Curr. Opin. Microbiol. 32, 103–112.

Godstime, O.C., Enwa, F.O., Augustina, J.O., Christopher, E.O., 2014. Mechanisms of antimicrobial actions of phytochemicals against enteric pathogens—a review. J. Pharm. Chem. Biol. Sci. 2, 77–85.

Gol, S., Pena, R.N., Rothschild, M.F., Tor, M., Estany, J., 2018. A polymorphism in the fatty acid desaturase-2 gene is associated with the arachidonic acid metabolism in pigs. Sci. Rep. 8, 1–9.

Gour, A., Jain, N.K., 2019. Advances in green synthesis of nanoparticles. Artif. Cells Nanomed. Biotechnol. 47, 844–851.

Han, J.I., Choi, H.K., Lee, S.W., Orwin, P.M., Kim, J., LaRoe, S.L., Kim, T.G., O'Neil, J., Leadbetter, J.R., Lee, S.Y., Hur, C.G., 2011. Complete genome sequence of the metabolically versatile plant growth-promoting endophyte *Variovorax paradoxus* S110. J. Bacteriol. 193, 1183–1190.

Hardoim, P.R., Van Overbeek, L.S., Berg, G., Pirttilä, A.M., Compant, S., Campisano, A., Döring, M., Sessitsch, A., 2015. The hidden world within plants: ecological and evolutionary considerations for defining functioning of microbial endophytes. Microbiol. Mol. Biol. Rev. 79, 293–320.

Harrison, J.J., Ceri, H., StrHarrison, J.J., Ceri, H., Stremick, C.A., Turner, R.J., 2004. Biofilm susceptibility to metal toxicity. Environ. Microbiol. 6 (12), 1220–1227.

Hassan, S.E.D., Salem, S.S., Fouda, A., Awad, M.A., El-Gamal, M.S., Abdo, A.M., 2018. New approach for antimicrobial activity and bio-control of various pathogens by biosynthesized copper nanoparticles using endophytic actinomycetes. J. Radiat. Res. Appl. Sci. 11, 262–270.

Hassan, S.E.D., Fouda, A., Radwan, A.A., Salem, S.S., Barghoth, M.G., Awad, M.A., Abdo, A. M., El-Gamal, M.S., 2019. Endophytic actinomycetes *Streptomyces* spp mediated biosynthesis of copper oxide nanoparticles as a promising tool for biotechnological applications. J. Biol. Inorg. Chem. 24, 377–393.

Hayat, R., Ali, S., Amara, U., Khalid, R., Ahmed, I., 2010. Soil beneficial bacteria and their role in plant growth promotion: a review. Ann. Microbiol. 60, 579–598.

Hu, X., Kandasamy Saravanakumar, T.J., Wang, M.H., 2019. Mycosynthesis, characterization, anticancer and antibacterial activity of silver nanoparticles from endophytic fungus *Talaromyces purpureogenus*. Int. J. Nanomedicine 14, 3427.

Ibrahim, E., Fouad, H., Zhang, M., Zhang, Y., Qiu, W., Yan, C., Li, B., Mo, J., Chen, J., 2019. Biosynthesis of silver nanoparticles using endophytic bacteria and their role in inhibition of rice pathogenic bacteria and plant growth promotion. RSC Adv. 9, 29293–29299.

Ibrahim, E., Zhang, M., Zhang, Y., Hossain, A., Qiu, W., Chen, Y., Wang, Y., Wu, W., Sun, G., Li, B., 2020. Green-synthesization of silver nanoparticles using endophytic bacteria isolated from garlic and its antifungal activity against wheat Fusarium head blight pathogen *Fusarium graminearum*. Nanomaterials 10, 219.

Iravani, S., 2014. Bacteria in nanoparticle synthesis: current status and future prospects. Int. Sch. Res. Notices 2014.

Iravani, S., Korbekandi, H., Mirmohammadi, S.V., Zolfaghari, B., 2014. Synthesis of silver nanoparticles: chemical, physical and biological methods. Res. Pharm. Sci. 9, 385.

Jo, Y.K., Kim, B.H., Jung, G., 2009. Antifungal activity of silver ions and nanoparticles on phytopathogenic fungi. Plant Dis. 93 (10), 1037–1043.

Jog, R., Pandya, M., Nareshkumar, G., Rajkumar, S., 2014. Mechanism of phosphate solubilization and antifungal activity of *Streptomyces* spp. isolated from wheat roots and rhizosphere and their application in improving plant growth. Microbiology 160, 778–788.

Joseph, B., Priya, R.M., 2011. Bioactive compounds from endophytes and their potential in. Am. J. Biochem. Mol. Biol. 1, 291–309.

Joshi, C.G., Danagoudar, A., Poyya, J., Kudva, A.K., Dhananjaya, B.L., 2017. Biogenic synthesis of gold nanoparticles by marine endophytic fungus-*Cladosporium cladosporioides* isolated from seaweed and evaluation of their antioxidant and antimicrobial properties. Process Biochem. 63, 137–144.

Kalaiselvam, M., 2013. Extracellular biosynthesis of silver nanoparticles by endophytic fungus *Aspergillus terreus* and its anti-dermatophytic activity. Int. J. Pharm. Biol. Arch. 4, 481–487.

Kang, J.W., Khan, Z., Doty, S.L., 2012. Biodegradation of trichloroethylene by an endophyte of hybrid poplar. Appl. Environ. Microbiol. 78, 3504–3507.

Katas, H., Lim, C.S., Azlan, A.Y.H.N., Buang, F., Busra, M.F.M., 2019. Antibacterial activity of biosynthesized gold nanoparticles using biomolecules from *Lignosus rhinocerotis* and chitosan. Saudi Pharm. J. 27 (2), 283–292.

Khan, A.L., Waqas, M., Kang, S.M., Al-Harrasi, A., Hussain, J., Al-Rawahi, A., Al-Khiziri, S., Ullah, I., Ali, L., Jung, H.Y., Lee, I.J., 2014. Bacterial endophyte *Sphingomonas* sp. LK11 produces gibberellins and IAA and promotes tomato plant growth. J. Microbiol. 52, 689–695.

Khandel, P., Yadaw, R.K., Soni, D.K., Kanwar, L., Shahi, S.K., 2018. Biogenesis of metal nanoparticles and their pharmacological applications: present status and application prospects. J. Nanostruct. Chem. 8, 217–254.

Khare, E., Mishra, J., Arora, N.K., 2018. Multifaceted interactions between endophytes and plant: developments and prospects. Front. Microbiol. 9, 2732.
Kumar, S., Kaushik, N., Edrada-Ebel, R., Ebel, R., Proksch, P., 2011. Isolation, characterization, and bioactivity of endophytic fungi of *Tylophora indica*. World J. Microbiol. Biotechnol. 27, 571–577.
Lata, R., Chowdhury, S., Gond, S.K., White Jr., J.F., 2018. Induction of abiotic stress tolerance in plants by endophytic microbes. Lett. Appl. Microbiol. 66, 268–276.
Le Ouay, B., Stellacci, F., 2015. Antibacterial activity of silver nanoparticles: a surface science insight. Nano Today 10 (3), 339–354.
Liao, S., Zhang, Y., Pan, X., Zhu, F., Jiang, C., Liu, Q., et al., 2019. Antibacterial activity and mechanism of silver nanoparticles against multidrug-resistant *Pseudomonas aeruginosa*. Int. J. Nanomedicine 14, 1469–1487. https://doi.org/10.2147/IJN.S191340.
McKnight, T.E., Melechko, A.V., Griffin, G.D., Guillorn, M.A., Merkulov, V.I., Serna, F., Hensley, D.K., Doktycz, M.J., Lowndes, D.H., Simpson, M.L., 2003. Intracellular integration of synthetic nanostructures with viable cells for controlled biochemical manipulation. Nanotechnology 14 (5), 551.
Messaoudi, O., Bendahou, M., 2020. Biological synthesis of nanoparticles using endophytic microorganisms: current development. In: Nanotechnology and the Environment. IntechOpen.
Messaoudi, O., Bendahou, M., Benamar, I., Abdelwouhid, D.E., 2015. Identification and preliminary characterization of non-polyene antibiotics secreted by new strain of actinomycete isolated from sebkha of Kenadsa, Algeria. Asian Pac. J. Trop. Biomed. 5, 438–445.
Miller, S.H., Browne, P., Prigent-Combaret, C., Combes-Meynet, E., Morrissey, J.P., O'Gara, F., 2010. Biochemical and genomic comparison of inorganic phosphate solubilization in *Pseudomonas species*. Environ. Microbiol. Rep. 2, 403–411.
Moghaddam, A.B., Namvar, F., Moniri, M., Azizi, S., Mohamad, R., 2015. Nanoparticles biosynthesized by fungi and yeast: a review of their preparation, properties, and medical applications. Molecules 20, 16540–16565.
Mukherjee, T., Banik, A., Mukhopadhyay, S.K., 2020. Plant growth-promoting traits of a thermophilic strain of the *Klebsiella* group with its effect on rice plant growth. Curr. Microbiol. 77, 2613–2622.
Musarrat, J., Dwivedi, S., Singh, B.R., Al-Khedhairy, A.A., Azam, A., Naqvi, A., 2010. Production of antimicrobial silver nanoparticles in water extracts of the fungus *Amylomyces rouxii* strain KSU-09. Bioresour. Technol. 101, 8772–8776.
Nair, D.N., Padmavathy, S., 2014. Impact of endophytic microorganisms on plants, environment and humans. Sci. World J. 2014. https://doi.org/10.1155/2014/250693, 250693.
Nies, D.H., 2000. Microbial heavy-metal resistance. Appl. Microbiol. Biotechnol. 51, 451–460.
Niño-Martínez, N., Salas Orozco, M.F., Martínez-Castañón, G.A., Torres Méndez, F., Ruiz, F., 2019. Molecular mechanisms of bacterial resistance to metal and metal oxide nanoparticles. Int. J. Mol. Sci. 20, 2808.
Olson, P.E., Castro, A., Joern, M., DuTeau, N.M., Pilon-Smits, E., Reardon, K.F., 2008. Effects of agronomic practices on phytoremediation of an aged PAH-contaminated soil. J. Environ. Qual. 37, 1439–1446.
Ovais, M., Khalil, A.T., Ayaz, M., Ahmad, I., Nethi, S.K., Mukherjee, S., 2018. Biosynthesis of metal nanoparticles via microbial enzymes: a mechanistic approach. Int. J. Mol. Sci. 19, 4100.
Park, H.J., Kim, S.H., Kim, H.J., Choi, S.H., 2006. A new composition of nanosized silica-silver for control of various plant diseases. Plant Pathol. J. 22 (3), 295–302.

Patel, M.V., Patel, R.K., 2014. Indole-3-acetic acid (IAA) production by endophytic bacteria isolated from saline dessert, the little Runn of Kutch. CIBTech J. Microbiol. 3, 17–28.

Piccoli, P., Bottini, R., 2013. Abiotic stress tolerance induced by endophytic PGPR. In: Symbiotic Endophytes. Springer, Berlin, Heidelberg, pp. 151–163.

Popli, D., Anil, V., Subramanyam, A.B., Namratha, M.N., Ranjitha, V.R., Rao, S.N., Rai, R.V., Govindappa, M., 2018. Endophyte fungi, *Cladosporium* species-mediated synthesis of silver nanoparticles possessing in vitro antioxidant, anti-diabetic and anti-Alzheimer activity. Artif. Cells Nanomed. Biotechnol. 46, 676–683.

Porter, A.L., Youtie, J., 2009. How interdisciplinary is nanotechnology? J. Nanopart. Res. 11, 1023–1041.

Potshangbam, M., Devi, S.I., Sahoo, D., Strobel, G.A., 2017. Functional characterization of endophytic fungal community associated with *Oryza sativa* L. and *Zea mays* L. Front. Microbiol. 8, 325.

Punja, Z.K., 2004. Fungal Disease Resistance in Plants: Biochemistry, Molecular Biology, and Genetic Engineering. Food Products Press.

Qais, F.A., Shafiq, A., Khan, H.M., Husain, F.M., Khan, R.A., Alenazi, B., et al., 2019. Antibacterial effect of silver nanoparticles synthesized using *Murraya koenigii* (L.) against multidrug-resistant pathogens. Bioinorg. Chem. Appl. 2019, 4649506. https://doi.org/10.1155/2019/4649506.

Qian, Y., Yu, H., He, D., Yang, H., Wang, W., Wan, X., Wang, L., 2013. Biosynthesis of silver nanoparticles by the endophytic fungus *Epicoccum nigrum* and their activity against pathogenic fungi. Bioprocess Biosyst. Eng. 36, 1613–1619.

Raheman, F., Deshmukh, S., Ingle, A., Gade, A., Rai, M., 2011. Silver nanoparticles: novel antimicrobial agent synthesized from an endophytic fungus *Pestalotia* sp. isolated from leaves of *Syzygium cumini* (L). Nano Biomed. Eng. 3, 174–178.

Rai, M., Deshmukh, S., Gade, A., 2012. Strategic nanoparticle-mediated gene transfer in plants and animals—a novel approach. Curr. Nanosci. 8, 170–179.

Rajabairavi, N., Raju, C.S., Karthikeyan, C., Varutharaju, K., Nethaji, S., Hameed, A.S.H., Shajahan, A., 2017. Biosynthesis of novel zinc oxide nanoparticles (ZnO NPs) using endophytic bacteria *Sphingobacterium thalpophilum*. In: Recent Trends in Materials Science and Applications. Springer, pp. 245–254.

Rajkumar, M., Ae, N., Freitas, H., 2009. Endophytic bacteria and their potential to enhance heavy metal phytoextraction. Chemosphere 77, 153–160.

Rajkumar, M., Ae, N., Prasad, M.N.V., Freitas, H., 2010. Potential of siderophore-producing bacteria for improving heavy metal phytoextraction. Trends Biotechnol. 28, 142–149.

Ramalingmam, P., Muthukrishnan, S., Thangaraj, P., 2015. Biosynthesis of silver nanoparticles using an endophytic fungus, *Curvularia lunata* and its antimicrobial potential. J. Nanosci. Nanoeng. 1, 241–247.

Rana, K.L., Kour, D., Yadav, N., Yadav, A.N., 2020. Endophytic microbes in nanotechnology: current development, and potential biotechnology applications. In: Microbial Endophytes. Woodhead Publishing, pp. 231–262.

Rani, P., Arya, V., 2017. Green synthesis of silver nanoparticles from endophytic fungus *Aspergillus niger*. In: Engineering Interventions in Agricultural Processing. 132. CRC Press, p. 131.

Ranjani, S., Adnan, M., Ruckmani, K., Hemalatha, S., 2020. Synthesis, characterization and applications of endophytic fungal nanoparticles. Inorg. Nano-Metal Chem. 51, 280–287. https://doi.org/10.1080/24701556.2020.1784231.

Roy, A., Bulut, O., Some, S., Mandal, A.K., Yilmaz, M.D., 2019. Green synthesis of silver nanoparticles: biomolecule-nanoparticle organizations targeting antimicrobial activity. RSC Adv. 9 (5), 2673–2702.

Rybakova, D., Cernava, T., Köberl, M., Liebminger, S., Etemadi, M., Berg, G., 2016. Endophytes-assisted biocontrol: novel insights in ecology and the mode of action of Paenibacillus. Plant and Soil 405, 125–140.
Santoyo, G., Moreno-Hagelsieb, G., del Carmen Orozco-Mosqueda, M., Glick, B.R., 2016. Plant growth-promoting bacterial endophytes. Microbiol. Res. 183, 92–99.
Sarkar, S., Sarkar, S.C., 2019. Chapter-4 application of nanotechnology in medicine. Med. Cal. Sci. 117, 49.
Shamaila, S., Zafar, N., Riaz, S., Sharif, R., Nazir, J., Naseem, S., 2016. Gold nanoparticles: an efficient antimicrobial agent against enteric bacterial human pathogen. Nanomaterials 6 (4), 71.
Shukla, S., Habbu, P., Kulkarni, V., Jagadish, K., Pandey, A., Sutariya, V., 2014. Endophytic microbes: a novel source for biologically/pharmacologically active secondary metabolites. Asian J. Pharm. Toxicol. 2, 1–6.
Singh, T., Jyoti, K., Patnaik, A., Singh, A., Chauhan, R., Chandel, S.S., 2017. Biosynthesis, characterization and antibacterial activity of silver nanoparticles using an endophytic fungal supernatant of *Raphanus sativus*. J. Genet. Eng. Biotechnol. 15, 31–39.
Singh, M., Srivastava, M., Kumar, A., Pandey, K.D., 2019. Biosynthesis of nanoparticles and applications in agriculture. In: Role of Plant Growth Promoting Microorganisms in Sustainable Agriculture and Nanotechnology. Woodhead Publishing, pp. 199–217.
Smith, S.A., Tank, D.C., Boulanger, L.A., Bascom-Slack, C.A., Eisenman, K., Kingery, D., Babbs, B., Fenn, K., Greene, J.S., Hann, B.D., Keehner, J., 2008. Bioactive endophytes warrant intensified exploration and conservation. PLoS One 3, e3052.
Soliman, H., Elsayed, A., Dyaa, A., 2018. Antimicrobial activity of silver nanoparticles biosynthesised by *Rhodotorula* sp. strain ATL72. Egypt. J. Basic Appl. Sci. 5, 228–233.
Stępniewska, Z., Kuźniar, A., 2013. Endophytic microorganisms—promising applications in bioremediation of greenhouse gases. Appl. Microbiol. Biotechnol. 97, 9589–9596.
Sudha, V., Govindaraj, R., Baskar, K., Al-Dhabi, N.A., Duraipandiyan, V., 2016. Biological properties of endophytic fungi. Braz. Arch. Biol. Sci. 59.
Sunkar, S., Nachiyar, C.V., 2012a. Biogenesis of antibacterial silver nanoparticles using the endophytic bacterium *Bacillus cereus* isolated from *Garcinia xanthochymus*. Asian Pac. J. Trop. Biomed. 2, 953–959.
Sunkar, S., Nachiyar, C.V., 2012b. Microbial synthesis and characterization of silver nanoparticles using the endophytic bacterium *Bacillus cereus*: a novel source in the benign synthesis. Glob. J. Med. Res. 12, 43–50.
Syed, B., Prasad, N.M., Satish, S., 2016. Endogenic mediated synthesis of gold nanoparticles bearing bactericidal activity. J. Microsc. Ultrastruct. 4, 162–166.
Syed, B., Prasad, M.N., Satish, S., 2019. Synthesis and characterization of silver nanobactericides produced by *Aneurinibacillus migulanus* 141, a novel endophyte inhabiting *Mimosa pudica* L. Arab. J. Chem. 12, 3743–3752.
Talekar, S., Ghodake, V., Ghotage, T., Rathod, P., Deshmukh, P., Nadar, S., Mulla, M., Ladole, M., 2012. Novel magnetic cross-linked enzyme aggregates (magnetic CLEAs) of alpha amylase. Bioresour. Technol. 123, 542–547.
Taniguchi, N., 1974. On the basic concept of nano-technology. In: Proceedings of the International Conference on Production Engineering Tokyo Part II. Japan Society of Precision Engineering.
Tao, C., 2018. Antimicrobial activity and toxicity of gold nanoparticles: research progress, challenges and prospects. Lett. Appl. Microbiol. 67 (6), 537–543.
Thomas, R., Jasim, B., Mathew, J., Radhakrishnan, E.K., 2012. Extracellular synthesis of silver nanoparticles by endophytic *Bordetella* sp. isolated from *Piper nigrum* and its antibacterial activity analysis. Nano Biomed. Eng. 4.

Timmusk, S., Paalme, V., Pavlicek, T., Bergquist, J., Vangala, A., Danilas, T., Nevo, E., 2011. Bacterial distribution in the rhizosphere of wild barley under contrasting microclimates. PLoS One 6, e17968.

Torney, F., Trewyn, B.G., Lin, V.S.Y., Wang, K., 2007. Mesoporous silica nanoparticles deliver DNA and chemicals into plants. Nat. Nanotechnol. 2, 295–300.

Velusamy, P., Kumar, G.V., Jeyanthi, V., Das, J., Pachaiappan, R., 2016. Bio-inspired green nanoparticles: synthesis, mechanism, and antibacterial application. Toxicol. Res. 32, 95–102.

Verma, V.C., Kharwar, R.N., Gange, A.C., 2010. Biosynthesis of antimicrobial silver nanoparticles by the endophytic fungus *Aspergillus clavatus*. Nanomedicine 5, 33–40.

Verma, S.K., Gond, S.K., Mishra, A., Sharma, V.K., Kumar, J., 2016. Biofabrication of antibacterial and antioxidant silver nanoparticles (Agnps) by an endophytic fungus *Pestalotia* Sp. isolated from *Madhuca longifolia*. J. Nanomater. Mol. Nanotechnol. 5, 3.

Xin, G., Zhang, G., Kang, J.W., Staley, J.T., Doty, S.L., 2009. A diazotrophic, indole-3-acetic acid-producing endophyte from wild cottonwood. Biol. Fertil. Soils 45, 669–674.

Yadav, A., Yadav, K., 2017. Exploring the potential of endophytes in agriculture: a minireview. Adv. Plants Agric. Res. 6, 102–106.

Yasmeen, R., 2021. Biological and clinical perspectives of nanobiotechnology. Res. J. Adv. Sci. 2, 1–8.

Yousef, N., Mawad, A., Abeed, A., 2019. Enhancement the cellulase activity induced by endophytic bacteria using calcium nanoparticles. Curr. Microbiol. 76, 346–354.

Metal stress removal and nanotechnology-driven solutions

6

Sougata Ghosh[a,c] *and Bishwarup Sarkar*[b]
[a]Department of Microbiology, School of Science, RK University, Rajkot, Gujarat, India, [b]College of Science, Northeastern University, Boston, MA, United States, [c]Department of Physics, Faculty of Science, Kasetsart University, Bangkok, Thailand

6.1 Introduction

Heavy metals (HMs) are naturally present in the earth's crust which is required by living organisms only in trace amounts (Mishra et al., 2019). These metals generally persist in the environment as contaminants and cannot be completely biodegraded, but it can be transformed into less severe toxic forms. However, HM contamination has become an emerging issue with rapid industrialization, excessive use of agrochemicals, unreasonable mining along with inappropriate waste management (Gautam et al., 2016). These anthropogenic activities have therefore severely contaminated the environment with several hazardous and toxic metals. Recently, HM contamination has been frequently detected in food and blood which in turn has attracted public attention (Zhou et al., 2021).

HM stress has also been reported to prominently affect plant growth and development, thereby indirectly affecting human health through the food chain (Naveedullah et al., 2013). In recent years, several studies have reported negative impact of HM contamination on important agricultural crops such as rice wherein toxicity of HMs is evident on a molecular, biochemical, physiological, and cellular level that ultimately results in decreased crop productivity (Arif et al., 2019). In addition, food quality as well as safety are also compromised due to cultivation of crops in excessive HM-contaminated soil which in turn increases the occurrence of health problems such as kidney and liver failure, infertility and reproductive disorders, several types of cancers, leukemia, mental illness, and others (Wei et al., 2019).

In recent times, nanotechnology has been extensively applied in industries, pharmaceuticals, cosmetics, and even in agricultural sectors (Zhou et al., 2021). Several studies have shown that nanoparticles assist in improvement in seed germination, plant photosynthesis, oxidative stress resistance, rhizome development, crop yield, and quality (Kah et al., 2019). Nanomaterials have even been applied as nanofertilizers, nanopesticides, and nanofungicides that are gradually released in environment as compared to conventional pesticides and can be easily adsorbed by the plants (Lowry et al., 2019). Furthermore, several nanoparticles have been shown to enhance activity of antioxidant enzymes in order to reduce reactive oxygen species (ROS)

Agricultural Nanobiotechnology. https://doi.org/10.1016/B978-0-323-91908-1.00004-3

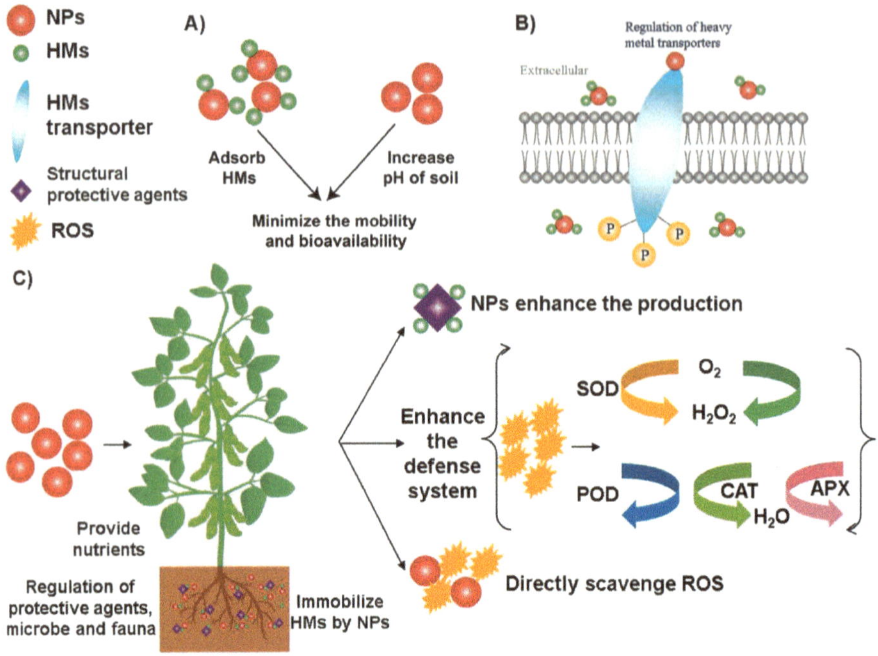

Fig. 6.1 Mechanisms of NPs to alleviate HM stress in soil and plant tissues. (A) NPs increase the soil pH and adsorb HM, which minimize the mobility and bioavailability of HM. (B) NPs can enhance the defense ability of plants by regulating the HM transport genes. (C) NPs enhance the plant defense system, promote the production of protective agents, and directly scavenge ROS, which ultimately improve the plant growth and nutritional content of fruits.
Reproduced with permission from Zhou, P., Adeel, M., Shakoor, N., Guo, M., Hao, Y., Azeem, I., Li, M., Liu, M., Rui, Y., 2021. Application of nanoparticles alleviates HMs stress and promotes plant growth: an overview. Nanomaterials (Basel) 11(1), 26. Copyright © 2020 by the authors. Licensee MDPI, Basel, Switzerland.

accumulation in plants, thus ameliorating plant stress that could improve yield and quality of crops (Ghosh et al., 2021a,b,c,d). In order to alleviate HM stress in plants, several strategies could be applied using nanoparticles as summarized in Fig. 6.1. Most of the nanoparticles either reduce the soil concentration of bioavailable HM, regulate expression of transport genes responsible for translocation of HM in plants, enhance antioxidant enzyme activities in plants, or promote production of phytoprotective agents (Ghosh, 2020). This chapter therefore explains, in detail, the application of different nanomaterials for alleviation of HM stress in plants.

6.2 Impact of metal stress in plants

Plants are continuously facing several abiotic and biotic stresses throughout their lifespan. Accumulation of HMs such as arsenic (As), iron (Fe), manganese (Mn), copper (Cu), nickel (Ni), cobalt (Co), cadmium (Cd), lead (Pb), mercury (Hg), and zinc (Zn) has become a common abiotic stress condition for plants nowadays due to increase in anthropogenic activities (Ghori et al., 2019). While some of these metals such as Zn, Cu, and Ni are important micronutrients and act as important cofactors for various enzymes in plants, other metals such as Cd and Pb are purely toxic and impart no beneficial role toward plants.

HM accumulation in plants leads to inactivation or denaturation of several important proteins which may alter membrane integrity and also interfere with basic metabolic reactions of plants such as photosynthesis, respiration, and homeostasis (Ghori et al., 2019). HM toxicity also stimulates ROS production that may lead to lipid peroxidation which in turn would result in membrane leakage, biomolecule damage, and DNA cleavage. Specific toxic effects of few HMs are summarized in Table 6.1. Although plants induce several molecular and physiological strategies in order to tolerate and defend against HM toxicity, some of which include sequestration of HMs in cytosol using phytochelatins, upregulation of certain stress pathways such as mitogen-activated protein kinase (MAPK) cascade pathway, expression of metal resistance genes stimulated by H_2O_2, detoxification of ROS using antioxidant enzymes, and many more. However, these mechanisms are not efficient enough to tolerate high concentrations of HMs in order to maintain proper yield and productivity of economically important plants.

Table 6.1 Toxic effects of HMs imparted on plants.

Heavy metal	Toxic effects
Zn	Growth retardation and senescence, chlorosis, root growth restriction, inhibition of photosystems, carbon fixation inhibition, increase in free calcium
Cd	Fe(II) deficiency, reduction in chlorophyll synthesis, water balance disturbance, inhibition in nutrient uptake, chromosomal fragmentation and aberration, reduction in respiration rate, accumulation of proline amino acid, decrease in plant biomass, and photosynthetic pigments
Cu	Chlorosis, oxidative stress induction, ion leakage, plant growth inhibition, retarded root growth, germination inhibition
Hg	Water uptake interference, oxidative stress induction, membrane degradation, biomolecules oxidation
Cr	Chlorosis, growth retardation, root injury, wilting, electron transport chain inhibition, carbon fixation inhibition, alteration in photosynthetic pigments, increased production of glutathione and ascorbate
Ni	Oxidative stress, nitrogen metabolism inhibition, mitotic inhibition, reduction in photosynthetic ability
Pb	Stunted growth, chlorosis, reduced root length, alteration in membrane permeability, hormonal imbalances, mineral nutrition disturbance, germination inhibition, stomatal pores closure

6.3 Nanoparticles for metal stress removal

Nanotechnology-driven solutions have come up as promising way to ameliorate metal stress in plants. This section gives an elaborate account of various nanoparticles that are listed in Table 6.2 with potential benefits for addressing metal stress in plants.

Table 6.2 Nanoparticles used for heavy metal stress removal in plants.

Nanoparticles (NPs)	Plant	Heavy metal (HM)	Mechanism of NPs in reducing HM stress	Reference
CeO_2NPs	*Oryza sativa* L.	Cd	CeO_2NPs alleviated Cd-induced reduction in shoot height and weight; increased chlorophyll content; reduced MDA and proline content; decreased oxidative DNA damage and reduced Cd content in plants	Wang et al. (2019)
CeO_2NPs and ZnONPs	*Oryza sativa* L.	As	CeO_2NPs increased shoot biomass in presence of As(III); ZnONPs decreased accumulation of As(III) and As(V)	Wang et al. (2018)
TiO_2NPs	*Zea mays* L.	Cd	Foliar exposure of TiO_2NPs protected loss of shoot and root dry weights as well as reduced MDA content and oxidative stress; root exposure of TiO_2NPs reduced loss of plasma membrane integrity	Lian et al. (2020)
TiO_2NPs	*Oryza sativa* L.	Cd	Increase in root length, plant height, fresh biomass, and photosynthesis rate in presence of Cd; decrease in Cd accumulation and organic acids content	Ji et al. (2017)
nZVI	*Oryza sativa* L.	Cd	Increase in biomass, shoot and root length under Cd stress; decrease in MDA and H_2O_2 levels; downregulation in expression of OsIRT1,2 and OsYSL2,15 genes	Guha et al. (2020)

Table 6.2 Continued

Nanoparticles (NPs)	Plant	Heavy metal (HM)	Mechanism of NPs in reducing HM stress	Reference
FeNPs	*Triticum aestivum*	Cd	FeNPs acted as growth promoter; foliar exposure increased photosynthetic pigments and stomatal conductance in presence of Cd while electrolytic leakage and Cd content were decreased	Hussain et al. (2019)
FeNPs	*Triticum aestivum*	Cd	FeNPs acted as growth promoter under drought and Cd stress; significantly reduced Cd content under drought stress; downregulated oxidative stress markers	Adrees et al. (2020)
Fe_3O_4NPs	*Triticum aestivum*	Pb, Zn, Cd, and Cu	Fe_3O_4NPs acted as growth promoter under Cd stress; decreased MDA content; and increased activity of antioxidant enzymes	Konate et al. (2017)
ZnONPs	*Oryza sativa* L.	As and Cd	ZNONPs reduced total As content and decreased Cd accumulation; facilitated microbial As methylation	Ma et al. (2020)
ZnONPs	*Triticum aestivum*	Cd	ZnONPs acted as growth promoter under drought and Cd stress; increased concentration of photosynthetic pigments; reduced Cd-induced oxidative stress; reduced Cd content in plants and Cd mobility in soil	Khan et al. (2019)
ZnONPs	*Triticum aestivum*	Cd	Foliar and soil application of ZnONPs increased plant growth in presence of Cd; reduced Cd concentrations; reduced electrolytic leakage	Hussain et al. (2018)

Continued

Table 6.2 Continued

Nanoparticles (NPs)	Plant	Heavy metal (HM)	Mechanism of NPs in reducing HM stress	Reference
ZnONPs	*Leucaena leucocephala*	Cd and Pb	Biogenic ZnONPs increased shoot and root length of plants in presence of Cd and Pb; enhanced photosynthetic pigments; declined MDA content; increased activity of antioxidant enzymes	Venkatachalam et al. (2017)
SNPs	*Brassica napus* L.	Hg	SNPs increased shoot and root growth in presence of Hg; reduced MDA contents and Hg accumulation; increased activity of antioxidant enzymes; higher accumulation of essential macro- and micronutrients under Hg stress	Yuan et al. (2021)
SNPs	*Helianthus annuus* L.	Mn	Biogenic SNPs mitigated Mn toxicity; increased pigment content; reduced electrolytic leakage; increased nutrient uptake and total amino acid content	Ragab and Saad-Allah (2020)
SeNPs	*Brassica napus* L.	Cd	Biogenic SeNPs ameliorated toxic effects of Cd; increased apical region viability; reduced oxidative stress; inhibited gene expression of NADPH oxidase and glycolate oxidase; maintained homeostasis of intracellular Ca concentration; helped in formation of disulfide bonds and restored waxy outer layer in leaves	Qi et al. (2021)

Table 6.2 Continued

Nanoparticles (NPs)	Plant	Heavy metal (HM)	Mechanism of NPs in reducing HM stress	Reference
SiO_2NPs	*Zea mays* L.	Al	SiO_2NPs mitigated Al toxicity in plant growth; increased photosynthetic, respiration rate, and stomatal conductance; reduced enzymes responsible for oxidative stress and flavonoids content	de Sousa et al. (2019)
PSINPs	*Zea mays* L.	Cu	PSINPs enhanced seed germination; increased growth parameters under Cu stress	Xin et al. (2020)
NHAP	*Lolium perenne* L.	Pb	NHAP increased plant biomass in presence of Pb; reduced Pb accumulation; reduced bioavailability of Pb in soil	Jin et al. (2016)

6.3.1 Cerium

Nanomaterials have been modeled in order to utilize it for elevated functions of plant organelles, tissues as well as whole organisms. Such approaches are also investigated in order to improve metal stress tolerance among plants. In one such study, Wang et al. (2019) showed alleviation of Cd stress from *Oryza sativa* L. (rice seedlings) after treatment with cerium oxide nanoparticles (CeO_2NPs). CeO_2NPs used in the study showed a cubic fluorite structure in transmission electron microscopy (TEM) images that had an average diameter of 620.73 ± 50.31 nm in deionized water along with a zeta potential value of -11.63 ± 0.32 mV. However, the nanoparticles were observed to be more stable when dispersed in Kimura nutrient solution as compared to deionized water. On the other hand, growth of rice seedlings was significantly inhibited in presence of 50 μM of $CdCl_2$ as well as 50 mM of NaCl as compared to control group. Shoot height was decreased by 28.22% in presence of $CdCl_2$; however, decrease in shoot height was reduced to 19.58% and 18.51% when $CdCl_2$-treated plants were in presence of NaCl and CeO_2NPs, respectively. Similar results were observed for the fresh shoot and root weights after application of CeO_2NPs in $CdCl_2$-treated rice seedlings. Moreover, salt stress was also reduced with respect to shoot height and shoot fresh weight after application of CeO_2NPs. Chlorophyll contents of rice seedlings were also drastically reduced in presence of 50 μM of $CdCl_2$ with 65.80% and 65.03% decrease in specific content of chlorophyll a and chlorophyll b, respectively. Furthermore, proline content in rice seedlings was investigated as proline accumulation is a known indicator of stress in plants. Level of proline was observed to increase when treated only with $CdCl_2$; however, coexposure of 50 μM of $CdCl_2$ with 200 mg/L of CeO_2NPs did not show elevation in proline content. Similarly, around 12.47% decline in proline content was observed after addition of CeO_2NPs in rice seedling under both Cd and salt stress. Amount of malondialdehyde (MDA) which is a lipid peroxidation product synthesized in presence of reactive oxygen species (ROS) was estimated. MDA content in shoots was increased by 2.43- and 2.96-folds after coexposure with $CdCl_2$ + NaCl and $CdCl_2$ + NaCl + CeO_2NPs, respectively. On the other hand, root MDA content was decreased by 18.89% after CeO_2NPs addition to plants that were coexposed with $CdCl_2$ and NaCl. Hence, it was speculated that CeO_2NPs might reduce MDA content in roots by inhibiting ROS production after exposure to $CdCl_2$ and NaCl. DNA damage by $CdCl_2$ was also evaluated by estimation of 8-hydroxydesoxiguanosine (8-OHdG) which is a hydroxylated product of guanine produced in presence of ROS. Significant reduction of 8-OHdG content was observed in presence of NaCl and $CdCl_2$ as compared to only NaCl treatment. Likewise, CeO_2NPs also alleviated oxidative DNA damage caused by NaCl. Cd content in rice seedlings was also found to be reduced by 8.38% after addition of CeO_2NPs while maximum reduction in Cd content of 15.96% was observed in plants that were exposed with NaCl. Therefore, it was suggested that CeO_2NPs could be used for removal of Cd metal stress in rice seedlings.

In another study, Wang et al. (2018) investigated the mutual effects of CeO_2NPs and zinc oxide nanoparticles (ZnONPs) on As uptake and its speciation in *Oryza sativa* (rice) plants. Both species of engineered nanoparticles were primarily spherical in shape while few particles had a triangular or irregular morphology as observed in TEM images. CeO_2NPs were found to have an average size of 10.5 nm while average diameter of ZnONPs was 68.1 nm. Hydrodynamic sizes of CeO_2NPs and ZnONPs were 961.83 ± 94.39 nm and 621.08 ± 7.63 nm, respectively, along with a zeta potential of -36.55 ± 4.72 and -28.80 ± 2.04 mV. Coexposure of CeO_2NPs with As(III) resulted in increased shoot biomass while ZnONPs in presence of As(III) or As(V) showed significant reduction in fresh root biomass of rice plants by 19% and 29%, respectively, as compared to seedlings that were treated with ZnONPs alone. Furthermore, accumulation of As(III) and As(V) in rice tissues was drastically reduced in presence of ZnONPs while no significant effect was observed in case of CeO_2NPs. After 6 days of ZnONPs coexposure, 72% and 68% decrease in As(III) and As(V) content were observed, respectively. Copresence of CeO_2NPs and As(III) exhibited negligible impact in rice roots while 20% reduction in As(V) species was observed when compared with plants that were exposed to As(III) only. ZnONPs showed 88% and 68% decrease of As(III) and As(V) in rice root tissues, respectively, which resulted in a smaller As(III)/As(V) ratio of 0.348 as compared to 0.942 ratio of As(III)/As (V) when rice seedlings were only exposed to As(III). In shoots of rice plants as well, 71% of As(III) was reduced in presence of ZnONPs. Moreover, presence of As(III) and As(V) resulted in accumulation of 6.5 and 2.3 times higher cerium in rice shoots, respectively, as compared to seedlings that were treated with CeO_2NPs alone.

6.3.2 Titanium dioxide

Among various engineered metal oxide nanoparticles, TiO_2NPs is considered as one of the most promising materials to alleviate oxidative stress-mediated cellular damage (Ghosh and Webster, 2021a, 2021b, 2021c). Hence, these materials can further be explored for the agricultural applications. Cd-mediated phytotoxicity in *Zea mays* L. (maize) was reported to be alleviated by Lian et al. (2020) using titanium dioxide nanoparticles (TiO_2NPs). The average size of nanoparticles used in this study was observed to be 6.5 ± 0.76 nm and had a surface area of $264.8\,m^2/g$. Single root exposure of 250 mg/L of TiO_2NPs showed no alteration on plant height and dry weight of roots; however, shoot dry weight and chlorophyll content were reduced by 26.1% and 36.0%, respectively. On the contrary, foliar exposure of 100 mg/L of TiO_2NPs showed no change in chlorophyll content but decreased dry weights of shoot and root by 23.1% and 38.3%, respectively. Further root exposure of 100 mg/L of TiO_2NPs along with Cd showed relative chlorophyll content, shoot and root dry weights to be reduced by 50.5%, 45.9%, and 43.1%, respectively. However, foliar exposure of TiO_2NPs and Cd showed no such reduction indicating possible protection of maize plants from Cd stress using TiO_2NPs.

Superoxide content of plants coexposed with Cd and TiO_2NPs was similar to that of plants having singular exposure to Cd. Additionally, loss of plasma membrane integrity in root tips was significantly reduced after coexposure of Cd with 250 mg/L of TiO_2NPs; however, foliar coexposure of Cd and TiO_2NPs showed heavy stains of Evans blue which is indicative of loss of plasma membrane integrity. Uptake as well as translocation of TiO_2NPs was also investigated through energy-dispersive X-ray spectroscopy (EDS) that showed 47.08% and 3.92% w/w of Ti were present in roots and leaves, respectively, after root exposure as seen in Fig. 6.2. Furthermore, TEM images of maize plant leaves showed penetration of TiO_2NPs into the cell membrane after root exposure which suggested presence of a symplastic pathway in roots for TiO_2NPs.

Foliar exposure of TiO_2NPs, on the other hand, did not show any Ti element in root tissues, while in leaf tissues, only 0.8% of Ti was observed. Concentration of Cd was influenced in presence of TiO_2NPs. Root exposure of 250 mg/L of TiO_2NPs showed 32.9% decrease of Cd content in roots and foliar exposure of 100 mg/L of TiO_2NPs resulted in 14.4%–17.8% decrease in Cd content from shoots along with 15.3% of Cd content from roots as well. Moreover, root exposure to Cd along with 100 mg/L of TiO_2NPs showed higher activity of peroxidase (POD), superoxide dismutase (SOD), and glutathione-*S*-transferase enzymes in leaves as compared to control, while foliar coexposure to Cd and TiO_2NPs significantly inhibited POD activity but showed no effect on superoxide and glutathione-*S*-transferase activity. MDA content after foliar exposure of TiO_2NPs was observed to be lowered while root exposure showed increased levels of MDA highlighting cell membrane injury. Hence, it was implied that foliar exposure of TiO_2NPs may suppress oxidative stress in maize. Analysis of metabolic profiles of maize leaves was also carried out using gas chromatography-mass spectrometry (GC-MS) that revealed TiO_2NPs exerted a stronger influence upon foliar exposure as compared to root exposure when maize plants

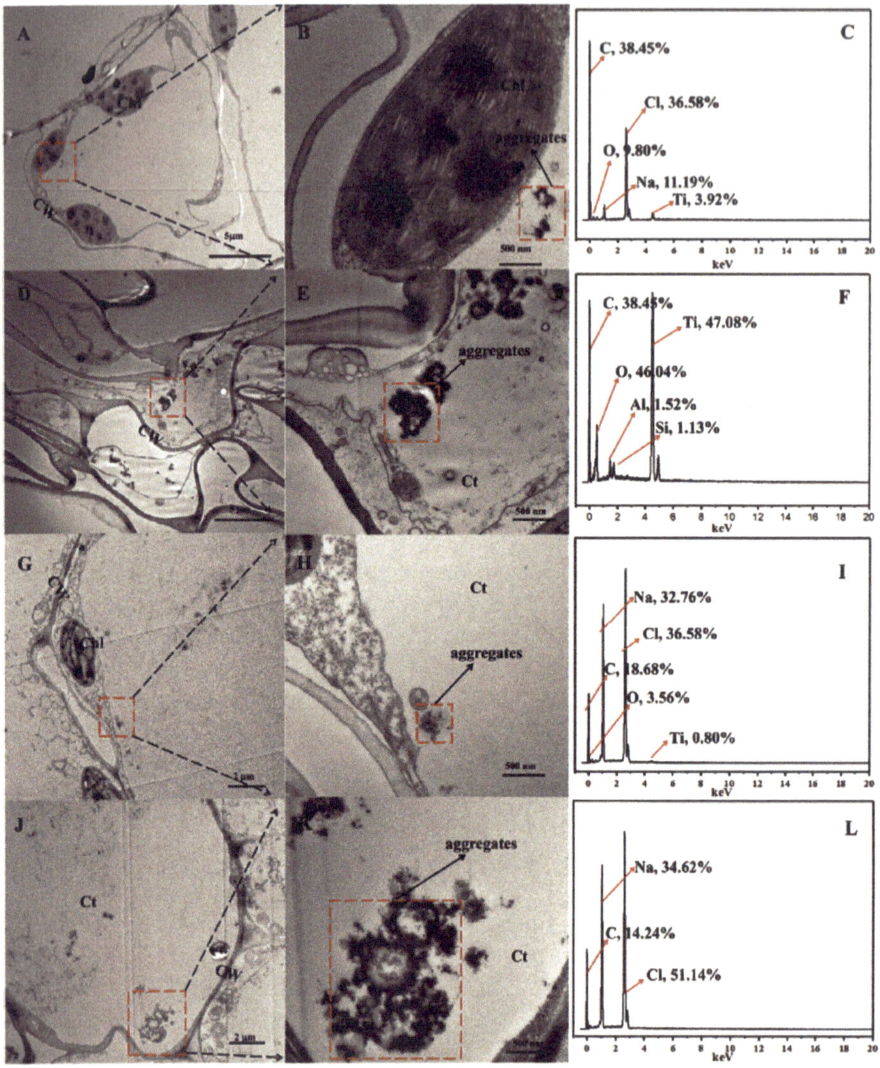

Fig. 6.2 TEM images of TiO_2 NPs in leaf (A, B, G, H) and root (D, E, J, K) sections of maize seedlings. Maize tissues were determined from CdTiO_2 NPs (250 mg/L RA) (A, B, G, H) treatment and CdTiO_2 NPs (250 mg/L FS) (D, E, J, K) treatment. The aggregates in red box from B, E, H, and K were further analyzed by EDS and shown in C, F, I, and L. *CW*, cell wall; *Ct*, cytoplasm; *Chl*, chloroplast.

were subjected to Cd stress. Out of 298 metabolites that were detected, 98 metabolites were identified to be involved in important metabolic pathways comprising of carbohydrate, amino acid, fatty acid, urea, and small-molecule acids. Around 60% of metabolites in leaves were upregulated after root exposure while 54% of metabolites were downregulated after foliar exposure of TiO_2NPs. Therefore, foliar exposure of TiO_2NPs was suggested to be an efficient approach for alleviating Cd stress from maize plants.

Ji et al. (2017) also reported similar findings of reducing Cd stress in rice seedlings after treatment with TiO_2NPs. Scanning electron microscopy (SEM) showed agglomeration of TiO_2NPs with an average size of 55 ± 42 nm. Root length was increased by 23.9% in presence of 10 mg/L of TiO_2NPs highlighting alleviation from 20 mg/L of Cd toxicity as compared to treatments with 10 and 20 mg/L of $CdCl_2$ that showed more than 45.5% and 37.9% reduction in root lengths, respectively. Similarly, TiO_2NPs enhanced plant height by 14.8% and 15.8% when seedlings were treated with 10 and 20 mg/L of $CdCl_2$, respectively. Moreover, fresh biomass of 20 mg/L $CdCl_2$-treated rice seedlings was increased by 35.6% in presence of 10 mg/L of TiO_2NPs while Cd and TiO_2NPs showed negligible alterations in the dry biomass of rice seedlings. Presence of 100 mg/L of $CdCl_2$ resulted in decrease of Ti from roots by 57.8% after treatment with 10 mg/L of TiO_2NPs while Ti content in leaves was observed to be increased by 123.6% in presence of $CdCl_2$ at a concentration of 20 mg/L. Simultaneous exposure of Cd and TiO_2NPs showed decrease in Cd accumulation in roots as well as leaves of rice plants. Furthermore, rice seedlings that were treated with 10 mg/L of TiO_2NPs only showed 35.8% increase in abscisic acid (ABA) content and gibberellic acid (GA) content was decreased after coexposure with 20 mg/L of Cd and 10 mg/L of TiO_2NPs. Indole-3-acetic acid (IAA) contents of rice plants were also decreased by 62.1% and 53.4% after coexposing with 20 mg/L of Cd along with 100 mg/L and 1000 mg/L of TiO_2NPs, respectively. Methyl jasmonate (JAME) content was increased by 25.6% in presence of 10 mg/L of Cd and 100 mg/L of TiO_2NPs while it decreased by 25.4% when TiO_2NPs concentration was increased to 1000 mg/L. MDA content was observed to be reduced by 9.1%, 22.0%, and 50.8% in copresence of 20 mg/L of Cd along with 10, 100, and 1000 mg/L of TiO_2NPs, respectively, as compared to plants that were only treated with Cd, thus indicating TiO_2NPs alleviated Cd toxicity. Additionally, POD activity was observed to increase by 10.8% and 13.8% when plants exposed with 10 mg/L of Cd were treated with 100 mg/L and 1000 mg/L of TiO_2NPs, respectively. SOD activity was also increased by 13.5% and 29.3% in presence of 100 mg/L and 1000 mg/L of TiO_2NPs, respectively. While net photosynthetic rate of rice seedlings was decreased by 13.3% and 29.9% when only treated with 10 and 20 mg/L of Cd, respectively, coexposure with TiO_2NPs showed comparative increase in rate of photosynthesis. Similar effects were seen with respect to chlorophyll content wherein 1000 mg/L of TiO_2NPs in presence of 20 mg/L of Cd increased chlorophyll content by 64.9% as compared to rice seedlings only treated with Cd. Finally, TEM images of rice seedling treated with TiO_2NPs revealed presence of spherical shaped particles that were confirmed EDS spectrum. The morphology of particles changed from spheres to flakes after coexposure with Cd.

6.3.3 Iron

Iron and its oxides are considered as attractive nanomaterials for magnetic-field-associated targeting potential. These exotic magnetic materials can be functionalized by various bioactive agents for diverse range of applications (Ghosh et al., 2015). Nanoscale zero-valent iron (nZVI) has been reported to be useful in HM remediation. In one such study, Guha et al. (2020) reported alleviation of Cd stress in rice plants using nZVI that had a mean hydrodynamic diameter of 2379 ± 20 nm along with a polydispersity index (PDI) and zeta potential value of 0.39 and -5.43 ± 4.51 mV, respectively. Treatment with 50–100 mg/L of nZVI was observed to enhance root length, shoot length, and biomass of rice plants while 200 mg/L of nZVI resulted in growth inhibition. Coexposure of 10 μM of Cd and 100 mg/L of nZVI also increased biomass, shoot length, and root length by 95.23%, 46.24%, and 88.58%, respectively. Moreover, MDA content of 17-day-old rice seedlings was increased when subjected to oxidative stress due to presence of 10, 50, and 250 μM of Cd that was significantly decreased when coexposed with 50 and 100 mg/L of nZVI. Principal content analysis showed combination of 10 μM Cd with 50 mg/L of nZVI had maximum Cd stress remediation; thus, it was considered to be most efficient dose. Maximum activities of SOD, catalase, glutathione peroxidase, and glutathione reductase enzymes were observed in seedlings when treated with the efficient dose. Activities of SOD, catalase, glutathione peroxidase, and glutathione reductase enzymes in root tissues were increased by 1.39-, 32.6-, 1.38-, and 2.27-folds, respectively, while shoot tissues showed a 14.5-, 5.75-, 2.07-, and 1.155-fold increase, respectively. Similarly, phytochelatin contents of root and shoot tissues were increased by 1.9- and 5.2-folds, respectively, while proline content of root and shoot tissues was enhanced by 1.4- and 1.34-folds as well. H_2O_2 levels for 17-day-old seedlings were also decreased by 27.92% and 14.15% in root and shoot tissues, respectively, after coexposure of nZVI and Cd as compared to seedlings that were only exposed to Cd. Intracellular production of ROS was confirmed using dichlorofluorescein dye that showed bright green fluorescence inferring high H_2O_2 levels. SEM images of root epidermis, root tip, and vascular bundles of 17-day-old seedlings revealed no alteration in morphology in presence of nZVI while exposure to Cd led to detachment and tearing in epidermal cells of root surface and root tip that was found to be minimized after treatment with nZVI as evident from Fig. 6.3. Furthermore, expression of OsIRT1,2 and OsYSL2,15 was significantly increased in roots and shoots after introduction of Cd. However, binary treatments with Cd and nZVI downregulated expression of these genes. nZVI was further observed to act as a growth promoter as it increased plant growth and biomass by 11.33% and 55%, respectively, as compared to plants that were treated only with Cd. Reduction in levels of important minerals such as N, P, K, Ca, Mg, and S as well as micronutrients such as Fe, Mn, and Zn due to Cd stress were ameliorated by nZVI. Bioconcentration factor (BCF) of Cd-treated plant shoots was significantly higher as compared to control that highlighted enrichment of the toxic metal within the plant. nZVI and Cd binary treatment, on the other hand, reduced BCF by 60.75% compared to only Cd-treated plants, thus preventing accumulation of Cd in rice plants.

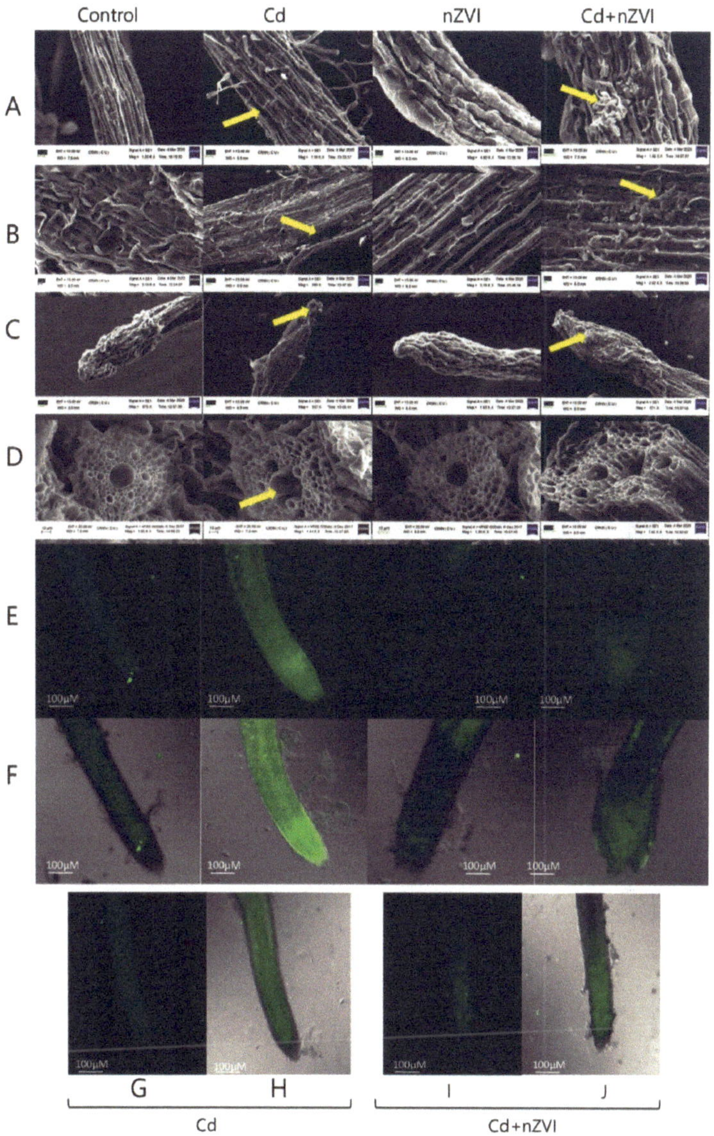

Fig. 6.3 Scanning electron micrograph of 17-day-old rice seedlings (A, B, C, and D) represents root surface, epidermal cells, root tips, and transverse section of root, respectively. *Arrows* indicate damaged cells and depositions on the root surfaces. Confocal microscopy (E) fluorescence (F) merged images with phase contrast showing cellular ROS tagged with DCF and (G, I) fluorescence (HeJ) merged images with phase contrast showing localization of Cd tagged with Leadmium Green AM dye in root tips of 17-day-old rice seedlings subjected to Cd and nZVI treatments.

Reproduced with permission from Guha, T., Barman, S., Mukherjee, A., Kundu, R., 2020. Nanoscale zero-valent iron modulates Fe/Cd transporters and immobilizes soil Cd for production of Cd free rice. Chemosphere 260, 127533. Copyright © 2020 Elsevier Ltd.

Availability of Cd in soil was also increased by 43.98% after nZVI application along with 62.89% Cd removal from carbonate fraction while 74.04% decrease was seen in level of soil-exchangeable Cd.

In another study, Hussain et al. (2019) investigated application of iron nanoparticles (FeNPs) on *Triticum aestivum* (wheat) plants that were grown in soil contaminated with Cd. Plants treated with FeNPs were healthy and were observed to have higher number of tillers as compared to control plants. The shoot length of plants treated with 20 mg/kg of FeNPs was 43% and 55% higher when applied over soil and leaves, respectively. Spike length of plants grown under Cd stress was also increased by 54% and 60% after soil and foliar application of 20 mg/kg of FeNPs, respectively. Dry weight of plants was observed to be slightly lower when FeNPs were applied on soil as compared to foliar application. Shoot dry weights after foliar application of 20 ppm of FeNPs were increased by 72% while soil amendments resulted in 52% enhancement of dry weight of shoots. Similarly, root dry weights were increased by 59% and 50% after foliar and soil application of 20 ppm of FeNPs, respectively. Chlorophyll a, b as well as carotenoid concentrations were increased by 70%, 139%, and 119%, respectively, as compared to control plants after foliar application of 20 mg/L of FeNPs. Photosynthetic rate was enhanced as well to around 102% after foliar spray of 20 ppm of FeNPs. Stomatal conductance showed a 116% increase after foliar spray of 20 mg/L of FeNPs while soil application resulted in stomatal conductance increase by 105% as compared to control. Electrolyte leakage (EL) of leaves was observed to be reduced after application of FeNPs. Both soil and foliar application of 20 mg/L of FeNPs showed similar decrease in EL of around 25% over control. On the other hand, 20 mg/kg of FeNPs enhanced SOD activity by 40% and 46% after foliar and soil spray, respectively, while POD activity of leaves was increased by 54% under same concentrations of FeNPs. Moreover, drastic reduction in Cd concentrations was observed in presence of FeNPs. Cd concentrations in shoots, roots, and grains of maize plants were minimized by 53%, 56%, and 84%, respectively, after foliar application of 20 mg/L of FeNPs while soil spray similar concentrations of FeNPs decreased Cd content in shoots, roots, and grains by 53%, 49%, and 81%, respectively. Fe concentrations were also found to be enhanced by 44%, 40%, and 169% in shoots, roots, and grains, respectively, after foliar use of 20 mg/L of FeNPs as compared to control.

Adrees et al. (2020) also reported similar observations on reducing Cd as well as drought stress in wheat plants after application of FeNPs in soil. FeNPs significantly enhanced morphology and growth of maize plants that were germinated in soil contaminated with Cd. Application of 100 mg/kg of FeNPs increased height of plant by 26% and 31% under normal water conditions and drought conditions, respectively. Spike length was also increased by 32% and 27% in drought and normal moisture contents, respectively, in presence of 100 mg/kg of FeNPs. Shoot dry weights were also improved by 43% and 64% after application of 100 mg/kg of FeNPs under drought stress and normal water conditions, respectively, along with increase in root dry weights by 52% and 78% under normal and limited water conditions, respectively. Chlorophyll a content was also enhanced by 65% under normal water conditions, while in presence of limited water supply, 66% increase was observed after treatment with 100 mg/kg of FeNPs. Moreover, concentrations of EL, H_2O_2, and MDA were increased in plants when grown under Cd stress. Concentrations of Cd in shoots ranged from 4.2 to 7.44 mg/kg while it was within 8.88–17.28 mg/kg for roots. While maximum amount of Cd was found under plants grown in drought and without any FeNPs, concentration of Cd was drastically reduced under normal moisture conditions and supplementation of 100 mg/kg of FeNPs. Amount of Cd in grains was found to lie within a range of 0.21–1.26 mg/kg with minimum Cd present after FeNPs treatment. Thus, it was indicated that accumulation of Cd was enhanced under drought conditions that, in turn, upregulated markers of oxidative stress in plants; however, application of FeNPs ameliorated soil bioavailable Cd.

Konate et al. (2017) also investigated mitigation of HM toxicity in wheat seedlings using magnetic nanoparticles (Fe_3O_4NPs). The average size of Fe_3O_4NPs used in this study was observed using TEM images and found to be 6.85 ± 1.70 nm with a negative zeta potential value of -12.7 ± 0.99 mV due to which particles were agglomerated in water. Pb, Zn, Cd, and Cu were used as toxic HMs that reduced the viability and growth of wheat seedlings. Rates of inhibition of seedling viability in presence of 10 mM of Cd, Cu, Pb, and Zn were 100%, 100%, 82.5%, and 22.5%, respectively, whereas seedlings treated with Fe_3O_4NPs showed 100% viability similar to that of control seedlings. Addition of 2000 mg/L of Fe_3O_4NPs resulted in decrease in root growth inhibition by 193.91%, 37.56%, 97.72%, and 31.89%, while inhibition of shoot growth was reduced by 65.75%, 25.06%, 87.35%, and 60.96% in plants growing under 1 mM of Pb, Zn, Cd, and Cu, respectively. However, addition of Fe_3O_4NPs did not result in decrease in growth inhibition when HMs were present at a concentration of 10 mM. Furthermore, administration of Fe_3O_4NPs reduced 54.04%, 23.95%, 65.51%, and 68.64% of Pb, Zn, Cd, and Cu accumulation in roots while 17.75%, 11.39%, 99.59%, and 52.94% of Pb, Zn, Cd, and Cu accumulation in shoots were reduced, respectively. Activities of SOD and POD enzymes in shoots of seedlings were reduced in presence of HMs whereas MDA content was significantly increased in roots which are indicative of oxidative stress induction in plants. However, addition of 2000 mg/L of Fe_3O_4NPs showed considerable increase in SOD activity in shoots by 26.35%, 38.13%, and 31.57% in plants that were treated with 1 mM of Pb, Cd, and Cu, respectively. Likewise, POD activity was enhanced by 4.39%, 39.05%, 27.97%, and 32.86% after addition of 2000 mg/L of Fe_3O_4NPs in plants treated with 1 mM of Pb, Zn, Cd, and Cu, respectively. Supplementation of Fe_3O_4NPs also exhibited decrease in MDA concentration of shoots by 51.46% in Cd-treated plants whereas MDA content in roots was reduced by 49.46% for plants grown in presence of 1 mM of Pb. Moreover, inductively coupled plasma-mass-spectrometry (ICP-MS) analysis revealed significant reduction in all HM concentrations after five days of addition of Fe_3O_4NPs, thus confirming the efficiency of Fe_3O_4NPs to act as HM adsorbent in order to alleviate stress in plants.

6.3.4 *Zinc oxide*

Zinc-oxide-based nanomaterials are not only biocompatible but also bioactive, thereby being used as antidiabetic, antimicrobial, and antioxidant agents (Adersh et al., 2015; Robkhob et al., 2020; Karmakar et al., 2020; Kitture et al., 2015). Hence, it is interesting to explore their role for induction of stress tolerance in the crops. Reduction in accumulation of As and Cd in rice plants was facilitated by zinc oxide nanoparticles (ZnONPs) as reported by Ma et al. (2020). ZnONPs used in this study were primarily spherical in shape as observed in TEM images; however, some of the particles were shown to exhibit triangular as well as irregular morphology. The mean diameter, hydrodynamic size, and zeta potential of ZnONPs were 68.1 nm, 621.08 ± 7.63 nm, and -28.80 ± 2.04 mV, respectively. Total As content in roots and shoots of rice plants was reduced by 39.5% and 60.2%, respectively, after application of ZnONPs as compared to control plants without any ZnONPs treatment. Zn^{2+} also lowered As levels by 83.3% and 80.0% in roots and shoots, respectively. While both ZnONPs and Zn^{2+} were able to reduce uptake of Cd by 61.8% and 51.7%, respectively, in roots of rice plants, only ZnONPs were able to lower Cd content by 26.3% and Zn^{2+} increased Cd concentration by 26.8% in shoots of rice plants. It was hypothesized that Zn may displace Cd from soil particles that could lead to increase in Cd bioavailability in case of high Zn^{2+} concentration. ZnONPs was proposed to lower such displacement by slow release of Zn^{2+} that may also reduce its toxicity toward plants. ZnONPs and Zn^{2+} amendments were further observed to only alter speciation of dimethylarsinic acid (DMA) in soil. Hence, higher content of DMA soils that were treated with ZnONPs was suggested to be due to stimulation of microbes that are able to methylate As. Moreover, Zn^{2+} significantly reduced concentration of inorganic As(V) by 77.9% while ZnONPs only showed 17.3% reduction in As(V) content. DMA and monomethylarsonic acid (MMA) content in roots of rice plants were also lowered by 75.8% and 86.3% using ZnONPs treatment while Zn^{2+} showed 94.2% and 96.1% reduction, respectively. As(III) content was not altered in presence of Zn amendments. It was also proposed that Zn^{2+} may facilitate reduction of As(V) by As reductase present in plant root cells which would result in formation of As(III) that may be either transported via *OsLsi2* xylem transporter to shoots of rice plant or they could be accumulated in root cells of the plant. As(III) content of rice shoots was also observed to be lowered by 46.8% and 86.3% after application of ZnONPs and Zn^{2+}, respectively. Therefore, ZnONPs were shown to significantly reduce As accumulation in rice tissues as well as lower Cd content in rice shoots.

ZnONPs were also reported by Khan et al. (2019) to ameliorate Cd accumulation in wheat plants. Height of plants was observed to be increased by 34% and 24% in presence of 100 mg/kg of ZnONPs under drought and normal water conditions, respectively. Likewise, enhancement in spike length by 27% and 37% was also seen without and with drought in presence of 100 mg/kg of ZnONPs, respectively. Furthermore, ZnONPs also enhanced dry biomass of roots, shoots, and grains of rice by 37%, 40%, and 52% under normal moisture content while it was increased by 54%, 46%, and 69% under drought conditions, respectively. In addition to Cd stress, drought stress was also observed to reduce chlorophyll a and chlorophyll b contents by 21% and 20%, respectively. However, after treatment with 100 mg/kg of ZnONPs, chlorophyll a and chlorophyll b contents were increased by 39% and 44% under normal water concentration while under drought conditions, chlorophyll a and chlorophyll b contents were increased by 59% and 49% by 100 mg/kg of ZnONPs, respectively. Addition of 100 mg/kg of ZnONPs further decreased oxidative stress and concentrations of EL, H_2O_2, and MDA by 33%, 35%, and 49% without drought stress, respectively. On the contrary, treatment with ZnONPs enhanced activities of SOD and POD enzymes in leaves of rice plants. Accumulation of Cd was observed to be highest in plants that were grown under drought stress without any ZnONPs amendments while lowest levels of Cd were observed in plants that were cultivated in normal moisture content and treated with 100 mg/kg of ZnONPs. Although plants grown with adequate water supply and treated with ZnONPs resulted in 78%, 64%, and 103% increase in Zn concentration in shoots, roots, and grains of rice plants as compared to plants without any nanoparticle treatment. Mobility of Cd in soil was also affected with increasing concentrations of ZnONPs with decrease in bioavailable Cd by 45% under normal moisture conditions and 37% under low water conditions after introduction of 100 mg/kg of ZnONPs.

Hussain et al. (2018) reported also reduction of Cd toxicity and accumulation in wheat plants using application of ZnONPs. Foliar spray of 100 mg/kg of ZnONPs increased shoot length by 55% while soil application of 100 mg/kg of ZnONPs showed a 43% increase as compared to control plants. Similarly, soil and foliar treatments increased spike length of 54% and 59%, respectively. Dry weights of shoots, roots, spikes, and grains were also enhanced after foliar application of 100 mg/L of ZnONPs by 72%, 59%, 90%, and 97%, respectively, while soil treatment of 100 mg/kg of ZnONPs increased the same by 52%, 50%, 69%, and 74%, respectively. Levels of chlorophyll a, b as well as carotenoids were increased by ZnONPs. Foliar spray of ZnONPs resulted in 60%, 139%, and 119% increase in chlorophyll a, b and carotenoids levels while soil amendments showed 53%, 138%, and 111% increase, respectively. Photosynthetic rate of plants was observed to be upregulated by 102% and 89% after foliar and soil application of 100 mg/kg of ZnONPs, respectively. EL levels in plants were observed to decrease in presence of 100 ppm of ZnONPs while SOD and POD enzyme activities were increased by 40% and 46% after foliar spray along with 46% and 54% increase when applied in soil, respectively. Cd concentrations in grains, shoots, and roots were decreased by 30%–77%, 10%–55%, and 25%–64% after foliar application of ZnONPs whereas soil treatments revealed 16%–78%, 17%–68%, and 23%–60% reduction, respectively, when compared with control plants. However, treatment with ZnONPs provided drastic increase in Zn concentration in various parts of wheat plants. Foliar treatment revealed 40%–185%, 10%–52%, and 7%–60% rise in levels of Zn in grains, shoots, and roots while soil treatment with 100 mg/kg of ZnONPs increased levels of Zn by 58%, 68%, and 190% in shoots, roots, and grains of wheat plants, respectively.

Leucaena leucocephala seedlings were ameliorated from Cd and Pb toxicity by Venkatachalam et al. (2017) using ZnONPs that was synthesized using extracts obtained from a marine green alga—*Ulva lactuca*. Fourier transform infrared (FTIR) spectroscopy revealed several strong peaks highlighting presence of bioactive molecules that may have been responsible for reduction of zinc. A prominent band at 3410 cm^{-1} was observed that indicated N-H stretching of amines, another peak at 2920 cm^{-1} represented C-H methylene stretching which was assigned to ammonium ions while other bands were also observed that corresponded with presence of other functional groups such as carboxylic acids, alcohols, phenols, esters, vinyl alkenes, and aliphatic amines. X-ray diffraction (XRD) results confirmed formation of crystalline ZnONPs. Field-emission scanning electron microscopy (FESEM) images revealed phytomolecule-coated ZnONPs were hexagonal, square as well as spherical in shape having a size range of 2–64 nm. Furthermore, energy-dispersive X-ray (EDX) spectroscopy results exhibited a strong peak at 1.4 and 8.6 keV which suggested presence of zinc element as a result of surface plasmon resonance in order to form ZnONPs. Later on, alterations in *L. leucocephala* plant growth were investigated in presence of 50 mg/L of Cd and 100 mg/L of Pb along with 25 mg/L of ZnONPs. Plants that were treated with Zn and ZnONPs were observed to have increased root and shoot length as compared to plants only treated with HMs. The biomass was also increased by 25.7% and 20.5% after ZnONPs treatment in plants grown under Cd and Pb stress, respectively. Photosynthetic pigments such as chlorophyll a, b and carotenoids were significantly increased in presence of ZnONPs. Chlorophyll a content was increased by 133.7% and 174.4% in plants grown under Cd and Pb stress and treated with ZnONPs, respectively, as compared to HM-stressed plants that were grown in absence of ZnONPs. Likewise, chlorophyll b content in Cd- and Pb-stressed plants was increased by 139.2% and 175.3% while carotenoids were enhanced by 132.4% and 160.0%, respectively, after application of ZnONPs. MDA content was also decreased from 115.4% to 87.2% after treatment with ZnONPs in Cd-stressed plants while plants grown in Pb stress showed decline from 102.6% to 79.5% after ZnONPs amendments. Furthermore, leaf tissues exhibited an increase in total soluble protein content by 115.8% and 113.2% after administration of ZnONPs. Activities of SOD, catalase, and POD enzymes were increased by 115.1%, 112.6%, and 110.2% in Cd-stressed plants after treatment with ZnONPs whereas Pb-stressed plants exhibited an increase of around 109.2%, 134.7%, and 127.1%, respectively. Moreover, four different POD isoforms were observed in native polyacrylamide gel electrophoresis (PAGE) wherein isoform 2 was significantly upregulated in presence of ZnONPs and HMs.

6.3.5 Sulfur

Certain nonmetallic elements can also be used to generate nanostructured materials for agricultural applications. Yuan et al. (2021) utilized sulfur nanoparticles (SNPs) to mitigate mercury (Hg) stress induced in *Brassica napus* L. plants. The SNPs were observed to be spherical in morphology with an average diameter of 48.8 nm and a zeta potential of 10.09 mV. Treatment of *B. napus* seedlings to 10 mg/L of Hg showed drastic reduction in shoot height, root length as well as fresh weight of plant. On the other hand, treatment of seeds with varying concentrations of SNPs resulted in 8.93%–11.2% increase in shoot height and 6.31%–12.8% increase in root length. Furthermore, fresh weight was increased by 16.3% in presence of 300 mg/L of SNPs. Cotreatment of plants treated with 300 mg/L of SNPs along with 10 mg/L of Hg demonstrated increased root length and shoot height by 148% and 69.1% along with 42.4% and 37.8% increase in dry weights of shoots and roots, respectively, as compared to plants that were treated with only Hg. Coexposure of Hg treated with plants with bulk sulfur particles (BSPs) also exhibited 28.3%, 36.2%, 19.7%, and 30.1% increase in shoot height, root length, dry weights of shoots, and roots, respectively. Hg accumulation was also increased wherein 286 and 601 ppm of Hg were found in shoots and roots of *B. napus* plants that were reduced by 6- and 8-folds after SNPs treatment, respectively. Chlorophyll content in plants only exposed to Hg was decreased by 9.7% whereas cotreatment with SNPs showed no such alteration as compared to control plants. Total protein contents in roots and shoots were also reduced to 39.6% and 31.6% in plants that were only treated with Hg whereas shoots and roots that were coexposed to Hg and 300 mg/L of SNPs showed a lower decrease in total protein content of around 16.6% and 22.9%, respectively. MDA content in shoots was increased by 53.6% after Hg treatment that was reduced down to 17.7% in presence of SNPs. Activities of antioxidant enzymes such as SOD in shoots and roots were drastically upregulated in shoots and roots by 1.8- and 2.6-folds when treated with Hg alone, respectively. Similarly, activity of glutathione reductase also increased by 70.4% while glutathione-*S*-transferase activity in shoots enhanced by 37.8% after administration of 10 mg/L of Hg. Other enzyme activities such as POD and ascorbate peroxidase were also increased by Hg treatment. Introduction of SNPs, however, reduced such enzyme activities; for example, POD activity in roots and shoots of plant was reduced by 16.9% and 43.4% while ascorbate POD activity was decreased by 26.8% and 10.9%, respectively. Concentrations of macronutrients such as K, Ca, P, and Mg were also reduced in shoots by 47.5%, 45.4%, 33.8%, and 16.1% while 28.5%, 31.9%, 24.7%, and 14.2% decrease in roots was observed, respectively. Level of micronutrients such as Zn and Mn was also reduced 59.1% and 24.2% in shoots while roots showed a reduction of 12% and 23%, respectively. On the other hand, addition of SNPs increased accumulation of essential macronutrients such as K, Ca, P, and Mg by 107%, 79.9%, 64.9%, and 25.6% in shoots whereas roots exhibited 66.2%, 79.8%, 10.5%, and 30.2% higher accumulation of macronutrients, respectively. Likewise, accumulation of micronutrients such as Mn, Zn, and Fe was also 0.7%, 182%, and 14.7% higher in shoot tissues, respectively.

In another study, Ragab and Saad-Allah (2020) fabricated SNPs using *Ocimum basilicum* L. leaf extract and reported its application in manganese (Mn) stress mitigation in *Helianthus annuus* L. seedlings. SNPs were synthesized by dissolving 0.2 M of sodium thiosulfate pentahydrate in leaf extract under stirring conditions followed by precipitation of sulfur particles using 10% HCl. The precipitate obtained was then purified and dried in a vacuum oven for 5 h at 60°C. Ultraviolet-visible (UV-VIS) spectroscopy results showed prominent peak at 272 nm indicating synthesis of SNPs. XRD pattern further demonstrated polycrystalline nature of SNPs while TEM images revealed spherical shape of SNPs having an average size of 25.9 nm. Moreover, *H. annuus* seedlings used in this study were treated with 100 mM $MnSO_4$ that resulted in considerable reduction in growth parameters of the plant such as shoot length, leaf area as well as relative water content of roots and shoots. Such inhibitory effects of Mn were alleviated in seeds that were presoaked in varying concentrations of SNPs. Maximum amelioration of Mn toxicity was observed in seeds treated with 50 and 100 μM SNPs that significantly increased shoot length as well as leaf area of *H. annuus* seedlings. Photosynthetic pigments such as chlorophyll a, b and carotenoids were also decreased by 45.96%, 39.02%, and 39.42% in presence of $MnSO_4$, respectively. Priming with 50 μM of SNPs comparatively increased content of chlorophyll a and b while treatment with 12.5 μM of SNPs showed maximum retrieval of carotenoids concentration. Drastic increase of 21.80% accumulation of Mn ions was observed in seedlings as compared to control that was markedly decreased when seeds were primed with SNPs. Furthermore, decline in N, P, K, Ca, and Mg contents due to Mn stress was also mitigated by SNPs addition. EL in leaves of *H. annuus* was increased by 65.26% under Mn stress conditions that were relieved in presence of 25 and 50 μM of SNPs. Crude protein content was also declined by 17.0% after Mn only treated plants whereas cotreatment with 200 μM of SNPs showed 23.57% enhancement in total protein content. Likewise, 25 and 100 μM of SNPs also increased total amino acid content by 81.41% and 82.05%, respectively.

6.3.6 Others

Various nanoparticles apart from those discussed in the earlier sections are also studied for agricultural applications. Selenium nanoparticles (SeNPs) were reported by Qi et al. (2021) to alleviate Cd toxicity in *B. napus* L. plants. SeNPs used in this study were biologically synthesized that exhibited spheroidal morphology. Cd was shown to reduce both root and shoot length of plants as compared to control along with 53% decrease in total dry weight. Root viability was also affected as well as stomatal closure was induced in presence of Cd as seen in Fig. 6.4. Both SeO_3^{2-} and SeNPs

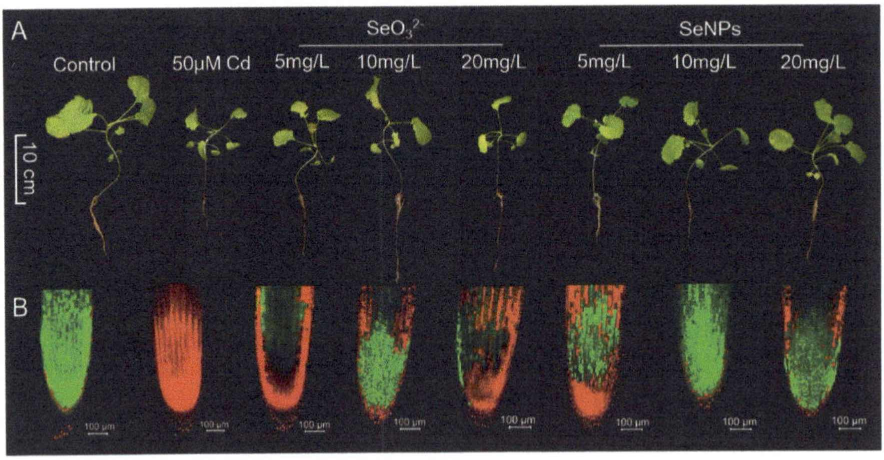

Fig. 6.4 (A) Effects of Se application on plant growth in the presence of 50 μM Cd and (B) laser scanning microscopy (LSM) imaging of root tips viability by fluorescein diacetate (FDA) and propidium iodide (PI) staining, where *green* and *red* fluorescence indicate living and dead cells, respectively.
Reproduced with permission from Qi, W.Y., Li, Q., Chen, H., Liu, J., Xing, S.F., Xu, M., Yan, Z., Song, C., Wang, S.G., 2021. Selenium nanoparticles ameliorate *Brassica napus* L. cadmium toxicity by inhibiting the respiratory burst and scavenging reactive oxygen species. J. Hazard. Mater. 417, 125900. Copyright © 2021 Elsevier B.V.

ameliorated such toxic effects of Cd wherein 10 mg/L of SeO_3^{2-} increased total dry weight of plants by 73% and 20 mg/L of SeNPs showed 106% increase in total dry weight as well. Accumulation of Cd was also reduced in presence of SeO_3^{2-} and SeNPs. Viability of apical meristem root was also drastically increased after application of 5 mg/L of SeO_3^{2-} or SeNPs while 10 mg/L of Se amendments increased cellular viability of apical region. Chlorophyll content of *B. napus* leaves was also declined in presence of Cd stress that was found to be increased significantly after 12 days of treatment with varying concentrations of SeNPs. ROS-mediated oxidative stress induced by Cd was also observed to be alleviated in presence of SeO_3^{2-} that decreased H_2O_2, O_2^-, and MDA levels by 44%, 49%, and 74%, respectively. SeNPs at a concentration of 10 mg/L also showed similar decrease of 39%, 61%, and 73% in levels of H_2O_2, O_2^-, and MDA while 5 mg/L of SeNPs also declined O_2^- and MDA content by 59% and 73%, respectively. Administration of 20 mg/L of SeNPs also enhanced catalase and ascorbate peroxidase activities by 72% and 75%, respectively, whereas 20 mg/L of SeO_3^{2-} showed no significant impact on such antioxidant enzymes. Transcriptional responses in presence of Cd and Se amendments were investigated as well using whole-genome RNA sequencing. Furthermore, oxidative stress detection assay and real-time quantitative PCR (RT-qPCR) were carried out that suggested underlying mechanisms of SeNPs in amelioration of Cd stress. SeNPs were proposed to inhibit *BnaRBOHC*, *BnaRBOHD1*, and *BnaRBOHF1* genes expression that are involved in translation of NADPH oxidases along with inhibition of *BnaGLO* gene expression which is responsible for glycolate oxidase production. Repression of these genes led to decreased ROS production, thereby decreasing oxidative protein as well as membrane lipid damage. Moreover, SeNPs were suggested to reduce Cd accumulation, maintain homeostasis of intracellular concentration of calcium, help in formation of disulfide bonds, and also restore waxy outer layer present on the surface of *B. napus* leaves.

Silicon dioxide nanoparticles (SiO_2NPs) were also applied by de Sousa et al. (2019) for amelioration of aluminum (Al) toxicity in maize plants. TEM images revealed spherical shape of SiO_2NPs that was used in this study with an average diameter of 20 nm. Concentration of Al was observed to increase in a dose-dependent manner in different components of maize plant. Administration of 24 mg/kg of Al resulted in 407-fold increase in Al levels in roots while a 119-fold increase was observed in leaves of maize plants. Polyphenol contents in soil were also increased by 8-fold after treatment with 24 mg/kg of Al. Moreover, concentrations of organic acids such as citrate, malate, and succinate were drastically increased by 78%, 145%, and 97%, respectively, in plants that were exposed to Al as well as SiO_2NPs. Exposure of Al also showed decrease in fresh and dry weights of plant that was mitigated after coexposure with SiO_2NPs. Photosynthetic rate was decreased by 82%, and respiration rate significantly increased by 200% in presence of Al stress while stomatal conductance was also decreased by 85% after application of 24 mg/kg of Al. Coexposure of Al-treated plants with SiO_2NPs, however, showed slight alterations in photosynthetic rates, respiration rates as well as stomatal conductance. Enzymes responsible for promoting oxidative stress such as glycolate oxidase and hydroxypyruvate reductase were increased by 148% and 167% in leaves treated with 24 mg/kg of Al that was attenuated after coapplication of SiO_2NPs. Flavonoids content in roots of maize plants was also increased by 73% in presence of severe Al stress that was controlled partially after cotreatment with SiO_2NPs. Activity of SOD was increased by 64% and 69% in roots and leaves of maize plants after Al exposure that was downregulated after SiO_2NPs amendments. Similar results were observed in case of other ROS-scavenging enzymes such as catalase, ascorbate peroxidase, catalase, guaiacol peroxidase, and others.

Polymeric nanoparticles were also implemented by Xin et al. (2020) to improve *Zea mays* (maize) growth under Cu stress. Polysuccinimide polymeric nanoparticles (PSINPs) were used in this study that had an average size of 14.2 ± 0.5 nm and a zeta potential of -20.7 ± 0.6 mV. Water absorption was increased in corn seeds that were treated with PSINPs with maximum water uptake efficiency of 82.3% observed in seeds treated with 200 mg/L of PSINPs. Seed germination percentage was also enhanced by 93.3% and 100% in presence of 100 and 200 mg/L of PSINPs whereas 800 mg/L of PSINPs exhibited 83.3% decrease in germination. Furthermore, mean germination time was also reduced by 18.8% and 31.2% after administration of 100 and 200 mg/L of PSINPs as well as higher growth rates were achieved in presence of PSINPs. Moreover, addition of more than 20 mg/L of Cu showed growth inhibition with 33.1% decrease in germination index value after treatment with 160 mg/L of Cu. However, treatment with PSINPs showed 13.4% and 10% higher growth percentages in plants grown under 80 and 160 mg/L of Cu as compared to plants without any PSINPs treatment. Similarly, PSINPs also alleviated reduction in growth rate of plants under Cu stress. Growth index was increased as well by 54.7% after cotreatment with 200 mg/L of PSINPs and 160 mg/L of Cu. Addition of 200 mg/L of PSINPs also mitigated inhibition of shoot and root length due to Cu stress. Additionally, fresh biomass of shoot and root was enhanced by 60.7% and 160% after plants grown in presence of 40 mg/L of Cu were treated with 200 mg/L of PSINPs. Although accumulation of Cu was increased upon addition of PSINPs, for instance, after addition of PSINPs, Cu uptake was increased by 2.3- and 1.7-folds in shoots and roots of plants grown in presence of 40 mg/L of Cu, respectively.

Nanohydroxyapatite (NHAP) was also investigated by Jin et al. (2016) for enhancing growth of *Lolium perenne* L. (ryegrass) in soil contaminated with Pb. Biomass of ryegrass was significantly reduced over time in presence of 500 and 800 mg/kg of Pb. Addition of NHAP resulted in increased biomass that suggested removal of toxic effects exhibited by HMs on ryegrass. Accumulation of Pb in aboveground components of ryegrass was also reduced from 31.85–155.09 to 6.60–119.82 mg/kg after NHAP treatment. It was hypothesized that such decrease in Pb levels could be due to formation of pyromorphite that reduces activity of Pb in soil as well as diminished absorptive capacity of plants. Moreover, it was also proposed that NHAP may increase total Pb extraction by ryegrass. Administration of NHAP was suggested to increase pH of soil, thus reducing bioavailability of Pb, whereas it was also suggested that NHAP may form stable complexes with Pb or it may adsorb Pb. Removal efficiency of NHAP ranged from 10.06% to 38.86%, thus suggesting efficient remediation of Pb-contaminated soil using NHAP and ryegrass.

6.4 Conclusion and future perspectives

HM stress is prominent in economically important plants, and therefore, efficient strategies are required in order to increase crop productivity and yield along with improving HM tolerance among plants. Application of nanoparticles could be one such beneficial approach that is shown to induce mechanisms to improve plant tolerance against HMs and increase productivity by increasing height and length of roots and shoots of plants under HM stress along with increasing activity of antioxidant enzymes such as SOD, POD, and catalase in order to alleviate excessive ROS production and regulate oxidative stress. Photosynthesis, respiration, and stomatal conductance can also be enhanced under specific concentrations of nanoparticles while it can also reduce accumulation of HM in roots as well as aboveground parts of plant. Further studies are therefore required on administration of nanoparticles in order to provide an effective solution for HM accumulation in plants.

References

Adersh, A., Kulkarni, A.R., Ghosh, S., More, P., Chopade, B.A., Gandhi, M.N., 2015. Surface defect rich ZnO quantum dots as antioxidant inhibiting α-amylase and α-glucosidase: a potential anti-diabetic nanomedicine. J. Mater. Chem. B 3, 4597–4606.

Adrees, M., Khan, Z.S., Ali, S., Hafeez, M., Khalid, S., Ur Rehman, M.Z., Hussain, A., Hussain, K., Chatha, S.A.S., Rizwan, M., 2020. Simultaneous mitigation of cadmium and drought stress in wheat by soil application of iron nanoparticles. Chemosphere 238, 124681.

Arif, N., Sharma, N.C., Yadav, V., Ramawat, N., Dubey, N.K., Tripathi, D.K., Chauhan, D.K., Sahi, S., 2019. Understanding HM stress in a rice crop: toxicity, tolerance mechanisms, and amelioration strategies. J. Plant Biol. 62, 239–253.

de Sousa, A., Saleh, A.M., Habeeb, T.H., Hassan, Y.M., Zrieq, R., Wadaan, M.A., Hozzein, W. N., Selim, S., Matos, M., AbdElgawad, H., 2019. Silicon dioxide nanoparticles ameliorate the phytotoxic hazards of aluminum in maize grown on acidic soil. Sci. Total Environ. 693, 133636.

Gautam, P.K., Gautam, R.K., Banerjee, S., Chattopadhyaya, M.C., Pandey, J.D., 2016. Heavy metals in the environment: fate, transport, toxicity and remediation technologies. In: Pathania, D. (Ed.), Heavy Metals: Sources, Toxicity and Remediation Techniques. Nova Science Publishers, Hauppauge, NY, pp. 101–130. Paperback ISBN: 978634847407.

Ghori, N.H., Ghori, T., Hayat, M.Q., Imadi, S.R., Gul, A., Altay, V., Ozturk, M., 2019. Heavy metal stress and responses in plants. Int. J. Environ. Sci. Technol. 16, 1807–1828.

Ghosh, S., 2020. Toxic metal removal using microbial nanotechnology. In: Rai, M., Golinska, P. (Eds.), Microbial Nanotechnology. CRC Press, Boca Raton, FL. eBook ISBN: 9780429276330.

Ghosh, S., Webster, T.J., 2021a. Metallic nanoscaffolds as osteogenic promoters: advances, challenges and scope. Metals 11, 1356.

Ghosh, S., Webster, T.J., 2021b. Mesoporous silica based nanostructures for bone tissue regeneration. Front. Mater. 8, 692309.

Ghosh, S., Webster, T.J., 2021c. Nanotechnology for water processing. In: Shah, M.P., Rodriguez-Couto, S., Mehta, K. (Eds.), The Future of Effluent Treatment Plants-Biological Treatment Systems. Elsevier, pp. 335–360. Paperback ISBN: 9780128229569.

Ghosh, S., More, P., Derle, A., Kitture, R., Kale, T., Gorain, M., Avasthi, A., Markad, P., Kundu, G.C., Kale, S., Dhavale, D.D., Bellare, J., Chopade, B.A., 2015. Diosgenin functionalized iron oxide nanoparticles as novel nanomaterial against breast cancer. J. Nanosci. Nanotechnol. 15 (12), 9464–9472.

Ghosh, S., Selvakumar, G., Ajilda, A.A.K., Webster, T.J., 2021a. Microbial biosorbents for heavy metal removal. In: Shah, M.P., Couto, S.R., Rudra, V.K. (Eds.), New Trends in Removal of Heavy Metals from Industrial Wastewater. Elsevier B.V., Amsterdam, Netherlands, pp. 213–262. eBook ISBN: 978-0-12-823108-1; Paperback ISBN: 978-0-12-822965-1.

Ghosh, S., Sharma, I., Nath, S., Webster, T.J., 2021b. Bioremediation—the natural solution. In: Shah, M.P., Couto, S.R. (Eds.), Microbial Ecology of Waste Water Treatment Plants (WWTPs). Elsevier, pp. 11–40. eBook ISBN: 978-0-12-822504-2; Paperback ISBN: 978-0-12-822503-5.

Ghosh, S., Joshi, K., Webster, T.J., 2021c. Removal of heavy metals by microbial communities. In: Shah, M.P., Couto, S.R. (Eds.), Waste Water Treatment Reactors: Microbial Community Structure. Elsevier, pp. 537–566. eBook ISBN: 978-0-12-824244-5; Paperback ISBN: 978-0-12-823991-9.

Ghosh, S., Khunt, N., Webster, T.J., 2021d. Arsenic removing prokaryotes as potential biofilters. In: Shah, M., Couto, S.R., Biswas, J.K. (Eds.), An Innovative Role of Biofiltration in Waste Water Treatment Plants. Elsevier (In Press).

Guha, T., Barman, S., Mukherjee, A., Kundu, R., 2020. Nano-scale zero valent iron modulates Fe/Cd transporters and immobilizes soil Cd for production of Cd free rice. Chemosphere 260, 127533.

Hussain, A., Ali, S., Rizwan, M., Ur Rehman, M.Z., Javed, M.R., Imran, M., Chatha, S.A.S., Nazir, R., 2018. Zinc oxide nanoparticles alter the wheat physiological response and reduce the cadmium uptake by plants. Environ. Pollut. 242 (Pt B), 1518–1526.

Hussain, A., Ali, S., Rizwan, M., Ur Rehman, M.Z., Qayyum, M.F., Wang, H., Rinklebe, J., 2019. Responses of wheat (*Triticum aestivum*) plants grown in a Cd contaminated soil to the application of iron oxide nanoparticles. Ecotoxicol. Environ. Saf. 173, 156–164.

Ji, Y., Zhou, Y., Ma, C., Feng, Y., Hao, Y., Rui, Y., Wu, W., Gui, X., Han, Y., Wang, Y., Xing, B., 2017. Jointed toxicity of TiO_2 NPs and Cd to rice seedlings: NPs alleviated Cd toxicity and Cd promoted NPs uptake. Plant Physiol. Biochem. 110, 82–93.

Jin, Y., Liu, W., Li, X.L., Shen, S.G., Liang, S.X., Liu, C., Shan, L., 2016. Nano-hydroxyapatite immobilized lead and enhanced plant growth of ryegrass in a contaminated soil. Ecol. Eng. 95, 25–29.

Kah, M., Tufenkji, N., White, J.C., 2019. Nano-enabled strategies to enhance crop nutrition and protection. Nat. Nanotechnol. 14, 532–540.

Karmakar, S., Ghosh, S., Kumbhakar, P., 2020. Enhanced sunlight driven photocatalytic and antibacterial activity of flower-like ZnO@MoS_2 nanocomposite. J. Nanopart. Res. 22, 11.

Khan, Z.S., Rizwan, M., Hafeez, M., Ali, S., Javed, M.R., Adrees, M., 2019. The accumulation of cadmium in wheat (*Triticum aestivum*) as influenced by zinc oxide nanoparticles and soil moisture conditions. Environ. Sci. Pollut. Res. Int. 26 (19), 19859–19870.

Kitture, R., Chordiya, K., Gaware, S., Ghosh, S., More, P.A., Kulkarni, P., Chopade, B.A., Kale, S.N., 2015. ZnO nanoparticles-red sandalwood conjugate: a promising anti-diabetic agent. J. Nanosci. Nanotechnol. 15, 4046–4051.

Konate, A., He, X., Zhang, Z., Ma, Y., Zhang, P., Alugongo, G.M., Rui, Y., 2017. Magnetic (Fe_3O_4) nanoparticles reduce HMs uptake and mitigate their toxicity in wheat seedling. Sustainability 9, 790.

Lian, J., Zhao, L., Wu, J., Xiong, H., Bao, Y., Zeb, A., Tang, J., Liu, W., 2020. Foliar spray of TiO_2 nanoparticles prevails over root application in reducing Cd accumulation and mitigating Cd-induced phytotoxicity in maize (*Zea mays* L.). Chemosphere 239, 124794.

Lowry, G.V., Avellan, A., Gilbertson, L.M., 2019. Opportunities and challenges for nanotechnology in the agri-tech revolution. Nat. Nanotechnol. 14, 517–522.

Ma, X., Sharifan, H., Dou, F., Sun, W., 2020. Simultaneous reduction of arsenic (As) and cadmium (Cd) accumulation in rice by zinc oxide nanoparticles. Chem. Eng. J. 384, 123802.

Mishra, S., Bharagava, R.N., More, N., Yadav, A., Zainith, S., Mani, S., Chowdhary, P., 2019. HM contamination: an alarming threat to environment and human health. In: Sobti, R., Arora, N., Kothari, R. (Eds.), Environmental Biotechnology: For Sustainable Future. Springer, Singapore, pp. 103–125.

Naveedullah, Hashmi, M.Z., Yu, C., Shen, H., Duan, D., Shen, C., Lou, L., Chen, Y., 2013. Risk assessment of heavy metals pollution in agricultural soils of siling reservoir watershed in Zhejiang province, China. Biomed. Res. Int. 2013, 590306.

Qi, W.Y., Li, Q., Chen, H., Liu, J., Xing, S.F., Xu, M., Yan, Z., Song, C., Wang, S.G., 2021. Selenium nanoparticles ameliorate *Brassica napus* L. cadmium toxicity by inhibiting the respiratory burst and scavenging reactive oxygen species. J. Hazard. Mater. 417, 125900.

Ragab, G.A., Saad-Allah, K.M., 2020. Green synthesis of sulfur nanoparticles using *Ocimum basilicum* leaves and its prospective effect on manganese-stressed *Helianthus annuus* (L.) seedlings. Ecotoxicol. Environ. Saf. 191, 110242.

Robkhob, P., Ghosh, S., Bellare, J., Jamdade, D., Tang, I.M., Thongmee, S., 2020. Effect of silver doping on antidiabetic and antioxidant potential of ZnO nanorods. J. Trace Elem. Med. Biol. 58, 126448.

Venkatachalam, P., Jayaraj, M., Manikandan, R., Geetha, N., Rene, E.R., Sharma, N.C., Sahi, S.V., 2017. Zinc oxide nanoparticles (ZnONPs) alleviate HM-induced toxicity in *Leucaena leucocephala* seedlings: a physiochemical analysis. Plant Physiol. Biochem. 110, 59–69.

Wang, X., Sun, W., Zhang, S., Sharifan, H., Ma, X., 2018. Elucidating the effects of cerium oxide nanoparticles and zinc oxide nanoparticles on arsenic uptake and speciation in rice (*Oryza sativa*) in a hydroponic system. Environ. Sci. Technol. 52 (17), 10040–10047.

Wang, Y., Wang, L., Ma, C., Wang, K., Hao, Y., Chen, Q., Mo, Y., Rui, Y., 2019. Effects of cerium oxide on rice seedlings as affected by co-exposure of cadmium and salt. Environ. Pollut. 252, 1087–1096.

Wei, J., Duan, M., Li, Y., Nwankwegu, A.S., Ji, Y., Zhang, J., 2019. Concentration and pollution assessment of heavy metals within surface sediments of the Raohe Basin, China. Sci. Rep. 9, 1–7.

Xin, X., Zhao, F., Rho, J.Y., Goodrich, S.L., Sumerlin, B.S., He, Z., 2020. Use of polymeric nanoparticles to improve seed germination and plant growth under copper stress. Sci. Total Environ. 745, 141055.

Yuan, H., Liu, Q., Guo, Z., Fu, J., Sun, Y., Gu, C., Xing, B., Dhankher, O.P., 2021. Sulfur nanoparticles improved plant growth and reduced mercury toxicity via mitigating the oxidative stress in *Brassica napus* L. J. Clean. Prod. 318, 128589.

Zhou, P., Adeel, M., Shakoor, N., Guo, M., Hao, Y., Azeem, I., Li, M., Liu, M., Rui, Y., 2021. Application of nanoparticles alleviates HMs stress and promotes plant growth: an overview. Nanomaterials (Basel) 11 (1), 26.

Role of nanoparticles in alleviation of drought stress in plants: Strategy to achieve sustainable agriculture system

7

Riti Thapar Kapoor[a], *Momina*[b], *and Mohd Rafatullah*[c]
[a]Amity Institute of Biotechnology, Amity University, Noida, India, [b]Civil Engineering Department, Jamia Millia Islamia, New Delhi, India, [c]School of Industrial Technology, Universiti Sains Malaysia, Penang, Malaysia

7.1 Introduction

Agriculture is one of the main pillars of nation's economy as it is a key to sustainable development. Global population growth will be 9–11 billion by 2050; however, crop production is not rising with a same speed (Roos et al., 2017). The increasing demand for food will need around 320–850 million hectares of extra farming land by the year 2050 (Frona et al., 2019). Hence, there is a challenge to increase crop production up to 70%–100% to fulfill world food requirement (Rodrigues et al., 2017). Approximately 40% of global arable field has deteriorated due to the change in climatic conditions and man-made interventions. This includes loss of soil fertility due to excessive use of synthetic pesticides and agrochemicals (Kale and Gawade, 2016). Two-thirds of the world population are facing food insecurity due to change in rainfall pattern as a result of changes in monsoon shifts (Aryal et al., 2020). The variation in climatic conditions is predicted to enhance recurrence of drought which will skeptically affect food security. Intergovernmental Panel on Climate Change stated if global warming continues, temperature of earth will rise up to 1.5°C between 2030 and 2052. Ray et al. (2019) reported that one-fifth of the global population will be affected by severe water deficit by the end of this century. Climate change catastrophe, that is, rise in global temperature and atmospheric CO_2 level, and scarcity of water resources will adversely affect plant development and pose menace to agriculture (Yin et al., 2018; Yang et al., 2019).

Agricultural activities require substantial amount of water. Water is essential for plant survival as it acts as reactant for vital activities of plant cells and is required for maintenance of turgor pressure and transport of nutrients. Drought is a multidimensional abiotic stress which poses serious threat to crop productivity due to unpredictable rainfall pattern and diminution of groundwater availability (Hussain et al., 2019). Drought is a main problem to food security and has catalyzed several famines in the past (Okorie et al., 2019). The unpredictable nature of drought depends on various factors such as rainfall, high temperature, light intensity, evapotranspiration, dry

Agricultural Nanobiotechnology. https://doi.org/10.1016/B978-0-323-91908-1.00011-0

wind, and water holding capacity around the rhizosphere of plants (Cohen et al., 2021). Plants as sessile organisms are often exposed to drought stress as they have no choice to escape from conditions. Drought stress is reported in the areas where crop production is mainly dependent on rainfall as compared to those areas which are mostly irrigated by rivers, canals, and water channels (Konapala et al., 2020). Drought stress affects plant-water relationship by reducing water content, inhibits transpiration, hampers photosynthesis, stomatal closure, and limits gaseous exchange (Schulze et al., 2019). Plants may dehydrate and die due to long exposure to water-deficit conditions (Takahashi et al., 2018). Drought causes crop losses in billions of dollars annually (Suzuki et al., 2014).

For ever-growing population, concurrent increase in agricultural productivity under meager water supply is required to fulfill world's food demand. Hence, there is an urgent need to develop and improve drought resistance in plants via safe, environmentally benign, and economic strategies. Many technological advancements in the past decade aim to address the challenges of crop production (Kou et al., 2018). Many approaches are used for the mitigation of drought stress such as use of omics technology like proteomics, genomics and metabolomics (Oliveira et al., 2020), application of microbes, hydrogel, polyamines and nanoparticles which may enhance plant resistance by regulating osmolytes and antioxidant defense system. Among different strategies adopted to combat drought-induced plant stress, application of nanomaterials has proved to be the most promising approach (Khan et al., 2017). Nanomaterials may show negative or positive impact on growth of plants as it depends on nanoparticles type, size, amount, mode of application, and plant species (Du et al., 2017). Thus, the present chapter provides an overview on pernicious impacts of drought on morphological and biochemical processes of plants, and highlights recent research findings with respect to the application of various types of nanomaterials in alleviation of drought-induced adverse effects on plants.

7.2 Approaches applied for the mitigation of drought stress

More than 80% of fresh biomass of plants is made up of water which shows significant role in different morphological and physiological phenomenon (Brodersen et al., 2019). Drought stress occurs if transpired water level in plants is more as compared to the water absorbed by the roots, and it may be due to reduced level of ground water or less water retention in soil particles (Salehi-Lisar and Bakhshayeshan-Agdam, 2016). Plants utilize different strategies like stress avoidance, escape, and tolerance for their survival against drought stress. Plants evolve to acquire various complex resistance mechanisms under the conditions of water scarcity which may vary from molecular to plant level (Galindo et al., 2020). The changes in growth pattern and structural dynamics, increase in length of roots, rolling of leaves, inhibition in transpiration, osmolytes accumulation, regulation of phytohormones, and delayed senescence are adopted by plants under drought. To escape adverse effects of water deficiency, plants show fast growth, shortening of the life cycle, self-pollination, and early flowering before the beginning of the dry season of the year

(Alvarez et al., 2018). The presence of hairy leaves, sunken stomata, and cuticle may also help in maintenance of water potential in plant tissues as the rate of water loss through leaf transpiration is reduced (Tiwari et al., 2020).

Application of plant growth regulators can enhance capacity of plants against drought. Adverse effect of water deficiency can be alleviated by application of gibberellic acid to maintain the internal water balance and protein synthesis in drought-stressed plants. The exogenous use of salicylic acid, abscisic acid, uniconazole, brassinolide, and jasmonic acid can improve crop yield under drought (Ahmad et al., 2018). Sedaghat et al. (2017) observed that salicylic acid utilization in wheat enhanced proline accumulation by the increase in abscisic acid content. Phytohormones such as ethylene and brassinolide also regulate plant defense system by scavenging reactive oxygen species under stress conditions in plants (Tanveer et al., 2019). The foliar application of proline on plants enhanced free proline content inside the plant cells and helped in escalation of drought tolerance capacity (Semida et al., 2020). The use of spermidine has also promoted tolerance to drought stress in barley and wheat (Sallam et al., 2019). Exposure to silicon increased relative water content in sorghum and sunflower (Hurtado et al., 2020). Selenium application accelerated growth, inhibited oxidative stress, and regulated water balance in the plants grown in water scarcity (Kaya et al., 2019). Its application in plants reduced cell membrane degradation and enhanced antioxidant enzymes activity (Jung et al., 2019). The potassium supplementation is also beneficial for physiological processes of plants as low grain production of crops can be enhanced via its utilization under drought conditions (Brestic et al., 2018).

The microbes significantly reduce injurious impact of drought by improving plant productivity (Khan et al., 2020). Utilization of growth-promoting rhizobacteria in plants can increase drought tolerance capacity of the crops as they form colonies in the root zones and enhance development of plants under stress (Khan et al., 2018). The inoculation of *Azospirillum brasilense*, *Bacillus amyloliquefaciens*, *Rhizobium leguminosarum*, and *Trichoderma harzianum* enhanced growth and drought tolerance index in plants (Hewedy et al., 2020; Kumar et al., 2020). Hydrogel is water-absorbing polymer, and its application in soil can reduce the need of frequent irrigation (Al-Jabari et al., 2019). Abobatta (2018) reported that hydrogel addition improved physico-chemical properties of soil with positive effects on the development of plants.

7.3 Application of nanoparticles in drought mitigation

Nanotechnology offers significant applications in various disciplines such as chemistry, pharmaceutical science, identification of plant diseases and protection of plants, food sector, fertilizers production, precision farming, and remediation of terrestrial environment (Monreal et al., 2016; Gil-Díaz et al., 2017). Nanoparticles are called as nanoscale particles. Nanoparticles can be considered as a link between bulk substances and atomic structures. They are becoming popular as novel materials for transformation of agricultural practices as they have a potential to increase food quality and quantity. According to Chhipa (2017), nanoparticles are considered as the most inspected materials of the century. Paul Ehrlich named nanomaterials as magic bullets

(Kreuter, 2007). European Commission defined nanomaterial as a natural or incidental or synthesized substance with half or more particles having 1–100nm size in unbound or aggregate state (Potocnik, 2011).

Nanoparticles are small molecular aggregates, may be inorganic, organic, or hybrid materials with outer interfacial layer having dimensions between 1 and 100nm. This layer is main component which determines its properties (Baby et al., 2018). Nanomaterials are produced from large materials, and small size of nanoparticles shows different chemical activities which are distinct as compared to their parent material.

The natural nanomaterials which are present in volcanic eruptions, forest fire, and mineral composites and also generated via anthropogenic activities such as combustion of coal, and fumes produced during welding work are also known as incidental nanomaterials (Dahoumane et al., 2017). The engineered nanomaterials can be categorized in four kinds such as metal- and carbon-based nanomaterials, metal oxides, dendrimers, and composites (Yu-Nam and Lead, 2008). Nanoparticles have diverse physico-chemical activities because of their minute particle size, tunable pore size, potential reactivity, enormous surface-to-volume ratio, high stability, multiple binding sites, electrical and optical properties, and spatial confinement (Seleiman et al., 2020). Nanoparticles show optical properties and quantum effects (Sudha et al., 2018). The engineered nanoparticles can easily alter their native properties, activity, and distribution inside the plant cells in comparison with nonmodified nanoparticles (Khan et al., 2017). The nanoparticles can cross cell wall and plasma membrane to assist efficacious absorption due to their small size and large surface area (Monica and Cremonini, 2009). They can penetrate easily into required locations in cells and helpings in adsorption and targeted delivery of specific molecules due to more surface area (Kashyap et al., 2015). Zhai et al. (2014) found intercellular transport of nanoparticles via plasmodesmata. Nanoparticles can be loaded or encapsulated with active ingredients or agrochemicals for their targeted and controlled release due to more surface area, rapid mass transfer ability, and easy binding (Ghormade et al., 2011). Nanoparticles act as superior carriers for biomolecules like plasmid deoxyribonucleic acid (DNA) and double-stranded ribonucleic acid (RNA) (Kwak et al., 2019). Nanoparticles act differently as their accessibility and reactivity in different environment is diversified (Ealias and Saravanakumar, 2017). The environmental conditions can affect oxidation state, stability, and precipitation properties of nanoparticles.

Nanotechnology enhances productivity of crops by increasing plant resistance under environmental stresses (Khan and Upadhyaya, 2019). To combat environmental stresses, plants develop molecular programs for fast adaptation but plant responses differ in same plant species (Kollist et al., 2019). The potential use of nanoparticles requires further studies, particularly their function as strain alleviator in plants (Vishwakarma et al., 2018). Nanoparticles can be applied as nanofertilizers, herbicides, pesticides, and nanosensors for farming (Yılmaz et al., 2017) which may decrease the consumption of synthetic agrochemicals and minimize environmental pollution (Pulimi and Subramanian, 2016). Several metal or metal oxides, metalloids and carbon-based nanoparticles are studied to assess their function in growth of plants and modulation of various processes in plants under drought stress (Table 7.1).

Table 7.1 Effect of nanoparticles in amelioration of drought-stress-incited damage in different plants.

S.No.	Nanoparticles	Plant	Mode of application	Concentration of nanoparticles	Size of nanoparticles (nm)	Effect on plants	References
(A) Carbon-based nanomaterials							
1.	Single-walled carbon nanotubes	*Hyoscyamus niger* L.	Seed socking	50–800 μg/mL	1.1 nm	Increase in water uptake, reduction in oxidative injury indices, activation of plant defense enzymes (SOD, POD, CAT, and APX), and biosynthesis of proteins, phenolic compounds, and proline	Hatami et al. (2017)
2.	Multiwalled carbon nanotubes	*Cucurbita pepo* L.	Seed socking	0.5–1 g/L	20–30 nm	Positive effect on water uptake, seed germination, and seedling growth	Karami and Sepehri (2017)
3.	Multiwalled carbon nanotubes	*Dodonaea viscosa* L.	Nanopriming	10, 30, 50, 100, and 200 mg/L	5–15 nm	Improved seed germination, seedling length, and biomass of root and stem	Yousefi et al. (2017)

Continued

Table 7.1 Continued

S.No.	Nanoparticles	Plant	Mode of application	Concentration of nanoparticles	Size of nanoparticles (nm)	Effect on plants	References
4.	Fullerenol	*Brassica napus* L.	Seed priming	10–100 mg/L	–	Repressing ROS accumulation by enhancing the regulatory mechanism of enzymatic and nonenzymatic antioxidants and ABA accumulation	Xiong et al. (2018)
5.	Carbon nanotubes and graphene	*Gossypium hirsutum*	Seed priming	50 and 200 μg/mL	13–18 nm	Enhanced fiber biomass and seedling growth	Pandey et al. (2019)
(B) Metalloids and nonmetallic nanoparticles							
1.	Phosphorus nanoparticles	*Gossypium barbadense* L.	Foliar spray	0.5–1.0 g/L	–	Improved macro- and micronutrients uptake	Hussien et al. (2015)
2.	Silicon dioxide	*Hordeum vulgare*	Foliar application	20 ppm	100 nm	Increased biomass and yield of barley	Ghorbanian et al. (2017)

3.	Silicon dioxide	*Cucumis sativus*	Amended soil	100, 200, 300, and 400 mg/kg	10 nm	Increased vegetative growth of plants, nutrient uptake, and fruit yield	Alsaeedi et al. (2019)
4.	Silicon dioxide	*Musa acuminata*	In vitro treatment	200, 400, and 600 mg/L	0.041–0.064 μm	Enhanced shoot growth and chlorophyll content, improved photosynthesis, maintained Na^+ and K^+ balance, reduced cell wall damage	Mahmoud et al. (2020)
(C) Metallic and metallic oxides nanoparticles							
1.	Titanium dioxide	*Triticum aestivum*	Spraying	0.01%, 0.02%, 0.03%	–	Increased growth and productivity, starch and gluten contents were also enhanced	Jaberzadeh et al. (2013)
2.	Zinc oxide	*Glycine max*	Petriplate exposure	0.5 and 1 g/L	–	Rate of seed germination was enhanced but dry weight was reduced	Sedghi et al. (2013)
3.	Titanium dioxide	*Ocimum basilicum*	Foliar application	0.01% and 0.03%	–	Improved negative effects of drought stress	Kiapour et al. (2015)

Continued

Table 7.1 Continued

S.No.	Nanoparticles	Plant	Mode of application	Concentration of nanoparticles	Size of nanoparticles (nm)	Effect on plants	References
4.	Silver nanoparticles	*Carum copticum*	Irrigation water	40 g/ha	20 nm	No positive effect on WUE and yield characteristics	Seghatoleslami et al. (2015)
5.	Potassium nanoparticles	*Cucurbita pepo*	Foliar spray	2.5 g/L	–	Increased growth traits and stomatal conductance	Gerdini (2016)
6.	Titanium dioxide	*Linum usitatissimum*	Leaf treatment	10, 100, and 500 mg/L	10–25 nm	Enhanced chlorophyll and carotenoids content and decreased MDA and H_2O_2 content	Aghdam et al. (2016)
7.	Titanium dioxide	*Dracocephalum moldavica*	Foliar application	10 ppm	10–25 nm	Higher accumulation of proline and reduction in MDA and H_2O_2 contents	Mohammadi et al. (2016))
8.	Copper and zinc oxide	*Triticum aestivum*	Addition in soil	CuO 300 mg/kg and ZnO 500 mg/kg	CuO 50 nm ZnO 70 nm	Nanoparticles interacted with root-colonizing microbes *Pseudomonas chlororaphis* O6 and enhanced plant growth	Yang et al. (2017)

9.	Cerium oxide nanoparticles	*Sorghum bicolor*	Foliar spraying	50 or 100 mg/L	15 nm	Increased leaf carbon assimilation rate, pollen germination, seed yield per plant, reduced free radicals accumulation, and lipid peroxidation	Djanaguiraman et al. (2018)
10.	Ferric oxide	*Fragaria × ananassa*	Murashige and Skoog medium	0.8 ppm	40–53 nm	Enhanced growth parameters of plant	Mozafari et al. (2018)
11.	Zinc oxide	*Sorghum bicolor*	Soil application	5 mg/kg	18 nm	Reduced the delay of flag leaf and grain head emergence, improved grain yield and nutrient translocation	Dimkpa et al. (2019)
12.	Nano-zero-valent iron	*Arabidopsis thaliana*	Soil application	500 mg/kg	54 nm	Beneficial effects on photosynthesis, sugar and starch synthesis, and increased uptake of nutrients such as phosphorus, zinc, manganese, and iron	Yoon et al. (2019)

Continued

Table 7.1 Continued

S.No.	Nanoparticles	Plant	Mode of application	Concentration of nanoparticles	Size of nanoparticles (nm)	Effect on plants	References
13.	Iron nanoparticles	*Triticum aestivum*	Soil application	100 mg/kg	50–100 nm	Increased photosynthesis, iron concentration, grain yield, and reduced oxidative stress	Adrees et al. (2020)
14.	Titanium dioxide	*Triticum aestivum*	Soil application	500, 1000, and 2000 mg/kg	10–25 nm	Increased seedling dry weight, relative water content, pigment content, improved stomatal conductance, transpiration rate, and antioxidative enzymes	Faraji and Sepehri (2020)
15.	Copper nanoparticles	*Zea mays*	Plants priming	3.33, 4.44, and 5.55 mg/L	30–40 nm	Enhanced leaf water content, plant biomass, grain yield, increased pigment, and carotenoid contents	Van Nguyen et al. (2021)

(D) Nanosize polymers and composites							
1.	Chitosan nanoparticles	*Hordeum vulgare*	Foliar application	30, 60, and 90 ppm	10–20 μm	Increased RWC, grain weight, protein, and proline contents, and SOD and CAT enzymes	Behboudi et al. (2018)
2.	Chitosan nanoparticles	*Triticum aestivum*	Foliar and soil application	30, 60, and 90 ppm	10–20 μm	Enhanced relative water content, pigment content, plant height, and biomass	Behboudi et al. (2019)
3.	Nanoclay	*Solanum tuberosum*	Nutrient solution	1 g/L	20–30 nm	Improved yield and oil percentage	Soltani et al. (2018)
4.	Analcite nanoparticles	*Triticum aestivum* and *Zea mays*	Soil application	500–1500 mg/L	100 μm	Increased photosynthesis, iron concentration, grain yield, and decreased oxidative stress	Zaimenko et al. (2014)
5.	ZnO, B_2O_3, and CuO nanocomposite	*Glycine max*	Foliar spray	1.77, 0.92, and 0.8 g/L	50–100 nm	Adaptation of strawberry plants to drought stress before field transplantation	Dimkpa et al. (2017)

There is a need to develop novel nanomaterials to promote crop yield under water-deficit conditions but their application in agriculture is still at primitive stage.

7.4 Uptake, translocation, and accumulation of nanoparticles in plants

Nanomaterials are taken from soil by plant roots via active transport mechanism through the xylem (Tripathi et al., 2017). Nanomaterials may change their structure and can form complex with membrane transporters (Dimkpa and Bindraban, 2018). The absorption, transport, and accretion of nanoparticles depend on their size, stability, type, composition, function of the nanoparticles, and species of plants. Sharif et al. (2013) stated that nanomaterials can easily penetrate via leaf cuticle and enter into the cytoplasm of cells. Nanomaterials can attach with cellular organelles and interfere with biochemical processes of cytoplasm (Zhang and Monteiro-Riviere, 2009). Unrine et al. (2012) found that 20-nm-size silver nanoparticles (AgNPs) entered in the cells by plasmodesmata. The nanoparticles react with sulfhydryl and carboxyl groups of biomolecules and may change activity of proteins in plant cells. Studies revealed that nanoparticles play an important function in plant's protection from drought stress via regulation of biochemical and molecular processes.

7.5 Mitigation of drought stress in plants by nanoparticles

Water scarcity is due to less or absence of rainfall resulting in low soil moisture content and water potential in different plant parts (Ristvey et al., 2019). The availability of less amount of water in rhizosphere hampers plant growth, thereby inhibiting water and nutrients uptake (Elemike et al., 2019). Drought stress inhibits water potential, stomatal conductance, and rate of transpiration with increasing leaf temperature that correlated linearly with increased drought severity as shown in Fig. 7.1. Reduced photosynthesis in drought-stressed plants is because of stomatal closure and increased leaf temperature leading to degradation of thylakoid membrane and disrupted Calvin cycle and impaired adenosine triphosphate (ATP) synthesis enzymes activities (Farooq et al., 2012). Plant productivity and quality are adversely affected under water deficit (Battaglia et al., 2018). The plant age, species, severity, and duration of drought stress are the main factors that influence development of plants against water deficiency (Gray and Brady, 2016). Drought stress enhances free radicals production, which imbalances cellular homeostasis (Sharma et al., 2020). Nanotechnology is a propitious technique to combat with the unfavorable impact of drought stress. Nanoparticles act as nanoregulators that can increase plant resistance to environmental stresses throughout their life cycle (Liu et al., 2016).

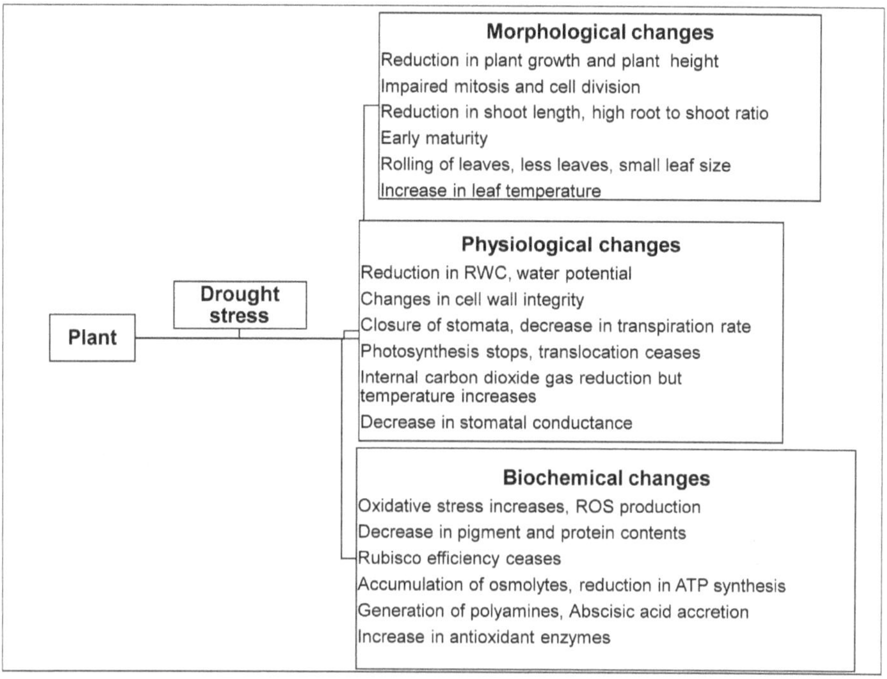

Fig. 7.1 Effect of drought stress on morphological, physiological, and biochemical processes of plants.

7.6 Nanoparticles-mediated regulation of morphological parameters under drought stress

7.6.1 Effect on growth parameters

There is a correlation between plant growth and water availability as cell enlargement is adversely affected compared to cell division under drought (Humplík et al., 2017). Growth of the plants can be reduced due to inhibition in the cell wall extensibility and turgor (Seleiman et al., 2019). Nanomaterials modulate growth and development of plants at multiple levels (Kwak and An, 2016).

They may show either beneficial or adverse effects on productivity of plants. Detrimental impact of nanomaterials depends on type, physico-chemical properties, size, amount, application mode, plant species, and soil condition (Du et al., 2017). Agathokleous et al. (2019) stated biphasic dose response (hormesis) of nanomaterials on plants, that is stimulation and inhibition with low and high doses, respectively. AgNPs inhibited root elongation in corn at 50–2500 mg/L concentration whereas the growth of watermelon and zucchini seedlings was significantly increased at the same concentration (Almutairi and Alharbi, 2015). Tiwari et al. (2017) observed better growth and photosynthetic performance of tomato plants with low concentration (0.5–2 g/L) of titanium dioxide nanoparticles (TiO_2 NPs) as compared to high concentration (4 g/L). Zinc oxide (ZnO) nanoparticles improved growth of beans but decreased the growth of wheat plants (Dimkpa and Bindraban, 2018).

The exposure of seeds to environmental stresses adversely affects establishment of seedlings and plant development. Lu et al. (2002) reported that nanomaterials can pierce seed coat and increase the absorption of water, which enhances activity of enzymes and finally promotes germination and growth of seedlings. Lu et al. (2002) observed increase in germination and growth of *Glycine max* seeds by TiO_2 and silicon dioxide (SiO_2) nanoparticles. TiO_2NPs enhanced plant growth by increasing photosynthesis and nitrogen metabolism (Yang et al., 2006). Application of carbon nanoparticles (CNPs) in combination with fertilizers can escalate rice yield by 10%, maize 11%, soybean 17%, wheat 29%, and vegetables production 12%–20% (Liu et al., 2009). SiO_2NPs-mediated increase in maize seed germination was due to more nutrients availability (Suriyaprabha et al., 2012). ZnONPs increased soybean seed germination under drought stress (Sedghi et al., 2013).

Adhikari et al. (2013) investigated promotive effects of SiO_2NPs on the seed germination and morphological parameters of rice. Foliar spraying of ZnONPs (1.5 mg/mL) on *Cicer arietinum* seeds exhibited escalation in seed germination and seedling biomass (Burman et al., 2013). Jaberzadeh et al. (2013) reported that utilization of TiO_2NPs showed positive effect on wheat productivity under drought stress conditions. Dhoke et al. (2013) found that ZnONPs significantly increased mung bean germination up to 80% and seedling vigor index. The iron nanoparticles (FeNPs) foliar supplementation enhanced safflower yield under water scarcity (Farinaz et al., 2014).

SiO_2NPs improved seedling biomass and vigor index of tomato (Siddiqui and Al-Whaibi, 2014). Foliar treatment of 0.5 g/L magnesium nanoparticles (MgNPs) on black-eyed pea significantly increased biomass and photosynthesis (Delfani et al., 2014). Carbon nanotubes (CNTs) enhanced seedling growth of wheat plants (Srivastava and Rao, 2014).

Laware and Raskar (2014) found that TiO_2NPs at less amount increased germination and seedling growth of onion with rise in amylase and protease activities but all the parameters reduced at high concentration. The increase in water utilization ability and yield of *Helianthus annuus* was reported by using ZnONPs under drought stress (Seghatoleslami and Forutani, 2015). Seed priming with SiO_2NPs enhanced seedling length, biomass, and vigor index in sunflower (Janmohammadi and Sabaghnia, 2015).

Rathore and Tarafdar (2015) observed foliar treatment of wheat plants with biosynthesized magnesium dioxide nanoparticles (MgO_2NPs) and observed significant increase in root length and biomass. Martínez-Fernández et al. (2015) observed that nanosized maghemite mitigated drought stress in *Helianthus annuus*. Saharan et al. (2016) reported an increase in corn seedlings growth with rise in α-amylase activity and starch content due to the application of Cu-chitosan nanoparticles. Liu et al. (2016) cultivated lettuce in different concentrations of iron oxide nanoparticles (Fe_3O_4NPs) which escalated shoot length by 12%–26%. Almutairi (2016) reported that application of nanomaterials elevated gene expression related to cell division and elongation. The seed germination, seedling length, and biomass of *Thymus kotschyanus* enhanced due to SiO_2NPs utilization (Khalaki et al., 2016). Hojjat and

Ganjali (2016) found positive impact of AgNPs on *Lens culinaris* under water-deficit conditions. The germination of seeds and level of antioxidants were significantly high in barley, corn, and soybean due to the treatment with CNTs (Liu et al., 2016).

Fitriani and Haryanti (2016) observed that SiO_2NPs of various amount enhanced plant height and leaf number in *Solanum lycopersicum*. The low doses of SiO_2NPs promoted radicle and plumule length and biomass in *Triticum aestivum* (Karimi and Mohsenzadeh, 2016). Shallan et al. (2016) revealed that under water deficiency, pretreatment of cotton plants with TiO_2 or SiO_2NPs showed rise in pigment content, free amino acids, sugar and protein contents, total phenolics, proline content, total antioxidant capacity, and enhancement in yield. The fullerenol nanoparticles ($C_{60}(OH)_{36}$NPs) can reduce the adverse effect of drought stress by maintaining supply of water to plant cells. These nanoparticles entered in tissues of root and leaves of plant and could bind with water molecules in different cells Borisev et al. (2016). ($C_{60}(OH)_{36}$NPs) function as intracellular binders of water and provide water availability for mitigation of drought stress in sugar beet (Borisev et al., 2016). ($C_{60}(OH)_{36}$NPs) utilization enhanced osmolyte contents in sugar beet roots and leaves under drought stress. The positive effect of silicon nanoparticles (SiNPs) on biomass of barley seedlings was linked with increase in water content and silicon accumulation in leaf tissues (Rastogi et al., 2017). Foliar spray of SiO_2NPs escalated *Cucumis sativus* productivity (Yassen et al., 2017). Zinc oxide (ZnONPs) and copper oxide (CuONPs) are rich sources of zinc and copper, respectively, and can be used as fertilizer. ZnONPs promoted lateral roots generation and CuONPs inspired root hairs elongation in seedlings of wheat under drought stress (Yang et al., 2017).

The treatment of fenugreek seeds with SiNPs enhanced 55% seedling length in comparison with control (Ivani et al., 2018). Palchoudhury et al. (2018) observed that low dose of ferric oxide nanoparticles (Fe_2O_3NPs) enhanced legumes root growth and presoaking of plants with nanoparticles increased root length up to 88%–366%. Sorghum showed an increased drought stress tolerance ability on treatment with ZnONPs through increased grain nitrogen transport and maintenance of nitrogen content (Dimkpa et al., 2019). The drought resistance was also enhanced by SiNPs in wheat plants (Rizwan et al., 2019). Hatami et al. (2019) observed presence of small pores in *Salvia mirzayanii* seedlings by the treatment of graphene oxide/polyaniline nanocomposite which may increase water absorption by roots. However, pore formation mechanism in root tissues due to nanoparticles treatment is not clear. The pore formation in the roots of nanoparticles-treated plants may be a response of plants against nanoparticles which act as external stressors. It is observed that smaller aggregates (less nanoparticle concentration) formed small pores whereas high nanoparticle concentration (large aggregates) formed large sized pores (Ghorbanpour et al., 2020).

The germination of seeds depends on reserve food availability and mobilization mediated by amylase and protease enzymes for plants survival until they can synthesize their own food by photosynthesis. The promotive impact of nanomaterials on germination of seeds can be explained by its role in regulation of aquaporins, which control permeability of water and indirectly plant growth (Heinen et al., 2009). Supplementation of plants with nanomaterials activates genes related to water channel protein which reflected better regulation of cell cycle (Khodakovskaya et al., 2012). Lahiani et al.

(2013) reported that multiwalled carbon nanotubes control water channel proteins gene expression in seed coat. It was further confirmed when nanotube accumulation was observed inside the treated seed coat. Navarro et al. (2008) stated that nanomaterials can sequester nutrients on their surface and serve as nutrient stock.

7.6.2 Effect on pigment content and photosynthesis

Photosynthesis is the basis for all the biochemical processes of plants and one of the most significant phenomena which affects productivity of plants. The decline in carbon dioxide conductance due to stomata and mesophyll limitations leads to decrease in photosynthesis in plants under water-deficit conditions (Singh and Thakur, 2018). Chowdhury et al. (2017) found changes in synthesis of chlorophyll *a* and *b* contents in soybean leaves under drought stress. Plants treated with nanomaterials exhibited improvement in pigment content, rate of photosynthesis, and efficiency of water utilization (Haghighi and Pessarakli, 2013). The photosynthesis is a sensitive phenomenon under abiotic stresses, although nanomaterials improved photosynthesis by repressing osmotic and oxidative stresses (Siddiqui et al., 2014). The decrease in photosynthetic activity under drought stress was due to reduced ability of the stomatal movement (Marcinska et al., 2013). The decline in chlorophyll content, photosystem II activity, changes in stomatal movement, and disruption in water supply resulted in decreased yield of plants (Xiang et al., 2013).

Bao-shan et al. (2004) reported that SiNPs enhanced chlorophyll synthesis. TiO_2NPs enhanced growth of spinach by rise in photosynthetic rate and metabolism of nitrogen (Yang et al., 2006). Gao et al. (2006) studied that nanoanatase TiO_2 promoted Rubsico and Rubisco-activase activities in spinach and enhanced photosynthesis.

The use of TiO_2NPs increased pigment content and Rubisco activity in spinach and resulted in the rise of biomass of plants (Gao et al., 2008). Cowpea yield was increased by foliar application of TiO_2NPs, and it also promoted photosynthesis (Owolade et al., 2008). SiNPs application escalated pigment amount in *Zea mays* and *Larix olgensis* (Yuvakkumar et al., 2011).

Ze et al. (2011) observed that TiO_2NPs enhanced LHCII II amount in *Arabidopsis thaliana* thylakoid membrane. They found that TiO_2 nanoparticles accelerated LHCII content by enhancing light absorption ability of chloroplast and regulating light energy distribution from PS I to PS II. Ghafariyan et al. (2013) found nanoparticles of ferric oxide (Fe_3O_4NPs) improved pigment content in soybean. Alidoust and Isoda (2013) studied effects of citric-acid-coated Fe_3O_4NPs on soybean seedlings given by root and leaves promoted root length and rate of photosynthesis. Pradhan et al. (2013) observed that manganese nanoparticles (MnNPs) enhanced pigment content by improving electron transport system. TiO_2NPs improved plant's ability to capture sunlight and increased photosynthetic efficiency in maize under drought stress (Akbari et al., 2014). SiO_2NPs increased photosynthesis, PSII activity, photochemical quenching, transpiration rate, and stomatal conductance in *Indocalamus barbatus* and *Cucurbita pepo* (Siddiqui et al., 2014). Zaimenko et al. (2014) investigated analcite nanoparticles ($AlSi_2O_6 \cdot H_2O$NPs) effect on wheat and corn which mitigated drought stress by increasing pigment and antioxidants content. Fe_2O_3NPs facilitate stomata opening via plasma membrane H^+-ATPase activation in *Arabidopsis thaliana* (Kim et al., 2015). The photosynthesis, stomatal conductance, and hawthorn seedlings biomass were elevated under drought stress due to SiNPs treatment (Ashkavand et al., 2015). SiNPs enhanced chlorophyll *a*, *b*, and carotenoid contents in *Zea mays* (Sharifi et al., 2016). Rise in pigment content and growth of plants due to SiNPs under drought stress enhanced N, K, and Si contents in plant tissues (Alsaeedi et al., 2019). Nanoparticles have the ability to enhance enzymes of C3 cycle and photosynthetic pathway which can promote carboxylation and plant growth (Lowry et al., 2019).

7.6.3 Nanoparticles-mediated mitigation of oxidative stress under drought condition

Drought generates oxidative stress by production of free radicals like superoxide, hydrogen peroxide, and hydroxyl radical. At low level, reactive oxygen species (ROS) promote defense system via activation of cell signaling mechanism and expression of genes. However, excessive ROS accumulation in plants under stress conditions shows degradation of biomolecules and cell membrane, and retards growth (Yadav et al., 2014). To combat oxidative stress, plants have both enzymatic and nonenzymatic defense system. The antioxidant enzymes are superoxide dismutase, catalase, ascorbate peroxidase, glutathione reductase, glutathione peroxidase, peroxidase, and nonenzymatic antioxidants (glutathione, ascorbate) which can remove free radicals. Thus, an efficient antioxidant system helps in removal of free radicals with reduction in electrolyte leakage and lipid peroxidation and maintains integrity of cell membrane (Hussain et al., 2019). Accumulation of trehalose, glycerol, proline, and glycine betaine preserves water level in plant cells.

Abid et al. (2017) found that antioxidant enzymes such as superoxide dismutase, catalase, and peroxidase were enhanced under water scarcity in *Vicia faba*. The antioxidants were increased in *Coleus plectranthus* under water-deficit condition (Prathyusha and Chaitanya, 2019).

Nanoparticles play significant function in amelioration of environmental stresses by osmolytes and free amino acids accretion and escalation in activities of antioxidant enzymes as illustrated in Fig. 7.2.

The nanomaterials application at low concentration was found effective in mitigation of stresses and promoted growth of plants (Soliman et al., 2015). Husen and Siddiqi (2014) reported that nanomaterials may show toxicological effects on plants under high concentration. Nanoparticles can mimic antioxidative enzymes activities and scavenge the ROS to protect plants under strain (Wei and Wang, 2013). Aghdam et al. (2015) reported effect on TiO_2NPs on morphological and biochemical mechanism of *Linum usitatissimum*, and results revealed lower dose showed better alleviation of drought-stress-inspired deterioration in comparison with high doses. CuO and ZnONPs reduced adverse effect of drought stress on winter wheat steppe ecotype Acveduc. CuO and ZnONPs increased 22% superoxide dismutase activity under

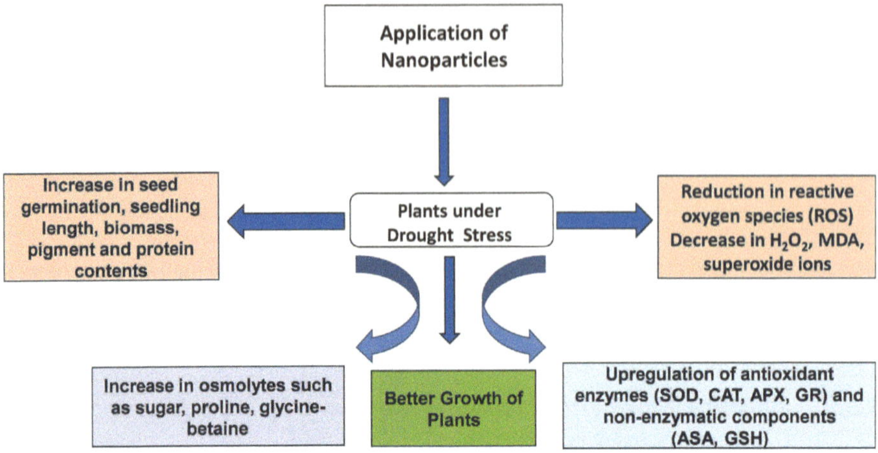

Fig. 7.2 Effect of nanoparticles in mitigation of drought stress in plants.

drought stress in seedlings of wheat steppe ecotype variety Acveduc in comparison with those which were grown without nanoparticles treatment. TiO_2NPs can increase utilization efficiency of plants for water and nitrogen and promote antioxidant enzymes activity in canola (Mahmoodzadeh et al., 2013).

Application of SiNPs increased catalase and peroxidase activities in faba bean leaves (Kalteh et al., 2014) and tomato (Siddiqui and Al-Whaibi, 2014) as compared to control plants. Malondialdehyde content is marker of oxidative damage to membrane lipids and enhances under drought stress. The treatment of plants with SiNPs reduced malondialdehyde amount (Ashkavand et al., 2015). Cerium oxide nanoparticles (CeO_2NPs) are significant scavengers of ROS (Walkey et al., 2015). Liu et al. (2016) reported amplification of superoxide dismutase, ascorbate peroxidase, and glutathione reductase in plants by fullerenol ($C_{60}(OH)_{36}$NPs) exposure which inhibited the unfavorable effect of drought stress. Siddiqi and Husen (2016) also found that activity of antioxidants enzymes was regulated by the application of nanoparticles. Palmqvist et al. (2017) found that Fe_2O_3NPs mitigate oxidative stress in *Brassica napus* under drought stress by decreasing hydrogen peroxide and malondialdehyde. The application of copper and zinc nanoparticles colloidal solution increased drought stress resistance in wheat by enhancing antioxidant enzymes and photosynthetic pigments (Taran et al., 2017). Djanaguiraman et al. (2018) observed that foliar spray of CeO_2NPs alleviated oxidative damage in sorghum under drought stress and showed 41% reduction in superoxide radicals, 36% H_2O_2 level, and lipid peroxidation (37%). Sturikova et al. (2018) reported abiotic stresses mitigation by nanoparticles was linked with increased activity of antioxidant enzymes.

Chen et al. (2018) synthesized molybdenum disulfide nanoparticles (MoS_2NPs), and they mimicked the function of antioxidant enzymes and improved plant stress resistance. Activity of SOD was increased in plants due to TiO_2NPs application (Ghasemlou et al., 2019). The drought resistance in wheat was also increased by CuO and ZnONPs utilization (Maswada et al., 2020).

Melatonin acts as an antioxidant or ROS scavenger or plant growth regulator, and it induces drought resistance mechanism in plants (Liang et al., 2019). Sharif et al. (2018) revealed that endogenous melatonin accretion can repair drought-stress-induced damage in mitochondria, increased amount of nitric oxide and antioxidants, and improved resistance of plants against water-deficit conditions. High concentration of melatonin increased drought resistance by controlling proline metabolism and improved plant growth (Fleta-Soriano et al., 2017). ZnONPs stimulated pathway for melatonin generation and promoted antioxidant defense system, which mitigated drought-stress-incited damage in mitochondria and chloroplast and promoted resistance against drought stress. Application of ZnONPs inspired drought tolerance ability by the overexpression of endogenous melatonin generation in corn (Sun et al., 2020). TEM images of chloroplast and mitochondria of corn leaves showed that they were affected by ZnONPs and drought stress as demonstrated in Fig. 7.3. As under drought stress, deformation of chloroplast membrane system, and starch grains degradation with appearance of vacuoles in the chloroplast lamella were observed. The mitochondria were also disintegrated under drought stress. No obvious changes were reported in ZnONPs-treated corn plants under drought stress.

7.7 Mechanism of action of nanomaterials under abiotic stress

ROS act as a strain indicator to promote defense system of plants (Dat et al., 2000). Nanomaterials at high concentration may cause oxidative strain in plants due to accretion of reactive oxygen and nitrogen species leading to damage in biomolecules and cell membrane (Khan et al., 2017). Nanoparticles induce ROS generation at one end (Simon et al., 2013); on the other hand, they can remove ROS by imitating function of the antioxidative enzymes (Wei and Wang, 2013). There is a need to analyze the dual nature of nanomaterials that how they provide protection against free radicals while they generate oxidative stress. Calcium (Ca^{2+}) helps in signaling responses to environmental stresses. Under stress conditions, Ca^{2+} transport to cytosol via Ca^{2+} channels leads to rise in Ca^{2+} content which initiates gene expression and plant resistance to stress (Khan et al., 2014). Nitric oxide promotes rise in cytosolic Ca^{2+} in plant cells due to exposure of abiotic stresses (Khan et al., 2012), and consequently, Ca^{2+} induces nitric oxide synthesis (Corpas et al., 2006). Mirzajani et al. (2014) stated that AgNPs responsive proteins were linked to Ca^{2+} regulation, cell division, and apoptosis in the roots of *Oryza sativa*. Miao et al. (2014) found that $C_{60}(OH)_{36}$NPs showed functional modulation of Ca^{2+}/calmodulin-dependent protein kinase II. Marmiroli et al. (2015) reported exposure of cadmium sulfide (CdS) quantum dots to *Arabidopsis thaliana* showed expression of calcium-binding protein CML45 and calcium-dependent protein kinase. The calcium-binding proteins regulated stress responses, and their upregulation escalated tolerance against abiotic stresses (Boudsocq and Sheen, 2013).

Nanoparticles increase nitrate reductase activity, which promotes synthesis of nitric oxide in plants (Chandra et al., 2015). NO can restrain nanoparticles-incited toxicity in plants and activates expression of antioxidant genes and inhibits free radical

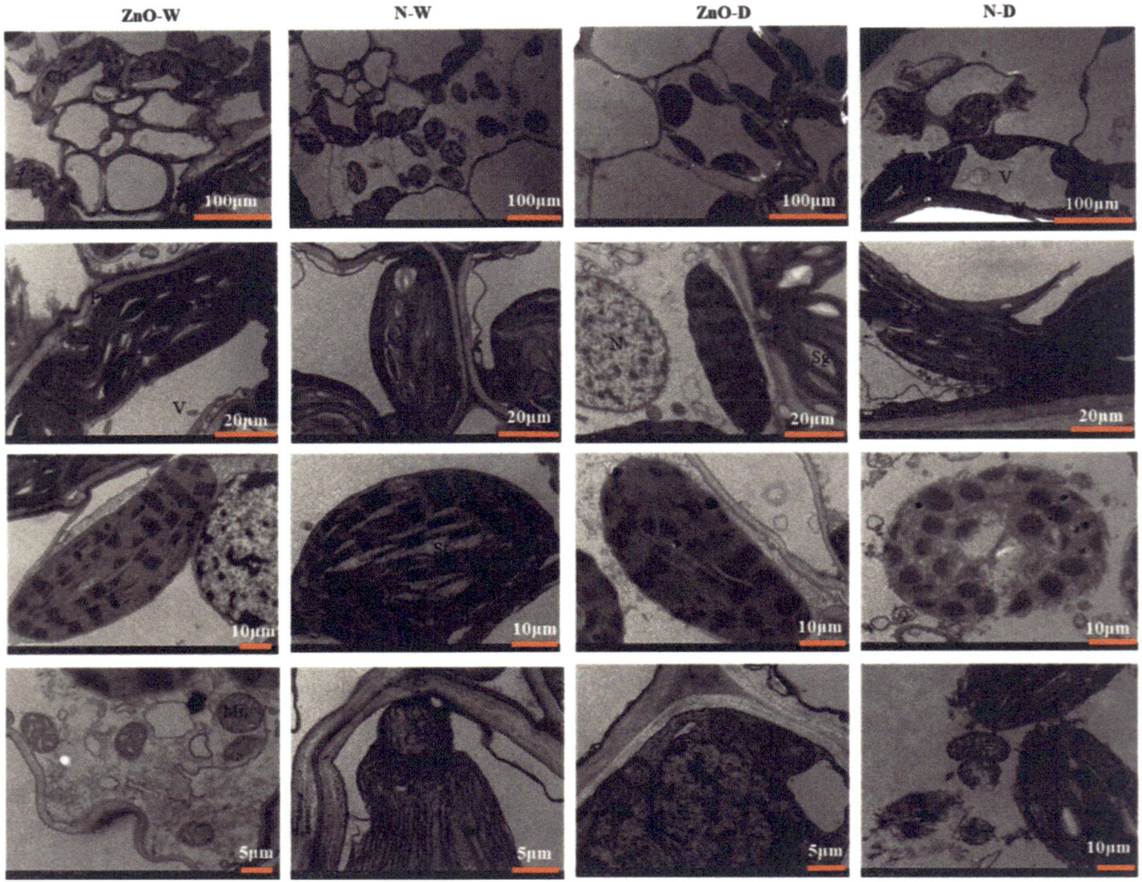

Fig. 7.3 TEM images of chloroplast and mitochondria of corn leaves affected by nano-ZnO and drought stress. *ZnO*, nano-ZnO treatment; *N*, non-ZnO treatment; *W*, normal water supply; *D*, drought stress; *Ch*, chloroplast; *Mi*, mitochondrion; *G*, granum; *SL*, stoma lamellae; *St*, stroma; *N*, nucleus; *V*, vacuolar; *Pm*, plasma membrane; *W*, cell wall; *Sg*, starch granule.

Reproduced with permission from Sun, L., Song, F., Guo, J., Zhu, X., Liu, S., Liu, F., Li, X., 2020. Nano-ZnO-induced drought tolerance is associated with melatonin synthesis and metabolism in maize. Int. J. Mol. Sci. 21, 782. Copyright © 2020 by the authors. Licensee MDPI, Basel, Switzerland.

production and lipid peroxidation (Chen et al., 2015). Nanoparticles mimic Ca^{2+}, bind with calcium-binding proteins, and induce stress responsive genes which may enhance tolerance under stress conditions (Mirzajani et al., 2014). Toxic impact of nanomaterials has been observed at high concentration whereas less amount of nanomaterials exhibits beneficial or no adverse impact on plants biochemical machinery. It indicates that less amount of nanomaterials controls normal functioning of antioxidant defense system which regulates free radicals production (Syu et al., 2014). However, these observations require intensive molecular level research to confirm nanomaterials' function as inducers or inhibitors of stress in plants.

7.8 Conclusion and future perspectives

Drought stress is a major future challenge for production of crops and food security. The present paper reveals that nanomaterials can ameliorate drought-stress-induced damages in plants by enhancing water and nutrients absorption and translocation, upregulation of aquaporins, reduction in water loss from leaves through stomatal closure due to abscisic acid accumulation, mitigation of osmotic stresses though accumulation of compatible solutes, increase in photosynthetic pigments and Rubisco activity, and boosting antioxidant defense system. Nanomaterials promoting drought resistance in plants under stress conditions need more studies to elucidate different plant responses such as phenological, anatomical, ecological, cytological, and molecular mechanisms. Nanoparticles assist site-targeted delivery of different nutrients required for the development of plants. However, there is a gap in our understanding about plant's absorption ability of nanoparticles and toxicity of nanomaterials. Hence, detailed investigations are needed to unravel exact mechanism of action, behavior and fate of nanoparticles, and their interactions with biomolecules in plant cells under water-deficit conditions. The nanoparticles utilization in agriculture may lead to sustainable development but its mode of application is still in its infancy. Therefore, understanding of interaction mechanism between nanoparticles and plants, optimization of size, and nanoparticles concentration and their role in gene expression in plants before their real application in the fields is required, so that their cost-effective preparation can be achieved by minimizing its possible negative impact on the crop plants and environment. The use of engineered nanomaterials has shown a completely new direction for sustainable agriculture which can reduce unpredictability in farming sector specifically under stress conditions.

Acknowledgments

The authors extend their appreciation to Amity University Noida and Universiti Sains Malaysia for providing all necessary facilities and valuable support.

References

Abid, G., Mhamdi, M., Mingeot, D., Aouida, M., Aroua, I., Muhovski, Y., Sassi, K., Souissi, F., Mannai, K., Jebara, M., 2017. Effect of drought stress on chlorophyll fluorescence, antioxidant enzyme activities and gene expression patterns in faba bean (*Vicia faba* L.). Arch. Agron. Soil Sci. 63, 536–552.

Abobatta, W., 2018. Impact of hydrogel polymer in agricultural sector. Adv. Agric. Environ. Sci. 1, 59–64.

Adhikari, T., Kundu, S., Rao, A.S., 2013. Impact of SiO_2 and Mo nano particles on seed germination of rice (*Oryza sativa* L.). Int. J. Agric. Food Sci. Technol. 4 (8), 809–816.

Adrees, M., Khan, Z.S., Ali, S., Hafeez, M., Khalid, S., Rehman, M.Z., Hussain, A., Hussain, K., Chatha, S.A.S., Rizwan, M., 2020. Simultaneous mitigation of cadmium and drought stress in wheat by soil application of iron nanoparticles. Chemosphere 238, 1–9.

Agathokleous, E., Feng, Z.Z., Ivo Iavicoli, I., Calabrese, E.J., 2019. The two faces of nanomaterials: a quantification of hormesis in algae and plants. Environ. Int. 131, 105044.

Aghdam, M.T.B., Mohammadi, H., Ghorbanpour, M., 2015. Effects of nanoparticulate anatase titanium dioxide on physiological and biochemical performance of *Linum usitatissimum* (Linaceae) under well-watered and drought stress conditions. Braz. J. Bot. 39, 139–146.

Aghdam, M.T.B., Mohammadi, H., Ghorbanpour, M., 2016. Effects of nanoparticulate anatase titanium dioxide on physiological and biochemical performance of *Linum usitatissimum* (Linaceae) under well-watered and drought stress conditions. Braz. J. Bot. 39, 139–146.

Ahmad, I., Kamran, M., Ali, S., Cai, T., Bilegjargal, B., Liu, T., Han, Q., 2018. Seed filling in maize and hormones crosstalk regulated by exogenous application of uniconazole in semi-arid regions. Environ. Sci. Pollut. Res. 25, 33225–33239.

Akbari, G.A., Morteza, E., Moaveni, P., Alahdadi, I., Bihamta, M.R., Hasanloo, T., 2014. Pigments apparatus and anthocyanins reactions of borage to irrigation, methylalcohol and titanium dioxide. Int. J. Biosci. 4 (7), 192–208.

Alidoust, D., Isoda, A., 2013. Effect of γFe_2O_3 nanoparticles on photosynthetic characteristic of soybean (*Glycine max* (L.) Merr.): foliar spray versus soil amendment. Acta Physiol. Plant. 35, 3365–3375.

Al-Jabari, M., Ghyadah, R.A., Alokely, R., 2019. Recovery of hydrogel from baby diaper wastes and its application for enhancing soil irrigation management. J. Environ. Manage. 239, 255–261.

Almutairi, Z.M., 2016. Effect of nanosilicon application on the expression of salt tolerance genes in germinating tomato (*Solanum lycopersicum* L.) seedlings under salt stress. Plant Omics 9 (1), 106–114.

Almutairi, Z.A., Alharbi, A., 2015. Effect of silver nanoparticles on seed germination of crop plants. J. Adv. Agric. 4, 280–285.

Alsaeedi, A., El-Ramady, H., Alshaal, T., El-Garawany, M., Elhawat, N., Al-Otaibi, A., 2019. Silica nanoparticles boost growth and productivity of cucumber under water deficit and salinity stresses by balancing nutrients uptake. Plant Physiol. Biochem. 139, 1–10.

Alvarez, S., Rodríguez, P., Broetto, F., Sanchez-Blanco, M.J., 2018. Long term responses and adaptive strategies of *Pistacia lentiscus* under moderate and severe deficit irrigation and salinity: osmotic and elastic adjustment, growth, ion uptake and photosynthetic activity. Agric. Water. Manag. 202, 253–262.

Aryal, J.P., Sapkota, T.B., Khurana, R., Khatri-Chhetri, A., Jat, M.L., 2020. Climate change and agriculture in South Asia: adaptation options in smallholder production systems. Environ. Dev. Sustain. 22, 5045–5075.

Ashkavand, P., Tabari, M., Zarafshar, M., Tomášková, I., Struve, D., 2015. Effect of SiO_2 nanoparticles on drought resistance in hawthorn seedlings. For. Res. Pap. 76 (4), 350–359.

Baby, A., Nazeerudeen, S., Ranganath, S., Samuel, R.S., 2018. Toxicological impacts of nanoparticles. In: Emerging Trends of Nanotechnology in Environment and Sustainability. Springer, Cham, pp. 77–85.

Bao-shan, L., Chun-hui, L., Li-jun, F., Shu-chun, Q., Min, Y., 2004. Effect of TMS (nanostructured silicon dioxide) on growth of Changbai larch seedlings. J. For. Res. 15 (2), 138–140.
Battaglia, M.L., Lee, C., Thomason, W., 2018. Corn yield components and yield responses to defoliation at different row widths. Agron. J. 110, 1–16.
Behboudi, F., Tahmasebi Sarvestani, Z., Zaman Kassaee, M., Modares Sanavi, S.A.M., Sorooshzadeh, A., 2018. Improving growth and yield of wheat under drought stress via application of SiO_2 nanoparticles. J. Agr. Sci. Tech. 20, 1479–1492.
Behboudi, F., Tahmasebi-Sarvestani, Z., Kassaee, M.Z., Mohammad Modarres-Sanavy, Sorooshzadeh, A., Mokhtassi-Bidgoli, A., 2019. Evaluation of chitosan nanoparticles effects with two application methods on wheat under drought stress. J. Plant. Nutr. 42 (13), 1439–1451.
Borisev, M., Borisev, I., Zupunski, M., Arsenov, D., Pajevic, S., Curcic, Z., Djordjevic, A., 2016. Drought impact is alleviated in sugar beets (*Beta vulgaris* L.) by foliar application of fullerenol nanoparticles. PLoS One 11 (11), e0166248.
Boudsocq, M., Sheen, J., 2013. CDPKs in immune and stress signaling. Trends Plant Sci. 18, 30–40.
Brestic, M., Zivcak, M., Hauptvogel, P., Misheva, S., Kocheva, K., Yang, X., Li, X., Allakhverdiev, S.I., 2018. Wheat plant selection for high yields entailed improvement of leaf anatomical and biochemical traits including tolerance to non-optimal temperature conditions. Photosynth. Res. 136, 245–255.
Brodersen, C.R., Roddy, A.B., Wason, J.W., McElrone, A.J., 2019. Functional status of xylem through time. Annu. Rev. Plant Biol. 70, 407–433.
Burman, U., Saini, M., Kumar, P., 2013. Effect of zinc oxide nanoparticles on growth and antioxidant system of chickpea seedlings. Toxicol. Environ. Chem. 95, 605–612.
Chandra, S., Chakraborty, N., Dasgupta, A., Sarkar, J., Panda, K., Acharya, K., 2015. Chitosan nanoparticles: a positive modulator of innate immune responses in plants. Sci. Rep. 5, 15195. https://doi.org/10.1038/srep15195.
Chen, J., Liu, X., Wang, C., Yin, S.S., Li, X.L., Hu, W.J., Simona, M., Shen, Z.J., Xiao, Q., Chu, C.C., Peng, X.X., Zheng, H.L., 2015. Nitric oxide ameliorates zinc oxide nanoparticles-induced phytotoxicity in rice seedlings. J. Hazard. Mater. 297, 173–182.
Chen, D., Wang, S., Yin, L., Deng, X., 2018. How does silicon mediate plant water uptake and loss under water deficiency? Front. Plant Sci. 9, 281.
Chhipa, H., 2017. Nanofertilizers and nanopesticides for agriculture. Environ. Chem. Lett. 15 (1), 15–22.
Chowdhury, J., Karim, M., Khaliq, Q., Ahmed, A., 2017. Effect of drought stress on biochemical change and cell membrane stability of soybean genotypes. Bangladesh J. Agric. Res. 42, 475–485.
Cohen, I., Zandalinas, S.I., Huck, C., Fritschi, F.B., Mittler, R., 2021. Meta-analysis of drought and heat stress combination impact on crop yield and yield components. Physiol. Plant. 171, 66–76.
Corpas, F.J., Barroso, J.B., Carreras, A., Valderrama, R., Palma, J.M., Leon, A.M., Sandalio, L. M., del Río, L.A., 2006. Constitutive arginine-dependent nitric oxide synthase activity in different organs of pea seedlings during plant development. Planta 224, 246–254.
Dahoumane, S.A., Jeffryes, C., Mechouet, M., Agathos, S.N., 2017. Biosynthesis of inorganic nanoparticles: a fresh look at the control of shape, size and composition. Bioengineering 4 (1), 14. https://doi.org/10.3390/bioengineering4010014.
Dat, J., Vandenabeele, S., Vranova, E., Van Montagu, M., Inze, D., Van Breusegem, F., 2000. Dual action of the active oxygen species during plant stress responses. Cell. Mol. Life Sci. 57 (5), 779–795.

Delfani, M., Baradarn Firouzabadi, M., Farrokhi, N., Makarian, H., 2014. Some physiological responses of black-eyed pea to iron and magnesium nanofertilizers. Commun. Soil Sci. Plant Anal. 45 (4), 530–540.

Dhoke, S.K., Mahajan, P., Kamble, R., Khanna, A., 2013. Effect of nanoparticles suspension on the growth of mung (*Vigna radiata*) seedlings by foliar spray method. Nanotechnol. Dev. 3, 1. https://doi.org/10.4081/nd.2013.e1.

Dimkpa, C., Bindraban, P., 2018. Nanofertilizers: new products for the industry? J. Agric. Food Chem. 66, 6462–6473.

Dimkpa, C.O., Bindraban, P.S., Fugice, J., Agyin-Birikorang, S., Singh, U., Hellums, D., 2017. Composite micronutrient nanoparticles and salts decrease drought stress in soybean. Agro. Sustain. Dev. 37 (1), 5.

Dimkpa, C.O., Singh, U., Bindraban, P.S., Elmer, W.H., Gardea-Torresdey, J.L., White, J.C., 2019. Zinc oxide nanoparticles alleviate drought-induced alterations in sorghum performance, nutrient acquisition, and grain fortification. Sci. Total Environ. 688, 926–934.

Djanaguiraman, M., Nair, R., Giraldo, J.P., Prasad, P.V.V., 2018. Cerium oxide nanoparticles decrease drought-induced oxidative damage in sorghum leading to higher photosynthesis and grain yield. ACS Omega 3 (10), 14406–14416.

Du, W., Tan, W., Peralta-Videa, J.R., Gardea-Torresdey, J.L., Ji, R., Yin, Y., Guo, H., 2017. Interaction of metal oxide nanoparticles with higher terrestrial plants: physiological and biochemical aspects. Plant Physiol. Biochem. 110, 110–225.

Ealias, A.M., Saravanakumar, M.P., 2017. A review on the classification, characterisation, synthesis of nanoparticles and their application. IOP Conf. Ser. Mater. Sci. Eng. 263, 032019.

Elemike, E.E., Uzoh, I.M., Onwudiwe, D.C., Babalola, O.O., 2019. The role of nanotechnology in the fortification of plant nutrients and improvement of crop production. Appl. Sci. 9, 499.

Faraji, J., Sepehri, A., 2020. Exogenous nitric oxide improves the protective effects of TiO_2 nanoparticles on growth, antioxidant system, and photosynthetic performance of wheat seedlings under drought stress. J. Soil Sci. Plant Nutr. 20, 703–714.

Farinaz, Z.D., Roozbahani, A., Hosnamidi, A., 2014. Evaluation the effect of water stress and foliar application of Fe nanoparticles on yield, yield components and oil percentage of safflower (*Carthamus tinctorius* L.). Int. J. Adv. Biol. Biomed. Res. 2, 1150–1159.

Farooq, M., Hussain, M., Wahid, A., Siddique, K.H.M., 2012. Drought stress in plants: an overview. In: Aroca, R. (Ed.), Plant Responses to Drought Stress. Springer-Verlag, Berlin, Heidelberg, pp. 1–33, https://doi.org/10.1007/978-3-642-32653-0-1.

Fitriani, H.P., Haryanti., 2016. Effect of the use of nanosilica fertilizer on the growth of tomato plant (*Solanum lycopersicum*). Bul. Anat. Dan. Fisiol. 24 (1), 34–41.

Fleta-Soriano, E., Diaz, L., Bonet, E., Munne-Bosch, S., 2017. Melatonin may exert a protective role against drought stress in maize. J. Agron. Crop Sci. 203, 286–294.

Frona, D., Szenderak, J., Harangi-Rakos, M., 2019. The challenge of feeding the world. Sustainability. 11, 5816. https://doi.org/10.3390/su11205816.

Galindo, A., Collado-Gonzalez, J., Griñán, I., Corell, M., Centeno, A., Martín-Palomo, M.J., Giron, I.F., Rodríguez, P., Cruz, Z.N., Memmi, H., 2020. Deficit irrigation and emerging fruit crops as a strategy to save water in Mediterranean semiarid agrosystems. Agric. Water Manag. 202, 311–324.

Gao, F., Hong, F., Liu, C., Zheng, L., Su, M., Wu, X., Yang, F., Wu, C., Yang, P., 2006. Mechanism of nano-anatase TiO_2 on promoting photosynthetic carbon reaction of spinach. Biol. Trace Elem. Res. 111, 239–253.

Gao, F., Liu, C., Qu, C., Zheng, L., Yang, F., Su, M., Hong, F., 2008. Was improvement of spinach growth by nano-TiO_2 treatment related to the changes of Rubisco activase? Biometals 21 (2), 211–217.

Gerdini, F.S., 2016. Effect of nano potassium fertilizer on some parchment pumpkin (Cucurbita pepo) morphological and physiological characteristics under drought conditions. Int. J. Farm. Alli. Sci. 5 (5), 367–371.

Ghafariyan, M.H., Malakouti, M.J., Dadpour, M.R., Stroeve, P., Mahmoudi, M., 2013. Effects of magnetite nanoparticles on soybean chlorophyll. Environ. Sci. Technol. 47, 10645–10652.

Ghasemlou, F., Amiri, H., Karamian, R., Mirzaie-asl, A., 2019. Alleviation of the effects of on drought stress *Verbascum nudicaule* by methyl jasmonate and titanium dioxide nanoparticles. Iran. J. Plant Physiol. 9 (8), 2911–2920.

Ghorbanian, H., Mohsen Janmohammadi, M., Ebadi-Segherloo, A., Sabaghnia, N., 2017. Genotypic response of barley to exogenous application of nanoparticles under water stress condition. Ann. Univ. Mariae Curie-Sklodowska Sect. C Biol. 72 (2), 1–13.

Ghorbanpour, M., Mohammadi, H., Kariman, K., 2020. Nanosilicon-based recovery of barley (*Hordeum vulgare*) plants subjected to drought stress. Environ. Sci. Nano 7 (2), 443–461.

Ghormade, V., Deshpande, M.V., Paknikar, K.M., 2011. Perspectives for nano-biotechnology enabled protection and nutrition of plants. Biotechnol. Adv. 29, 792–803.

Gil-Díaz, M., Pinilla, P., Alonso, J., Lobo, M.C., 2017. Viability of a nanoremediation process in single or multi-metal(loid) contaminated soils. J. Hazard. Mater. 321, 812–819.

Gray, S.B., Brady, S.M., 2016. Plant developmental responses to climate change. Dev. Biol. 419, 64–77.

Haghighi, M., Pessarakli, M., 2013. Influence of silicon and nano-silicon on salinity tolerance of cherry tomatoes (*Solanum lycopersicum* L.) at early growth stage. Sci. Hortic. 161, 111–117.

Hatami, M., Hadian, J., Ghorbanpour, M., 2017. Mechanisms underlying toxicity and stimulatory role of single-walled carbon nanotubes in Hyoscyamus niger during drought stress simulated by polyethylene glycol. J. Hazard. Mater. 15 (324), 306–320.

Hatami, M., Hosseini, S.M., Ghorbanpour, M., Kariman, K., 2019. Physiological and antioxidative responses to GO/PANI nanocomposite in intact and demucilaged seeds and young seedlings of *Salvia mirzayanii*. Chemosphere 233, 920–935.

Heinen, R.B., Ye, Q., Chaumont, F., 2009. Role of aquaporins in leaf physiology. J. Exp. Bot. 11, 2971–2985.

Hewedy, O.A., Abdel Lateif, K.S., Seleiman, M.F., Shami, A., Albarakaty, F.M., El-Meihy, R. M., 2020. Phylogenetic diversity of *Trichoderma* strains and their antagonistic potential against soil-borne pathogens under stress conditions. Biology 9, 189.

Hojjat, S.S., Ganjali, A., 2016. The effect of silver nanoparticle on lentil seed germination under drought stress. Int. J. Farm Allied Sci. 5 (3), 208–212.

Humplík, J.F., Bergougnoux, V., Van Volkenburgh, E., 2017. To stimulate or inhibit? That is the question for the function of abscisic acid. Trends Plant Sci. 22, 830–841.

Hurtado, A.C., Chiconato, D.A., de Mello Prado, R., Junior, G.D.S.S., Gratão, P.L., Felisberto, G., Viciedo, D.O., Dos Santos, D.M.M., 2020. Different methods of silicon application attenuate salt stress in sorghum and sunflower by modifying the antioxidative defense mechanism. Ecotoxicol. Environ. Saf. 203. https://doi.org/10.1016/j.ecoenv.2020.110964, 110964.

Husen, A., Siddiqi, K.S., 2014. Phytosynthesis of nanoparticles: concept, controversy and application. Nanoscale Res. Lett. 9, 229.

Hussain, A., Rizwan, M., Ali, Q., Ali, S., 2019. Seed priming with silicon nanoparticles improved the biomass and yield while reduced the oxidative stress and cadmium concentration in wheat grains. Environ. Sci. Pollut. Res. 26, 1–10.

Hussien, M.M., El-Ashry, S.M., Haggag, W.M., Mubarak, D.M., 2015. Response of mineral status to nano-fertilizer and moisture stress during different growth stages of cotton plants. Int. J. Chem. Tech. Res. 8, 643–650.

Ivani, R., Sanaei Nejad, S.H., Ghahraman, B., Astaraei, A.R., Feizi, H., 2018. Role of bulk and nanosized SiO_2 to overcome salt stress during fenugreek germination (*Trigonella foenum-graceum* L.). Plant Signal. Behav. 13 (7), e1044190.

Jaberzadeh, A., Moaveni, P., Moghadam, H.R.T., Zahedi, H., 2013. Influence of bulk and nanoparticles titanium foliar application on some agronomic traits, seed gluten and starch contents of wheat subjected to water deficit stress. Not. Bot. Horti. Agrobot. Cluj Napoca 41, 201–207.

Janmohammadi, M., Sabaghnia, N., 2015. Effect of pre-sowing seed treatments with silicon nanoparticles on germinability of sunflower (*Helianthus annuus*). Bot. Lith. 21 (1), 13–21.

Jung, H.I., Kong, M.S., Lee, B.R., Kim, T.H., Chae, M.J., Sung, J.K., Lee, C.H., Lee, E.J., Jung, G.B., Kim, Y.H., 2019. Exogenous glutathione increases arsenic translocation into shoots and alleviates arsenic-induced oxidative stress by sustaining ascorbate-glutathione homeostasis in rice seedlings. Front. Plant Sci. 10, 1089. https://doi.org/10.3389/fpls.2019.01089.

Kale, A.P., Gawade, S.N., 2016. Studies of nanoparticle induced nutrient use efficiency of fertilizer and crop productivity. Green Chem. Technol. Lett. 2 (2), 88–92.

Kalteh, M., Zarrin, T.A., Shahram, A., Maryam, M.A., Alireza, F.N., 2014. Effect of silica nanoparticles on basil (*Ocimum basilicum*) under salinity stress. J. Chem. Health Risks. 4 (3), 49–55.

Karami, A., Sepehri, A., 2017. Multiwalled carbon nanotubes and nitric oxide modulate the germination and early seedling growth of barley under drought and salinity. Agric. Conspec. Sci. 82, 331–339.

Karimi, J., Mohsenzadeh, S., 2016. Effects of silicon oxide nanoparticles on growth and physiology of wheat seedlings. Russ. J. Plant Physiol. 63 (1), 119–123.

Kashyap, P.L., Xiang, X., Heiden, P., 2015. Chitosan nanoparticle based delivery systems for sustainable agriculture. Int. J. Biol. Macromol. 77, 36–51.

Kaya, C., Okant, M., Ugurlar, F., Alyemeni, M.N., Ashraf, M., Ahmad, P., 2019. Melatonin-mediated nitric oxide improves tolerance to cadmium toxicity by reducing oxidative stress in wheat plants. Chemosphere 225, 627–638.

Khalaki, M., Ghorbani, A., Moameri, M., 2016. Effects of silica and silver nanoparticles on seed germination traits of *Thymus kotschyanus* in laboratory conditions. J. Rangel. Sci. 6 (3), 221–231.

Khan, Z., Upadhyaya, H., 2019. Impact of nanoparticles on abiotic stress responses in plants: an overview. In: Nanomaterials in Plants, Algae and Microorganisms. Academic Press, Cambridge, pp. 305–322.

Khan, N.A., Nazar, R., Iqbal, N., Anjum, N.A., 2012. Phytohormones and Abiotic Stress Tolerance in Plants. Springer, Berlin, Heidelberg, https://doi.org/10.1007/978-3-642-25829-9.

Khan, M.N., Mohammad, F., Mobin, M., Saqib, M.A., 2014. Tolerance of plants to abiotic stress: a role of nitric oxide and calcium. In: Khan, M.N., Mobin, M., Mohammad, F., Corpas, F.J. (Eds.), Nitric Oxide in Plants: Metabolism and Role. Springer International Publishing, Switzerland.

Khan, I., Saeed, K., Khan, I., 2017. Nanoparticles: properties, applications and toxicities. Arab. J. Chem. 12 (7), 908–931.

Khan, N., Bano, A., Shahid, M.A., Nasim, W., Babar, M.A., 2018. Interaction between PGPR and PGR for water conservation and plant growth attributes under drought condition. Biologia 73, 1083–1098.

Khan, N., Bano, A., Ali, S., Babar, M.A., 2020. Crosstalk amongst phytohormones from planta and PGPR under biotic and abiotic stresses. Plant Growth Regul. 90, 189–203.

Khodakovskaya, M.V., de Silva, K., Biris, A.S., Dervishi, E., Villagarcia, H., 2012. Carbon nanotubes induce growth enhancement of tobacco cells. ACS Nano 6, 2128–2135.

Kiapour, H., Moaveni, P., Habibi, D., Sani, B., 2015. Evaluation of the application of gibbrellic acid and titanium dioxide nanoparticles under drought stress on some traits of basil (*Ocimum basilicum* L.). Int. J. Agron. Agric. Res. 6 (4), 138–150.

Kim, J.H., Oh, Y., Yoon, H., Hwang, I., Chang, Y.S., 2015. Iron nanoparticle-induced activation of plasma membrane H^+-ATPase promotes stomatal opening in *Arabidopsis thaliana*. Environ. Sci. Technol. 4, 1113–1119.

Kollist, H., Zandalinas, S.I., Sengupta, S., Nuhkat, M., Kangasjärvi, J., Mittler, R., 2019. Rapid responses to abiotic stress: priming the landscape for the signal transduction network. Trends Plant Sci. 24 (1), 25–37.

Konapala, G., Mishra, A.K., Wada, Y., Mann, M.E., 2020. Climate change will affect global water availability through compounding changes in seasonal precipitation and evaporation. Nat. Commun. 11, 1–10.

Kou, T.J., Yu, W.W., Lam, S.K., Chen, D.L., Hou, Y.P., Li, Z.Y., 2018. Differential root responses in two cultivars of winter wheat (*Triticum aestivum* L.) to elevated ozone concentration under fully open-air field conditions. J. Agron. Crop Sci. 204, 325–332.

Kreuter, J., 2007. Nanoparticles—a historical perspective. Int. J. Pharm. 331 (1), 1–10. https://doi.org/10.1016/j.ijpharm.2006.10.021.

Kumar, A., Singh, S., Gaurav, A.K., Srivastava, S., Verma, J.P., 2020. Plant growth-promoting bacteria: biological tools for the mitigation of salinity stress in plants. Front. Microbiol. 11, 1216. https://doi.org/10.3389/fmicb.2020.01216.

Kwak, J.I., An, Y.J., 2016. The current state of the art in research on engineered nanomaterials and terrestrial environments: different-scale approaches. Environ. Res. 151, 368–382.

Kwak, S.Y., Lew, T.T.S., Sweeney, C.J., Koman, V.B., Wong, M.H., Bohmert-Tatarev, K., Snell, K.D., Seo, J.S., Chua, N.H., Strano, M.S., 2019. Chloroplast-selective gene delivery and expression in planta using chitosan-complexed single-walled carbon nanotube carriers. Nat. Nanotechnol. 14 (5), 447–455.

Lahiani, M.H., Dervishi, E., Chen, J., Nima, Z., Gaume, A., Biris, A.S., Khodakovskaya, M.V., 2013. Impact of carbon nanotube exposure to seeds of valuable crops. ACS Appl. Mater. Interfaces 5, 7965–7973.

Laware, S.L., Raskar, S., 2014. Effect of titanium dioxide nanoparticles on hydrolytic and antioxidant enzymes during seed germination in onion. Int. J. Curr. Microbiol. App. Sci. 3 (7), 749–760.

Liang, D., Ni, Z., Xia, H., Xie, Y., Lv, X., Wang, J., Lin, L., Deng, Q., Luo, X., 2019. Exogenous melatonin promotes biomass accumulation and photosynthesis of kiwifruit seedlings under drought stress. Sci. Hortic. 246, 34–43.

Liu, J., Zhang, Y., Zhang, Z., 2009. The application research of nano-biotechnology to promote increasing of vegetable production. Hubei Agric. Sci. 1, 041.

Liu, F.Y., Xiong, F.X., Fan, Y.K., Li, J., Wang, H.Z., Xing, G.M., He, R., 2016. Facile and scalable fabrication engineering of fullerenol nanoparticles by improved alkaline-oxidation approach and its antioxidant potential in maize. J. Nanopart. Res. 18 (11), 338. https://doi.org/10.1007/s11051-016-3642-4.

Lowry, G.V., Avellan, A., Gilbertson, L.M., 2019. Opportunities and challenges for nanotechnology in the agri-tech revolution. Nat. Nanotechnol. 14 (6), 517–522.

Lu, C.M., Zhang, C.Y., Wu, J.Q., Tao, M.X., 2002. Research of the effect of nanometer on germination and growth enhancement of *Glycine max* and its mechanism. Soybean Sci. 21, 168–172.

Mahmoodzadeh, H., Nabavi, M., Kashefi, H., 2013. Effect of nanoscale titanium dioxide particles on the germination and growth of canola (*Brassica napus*). J. Ornam. Hortic. Plants 3, 25–32.

Mahmoud, L.M., Dutt, M., Shalan, A.M., El-Kadya, M.E., El-Boraya, M.S., Shabana, Y.M., Grosser, J.W., 2020. Silicon nanoparticles mitigate oxidative stress of in vitro-derived banana (Musa acuminata 'Grand Nain') under simulated water deficit or salinity stress. South Afr. J. Bot. 132, 155–163.

Marcinska, I., Czyczyło-Mysza, I., Skrzypek, E., Filek, M., Grzesiak, S., Grzesiak, M.T., Janowiak, F., Hura, T., Dziurka, M., Dziurka, K., 2013. Impact of osmotic stress on physiological and biochemical characteristics in drought-susceptible and drought-resistant wheat genotypes. Acta Physiol. Plant. 35, 451–461.

Marmiroli, M., Imperiale, D., Pagano, L., Villani, M., Zappettini, A., Marmiroli, N., 2015. The proteomic response of *Arabidopsis thaliana* to cadmium sulphide quantum dots, and its correlation with the transcriptomic response. Front. Plant Sci. 6, 1104. https://doi.org/10.3389/fpls.2015.01104.

Martínez-Fernández, D., Vítkova, M., Bernal, M.P., Komárek, M., 2015. Effects of nano-maghemite on trace element accumulation and drought response of *Helianthus annuus* L. in a contaminated mine soil. Water Air Soil Pollut. 226, 1–9.

Maswada, H.F., Mazrou, Y.S., Elzaawely, A.A., Alam-Eldein, S.M., 2020. Nanomaterials. Effective tools for field and horticultural crops to cope with drought stress: a review. Span. J. Agric. Res. 18, 15. https://doi.org/10.5424/sjar/2020182-16181.

Miao, Y., Xu, J., Shen, Y., Chen, L., Bian, Y., Hu, Y., Zhou, W., Zheng, F., Man, N., Shen, Y., Zhang, Y., Wang, M., Wen, L., 2014. Nanoparticle as signaling protein mimic: robust structural and functional modulation of CaMKII upon specific binding to fullerene C60 nanocrystals. ACS Nano 8, 6131–6144.

Mirzajani, F., Askari, H., Hamzelou, S., Schober, Y., Reompp, A., Ghassempour, A., Spengler, B., 2014. Proteomics study of silver nanoparticles toxicity on *Oryza sativa* L. Ecotoxicol. Environ. Saf. 108, 335–339.

Mohammadi, H., Esmailpour, M., Gheranpaye, A., 2016. Effects of TiO_2 nanoparticles and water-deficit stress on morpho-physiological characteristics of dragonhead (*Dracocephalum moldavica* L.) plants. Acta Agric. Slov. 107 (2), 385–396.

Monica, R.C., Cremonini, R., 2009. Nanoparticles and higher plants. Caryologia 62 (2), 161–165.

Monreal, C.M., DeRosa, M., Mallubhotla, S.C., Bindraban, P.S., Dimkpa, C., 2016. Nanotechnologies for increasing the crop use efficiency of fertilizer-micronutrients. Biol. Fertil. Soils 52, 423–437.

Mozafari, A., Dedejani, A.S., Ghaderi, N., 2018. Positive responses of strawberry (Fragaria × ananassa Duch.) explants to salicylic and iron nanoparticle application under salinity conditions. Plant Cell Tissue Organ Cult 134 (2), 267–275.

Navarro, E., Baun, A., Behra, R., Hartmann, N.B., Filser, J., Miao, A., Quigg, A., Santschi, P.H., Sigg, L., 2008. Environmental behavior and ecotoxicity of engineered nanoparticles to algae, plants, and fungi. Ecotoxicology 17, 372–386.

Okorie, V.O., Mphambukeli, T.N., Amusan, S.O., 2019. Exploring the political economy of water and food security nexus in BRICS. Afr. Insight 48, 21–38.

Oliveira, I.C.M., Guilhen, J.H.S., de Oliveira Ribeiro, P.C., Gezan, S.A., Schaffert, R.E., Simeone, M.L.F., Pastina, M.M., 2020. Genotype-by-environment interaction and yield stability analysis of biomass sorghum hybrids using factor analytic models and environmental covariates. Field Crop Res. 257, 107929.

Owolade, O.F., Ogunleti, D.O., Adenekan, M.O., 2008. Titanium dioxide affects diseases, development and yield of edible cowpea. Elec. J. Env. Agricult. Food Chem. 7 (5), 2942–2947.

Palchoudhury, S., Jungjohann, K.L., Weerasena, L., Arabshahi, A., Gharge, U., Albattah, A., Miller, J., Patel, K., Holler, R.A., 2018. Enhanced legume root growth with pre-soaking in α-Fe_2O_3 nanoparticle fertilizer. RSC Adv. 8 (43), 24075–24083.

Palmqvist, N.G.M., Seisenbaeva, G.A., Svedlindh, P., Kessler, V.G., 2017. Maghemite nanoparticles acts as nanozymes, improving growth and abiotic stress tolerance in *Brassica napus*. Nanoscale Res. Lett. 12 (1), 631. https://doi.org/10.1186/s11671-017-2404-2.

Pandey, K., Anas, M., Hicks, V.K., Green, M.J., Khodakovskaya, M.V., 2019. Improvement of commercially valuable traits of industrial crops by application of carbon-based nanomaterials. Sci. Rep. 9, 19358.

Potocnik, J., 2011. Off. J. Eur. Communities: Legis L275, 38–40. https://doi.org/10.3000/19770677.L_2011.275.eng.

Pradhan, S., Patra, P., Das, S., Chandra, S., Mitra, S., Dey, K.K., Akbar, S., Palit, P., Goswami, A., 2013. Photochemical modulation of biosafe manganese nanoparticles on *Vigna radiata*: a detailed molecular, biochemical, and biophysical study. Environ. Sci. Technol. 47, 13122–13131.

Prathyusha, I.V.S.N., Chaitanya, K.V., 2019. Effect of water stress on the physiological and biochemical responses of two different *Coleus* (*Plectranthus*) species. Biol. Fut. 70, 312–322.

Pulimi, M., Subramanian, S., 2016. Nanomaterials for soil fertilisation and contaminant removal. In: Ranjan, S., et al. (Eds.), Nanoscience in Food and Agriculture 1. Sustainable Agriculture Reviews, vol. 20. Springer International Publishing, Switzerland, pp. 229–246, https://doi.org/10.1007/978-3-319-39303-2_8.

Rastogi, A., Zivcak, M., Sytar, O., Kalaji, H.M., He, X., Mbarki, S., Brestic, M., 2017. Impact of metal and metal oxide nanoparticles on plant: a critical review. Front. Chem. 5, 78. https://doi.org/10.3389/fchem.2017.00078.

Rathore, I., Tarafdar, J., 2015. Perspectives of biosynthesized magnesium nanoparticles in foliar application of wheat plant. J. Bionanosci. 9, 209–214.

Ray, D.K., West, P.C., Clark, M., Gerber, J.S., Prishchepov, A.V., Chatterjee, S., 2019. Climate change has likely already affected global food production. PLoS One 14. https://doi.org/10.1371/journal.pone.0217148, 217148.

Ristvey, A.G., Belayneh, B.E., Lea-Cox, J.D., 2019. A comparison of irrigation-water containment methods and management strategies between two ornamental production systems to minimize water security threats. Water 11, 2558. https://doi.org/10.3390/w11122558.

Rizwan, M., Ali, S., ur Rehman, M.Z., Malik, S., Adrees, M., Qayyum, M.F., Alamri, S.A., Alyemeni, M.N., Ahmad, P., 2019. Effect of foliar applications of silicon and titanium dioxide nanoparticles on growth, oxidative stress, and cadmium accumulation by rice (*Oryza sativa*). Acta Physiol. Plant. 41, 35. https://doi.org/10.1007/s11738-019-2828-7.

Rodrigues, S.M., Demokritou, P., Dokoozlian, N., Hendren, C.O., Karn, B., Mauter, M.S., Sadik, O.A., Safarpour, M., Unrine, J.M., Viers, J., Welle, P., White, J.C., Wiesner, M. R., Lowry, G.V., 2017. Nanotechnology for sustainable food production: promising opportunities and scientific challenges. Environ. Sci. Nano 4 (4), 767–781.

Roos, E., Bajzelj, B., Smith, P., Patel, M., Little, D., Garnett, T., 2017. Greedy or needy? Land use and climate impacts of food in 2050 under different livestock futures. Glob. Environ. Chang. 47, 1–12.

Saharan, V., Kumaraswamy, R.V., Choudhary, R.C., Kumari, S., Pal, A., Raliya, R., Biswas, P., 2016. Cu-chitosan nanoparticle mediated sustainable approach to enhance seedling growth in maize by mobilizing reserved food. J. Agric. Food Chem. 64 (31), 6148–6155.

Salehi-Lisar, S.Y., Bakhshayeshan-Agdam, H., 2016. Drought stress in plants: causes, consequences, and tolerance. In: Drought Stress Tolerance in Plants. Springer, Cham, Switzerland.

Sallam, A., Alqudah, A.M., Dawood, M.F., Baenziger, P.S., Borner, A., 2019. Drought stress tolerance in wheat and barley: advances in physiology, breeding and genetics research. Int. J. Mol. Sci. 20, 3137. https://doi.org/10.3390/ijms20133137.

Schulze, E.D., Beck, E., Buchmann, N., Clemens, S., Müller-Hohenstein, K., Scherer-Lorenzen, M., 2019. Water deficiency (drought). In: Plant Ecology. Springer, Berlin, pp. 165–202.

Sedaghat, M., Tahmasebi-Sarvestani, Z., Emam, Y., Mokhtassi-Bidgoli, A., 2017. Physiological and antioxidant responses of winter wheat cultivars to strigolactone and salicylic acid in drought. Plant Physiol. Biochem. 119, 59–69.

Sedghi, M., Hadi, M., Toluie, S.G., 2013. Effect of nano zinc oxide on the germination parameters of soybean seeds under drought stress. Ann. West Univ. Timişoara Ser. Biol. 16, 73–78.

Seghatoleslami, M., Feizi, H., Mousavi, G., Berahmand, A., 2015. Effect of magnetic field and silver nanoparticles on yield and water use efficiency of *Carum copticum* under water stress conditions. Pol. J. Chem. Technol. 17 (1), 110–114.

Seghatoleslami, M., Forutani, R., 2015. Yield and water use efficiency of sunflower as affected by nano ZnO and water stress. J. Adv. Agric. Technol. 2, 34–37.

Seleiman, M.F., Refay, Y., Al-Suhaibani, N., Al-Ashkar, I., El-Hendawy, S., Hafez, E.M., 2019. Integrative effects of rice-straw biochar and silicon on oil and seed quality, yield and physiological traits of *Helianthus annuus* L. grown under water deficit stress. Agronomy 9, 637. https://doi.org/10.3390/agronomy9100637.

Seleiman, M.F., Almutairi, K.F., Alotaibi, M., Shami, A., Alhammad, B.A., Battaglia, M.L., 2020. Nano fertilization as an emerging fertilization technique: why modern agriculture can benefit from its use? Plan. Theory 10, 2. https://doi.org/10.3390/plants10010002.

Semida, W.M., Abdelkhalik, A., Rady, M.O., Marey, R.A., Abd El-Mageed, T.A., 2020. Exogenously applied proline enhances growth and productivity of drought stressed onion by improving photosynthetic efficiency, water use efficiency and up-regulating osmoprotectants. Sci. Hortic. 272, 109580.

Shallan, M.A., Hassan, H.M., Namich, A.A., Ibrahim, A.A., 2016. Biochemical and physiological effects of TiO_2 and SiO_2 nanoparticles on cotton plant under drought stress. Res. J. Pharm. Biol. Chem. Sci. 7 (4), 1540–1551.

Sharif, F., Westerhoff, P., Herckes, P., 2013. Sorption of trace organics and engineered nanomaterials on to wet land plant material. Environ Sci Process Impacts 15 (1), 267–274.

Sharif, R., Xie, C., Zhang, H., Arnao, M., Ali, M., Ali, Q., Muhammad, I., Shalmani, A., Nawaz, M.A., Chen, P., 2018. Melatonin and its effects on plant systems. Molecules 23, 2352. https://doi.org/10.3390/molecules23092352.

Sharifi, R.J., Sharifirad, M., Teixeira, D.S.J., 2016. Morphological, physiological and biochemical responses of crops (*Zea mays* L., *Phaseolus vulgaris* L.), medicinal plants (*Hyssopus officinalis* L., *Nigella sativa* L.), and weeds (*Amaranthus retroflexus* L., *Taraxacum officinale* FH Wigg) exposed to SiO_2 nanoparticles. J. Agric. Sci. Technol. 18, 1027–1040.

Sharma, A., Wang, J., Xu, D., Tao, S., Chong, S., Yan, D., Li, Z., Yuan, H., Zheng, B., 2020. Melatonin regulates the functional components of photosynthesis, antioxidant system, gene expression, and metabolic pathways to induce drought resistance in grafted *Carya cathayensis* plants. Sci. Total Environ. 713. https://doi.org/10.1016/j.scitotenv.2020.136675, 136675.

Siddiqi, K.S., Husen, A., 2016. Engineered gold nanoparticles and plant adaptation potential. Nanoscale Res. Lett. 11, 400. https://doi.org/10.1186/s11671-016-1607-2.

Siddiqui, M.H., Al-Whaibi, M.H., 2014. Role of nano-SiO_2 in germination of tomato (*Lycopersicum esculentum* seeds Mill.). Saudi J. Biol. Sci. 21, 13–17.

Siddiqui, M.H., Al-Whaibi, M.H., Faisal, M., Al Sahli, A.A., 2014. Nano-silicon dioxide mitigates the adverse effects of salt stress on *Cucurbita pepo* L. Environ. Toxicol. Chem. 33, 2429–2437.

Simon, D.F., Domingos, R.F., Hauser, C., Hutchins, C.M., Zerges, W., Wilkinson, K.J., 2013. Transcriptome sequencing (RNA-seq) analysis of the effects of metal nanoparticle exposure on the transcriptome of *Chlamydomonas reinhardtii*. Appl. Environ. Microbiol. 79, 4774–4785.

Singh, J., Thakur, J.K., 2018. Photosynthesis and abiotic stress in plants. In: Biotic and Abiotic Stress Tolerance in Plants. Springer, Singapore, pp. 27–46.

Soliman, A.S., El-feky, S.A., Darwish, E., 2015. Alleviation of salt stress on Moringa peregrine using foliar application of nanofertilizers. J. Hortic. For. 7 (2), 36–47.

Soltani, M., Kafi, M., Nezami, A., Taghiyari, H.R., 2018. Effects of silicon application at nano and micro scales on the growth and nutrient uptake of potato minitubers (*Solanum tuberosum* var. Agria) in greenhouse conditions. Bio Nano Sci. 8, 218–228.

Srivastava, A., Rao, D., 2014. Enhancement of seed germination and plant growth of wheat, maize, peanut and garlic using multiwalled carbon nanotubes. Eur. Chem. Bull. 3, 502–504.

Sturikova, H., Krystofova, O., Huska, D., Adam, V., 2018. Zinc, zinc nanoparticles and plants. J. Hazard. Mater. 349, 101–110.

Sudha, P.N., Sangeetha, K., Vijayalakshmi, K., Barhoum, A., 2018. Nanomaterials history, classification, unique properties, production and market. In: Emerging Applications of Nanoparticles and Architecture Nanostructures. Elsevier, Amsterdam, pp. 341–384.

Sun, L., Song, F., Guo, J., Zhu, X., Liu, S., Liu, F., Xiangnan Li, X., 2020. Nano-ZnO-induced drought tolerance is associated with melatonin synthesis and metabolism in maize. Int. J. Mol. Sci. 21, 782. https://doi.org/10.3390/ijms21030782.

Suriyaprabha, R., Karunakaran, G., Yuvakkumar, R., Prabu, P., Rajendran, V., Kannan, N., 2012. Growth and physiological responses of maize (*Zea mays* L.) to porous silica nanoparticles in soil. J. Nanopart. Res. 14 (12), 1294. https://doi.org/10.1007/s11051-012-1294-6.

Suzuki, N., Rivero, R.M., Shulaev, V., Blumwald, E., Mittler, R., 2014. Abiotic and biotic stress combinations. New Phytol. 203, 32–43.

Syu, Y.Y., Hung, J.H., Chen, J.C., Chuang, H.W., 2014. Impact of size and shape of silver nanoparticles on *Arabidopsis* plant growth and gene expression. Plant Physiol. Biochem. 83, 57 61.

Takahashi, F., Kuromori, T., Sato, H., Shinozaki, K., 2018. Regulatory gene networks in drought stress responses and resistance in plants. In: Survival Strategies in Extreme Cold and Desiccation. Springer, Singapore, pp. 189–214.

Tanveer, M., Shahzad, B., Sharma, A., Khan, E.A., 2019. 24-Epibrassinolide application in plants: an implication for improving drought stress tolerance in plants. Plant Physiol. Biochem. 135, 295–303.

Taran, N., Storozhenko, V., Svietlova, N., Batsmanova, L., Shvartau, V., Kovalenko, M., 2017. Effect of zinc and copper nanoparticles on drought resistance of wheat seedlings. Nanoscale Res. Lett. 12, 60. https://doi.org/10.1186/s11671-017-1839-9.

Tiwari, M., Sharma, N.C., Fleischmann, P., Burbage, J., Venkatachalam, P., Sahi, S.V., 2017. Nanotitania exposure causes alterations in physiological, nutritional and stress responses in tomato (*Solanum lycopersicum*). Front. Plant Sci. 8, 633. https://doi.org/10.3389/fpls.2017.00633.

Tiwari, P., Srivastava, D., Chauhan, A.S., Indoliya, Y., Singh, P.K., Tiwari, S., Fatima, T., Mishra, S.K., Dwivedi, S., Agarwal, L., Singh, P.C., Asif, M.H., Tripathi, R.D., Shirke, P.A., Chakrabarty, D., Chauhan, P.S., Nautiyal, C.S., 2020. Root system architecture, physiological analysis and dynamic transcriptomics unravel the drought-responsive traits in rice genotypes. Ecotoxicol. Environ. Saf. 207. https://doi.org/10.1016/j.ecoenv.2020.111252, 111252.

Tripathi, D.K., Singh, S., Singh, S., Pandey, R., Singh, V.P., Sharma, N.C., Chauhan, D.K., 2017. An overview on manufactured nanoparticles in plants: uptake, translocation, accumulation and phytotoxicity. Plant Physiol. Biochem. 110, 2–12.

Unrine, J.M., Colman, B.P., Bone, A.J., Gondikas, A.P., Matson, C.W., 2012. Biotic and abiotic interactions in aquatic microcosms determine fate and toxicity of Ag nanoparticles. Part 1. Aggregation and dissolution. Environ. Sci. Technol. 46 (13), 6915–6924.

Van Nguyen, D., Nguyen, H.M., Le, N.T., Nguyen, K.H., Nguyen, H.T., Le, H.M., Nguyen, A. T., Dinh, N.T.T., Hoang, S.A., Van Ha, C., 2021. Copper nanoparticle application enhances plant growth and grain yield in maize under drought stress conditions. J. Plant Growth Regul. 40, 1–12.

Vishwakarma, K., Upadhyay, N., Kumar, N., Tripathi, D.K., Chauhan, D.K., Sharma, S., Sahi, S., 2018. Potential applications and avenues of nanotechnology in sustainable agriculture. In: Nanomaterials in Plants, Algae, and Microorganisms. Academic Press, Cambridge, pp. 473–500.

Walkey, C., Das, S., Seal, S., Erlichman, J., Heckman, K., Ghibelli, L., Traversa, E., McGinnis, J.F., Self, W.T., 2015. Catalytic properties and biomedical applications of cerium oxide nanoparticles. Environ. Sci. Nano 2 (1), 33–53.

Wei, H., Wang, E., 2013. Nanomaterials with enzyme-like characteristics (nanozymes): next-generation artificial enzymes. Chem. Soc. Rev. 42, 6060–6093.

Xiang, D.B., Peng, L.X., Zhao, J.L., Zou, L., Zhao, G., Song, C., 2013. Effect of drought stress on yield, chlorophyll contents and photosynthesis in tartary buckwheat (*Fagopyrum tataricum*). J. Food Agric. Environ. 11, 1358–1363.

Xiong, J.L., Li, J., Wang, H.C., Zhang, C.L., Naeem, M.S., 2018. Fullerol improves seed germination, biomass accumulation, photosynthesis and antioxidant system in *Brassica napus* L. under water stress. Plant Physiol. Biochem. 129, 130–140.

Yadav, T., Mungray, A.A., Mungray, A.K., 2014. Fabricated nanoparticles: current status and potential phytotoxic threats. In: Whitacre, D.M. (Ed.), Reviews of Environmental Contamination and Toxicology. Springer International Publishing, Switzerland.

Yang, F., Hong, F., You, W., Liu, C., Gao, F., Wu, C., Yang, P., 2006. Influence of nano-anatase TiO_2 on the nitrogen metabolism of growing spinach. Biol. Trace Elem. Res. 110 (2), 179–190.

Yang, K.Y., Doxey, S., McLean, J.E., Britt, D., Watson, A., Al Qassy, D., Anderson, A.J., 2017. Remodeling of root morphology by CuO and ZnO nanoparticles: effects on drought tolerance for plants colonized by a beneficial pseudomonad. Botany 96 (3), 175–186.

Yang, H., Huntingford, C., Wiltshire, A., Sitch, S., Mercado, L., 2019. Compensatory climate effects link trends in global runoff to rising atmospheric CO_2 concentration. Environ. Res. Lett. 14. https://doi.org/10.1088/1748-9326/ab5c6f, 124075.

Yassen, A., Abdallah, E., Gaballah, M., Zaghloul, S., 2017. Role of silicon dioxide nano fertilizer in mitigating salt stress on growth, yield and chemical composition of cucumber (*Cucumis sativus* L.). Int. J. Agric. Res. 22, 130–135.

Yılmaz, E., Ozgur, E., Bereli, N., Turkmen, D., Denizli, A., 2017. Plastic antibody based surface plasmon resonance nanosensors for selective atrazine detection. Mater. Sci. Eng. C 73, 603–610.

Yin, J., Gentine, P., Zhou, S., Sullivan, S.C., Wang, R., Zhang, Y., Guo, S., 2018. Large increase in global storm runoff extremes driven by climate and anthropogenic changes. Nat. Commun. 9, 1–10.

Yoon, H., Kang, Y.G., Chang, Y.S., Kim, J.H., 2019. Effects of zerovalent iron nanoparticles on photosynthesis and biochemical adaptation of soil-grown *Arabidopsis thaliana*. Nanomaterials 9, 1543.

Yousefi, S., Kartoolinejad, D., Naghdi, R., 2017. Effects of priming with multi-walled carbon nanotubes on seed physiological characteristics of Hopbush (*Dodonaea viscosa* L.) under drought stress. Int. J. Environ. Stud. 74 (4), 528–539.

Yu-Nam, Y., Lead, R., 2008. Manufactured nanoparticles: an overview of their chemistry, interactions and potential environmental implications. Sci. Total Environ. 400, 396–414.

Yuvakkumar, R., Elango, V., Rajendran, V., Kannan, N.S., Prabu, P., 2011. Influence of nanosilica powder on the growth of maize crop (*Zea mays* L.). Int. J. Green Nanotechnol. 3 (3), 180–190.

Zaimenko, N.V., Didyk, N.P., Dzyuba, O.I., Zakrasov, O.V., Rositska, N.V., Viter, A.V., 2014. Enhancement of drought resistance in wheat and corn by nanoparticles of natural mineral analcite. Ecol. Balk. 6, 1–10.

Ze, Y., Liu, C., Wang, L., Hong, M., Hong, F., 2011. The regulation of TiO_2 nanoparticles on the expression of light-harvesting complex II and photosynthesis of chloroplasts of *Arabidopsis thaliana*. Biol. Trace Elem. Res. 143, 1131–1141.

Zhai, G., Walters, K.S., Peate, D.W., Alvarez, P.J., Schnoor, J.L., 2014. Transport of gold nanoparticles through plasmodesmata and precipitation of gold ions in woody poplar. Environ. Sci. Technol. Lett. 1 (2), 146–151.

Zhang, L.W., Monteiro-Riviere, N.A., 2009. Mechanisms of quantum dot nanoparticle cellular uptake. Toxicol. Sci. 110 (1), 138–155.

Nanoparticles for effective management of salinity stress in plants

8

Rahul Nitnavare[a,b], *Joorie Bhattacharya*[c,d], *and Sougata Ghosh*[e,f]
[a]Division of Plant and Crop Sciences, School of Biosciences, University of Nottingham, Leicestershire, United Kingdom, [b]Department of Plant Sciences, Rothamsted Research, Hertfordshire, United Kingdom, [c]Genetic Gains, International Crops Research Institute for the Semi-Arid Tropics, Hyderabad, Telangana, India, [d]Department of Genetics, Osmania University, Hyderabad, Telangana, India, [e]Department of Microbiology, School of Science, RK University, Rajkot, Gujarat, India, [f]Department of Physics, Faculty of Science, Kasetsart University, Bangkok, Thailand

8.1 Introduction

Salt stress is one of the most severe abiotic factors which impact the agricultural productivity. High concentrations of salt in the soil limit crop yield and also the usage of uncultivated land. The prevalence of salinization is more in arid and semiarid regions where due to low rainfall the leaching of soil is deprived and salt buildup is observed in the root zone. According to the United Nations Environment Program, the percentage of agricultural land and crop land which is salt-stressed is approximately 20% and 50%, respectively. It is also estimated that the average per annum infrastructural expenses for removal of salt in order to ensure sustainable agriculture are $US 12 billion (Pitman and Läuchli, 2002). This surmounts to huge economical investments. The water and soil management practices have allowed the cultivation of crops in salt-stressed land as well. The response to salinization is carried out by irrigation of such high-saline soil. However, these techniques come with the added disadvantage of comparatively lower yield and high use of resources which makes it cost-ineffective.

Considering the fact that the 96.5% of available global water is seawater, and saline groundwater accounts for 1%; this leaves only 2.5% of the global reserve as freshwater (Pitman and Läuchli, 2002). Therefore, the usability of water for agriculture purposes is highly limited. In addition to this, the scarcity of water in arid and semiarid regions creates challenges toward irrigated agriculture. Keeping this in mind, progress has been made in various aspects of crop improvement to deal with salinity stress. The first of its kind would be the progress made in genetics in terms of development of molecular markers. Certain plants possess inherent mechanisms to deal with salt stress in the form of osmolytes, osmoprotectant, ion transport systems, ion compartmentalization, and signaling molecules (Yokoi et al., 2002). These determinants regulate homeostasis which can be studied further using biotechnological techniques to obtain salt tolerant crops.

Agricultural Nanobiotechnology. https://doi.org/10.1016/B978-0-323-91908-1.00013-4

Markers in combination with conventional techniques provide a platform for generation of cultivars from wild varieties. These cultivars might be further used to comprehend the mechanism related to plant tolerance. The other aspect would be specifically engineering the plants to increase salt tolerance by identification of genes associated with salinity regulation. Various studies have been already carried out based on genetic engineering approaches including those in model crops such as *Arabidopsis* (Pitman and Läuchli, 2002; Yokoi et al., 2002; Acosta-Motos et al., 2020). All of these techniques have given significant results in understanding the salt-tolerance mechanism in plant systems. However, the limitations associated with such techniques allow room for other technologies to contribute in crop improvement. One of these technologies is the generation of nanoparticles (NPs) for coping with salt stress. Due to their unique nature, NPs are widely used in various facets of crop improvement and salinity tolerance likewise.

Metal- and nonmetal-based NPs have been found to have amelioration effects on salt-stressed seedlings and in turn reduce the salinity stress on germination on new seeds. Silicon (Si)-based NPs demonstrate enhanced abiotic stress tolerance and have been observed to also impart mechanical strength to plants (Sabaghnia and Janmohammadi, 2015). However, higher concentrations of Si can be toxic to the environment. On the other hand, colloidal NPs which are smaller in size and effective even in lower concentrations are less toxic than the salt forms. Similarly, metal-based nanoparticles such as gold, silver, copper, titanium, platinum, and palladium along with other nonmetallic NPs like carbon, graphene, and polymeric NPs have better scope for crop improvement and productivity enhancement. However, in this case as well, the concentration of the NPs plays a crucial role in determining the efficiency and effectiveness of the NPs (Poddar et al., 2018). Such NPs or engineered nanomaterials (ENMs) hold the potential to improve tolerance of plants to salt stress which in turn have results in other effects on the plant such as enhanced management toward oxidative stress (Lowry et al., 2019). While the integration of nanotechnology with sustainable agriculture has immense relevance in today's agricultural scenario, certain efforts still need to be made on aspects such as studying the relationship between NPs and different molecules of plant system, validation of permissible concentrations of NPs, and environmental safety. As nanotechnology continues to improve, more refined and specific forms of ENMs may be developed in the near future which would have a cost-effective delivery system (Jasrotia et al., 2018; Lowry et al., 2019).

8.2 Nanostructures for management of salt stress

Due to their unique physicochemical properties, nanoparticles are extensively used in increasing plant metabolism. ENMs can enter plant cells and target specific genes for regulation of traits. The size, shape and chemical composition altogether define the adequacy of NP (Lowry et al., 2019). They usually are fine particles of the diametric range of 1–100 nm. However, larger nanostructures such as nanotubes and graphene sheets have also been seen to have size of 1–2 μm.

A variety of metal-based NPs such as gold, silver, titanium, copper, iron, and zinc nanoparticles have been tested in controlled environments to study their effects on plant growth and development. These are usually spherical, triangular, hexagonal, oval, and cubic in shape (Ishak et al., 2019). Apart from these, nonmetallic NPs essentially include carbon-based, polymer-based, and graphene-based nanocomposites. These nanoparticles have a more nontoxic origin, and studies are still at the preliminary stage. More research is required on the interaction of these kinds of nanoparticles with the biomolecules and their impact on regulation of gene expression.

8.3 Metal nanoparticles

Several metal nanoparticles have shown promising applications for ameliorating salt stress in plants which are enlisted in Table 8.1.

8.3.1 Copper

In plants, copper as an element forms an important part of essential micronutrient and takes part in vital activities such as protein regulation, cellular metabolism, signaling, and oxidative stress response. Further, copper nanoparticles (CuNPs) are one of the most studied nanoparticles in plants. CuNPs are known to induce antioxidant properties in plants by interacting with intracellular structures which stimulate the formation of reactive oxygen species (ROS) and/or reactive nitrogen species (RNS). Additionally, under salt-stress conditions, plants have two major responses: a water-deficit phase and a salt accumulation phase. In response to this, the oxidative stress increases as a result of low stomatal conductance and transpiration along with deceleration of photosynthetic activity. This in turn activates the plant's defense system and synthesizes antioxidant compounds (Pérez-Labrada et al., 2019).

While several studies have been partaken in studying the effect of NPs on salinity stress, its effect on postharvest fruit has seldom been explored. In one such study, the effect of salinity on fruit quality has been studied in tomato. Plants under salt stress were seen to produce reduced concentrations of bioactive compounds such as total phenols, β-carotene and vitamin C, and antioxidant properties in the fruit. Upon the application of CuNPs, the accumulation of bioactive compounds was maintained. For treatment under saline-stress conditions, 50 mM concentration of NaCl was used and the applied CuNPs were spherical in shape with an average size of 50 nm (Hernández-Fuentes et al., 2017).

One of the major drawbacks of NPs is their insolubility in aqueous solutions which hinders assessing its toxicity. Therefore, an effective alternative is encapsulation of the NP. Chitosan is a widely used polymer due to its availability, biocompatibility, biodegradability, and nontoxic nature. Encapsulation of CuNPs has demonstrated additional positive effects such as increase in growth of plants as well as exhibition of antimicrobial properties. In a study by Hernández-Hernández et al. (2018a), chitosan-PVA (polyvinyl-alcohol) encapsulated CuNPs demonstrated enhanced salinity tolerance in tomato plants. Under salt-stress conditions, the yield of the crop was

Table 8.1 Metallic nanostructures for ameliorating salinity stress in plants.

Type of nanoparticles	Shape	Size (nm)	Plant/crop	Maximum salt (NaCl) concentration tolerated	References
CuNPs	Spherical	50	*Solanum lycopersicum*	50 mM	Hernández-Fuentes et al. (2017)
Cs-PVA +CuNPs	Spherical	25	*Solanum lycopersicum*	100 mM	Hernández-Hernández et al. (2018a,b)
CuNPs	Spherical	48.3	*Solanum lycopersicum*	50 mM	Pérez-Labrada et al. (2019)
CuNPs	Irregular	22.44–44.26	*Solanum lycopersicum*	0.5 mM	Noman et al. (2021)
AuNPs	Spherical	5–15	*Williopsis saturnus*	0%–4%	Mohite et al. (2017)
AgNPs	–	–	*Solanum lycopersicum*	100 nm and 150 nm	Almutairi (2016)
AgNPs	Spherical	15–29	*Triticum aestivum*	150 mM	Mohamed et al. (2017)
AgNPs	Spherical and semispherical	15–30	*Triticum aestivum*	25 and 100 mM	Ismail and Abou-Zeid (2018)
AgNPs	–	–	*Thymus vulgaris* and *Thymus daenensis*	0, 100, 200, and 300 mM	Ghavam (2018)
AgNPs	Spherical	100	*Satureja hortensis*	0, 30, 60, 90, and 120 $mM L^{-1}$	Nejatzadeh (2021)
AgNPs	–	–	*Allium cepa*	50 mM	Jahangir et al. (2020)
AgNPs	Spherical	75	*Linum usitatissimum*	0, 25, 50, and 100 mM	Khalofah et al. (2021)
AgNPs	–	50–100	*Pennisetum glaucum*	0, 120, 150 mM	Khan et al. (2020)
n-TiO_2	Spherical	10–25	*Hordeum vulgare* L.	1, 100, and 200 mM	Karami and Sepehri (2018)
n-TiO_2	–	20–25	*Vicia faba cv. Misr-1*	180 mM	Abdel Latef et al. (2018)
n-TiO_2	Oval, cubic, and rod-shaped	68	*Phaseolus vulgaris* L.	100 mM	El-Gazzar et al. (2020)
n-TiO_2	–	10–25	*Zea mays*	200 mM	Shah et al. (2021)

n-TiO_2	Spherical	70–90	*Dracocephalum moldavica* L.	100 mM	Gohari et al. (2020)
Fe_2O_3NPs	–	27.3–34.62	*Mentha piperita*	0, 50, 100, and 150 mM	Askary et al. (2017)
Fe_2O_3NPs and ZnONPs	–	20	*Triticum aestivum*	0, 75, and 150 mM	Fathi et al. (2017)
ZnONPs	Cylindrical	21.3	*Lupinus termis*	150 mM	Abdel Latef et al. (2017)
ZnONPs	Hexagonal and spherical	16–35	*Abelmoschus esculentus*	0%, 10%, 25%, 50%, 75%, and 100% SW	Alabdallah and Alzahrani (2020a)
ZnONPs	Hexagonal and spherical	16–35	*Vigna unguiculata*	0%, 10%, 25%, 50%, 75%, and 100% SW	Alabdallah and Alzahrani (2020b)
ZnNPs	–	17–40	*Abelmoschus esculentus*	10 dS/m	Zafar et al. (2021)
MnNPs	Spherical	50	*Capsicum annuum* L.	100 mM	Ye et al. (2020)
SeNPs	Spherical	2–20	*Solanum lycopersicum*	50 mM	Morales-Espinoza et al. (2019)
Cs-SeNPs	Spherical	50	*Momordica charantia*	0, 50, and 100 mM	Sheikhalipour et al. (2021)

also found to increase when exposed to chitosan-PVA(Cs-PVA)+CuNPs. This was due to increase in content of phenols, β-carotene, vitamin C, and lycopene in stressed tomato plants. Jasmonic acid (JA) is one of the key hormones involved in the salt-stress pathway. Additionally, antioxidant enzyme activation is another mechanism employed by plants when under salt stress. Superoxide dismutase (SOD) enzyme converts the toxic ROS into less toxic H_2O_2. The hydrogen peroxide is then scavenged by enzymes like catalase (CAT), peroxidase ascorbate (APX), and peroxidase glutathione (GPX). The application of Cs-PVA+CuNPs increased the expression of SOD and JA both in the presence and in the absence of salt stress. This indicated that the NPs stimulated the first line of defense in saline conditions. Further, under salt stress, application of the NPs increased the diameter and height of the tomato plant. This was indicative of tolerance of plants to salinity stress (Hernández-Hernández et al., 2018b).

Several data report that NPs induce salinity tolerance in plants by regulating ROS and RNS. As mentioned previously, CuNPs are capable of stimulating the antioxidant system in plants which in turn leads to salinity stress tolerance. In addition to this, several other biochemical responses have been reported in plants exposed to foliar application of CuNPs and subjected to stress tolerance. When treated with 50 mM NaCl, tomato plants underwent severe physiological damage. However, when foliar application of CuNPs was introduced, an increase in components like phenols, vitamin C, flavonoids, lycopene, and glutathione in leaves and fruits was observed. Further, enzyme activities of phenylalanine ammonia lyase (PAL), glutathione peroxidase (GPX), APX, SOD, and CAT were also enhanced in the leaf tissues. All of these components are bioactive compounds involved in the plant's antioxidant mechanism. The findings suggest that foliar application of CuNPs has a positive impact on the plant's salt-stress tolerance and therefore is a promising strategy to reduce salt-stress damage in tomato plants (Pérez-Labrada et al., 2019).

In recent time, bioengineered NPs (BNPs) have gained an upper hand over chemically produced NPs and are being used efficiently in various fields such as agriculture, pharmaceuticals, biomedical, and food. Such BNPs have been utilized to elucidate their application in maize seedlings under saline stress. CuNPs were generated from bacterial strain *Klebsiella pneumoniae* (NST2). The biogenic CuNPs were crystalline and ranging from 22.44 to 44.26 nm in size which was stabilized by certain functional groups like carbonyl and amine. Transmission electron microscopy (TEM) analysis also revealed that the BNPs were mostly irregular in shape and existed in spherical and aggregated condition. When the synthesized CuNPs were mixed in saline soil with salt concentration of approximate 0.5 mM, the maize seedlings exhibited enhanced root and shoot length along with dry and fresh weight. Additionally, at high concentrations of CuNPs (100 mg/kg soil), the lipid peroxidation and the oxidative damage in the seedling were reduced significantly promoting antioxidant activities as depicted in Fig. 8.1. Also, the ionic content of Na^+ and Cl^- declined with reduction in concentration of ROS. Thus, BNPs can be a more sustainable alternative to traditional forms of salinity stress management in plants (Noman et al., 2021). In a nutshell, the fundamental studies performed to explore the efficacy of CuNPs to induce plant saline stress tolerance have been quite comprehensive.

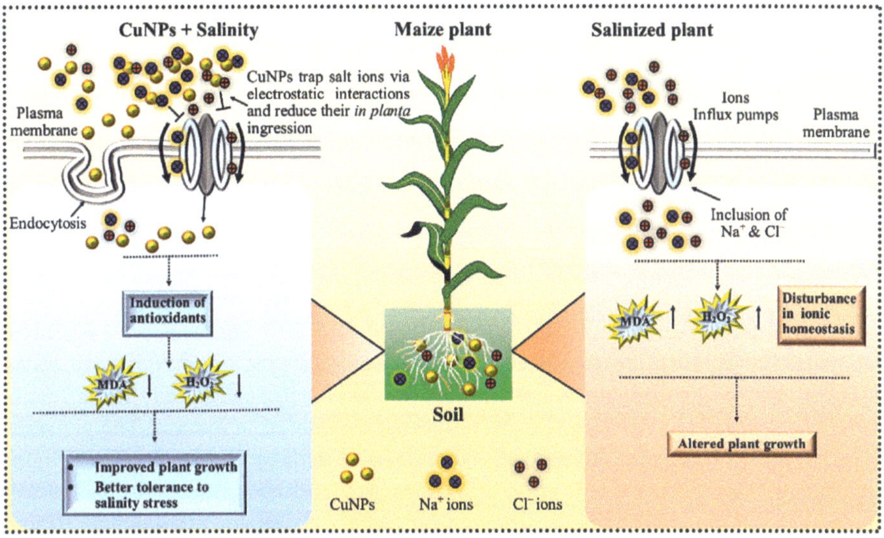

Fig. 8.1 Mechanism of action of biogenic CuNPs to relieve plants from adverse effects of salinity stress. Biogenic CuNPs released salinized plants from stress by activating of stress-defensive mechanisms (such as capturing salt ions via electrostatic interactions and activation of antioxidative machinery) in plants. CuNPs together with activated stress-protecting mechanisms helped plants to re-gain its ionic homeostasis and normal growth with improved stress-tolerating potential. *Note*: In figure, *down- and upward arrows* denote the decrease and increase in concentrations of MDA and H_2O_2, respectively.
Reproduced with permission from Noman, M., Ahmed, T., Shahid, M., Bilal, M., Niazi, K., Qasim, M., Kouadri, F., Abdulmajeed, A.M., Mohammed, S., Ahmad, N., Zafar, M., Ali, S., 2021. Biogenic copper nanoparticles produced by using the *Klebsiella pneumoniae* strain NST2 curtailed salt-stress effects in maize by modulating the cellular oxidative repair mechanisms. Ecotoxicol. Environ. Saf. 217, 112264. Copyright © 2021 The Authors.

8.3.2 Gold

Salt tolerance in plants is based on various additional mechanisms which contribute collectively to the inherent tolerance of the plant system. These can be in the form of antioxidant elements, ionic homeostasis as well as the nitrogen use efficiency of plants. In a study pertaining to salt tolerance in wheat (*Triticum aestivum*), the effect of AuNPs has been studied on nitrogen assimilation, K^+/Na^+ ration, chlorophyll concentration, and defense system of salt-stressed wheat. The AuNPs were chemically synthesized and were characterized accordingly using DLS and TEM, and the particles were found to be mostly spherical with a diameter of 16 nm. At salt concentrations of 100 mM, enhanced levels of oxidative stress markers, viz. reactive oxygen species (ROS) and thiobarbituric acid reactive substances (TBARSs), were observed, which was significantly reduced upon treatment with AuNPs. Additionally, salt stress led to drastic reduction of nitrate reductase (27.73%) as well as nitrogen assimilation ability (30.24%) of the plant which increased upon exposure to AuNPs at 58.81% and 58.23%, respectively. Apart from this, AuNPs were also seen to improve the chlorophyll content, relative water content, and dry mass accumulation of salt-stressed wheat. In conclusion, regulation of these traits upon treatment with AuNPs was found to have a direct effect on the salt tolerance of wheat (Wahid et al., 2022).

8.3.3 Silver

Various genes are involved in determination of salt uptake in plants and activation of defense mechanisms. These genes govern the soil salt uptake of the plant and also are involved in maintaining osmotic balance throughout the plant body. The concentration of salt in the environment of the growing plant also determines the germination and growth phase. Excessive salt is known to reduce leaf and root growth. Silver nanoparticles (AgNPs) are known to be quite effective in plant improvement in terms of enhancing seed germination, growth, and photosynthetic efficiency. The effect of AgNPs in regulating tolerance to salinity has been explored in tomato. The germination efficiency of salt-stress tomato seeds was studied in the presence of AgNPs. In the presence of high salt concentrations (100 mM and 150 mM), the germination was observed to be hindered while treatment with AgNPs under salt stress enhanced germination rate, root length, and dry/fresh weight of tomato. In addition to this, the genes involved were investigated. AgNPs upregulated all four major salt-stress genes, *AREB*, *MAPK2*, *P5CS*, and *CRK1*. However, three other genes, *AREB*, *MAPK2*, *P5CS* and *CRK1* were downregulated under both salt concentration levels. The differentially regulated genes are involved in regulating abscisic acid (ABA) which in turn mediates salt-responsive genes of the plant. The gene expression pattern provides promising foundation for studying the impact of AgNP in salt-stress regulatory genes (Almutairi, 2016).

One of the first studies for salt tolerance in wheat seedling was done by Mohamed et al. (2017). In this, the effect of AgNPs was evaluated on salt stress in wheat and also to examine the various physiological and biochemical changes occurring due to NaCl exposure. The AgNPs isolated from *Aspergillus terreus* were found to be spherical and 15–29 nm in size. When grown under salt stress (150 mM of NaCl), the wheat seedlings were found to increase proline, soluble sugar, and hydrogen peroxide concentration with decreased shoot biomass. Seed priming with AgNPs at low concentrations significantly increased the seedling biomass. This also increased chlorophyll content which was reduced during salt stress. Further, antioxidant enzymes such as peroxidases (PODs) increased in wheat seedling. The study, however, also implied the dose dependency on salt-stress alleviation. Priming with low concentrations of AgNPs might thus be an effective strategy to enhance salt tolerance in wheat (Mohamed et al., 2017).

Seed priming enhances seed germination and is used as a strategy for plants growing in salt-stressed soils. It allows basic metabolic processes to complete until seed germination. Nanoparticles for seed priming are therefore widely used successfully. AgNPs extracted from *Capparis spinosa* were utilized to study its effect on salt stress in *T. aestivum* L. (wheat). Wheat grown in salt-stressed environment (25 and 100 mM NaCl) was primed with AgNPs that induced germination. Additionally, it regulated phytohormones such as IBA (indole-3-butyric acid), BAP (6-benzylaminopurine),

NAA (1-naphthalene acetic acid), and ABA. Salt stress inhibited the plant growth and hormone stability which could be improved by priming with AgNPs. Further studies would also shed light on the physiological and molecular aspects of the effects due to AgNPs priming (Ismail and Abou-Zeid, 2018). On similar lines, the seed germination rate was regulated using AgNPs in two thyme species, viz. *Thymus vulgaris* and *Thymus daenensis*, under variable salinity. The plants were subjected to four concentrations of NaCl, 0, 100, 200, and 300 nM, and their effects on the thyme species were observed. Increased salt stress notably decreased the germination rate which was attributed to low osmotic pressure and water intake due to hindrance by Na^+ and Cl^- ions. Upon treatment with AgNPs, the germination rate and root vigor increased when treated with 200 mM salt concentration while the shoot length and seed vigor were enhanced at 100 mM salt concentration (Ghavam, 2018).

While studies have elucidated the influence of salinity and AgNPs on seed germination, the effect varies from crop to crop mainly due to the morphological diversity of nanoparticles. Similar to thyme, another medicinal plant which was explored for studying the effect of AgNPs on salt-treated plants was *Satureja hortensis* L. Priming of the plant with AgNPs before planting induced germination and improved seed emergence. Varying salt concentrations (0, 30, 60, 90, and 120 mML^{-1}) were utilized along with AgNPs concentrations (0, 40, 60, and 80 ppm). Results demonstrated that significant increase in salt tolerance was induced after treatment with AgNPs apart from improved seed germination. The morphology of the synthesized AgNPs was studied using TEM and was found to be spherical in nature with average size of 100 nm. Overall, the application of AgNPs showed promising results in salinity tolerance in *S. hortensis*, and further field experiments would unravel its efficacy in inducing stimulating mechanisms in plants (Nejatzadeh, 2021).

Another crop wherein effects of seed priming using AgNPs under salt stress have been explored is in pearl millet (*Pennisetum glaucum* L.). Three different concentrations of salt levels (0, 120, and 150 mM NaCl) have been taken to assess the variations in physiology and morphology of the crop. The salt stress was found to have adverse effects on the plant growth showing restricted growth and production of hydrogen peroxide and malondialdehyde (MDA). Seed priming with 20 mM AgNPs led to increased plant growth, improved proline levels as well as antioxidant enzyme activities. The commercially available AgNPs had a diametric range of 50–100 nm. Thus, seed priming in pearl millet under salinity conditions enhanced physiological and biochemical responses which can attribute to enhanced salt tolerance (Khan et al., 2020).

Plant-growth-promoting rhizobacteria (PGPR) contribute toward the overall growth and yield of a plant and negate the detrimental effects of high salinity. PGPRs are also known to induce phenols which are important components of antioxidants in plants. Since they reside in the soil, their relation with plant root system is highly probable which in turn influences the plant physiology. A combination of PGPRs and AgNPs offers a robust solution for curbing salt stress in plants. In a study using onion seedling, two PGPR strains *Bacillus pumilus* and *Pseudomonas moraviensis* were used to inoculate onion seedling which were treated with 50 mM of NaCl. After transplantation of the onion seedlings, the rhizosphere soil was treated with 5 ppm of AgNPs. It was observed that the combination of PGPR and AgNPs increased the total chlorophyll *and* carotenoid content as well as protein content of onion bulb. Also, sugar and proline contents were significantly increased in the plant. It was also observed that *B. pumilus* demonstrated greater efficiency to salt-stress tolerance as compared to *P. moraviensis*. PGPRs appear to facilitate the enhanced fresh weight in the onion bulb along with AgNPs. The study provided foundation for salt-stress tolerance using PGPRs in assistance with AgNPs as well as enhanced growth parameters (Jahangir et al., 2020).

Comamonas testosteroni along with AgNPs improved salt-stress tolerance in *Linum usitatissimum* L. (linseed). The AgNPs were globular in shape with an average diameter of 75 nm. Under saline conditions, concentration of pigments, carotenoid, anthocyanin, sugar, proline, H_2O_2, malondialdehyde, and proteins varied. When treated with *C. testosteroni* and AgNPs, H_2O_2 and malondialdehyde were reduced while antioxidants increased in linseed. Additionally, elevated photosynthesis pigmentation levels were observed which implied toward stimulation of inherent defense system in plant. The synergistic combination of the bacteria and NPs resulted in alleviated tolerance of linseed against salinity and also increase in plant growth and yield (Khalofah et al., 2021).

8.3.4 Titanium

In a study using salt-stressed barley, the impact of treatment with nanotitanium dioxide (n-TiO_2) was assessed. Commercially available spherical n-TiO_2 with a size of 10–25 nm were used for treating salt-stressed barley at variable concentrations (500, 1000, and 2000 mg/kg). The barley plants demonstrated stunted root and shoot growth, reduced chlorophyll, and photosynthetic content under high salinity conditions. Treatment with n-TiO_2 increased root and shoot length along with chlorophyll content and photosynthetic activity. Further, sodium nitroprusside (nitric oxide donor) along with n-TiO_2 enhanced the salt tolerance by enhancing the antioxidant activities. Salt stress causes heightened levels of malondialdehyde (MDA) and hydrogen peroxide (H_2O_2). Using n-TiO_2 in combination with sodium nitroprusside decreased the level of these enzymes (Karami and Sepehri, 2018).

Apart from cereals, the effect of n-TiO_2 was explored in the salt-stressed legume, broad bean (*Vicia faba* cv. Misr-1). High salt conditions have turned cultivable land into wastelands that have rendered the farming of broad bean difficult. Owing to this, the effect of n-TiO_2 on mitigating the impact of soil salinity on the growth and development of broad bean was examined. The effect of n-TiO_2 was studied in a concentration-dependent manner where broad beans were treated with 180 mM of NaCl. The optimum n-TiO_2 concentration that could ameliorate salt stress was 0.01%, while at higher concentrations, the nanoparticles formed agglomerates which essentially hindered their uptake by the legume. At optimum n-TiO_2 concentrations under salt stress, broad beans exhibited enhanced photosynthetic and antioxidant activity. The size of n-TiO_2 was in the range of 20–25 nm (Abdel Latef et al., 2018).

The arbuscular mycorrhizal fungi (AMF) are symbionts which aid plants to thrive under environmentally challenging conditions. In one such study, the effect of n-TiO_2 on the activity of AMF in common bean (*Phaseolus vulgaris* L.) under salt conditions was studied. The n-TiO_2 were oval, cubic, and rod-shaped with an average diameter of 68 nm. At low saline conditions, the AMF activity contributing to plant growth increased from 180% to 224%, while at high saline conditions, an increase from 48% to 130% was seen in the presence of n-TiO_2. Further, the AMF activity for plant dry biomass increased from 277% to 465% and 883% at low and high saline conditions, respectively, when treated with n-TiO_2. The tolerance of the AMF to 100 mM NaCl plant saline conditions reached 110% which is quite significant. Thus, n-TiO_2 was seen to substantially impart salt tolerance to AMF. However, this effect was only toward the AMF, which in turn contributed to the plants growth and development, and not the common bean in itself (El-Gazzar et al., 2020).

The impact of n-TiO_2 on the overall germination and physiology of maize under salt stress has been explored wherein maize (*Zea mays*) seeds were soaked in 40, 60, and 80 ppm of n-TiO_2 and treated with 200 mM NaCl solution. About 60-ppm n-TiO_2 was observed to be the optimum concentration for efficient germination. The n-TiO_2 used for priming the maize seeds were characterized using scanning electron microscopy (SEM) and X-ray diffraction (XRD), and were found to be of the size range of 10–25 nm. The salt-stress tolerance has been examined by using the relative water content (RWC) of leaf technique which behaves as an indicator of the water status in plants. Under high salt conditions, the RWC of maize was seen to decrease. When treated with n-TiO_2, the RWC of the maize leaf significantly increased. Apart from this, the n-TiO_2-treated maize also enhanced the shoot length, root length, seedling weight, and seedling dry weight of the cereal. Further, n-TiO_2 had a positive effect on proline and phenolic content along with improving superoxide dismutase, phenylalanine ammonia lyase, malondialdehyde, catalase, and electrolytes like K^+ and Na^+ levels. Thus, seed priming with n-TiO_2 is an effective means of achieving salt tolerance in maize (Shah et al., 2021).

The role of n-TiO_2 has been studied in salt-stressed Moldavian balm (*Dracocephalum moldavica* L.) grown under salt-stress conditions of 0, 50, and 100 mM NaCl. Varying concentrations of n-TiO_2 were utilized for the experiment; however, 100 mg/L of n-TiO_2 was found to be the most effective compared to the other concentrations. Under salinity conditions, the n-TiO_2 was seen to improve antioxidant enzyme activity and essential oil content (1.19%) such as geranial, z-citral, geranyl acetate, and geraniol. SEM and TEM analyses demonstrated that the n-TiO_2 were spherical in shape with a diameter of 70–90 nm as seen in Fig. 8.2 (Gohari et al., 2020).

8.3.5 Others

Iron (Fe) is the fourth most abundant metal found in the earth's crust and is one of the most widely used metals for nanoparticle synthesis due to its availability. Additionally, FeNPs production is cost-effective and less toxic along with possessing a surface cover of magnetic nanoparticles. All of these properties make FeNPs unique and sought after for use in the field of medicine, agriculture, and pharmaceuticals. Therefore, its effect on abiotic stress such as high salinity has been studied in different crop species. In one such study, the impact of Fe_2O_3NPs has been studied in *Mentha piperita* (peppermint) under different NaCl concentrations (0, 50, 100, and

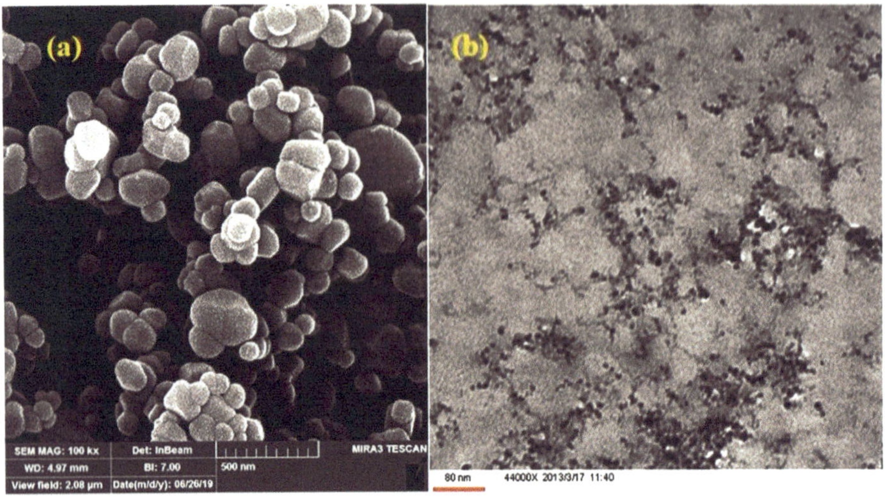

Fig. 8.2 (A) SEM and (B) TEM images of TiO_2 NPs.
Reproduced with permission from Gohari, G., Mohammadi, A., Akbari, A., Panahirad, S., Dadpour, M.R., Fotopoulos, V., Kimura, S., 2020. Titanium dioxide nanoparticles (TiO_2 NPs) promote growth and ameliorate salinity stress effects on essential oil profile and biochemical attributes of *Dracocephalum moldavica*. Sci Rep 10(1), 1–14.

150 mM). XRD and SEM images revealed that the magnetite NPs were of the average size of 27.3–34.62 nm. It was observed that salt stress results in increased proline content and lipid peroxidation along with decreased dry/fresh weight (DW and FW). Upon application of Fe_2O_3NP, the lipid peroxidation and proline levels were seen to drastically reduce. Additionally, DW and FW enhanced in the presence of NPs. Maximum activities of antioxidant enzymes (I%) like CAT, SOD, and guaiacol peroxidase (GPOX) increased when under 150 mM of salt concentration. However, exposure to Fe_2O_3NPs caused reduced antioxidant activities which implied toward the need for appropriate concentrations of nanoparticles for stress resistance (Askary et al., 2017).

Combinations of multiple metal oxide nanoparticles have also been carried out to study their effects on salinity. Fe_2O_3NPs and ZnONPs were treated with wheat genotypes under three levels of NaCl concentrations. Foliar application was done to wheat, and it was observed that the combination of the two significantly increased plant height, shoot dry weight, leaf area, chlorophyll content as well as that of potassium (K), Fe, and Zn. The Na concentration along with Na/K ratio also decreased which implied toward. While the study demonstrated promising results toward foliar application of combinational NPs, the antagonistic effects of the two metal oxides need to be explored further. The competitive uptake of the metal oxides is a possibility, and effects of varying concentrations of metal NPs could be explored further to define the pros and cons of this technique (Fathi et al., 2017).

As a combination, chitosan-selenium nanoparticles (Cs-Se-NPs) have been synthesized to examine their ability to reduce salt stress in bitter melon (*Momordica charantia*) under NaCl concentrations of 0, 50, and 100 mM. Foliar application of Cs-Se-NPs increased the growth of bitter melon and was also observed to improve enzymatic activity of antioxidants and intrinsic antioxidant compounds like flavonoids, ascorbate, and anthocyanin. The fruit quality of bitter melon was enhanced due to increased levels of oils after treatment with Cs-Se-NPs. Additionally, MDA and H_2O_2 levels were decreased upon NP application. The Cs-Se-NPs were spherical in shape with an average size of 50 nm. Hence, this strategy can be employed for the improvement of dietary quality of vegetables (Sheikhalipour et al., 2021).

Zinc (Zn) in itself is an essential element for plant growth and facilitates development of plant. One of the primary responses of plants under salt stress is reduced growth as well as physiological and biochemical imbalance. Seed priming with zinc nanoparticles (ZNPs) positively impacts growth in salt-stressed plants as it increases auxin levels, protein synthesis, and phospholipid accumulation. This helps to sustain the integrity of biological membranes and reduces uptake of Na and Cl ions. *Lupinus termis*, lupine, were grown in pots under salt-stress conditions (150 mM NaCl). The seeds of lupine were primed with increasing concentrations of ZnNPs (20, 40, and 60 mg/L). The NPs were elongated and cylindrical with an average particle size of 21.3 nm. While the salinized plants predictably showed stunted growth, decreased pigment levels, and enhanced synthesis of organic solutes like sugar, proteins as well as MDA, SOD, POD, and APX, seed primed plants exhibited stimulated growth. Enhanced levels of pigment along with reduction in MDA and Na were also observed. The most effective changes were observed at 60 mg/L ZnO primed plants, and therefore, this strategy can be used for inducing salt tolerance in lupine plants (Abdel Latef et al., 2017).

The effects of foliar application of ZnONPs have also been demonstrated in salt-stressed *Abelmoschus esculentus* L. Moench cv. Hasawi (Okra). Okra plants were treated with seawater (SW) with varying salt concentrations (0%, 10%, 25%, 50%, 75%, and 100%). With increase in salt concentration, decrease in chlorophyll *a* and *b* was observed, and between 0% and 75% SW, an increase in proline and organic solute content was observed. Additionally, the levels of SOD and CAT decreased at 100% SW. Green synthesis of ZnONPs led to alleviation of salt stress. The NPs were hexagonal and spherical in shape with a diametric size range of 16–35 nm. The application of ZnONPs increased pigment levels and antioxidant enzymes (SOD and CAT) along with decreasing proline and organic solute concentrations. Thus, the data suggest that ZnONPs can be utilized as a strategy to enhance plants' tolerance to salt conditions in an environment-friendly manner (Alabdallah and Alzahrani, 2020a). An extension of this study was performed with *Vigna unguiculata* L. var. California blackeye NO.46 (cowpea). Similar experimental backgrounds were undertaken, and it was observed that shoot and root lengths, FW and DW of leaf and shoot decreased significantly under salt stress along with reduction in relative growth rate (RGR) with increasing salt concentrations. Foliar application of ZnONPs enhanced the growth parameters of cowpea.

Thus, foliar applications can possibly be a much more stable alternative to bulk applications as they would be relatively more environment friendly and less toxic (Alabdallah and Alzahrani, 2020b).

The type of NPs also plays a vital role in determining the efficacy of NP-based induction of salt-stress tolerance in plants. In a study by Zafar et al., 2021, ZnNPs were synthesized using chemical and green methods. Chemical means using coprecipitation was utilized to develop Zn-CNPs while green synthesis of ZnNPs for synthesis of Zn-GNPs was done via using *Sorghum bicolor* L. as a substrate. XRD analysis determined that both the NPs were of the average size ranging from 17 to 40 nm. FTIR analysis revealed the presence of functional groups in the NPs such as amines, ketones, aldehydes, and carboxylic acid. Foliar application of the ZnNPs led to enhanced growth of salt-stressed okra. It was also determined that due to their size, the ZnNPs were able to penetrate the cell wall of plants and regulate metabolic activities of the cell resulting in higher plant production. Additionally, foliar application leads to gradual uptake of the NPs which essentially reduces chances of toxicity in the plant. Treatment with Zn-CNPs and Zn-GNPs enhanced the activities of antioxidants such as SOD, CAT, and POD resulting in improved salt tolerance. Inhibition of ROS synthesis was also observed which is vital for improved antioxidant properties. However, Zn-GNPs demonstrated slightly superior regulation of antioxidant properties under saline environment and therefore can be concluded to be a relatively better approach for reducing salinity stress in plants (Zafar et al., 2021).

Another metal which has been identified for synthesis of NPs is manganese (Mn). Manganese (Mn) oxide as a nanopriming agent for improving salt-stress tolerance in *Capsicum annuum* L. Capsicum plants under 100 mM NaCl was treated with MnNPs. The salt stress caused reduced plant growth and altered the protein and lignin patterns. Further, macronutrients like Na, Mg, Ca, Mn, Zn, K, and S were redistributed in root and shoot under salt stress. The MnNPs penetrated the seed coat and formed a nanoparticle-corona complex. The MnNPs were spherical with an average size of 50 nm. Also, priming with MnNPs enhanced SOD expression which in turn assisted in regulation of oxidative stress. Nanopriming as a technique can be explored further to comprehend biochemical responses of plants to abiotic stress and also to study the interaction of nanomaterials with plant biomolecules (Ye et al., 2020).

Selenium (Se), as an element, is also essential to human health and plant metabolism. Se functions as a cofactor for antioxidant enzyme metabolism along with playing a role in determining antioxidant capacity. SeNPs can enhance antioxidant properties and consequently the tolerance to oxidative stress. When tomato plant was subjected to 50 mM NaCl stress, they exhibited reduced fruit yield and size along with stunted growth of the plant as seen in Fig. 8.3. Four doses of SeNPs (1, 5, 10, and 20 mg/L) were used to mitigate the salt stress. The application led to increase in yield and growth of the tomato plant and also in enzymatic and nonenzymatic properties. An increase in antioxidant content such as lycopene, flavonoids, phenols, and β-carotene provided an additional advantage for the growth of tomato under salt stress. The SeNPs were spherical in shape with an average size range of 2–20 nm. Future studies can be carried out to further enhance the nutritional value of tomato and other related crops under salinity conditions (Morales-Espinoza et al., 2019).

Fig. 8.3 A tomato crop under NaCl stress and an application of Se NPs. T0: control. NaCl: NaCl-only treatment. 1, 5, 10, and 20+NaCl: represent the amount of Se NPs applied (mg/L) plus 50 mM of NaCl.
Reproduced with permission from Morales-Espinoza, M.C., Cadenas-Pliego, G., Pérez-Alvarez, M., Hernández-Fuentes, A.D., De La Fuente, M.C., Benavides-Mendoza, A., Valdés-Reyna, J., Juárez-Maldonado, A., 2019. Se nanoparticles induce changes in the growth, antioxidant responses, and fruit quality of tomato developed under NaCl stress. Molecules 24, 3030.

8.4 Nonmetallic nanoparticles

Various nonmetallic nanoparticles are also reported to improve salt tolerance of the plants which are presented in Table 8.2.

8.4.1 Carbon nanotubes and nanohorns

Carbon nanotubes (CNTs) have been used extensively in material science, agriculture, and pharmaceutical industries due to their high surface-to-volume ratio. In a study pertaining to 100-mM salt-stressed broccoli (*Brassica oleracea*), multiwalled CNTs (MWCNTs), of concentration of 10 mg/L, were employed to impart salt tolerance. Under high saline conditions, the germination and growth efficiency of the plants were compromised. Upon exposure to MWCNTs, the negative effects of salt were alleviated. The MWCNTs entered the cells of adult plants as seen in Fig. 8.4 and eventually caused enhanced water uptake, net assimilation of CO_2, and aquaporin transduction which improved the overall water transport of the plant. The MWCNTs were larger than 100–500 nm in size (Martínez-Ballesta et al., 2016).

Under salt stress, reactive oxygen species (ROS) such as hydrogen peroxide (H_2O_2) have been known to cause oxidative damage to the cells. Therefore, monitoring and detection of H_2O_2 is crucial for examining plants under salt stress. Electrochemical detection methods are simple, rapid, and efficient, due to which they are the most preferred.

Table 8.2 Nonmetallic nanoparticles for improvement of salt tolerance in plants.

Type of nanoparticles	Shape	Size (nm)	Plant/crop	Maximum salt (NaCl) concentration tolerated	References
MWCNTs	Cylindrical	200	*Brassica oleracea*	100 mM	Martínez-Ballesta et al. (2016)
SWCNTs	Tangled SWCNTs	–	*Aloe vera*	0.3 M	Ren et al. (2013)
MWCNTs	Cylindrical	1000–12,000	*Sorghum bicolor* L. Moench and *Panicum virgatum* L.	100 mM	Pandey et al. (2018)
SWCNHs	Horn-shaped	80–100	*Sophora alopecuroides*	100 mM	Wan et al. (2020)
Graphene	Sheets	1000–2000	*Sorghum bicolor* L. Moench and *Panicum virgatum* L.	100 mM	Pandey et al. (2018)
Graphene oxide	Crumpled thin flake	–	*Pennisetum glaucum* L.	3.16 dS/m and 10.29 dS/m	Mahmoud and Abdelhameed (2021)
Graphene oxide-glycine betaine	Sheet-like structures	–	*Ocimum basilicum* L.	0, 50, and 100 mM	Ganjavi et al. (2021)
S-nitroso-MSA-CsNPs	–	38.81 ± 18.10	*Zea mays*	300 mM	Oliveira et al. (2016)
CsNPs	–	46.32	*Phaseolus vulgaris* L.	100 mM	Zayed et al. (2017)
CsNPs	Spherical	500	*Vigna radiata* L.	4 dS/m and 8 dS/m	Sen et al. (2020)
CsNPs	–	–	*Silybum marianum* (L.) Gaertn.	0, 50, 100, and 150 mM	Mosavikia et al. (2020)
Hydrogels	–	–	*Festuca rubra* L., *Lolium perenne* L., and *Poa pratensis* L.	0, 5, and 10 g dm^{-3}	Anna et al. (2011)

NSi	–	40	*Vicia faba* L.	0, 50, 100, and 200 mM	Qados and Moftah (2015)
NSi	–	7	*Ocimum basilicum*	1, 3, and 6 dS/m	Kalteh et al. (2014)
SiO_2NPs	–	10–20 nm	*Fragaria* × *ananassa* Duch.	0, 25, and 50 mM of NaCl	Avestan et al. (2019)
Nanocomposite-Bent-pro	Pro intercalated between Bent monolayers	–	*Arabidopsis thaliana*	200 mM	Merino et al. (2021)
SiO_2NPs	–	–	*Solanum tuberosum* L.	10 dS/m	Ali et al. (2021)
Polyhydroxy fullerol	Spherical cage-like	1.3 nm	*Triticum aestivum* L.	150 mM	Shafiq et al. (2019)
Polyhydroxy fullerol	Spherical cage-like	1.3 nm	*Triticum aestivum* L.	150 mM	Shafiq et al. (2020)

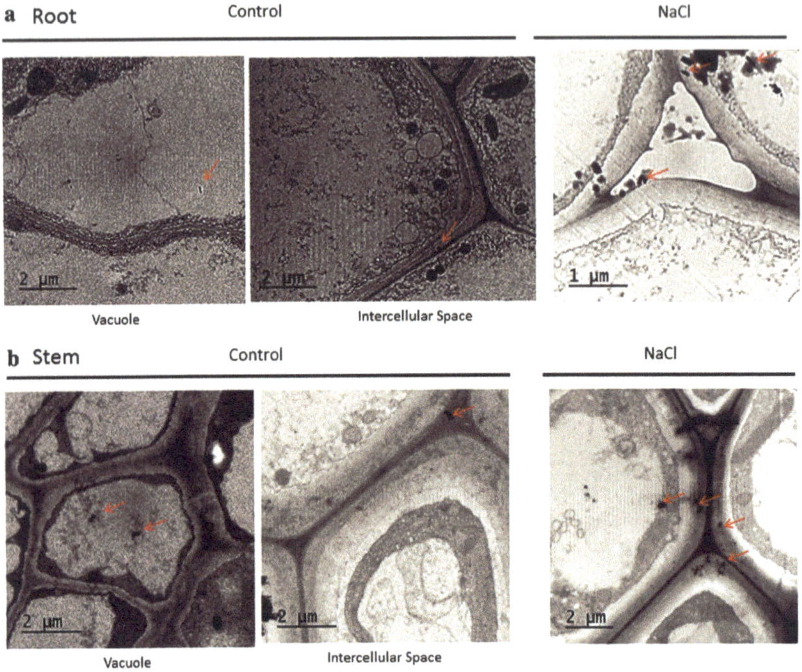

Fig. 8.4 TEM characterization of MWCNTs uptake when adult broccoli plants were grown with MWCNTs (10 mg/L) in the nutrient solution during 7 days. MWCNTs were allocated in different organs, (A) root and (B) stem of control and 100 mM NaCl-treated plants. The *arrows* indicate the MWCNTs in intercellular space, vacuole, and cytoplasm.
Reproduced with permission from Martínez-Ballesta, M.C., Zapata, L., Chalbi, N., Carvajal, M., 2016. Multiwalled carbon nanotubes enter broccoli cells enhancing growth and water uptake of plants exposed to salinity. J. Nanobiotechnol. 14 (1),1–14.

For this purpose, single-walled CNTs (SWCNTs)-modified electrochemical carbon fiber ultramicroelectrode (CFUME) was developed for H_2O_2 sensing. In vivo monitoring of H_2O_2 in aloe leaves irrigated with 0.3 M salt solution was achieved. After 12.5 h, an increase in H_2O_2 was observed. To further validate the study, CAT was injected which leads to rapid decrease in H_2O_2 suggesting it's scavenging. Thus, CFUMEs can be used as an efficient tool for in vivo studies pertaining to ROS generation under salt stress (Ren et al., 2013).

MWCNTs have also been utilized to investigate their effect on bioenergy crops, viz. sorghum (*S. bicolor* L. Moench) and switchgrass (*Panicum virgatum* L.). High level of soil salinity significantly hinders seed germination and plant growth, and also reduces biomass accumulation of energy. MWCNTs (of length 1–12 µm) treated seeds of sorghum and switchgrass at concentrations of 50 and 200 µg/mL. Addition of 100 mM NaCl to the seeds caused reduction in root, shoot length, and seed germination as compared to seeds which were untreated. Highest germination efficiency of seeds exposed to NaCl was seen at 200 µg/mL of MWCNTs. Expression profiles of the two genes under 100 mM salt stress exhibited that the plasma membrane intrinsic proteins (PIPs) *1;5* and tonoplast intrinsic proteins (PIPs) *1;1* were downregulated in root and shoot. Supplementation with MWCNTs enhanced the expression of the aquaporin genes. Since aquaporins are known to play an important role in regulation of abiotic stress, the MWCNTs regulate plant stress response gene (Pandey et al., 2018).

Apart from nanotubes, carbon nanohorns also possess application in biomedical and agriculture industry. Under salt-stress conditions, the overall physiology of the plant *Sophora alopecuroides* was severely affected. Foliar spraying with single-walled carbon nanohorns (SWCNHs) increased PSII (photosystem II) photosynthetic activity along with enhancing plant height, protein, and soluble sugar content for plants exposed to 100 mM NaCl. The nanohorns had a pipe diameter of 80–100 nm. Metabolomics study suggested that SWCNHs bring about salt tolerance by regulating carbon and nitrogen processes along with enhancing glycolysis and TCA cycle. They also increase the unsaturated fatty acid levels in the cell (Wan et al., 2020).

8.4.2 Graphene and graphene oxide

Among the carbon-based nanomaterials (CBNs), graphene nanoplates were explored for salt tolerance in switchgrass (*P. virgatum* L.) and sorghum (*S. bicolor* L. Moench). For examining the sensitivity of sorghum and switchgrass to NaCl, the seeds of the plants were exposed to 100 mM salt and then treated with 50, 100, 200, 500, and 1000 μg/mL graphene. The 500 μg/mL concentration of graphene was most effective for salt tolerance in switchgrass, while in case of sorghum seeds, the germination efficiency was maximum within 100–1000 μg/mL. Aquaporins, which are essential for maintaining the water balance in the plant, also were studied for sorghum under salt stress. The *PIP 1;5* and *TIP1;1* are two major aquaporin genes in sorghum, and their expression levels were analyzed for plants under 100 mM NaCl stress. About 50–200 μg/mL of graphene was seen to be most efficient for the activation of expression of the two genes (Pandey et al., 2018).

Soil salinity is known to induce accumulation of ROS which in turn stimulates the production of MDA, H_2O_2, and leads to decrease in enzymatic activity of SOD and peroxidase (POD) reducing the crop yield. Pearl millet (*P. glaucum* L.), a semiarid crop, is being recently explored due to its high protein content. The effect of salt stress to this crop has been studied, and the concomitant effect of graphene oxide has also been explored. Salt concentrations of 3.16 dS/m and 10.29 dS/m have been considered in soil and water, and the effect of graphene oxide treatment has been observed. The foliar application of graphene oxide induced salinity stress tolerance by regulating the morphological and biophysical traits of pearl millet. Further, the level of ROS reduced in the crop along with substantial improvement in plant growth, protein content, yield, and photosynthetic pigment content. The graphene oxide exhibited a tendency to unfold and form a crumpled thin flake. Foliar application of graphene oxide increased the weight of pearl millet plants by 2.4-folds as compared to salt-stressed plants. Additionally, the grain yield increased by 78.2% when salt-stressed plants were sprayed with graphene oxide. The ideal concentration of graphene oxide for optimum growth of pearl millet under salt-stress conditions was 20 mg/L (Mahmoud and Abdelhameed, 2021).

The application of graphene oxide in salt-stress tolerance in plants was also explored in sweet basil (*Ocimum basilicum* L.) wherein graphene oxide nanoparticles were utilized as a carrier for glycine betaine (GB). GB, which is a free amino acid, has its role in cellular functions and is known to be actively involved against salinity stress. Under such osmotic stress conditions, GB protects enzymes involved in stress such as Rubisco and malate dehydrogenase. Sweet basil plants under salt-stress conditions of 0, 50, and 100 mM were exposed to graphene oxide, GB and graphene oxide-GB of concentrations of 0, 50, and 100 mg/L. Graphene oxide and GB were collectively seen to negate the effects of salt stress by increasing photosynthetic pigments, membrane stability index, proline, phenols, and antioxidant activity along with reducing MDA and H_2O_2 levels (Ganjavi et al., 2021).

8.4.3 Polymeric nanoparticles

Polymeric nanoparticles comprise of a wide range of multifunctional polymers such as chitosan which plays a significant role in abiotic stress tolerance. They are known to regulate enzymatic activities and antioxidant activity which aids the plants during such stress conditions.

Nitric oxide (NO) molecules are involved in plant abiotic stress responses; however, owing to their instability, their application in agricultural purposes is limited. Due to this, nitric-oxide-releasing nanoparticles are being explored. Chitosan nanoparticles (CsNPs) containing the NO donor S-nitroso-mercaptosuccinic (S-nitroso-MSA) acid were synthesized and characterized for their salt tolerance imparting ability in maize (*Z. mays*). The NO donor was encapsulated with the CsNPs, and the free group of MSA-CsNPs was nitrosated generating NO-releasing NPs (S-nitroso-MSA-CsNPs). Maize plants exposed to 300 mM NaCl when treated with S-nitroso-MSA-CsNPs showed enhanced photosystem II activity and electron transport rate along with increased root and shoot growth. The hydrodynamic size of the MSA-CsNPs was found to be 38.81 ± 18.10 nm. The encapsulation efficiency of the MSA in the CsNPs was observed to be almost 91%. The study implied toward the overall impact of MSA nanoencapsulation in enhancing the bioactivity of NO in salt-stressed maize plants (Oliveira et al., 2016).

The effect of CsNPs on salt stress was also studied in bean (*P. vulgaris* L.). A wide range of NaCl concentrations from 50 to 150 mM was selected for the study and, 100 mM was finally used as optimum salt concentration for stress studies owing to reduced salt-tolerance index. The salt-stressed bean plants were then treated with CsNPs of concentration of 0.1%, 0.2%, and 0.3% among which 0.3% showed optimum activity in enhancing germination rate along with other growth variables like plant height, root and shoot length, and leaf area. Additionally, even at 0.1% of CsNPs, positive effect was seen for antioxidant enzymes, catalase, proline and carotenoids, relative water content, and membrane stability index. The average size of the CsNPs was studied using a zetasizer, and the mean diameter was found to be 46.32 nm (Zayed et al., 2017).

While metal nanoparticles, such as CuNPs, are effectively utilized for plants studies, they might be toxic to plants at higher concentrations. Therefore, CuNPs coupled with polymers like chitosan are being investigated which are known to be less toxic as compared to free CuNPs. The application of chitosan-polyvinyl alcohol (Cs-PVA) and CuNPs on the growth of tomato (*Solanum lycopersicum* L.) under salt stress was examined. The CuNPs-Cs-PVA complex was developed by absorption in hydrogels. Cs-PVA and CuNPs were seen to promote plant growth as well as key osmotic balance components such as jasmonic acid and superoxide dismutase (SOD) under 100 mM of salt-stress conditions. SOD is the primary enzyme against ROS-forming H_2O_2 which in turn is scavenged by CAT, peroxidase ascorbate, and peroxidase glutathione. The CuNPs were roughly spherical in morphology and 25 nm in diameter (Hernández-Hernández et al., 2018a).

The effect of CsNPs after solid matrix priming (SMP) was analyzed in salt-stressed mung bean (*Vigna radiata* L.). Priming utilizes the characteristic features of the material to regulate water uptake of the seed and is an important method for enhancement of seed germination under saline conditions. Solid matrix is preferred over liquid priming or hydropriming as it uses restricted amount of liquid per unit of seed. The ability of elicitors like CsNPs on oxidative stress markers like MDA, H_2O_2, proline, chlorophyll, protein content, and carotenoids was also investigated. Mung bean seeds treated with 4 dS/m and 8 dS/m of salt stress were exposed to CsNPs which revealed the antioxidant activities of SOD and CAT along with reduction in MDA and H_2O_2. This also resulted in enhanced chlorophyll content and photosynthesis. The CsNPs showed spherical morphology of mean diameter of 500 nm. The study conclusively suggested the potential of SMP with CsNPs to overcome salt-stress tolerance (Sen et al., 2020).

On similar lines, the potential of CsNPs seed priming in salt-stress milk thistle (*Silybum marianum* (L.) Gaertn.) seeds was studied. Salt stress of concentrations of 0, 50, 100, and 150 mM was employed with CsNPs concentration of 0%, 0.25%, 0.5%, and 1%. At high concentrations of salt, an astounding decrease in germination percentage, seedling length, seedling vigor index, and increase in SOD activity were observed. The salt stress also decreased chlorophyll content. The optimum application of 0.25% of CsNPs to salt-stressed milk thistle seeds demonstrated improvement in physiological mechanisms such as chlorophyll content and pigment synthesis along with enhanced antioxidant activities. Therefore, seed priming by CsNPs can be a potential method for achieving optimum growth in salt-stressed environments (Mosavikia et al., 2020).

Other polymeric substances which are utilized to study the effect of salt stress are hydrogels. Owing to their large surface area, hydrogels are used extensively in drought-prone areas as they facilitate water uptake and decrease evapotranspiration in plants. Apart from water, hydrogels are also capable of absorption of cations which suggests its role in saline conditions. Hydrogels have been utilized for exploring their role in salt-stressed road-side turf grass species, viz. red fescue (*Festuca rubra* L.), kentucky bluegrass (*Poa pratensis* L.), and perennial ryegrass (*Lolium perenne* L.). The plants were treated with salt solutions of concentrations of $0\,g\,dm^{-3}$, $5\,g\,dm^{-3}$, and $10\,g\,dm^{-3}$, and cultivated with hydrogel. The effectiveness of the hydrogel was examined based on the turf height, above-ground fresh matter, and extent of membrane injury. The efficacy of hydrogel was found to be optimum at 5 $g\,dm^{-3}$, while at higher concentrations, it decreased (Anna et al., 2011).

8.4.4 Others

Nanosilicon (NSi) is one of the popular nanomaterials used in agriculture as they have beneficial effects on plant germination, growth, and overall physiological development of the plant. The impact of NSi on salt-stressed faba beans (*V. faba* L.) was studied. Faba beans are widely cultivated due to their high nutritional content and are found mostly in the Asian and African regions. Most of these regions have high salt in soil and therefore require a robust salt-stress ameliorating technique. Faba bean could also be used as a model for other leguminous crops which are grown widely in these regions. Faba beans salt-stressed at concentration of 0, 50, 100, and 200 mM were treated with NSi (diameter 40 nm) at concentrations of 0, 1, 2, and 3 mM. NSi treatment improved germination attributes by enhancing germination rate and percentage along with reducing mean germination time. Further, NSi was also seen to boost the vegetative traits like plant height, number of leaves, number of pods, leaf area, nutrient element levels, and cumulative yield parameters of faba beans. The optimum concentration of NSi was 2 mM which showed significant efficiency (Qados and Moftah, 2015).

Effect of NSi on salt-stressed basil (*O. basilicum*) was assessed at salt concentrations of 1, 3, and 6 dS/m. Physiological and morphological traits such as fresh/dry weight along with chlorophyll and proline content were analyzed after treatment with NSi. The NSi particles possessed a surface area of 400 m^2 and diameter of 7 nm. While parameters like fresh/dry weight and chlorophyll content decreased under stress conditions, proline levels increased as a response to salt stress. Upon exposure to NSi, the fresh and dry weight of the shoot was improved which was attributed to cell wall stability conferred by the NSi. NSi also improved water translocation and efficiency in xylem. Similar enhancement was observed with chlorophyll content in the basil plant. Proline content, however, remained high due to tolerance induction in the plant (Kalteh et al., 2014).

Response of strawberry (*Fragaria × ananassa* Duch.) plants to salt stress and their subsequent exposure to nanosilicon dioxide (SiO_2NPs) were examined. Strawberry plants were treated with 0, 25, and 50 mM of NaCl which reduced plant biomass, chlorophyll content, and the RWC. The salt-treated plants also had lower membrane stability index (MSI) (10%) and high proline content (81%). MSI allows for the evaluation of water deficit in plant tissues as it causes structural impairment of the cell membrane. The plants exposed to SiO_2-NPs were able to overcome the decrease in chlorophyll content and epicuticular structure, and SEM images implied that nanoparticles-treated plants had crystal wax deposits on the epicuticular layer. The plants also accumulated less proline relative to salt-stressed plants. The SiO_2NPs were obtained commercially and had a particle size of 10–20 nm (Avestan et al., 2019).

Nanoclay materials which belong to the smectite group, an expansible form of clay, have multiple applications. Nanoclays are essentially crystalline oxide sheets of aluminum and silicon (aluminosilicates) stacked on top of each other. Montmorillonite (Mnt) is the most commonly utilized nanoclay and is the main component of bentonite which is naturally found. As discussed earlier, under saline conditions, plants response mechanism is induced and the primary response is the production of the amino-acid, proline (Pro). Pro behaves as an osmoprotectant maintaining turgor pressure and cellular membrane balance and structure. Pro also functions in removing free radicals such as ROS by scavenging activity. The incorporation of proline into bentonite could therefore open avenues for protection against stress conditions. For this, a nanocomposite, Bent-pro, was developed to assess their role in plant stress response. The nanocomposite was present in a layered arrangement with the pro in its zwitterionic form intercalated between the bent monolayers. *Arabidopsis thaliana* seeds treated with Bent-pro were grown in 200 mM NaCl-supplemented media. They were protected against seed germination inhibition and low chlorophyll synthesis (Merino et al., 2021).

The effect of SiO_2NPs was checked on the growth and yield of potato (*Solanum tuberosum* L.) plants growing under salt stress. Irrigation of the cultivars with salinity of 10 dS/m demonstrated reduction in vegetative growth and yield parameters like height, number of tubers, number of branches, and leaf area. High-saline soil causes accumulation of sodium ions which hinders plant function such as respiration and photosynthesis. Additionally, salinity also induces the production of ROS causing oxidative imbalance. Treatment with 50 mg/L of SiO_2NPs resulted in enhanced vegetative growth and improved yield parameters (Ali et al., 2021).

Among carbon nanoparticles, fullerenes (C_{60}) are widely known. However, C_{60} prepared in organic solvents cannot be used directly on plants due to their toxicity, and therefore, a derivative of C_{60} which is fullerol was explored. Fullerol [$C_{60}(OH)_x$] or polyhydroxy fullerenes (PHF) are derivatives of C_{60} that behave as radical sponges along with possessing higher electron affinities. They are spherical carbon cages with hydroxyl groups on the surface having an average size of 1.3 nm. Additionally, since they are water soluble, they are nontoxic to plants. The effect of PHF in salt-stressed wheat (*T. aestivum* L.) through seed priming was studied. Treatment with 150 mM NaCl solution led to decreased root and shoot length, and chlorophyll content along with increased SOD, CAT, and POD activities. Oxidative stress also increased which was determined by enhanced levels of H_2O_2 and MDA. The PHF-treated plants showed recovery in root and shoot length along with chlorophyll levels. Also the oxidative stress was significantly reduced as H_2O_2 and MDA levels decreased (Shafiq et al., 2019).

In a subsequent experiment, the influence of foliar application of PHF on salt-stressed plants was investigated. Wheat plants grown at 150 mM NaCl-supplemented soil were treated with 0, 10, 40, 80, and 100 nM concentrations of foliar sprayed PHF. The fullerol caused evident increase in shoot growth and least effect on root development. At 40 and 80 nM concentrations, the improved antioxidant activity was observed along with reduced H_2O_2. However, MDA levels remained unchanged. PHF also improved sugar, protein, and free amino acid content of the salt-stressed wheat. The photosynthetic activity, osmolyte accumulation, and tissue ion compartmentalization together contributed to the grain yield that was notably enhanced after fullerol foliar spray. In both the cases, the differential response of the plant could be due to difference in exposure to salt stress. In case of foliar spraying, the plants are already exposed to stress which affects their biochemistry which is not the case in PHF seed-primed wheat which is exposed to salt later (Shafiq et al., 2020).

8.4.5 Conclusions and future perspectives

Like other abiotic stresses, salt stress is a major cause of crop loss that affects a wide range of crop species globally. About 20% of the cultivable land contains high salt concentrations which is one of the primary stress conditions studied (Khan and Upadhyaya, 2019). Salinity drastically affects the osmotic and nutritional balance of a crop and influences the stomatal conductance and production of photosynthetic pigments. This consequently leads to decreased CO_2 assimilation causing stunted growth and lowered productivity. Apart from this, salinity stress also affects proteins and genes, such as aquaporins, especially in the root system. Salt hinders the uptake of water from the soil causing reduced water transportation. The use of different kinds of nanoparticles is gradually becoming extensive and integral for agriculture. Nanoparticles are known to regulate the physiological and biochemical aspects of the plant including the antioxidant mechanism, photosynthetic and respiratory processes, yield, and overall development. Nanoparticles are therefore being explored as a means to overcome the deleterious effects of salinity stress on different plant species. Due to their unique physiochemical properties, nanoparticles are known to boost plant metabolism (Ahmad and Akhtar, 2019). Various metal nanoparticles such as gold, silver, copper, zinc, iron, and titanium are successfully utilized for alleviating salt stress in various plant species. Similarly, nonmetallic nanoparticles like polymers, nanotubes, nanohorns, fullerenes, and nanocomposites are developed which can significantly impart tolerance to salt-stressed plants.

Although nanoparticles are successfully utilized for the agriculture, pharmaceutical, and biomedical purposes, they pose certain drawbacks and shortcomings. They might render environmental pollution posing a threat to human and animal health. Additionally, higher concentrations of nanoparticle can be toxic to plant systems. Hence, an optimum and minimum concentration of nanoparticles should be standardized for incorporating into the plant system (Zulfiqar and Ashraf, 2021). Reports have also demonstrated that phytotoxic effects of nanoparticles might cause oxidative stress damage to the plant (Rastogi et al., 2019). Therefore, it is crucial to overcome these challenges faced by application of nanoparticles. One such technique would be the employment of polymer-based nanoparticles, such as CsNPs, which are nontoxic to the plant and environment due to their natural origin. Also, toxicity to the plants is highly dependent on the plants species as well as salt levels.

On the other hand, traditional methods of abiotic stress management are time-consuming and labor-intensive while genetically modified crops, which are even though highly effective, are not yet accepted. This calls for an effective alternative and therefore nanotechnology can be a potential alternative strategy to adopt at the farm level. The chapter has tried to summarize the various types of nanomaterials employed for assessing the biochemical and morphological changes of salt-stressed plants. Also, as seen above, most of the studies are still at their preliminary stage and would require further research to form a robust conclusion. Another aspect worth mentioning is that nanoparticle-based research has mainly been performed in controlled environment which overlooks the effects of natural environmental effects on the plant system. Thus, field trials are crucial to have a complete comprehension of influence of nanoparticles on salt-stressed plants. A collective study on multiple stresses would also be useful which will mimic the field environment. Such exploratory studies will lead to value addition to the ongoing nanotechnological applications for revolutionizing agriculture.

References

Abdel Latef, A.A.H., Abu Alhmad, M.F., Abdelfattah, K.E., 2017. The possible roles of priming with ZnO nanoparticles in mitigation of salinity stress in lupine (*Lupinus termis*) plants. J. Plant Growth Regul. 36 (1), 60–70.

Abdel Latef, A.A.H., Srivastava, A.K., El-sadek, M.S.A., Kordrostami, M., Tran, L.S.P., 2018. Titanium dioxide nanoparticles improve growth and enhance tolerance of broad bean plants under saline soil conditions. Land Degrad. Dev. 29 (4), 1065–1073.

Acosta-Motos, J.R., Penella, C., Hernández, J.A., Díaz-Vivancos, P., Sánchez-Blanco, M.J., Navarro, J.M., Gómez-Bellot, M.J., Barba-Espín, G., 2020. Towards a sustainable agriculture: strategies involving phytoprotectants against salt stress. Agronomy 10 (2), 1–32.

Ahmad, I., Akhtar, M.S., 2019. In: Akhtar, M.S. (Ed.), Salt stress, microbes, and plant interactions: causes and solution. Springer International Publishing, Switzerland, pp. 199–215.

Alabdallah, N.M., Alzahrani, H.S., 2020a. The potential mitigation effect of ZnO nanoparticles on [*Abelmoschus esculentus* L. Moench] metabolism under salt stress conditions. Saudi J. Biol. Sci. 27 (11), 3132–3137.

Alabdallah, N.M., Alzahrani, H.S., 2020b. Impact of ZnO nanoparticles on growth of cowpea and okra plants under salt stress conditions. Biosci. Biotechnol. Res. Asia 17 (2), 329–340.

Ali, N.M., Altaey, D.K.A., Altaee, N.H., 2021. The impact of selenium, nano (Sio2) and organic fertilization on growth and yield of potato *Solanum tuberosum* L. under salt stress conditions. IOP Conf. Ser. Earth Environ. Sci. 735 (1), 012042.

Almutairi, Z.M., 2016. Influence of silver nano-particles on the salt resistance of tomato (*Solanum lycopersicum*) during germination. Int. J. Agric. Biol. 18 (2), 449–457.

Anna, H., Mariola, W., Zbigniew, K., 2011. Effect of hydrogel on the turf grass species growing under salt stress. Ann. Warsaw Univ. Life Sci. SGGW Land Reclam. 43 (1), 47–55.

Askary, M., Talebi, S.M., Amini, F., Bangan, A.D.B., 2017. Effects of iron nanoparticles on *Mentha piperita* L. under salinity stress. Biologija 63 (1), 65–75.

Avestan, S., Ghasemnezhad, M., Esfahani, M., Byrt, C.S., 2019. Application of nano-silicon dioxide improves salt stress tolerance in strawberry plants. Agronomy 9 (5), 1–17.

El-Gazzar, N., Almaary, K., Ismail, A., Polizzi, G., 2020. Influence of *Funneliformis mosseae* enhanced with titanium dioxide nanoparticles (TiO_2NPs) on *Phaseolus vulgaris* L. under salinity stress. PLoS One 15 (8), 1–23.

Fathi, A., Zahedi, M., Torabian, S., Khoshgoftar, A., 2017. Response of wheat genotypes to foliar spray of ZnO and Fe_2O_3 nanoparticles under salt stress. J. Plant Nutr. 40 (10), 1376–1385.

Ganjavi, A.S., Oraei, M., Gohari, G., Akbari, A., Faramarzi, A., 2021. Glycine betaine functionalized graphene oxide as a new engineering nanoparticle lessens salt stress impacts in sweet basil (*Ocimum basilicum* L.). Plant Physiol. Biochem. 162, 14–26.

Ghavam, M., 2018. Effect of silver nanoparticles on seed germination and seedling growth in *Thymus vulgaris* L. and *Thymus daenensis* Celak under salinity stress. J. Rangel. Sci. 8 (1), 93–100.

Gohari, G., Mohammadi, A., Akbari, A., Panahirad, S., Dadpour, M.R., Fotopoulos, V., Kimura, S., 2020. Titanium dioxide nanoparticles (TiO_2 NPs) promote growth and ameliorate salinity stress effects on essential oil profile and biochemical attributes of *Dracocephalum moldavica*. Sci. Rep. 10 (1), 1–14.

Hernández-Fuentes, A.D., López-Vargas, E.R., Pinedo-Espinoza, J.M., Campos-Montiel, R.G., Valdés-Reyna, J., Juárez-Maldonado, A., 2017. Postharvest behavior of bioactive compounds in tomato fruits treated with Cu nanoparticles and NaCl stress. Appl. Sci. 7 (10), 980.

Hernández-Hernández, H., González-Moralesto, S., Benavides-Mendoza, A., Ortega-Ortiz, H., Cadenas-Pliego, G., Juárez-Maldonado, A., 2018a. Effects of chitosan—PVA and Cu nanoparticles on the growth and antioxidant capacity of tomato under. Molecules 23 (178), 1–15.

Hernández-Hernández, H., Juárez-Maldonado, A., Benavides-Mendoza, A., Ortega-Ortiz, H., Cadenas-Pliego, G., Sánchez-Aspeytia, D., González-Morales, S., 2018b. Chitosan-PVA and copper nanoparticles improve growth and overexpress the SOD and JA genes in tomato plants under salt stress. Agronomy 8 (175), 2–10.

Ishak, N.M., Kamarudin, S.K., Timmiati, S.N., 2019. Green synthesis of metal and metal oxide nanoparticles via plant extracts: an overview. Mater. Res. Exp. 6 (11), 112004.

Ismail, G., Abou-Zeid, H., 2018. The role of priming with biosynthesized silver nanoparticles in the response of *Triticum aestivum* L. to salt stress. Egypt. J. Bot. 58 (1), 73–85.

Jahangir, S., Javed, K., Bano, A., 2020. Nanoparticles and plant growth promoting rhizobacteria (PGPR) modulate the physiology of onion plant under salt stress. Pak. J. Bot. 52 (4), 1 473–1480.

Jasrotia, P., Kashyap, P.L., Bhardwaj, A.K., Kumar, S., Singh, G.P., 2018. Nanotechnology scope and applications for wheat production and quality enhancement: a review of recent advances. Wheat Barley Res. 10 (1), 1–14.

Kalteh, M., Alipour, Z.T., Ashraf, S., Aliabadi, M.M., Nosratabadi, A.F., 2014. Effect of silica nanoparticles on basil (*Ocimum basilicum*) under salinity stress. J. Chem. Health Risks 4 (3), 49–55.

Karami, A., Sepehri, A., 2018. Nano titanium dioxide and nitric oxide alleviate salt induced changes in seedling growth, physiological and photosynthesis attributes of barley. Zemdirbyste-Agriculture 105 (2), 123–132.

Khalofah, A., Kilany, M., Migdadi, H., 2021. Phytostimulatory influence of *Comamonas testosteroni* and silver nanoparticles on *Linum usitatissimum* l. Under salinity stress. Plants 10 (4), 790.

Khan, Z., Upadhyaya, H., 2019. Impact of nanoparticles on abiotic stress responses in plants: an overview. In: Tripathi, D.K., Ahmad, P., Sharma, S., Chauhan, D.K., Dubey, N.K. (Eds.), Nanomaterials in Plants, Algae and Microorganisms, Concepts and Controversies. 2. Academic Press, Elsevier, ISBN: 978-0-12-811488-9, pp. 305–322.

Khan, I., Raza, M.A., Awan, S.A., Shah, G.A., Rizwan, M., Ali, B., Tariq, R., Hassan, M.J., Alyemeni, M.N., Brestic, M., Zhang, X., Ali, S., Huang, L., 2020. Amelioration of salt induced toxicity in pearl millet by seed priming with silver nanoparticles (AgNPs): the oxidative damage, antioxidant enzymes and ions uptake are major determinants of salt tolerant capacity. Plant Physiol. Biochem. 156, 221–232.

Lowry, G.V., Avellan, A., Gilbertson, L.M., 2019. Opportunities and challenges for nanotechnology in the agri-tech revolution. Nat. Nanotechnol. 14 (6), 517–522.

Mahmoud, N.E., Abdelhameed, R.M., 2021. Superiority of modified graphene oxide for enhancing the growth, yield, and antioxidant potential of pearl millet (*Pennisetum glaucum* L.) under salt stress. Plant Stress 2, 100025.

Martínez-Ballesta, M.C., Zapata, L., Chalbi, N., Carvajal, M., 2016. Multiwalled carbon nanotubes enter broccoli cells enhancing growth and water uptake of plants exposed to salinity. J Nanobiotechnol. 14 (1), 1–14.

Merino, D., Iglesias, M.J., Mansilla, A.Y., Casalongué, C.A., Alvarez, V.A., 2021. Fighting against plant saline stress: development of a novel bioactive composite based on bentonite and l-proline. Clay Clay Miner. 69 (2), 232–242.

Mohamed, A.K.S.H., Qayyum, M.F., Abdel-Hadi, A.M., Rehman, R.A., Ali, S., Rizwan, M., 2017. Interactive effect of salinity and silver nanoparticles on photosynthetic and biochemical parameters of wheat. Arch. Agron. Soil Sci. 63 (12), 1736–1747.

Mohite, P., Kumar, A.R., Zinjarde, S., 2017. Relationship between salt tolerance and nanoparticle synthesis by Williopsis saturnus NCIM 3298. World J. Microbiol. Biotechnol. 33, 163.

Morales-Espinoza, M.C., Cadenas-Pliego, G., Pérez-Alvarez, M., Hernández-Fuentes, A.D., De La Fuente, M.C., Benavides-Mendoza, A., Valdés-Reyna, J., Juárez-Maldonado, A., 2019. Se nanoparticles induce changes in the growth, antioxidant responses, and fruit quality of tomato developed under NaCl stress. Molecules 24 (17), 3030.

Mosavikia, A.A., Mosavi, S.G., Seghatoleslami, M., Baradaran, R., 2020. Chitosan nanoparticle and pyridoxine seed priming improves tolerance to salinity in milk thistle seedling [*Silybum marianum* (L.) Gaertn.]. Not. Bot. Horti Agrobot. Cluj-Napoca 48 (1), 221–233.

Nejatzadeh, F., 2021. Effect of silver nanoparticles on salt tolerance of *Satureja hortensis* l. during in vitro and in vivo germination tests. Heliyon 7 (2), e05981.

Noman, M., Ahmed, T., Shahid, M., Bilal, M., Niazi, K., Qasim, M., Kouadri, F., Abdulmajeed, A.M., Mohammed, S., Ahmad, N., Zafar, M., Ali, S., 2021. Biogenic copper nanoparticles produced by using the *Klebsiella pneumoniae* strain NST2 curtailed salt stress effects in maize by modulating the cellular oxidative repair mechanisms. Ecotoxicol. Environ. Saf. 217, 112264.

Oliveira, H.C., Gomes, B.C.R., Pelegrino, M.T., Seabra, A.B., 2016. Nitric oxide-releasing chitosan nanoparticles alleviate the effects of salt stress in maize plants. Nitric Oxide Biol. Chem. 61, 10–19.

Pandey, K., Lahiani, M.H., Hicks, V.K., Keith Hudson, M., Green, M.J., Khodakovskaya, M., 2018. Effects of carbon-based nanomaterials on seed germination, biomass accumulation and salt stress response of bioenergy crops. PLoS One 13 (8), 1–17.

Pérez-Labrada, F., López-Vargas, E.R., Ortega-Ortiz, H., Cadenas-Pliego, G., Benavides-Mendoza, A., Juárez-Maldonado, A., 2019. Responses of tomato plants under saline stress to foliar application of copper nanoparticle. Plants 8 (6), 1–17.

Pitman, M.G., Läuchli, A., 2002. Global impact of salinity and agricultural ecosystems. In: Läuchli, A., Lüttge, U. (Eds.), Salinity: Environment—Plants—Molecules. Springer, Dordrecht, pp. 3–20.

Poddar, K., Vijayan, J., Ray, S., Adak, T., 2018. Nanotechnology for sustainable agriculture. In: Biotechnology for Sustainable Agriculture. Woodhead Publishing, pp. 281–303.

Qados, A., Moftah, A., 2015. Influence of silicon and nano-silicon on germination, growth and yield of faba bean (*Vicia faba* l.) under salt stress conditions. Am. J. Exp. Agric. 5 (6), 509–524.

Rastogi, A., Zivcak, M., Tripathi, D.K., Yadav, S., Kalaji, H.M., Brestic, M., 2019. Phytotoxic effect of silver nanoparticles in *Triticum aestivum*: improper regulation of photosystem I activity as the reason for oxidative damage in the chloroplast. Photosynthetica 57, 209–216.

Ren, Q.Q., Yuan, X.J., Huang, X.R., Wen, W., Zhao, Y.D., Chen, W., 2013. *In vivo* monitoring of oxidative burst on aloe under salinity stress using hemoglobin and single-walled carbon nanotubes modified carbon fiber ultramicroelectrode. Biosens. Bioelectron. 50, 318–324.
Sabaghnia, N., Janmohammadi, M., 2015. Effect of nano-silicon particles application on salinity tolerance in early growth of some lentil genotypes. Ann. UMCS Biol. 69 (2), 39–55.
Sen, S.K., Chouhan, D., Das, D., Ghosh, R., Mandal, P., 2020. Improvisation of salinity stress response in mung bean through solid matrix priming with normal and nano-sized chitosan. Int. J. Biol. Macromol. 145, 108–123.
Shafiq, F., Iqbal, M., Ali, M., Ashraf, M.A., 2019. Seed pre-treatment with polyhydroxy fullerene nanoparticles confer salt tolerance in wheat through upregulation of H_2O_2 neutralizing enzymes and phosphorus uptake. J. Soil Sci. Plant Nutr. 19 (4), 734–742.
Shafiq, F., Iqbal, M., Ashraf, M.A., Ali, M., 2020. Foliar applied fullerol differentially improves salt tolerance in wheat through ion compartmentalization, osmotic adjustments and regulation of enzymatic antioxidants. Physiol. Mol. Biol. Plants 26 (3), 475–487.
Shah, T., Latif, S., Saeed, F., Ali, I., Ullah, S., Abdullah Alsahli, A., Jan, S., Ahmad, P., 2021. Seed priming with titanium dioxide nanoparticles enhances seed vigor, leaf water status, and antioxidant enzyme activities in maize (*Zea mays* L.) under salinity stress. J. King Saud Univ. Sci. 33 (1), 101207.
Sheikhalipour, M., Esmaielpour, B., Behnamian, M., Gohari, G., Giglou, M.T., Vachova, P., Rastogi, A., Brestic, M., Skalicky, M., 2021. Chitosan–selenium nanoparticle (Cs–Se NP) foliar spray alleviates salt stress in bitter melon. Nanomaterials 11 (3), 1–23.
Wahid, I., Rani, P., Kumari, S., Ahmad, R., Hussain, S.J., Alamri, S., Tripathy, N., Khan, M.I. R., 2022. Biosynthesized gold nanoparticles maintained nitrogen metabolism, nitric oxide synthesis, ions balance, and stabilizes the defense systems to improve salt stress tolerance in wheat. Chemosphere 287, 132142.
Wan, J., Wang, R., Bai, H., Wang, Y., Xu, J., 2020. Comparative physiological and metabolomics analysis reveals that single-walled carbon nanohorns and ZnO nanoparticles affect salt tolerance in: *Sophora alopecuroides*. Environ. Sci. Nano 7 (10), 2968–2981.
Ye, Y., Cota-Ruiz, K., Hernández-Viezcas, J.A., Valdés, C., Medina-Velo, I.A., Turley, R.S., Peralta-Videa, J.R., Gardea-Torresdey, J.L., 2020. Manganese nanoparticles control salinity-modulated molecular responses in *Capsicum annuum* L. through priming: a sustainable approach for agriculture. ACS Sustain. Chem. Eng. 8 (3), 1427–1436.
Yokoi, S., Bressan, R., Hasegawa, P., 2002. Salt stress tolerance of plants. JIRCAS Work Rep. 23 (01), 25–33.
Zafar, S., Hasnain, Z., Aslam, N., Mumtaz, S., Jaafar, H.Z.E., Megat Wahabm, P.E., Qayum, M., Ormenisan, A.N., 2021. Impact of Zn nanoparticles synthesized via green and chemical approach on okra (*Abelmoschus esculentus* L.) growth under salt stress. Sustainability 13 (7), 3694.
Zayed, M., Elkafafi, S., Zedan, A., Dawoud, S., 2017. Effect of nano chitosan on growth, physiological and biochemical parameters of *Phaseolus vulgaris* under salt stress. J. Plant Prod. 8 (5), 577–585.
Zulfiqar, F., Ashraf, M., 2021. Nanoparticles potentially mediate salt stress tolerance in plants. Plant Physiol. Biochem. 160, 257–268.

An overview of application of carbon nanotubes in various agricultural practices

9

Momina[a], *Kafeel Ahmad*[a], *Riti Thapar Kapoor*[b], *and Mohd Rafatullah*[c]
[a]Civil Engineering Department, Jamia Millia Islamia, New Delhi, India, [b]Amity Institute of Biotechnology, Amity University, Noida, India, [c]School of Industrial Technology, Universiti Sains Malaysia, Penang, Malaysia

9.1 Introduction

Agriculture is a main building block of national economy of developing countries. The gross domestic production (GDP) of any nation can be enhanced by increase in food production. The demand for food and energy has been rising due to the growth in the global population, which is estimated to cross 9.6 billion by 2050 (FAO, 2017). Approximately 815 million people are currently undernourished, and it is expected that by the year 2050, two billion people would come under this category. This reason pushes the food and agriculture sector to develop efficient agricultural practices for the reduction of poverty and hunger (Zulfiqar et al., 2019). According to Intergovernmental Panel on Climate Change, emissions of greenhouse gases (GHGs), climate change, food security, soil health, and availability of water have a direct influence on agricultural sector which will reduce yields of the crops. Therefore, an adaptation of efficient practices and regulation of agricultural environment can improve food production for the increasing global population in a sustainable manner (FAO, 2017).

Nanotechnology has shown its application in different fields including solar, environment, medicine, pharmaceuticals, and also agriculture. It has great potential to modernize various agricultural fields such as food additives, agricultural production, animal feed, food processing, and crop improvement (Singh, 2014). Nanotechnology comprises the manufacturing manipulation, and characterization of nanosized materials possessing exclusive properties such as large surface-to-mass ratio and enhanced surface reactivity to improve crop and livestock productivity (Peters et al., 2016). Productivity of agriculture can be amended by development of nanomaterial-inspired genetically modified plants, site-specific drug and gene delivery at cellular level, and nano-array-based genetic modification of plants under stress conditions. Nanotechnology can deliver agrochemicals in a precise way to increase plant growth, disease resistance, and nutrient use (Pramanik et al., 2020). In this regard, various nanomaterials are used in the application of productivity of agriculture and food such as metal oxide, liposomes, dendrimers, and carbon-based nanomaterials.

Agricultural Nanobiotechnology. https://doi.org/10.1016/B978-0-323-91908-1.00002-X

On the contrary, nanotoxicity is causing risk to agriculture, health, and the environment. Nanoparticles can translocate into the body tissues/organs and lymphatic system due to their small sizes after their internalization. However, some nanoparticles can cause irreversible damage to cells and organelles depending on the composition and size. Parameters that classified the effect of nanoparticle toxicity in plants are still not well defined. The reason can be the different characteristics of nanoparticles because some of them are toxic to plants and some are nontoxic. Therefore, we will discuss application of nanomaterials in different agricultural practices. Moreover, the effect of carbon-based nanomaterials on plant growth, plant-associated soil microbes, and their biotic and abiotic stress tolerance has been discussed. The chapter also highlights the toxicological effects of carbon-based nanomaterials.

9.2 Overview of the application of nanomaterials in agriculture and food sector

Nanomaterials have at least one dimension and range from 1 to 100 nm. Properties of nanomaterials at this nanoscale differ in terms of physical, biological, and chemical as compared to larger particles. Due to minute size, they can be used for revolutionizing various fields such as food, environmental sciences, and agriculture (Kaphle et al., 2017). The applications of nanotechnology in agriculture are mentioned in Fig. 9.1. Nanomaterials are employed to develop nanosensors to enhance the detection of contaminants, pests, nutrients, biomolecules, etc. (Sharon et al., 2010). They also used to provide industries with devices to alter genes and generate novel new organisms (Torney et al., 2007). In one of the recent research works, complete replacement of genes was carried out from one bacterium to another (ETC group, 2007). These materials improve the quality of food, safety of food, fortification, and act as biosensors for detection of adulterated food (Ranjan et al., 2014). Various nanomaterials are also

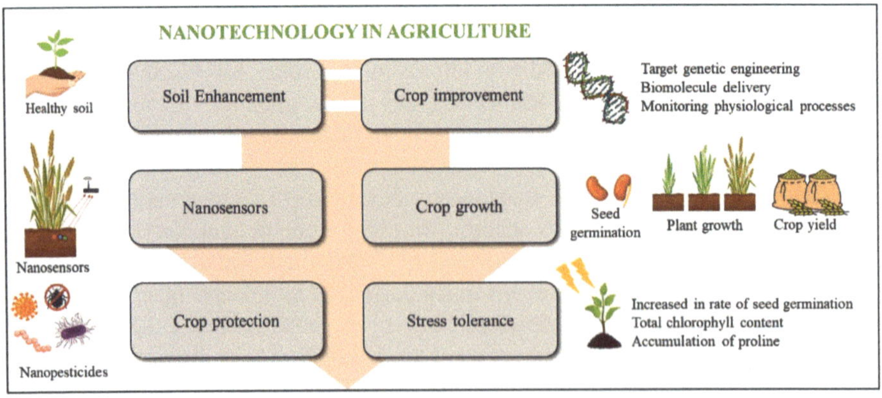

Fig. 9.1 Application of nanotechnology in different agricultural practices.

employed for crop growth and its improvements such as absorption, translocation, accumulation, and overall development of crop plants. The small size and large surface area of nanomaterials have the advantage of penetrating the seed coats, afterward emerging roots, and provide better availability of molecules to the radicals of seeds. It also enhances the rate of germination, length of roots and shoot, yield and production of seedlings in several crop plants. Other than the size of nanomaterials, zeta potential, shape, and aggregation of nanomaterials are also considered as important physico-chemical properties for the enhancement of photosynthetic activity and nitrogen metabolism (Peters et al., 2016). Various nanomaterials such as metal, metal oxide, dendrimers, liposomes, carbon-based, and nanocomposite materials are stated commonly for the application of agricultural activities.

9.2.1 Metal and metal oxide nanoparticles

The utilization of metal-based nanoparticles in agricultural activities is increasing due to their physico-chemical and biological properties. The optical, thermal, surface, and electrical properties of metal-based nanoparticles are better than their bulk native compounds. Copper-based nanoparticles are recognized as antifungal agents which eliminates the fungal species causing the disease to various plants (Kanhed et al., 2014). It shows good activity against the plant pathogenic species of fungi like *Phoma destructiva, Alternaria alternata,* and *Fusarium oxysporum.* Magnesium nanoparticles can enhance the absorption of solar light by leaves. It increases the length of plant root, the number of tips, and vegetative biomass which eventually increases crop production and is considered as perks of agriculture purposes (Rathore and Tarafdar, 2015). Manganese nanoparticles played a significant function in photosynthesis to enhance growth of plants (Pradhan et al., 2013). Iron and iron oxide nanoparticles were found to be effective for elimination of harmful pesticides (Siddiqi et al., 2016). However, some reports revealed the adverse impact of metal-based nanoparticles such as use of silica nanoparticles was found harmful against plants due to decrement in pH value after the addition of nanoparticles (Tripathi et al., 2017). The lack of understanding of a mechanism of nanoparticle interaction with plants at a large scale and lack of standardization for nanoparticles phytotoxic assay are limiting their full utility in the agricultural sector. Therefore, more research is required to understand the effect of nanoparticles on the plant's physiology and biology. Fate and behavior of metal-oxide nanoparticles in agriculture system are also needed to be investigated due to the aggregation behavior of the nanoparticles. Understanding and explaining the aggregation nature of metal-based NPs requires well-established and convincing colloid science theories. The nanoscale properties such as composition, structure, and shape make it difficult to understand metal-based nanoparticles assembly within the biological systems like plant tissues (Hotze et al., 2010). One of the challenges that studies have highlighted in toxicity evaluation of metal-based nanoparticles is that the toxicity is either caused by metal ions that are liberated or due to nanoparticles itself or combination of both.

9.2.2 Liposomes

Liposomes act as a delivery agent, which is able to transport any type of molecules such as pesticides, drugs, enzymes, fertilizers, and antimicrobials. These lipid vesicles are mainly formed by dispersing amphiphilic molecules in aqueous solvents (Kaphle et al., 2018). Vesicles contain a bi-layer membrane that consists of various functional molecules in the internal cavities of the membrane which are beneficial for their application in various fields including the agriculture and the food sector. Liposomes are circular which may contain hydrophilic and lipophilic and made up of the single membrane of size <30nm whereas bi-layer membrane has a size of 30–100nm (Saneja et al., 2020) as shown in Fig. 9.2. These nano-based carriers are one of the most favorable encapsulated agents which increase their usage in the food processing and dairy industries. They are able to carry vitamins, enzymes, and flavoring agents for food. Further, lipid-based microencapsulation techniques improve the entrapment ability of materials with a wide range of solubility, which eventually increase the production of food with minimum cost. Other functional properties such as stability, biocompatibility with food constituents, low manufacturing cost, and their feasibility as antimicrobials and nutraceuticals are advantageous (Taylor et al., 2005). Liposomes are applied in the ripening and hardening of cheese by the addition of edible enzymes before isolation of curds, which can reduce time and cost. However, the disadvantage

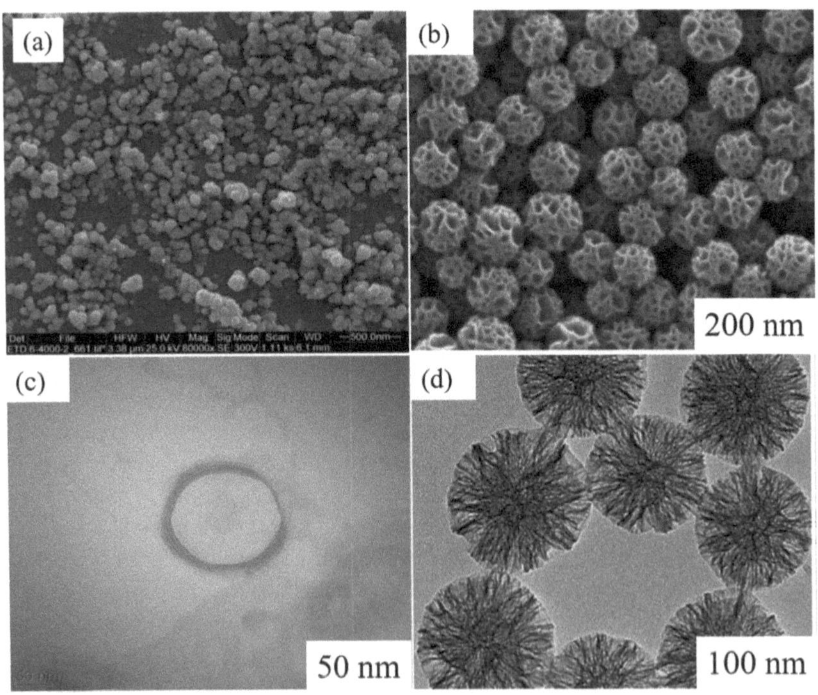

Fig. 9.2 SEM images of (A) liposome and (B) dendrimer, and TEM images of (C) liposome and (D) dendrimer (Du et al., 2014; Sarabandi et al., 2019; Sebaaly et al., 2015).

of the addition of enzymes is the rapid breakdown of proteins of casein. Therefore, proteinases were encapsulated in liposome nanoparticles to make firm cheese that prevented any offensive enzymatic effect (Ahn-Yoon et al., 2003). In another study, lipids encapsulated in liposome nanoparticles decreased the cheese hardness and enhanced elasticity and tackiness of cheddar cheese (Kheadr et al., 2002). Few reports on the encapsulation of essential oils in liposome nanoparticles found them stable which did not degrade in the natural environment in presence of light, oxygen, and temperature. Liposome protects the fluidity and stability of essential oil at low temperature (3–4°C) for at least 6 months (Sherry et al., 2013). Further, the encapsulation of essential oil in liposomes enhances the antimicrobial activity (Khosravi-Darani et al., 2016) and protects the plants from degradation after exposure to ultra-violet light (Sebaaly et al., 2015). It has also found its application in protecting and preserving vitamins in various types of food. The stability of liposome-encapsulated vitamin C and antioxidant efficacy has been reported (Taylor et al., 2005). However, some studies stated toxic effect of cationic liposomes such as proliferation, inflammation, and carcinogenic activity. Other issues such as lack of potential sterilization techniques, nonstability, and scale-up problems need to be considered for better food processing.

9.2.3 Dendrimers

Dendrimer's central core consists of the branched, tree structure, and attached with functional groups as shown in Fig. 9.2. Basically, they are synthetic polymers that consist of branched topologies that make them different from linear polymer. Dendrimer properties such as degradability, biological activity, and solubility are influenced by the type of molecules that forms the polymer. Commonly used dendrimers are polyamines, polyamidoamines, polyester, poly(aryl ethers), deoxyribonucleic acid (DNA), and carbohydrates (Kaphle et al., 2018). They have multiple applications but they are widely used as delivery agents for transport of drugs, nucleotides, and other molecules in agricultural practices. They are considered as an important vehicle for transporting materials to tissues, enhancement of the solubility, adhesion, improves penetration of active agent in soil, and help in protecting active agents from enzymatic degradation caused by seeds or plants (Elmer and White, 2018). However, some investigations found that the presence of the amine group and other cationic groups in dendrimer exerts toxicity (Agashe et al., 2006; Jevprasesphant et al., 2003). Some other effects such as hematological, hemolytic, and immunological effects were also observed while using dendrimer.

9.2.4 Carbon-related nanomaterials

Carbon-based nanomaterials are diverse in shape, function, size, and type of synthesis. Commonly used carbon nanomaterials are graphene oxides (reduced and oxidized form), fullerene (C_{60}; buckyballs), and carbon nanotubes (single-walled or multiwalled). They can be in the form of tube-shaped, horn-shaped, spherical, or ellipsoidal. They are used in different applications like textiles, paints, medicine, and electronics due to their extraordinary optical, thermal, electrical, and mechanical properties (Srivastava et al., 2015). The application of carbon-based materials in agricultural practices has been explored as smart drug delivery agents and vehicles. They can be loaded with appropriate drugs and can transfer drugs or active agents to particular target sites within the cell. Among various carbon-based nanomaterials, carbon nanotubes (CNTs) have gained interest among researchers for different applications in agricultural practices. The number of papers published for CNTs from the last 9 years is enhanced due to their extra-ordinary physical and chemical qualities as compared to fullerene and graphene oxide (Fig. 9.3).

9.3 Carbon nanotubes (CNTs): Types and structure

The structure of CNTs is very similar to graphite, with sp^2 bonded carbon. The CNTs structure has an impact on its properties such as electrical conductivity, thermal conductivity, density, and structure of the lattice. As their name "Carbon Nanotube" implies, their diameter is in nano range (generally 1–100 nm). CNTs are of two types: single-walled carbon nanotubes (SWNTs) and multiwalled carbon nanotubes (MWNTs) (Iijima and Ichihashi, 1993). MWNTs are cylindrical which are made up of single graphene sheets and have a vacant center. External diameter of the cylinder varies from 2 to 100 nm, and inner diameter varies from 1 to 3 nm. Each cylinder is placed coaxially with other cylinder, and it is made up of graphene sheet that covers empty central part. The length of the cylinder ranges from 1 to several micrometer (Dresselhaus et al., 1996). SWNTs consist of single cylinder made up of graphene sheet and having a diameter of 0.4–2 nm (Qin et al., 2000; Wang et al., 2000). Nanotubes are hexagonally close-packed bundles that are linked by Van der Waals forces. Based on the diameter and helicity, SWNTs can be either metallic or semiconducting.

9.4 Modification approaches for CNTs using biological molecules

Modification or functionalization using biological molecules is a procedure of immobilizing biomolecules on the surface of the material which provides specific role like catalytic activity or biospecificity (Zhou et al., 2019). Various studies are performed to immobilize biological molecules (enzymes, nucleic acid, proteins, and antibodies), on CNTs for their application in agriculture, primarily as biosensors (Besteman et al., 2003; Blanford et al., 2008; Tîlmaciu and Morris, 2015). Functionalization approaches of CNTs can be classified into two types: covalent and noncovalent functionalization.

Covalent functionalization involves the introduction of chemical groups on the CNTs sidewalls to form amine-functionalized CNTs, carboxylated functionalized CNTs, sulfhydryl functionalized CNTs, etc. The functional groups of CNTs covalently attach with immobilized biomolecules on CNTs surface. Carboxyl groups of CNTs once activated, they can react directly with primary amines of biomolecules

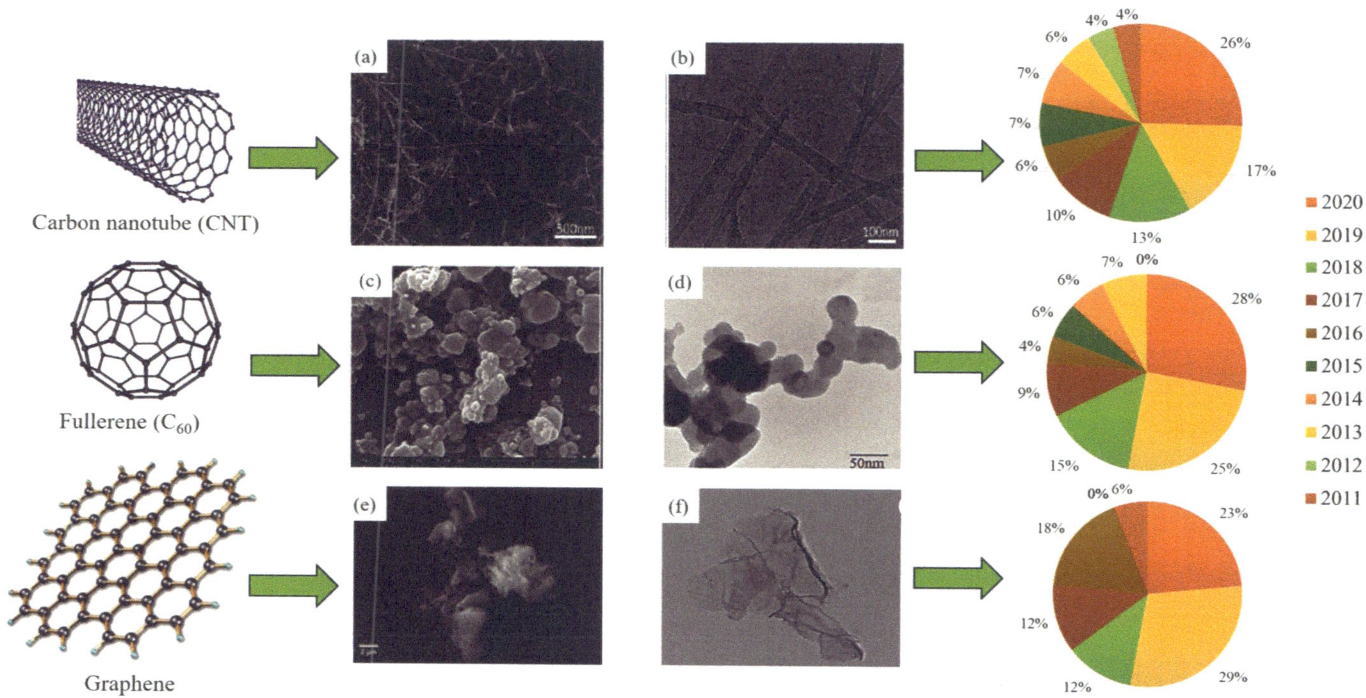

Fig. 9.3 Structure of carbon-based nanomaterials (CNT, fullerene, and graphene), SEM and TEM images of (A, B) CNT (Ma et al., 2016), (C, D) fullerene (Ding et al., 2018), and (E, F) graphene oxide (Kumari et al., 2020), and papers published in the last 9 years (2011 – 2020) in the application of carbon-based nanomaterials (CNT, fullerene, and graphene oxide) in agricultural practices.
Data taken from Scopus.

using carbodiimide compounds. One such example of carbodiimide compound is N-ethyl-N′-(3-dimethyl aminopropyl) carbodiimide hydrochloride (EDC) which can be easily replaced by amine groups in biomolecules increasing its dispersibility in an aqueous medium (Singh et al., 2016). Several proteins, antibodies, and enzymes are studied for the immobilization of molecules on carboxylated CNTs (Singh et al., 2013; Zeng et al., 2007). Covalent functionalization can affect the sp^2 structure of CNTs which could impact its electrical and mechanical properties. However, *noncovalent functionalization* retains the intrinsic properties of CNTs. Noncovalent functionalization is performed by electrostatic interactions, π-π interaction, and CH-π interaction between biomolecules and CNTs. One such example of noncovalent functionalization of CNTs applying biological molecule is 1-pyrenebutanoic acid succinate succinimidyl ester on hydrophobic surface of SWNTs (Chen et al., 2001). Succinimidyl ester groups were replaced by a primary or secondary amine group of protein and attached by a nucleophilic substitution reaction. In another study, poly(methacrylic acid-*co*-acrylamide) was utilized to immobilize myoglobin on MWNTs (Zhan et al., 2014) which increased the stability and solubility of MWCNTs. In this study, myoglobin was immobilized on poly(methacrylic acid-co-acrylamide)/MWCNT via electrostatic interaction. In all noncovalent functionalization, the biocompatibility of CNTs in different applications including agriculture was improved. However, the bond between the coupling agents and CNTs in noncovalent functionalization is not as strong as in covalent functionalization, which may affect the stability of the system (Zhou et al., 2019).

9.5 Agriculture and environmental applications of CNTs

According to Food and Agriculture Organization, United Nations (FAO), production of enough food and agricultural products has been a challenge due to the degradation and depletion of land and water. Nanotechnology has been considered an effective tool to enhance agricultural yield by using less farming land. Precision farming develops sustainable agriculture by minimizing energy and waste (Raliya et al., 2018). Nanofertilizers provide higher nutrients having high plant uptake efficiencies and allow less release of active components. Several studies are carried out on the nutrients absorption and transport in various parts of plants. The high surface area and solubility in different solvents and minute particle size of nanofertilizers can fulfill the requirement of plant roots as compared to conventional fertilizers (Mahanta et al., 2019). Plant protection products such as herbicides and pesticides based on nanomaterials can increase the efficacy of pesticides, stability of payloads, and reduce the requirement of high dose of application to plants (Hayles et al., 2017). Use of nanoencapsulation of pesticides is a promising strategy that improves the conventional formulations of pesticides and developed a new delivery system for plant protection. Mancozeb and Zineb pesticides were encapsulated in MWCNT-graft poly(citric acid) composite to study their effect on *Alternaria alternata* fungi. The encapsulation of Zineb pesticides showed extraordinary effect as compared to Mancozeb, and its encapsulation in MWCNT-g-PCA depends on the pH and time of stirring (Sarlak et al., 2014). The carbon-based nanomaterials provide around 40% contribution in farming sector either as active components, sensors, or as additives (Zaytseva and Neumann, 2016).

Environmental and agricultural pollution is the main problem of world. Various types of contaminants are increasing in an environment that requires proper and efficient control methods. The efficiency of traditional techniques used nowadays is causing risk to sustainability of the environment. Activated carbon is widely employed in treatment of wastewater due to its large surface area and its capability to adsorb various types of pollutants with good efficiency. However, their adsorption efficiency is low and ineffective for various types of microorganisms (Majeed et al., 2020). Therefore, CNTs are used to improve wastewater management. Many examples are available in the literature; for example, the adsorption capacity of CNTs for microcystins (cyanobacterial toxins) is higher than that of activated carbon. Various herbicides, insecticides, and pesticides can be adsorbed from wastewater (Srivastava et al., 2011; Wang et al., 2013) by CNTs due to the high chemical affinity, mechanical and thermal stability, high surface area, and possess antimicrobial activities.

9.5.1 Effect of CNTs in plant growth and development

Most of studies performed on plant-nanomaterial interactions are not only focused on economic risk assessment but also concentrated on how to increase crop yields and decrease agrochemicals utilization. The studies are performed on analyzing effect of plant growth, seed germination, cell culture, and gene expression using CNTs. Table 9.1 shows how multiple CNTs have varying effects on different plant species, their varieties, and stage of their developmental growth.

The three phases of germination are imbibition, lag phase, and radicle protrusion by the testa. Seed germination increases with an increase in the hydration of the seed, where hydration increases the RNA and protein synthesis, faster embryo growth, high ATP availability, and heal deteriorated seeds (Pourkhaloee et al., 2011). Effect of SWNTs on seed germination and seedling growth of pepper, salvia, and tall fescue was reported. The highest germination rate was observed for a medium concentration of SWNTs (Pourkhaloee et al., 2011). Seeds of *Phaseolus mungo* and *Brassica juncea* were reported to germinate in 10, 20, and 40 mg/L of MWCNTs, and nontoxic nature of CNTs was reported (Ghodake et al., 2010). The positive effect of single-walled carbon nanohorns (SWCNHs) at 0.025, 0.05, and 0.10 mg/L concentration was observed on corn, barley, rice, soybean, and tomato roots and cells. It was also nontoxic to plant cell culture (tobacco callus) (Lahiani et al., 2015). In the case of MWCNTs, a lot of studies have reported on improved germination of tomato, rice, barley, soybean, and maize (Khodakovskaya et al., 2009; Lahiani et al., 2013; Nair et al., 2012; Zhai et al., 2015). An increase in germination rate can be due to increase in water uptake inside the seeds. It was due to penetration of nanoparticles inside the seed coat which supported allowance of water uptake in seeds. Further, MWCNT was also studied to activate the expression of genes in seed-located channel (aquaporin) of soybean, barley, and maize (Lahiani et al., 2013).

Table 9.1 Effect of CNTs on the growth of various plant species.

CNT	Plant species	Concentration	Effects	References
SWNTs	Salvia, pepper, and tall fescue	10 mg/L, 30 mg/L	Increase in germination of seeds	Pourkhaloee et al. (2011)
o-MWCNTs	Wheat (*Triticum aestivum*)	40, 80, and 160 mg/L	Increase in root length of wheat seedlings	Wang et al. (2012)
SWNTs	Maize (*Zea mays*)	20 mg/L	Seed germination is not affected	Yan et al. (2013)
SWCNHs	Barley, corn, soybean, rice, switchgrass, and tomato roots	0.025, 0.05, and 0.10 mg/L	Activated seed germination and plant growth	Lahiani et al. (2015)
MWCNTs	Tomato (*Solanum lycopersicum*)	10–40 μg/mL	Rate of germination of seeds enhanced	Khodakovskaya et al. (2009)
MWCNTs	Rice (*Oryza. sativa*)	25, 75 and 100 μg/mL	Faster germination for seeds in media with MWCNT	Nair et al. (2012)
MWCNTs	Barley, corn, soybean	50, 100, and 200 μg/mL	Accelerated seed germination	Lahiani et al. (2013)
MWCNTs	Maize (*Z. mays*) and soybean (*Glycine* max)	10.0 to 50.0 mg/L	Increase in seed germination	Zhai et al. (2015)
SWNTs	Fig plants (*Ficus carica*)	2 μg/mL and 6 μg/mL	Development of roots and stem of a plant	Flores et al. (2013)
CNTs	Tomato (*S. lycopersicum*)	50, 100, and 200 mg/L	Produces 2–4 times more flowers and fruit	Khodakovskaya et al. (2013)
SWCNTs	Salvia (*Salvia officinalis*), pepper (*Capsicum annuum*), and tall fescue (*Festuca arundinacea*)	10–30 mg/L	Seed germination increased	Pourkhaloee et al. (2011)
Pristine and oxidized MWCNTs	Mustard (*Brassica juncea*)	2.3–46 mg/L	Increases root and shoot, improves seed germination	Mondal et al. (2011)

SWNT-COOH	Blackberry (*Rubus adenotrichos*)	4, 8, and 12 μg/mL	Improved plant growth and roots	Flores et al. (2014)
MWCNTs	Broccoli	10, 20, 40 and 60 mg/L	Improved growth of plants	Martínez-Ballesta et al. (2016)
MWCNTs-COOH	Maize (*Z. mays*), wheat (*T. aestivum*), peanut (*Arachis hypogaea*), and garlic bulb (*Allium sativum*)	20 and 50 μg/mL	Positive effect on elongation of roots and shoot	Srivastava and Rao (2014)
CNTs	Gram (*Cicer arietinum*) plants	6 μg/mL	Positive effect on elongation of roots and shoot	Tripathi et al. (2011)
CNTs	Tomato (*Lycopersicum esculentum*), onion (*Allium cepa*), turnip (*Brassica rapa*), and radish (*Raphanus sativus*)	0, 10, 20, and 40 mg/L	Tomato and radish shoots increased	Haghighi and Teixeira Da Silva (2014)
Hollow CNTs	*Oryza sativa* L.	0, 10, 50, and 300 mg/L	Decreases rice growth	Hao et al. (2016)
MWCNTs	*Cucurbita pepo* L	0, 125, 250, 500, and 1000 μg/mL	Toxic effect on seed germination	Hatami (2017)
MWCNTs-COOH	*Ocimum basilicum*	0, 25, 50, and 100 mg/L	Causes toxic effect on plant growth	Gohari et al. (2020b)
MWCNTs and SWCNTs	*S. lycopersicum*	1 and 10 mg/kg	Delay in growth and flowering, SWCNT increases salicylic acid.	Jordan et al. (2020)

Interaction of young seedlings with SWNTs was reported on various plant species such as fig plants (*Ficus carica*) (Flores et al., 2013), tomato (*Solanum lycopersicum*) (Khodakovskaya et al., 2011), and maize (*Zea mays*) (Yan et al., 2013). It was found that the application of CNTs to tomato plant produced 2–4 times more fruits per plant endorsing it as an effective approach for the development of plants (Khodakovskaya et al., 2013). On the contrary, MWCNTs also activated the expression of genes in tomato plants for its development and division of cells (Villagarcia et al., 2012). Pourkhaloee et al. (2011) observed effect of SWCNTs on salvia, pepper, and tall fescue, as well. The development in the formation of seedling biomass increases from 10 to 30 mg/L of SWNTs while a negative response was observed at 40 mg/L. Similar results were reported by functionalized SWNTs (SWNT-COOH) on blackberry (*Rubus adenotrichos*) (Flores et al., 2014). Studies have shown the growth of broccoli plants under salinity as of the extending applicability of nanobiotechnology (Martínez-Ballesta et al., 2016). CNTs also have a positive impact on the root and shoot lengths of maize, wheat, peanut, and garlic bulb (Srivastava and Rao, 2014). It was reported that the shoot and roots of plants were fully germinated in less time. In general, CNTs are acknowledged to promote growth and development of plants; however, their physiological functions are still unknown.

Further, effect of functionalized and nonfunctionalized CNTs on the growth of the plant is mainly dependent on the plant type as reported by Cañas et al. (2008). The length of roots of onion and cucumber plants was increased by nonfunctionalized CNTs, whereas roots of tomatoes were inhibited by the same. Moreover, roots of carrots and cabbage were not affected much by either functionalized or nonfunctionalized CNTs (Cañas et al., 2008). Current research shows positive effect of MWCNTs on the growth of plants is mainly due to alteration in lipid composition, permeability of plasma membrane, stiffness in roots, and enhancement in gibberellin amount (Zhang et al., 2017). The synthesis and formation CNTs as transporters for the plant cell is a crucial step for labeling/imaging of plant cells and genetic transformation. These are also essential for our further understanding of plant cell biology and crop production through extensive researches on the transporting capability of the synthesized CNTs (Vera-Reyes et al., 2018).

9.5.2 Effect of CNTs in plant-associated soil microbes

The protection of plant-associated soil microbes is important because they have a direct effect on the quality of soil due to several processes such as decomposition of organic wastes, nutrient cycling, and their interactions with terrestrial plant species. CNTs showed great application potential in various fields of agricultural practices including soil microbes. However, generally, most of the carbon-related nanomaterials show a toxic effect on soil microbes. Very few studies discovered neutral and positive impact of CNTs on soil microbes. Shrestha et al. (2013) stated no toxic effect of MWCNTs on soil microbial activities after exposure of three months. Authors observed no effect for low concentration of CNTs (10, 100, and 1000 mg/kg), but, at 10000 mg/kg, bacterial species (*Rhodococcus, Cellulomonas, Nocardioides,* and *Pseudomonas*) increased. CNTs can be toxic to soil microorganisms because they can either increase or decrease the concentration of organic compounds and can alter the availability of nutrients. Moreover, the diffusion of modified toxic metal into cell membranes can disturb the soil/plant environmental balance (Wang et al., 2014). Therefore, a study on the impact of CNTs on soil is required to understand its effect on the soil ecology and cycling, and the bioavailability of nutrients (Vaishlya et al., 2015).

A decrease in protozoan, bacterial, and enzyme activities was also reported by CNTs. For example, the short-duration impact of MWCNTs on soil microbes such as extracellular enzymes (cellobiohydrolase, 1,4-*N*-acetylglucosaminidase, 1,4-b-glucosidase, xylosidase, phosphatase) and microbial biomass C and N in soils were studied by Chung et al. (2011). It was observed that more amount of nanoparticles reduced microbial biomass in soil. Similar behavior was observed (Jin et al., 2013; Rodrigues et al., 2013) with SWNTs on soil microorganisms. Moreover, CNTs also cause cytotoxic effects because they might produce reactive oxygen species (ROS) which can damage proteins, DNA, and membrane of the cell apart from the disruption of the bacterial cell wall. CNTs were also reported to have adverse effects on the soil microbial community. Rodrigues et al. (2013) reported that number of colony-forming units (CFUs) decreased in 72 h and fungal community was not able to make progress in fourteen days after the exposure of soil to SWCNTs. Further, the prolonged effect of CNTs and functionalized CNTs on soil microbial activities is still lacking. According to Ge et al. (2016), exposure to MWCNTs for one year reduced the DNA content and altered the bacterial communities in soil. Kerfahi et al. (2015) studied that bacterial diversity of soil does not get affected by the raw and acid-functionalized MWCNTs; however, a higher concentration of functionalized CNTs can alter the soil bacterial communities' composition. Effect of various types of CNTs on soil microbial community composition and structure was investigated by Wu et al. (2019). The type of CNTs is a key determinant of the soil microbial activity; for example, MWCNTs cause great alteration in number of soil bacteria and alpha diversity, whereas SWCNTs change the beta diversity more intensively than MWCNTs.

9.5.3 Modified CNTs as biosensors

The concept of CNTs-based biosensors was developed in 1962 by Clarke. Biosensors mainly have two components: one is a biologically sensitive element and the second is transducers as illustrated in Fig. 9.4. Generally used bioreceptors or biomolecules are proteins (e.g., enzymes, antibodies, cell receptors), nucleic acid, microorganisms, or biological tissues (Allen et al., 2007; Guo, 2013). Commonly used transducers are optical, piezoelectric, thermal, or electrochemical which are used to change concentration of samples to detectable physical signals (current, absorbance, mass, etc.). The large surface area/volume ratio and good conductance make CNTs superior biosensors and interact with biological molecules. The hollow structure of CNTs gives an advantage of good adsorption of enzymes. These enzymes-functionalized CNTs were used in the amperometric biosensors. Different enzymes were used for different analytes for catalysis or reduction of biological molecules. Glucose oxidase (GOD) (Yang et al., 2015), aflatoxin oxidase (Li et al., 2011), lactic acid oxidase

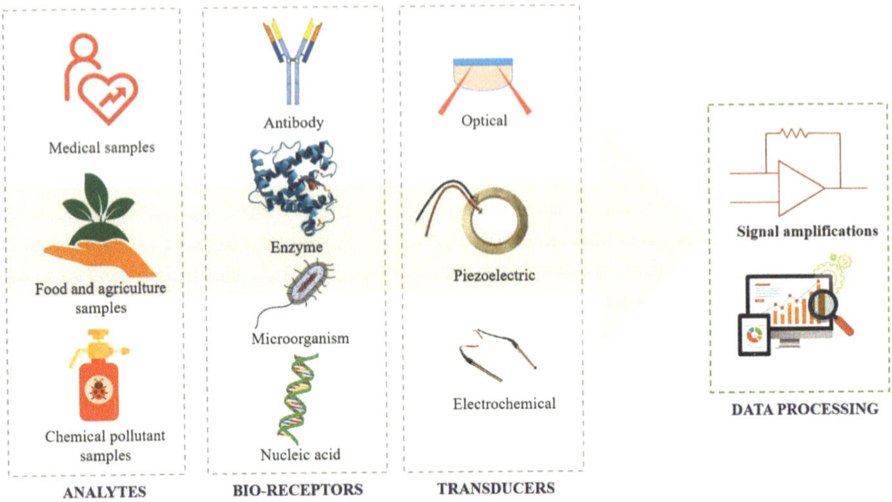

Fig. 9.4 Different components of CNTs-based biosensors.

(Pavlidis et al., 2010), reduced nicotinamide adenine dinucleotide (NADH) (Liang et al., 2013; Meng et al., 2009), hydrogen peroxide (Wang et al., 2003), etc., are some of the reported enzymes used for different analytes.

CNTs have strong sensitivity toward food adulterants and contaminants. Food-contaminating pathogenic bacteria can be detected by identifying a whole bacterial cell or bacterial genetic materials. SWCNTs were used as a DNA sensor for identification of *Salmonella* using N-ethyl-N′-(3-dimethyl aminopropyl) carbodiimide hydrochloride (3-dimethyl aminopropyl) with electrochemical transducers (Weber et al., 2011). The sensor sensitivity was observed at a target concentration of 1×10^{-9} mol/L DNA. Furthermore, MWCNTs-based aptasensor was developed by indium tin oxide (ITO) for identification of *Salmonella enteritidis* and *Salmonella typhimurium*. The sensor showed sensitivity at 5.5×10^1 and 6.7×10^1 CFU/mL, respectively (Hasan et al., 2018). CNTs have superb luminescence properties, luminous intensity, and excellent wavelength conversion function, which make them effective optical biosensors (Momina and Ahmad, 2021; Vithanage et al., 2017; Yang et al., 2015). Aflatoxins, deoxynivalenol, fumonisin, zearalenone, ochratoxin, and patulin are carcinogenic and toxic compounds found in contaminated food (Kumar et al., 2017; Malhotra et al., 2015). Enzyme-functionalized CNTs have great potential to monitor and detect food-contaminating toxins. CNTs were also reported to detect pesticides, fertilizers, and other toxic chemicals; for example, MWCNTs-silica-based composite was used for the detection of Cd^{2+} ions (Bagheri et al., 2013). Sunset yellow and tetrazine are synthetic food dyes that are found in food products like bakery products, beverages, and dairy products. Presence of dyes more than their permissible limits in food products can show an adverse effect on neurological behavior. Carbon ceramic electrode modified with MWCNTs ionic liquid nanocomposites was used to detect sunset yellow and tetrazine dye with a detection limit of

0.045 mg/L and 0.59 mg/L (Majidi et al., 2013). Neonicotinoid-based pesticides, such as acetamiprid, can cause agonistic effects against nicotinic acetylcholine receptors (Jampílek and KráL'Ová, 2015). MWCNTs-graphene oxide nanocomposite decorated with gold nanoparticles was developed as an ultrasensitive label-free electrochemical impedimetric aptasensor for identification of acetamiprid with a detection limit of 1.7×10^{-14} mol/dm^3 (Jampílek and KráL'Ová, 2015).

Quinoxaline-2-carboxylic acid is a constituent of food substances which shows carcinogenic and mutagenic effects. The MWCNTs functionalized with chitosan, and fabricated on glassy carbon electrode gel, were used as an electrochemical sensor to detect quinoxaline-2-carboxylic acid at very less concentration up to 4.4×10^{-7} mol/L (Yang et al., 2013). The immobilized fructose dehydrogenase enzyme-based biosensor was developed for D-fructose detection (Antiochia et al., 2013). The biosensor is based on a CNTs-based electrode modification with an osmium redox polymer which was applied as a redox mediator to shuttle electrons between SWCNTs and immobilized enzyme. Biosensor showed a less detection limit of 1 μM with good stability and rapid response.

Molecules such as ROS, glucose, nitric oxide (NO), and nitroaromatics can be detected and monitored in some parts of the leaf using SWCNTs sensors with high fluorescent intensity and high spatiotemporal resolutions (Giraldo et al., 2014; Wong et al., 2017). Varied concentration of analytes in plant cells such as ROS and calcium binds with the SWCNTs sensor, changes the shifting wavelength, and rapidly changes pattern with time and subcellular positions (Kruss et al., 2014; Zhang et al., 2013). In recent decades, wearable sensing technology is increasing for skin application and human clothing. It brings new ideas in agricultural practices that offer wearable nanoelectronic circuits on the surface of plants to monitor volatile compounds at low concentration (Heikenfeld et al., 2018). An electronic device based on SWCNTs channels and graphitic electrodes was developed and used to trace the levels of gaseous compounds up to 5 ppm concentration on the leaf surface of plants (Lee et al., 2014). The CNTs-based biosensor is also used for the detection of plant hormone (ethylene) which is a beneficial indicator for the ripening of fruits (Mittal et al., 2020).

9.5.4 CNTs in abiotic and biotic stress tolerance

Any unfavorable conditions or substances that affect growth and development of plant are known as plant stress. Plant organisms can reach stress conditions because of various aspects which are divided into two groups: abiotic and biotic stresses. Effects of stresses on the plants growth are mentioned in Fig. 9.5. In addition to the function of carbon-based nanoparticles in the growth of the plants, they also play important role in the plant protection from biotic and abiotic stresses. Both abiotic and biotic stresses can give rise to various issues that can affect development of plants. To improve the survival rate of plants, they need to adapt in such adverse environmental conditions (Hakim et al., 2018).

Drought stress can reduce the water content and alter the various physiological and molecular behaviors of plants such as reduced photosynthesis which further results in

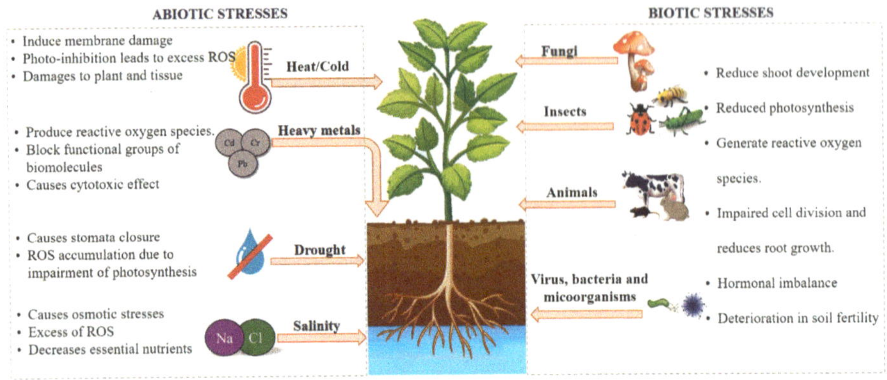

Fig. 9.5 Effect of abiotic and biotic stresses on plants development and growth.

generation of ROS. Cell membrane, DNA, and proteins get damaged due to ROS production (Duhan et al., 2017). Therefore, maintaining and controlling the level of ROS to establish an equilibrium is crucial, which can be addressed by the use of various carbon-based nanomaterials. However, very few studies are reported to date on the modulation of stress tolerance of plants employing different types of CNTs. Hatami (2017) investigated the impact of MWCNTs on *Cucurbita pepo* L under two levels of water potential: well-watered (0 MPa) and water stress (−0.1 MPa) for 14 days. MWCNTs showed a negative effect on the germination and growth of *C. pepo* L under both water potentials. Effect of MWCNTs in vegetable crop broccoli under saline regimes improved the photosynthesis rate and water absorption (Martínez-Ballesta et al., 2016). Moreover, in this study, MWCNTs altered the lipid composition and improved the aquaporin transduction which enhanced the water uptake and reduced the salinity. In another study, exposure of MWCNTs to salt stress reduced the ROS level, thiobarbituric acid production, and Na^+/K^+ ratio in rapeseed (*Brassica napus* L.) (Zhao et al., 2019). Recently, Gohari et al. (2020a,b) showed that low concentration of MWCNTs had an ameliorating effect on salt-stressed *Ocimum basilicum* plants while high concentration shows phytotoxic effects.

9.6 Conclusions and future perspectives

Apart from having various advantages, a lot of questions are still unanswered on the toxicity of CNTs. Toxicity of CNTs is dependent on its concentration, plant species on which it is applied, and their growth/exposure conditions. Generally, the cationic surface of CNTs is believed to be more toxic in comparison with the anionic and neutral surface. Acid-functionalized CNTs can cause more cytotoxicity as compared to neutral tubes. Further, the smaller particles of SWCNTs can be easily translocated into lymph nodes due to phagocytosis by macrophages and larger particles may aggregate into organs (Kaphle et al., 2018). The process of CNTs toxicity toward soil microbes is not clear. According to Jackson et al. (2013), CNTs interact strongly with bacteria via electrostatic attraction which may disrupt or damage the cell membrane. Another possible mechanism is that CNTs can interact with organelles to damage the protein and degrade the DNA resulting in cell death (Jackson et al., 2013). Therefore, it is necessary to design different realistic exposures to observe the toxic impact of CNTs on plants and soil. There are many knowledge gaps toward effective use of CNTs in agricultural products which are as follows:

- CNTs play a significant role in water uptake during seed germination. But proper mechanism of uptake of water by nanoparticles is not clear and requires in-depth investigations.
- Translocation of SWCNTs causes toxicity to plants that include damage to DNA and protein, aggregation into organs, and alteration of gene expression in plants. Therefore, the behavior and mechanism of transmission and translocation of SWCNTs in different parts of plants require further research.
- Various studies reported the positive effect of accumulation of CNTs in plants, but, post-uptake behavior of CNTs in plants also needs to be investigated.
- CNTs-based biosensor has shown unique properties and has been researched extensively over the past. However, the prevailing challenges in utilizing biosensor CNTs in agriculture still require improvements in surface functionalization techniques to improve their overall performance. The current practices that are utilized for surface functionalization such as covalent funtionalization are still not matured enough from its applicability and reproducibility standpoint. Further research efforts are required to find more reliable and simple methods for surface functionalization of CNTs-based biosensors.
- MWCNTs show positive effects in plants under salt stress by refining the capacity of water uptake. However, these biochemical changes are also needed to be investigated for movement of MWCNTs in different plant parts such as fruits and leaves in adverse environmental conditions.
- Environmental stresses such as extreme temperatures, water stress, and salinity really play a crucial role in removing physiological effects of CNTs in the plant system. However, there is still a need for the study, in detail, of the behavior of CNTs plant cell interface under stress conditions for the high-value crops such as their tolerance to water deficit, especially in semiarid and arid regions.
- Fate and behavior of CNTs in plants should be studied by high-precision analytical techniques such as synchrotron-based techniques to identify presence of small-size nanoparticles in plants.

The nanotechnology field is developing very rapidly over the last decade and bringing positive impacts across multiple fields of science and technology including biotechnology and agriculture. A variety of nanoparticles such as metal and metal oxide nanoparticles, liposomes, dendrimers, and carbon-related nanomaterials are limelight and developing for their applicability in agriculture practices. The size of CNTs is of utmost significance because it can pierce the plant cell wall and improve the biomass and crop yield by changing their metabolic functions. CNTs also exhibit the vast potential for their use in nanosensors and nanobiosensor due to their specific optical, thermal, chemical, and mechanical properties. However, challenge of phytotoxicity due to the small size of CNTs needs to be carefully addressed and requires extensive research. Further progress in the field of CNTs will be able to address the above-mentioned challenges that would bring sustainable and environmentally friendly solutions.

Acknowledgments

The authors extend their appreciation to Jamia Millia Islamia and Universiti Sains Malaysia through the Fundamental Research Grant Scheme (FRGS) (grant number, 203/PTEKIND/6711822) for providing facilities.

References

Agashe, H.B., Dutta, T., Garg, M., Jain, N.K., 2006. Investigations on the toxicological profile of functionalized fifth-generation poly(propylene imine) dendrimer. J. Pharm. Pharmacol. 58, 1491–1498. https://doi.org/10.1211/jpp.58.11.0010.

Ahn-Yoon, S., DeCory, T.R., Baeumner, A.J., Durst, R.A., 2003. Ganglioside-liposome immunoassay for the ultrasensitive detection of cholera toxin. Anal. Chem. 75, 2256–2261. https://doi.org/10.1021/ac026428t.

Allen, B.L., Kichambare, P.D., Star, A., 2007. Carbon nanotube field-effect-transistor-based biosensors. Adv. Mater. 19, 1439–1451. https://doi.org/10.1002/adma.200602043.

Antiochia, R., Vinci, G., Gorton, L., 2013. Rapid and direct determination of fructose in food: a new osmium-polymer mediated biosensor. Food Chem. 140, 742–747. https://doi.org/10.1016/j.foodchem.2012.11.023.

Bagheri, H., Afkhami, A., Shirzadmehr, A., Khoshsafar, H., Khoshsafar, H., Ghaedi, H., 2013. Novel potentiometric sensor for the determination of Cd^{2+} based on a new nano-composite. Int. J. Environ. Anal. Chem. 93, 578–591. https://doi.org/10.1080/03067319.2011.649741.

Besteman, K., Lee, J.-O., Wiertz, F.G.M., Heering, H.A., Dekker, C., 2003. Enzyme-coated carbon nanotubes as single-molecule biosensors. Nano Lett. 3, 727–730. https://doi.org/10.1021/nl034139u.

Blanford, C.F., Foster, C.E., Heath, R.S., Armstrong, F.A., 2008. Efficient electrocatalytic oxygen reduction by the "blue" copper oxidase, laccase, directly attached to chemically modified carbons. Faraday Discuss. 140, 319–335. https://doi.org/10.1039/b808939f.

Cañas, J.E., Long, M., Nations, S., Vadan, R., Dai, L., Luo, M., Ambikapathi, R., Lee, E.H., Olszyk, D., 2008. Effects of functionalized and nonfunctionalized single-walled carbon nanotubes on root elongation of select crop species. Environ. Toxicol. Chem. 27, 1922. https://doi.org/10.1897/08-117.1.

Chen, R.J., Zhang, Y., Wang, D., Dai, H., 2001. Noncovalent sidewall functionalization of single-walled carbon nanotubes for protein immobilization. J. Am. Chem. Soc. 123, 3838–3839. https://doi.org/10.1021/ja010172b.

Chung, H., Son, Y., Yoon, T.K., Kim, S., Kim, W., 2011. The effect of multi-walled carbon nanotubes on soil microbial activity. Ecotoxicol. Environ. Saf. 74, 569–575. https://doi.org/10.1016/j.ecoenv.2011.01.004.

Ding, X., Chen, X., Chen, X., Zhao, X., Li, N., 2018. A QCM humidity sensor based on fullerene/graphene oxide nanocomposites with high quality factor. Sens. Actuators B 266, 534–542. https://doi.org/10.1016/j.snb.2018.03.143.

Dresselhaus, M.S., Dresselhaus, G., Eklund, P.C., 1996. Science of Fullerenes and Carbon Nanotubes: Their Properties and Applications. Elsevier, San Diego.

Du, X., Shi, B., Tang, Y., Dai, S., Qiao, S.Z., 2014. Label-free dendrimer-like silica nanohybrids for traceable and controlled gene delivery. Biomaterials 35, 5580–5590. https://doi.org/10.1016/j.biomaterials.2014.03.051.

Duhan, J.S., Kumar, R., Kumar, N., Kaur, P., Nehra, K., Duhan, S., 2017. Nanotechnology: the new perspective in precision agriculture. Biotechnol. Rep. https://doi.org/10.1016/j.btre.2017.03.002.

Elmer, W., White, J.C., 2018. The future of nanotechnology in plant pathology. Annu. Rev. Phytopathol. 56, 111–133. https://doi.org/10.1146/annurev-phyto-080417-050108.

ETC group, 2007. Extreme Genetic Engineering: An Introduction to Synthetic Biology. Ottawa, Canada.

FAO, 2017. The Future of Food and Agriculture- Trends and Challenges. Rome.

Flores, D., Chaves, J.S., Chacón, R., Schmidt, A., 2013. A novel technique using SWCNTs to enhanced development and root growth of fig plants (*Ficus carica*). In: Technical Proceedings of the 2013 NSTI Nanotechnology Conference and Expo (NSTI-Nanotech'13), pp. 167–170.

Flores, D., Chacón, R., Alvarado, L., Schmidt, A., Alvarado, C., Chaves, J., 2014. Effect of using two different types of carbon nanotubes for blackberry (Rubus adenotrichos) in vitro plant rooting, growth and histology. Am. J. Plant Sci. 05, 3510–3518. https://doi.org/10.4236/ajps.2014.524367.

Ge, Y., Priester, J.H., Mortimer, M., Chang, C.H., Ji, Z., Schimel, J.P., Holden, P.A., 2016. Long-term effects of multiwalled carbon nanotubes and graphene on microbial communities in dry soil. Environ. Sci. Technol. 50, 3965–3974. https://doi.org/10.1021/acs.est.5b05620.

Ghodake, G., Seo, Y.D., Park, D., Lee, D.S., 2010. Phytotoxicity of carbon nanotubes assessed by brassica juncea and phaseolus Mungo. J. Nanoelectron. Optoelectron. 5, 157–160. https://doi.org/10.1166/jno.2010.1084.

Giraldo, J.P., Landry, M.P., Faltermeier, S.M., McNicholas, T.P., Iverson, N.M., Boghossian, A.A., Reuel, N.F., Hilmer, A.J., Sen, F., Brew, J.A., Strano, M.S., 2014. Plant nanobionics approach to augment photosynthesis and biochemical sensing. Nat. Mater. 13, 400–408. https://doi.org/10.1038/nmat3890.

Gohari, G., Mohammadi, A., Akbari, A., Panahirad, S., Dadpour, M.R., Fotopoulos, V., Kimura, S., 2020a. Titanium dioxide nanoparticles (TiO_2 NPs) promote growth and ameliorate salinity stress effects on essential oil profile and biochemical attributes of *Dracocephalum moldavica*. Sci. Rep. 10, 1–14. https://doi.org/10.1038/s41598-020-57794-1.

Gohari, G., Safai, F., Panahirad, S., Akbari, A., Rasouli, F., Dadpour, M.R., Fotopoulos, V., 2020b. Modified multiwall carbon nanotubes display either phytotoxic or growth promoting and stress protecting activity in *Ocimum basilicum* L. in a concentration-dependent manner. Chemosphere 249. https://doi.org/10.1016/j.chemosphere.2020.126171, 126171.

Guo, X., 2013. Single-molecule electrical biosensors based on single-walled carbon nanotubes. Adv. Mater. 25, 3397–3408. https://doi.org/10.1002/adma.201301219.

Haghighi, M., Teixeira Da Silva, J.A., 2014. The effect of carbon nanotubes on the seed germination and seedling growth of four vegetable species. J. Crop. Sci. Biotechnol. 17, 201–208. https://doi.org/10.1007/s12892-014-0057-6.

Hakim, Ullah, A., Hussain, A., Shaban, M., Khan, A.H., Alariqi, M., Gul, S., Jun, Z., Lin, S., Li, J., Jin, S., Munis, M.F.H., 2018. Osmotin: a plant defense tool against biotic and abiotic stresses. Plant Physiol. Biochem. 123, 149–159. https://doi.org/10.1016/j.plaphy.2017.12.012.

Hao, Y., Yu, F., Lv, R., Ma, C., Zhang, Z., Rui, Y., Liu, L., Cao, W., Xing, B., 2016. Carbon nanotubes filled with different ferromagnetic alloys affect the growth and development of rice seedlings by changing the C:N ratio and plant hormones concentrations. PLoS One 11, 1–19. https://doi.org/10.1371/journal.pone.0157264.

Hasan, M.R., Pulingam, T., Appaturi, J.N., Zifruddin, A.N., Teh, S.J., Lim, T.W., Ibrahim, F., Leo, B.F., Thong, K.L., 2018. Carbon nanotube-based aptasensor for sensitive electrochemical detection of whole-cell Salmonella. Anal. Biochem. 554, 34–43. https://doi.org/10.1016/j.ab.2018.06.001.

Hatami, M., 2017. Toxicity assessment of multi-walled carbon nanotubes on *Cucurbita pepo* L. under well-watered and water-stressed conditions. Ecotoxicol. Environ. Saf. 142, 274–283. https://doi.org/10.1016/j.ecoenv.2017.04.018.

Hayles, J., Johnson, L., Worthley, C., Losic, D., 2017. Nanopesticides: a review of current research and perspectives. In: New Pesticides and Soil Sensors, pp. 193–225, https://doi.org/10.1016/b978-0-12-804299-1.00006-0.

Heikenfeld, J., Jajack, A., Rogers, J., Gutruf, P., Tian, L., Pan, T., Li, R., Khine, M., Kim, J., Wang, J., Kim, J., 2018. Wearable sensors: modalities, challenges, and prospects. Lab Chip 18, 217–248.

Hotze, E.M., Phenrat, T., Lowry, G.V., 2010. Nanoparticle aggregation: challenges to understanding transport and reactivity in the environment. J. Environ. Qual. 39, 1909–1924. https://doi.org/10.2134/jeq2009.0462.

Iijima, S., Ichihashi, T., 1993. Single-shell carbon nanotubes of 1-nm diameter. Nature 363, 603–605. https://doi.org/10.1038/363603a0.

Jackson, P., Jacobsen Raun, N., Baun, A., Birkedal, R., Kühnel, D., Jensen Alstrup, K., Vogel, U., Wallin, H., 2013. Bioaccumulation and ecotoxicity of carbon nanotubes. Chem. Cent. J. 7, 154. https://doi.org/10.1186/1752-153X-7-154.

Jampílek, J., KráL'Ová, K., 2015. Application of nanotechnology in agriculture and food industry, its prospects and risks. Ecol. Chem. Eng. S. 22, 321–361. https://doi.org/10.1515/eces-2015-0018.

Jevprasesphant, R., Penny, J., Jalal, R., Attwood, D., McKeown, N.B., D'Emanuele, A., 2003. The influence of surface modification on the cytotoxicity of PAMAM dendrimers. Int. J. Pharm. 252, 263–266. https://doi.org/10.1016/S0378-5173(02)00623-3.

Jin, L., Son, Y., Yoon, T.K., Kang, Y.J., Kim, W., Chung, H., 2013. High concentrations of single-walled carbon nanotubes lower soil enzyme activity and microbial biomass. Ecotoxicol. Environ. Saf. 88, 9–15. https://doi.org/10.1016/j.ecoenv.2012.10.031.

Jordan, J.T., Oates, R.P., Subbiah, S., Payton, P.R., Singh, K.P., Shah, S.A., Green, M.J., Klein, D.M., Cañas-Carrell, J.E., 2020. Carbon nanotubes affect early growth, flowering time and phytohormones in tomato. Chemosphere 256. https://doi.org/10.1016/j.chemosphere.2020.127042, 127042.

Kanhed, P., Birla, S., Gaikwad, S., Gade, A., Seabra, A.B., Rubilar, O., Duran, N., Rai, M., 2014. In vitro antifungal efficacy of copper nanoparticles against selected crop pathogenic fungi. Mater. Lett. 115, 13–17. https://doi.org/10.1016/j.matlet.2013.10.011.

Kaphle, A., Navya, P.N., Umapathi, A., Chopra, M., Daima, H.K., 2017. Nanomaterial impact, toxicity and regulation in agriculture. In: Food and Environment. Springer, Cham, pp. 205–242, https://doi.org/10.1007/978-3-319-58496-6_8.

Kaphle, A., Navya, P.N., Umapathi, A., Daima, H.K., 2018. Nanomaterials for agriculture, food and environment: applications, toxicity and regulation. Environ. Chem. Lett. 16, 43–58. https://doi.org/10.1007/s10311-017-0662-y.

Kerfahi, D., Tripathi, B.M., Singh, D., Kim, H., Lee, S., Lee, J., Adams, J.M., 2015. Effects of functionalized and raw multi-walled carbon nanotubes on soil bacterial community composition. PLoS One 10. https://doi.org/10.1371/journal.pone.0123042, e0123042.

Kheadr, E.E., Vuillemard, J.C., El-Deeb, S.A., 2002. Acceleration of cheddar cheese lipolysis by using liposome-entrapped lipases. J. Food Sci. 67, 485–492. https://doi.org/10.1111/j.1365-2621.2002.tb10624.x.

Khodakovskaya, M., Dervishi, E., Mahmood, M., Xu, Y., Li, Z., Watanabe, F., Biris, A.S., 2009. Carbon nanotubes are able to penetrate plant seed coat and dramatically affect seed germination and plant growth. ACS Nano 3, 3221–3227. https://doi.org/10.1021/nn900887m.

Khodakovskaya, M.V., De Silva, K., Nedosekin, D.A., Dervishi, E., Biris, A.S., Shashkov, E.V., Galanzha, E.I., Zharov, V.P., 2011. Complex genetic, photothermal, and photoacoustic analysis of nanoparticle-plant interactions. Proc. Natl. Acad. Sci. U. S. A. 108, 1028–1033. https://doi.org/10.1073/pnas.1008856108.

Khodakovskaya, M.V., Kim, B.-S., Kim, J.N., Alimohammadi, M., Dervishi, E., Mustafa, T., Cernigla, C.E., 2013. Carbon nanotubes as plant growth regulators: effects on tomato growth, reproductive system, and soil microbial community. Small 9, 115–123. https://doi.org/10.1002/smll.201201225.

Khosravi-Darani, K., Khoosfi, M.E., Hosseini, H., 2016. Encapsulation of Zataria multiflora Boiss. Essential oil in liposome: antibacterial activity against *E. Coli* O157:H7 in broth media and minced beef. J. Food Saf. 36, 515–523. https://doi.org/10.1111/jfs.12271.

Kruss, S., Landry, M.P., Vander Ende, E., Lima, B.M.A., Reuel, N.F., Zhang, J., Nelson, J., Mu, B., Hilmer, A., Strano, M., 2014. Neurotransmitter detection using corona phase molecular recognition on fluorescent single-walled carbon nanotube sensors. J. Am. Chem. Soc. 136, 713–724. https://doi.org/10.1021/ja410433b.

Kumar, V., Guleria, P., Mehta, S.K., 2017. Nanosensors for food quality and safety assessment. Environ. Chem. Lett. 15, 165–177. https://doi.org/10.1007/s10311-017-0616-4.

Kumari, S., Sharma, P., Yadav, S., Kumar, J., Vij, A., Rawat, P., Kumar, S., Sinha, C., Bhattacharya, J., Srivastava, C.M., Majumder, S., 2020. A novel synthesis of the graphene oxide-silver (GO-Ag) nanocomposite for unique physiochemical applications. ACS Omega 5, 5041–5047. https://doi.org/10.1021/acsomega.9b03976.

Lahiani, M.H., Dervishi, E., Chen, J., Nima, Z., Gaume, A., Biris, A.S., Khodakovskaya, M.V., 2013. Impact of carbon nanotube exposure to seeds of valuable crops. ACS Appl. Mater. Interfaces 5, 7965–7973. https://doi.org/10.1021/am402052x.

Lahiani, M.H., Chen, J., Irin, F., Puretzky, A.A., Green, M.J., Khodakovskaya, M.V., 2015. Interaction of carbon nanohorns with plants: uptake and biological effects. Carbon N. Y. 81, 607–619. https://doi.org/10.1016/j.carbon.2014.09.095.

Lee, K., Park, J., Lee, M.-S., Kim, J., Gwan Hyun, B., Jun Kang, D., Na, K., Young Lee, C., Bien, F., Park, J.-U., 2014. In-situ synthesis of carbon nanotube – graphite electronic devices and their integrations onto surfaces of live plants and insects. Am. Chem. Soc. 14, 2647–2657. https://doi.org/10.1021/nl500513n.

Li, S.c., Chen, J.h., Cao, H., Yao, D.s., Liu, D.l., 2011. Amperometric biosensor for aflatoxin B1 based on aflatoxin-oxidase immobilized on multiwalled carbon nanotubes. Food Control 22, 43–49. https://doi.org/10.1016/j.foodcont.2010.05.005.

Liang, B., Li, L., Tang, X.J., Lang, Q., Wang, H., Li, F., Shi, J., Shen, W., Palchetti, I., Mascini, M., Liu, A., 2013. Microbial surface display of glucose dehydrogenase for amperometric glucose biosensor. Biosens. Bioelectron. 45, 19–24. https://doi.org/10.1016/j.bios.2013.01.050.

Ma, Z., Wang, Y., Yang, Y., Yousaf, M., Zou, M., Cao, A., Han, R.P.S., 2016. Flexible hybrid carbon nanotube sponges embedded with SnS_2 from tubular nanosheaths to nanosheets as free-standing anodes for lithium-ion batteries. RSC Adv. 6, 30098–30105. https://doi.org/10.1039/c6ra01143h.

Mahanta, N., Dambale, A., Rajkhowa, M., 2019. Nutrient use efficiency through Nano fertilizers. Int. J. Chem. Stud. 7, 2839–2842.

Majeed, N., Panigrahi, K.C.S., Sukla, L.B., John, R., Panigrahy, M., 2020. Application of carbon nanomaterials in plant biotechnology. Mater. Today Proc. 30, 340–345. https://doi.org/10.1016/j.matpr.2020.01.618.

Majidi, M.R., Fadakar, B.B.R., Naseri, A., 2013. Carbon nanotube-ionic liquid (CNT-IL) nanocomposite modified sol-gel derived carbon-ceramic electrode for simultaneous determination of sunset yellow and tartrazine in food samples. Food Anal. Methods 6, 1388–1397. https://doi.org/10.1007/s12161-012-9556-6.

Malhotra, B.D., Srivastava, S., Augustine, S., 2015. Biosensors for food toxin detection: carbon nanotubes and graphene. Mater. Res. Soc. Symp. Proc. 1725, 33–43. https://doi.org/10.1557/opl.2015.165.

Martínez-Ballesta, M.C., Zapata, L., Chalbi, N., Carvajal, M., 2016. Multiwalled carbon nanotubes enter broccoli cells enhancing growth and water uptake of plants exposed to salinity. J. Nanobiotechnol. 14, 42. https://doi.org/10.1186/s12951-016-0199-4.

Meng, L., Wu, P., Chen, G., Cai, C., Sun, Y., Yuan, Z., 2009. Low potential detection of glutamate based on the electrocatalytic oxidation of NADH at thionine/single-walled carbon nanotubes composite modified electrode. Biosens. Bioelectron. 24, 1751–1756. https://doi.org/10.1016/j.bios.2008.09.001.

Mittal, D., Kaur, G., Singh, P., Yadav, K., Ali, S.A., 2020. Nanoparticle-based sustainable agriculture and food science: recent advances and future outlook. Front. Nanotechnol. 2, 1–37. https://doi.org/10.3389/fnano.2020.579954.

Momina, Ahmad, K., 2021. Study of different polymer nanocomposites and their pollutant removal efficiency: review. Polymer (Guildf) 217. https://doi.org/10.1016/j.polymer.2021.123453, 123453.

Mondal, A., Basu, R., Das, S., Nandy, P., 2011. Beneficial role of carbon nanotubes on mustard plant growth: an agricultural prospect. J. Nanopart. Res. 13, 4519–4528. https://doi.org/10.1007/s11051-011-0406-z.

Nair, R., Mohamed, M.S., Gao, W., Maekawa, T., Yoshida, Y., Ajayan, P.M., Kumar, D.S., 2012. Effect of carbon nanomaterials on the germination and growth of rice plants. J. Nanosci. Nanotechnol. 12, 2212–2220. https://doi.org/10.1166/jnn.2012.5775.

Pavlidis, I.V., Tsoufis, T., Enotiadis, A., Gournis, D., Stamatis, H., 2010. Functionalized multi-wall carbon nanotubes for lipase immobilization. Adv. Eng. Mater. 12, B179–B183. https://doi.org/10.1002/adem.200980021.

Peters, R.J.B., Bouwmeester, H., Gottardo, S., Amenta, V., Arena, M., Brandhoff, P., Marvin, H.J.P., Mech, A., Moniz, F.B., Pesudo, L.Q., Rauscher, H., Schoonjans, R., Undas, A.K., Vettori, M.V., Weigel, S., Aschberger, K., 2016. Nanomaterials for products and application in agriculture, feed and food. Trends Food Sci. Technol. 54, 155–164. https://doi.org/10.1016/j.tifs.2016.06.008.

Pourkhaloee, A., Haghigh, M., Saharkhiz, M.J., Jouzi, H., Doroodmand, M.M., 2011. Carbon nanotubes can promote seed germination via seed coat penetration. Carbon nanotubes can promote seed germination via seed coat penetration. Seed Technol. 33, 155–169.

Pradhan, S., Patra, P., Das, S., Chandra, S., Mitra, S., Dey, K.K., Akbar, S., Palit, P., Goswami, A., 2013. Photochemical modulation of biosafe manganese nanoparticles on *Vigna radiata*: a detailed molecular, biochemical, and biophysical study. Environ. Sci. Technol. 47, 13122–13131. https://doi.org/10.1021/es402659t.

Pramanik, P., Krishnan, P., Maity, A., Mridha, N., Mukherjee, A., Rai, V., 2020. Application of nanotechnology in agriculture. Environ. Nanotechnol., 317–348. https://doi.org/10.1007/978-3-030-26668-4_9.

Qin, L.C., Zhao, X., Hirahara, K., Miyamoto, Y., Ando, Y., Iijima, S., 2000. Materials science: the smallest carbon nanotube. Nature 408, 50. https://doi.org/10.1038/35040699.

Raliya, R., Saharan, V., Dimkpa, C., Biswas, P., 2018. Nanofertilizer for precision and sustainable agriculture: current state and future perspectives. J. Agric. Food Chem. 66, 6487–6503. https://doi.org/10.1021/acs.jafc.7b02178.

Ranjan, S., Dasgupta, N., Chakraborty, A.R., Melvin Samuel, S., Ramalingam, C., Shanker, R., Kumar, A., 2014. Nanoscience and nanotechnologies in food industries: opportunities and research trends. J. Nanopart. Res. 16, 1–23. https://doi.org/10.1007/s11051-014-2464-5.

Rathore, I., Tarafdar, J.C., 2015. Perspectives of biosynthesized magnesium nanoparticles in foliar application of wheat plant. J. Bionanosci. 9, 209–214. https://doi.org/10.1166/jbns.2015.1296.

Rodrigues, D.F., Jaisi, D.P., Elimelech, M., 2013. Toxicity of functionalized single-walled carbon nanotubes on soil microbial communities: implications for nutrient cycling in soil. Environ. Sci. Technol. 47, 625–633. https://doi.org/10.1021/es304002q.

Saneja, A., Panda, A.K., Lichtfouse, E., 2020. Pharmaceutical technology for natural products delivery. In: Fundamentals and Applications, Sustainable Agriculture Reviews. vol. 1. Springer International Publishing, Cham, https://doi.org/10.1007/978-3-030-41838-0.

Sarabandi, K., Jafari, S.M., Mohammadi, M., Akbarbaglu, Z., Pezeshki, A., Khakbaz Heshmati, M., 2019. Production of reconstitutable nanoliposomes loaded with flaxseed protein hydrolysates: stability and characterization. Food Hydrocoll. 96, 442–450. https://doi.org/10.1016/j.foodhyd.2019.05.047.

Sarlak, N., Taherifar, A., Salehi, F., 2014. Synthesis of nanopesticides by encapsulating pesticide nanoparticles using functionalized carbon nanotubes and application of new nanocomposite for plant disease treatment. J. Agric. Food Chem. 62, 4833–4838. https://doi.org/10.1021/jf404720d.

Sebaaly, C., Jraij, A., Fessi, H., Charcosset, C., Greige-Gerges, H., 2015. Preparation and characterization of clove essential oil-loaded liposomes. Food Chem. 178, 52–62. https://doi.org/10.1016/j.foodchem.2015.01.067.

Sharon, M., Choudhary, A.K., Kumar, R., 2010. Nanotechnology in agricultural diseases and food safety. J. Phytology 2010, 83–92.

Sherry, M., Charcosset, C., Fessi, H., Greige-Gerges, H., 2013. Essential oils encapsulated in liposomes: a review. J. Liposome Res. 23, 268–275. https://doi.org/10.3109/08982104.2013.819888.

Shrestha, B., Acosta-Martinez, V., Cox, S.B., Green, M.J., Li, S., Cañas-Carrell, J.E., 2013. An evaluation of the impact of multiwalled carbon nanotubes on soil microbial community structure and functioning. J. Hazard. Mater. 261, 188–197. https://doi.org/10.1016/j.jhazmat.2013.07.031.

Siddiqi, K.S., ur Rahman, A., Tajuddin, Husen, A., 2016. Biogenic fabrication of iron/iron oxide nanoparticles and their application. Nanoscale Res. Lett. 11, 1–13. https://doi.org/10.1186/s11671-016-1714-0.

Singh, S.B., 2014. Nanotechnology in Agri-food production: an overview. Nanotechnol. Sci. Appl. 7, 31–53. https://doi.org/10.2147/NSA.S39406.

Singh, C., Srivastava, S., Ali, M.A., Gupta, T.K., Sumana, G., Srivastava, A., Mathur, R.B., Malhotra, B.D., 2013. Carboxylated multiwalled carbon nanotubes based biosensor for aflatoxin detection. Sens. Actuators B 185, 258–264. https://doi.org/10.1016/j.snb.2013.04.040.

Singh, B., Lohan, S., Sandhu, P.S., Jain, A., Mehta, S.K., 2016. Functionalized carbon nanotubes and their promising applications in therapeutics and diagnostics. In: Nanobiomaterials in Medical Imaging: Applications of Nanobiomaterials. Elsevier Inc, pp. 455–478, https://doi.org/10.1016/B978-0-323-41736-5.00015-7.

Srivastava, A., Rao, D., 2014. Enhancement of seed germination and plant growth of wheat, maize, peanut and garlic using multiwalled carbon nanotubes. Eur. Chem. Bull. 3, 502–504.

Srivastava, M., Abhilash, P.C., Singh, N., 2011. Remediation of lindane using engineered nanoparticles. J. Biomed. Nanotechnol. 7, 172–174. https://doi.org/10.1166/jbn.2011.1255.

Srivastava, V., Gusain, D., Sharma, Y.C., 2015. Critical review on the toxicity of some widely used engineered nanoparticles. Ind. Eng. Chem. Res. 54, 6209–6233. https://doi.org/10.1021/acs.iecr.5b01610.

Taylor, T.M., Davidson, P.M., Bruce, B.D., Weiss, J., 2005. Liposomal nanocapsules in food science and agriculture. Crit. Rev. Food Sci. Nutr. 45, 587–605. https://doi.org/10.1080/10408390591001135.

Tîlmaciu, C.M., Morris, M.C., 2015. Carbon nanotube biosensors. Front. Chem. 3, 59. https://doi.org/10.3389/fchem.2015.00059.

Torney, F., Trewyn, B.G., Lin, V.S.Y., Wang, K., 2007. Mesoporous silica nanoparticles deliver DNA and chemicals into plants. Nat. Nanotechnol. 2, 295–300. https://doi.org/10.1038/nnano.2007.108.

Tripathi, S., Sonkar, S.K., Sarkar, S., 2011. Growth stimulation of gram (*Cicer arietinum*) plant by water soluble carbon nanotubes. Nanoscale 3, 1176–1181. https://doi.org/10.1039/c0nr00722f.

Tripathi, D.K., Singh, S., Singh, S., Srivastava, P.K., Singh, V.P., Singh, S., Prasad, S.M., Singh, P.K., Dubey, N.K., Pandey, A.C., Chauhan, D.K., 2017. Nitric oxide alleviates silver nanoparticles (AgNps)-induced phytotoxicity in *Pisum sativum* seedlings. Plant Physiol. Biochem. 110, 167–177. https://doi.org/10.1016/j.plaphy.2016.06.015.

Vaishlya, O.B., Osipov, N.N., Guseva, N.V., 2015. Carbon nanotubes influence the enzyme activity of biogeochemical cycles of carbon, nitrogen, phosphorus and the pathogenesis of plants in annual agroecosystems. IOP Conf. Ser.: Mater. Sci. Eng., 012082. https://doi.org/10.1088/1757-899X/91/1/012082.

Vera-Reyes, I., Vázquez-Núñez, E., Lira-Saldivar, R., Méndez-Argüello, B., 2018. Effects of nanoparticles on germination, growth, and plant crop development. Agric. Nanobiotechnol., 77–110. https://doi.org/10.1007/978-3-319-96719-6_5.

Villagarcia, H., Dervishi, E., de Silva, K., Biris, A.S., Khodakovskaya, M.V., 2012. Surface chemistry of carbon nanotubes impacts the growth and expression of water channel protein in tomato plants. Small 8, 2328–2334. https://doi.org/10.1002/smll.201102661.

Vithanage, M., Seneviratne, M., Ahmad, M., Sarkar, B., Ok, Y.S., 2017. Contrasting effects of engineered carbon nanotubes on plants: a review. Environ. Geochem. Health 39, 1421–1439. https://doi.org/10.1007/s10653-017-9957-y.

Wang, N., Tang, Z.K., Li, G.D., Chen, J.S., 2000. Materials science: single-walled 4Å carbon nanotube arrays. Nature 408, 50–51. https://doi.org/10.1038/35040702.

Wang, J., Musameh, M., Lin, Y., 2003. Solubilization of carbon nanotubes by Nafion toward the preparation of amperometric biosensors. J. Am. Chem. Soc. 125, 2408–2409. https://doi.org/10.1021/ja028951v.

Wang, X., Han, H., Liu, X., Gu, X., Chen, K., Lu, D., 2012. Multi-walled carbon nanotubes can enhance root elongation of wheat (Triticum aestivum) plants. J. Nanopart. Res. 14, 1–10. https://doi.org/10.1007/s11051-012-0841-5.

Wang, L., Fortner, J.D., Hou, L., Zhang, C., Kan, A.T., Tomson, M.B., Chen, W., 2013. Contaminant-mobilizing capability of fullerene nanoparticles (nC_{60}): Effect of solvent-exchange process in nC_{60} formation. Environ. Toxicol. Chem. 32, 329–336. https://doi.org/10.1002/etc.2074.

Wang, C., Liu, H., Chen, J., Tian, Y., Shi, J., Li, D., Guo, C., Ma, Q., 2014. Carboxylated multi-walled carbon nanotubes aggravated biochemical and subcellular damages in leaves of broad bean (*Vicia faba* L.) seedlings under combined stress of lead and cadmium. J. Hazard. Mater. 274, 404–412. https://doi.org/10.1016/j.jhazmat.2014.04.036.

Weber, J.E., Pillai, S., Ram, M.K., Kumar, A., Singh, S.R., 2011. Electrochemical impedance-based DNA sensor using a modified single walled carbon nanotube electrode. Mater. Sci. Eng. C 31, 821–825. https://doi.org/10.1016/j.msec.2010.12.009.

Wong, M.H., Giraldo, J.P., Kwak, S.Y., Koman, V.B., Sinclair, R., Lew, T.T.S., Bisker, G., Liu, P., Strano, M.S., 2017. Nitroaromatic detection and infrared communication from wild-type plants using plant nanobionics. Nat. Mater. 16, 264–272. https://doi.org/10.1038/nmat4771.

Wu, F., You, Y., Zhang, X., Zhang, H., Chen, W., Yang, Y., Werner, D., Tao, S., Wang, X., 2019. Effects of various carbon nanotubes on soil bacterial community composition and structure. Environ. Sci. Technol. 53, 5707–5716. https://doi.org/10.1021/acs.est.8b06909.

Yan, S., Zhao, L., Li, H., Zhang, Q., Tan, J., Huang, M., He, S., Li, L., 2013. Single-walled carbon nanotubes selectively influence maize root tissue development accompanied by the change in the related gene expression. J. Hazard. Mater. 246-247, 110–118. https://doi.org/10.1016/j.jhazmat.2012.12.013.

Yang, Y., Fang, G., Liu, G., Pan, M., Wang, X., Kong, L., He, X., Wang, S., 2013. Electrochemical sensor based on molecularly imprinted polymer film via sol-gel technology and multi-walled carbon nanotubes-chitosan functional layer for sensitive determination of quinoxaline-2-carboxylic acid. Biosens. Bioelectron. 47, 475–481. https://doi.org/10.1016/j.bios.2013.03.054.

Yang, N., Chen, X., Ren, T., Zhang, P., Yang, D., 2015. Carbon nanotube based biosensors. Sens. Actuators B 207, 690–715. https://doi.org/10.1016/j.snb.2014.10.040.

Zaytseva, O., Neumann, G., 2016. Carbon nanomaterials: production, impact on plant development, agricultural and environmental applications. Chem. Biol. Technol. Agric. 3, 17. https://doi.org/10.1186/s40538-016-0070-8.

Zeng, Y.-L., Huang, Y.-F., Jiang, J.-H., Zhang, X.-B., Tang, C.-R., Shen, G.-L., Ru-Qin, Y., 2007. Functionalization of multi-walled carbon nanotubes with poly (amidoamine) dendrimer for mediator-free glucose biosensor. Electrochemistry 9, 185–190.

Zhai, G., Gutowski, S.M., Walters, K.S., Yan, B., Schnoor, J.L., 2015. Charge, size, and cellular selectivity for multiwall carbon nanotubes by maize and soybean. Environ. Sci. Technol. 49, 7380–7390. https://doi.org/10.1021/acs.est.5b01145.

Zhan, K., Chen, Y., Liu, H., Zhang, H., Chen, Y., Ni, H., Wu, M., Sun, D., 2014. A facile method for the immobilization of myoglobin on multi-walled carbon nanotubes: poly(methacrylic acid-co-acrylamide) nanocomposite and its application for direct bio-detection of H2O2. J. Electroanal. Chem. 724, 80–86. https://doi.org/10.1016/j.jelechem.2014.04.012.

Zhang, J., Landry, M.P., Barone, P.W., Kim, J.H., Lin, S., Ulissi, Z.W., Lin, D., Mu, B., Boghossian, A.A., Hilmer, A.J., Rwei, A., Hinckley, A.C., Kruss, S., Shandell, M.A., Nair, N., Blake, S., Şen, F., Şen, S., Croy, R.G., Li, D., Yum, K., Ahn, J.H., Jin, H., Heller, D.A., Essigmann, J.M., Blankschtein, D., Strano, M.S., 2013. Molecular recognition using corona phase complexes made of synthetic polymers adsorbed on carbon nanotubes. Nat. Nanotechnol. 8, 959–968. https://doi.org/10.1038/nnano.2013.236.

Zhang, H., Yue, M., Zheng, X., Xie, C., Zhou, H., Li, L., 2017. Physiological effects of single- and multi-walled carbon nanotubes on rice seedlings. IEEE Trans. Nanobioscience 16, 563–570. https://doi.org/10.1109/TNB.2017.2715359.

Zhao, G., Zhao, Y., Lou, W., Su, J., Wei, S., Yang, X., Wang, R., Guan, R., Pu, H., Shen, W., 2019. Nitrate reductase-dependent nitric oxide is crucial for multi-walled carbon nanotube-induced plant tolerance against salinity. Nanoscale 11, 10511–10523. https://doi.org/10.1039/c8nr10514f.

Zhou, Y., Fang, Y., Ramasamy, R.P., 2019. Non-covalent functionalization of carbon nanotubes for electrochemical biosensor development. Sensors 19, 1–29. https://doi.org/10.3390/s19020392.

Zulfiqar, F., Navarro, M., Ashraf, M., Akram, N.A., Munné-Bosch, S., 2019. Nanofertilizer use for sustainable agriculture: advantages and limitations. Plant Sci. 289. https://doi.org/10.1016/j.plantsci.2019.110270, 110270.

Quantum dots as promising nanomaterials in agriculture

10

Archita Gupta, Sanjay Kumar Mehta, Kanishka Kunal, Kunal Mukhopadhyay, and Sneha Singh
Department of Bioengineering and Biotechnology, Birla Institute of Technology, Mesra, Ranchi, Jharkhand, India

10.1 Introduction

Nanomaterials and their extraordinary properties have revolutionized the scientific community and continuing to flabbergast with every new finding. Recent nanotechnological developments have offered enormous knowledge and technological platforms with applications ranging from medicine, agriculture, electronics, optics, and many others. Nanomaterials are present in different dimensions based on their electronic arrangement in different axes like zero-dimensional (nanoclusters and metallic nanoparticles), one-dimensional (nanotubes and nanowires), two-dimensional (graphene sheets, and nanoplates), and three-dimensional (nanocrystals). Nowadays, intensive studies into a new dimensional confinement of electrons are upsurging leading to the developments entailing quantum dots (QDs). These are zero-dimensional dot-sized (less than 10 nm) nanomaterials equivalent to exciton Bohr radius, thus instigated as "artificial atoms." QDs are highly luminescent crystalline nanostructures that comprise elements from II–VI (CdSe, ZnS, and CdTe), III–V (InP and InAs), and IV–VI (PbSe and PbS) groups (Bonilla and Kouznetsov, 2016).

Significant advances in the field of QDs are achieved due to their eminent physiochemical properties varied from optical to electrical to biological. These properties are greatly influenced by the size, shape, and composition of QDs. Architecturally, QDs are developed in the combinations of core, shell, and capping ligands. Surface chemistry showcased by the core/shell/ligand makes QDs highly stable, water soluble, efficient, sensitive, robust, and biocompatible. Different synthesis procedures and biofunctionalizations allow improvement in different physiochemical properties of QDs.

There is an increasing utilization of QDs in plant sciences wherein they are used as biomarkers of cells or cell walls for plant bioimaging, increasing plant yield due to different light excitation wavelengths, eradication of plant diseases due to eminent antibacterial potency of small-sized QDs, and many more. Optical properties of QDs, their conjugation with other moieties, and surrounding interface are under extensive study for agricultural production (Ung et al., 2012). Pesticides and/or herbicides are frequently used in agriculture which are not degraded easily, thereby posing concerns to environment, animals, and human beings. Organophosphorus-based

Agricultural Nanobiotechnology. https://doi.org/10.1016/B978-0-323-91908-1.00016-X

chemicals are used for a long time in agriculture and tend to get accumulated in the field, thereby affecting the soil quality and productivity. Photoluminescence property of QDs plays a crucial role in highly sensitive and selective detection of even very low concentration of such chemicals easily (Du et al., 2019). QDs are also reported to significantly influence plant physiological processes. Even the low doses of QDs can activate photosynthesis resulting in enhanced growth of plants and subsequent increase in productive yield (Chakravarty et al., 2015). The quantum size of QDs allows their easy penetration into plant cells for smart delivery. QDs are also effective in detecting and eradicating plant disease caused by bacteria and fungi without affecting normal plant growth (Kasibabu et al., 2015). An electric light source with spectral composition is the most widely used artificial light source in greenhouse cultivation. QDs provide an efficient and cost-effective light source alternative with proper spectral composition to intensify the photosynthetic rate of plants (Pavlov et al., 2017). However, the frequent use and very high surface charge possessed by QDs can easily interact with biomolecules rendering dose-dependent toxic effects, thereby still limiting their clinical use.

In toto, utilization of QDs in agricultural applications requires understanding their properties and surface chemistry. Therefore, the chapter focuses onto determining the properties of QDs, their synthesis mechanisms, and passivation techniques to make highly stable QDs in the first section. The second section of the chapter elaborates on different types of QDs researched so far for efficient utilization in agricultural prospects. Final part of the chapter discusses different avenues explored in enhancing productivity of agricultural products by QDs *hitherto* and associated toxicity profile of QDs.

10.2 Unraveling the potential of quantum dots (QDs)

The field of nanoscience has recently become the most researched area of the technological development. For the past two decades, nanoscience attained rapid development opening up new avenues for designing nanomaterials for different applications (Bonilla and Kouznetsov, 2016). Nanoparticles are generally divided into four classes based on their dimensionality: three-dimensional, two-dimensional, one-dimensional, and zero-dimensional, comprising, for instance, nanocrystals, graphene sheets, nanotubes, and metallic nanoparticles, respectively. The dimension-based classification of nanoparticles is determined through their electronic arrangement in directional axis where dimension in all axis is more than 100 nm for three-dimensional, one axis is less than 100 nm in two-dimensional, two axes under 100 nm for one-dimensional and all axis less than 100 nm for zero-dimensional nanoparticles (Jeevanandam et al., 2018). Currently, a new dimensional confinement of electron led to the development of novel nanoparticles where electrons are confined in each dimension fostering zero-dimensional nanoparticle quantum dots (QDs) (Zeng et al., 2018). QDs were first theoretically introduced in the 1970s and were first synthesized in the early 1980s. QDs are a novel class of inorganic fluorophores that are recently been accentuated due to their exceptional properties. They are nanometric semiconductors composed of elements of mainly II–VI or III–V groups having dimensions less than an exciton Bohr radius (Cassidy and Zamkov, 2020). QDs are non-natural semiconducting nanoparticles that can transport electrons and emit light under different wavelengths. When a photon of light falls on the surface of QDs, the electrons are excited to a higher energy state and the material emits light when the electron return to ground state. Depending on the bandgap of the material, reemission of fluorescence of different colors takes place. These quantum-sized particles are more of considerable interest due to their dimensional similarities to biomacromolecules (Abbasi et al., 2016). QDs are defined as artificial atoms as they exhibit discrete energy states and tunable bandgaps which can be modulated by changing the size, shape, composition, and structure. QDs are less than 10-nm-sized nanoparticles exhibiting unique quantum effects like inelastic electron coupling, quantum tunneling, and quantum gravitation (Jeevanandam et al., 2021). These effects are mainly responsible for the eminent properties possessed by QDs.

10.2.1 Physiochemical properties of QDs

QDs are highly luminescent crystalline nanostructures within the size range of 1 to 10 nm and comprises elements from II–VI (CdSe, ZnS, and CdTe), III–V (InP and InAs), and IV–VI (PbSe and PbS) groups (Bonilla and Kouznetsov, 2016). At such small size, physiochemical properties are dramatically influenced due to quantum confinement defining the electronic rearrangement and movement within the nanocrystals. After the absorption of light, holes, and electrons are attracted toward each other generating a pair of electron-hole also known as exciton (Fig. 10.1A). Average distance between the electron and a hole within an exciton represents the exciton Bohr radius. If the size of material is increasingly getting small it reflects, it is shifting toward the exciton Bohr radius or presents continuous change in energy levels.

QDs represent physiochemical properties like high photostability, broad absorption and narrow emission spectra, slow decay rate, and broad energy gaps. The physiochemical properties of QDs are governed by their shape, size, and structural composition which particularly define crystallinity and electronic configuration (Lee et al., 2018). Different applications rendered by QDs are mainly due to their exclusive physiochemical properties. The extraordinary properties of QDs are reflected through the size that showcases inverse relationship with the bandgap energy as shown in Fig. 10.1B. As the size of dots decreases, the bandgap energy increases and will simultaneously decrease the excitation and emission wavelengths (Singh et al., 2018). This is known as the quantum size effect wherein the fluorescent color emitted by QDs after UV irradiation can be tailored through the size of the particles.

The distribution of QDs along with the size plays a significant role in enabling the wide absorbance range with narrow and symmetric emission spectrum. With respect to the architecture of a QD, it consists of a core, a shell, and a capping ligand. The core having the low bandgap energy is responsible for providing the primary physiochemical properties, whereas the shell with high bandgap energy is responsible for passivating the bare QD surface. The core QD represents loss in quantum yield due to highly exposed surface and instability, whereas shell over the cores provides

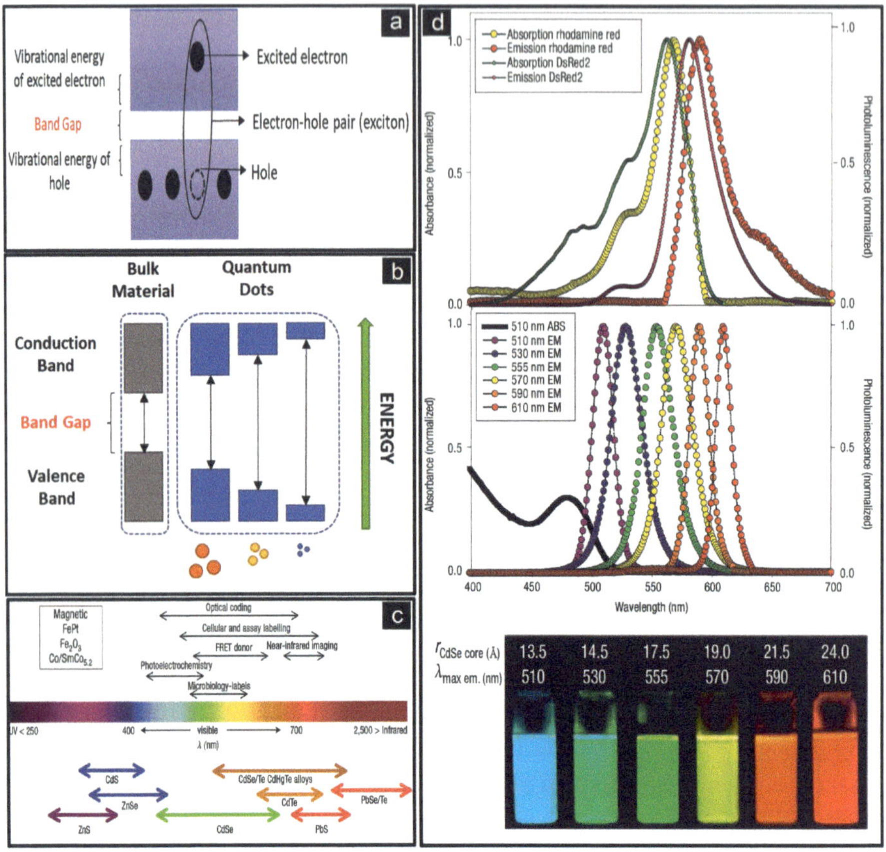

Fig. 10.1 Physiochemical properties leading to varied applications of QDs: (A) representative image defining the confinement energies by electron, hole, and their pairs, (B) variation in bandgap and energies as a function of QD sizes, (C) QD core material as a function of superimposed emission wavelength with the spectrum and possible activities, and (D) comparison of rhodamine and genetically modified dye with QDs in terms of their fluorescence along with size-dependent fluorescence properties (C–D).
Reprinted with permission from Medintz, I.L., Uyeda, H.T., Goldman, E.R. and Mattoussi, H., 2005. Quantum dot bioconjugates for imaging, labelling and sensing. Nat. Mater. 4(6), 435–446.

stability and avoids quantum loss, blinking, and photobleaching (Neo et al., 2014). The shell possesses two important characteristics: first, it provides physical barrier between the core and the surrounding medium; second, it provides saturation of defect state reducing the crystalline defect generated by dangling bonds on core surface. Capping ligands like alcohols, carboxylic acids, thiols, primary amines, and organophosphates are equally important for providing stability, controlled morphological distribution, and preventing agglomeration of QDs (Grandhi and Viswanatha, 2016; Xie et al., 2019). They represent the key parameters as the solubilization characteristics are greatly influenced by the nature of ligand and hence the physiochemical profiles (Karakoti et al., 2015). Therefore, the properties of QDs are completely dependent on the composition and final arrangement of core/shell/ligand in the QDs (Fig. 10.4C).

10.2.1.1 Optical properties

The quantum confinement defines the optical properties of QDs making them ideal candidates for bioimaging and biosensing pesticides/herbicides (Schiffman and Balakrishna, 2018; Jafari et al., 2019). As shown in Fig. 10.1C, most of the QDs are excited under various energy sources providing different colors under visible to infrared region (Rossi et al., 2018). When the electrons are excited, it leaves a hole behind and releases photons in the form of energy known as fluorescence. It is easy and possible to generate a continuous bandgap through QDs as compared to bulk materials, thereby making them more fluorescent (Singh et al., 2018). Multiple QDs with same composition but different sizes emit different colors. Smaller QDs have larger bandgap energy yielding blueshifted and thereby provide blue color, whereas larger sized QDs have low bandgap energy, thereby rendering redshifted and emitting red color (Fig. 10.1D) (Bonilla and Kouznetsov, 2016). This important optical property features significant biological applications due to their vivid spectra contributed by size of QDs where simultaneous emission of different colors by different-sized QDs can be obtained through same light source.

High molar extinction coefficient and narrow emission spectra of QDs are very important characteristics for their applications in biosensing (Martynenko et al., 2017). Fu et al. (2020) synthesized carbon QDs coated with polyethylene glycol for live cell imaging in mice. The QDs were easily visible in bone marrow and lymph nodes even after a long time suggesting the stability of QDs rendered by coating through ligand. In a study, it was reported that the photostability of QDs is very high as compared to the conventional organic dyes used for bioimaging (Bajwa et al., 2016). Conventional dyes have a broad emission spectrum leading to overlapping between subsequent dyes with same spectral range, and it becomes highly difficult to differentially use them as fluorescent probes (Fig. 10.1D). On the contrary, QDs are highly preferred due to their narrow spectrum and can easily be controlled by the size, composition, and surface passivation of QDs. Reports have suggested that QDs are more stable than Alexa488 which is considered the most stable organic dye (Wu et al., 2003; Karakoti et al., 2015). CdSe/ZnS QDs coated with dihydrolipoic acid showed no loss in fluorescence even after 14h and considered 100 times more stable and 20 times brighter than the rhodamine 6G dye (Debruyne et al., 2015; Abbasi et al., 2016). Based on their optical properties, QDs are also used in photothermal therapy for treating cancer. Codoped QDs with nitrogen and phosphorus were prepared with the help of ATP biomolecules and evidenced to emit brighter light due to small size and was able to successfully prevent cancer growth (Ananthanarayanan et al., 2015). Zhang et al. (2019) synthesized a hybrid QD structure with zero- and two-dimensional for very fast charge transfer in optoelectronic devices. A study reported that the photoluminescence property of QDs can be improved by coarse adjustment of quantum confinement (Liu et al., 2018). This property of QDs will make them potential candidate for long-term cell imaging or where high and controlled intensities of light are desired for longer durations.

10.2.1.2 Thermal properties

The emission band of QDs is also dependent on temperature. The electrical conductivity of QDs increases linearly with an increase in temperature. QDs exploit this property by their use in LEDs where thermal stability of LEDs is monitored through the luminescence which was maintained to 97% at higher temperature as compared to conventional LEDs showing luminescence to 65% (Vu et al., 2016). In a report, InAs QDs were designed at various temperatures to study the effect of temperature on photoluminescence (Korsch et al., 2019). It was observed that with increasing temperature photoluminescence increases but it significantly decreases at temperature above 850°C due to the degradation of QDs. Maestro et al. (2011) researched developing nanothermometers by exploiting the thermal sensitivity of CdTe and CdSe QDs. It was observed that QDs showed very high sensitivity of about 0.2 °C and CdTe QDs proved to be more sensitive than CdSe QDs. Biomedical applications require thermal imaging to study tissue dynamics. Temperature plays a significant role in determining the cell metabolism and mechanism. Hence, it is necessary to determine the temperature change during cell division and cellular death to diagnose various diseases of concern in humans. High thermal sensitivity of CdTe QDs can be utilized for detecting early diseases and their progression within the human body (Haro-González et al., 2012). Recently, QDs synthesized under different thermal conditions were tested for their thermal conductivity. QDs synthesized at supercritical and subcritical temperatures showed higher conductivity and their safe use at high temperatures make them potential material for thermal therapies as compared to graphene oxide sheets (Tayyebi et al., 2018).

10.2.1.3 Electrical properties

In QDs, electrons and holes are confined in all three directions allowing them to have distinct electrical properties. Studies have reported that molecular orbital theory of QDs defines the electronic structure. Carbon QDs show high energy by transition of electrons at $n \rightarrow \pi^*$ and $\pi \rightarrow \pi^*$. Imran et al. (2019) used CdSe and CdTe in applications where energy harvesting is required and observed that high atomic number significantly changes the properties of QDs. It is also observed that electrons flow with higher rate in graphene QDs as compared to silicon materials, thereby rendering their potential in the next-generation electronic devices.

10.2.1.4 Photocatalytic properties

Numerous QDs are used for water treatment and removal of pollutants due to their catalytic properties upon exposure to light. Recently, it was reported that cadmium oxide QDs enhance photocatalytic properties in visible region (Reshak, 2017). Moreover, another report suggested that functionalization of silicon QDs allows better photocatalytic potential as compared to nonfunctionalized moieties.

Kumar et al. (2018) incorporated carbon QDs onto TiO_2 nanoparticles to improve their photocatalytic capability in removing organic pollutants. These exclusive and vast properties of QDs allow their applications in a variety of fields ranging from electrical to biological.

10.2.2 Synthesis procedures and surface passivation for different QDs

QDs render unique electrical and fluorescence properties like high photostability, narrow emission spectra, and wide absorption spectra (Yuan et al., 2019). QDs are the first nanotechnological intervention incorporated into biological products for commercialization. CdSe/ZnS QDs are the most commonly available commercial product as secondary antibody conjugates comprising core of CdSe of 10–50 atoms in diameter (Vo et al., 2015). There is an increasing utilization of QDs in plant sciences wherein they are used as biomarkers of cells or cell walls for plant bioimaging, increasing plant yield due to different light excitation wavelengths, eradication of plant diseases due to eminent antibacterial potency of small-sized QDs and many more. Different synthesis procedures and biofunctionalizations allow improvement in different physiochemical properties of QDs. Therefore, it becomes the prime need to explore various synthesis procedures and surface passivation of QDs to make them application-based tunable nanomaterials.

10.2.2.1 Synthesis methods

QDs are synthesized broadly through two basic processing methods: top-down and bottom-up approaches. Top-down approach includes molecular beam epitaxy, electron beam lithography, ion implantation, and X-ray lithography. Under this method, bulk material is reduced down to quantum-sized materials to render unique properties. To achieve QD size of 30 nm and less, mainly electron beam lithography, wet and reactive ion etching are prominently used. Moreover, laser and ion beams are also used to fabricate zero-dimensional dots. Structural imperfection and the presence of impurities limit the use of this processing method (Ding et al., 2019). Also, majorly explored optical property of QDs shifts the extensive use of synthesis process toward bottom-up approach. In this alternative synthesis approach, colloidal QDs are synthesized by chemical reduction followed by self-assembly (Namdari et al., 2017). Wet chemical and vapor phase methods are the commonly used bottom-up approaches for synthesizing QDs. In wet chemical method, a conventional precipitation in solution involving nucleation and growth of nanoparticles under controlled parameters is achieved. This method incorporates sol-gel, microemulsion, hot-solution decomposition, competitive or electrochemistry, and sonic or microwaves (Moghaddam et al., 2015; Chen et al., 2016; Gomase et al., 2019; Singh et al., 2019). In vapor phase method, the layers of QD are grown in atom-over-atom manner wherein self-assembly on substrate takes place without patterning (Pu et al., 2018). Techniques like sputtering, molecular beam epitaxy, liquid metal ion sources, and aggregation of gaseous monomers are included within vapor phase method. Molecular beam epitaxy is generally used for the self-assembly of QDs from III-V and II-VI groups with large lattice mismatch (Valizadeh et al., 2012).

Synthesis of QDs significantly influences their properties; therefore, core strategies involved in QD fabrication need to be understood. By tuning the size and shape of QDs, the properties associated with different biological applications ranging from electronics, photocatalysis, biomedicine, agriculture, and pharmaceutics are enhanced by using different synthesis approaches. Through various synthesis mechanisms like physical, chemical, and biological, it is possible to obtain QDs with enhanced spectral emission and absorption (Fig. 10.2). Mostly, chemical and biological methods are used for synthesis and physical method is used for their coating onto bulk material or substrates (Panagiotopoulou et al., 2016). Some QDs are also synthesized through physical method but they lack stability and are cost-intensive (Wang et al., 2012). Chemical method of synthesis has drawbacks like toxicity to human health and environment (Oh et al., 2016). Biological synthesis on the contrary enhances biocompatibility and bioavailability of QDs.

Physical approaches: Numerous synthetic methods are used for the synthesis of QDs involving the requirements of material and size. In physical method of synthesis, growth of QDs along with nucleation takes place within vapor phase (Brichkin and Razumov, 2016). Fabrication of QDs through physical method is advantageous due to less cumbersome technique and uniform deposition on substrate with controlled parameters (Gomase et al., 2019). In vapor deposition technique, QDs are synthesized through atom-by-atom growth process. Molecular beam epitaxy is one of such method wherein thin films of semiconductor elements are grown epitaxial under high vacuum with 3000 nm/h of deposition rate (Borovaya et al., 2015). The reduction of epilayers to dot size is achieved by varying the deposition thickness onto the substrate. A well-organized quantum-sized PbTe/CdTe layers were attained by depositing PbTe (1–10 nm thickness) on CdTe (50 nm thickness) layers under annealing temperature of 340°C. The size and shape variations in layer thickness were determined through the cross-sectional analysis by dark-field transmission electron microscopy. Such changes in size allow tuning of QD luminescence in spectral range of 2.2 to 3.7 μm, allowing their potential use in luminescent diodes (Ding et al., 2016).

Physical vapor deposition is the most common vapor phase deposition method comprising laser ablation, sputtering, and thermal vapor deposition. In laser ablation technique, the high laser beams upon hitting the target surface allow the release of substrate material by continuous increment in temperature onto the substrate surface (Kim et al., 2017). CdSe QDs were synthesized through laser ablation in aqueous medium as indicated by the color change in the suspension (Fig. 10.2). Quantum-sized CdSe dots of 5 nm were confirmed through X-ray diffraction and electron microscopy. The integration of CdSe QDs with ZnO nanowires allows improved solar cell efficiency under quantum sensitization. The QDs synthesized were ligand free, thereby reducing the restricted transferring of electrons between QD and nanowire (Qi et al., 2015). Similar to laser ablation is sputtering; the substrate is bombarded with gaseous ions between target and substrate to deposit molecules under high-accelerated voltage within vacuum chamber (Baptista et al., 2018).

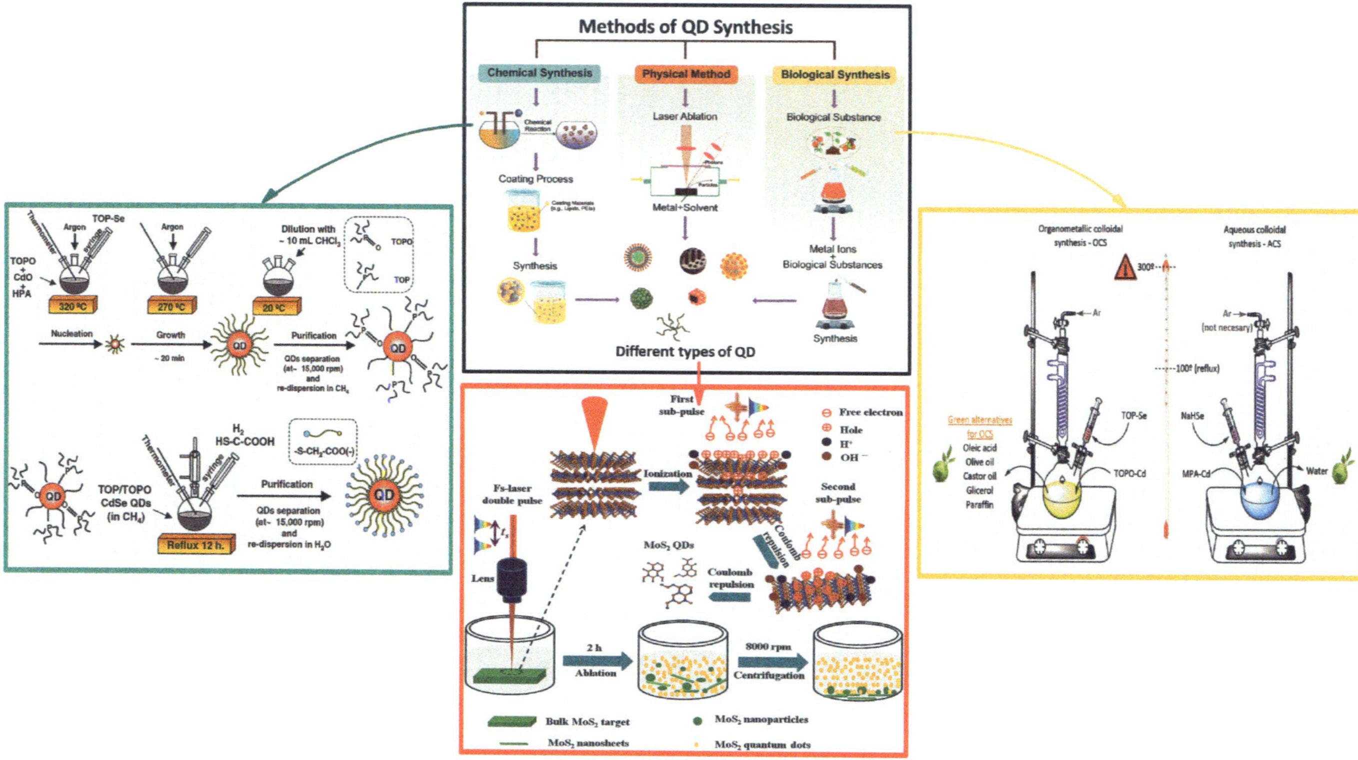

Fig. 10.2 Different methods of synthesis of QDs via physical, chemical, and biological approaches.
Reprinted with permission from Bonilla, C.M. and Kouznetsov, V.V., 2016. Green quantum dots: basics, Green synthesis, and nanotechnological applications. Green Nanotechnology-Overview and Further Prospects, pp. 174–192; Pang, C. and Gong, Y., 2019. Current status and future prospects of semiconductor quantum dots in botany. J. Agric. Food Chem. 67(27), 7561–7568; Jeyaraj, M., Gurunathan, S., Qasim, M., Kang, M.H. and Kim, J.H., 2019. A comprehensive review on the synthesis, characterization, and biomedical application of platinum nanoparticles. Nanomaterials 9(12), 1719; Li, B., Jiang, L., Li, X., Ran, P., Zuo, P., Wang, A., Qu, L., Zhao, Y., Cheng, Z. and Lu, Y., 2017a. Preparation of monolayer MoS 2 quantum dots using temporally shaped femtosecond laser ablation of bulk MoS 2 targets in water. Sci. Rep. 7(1), 1–12.

Recently, CdSe QDs are synthesized through radio frequency magnetron sputtering with controlled size of 7.2 nm in substrate distance of 13.5 cm under 2.2×10^{-1} mbar pressure (Dahi et al., 2016). The size of CdSe QD significantly increases by increasing the radio frequency rate and deposition rate. In thermal deposition method, solid material is heated in vacuum chamber, vaporize, and hit the substrate. It was reported that Nb_2O_5 QDs of definitive sizes were synthesized through thermal deposition by changing the deposition rate (Dhawan et al., 2014). Another important technique is chemical vapor deposition that is different from other physical deposition techniques as deposition is mediated through the chemical reactions. Graphene QDs were deposited onto copper substrates through this method with uniform size distribution of 5–15 nm attained through controlling the growth time, flow rate of gas, and growth cycles (Fan et al., 2013). Kumar et al. (2018) synthesized nitrogen-doped graphene QDs where chitosan served as the carbon source. This method fabricated QDs of 10–15 nm size range. Although physical methods have been successfully used for deposition of QDs, they have certain limitations like involvement of expensive instruments and difficulty in separation of QDs from the substrates (Pu et al., 2018).

Chemical approaches: Chemical synthesis approaches are broadly used in place of physical methods to obtain QDs with tunable size, colloidal characteristics, and properties by varying the concentration of precursor used in synthesis, solvents, reaction time, and temperature (Karmakar, 2015). Synthesis of QDs through chemical approach involves two main steps—nucleation and growth. In a study, the effect of temperature and solvent polarity onto QD nucleation and growth was determined (Li et al., 2016). It was observed that QD size is significantly influenced by the aforementioned parameters. The increase in solvent polarity decreases the QD size, whereas the increase in temperature increases the size of QD (Li et al., 2016). Colloidal and water-soluble QDs are synthesized exclusively through hot injection, template-assisted synthesis, and solvothermal methods. During hot injection method, cold precursor with hot organometallic precursor at very high temperature led to instantaneous nuclei formation and size reduction of QD (Salaheldin et al., 2017). In a study, highly luminescent multicolored CdSe QDs were synthesized in single step through hot injection method. QDs ranging in spectral variations of 445 to 643 nm were fabricated by varying molar ratios of selenium to sulfur and cadmium to zinc (Osman et al., 2019). In another study, bandgap in nanocrystals was reduced by incorporation of sulfur into CdSe semiconductor crystals through hot injection method (Mahajan et al., 2013). This method is frequently used but requires harsh conditions during synthesis, thereby hindering scaled-up synthesis of QDs.

To overcome the limitations of hot injection method, noninjection method was introduced wherein reagents were loaded at optimum temperature and refluxed under heat for initiating nucleation leading to easy large-scale production (Hu and Zhu, 2015). Chen et al. (2016) synthesized size-tunable perovskite cesium-lead halide QDs through noninjection method by varying the reaction temperature. The emission spectra were also tuned to the range of UV-infrared for effective luminescence. Solvothermal method on the contrary is cost-effective, simple, and ecofriendly for the development of QDs. In this method, precursors are heated under controlled pressure and temperature maintained via jacketed autoclave for enhanced QD synthesis

(Xu et al., 2016). Recently, biocompatible 2–5-nm-sized carbon dots are synthesized using solvothermal technique rendering high cell viability even at very high concentration (Zhao et al., 2018). Qi et al. (2018) reported two different groups of graphene QDs for tuning photoluminescence. First group comprises of QDs of similar sizes but different oxidation states, whereas the other group consists of different sizes and similar oxidation states. It was observed that large size of QDs and high oxidation surfaces produce a redshifted photoluminescence.

Template-assisted synthesis of QDs is prominently used in controlling the morphology of QDs easily wherein the growth of nuclei takes place within the holes and defects of the substrate (Li et al., 2017). Porous polymeric membranes and anodic aluminum oxides are widely used as templates for synthesizing quantum nanocrystals. Faucher et al. (2018) have synthesized polymer-encapsulated CdSe/ZnS QDs wherein the surface properties are modulated through the polymer coatings. Another study demonstrated the synthesis of nanoporous CdSe QDs as nanostructured films onto graphene sheets. A hexagonal array with a dot size of 10 nm was obtained indicating the importance of graphene template for structural arrangement of QDs with enhanced electromechanical properties (Kim et al., 2010). Conventionally, synthesis of colloidal QDs through II to VI groups was similar to that of colloidal gold nanoparticles, but resulted in larger nanocrystals with very high polydispersity and low fluorescence. Later Cha et al. (2017) explored the synthesis through organic medium and organometallic precursors, thereby generating small-sized, monodisperse, crystalline, and highly fluorescent QDs with no structural defects. High-temperature pyrolytic method of QD synthesis combines organometallic precursors, like cadmium, zinc, mercury, and chalcogen precursors, like sulfur, tellurium, and selenium, in appropriate solvents at high temperatures.

Highly stable QDs are synthesized through solvents which are stable at very high temperatures and render stabilization of QDs from aggregation (Gomase et al., 2019). Specifically, tri-*n*-octylphosphine oxide (TOPO) is the most commonly used solvent which can easily combine both groups of precursors. Under this process, rapid nucleation with epitaxial growth takes place at comparatively lower temperature (Fig. 10.2). Finally, the reaction is preceded at lower temperature to cease any further growth of the particles. The size of QD is controlled through the precursor concentration, temperature, and reaction duration (Zhang et al., 2015). The reactions are required to be performed under inert environment to avoid oxidation of precursors; however, QDs after synthesis can be processed under ambient conditions making them easy and stable nanocrystals. To passivate QD cores, capping agents are added slowly to the reaction mixture during annealing phase allowing epitaxial deposition of material onto the core rather than any further nucleation (Karakoti et al., 2015). The growth of the shell can be optimized by observing further fluorescence quenching. QDs synthesize through this method are highly fluorescent, stable, and monodisperse but are not soluble in water neither can be used for bioconjugation of biomacromolecules. To achieve this, TOPO must be modified with some surface modulators or linkers.

Both physical and chemical approaches are dependent on the precursor concentration, reducing and stabilizing potentials of precursor in tailoring QD size and properties. Chemical process also possesses limitations of use of toxic chemicals as precursor and stabilizing agents that hinder the use of QDs for biomedical applications (Jeevanandam et al., 2020). However, recently new surface modification techniques are adopted to synthesize biocompatible QDs by displacing surface moieties with more biocompatible monomers like polyethylene glycol and ammonium containing silane (Fu et al., 2020; Xie et al., 2020). Biocompatibility can also be enhanced by coating the surfaces with biomolecules via different linker molecules like histidine, glutathione, mercaptoacetic acid, dithiothretol, and mercaptosuccinic acid (Foubert et al., 2016).

Biological approaches: QDs have been proposed to have great potential in biomedical applications due to targeted drug delivery and bioimaging (Abbasi et al., 2016). Therefore, QDs are expected to be stable, bioactive, biocompatible, and bioavailable which are not achieved by conventionally used physical and chemical approaches of QD synthesis (Huang et al., 2019). Hence, this newly emerging biological synthesis of QDs is recently been upsurged to overcome the challenges faced so far.

Biological method includes plant- and microbe-mediated synthesis of QDs. Plant-based synthesis of QD is the most prominently used biological method due to high availability, safety, and the presence of phytochemicals that act as reducing and stabilizing agents simultaneously (Andra et al., 2019). The most commonly and easily synthesized QDs through biological methods are carbon QDs. Recently, carbon QDs made of sugarcane bagasse pulp resulted in very small QDs of approximately 4.1 ± 0.17 nm (Thambiraj and Shankaran, 2016). In other study, carbon QDs doped with nitrogen were synthesized using extract of *Actinidia deliciosa* fruits which worked as a carbon source while ammonium worked as a nitrogen dopant (Arul and Sethuraman, 2018). These highly fluorescent and spherical dots with an average particle size of 3.59 nm emitted blue color. These dots were tested against cell lines for biocompatibility where they demonstrated no cytotoxicity toward L-929 and MCF-7 cell lines. The effect of these doped carbon QDs was also determined to catalytically remove Rhodamine B dye in the presence of borohydride. It was observed that due to the presence of dots, the intensity of dye reduced significantly within 10 min as compared to no-dot-treated group (Chen et al., 2016). Vandarkuzhali et al. (2017) synthesized small-sized carbon dots using pseudostem of banana using hydrothermal treatment. The study represented that 2.5-nm dots render low toxicity against HeLa cells even at high concentrations. These dots were further used for bioimaging and for the detection of iron. Some research groups synthesized hairy root culture of *Linaria maroccana* L.-mediated CdS QDs with the particle size range of 5 to 7 nm (Borovaya et al., 2015, b).

Plant-mediated synthesis of QDs poses less toxicity to human cells sequestering potential application in cell biology, drug delivery, and bioimaging. However, the stability of QDs is the major concern which is recently been researched onto overcoming via incorporating hybrid synthesis approaches like simultaneously using microwave or ultrasound during synthesis (Jeevanandam et al., 2021). Biogenic synthesis approach can be mediated through two synthesis variations based on the use of solvents naming organometallic colloidal synthesis and aqueous colloidal synthesis with respect to the use of organic solvent and aqueous solvent, respectively, during the synthesis as presented in Fig. 10.2.

Microbial-based synthesis approach is also adopted to fabricate more stable QDs. Prokaryotes have an advantage of faster growth rate which potentially develop biocompatible QDs. Recently, CdTe QDs are synthesized through *E. coli* and compared the toxicity profile to the QDs synthesized through microwave irradiation (Kominkova et al., 2017). The results suggested that *E. coli*-based QDs maintained cell viability to about 92.9% compared to other counterpart synthesis method. Bruna et al. (2019) synthesized 2–5-nm-sized CdS from *Halobacillus* sp. DS2 which maintained stability and fluorescence even at very high concentration of salt. Researchers reported the synthesis of CdS QDs from *Pleurotus ostreatus* where QDs of wurtzite crystalline structure and 4–5 nm in size were synthesized from $CdSO_4$ upon incubation at 26°C for 10 days (Borovaya et al., 2015). The synthesis of 30–40-nm-sized ZnS QDs was optimized by varying the concentration of $ZnSO_4$ and *Saccharomyces cerevisiae* under different time frames (Mala and Rose, 2014). Microalgal biomass like eutrophic algal blooms has also been utilized for synthesizing carbon QDs. These 8-nm-sized QDs were proved to be biocompatible with MCF-7 cell lines and were also able to bioimage cell lines in vitro (Guo et al., 2017). The cell penetration capability of QDs was demonstrated via photoluminescence potential of the QDs (Ramanan et al., 2016). The concentration of precursors, volume of the extracts, temperature, and reaction time are the crucial parameters in optimizing the biogenic synthesis of QDs (Singh et al., 2018; Brooks and Lefebvre, 2017). Biosynthesized QDs are found to be nontoxic with enhanced cellular uptake and bioimaging potential. However, this method has inherent limitations pertaining to the scaled-up production of QDs due to instability and long reaction time.

10.2.2.2 Surface passivation

The small size and high surface-to-volume ratio of QDs are highly advantageous in applications but the dangling electronic band onto their surfaces quenches the photoluminescence and decreases the quantum yield (Dimos, 2016). Therefore, passivation of core QDs to provide a wider bandgap is necessary to maintain the surface chemistry. The synthesis of QD can be tailored with core, shell, and/or coatings to significantly affect photochemical properties. The choice of shell and/or coatings is highly critical as the shell stabilizes the dots, whereas the coatings render photophysical properties for desired applications. Bare core of QDs is not appropriately utilized due to certain shortcomings (Badıllı et al., 2020). First, the crystallinity of QDs is not maintained and they tend to switch between fluorescent and nonfluorescent states even after continuous excitation. Second, the cores have very high surface-to-volume ratio making them highly reactive and unstable leading to their photochemical degradation. Therefore, capping these bare QDs is highly desired to enhance their performance. A study reported capping of core QDs with ZnS leading to higher quantum yield, stability, and luminescence at room temperature (Liu et al., 2015). The shell of ZnS also prevents oxidation of the particles and enhances their colloidal stability and quantum yield as compared to bare dots (Luo et al., 2015). The shell of ZnS can further be improved by the addition of Cd to provide even coating across QD surface leading to the quantum yield of about 95% (Vasudevan et al., 2015). However, this stand-alone coating is not sufficient for biological applications; therefore, more biocompatible modifier like polyethylene glycol is the first choice to increase further stability and nonspecific adsorption (Abbasi et al., 2016).

Solubilization of QD is also an important factor for its successful utilization in biological applications. Nonaqueous dispersed QDs can easily be prepared in hydrophobic solvents but to make them aqueous dispersed moieties, surface chemistry plays a significant role. For successful modification of surfaces without affecting the properties of QD, the key steps involved require the exchange of hydrophobic layer with the hydrophilic one or an additional layer of hydrophilic or amphiphilic groups (Fig. 10.3) (Zhan et al., 2015; Wang et al., 2018). The addition of an extra layer is considered more advantageous than the replacement strategy as the native surfactant molecules appeared to maintain the stability and fluorescence efficiency compared to the cores/shells where native layer has been completely removed (Karakoti et al., 2015). Chitosan and cyclodextrin are the most prominently used polymers for passivation and intracellular delivery of QDs (Dogan et al., 2016; Zhao et al., 2018). Other modifiers like phospholipids, dithiothreitol, dendrons, oligomeric ligands, silica, poly(maleic anhydride alt-1-tetradecene), and mercaptopropionic acid can also be used for successful passivation (Karakoti et al., 2015; Vardar et al., 2018; Beloglazova et al., 2015). The hydroxyl and amine groups of chitosan, carboxyl groups of mercaptopropionic acid, and thiol groups of silica are the important groups to play role in surface chemistry (Abbasi et al., 2016). In a study, graphene QDs are surface-passivated to generate two different groups varying in size and oxidation states depending on the level of surface coatings (Qi et al., 2018). The mechanism of photoluminescence and surface chemistry was monitored through electrochemiluminescence to investigate the sizes of dots with respect to surface chemistry adopted. It was observed that large-sized dots with higher oxidation states shifted their photoluminescence toward red wavelength. The technique also revealed that the coating provided negligible defects to the surfaces that are responsible for phenomenal photoluminescence. The surface modification strategies like encapsulation, ligand exchange, and bioconjugation are required for improving solubilization, stability, and specificity for the QDs (Fig. 10.3).

Encapsulation: Main challenge faced by QDs is their solubility in water, to enhance this, researchers have focused onto encapsulating with the hydrophilic materials like polymers or hydrogels. Ligand exchanging or surface coating can also be used for substituting hydrophobic molecules. The choice of encapsulating agent is highly dependent on the final application. Amphiphilic polymer chains can also be utilized to coat QDs. For a better yield of QDs, the original ligand needs to be left on the surface so that the hydrophobic regions of the ligand allow interaction during cross-linking the polymers for QD encapsulation (Lv et al., 2019). Designing QDs with amphiphilic layer is much better than any other method and adding functional groups to the polymer can impart more efficiency to QDs during final functionalization. Recently, "click chemistry" has come up using amphiphilic polymer with anhydride backbone and acetylene group functionalization to coat CdSe QDs (Nxele and Nyokong, 2016). This method includes polymeric precursors with simple functionalization of polymer in common which yields stable, robust, and water-soluble QDs requiring no additional cross-linking. Another strategy of coating QDs

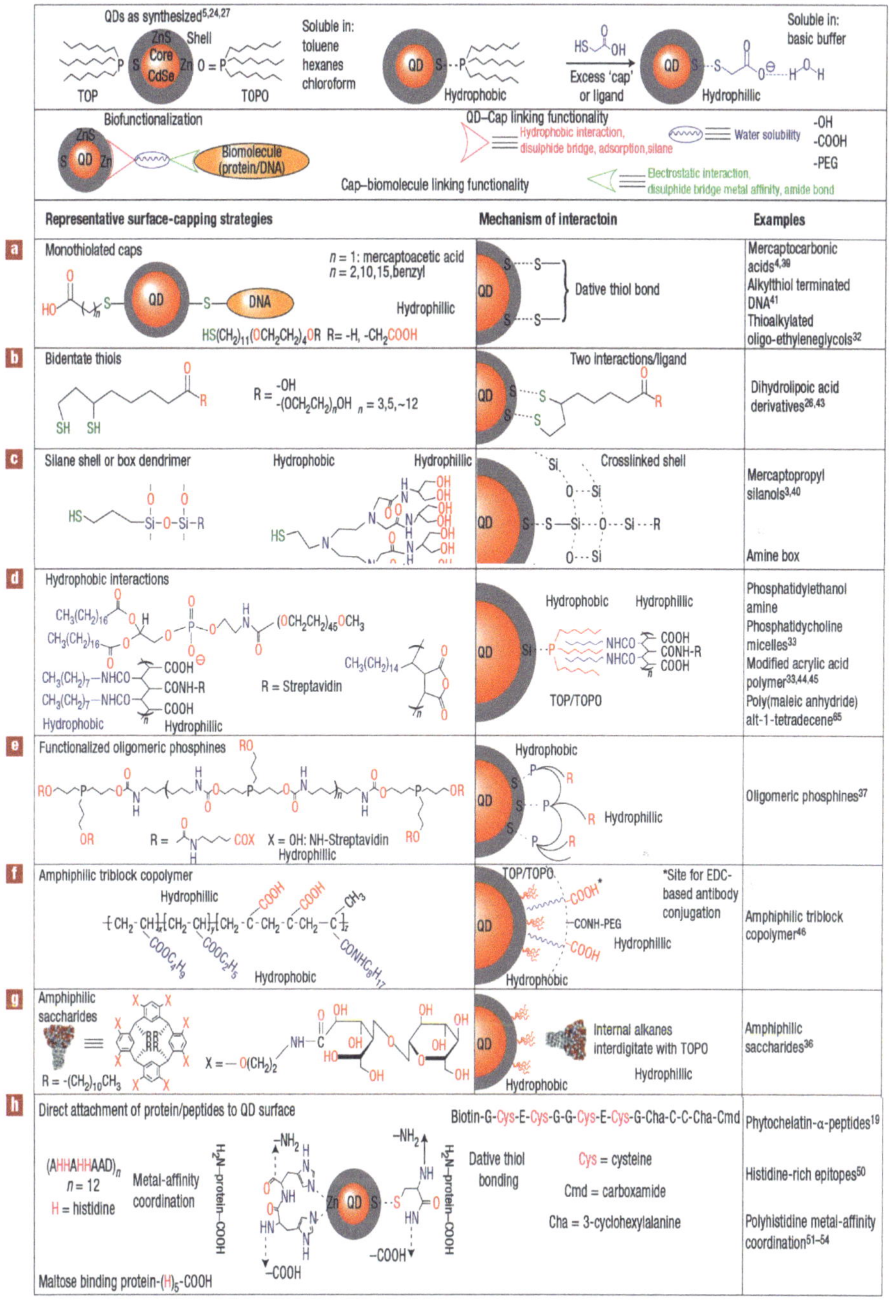

Fig. 10.3 Different ways of QD passivation to generate a stable nanomaterial for efficient end application.

Reprinted with permission from Medintz, I.L., Uyeda, H.T., Goldman, E.R. and Mattoussi, H., 2005. Quantum dot bioconjugates for imaging, labeling and sensing. Nat. Mater. 4(6), 435–446.

is using multidentate polymers like oligomeric alkyl phosphine-, bidentate thiol-, amine-, and thiol-containing polymers (Mai et al., 2020; Ma et al., 2016). Encapsulating QDs with multidentate significantly help in reducing the size, increasing the aqueous solubility, and making robust QDs for any further modifications. Such polymers are capable of interacting QD surface from multiple sites wherein the multivalent interactions will increase the overall stability between the interface of QDs and ligands. The structural analysis and control over the distribution of functional groups onto QDs are highly desired to achieve clinical applications.

Ligand exchange: Another prominently used method to increase solubilization is ligand exchange. This technique requires the replacement of organic ligand with the bifunctional hydrophilic polymer like mercaptoundecanoic acid, dihydrolipoic acid, and thiolated molecules that can bind to QDs from their one end and provide solubility through the other end (Kim et al., 2020; Santana et al., 2019; Li et al., 2019). By this technique, the size is not increased, but the photoluminescence yield is highly increased due to direct binding to QD surface. This method is very simple but requires large number of ligands and lack long-term stability, especially at highly acidic and basic buffers (Bi et al., 2019). Adding multiple coordinating groups add an extra advantage of stability at vast ionic strengths. Major success of this method lies in the affinity of groups onto the surface and the degree of colloidal stability imparted by the hydrophilic polymers. For example, adding polyethylene glycol increases stability, solubility, and reduces nonspecific interactions. In a study, an amino acid histidine is used to exchange the native ligand TOPO from the surface and further functionalization with the histidine intermediate takes place (Ruiyi et al., 2017). This method resulted in the high surface transfer of the ligand with increased quantum yield as compared to encapsulation technique. Ligands can further be exchanged through self-assembled monolayers (dihydrolipoic acid and polyethylene glycol) in a homogenous mixture (Sun et al., 2017).

Bioconjugation: QDs can be conjugated to proteins, antibodies, and oligonucleotides to redirect them to in vitro and in vivo cellular uptake, therapeutic targeting, and bioimaging (Foubert et al., 2016). Bioconjugation capability mainly depends on the ligand chemistries to provide appropriate functional moieties on the surface for covalent or noncovalent interactions. EDC/NHS is one of them which allow conjugation of carboxyl- and amine-terminated QDs to amine- and carboxyl-terminated biomolecules, respectively. Similarly, sulfosuccinimidyl 6-(3′-[2-pyridyldithio]-propionamido) hexanoate has been exploited to link amine-terminated QDs with thiol-terminated biomolecules (Omar et al., 2019). The carboxy-terminated CdSe/ZnS QDs were conjugated with the amine-terminated IgG using EDC/NHS chemistry yielding 25–31-nm-sized particles, allowing better detection of microbes onto the membrane surface as compared to dye-labeled IgGs with 100 times more sensitivity (Pisanicli et al., 2014). However, this carbodiimide chemistry poses some limitations of NHS esters to be hydrolyzed in conjugation conditions. To overcome this, QD ligands are preferred to be replaced by oligonucleotides enabling electrostatic attachment (Sedighi and Krull, 2016). But this technique will require stringent characterization and purification method to avoid any toxicity through heavy metal ligands, is labor-intensive, and results in lower yields. Apart from this, thiol-coated biomolecules were also used in phase transfer but their use is limited due to easy photooxidation of ligand. Other common method is biotin-avidin interactions wherein avidin-coated QDs are attached to biotinylated biomolecules (Jin et al., 2019). This allows increased size of QD and limits their use to only ex vivo or superficial tissue imaging. QDs are also functionalized through haloalkane dehalogenase ligand (HaloTag) via site-specific conjugation (Chen et al., 2018; Liu et al., 2016). This interaction with protein is highly sensitive and efficiently used in cellular processes without posing any toxic effects on mediator proteins (Hatakeyama et al., 2017). A study reported that QDs are developed using *E. coli* lipoic acid ligase and HaloTag for targeting cellular proteins for dynamic bioimaging (Wagner et al., 2019). These methods have been overcome by newly emerging techniques like bio-orthogonal reaction and click chemistry to subdue the limitations of cross-reactivity and inefficient binding. The terminal azides and alkynes used during click chemistry suggest their easy utilization in binding to biomolecules and easy cross-linking through standard coupling reactions like esterification (Blanco-Canosa et al., 2014). These methods can be exploited in a high-throughput manner to generate QD libraries for further characterization.

10.3 Types of QDs

QDs can be classified into two categories depending on the material used and procedure to fabricate them: (i) composition-based QDs and (ii) structure-based QDs. Researchers have developed many types of QDs depending on various applications. The material and structure of a QD define the various properties that it carries. Depending on the material and structure, QDs can emit radiation in between ultraviolet (UV) and infrared (IR) region of spectra. In the past years, researchers have experimented on vast number of materials as well as combination of materials in different shapes, sizes, and structures to enhance the mechanism and properties of QDs for application in various fields.

10.3.1 Composition-based QDs

Generally, QDs are fabricated using III–V or II–IV elements. Semiconductor QDs were the first to be developed, but due to the toxic nature and high cost of the heavy metals, different materials were used with enhanced optical and electrical properties for the development of QDs. Materials such as cadmium (Cd), zinc (Zn), telluride (Te), selenide (Se), mercury (Hg), lead (Pb), graphene (Ge), silicon (Si), carbon (C), and sulfur (S) are used for the synthesis of QDs. Fabrication of single material QD or colloidal QD depends on the optoelectronic properties of the materials and the applications, in which the QD is to be used.

10.3.1.1 Carbon QDs

Carbon nanomaterials are of most importance in the technology field due to their wide range of properties and applications. Owing to their unique electrical, optical, thermal, and mechanical properties, they have found their application in the form of carbon nanotubes (CNT), nanodiamonds, graphene sheets, and carbon QDs (CQDs). Carbon QDs are a new type of carbon nanomaterials of size less than 10 nm, which has attracted interest in recent years because of their photochemical, electrical, optical, and fluorescent emission properties (Lim et al., 2015). Semiconductor-based QDs were studied extensively for years because of their optical and tunable fluorescent emission properties and their application in a wide range of areas such as solar cells, biosensors, bioimaging, quantum computing, lasers, and microscopy. However, the use of semiconductor-based QDs is limited, especially in the field of biomedical and environment due to their high toxicity. The use of heavy metals for the production of semiconductor-based QDs is the main reason for the highly toxic property of the QDs. In the year 2004, carbon QDs were accidentally discovered while purification and separation of single-walled carbon nanotubes (Molaei, 2019). Since then, the CQDs have attracted extensive interest and study because of their properties such as low toxicity, biocompatibility, good water solubility, low cost, chemical inertness, and eco-friendly nature. These properties with similar fluorescent properties provide a tremendous potential advantage to CQDs over semiconductor-based QDs over a wide range of fields. First, biocompatibility, good water solubility, and eco-friendly nature of CQDs make them a better candidate to be used in the area of biosensing, bioimaging, biomedicine, drug delivery, and environmental application. Second, carbon as an element is present in abundance in the environment, which makes the production of CQDs cost-effective. And third, the property of chemical inertness provides stability and better storage to CQDs.

In recent years, several methods have been developed for the synthesis of CQDs. CQDs can be synthesized by both top-down and bottom-up approaches (Wang et al., 2019). In top-down method, large carbon compounds such as carbon nanotube, nanodiamond, graphite oxide, graphite, and activated carbon are broken down using laser ablation, electrochemical oxidation, or arc discharge for the preparation of CQDs. Moreover, in bottom-up approach, carbon precursors such as silica-polymer nanocomposites, citrate, and carbohydrate are used for the synthesis of CQDs via combustion, thermal decomposition, and microwave-assisted methods.

10.3.1.2 Graphene QDs

Graphene QDs (GQDs) are a type of zero-dimensional carbon QDs that derive their characteristics from both graphene and carbon. GQDs are obtained by converting 2D graphene sheets, graphene, and graphene oxide into GQDs. GQDs possess unique properties because of the presence of graphene structure inside the QDs and resemble the crystalline structure of single- or multiple-layered graphene (Facure et al., 2020). The GQDs due to their unique structure provide better crystallinity compared to the CQDs. Better crystalline structure and presence of graphene characteristics in GQDs result in superior properties such as high photostability against photobleaching, strong elasticity, high electrical conductivity, and high thermal stability. Apart from the mentioned characteristics, GQDs also consist of carbon-based properties—low toxicity, biocompatibility, and eco-friendly nature. However, compared to their other carbon counterparts, GQDs are easily soluble in water, and different solvents as preferred (Choi, 2017). The superior properties of GQDs compared to CQDs and semiconductor

QDs have attracted the attention of researchers from around the globe. Though belonging to the same category, GQDs have different structures compared to CQDs. Due to the presence of graphene lattice inside GQDs, a large number of sp^2 hybridization carbon structures is present which results in high crystallinity. In contrast, CQDs are quasi-spherical particles consisting of both crystalline and amorphous structures (Bacon et al., 2014). Generally, CQDs consist of less sp^2 hybridization, which results in poor crystallinity compared to GQDs. The unique property and structure of GQDs have found their application in numerous fields such as photodetectors, light-emitting diodes, solar cells, photocatalysis, bioimaging, drug delivery, and biomedical and agricultural applications (Tian et al., 2018).

10.3.1.3 Silicon QDs

Silicon is one of the most abundant elements present on the Earth and is of great importance in the production of microelectronic and photovoltaic devices, but due to the presence of indirect bandgap (bandgap energy 1.2 eV) in bulk silicon, its use is limited in optoelectronic devices (Dutta et al., 2019). The indirect bandgap makes silicon poor light emitter and absorber, but the electronic and optical property of silicon can be modified by reducing the size of the crystal silicon in the nano regime less than Bohr's excitation radius (4.3 nm). Thus, the optical properties of silicon QDs (SiQDs) are tunable depending upon the size and chemical environment of the silicon crystal. Size control fabrication of SiQDs and with the help of surface chemistry, the optical property of SiQDs can be tunable to a broad range. The SiQDs can emit light in the region of UV to near-IR wavelength, based on the modulations in size and surface chemistry of SiQDs (Dohnalová et al., 2014). The electronic property of SiQDs can also be enhanced by confining silicon into an insulating material that has a higher bandgap than silicon such as silicon nitride (Si_3N_4) and silicon carbide (SiC) (Chinnathambi et al., 2014). Silicon, which is already an important material for the production of microelectronics and photovoltaic devices, with its enhanced optical and electrical property, it can be used for the development of Si optoelectronics and photonics devices. Given the nontoxic and biodegradable nature of SiQDs and the wide range of emission spectrum, SiQDs can substitute semiconductor-based QDs in a wide area of application. SiQDs can be used for deposition at low temperatures on any substrate with the help of various techniques such as painting, spraying, printing, and stamping. The nontoxic and biodegradable nature of SiQDs allows it to be suitable for application in the area of health and environment, optoelectronic toys, fluorescent imaging, drug delivery, waste sterilization, cosmetics, phytotherapeutics, biosensors, and many more. SiQDs have great potential but still lots of research are needed to be carried out to fully appreciate the application of SiQDs.

10.3.1.4 II to IV semiconductor QDs

Semiconductor QDs have been the focus of research for many decades because their optoelectronic properties are tunable depending upon their size and shape. Semiconductor QDs are mostly fabricated by using elements from group II–IV; these elements consist of cadmium, zinc, selenide, sulfide, and telluride (Pu et al., 2018). In the last few decades, several methods have been developed for the fabrication of colloidal semiconductors such as cadmium selenide (CdSe), cadmium telluride (CdTe), cadmium sulfide (CdS), zinc sulfide (ZnS), zinc telluride (ZnTe), and zinc selenide (ZnSe). II–IV semiconductor QDs have shown promising results in various applications due to their direct bandgap and properties comprising of both insulator and conductor. II–IV semiconductors are known for their highly luminescent property, size-tunable emission, and absorption (Chistyakov et al., 2017). Moreover, most of the II–IV semiconductor QDs emit light in the visible spectrum with the exception being only CdTe QDs which emit light between the visible spectrum and near-IR spectrum, due to which the II–IV semiconductor QDs have found a wide range of applications in optoelectronics and biolabeling (Martynenko et al., 2017). Among II–IV semiconductors, CdTe QDs with bandgap energy of 1.5 eV emit in between visible and near-IR regions, depending upon the different sizes and shapes corresponding to quantum confinement of the charge carriers, because of which CdTe is highly famous among the researchers compared to CdSe (Imran et al., 2019). This property of the CdTe quantum dot has proved highly beneficial in detecting antigens of different sizes. Moreover, CdTe QDs can be easily synthesized in an aqueous phase that makes CdTe QDs highly desirable for the use of biolabeling, drug delivery, detection of tumor cells, pesticides, and clenbuterol detection (Abbasi et al., 2016). In contrast, CdSe QDs are synthesized using ligand exchange in an organic solvent at the high boiling point; this makes the production methodology of CdSe QDs more complex compared to CdTe QDs. It should be noted that even having such high potential, the use of Cd-based QDs is limited because of their toxicity that has proved hazardous for biological and environmental use. Group II–IV semiconductor is characterized by their holes and electrons being three-dimensionally confined to the materials' Bohr radius. This characteristic results in a spherical shape, good crystallinity, high optoelectronic property of Cd-based QDs. By shelling CdSe and CdTe QDs with bandgap materials such as ZnS, ZnSe, and CdS result in the fabrication of core/shell or core/shell/shell structure, which helps in enhancing the properties of Cd-based QDs (Cassidy and Zamkov, 2020).

10.3.2 Structure-based QDs

QDs are known for their size-tunable optoelectronic properties but to some extent only due to instability and small-sized toxicity profile. So, to further improve the property of a quantum dot, different types of structural alignments are developed. Generally, bare core is synthesized from the aforementioned quantum materials. To increase the stability, cores are further encapsulated through the shell made of other quantum materials to generate a core/shell or alloyed form of QDs (Fig. 10.4A).

10.3.2.1 Core type QDs

These types of QDs are fabricated using a single component and have uniform internal composition. The material carrying the property of size-tunable emission is suitable for core type structure. The best examples of such structures are CQDs, GQDs, and PbS QDs.

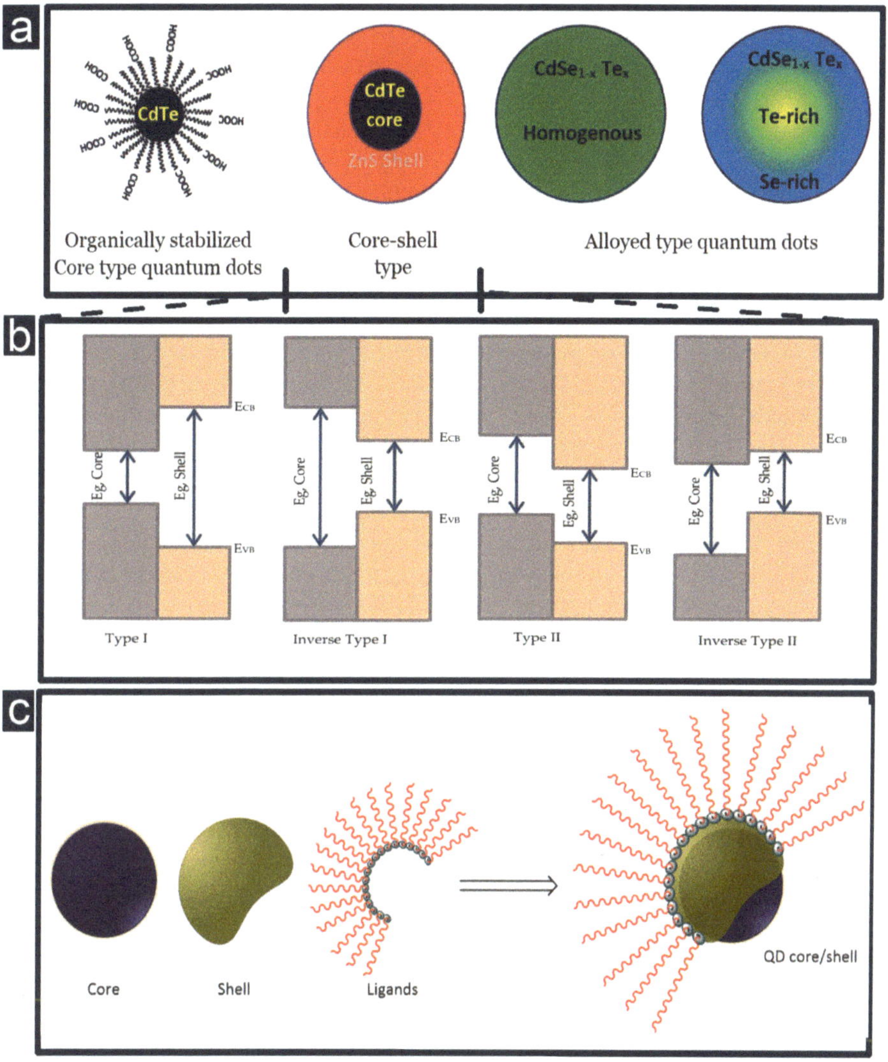

Fig. 10.4 (A) Different types of structure-based QDs representing core, core/shell, and alloyed type of QDs; (B) different types of core/shell type QDs representing type I, inverse type I, type II, and inverse type II; (C) architectural arrangement of QDs representing core, shell, and ligand to generate a stable core/shell/ligand complex.
Panel (A) reprinted with permission from Munasinghe, E., Aththapaththu, M. and Jayarathne, L., 2019. Magnetic and Quantum Dot Nanoparticles for Drug Delivery and Diagnostic Systems. In Colloid Science in Pharmaceutical Nanotechnology (pp.17–32). IntechOpen; (B) from Vasudevan, D., Gaddam, R.R., Trinchi, A. and Cole, I., 2015. Core-shell quantum dots: Properties and applications. J. Alloys Compd. 636, 395–404; (C) from Bonilla, C.M. and Kouznetsov, V.V., 2016. Green quantum dots: basics, Green synthesis, and nanotechnological applications. Green Nanotechnology-Overview and Further Prospects, pp. 174–192.

10.3.2.2 Core/shell type QDs

When one material is embedded inside another material that acts as a shell, then this type of structure is known as core/shell QDs. The electron-hole pairs (exciton decay) provide the luminescent property to the QDs by radiative pathway. Sometimes non-radiative pathways can also result in exciton decay, which affects the emission efficiency or optoelectronic property of the quantum dot (Vasudevan et al., 2015). To overcome such kind of limitation and to increase the brightness and emission efficiency of the QDs, a core/shell type of structure is developed. In core/shell-type structure, a quantum dot is shelled with another material having a wider bandgap. Core/shell-type QDs are differentiated by the position of their valence and conduction band and the presence of essential energy gap between core and shell in the semiconductor. The core/shell-type QDs are classified into four different types (Fig. 10.4B): (i) type I: core can have small bandgap than a shell, (ii) inverse type I: shell can have a small bandgap than the core, (iii) type II: the valence band of the core can be found within the bandgap of shell, and (iv) inverse type II: the conduction band of the core can be found within the bandgap of shell (AbouElhamd et al., 2019). In type I core/shell QDs, the overall emitted wavelength is controlled by the thickness of the shell and delocalization of charge carriers in the shell. The examples of type I QDs are (CdS)HgS, (ZnSe)CdSe, and (CdS)CdSe. In inverse type I core/shell QDs, the shell has a wider bandgap compared to the core. This characteristic helps in the passivation of non-radiative recombination sites present on the core, which results in the improvement of the quantum yield of the QD. The typical example of inverse type I core/shell QDs are (CdSe)CdS, (ZnS)CdSe, and (CdSe)InAs. The type II core/shell QDs are bestowed with a number of properties as compared to type I because of the spatial separation between the charge carriers of type II QDs (Nandan and Mehata, 2019). In type II, one charge carrier is positioned in the core and the other is primarily positioned in the shell; this causes spatial separation between the charged particles. Common examples for type II QDs are (CdTe)CdSe and (CdSe)ZnTe. The inverse type II QDs comprise(*InP*)CdS and (PbS)CdS QDs wherein one of the excited electrons or holes is confined in shell/core structure, whereas the other one remains localized to the core.

10.3.2.3 Alloyed-type QDs

The ability to tune the optoelectronic properties of QDs by controlling the size and surface modification has become a common method. Still, there are limitations to tune the optoelectronic properties of QDs by using such methods. To overcome such limitations, alloyed-type QDs have been discovered. Alloyed type of QDs enhances the optoelectronic properties of QDs without altering the size, internal structure, and surface chemistry. For example, CdS_xSe_{1-x}/ZnS, the quantum dot of diameter of 6 nm, emits different wavelengths of lights without changing the size of the quantum dot (Prusty et al., 2021). Similarly, many alloyed semiconductor QDs of different bandgaps have exhibited distinct characteristics compared to their parent counterparts. Tuning of optoelectronic properties of QDs by changing the particle size has caused problems in different applications such as biolabeling, superlattice structure, and nanoelectronics; due to this, alloyed QDs have attracted interest in recent times.

10.4 Role of QD in agriculture

As evident from the history of human civilization, the production of crops for food and sustenance has been the most critical and essential thing to achieve. For the past two decades, the advancements in the new agricultural technologies have significantly reduced the number of undernourished populations. But still, people are suffering from the problem of insufficient productive yield, pesticide/herbicide accumulation, decreased soil quality, decreased nutrition, moisture content, and prevalence of plant-related disease (Li et al., 2020). Productive growth of crops requires surplus number of fertilizers, pesticides, herbicides, and water. Inefficient utilization of herbicides and pesticides has strategically influenced threats to biodiversity and ecosystem. Climatic change, pests, and diseases enormously increase the stress onto the agriculture sector (Zhao et al., 2017). Therefore, there is an urgent requirement for eco-friendly technologies to overcome the challenges faced by the agricultural industries to improve production and distribution of crops. Nanotechnological interventions have already impacted various facets of human existence. Use of various nanomaterials to address issues faced by the agricultural industry has engrossed reasonable attention in recent times. Considerable efforts to monitor crop health, enhanced plant growth, management of stress, eradication of diseases, determining environmental pollutants, and improving efficiency of fertilizers are being explored with the help of nanomaterials (Giraldo et al., 2019; Zaytseva and Neumann, 2016; Prasad et al., 2017; Nuruzzaman et al., 2016; Li et al., 2020; Athanassiou et al., 2018; Wu et al., 2017). QDs belonging to the novel class of nanomaterial family are the focal point of research due to superior physiochemical properties. Small size, photostability, biocompatibility, aqueous solubility, photoluminescence, noncross-reactivity, and easy functionalization with biomolecules have allowed reasonable escalation toward the utilization of QDs in agriculture. QDs play an essential role in regulating the plant physiological processes, controlling stress, detecting pollutants from soil and water, eradicating diseases, and enhancing the susceptibility of greenhouses for efficient crop growth and high production yield.

10.4.1 Effect of QDs on plant physiology

Crops maintain ecological balance, provide food for human beings, and feed for livestock. But increasing population is upsurging the demand leading to huge challenges faced by agricultural industry (Zhao et al., 2017). Materials are required to enhance plants' growth and productivity. QDs have played an essential role in augmenting the growth of stem and root, increasing the photosynthesis and nutrient contents, respectively. *Vigna radiata* are the simplest and commonly used plant models that are preferred most in the studies due to their high growth rate (Li et al., 2020). When Wang et al. (2018) treated the plants with CQDs, they increased root length, shoot length, and plant weight as compared to nontreated group (Fig. 10.5A). In another study, researchers concluded that total chlorophyll percentage enhanced in hydroponically grown *V. radiata* and tomato when their seeds were treated with GQDs (Feng et al., 2019). Seed treatment with a low concentration of GQDs resulted in an increased amount of chlorophyll, whereas that with a high concentration showed adverse effects on growth in both crops.

Proline, an amino acid, is used by plants to overcome salinity-related stress causing problems like morphological, physiological, productivity, and biochemical attributes of plants by altering the osmotic properties of cytoplasm. Due to salinity stress, there is an increment in programmed cell death and ROS generation, which is controlled by proline (Wani et al., 2020; Khanna-Chopra et al., 2019). Therefore, regarding these issues, researchers tried to synthesize nanosized proline-coated CQDs by hydrothermal methods. They checked the prepared salinity effect of CQDs on the grapevine and speculated a significant increase in an amount of proline, fresh and dry weight of leaf, photosynthetic pigments, and antioxidant enzyme activities. However, there is a decrease in the content of malondialdehyde and hydrogen peroxide electrolyte leakage leading to salinity stress reduction. It was observed that a higher concentration of proline is required for the effective results in free form, whereas with CQD, less concentration is enough for generating prominent results (Gohari et al., 2021). CQDs synthesized through L-cysteine and glucose via hydrothermal method were investigated as a hydroponic nutrient for the development of lettuce and tomatoes. The results evidenced that CQDs enhanced germination of seeds, root development, and elongation of hypocotyl by overexpression of aquaporin proteins. CQDs also increased the yield and nutritional content of plants by accelerating the mineral absorption capability and photosynthesis of plants (Kou et al., 2021). In a different research, it was investigated that CdSe/ZnS QDs passivated with Silwet L-77 surfactant and mercaptoacetic acid can easily penetrate maize root tissues and help in tracking the live physiological processes in the maize root system (Hu et al., 2010). It was seen that QDs were stable for several days when reached in the stele of root seedlings through the apoplastic pathway.

Researchers also concluded that the germination of seed and growth of root were not affected under different concentrations of QDs. CQDs can play an important role in increasing the rate of photosynthesis and therefore enhance the productivity of agricultural products. One of the research groups showed that the treatment of CQDs to *Zea mays* and *Oryza sativa* crops increased the rate of electron transport and subsequently enhanced the photosynthesis (Tan et al., 2021). Also, nitrogen-doped CQDs affected the growth of the *V. radiata* sprouts, and researches observed via photoluminescence the presence of CQDs into different plant parts and their intensity at different incubation periods (Fig. 10.5E). In a study, the development of SiQDs by a biomimetic approach was done and their impact was studied on cucumber seedlings. The SiQDs consist of elements, such as carbon, hydrogen, nitrogen, and oxygen, which are important for plant physiology. Seedling treated with SiQDs showed effective results after the incubation period as compared to control. It also observed that the rate of water absorbed by the root system was enhanced by 74.6% due to the expression of aquaporins (Li et al., 2019). This protein also participates in the development

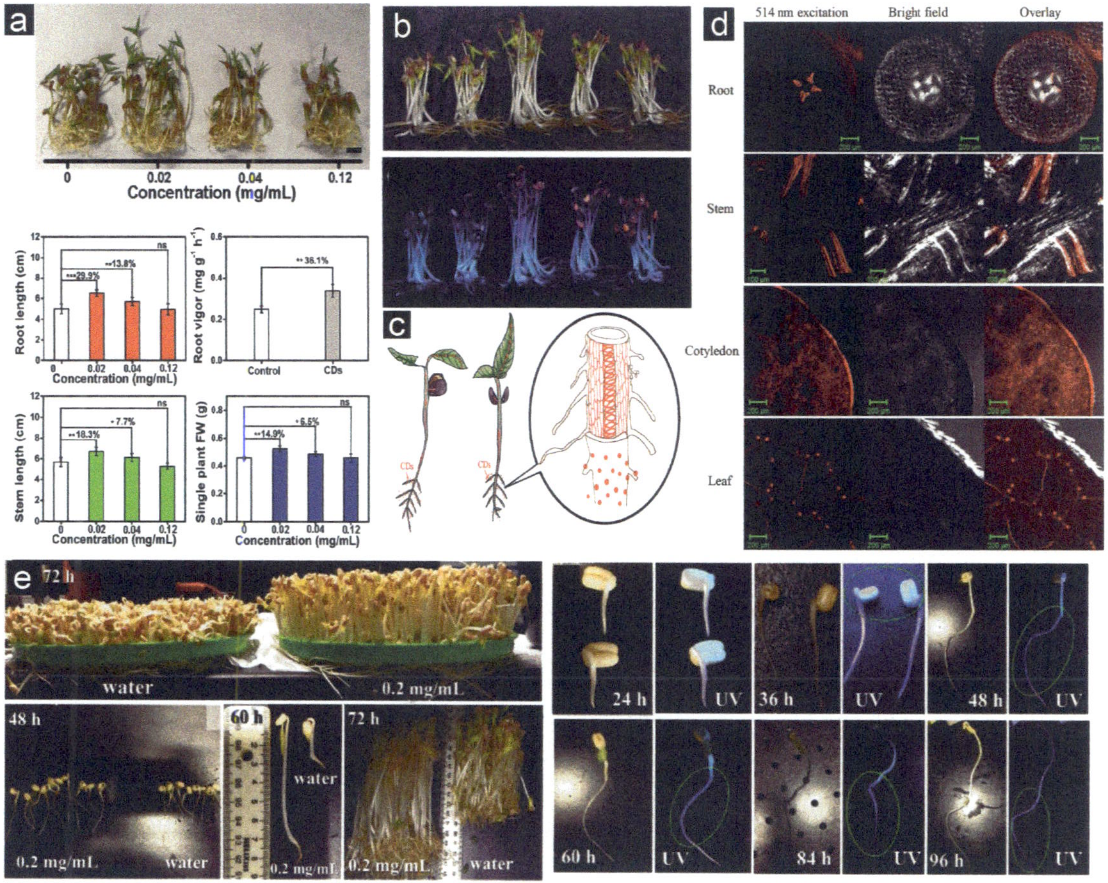

Fig. 10.5 See figure legend on next page

Fig. 10.5 Effects of QDs on physiology of *Vigna radiata* plants (A) growth of *V. radiata* sprouts under different concentrations of CQD and physiological changes after treatment with CQDs; (B) representative images of CQD (0, 0.1, 0.4, 0.7, and 1.0 mg/mL)-treated *V. radiata* under bright light and 365-nm UV light; (C) CQD uptake and accumulation by *V. radiata* plants; (D) laser scanning microscopic images representing transverse sections of different parts of the *V. radiata* plant after 5 days of treatment with CQDs; (E) *V. radiata* exposed to 0.2 mg/mL of nitrogen-doped CQDs, photoluminescence of CQDs under 365-nm UV exposure at different incubation times.

Panel (A) reprinted with permission from Wang, H., Zhang, M., Song, Y., Li, H., Huang, H., Shao, M., Liu, Y. and Kang, Z., 2018. Carbon dots promote the growth and photosynthesis of mung bean sprouts. Carbon 136, 94–102; (B–D) from Li, W., Zheng, Y., Zhang, H., Liu, Z., Su, W., Chen, S., Liu, Y., Zhuang, J. and Lei, B., 2016. Phytotoxicity, uptake, and translocation of fluorescent carbon dots in mung bean plants. ACS Appl. Mater. Interfaces 8(31), 19939–19,945; (E) from Wang, A., Kang, F., Wang, Z., Shao, Q., Li, Z., Zhu, G., Lu, J. and Li, Y.Y., 2019b. Facile Synthesis of Nitrogen-Rich Carbon Dots as Fertilizers for Mung Bean Sprouts. Adv. Sustain. Syst. 3(3), 1800132.

of seed, photosynthesis process, and lengthening of cells. They demonstrated that SiQDs can transport from the root system to leaves through vascular bundles. In another study, CQDs were also tested for the growth of *V. radiata* and reported to significantly increase the plant biomass (Li et al., 2016). Upon detailed investigations, it was revealed that the dots were uptake by the roots, distributed to different parts of the plant where they accumulated for bringing the required physiological and cellular changes promoting growth of *V. radiata* as presented in Fig. 10.5B–D. Similarly, another research group used GQDs as a nanofertilizer to check the growth of *Coriandrum sativum* L. and *Allium sativum* plants. Both plant seeds were treated with 0.2 mg/mL of GQDs and resulted in increased growth rate of root, leaf length, number of flowers, shoots, and fruits (Chakravarty et al., 2015). Untreated groups of plants showed less healthy growth and development of early chlorosis.

10.4.2 QD for determining soil quality

For enhancing the productivity of agricultural products, it is necessary to monitor soil moisture because it indicates the available water in the soil for the crops. At present times, existing equipment used for measuring soil moisture is costly and unaffordable for the farmers (Su et al., 2014). So, researchers are focusing to develop a rapid and accurate moisture measurement alternative using QDs. In a study, GQD-based sensor was prepared from graphene oxide and electrolyte (propylene carbonate) by electrochemical method for sensing soil moisture as presented in Fig. 10.6A (Kalita et al., 2020). Different size ranges of GQDs were obtained by varying the concentration of electrolytes and reaction time. Black and red soils were used as test samples to check the potential of GQDs to measure moisture content. Rapid detection was obtained using QDs with only 180 s of detection time. Moreover, the stability of QD-based sensor is also for 5 months and great reproducibility as observed. Hence, QD-based sensors for the detection of moisture content are highly beneficial in agricultural fields.

Herbicides like alachlor are used in the agriculture field to control the growth of weeds and hence the better production of crops (Rose et al., 2016). But due to its excessive use and nondegradability, its accumulation can pollute the soil and water. This carcinogenic compound can damage liver and kidneys upon the consumption through polluted water or the crops with an accumulated amount of alachlor (Nykiel-Szymańska et al., 2018; Elshafey and Radi, 2018; Segal et al., 2019). Hence, we have to be concerned about these problems and need a suitable strategy for the measurement of the residue level of herbicides. Highly stable CQDs doped with nitrogen were synthesized through a microwave-assisted hydrothermal technique using phenylenediamine for the fluorescent-based detection of the alachlor (Du et al., 2019). The presence of trace amounts of alachlor in the soil can be selectively detected with high specificity (Fig. 10.6C). The red-emissive QDs show decline in fluorescence as the concentration of alachlor increases in the soil. Therefore, such doped QDs can prove to be promising fluorescent probes for detecting the presence and concentration of herbicides in the soil.

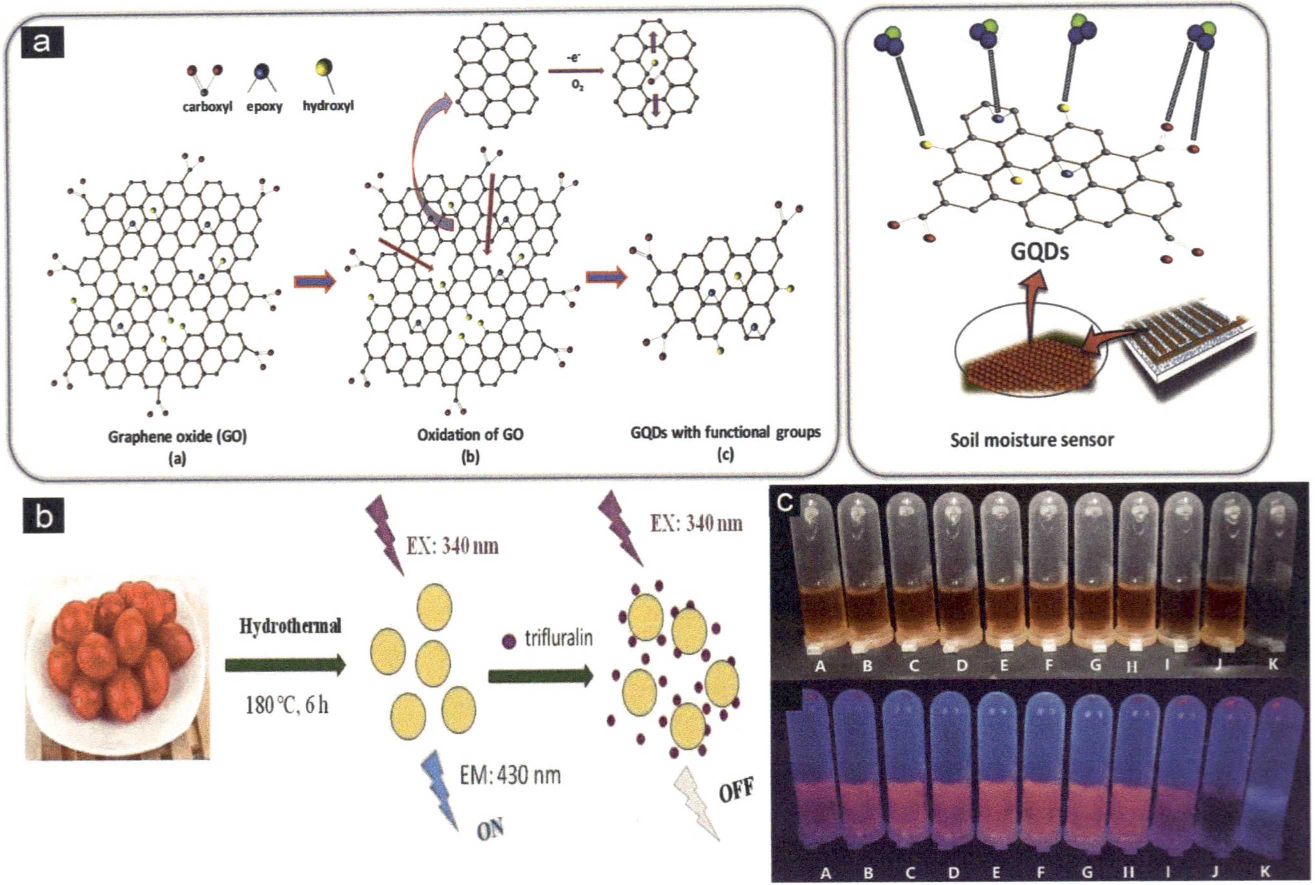

Fig. 10.6 See figure legend on opposite page

Fig. 10.6 Effects of QDs on determining soil quality (A) synthesis of GQD by oxidation of graphene oxide and passivation with different functional groups to develop fluorescence biosensor for the detection of moisture content in soil; (B) photoluminescence-based detection of trifluralin by green-synthesized CQDs wherein the fluorescence is quenched after addition of herbicide; and (C) selectivity of nitrogen-doped CQDs in detecting herbicides trifluralin (J) and alachlor (K) by fluorescence quenching.

Panel (A) reprinted with permission from Kalita, H., Palaparthy, V.S., Baghini, M.S. and Aslam, M., 2020. Electrochemical synthesis of graphene quantum dots from graphene oxide at room temperature and its soil moisture sensing properties. Carbon, 165, 9–17; (B) from Lai, Z., Guo, X., Cheng, Z., Ruan, G. and Du, F., 2020. Green synthesis of fluorescent carbon dots from cherry tomatoes for highly effective detection of trifluralin herbicide in soil samples. ChemistrySelect, 5(6), pp.1956–1960; (C) from Du, F., Cheng, Z., Lai, Z., Ruan, G. and Zhao, C., 2019. Red-emissive nitrogen doped carbon quantum dots for highly selective and sensitive fluorescence detection of the alachlor herbicide in soil samples. New J. Chem. 43(47), pp.18695–18701.

Trifluralin herbicides are used in the soil to control the growth of unusual grasses and weeds mainly in the vegetable field, grain, and fruit crops (Epp et al., 2018; Milhome et al., 2018). The mode of action of these herbicides is selective and impactful to the crops. Its residue present in soil or water can be detected via electrochemical, chromatographic, and fluorescent techniques (Karasali et al., 2017; Zhang et al., 2019; Irandoust et al., 2018). Recently, fluorescent methods gained attention for herbicide detection also because they are simple, rapid, and cost-effective. So, researchers focused on developing sensors that are based on CQDs having fluorescence properties and can be utilized to check the cutoff limit of trifluralin residue (Jafari et al., 2019). Along these, CQDs have no toxicity if present in lower concentrations, showcase good physicochemical properties, and high sensing accuracy. Green synthesis approach was exploited to synthesize CQDs using cherry tomato for the fluorescent-based detection of trifluralin herbicide. The presence of low amounts of trifluralin in the soil can be detected with high precision (Fig. 10.6B). Researchers demonstrated positive results of sensing wherein CQDs were able to detect even 0.5 nM of trifluralin in the soil (Lai et al., 2020). Hence, green synthesis approach-based CQDs can prove to be effective fluorescent probes for detecting the residue and concentration of herbicides in the soil.

10.4.3 Diagnosing plant diseases using QD

Antibiotic like kasugamycin (KAS) has been frequently used to cure plant diseases such as rice blast, apple fire blight, olive knot, and bacterial fruit blotch caused by *Pyricularia grisea*, *Erwinia amylovora*, *Pseudomonas savastanoi*, and *Acidovorax citrulli*, respectively (Nguyen et al., 2018; Sundin and Wang, 2018). KAS-conjugated ZnO QDs (KAS-ZnO QDs) were found to prevent the growth of *A. citrulli* bacteria (Liang et al., 2019). Humidity favors the development of these bacteria, at the site of disease with symptoms of lesions, that contain low pH water (Langemeier et al., 2017; Yan et al., 2017). Researchers investigated that KAS-ZnO QDs gave effective antimicrobial activity in the case of bacterial fruit blotch by sustained release of antibiotic from the QD surface with no reportable toxic effect on the watermelon plant (Fig. 10.7A). Also, conjugation to QDs surface prevents photodegradation of KAS due to the UV shielding effect of QDs (Zheng et al., 2016). Synergistic antimicrobial property can be due to the release of KAS and Zn^{2+} from KAS-ZnO QDs in an acidic environment. However, it was also reported that the release of Zn^{2+} ion from ZnO QDs has antimicrobial effects against many bacteria like *Escherichia coli*, *Proteus mirabilis*, *Staphylococcus aureus*, and *Klebsiella pneumonia* (Redfern et al., 2018).

Apart from these, *Pseudomonas aeruginosa*, a bacterium, and the fungi, *Fusarium avenaceum*, also infect the plants and cause severe damage to them. Detection of these two microorganisms can be possible through fluorescence imaging (Wang et al., 2014). Therefore, researchers developed multicolor emission fluorescent probes that target the bacteria and fungus (Kasibabu et al., 2015). Green-synthesized and nontoxic carbon dots were prepared by a hydrothermal method which have fluorescence property and hence can be used for cell imaging of bacteria and fungus. A plant virus known as beet necrotic yellow vein virus (BNYVV), transmitted by a vector *Polymyxa betae*, causes rhizomania disease in sugar beet. Infection of plants indicates necrosis,

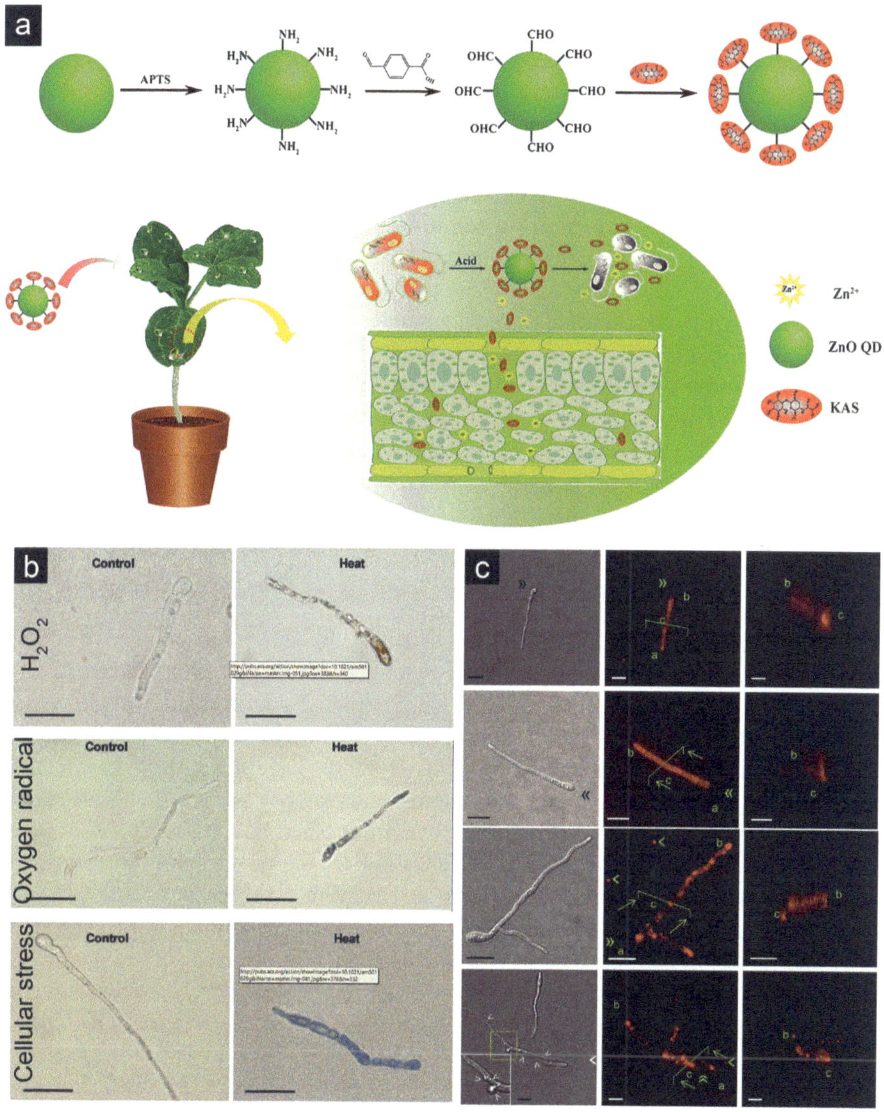

Fig. 10.7 Effects of QDs in detection and eradication of plant diseases (A) synthesis and functionalization of kasugamycin on ZnO QDs for controlled release of antibiotic for antimicrobial activity against *Acidovorax citrulli*; (B) optical microscopic images representing effect of MPA-QDs against *F. oxysporum* via H_2O_2 generation, oxygen radical generation, and cellular stress; and (C) confocal micrographic images representing effect of MPA-QDs in a suspension of growing *F. oxysporum* after different incubation times.

Panel (A) Reprinted with permission from Liang, Y., Duan, Y., Fan, C., Dong, H., Yang, J., Tang, J., Tang, G., Wang, W., Jiang, N. and Cao, Y., 2019. Preparation of kasugamycin conjugation based on ZnO quantum dots for improving its effective utilization. Chem. Eng. J. 361, 671–679; (B and C) from Rispail, N., De Matteis, L., Santos, R., Miguel, A.S., Custardoy, L., Testillano, P. S., Risueño, M.C., Pérez-de-Luque, A., Maycock, C., Fevereiro, P. and Oliva, A., 2014. Quantum dot and superparamagnetic nanoparticle interaction with pathogenic fungi: internalization and toxicity profile. ACS Appl. Mater. Interfaces 6(12), 9100–9110.

yellow vein, and yellow spots on leaves (Chiodi et al., 2021). In a study, antibody-functionalized CdTe QDs were formed for the detection of *Polymyxa betae*. The antiglutathione-*S*-transferase proteins conjugated with QDs and rhodamine dye were attached with glutathione-*S*-transferase proteins. Further, an antigen-antibody complex was formed which results the FRET mechanism (Safarpour et al., 2012). In another study, immunosensors based on CdTe QDs are highly sensitive, specific, and stable. *Fusarium oxysporum* is a soilborne plant pathogen. Its chlamydospores reach a vascular bundle via root penetration and cause wilting and death in plants (Arie, 2019). Therefore, it is necessary to detect this fungus to save fruitful agricultural plants. In a study, 3-mercaptopropionic acid-coated QDs (MPA-QDs) were prepared and their interaction with*Fusarium oxysporum* hyphal cells was checked (Rispail et al., 2014). Interestingly, nanomaterials quickly interacted with the hyphal cells labeling the presence of the fungus and internalization of MPA-QDs by the hyphae were observed (Fig. 10.7B and C). Although low toxicity was seen, they can be used for detection and control approaches based on nanotechnology.

10.4.4 QD for detection of pesticides in agricultural products

A lot of pesticides like parathion, malathion, and methyl parathion are used to control the pests in the field of agriculture. These belong to organophosphorus class of pesticides which is the most common and toxic group of pesticides in the world. When crops are treated with the pesticides, they remain on the surface of the crops and also pollute the environment. Once these pesticides enter the food chain, they show serious health concerns as they significantly inhibit the activity of acetylcholinesterase enzyme, thereby interfering the neurotransmission (Chen and Wang, 2020). Therefore, by the Environmental Protection Agency (EPA), the organophosphorus insecticides are replaced by acetamiprid, a broad-spectrum insecticide that causes the death of the insects. However, the excess use of this may also accumulate in the soil gradually and further contaminate the groundwater and air.

For the detection of pesticides/insecticides in agricultural products as well as in environmental samples, various researches are going on Researchers prepared QDs conjugated with antibody, and immunochromatographic strip (ICS) assay was developed for quick and sensitive detection of acetamiprid residues in agricultural products and environmental specimens (Liu et al., 2019). Its extraordinary benefits like being fast, sensitive, affordable, real-time screening, and size-tunable fluorescence allow ICS assay as an effective tool in determining food safety. Similarly, in other investigations, researchers built up an immunochromatographic assay (ICA) utilizing QD submicrobeads (QBs) as fluorescent probes for the quick and quantitative on-site detection of parathion in agrarian products (Fig. 10.8A). Under ideal conditions, the created QB-ICA had a direct reach from 0.7 to 13.0ng/mL, limit of detection at 10% inhibitory concentration worth of 0.411 ng/mL, and visual identification limits are 6.25 ng/mL (Liu et al., 2020). This proposed methodology was effectively applied for parathion detection in spiked agricultural samples demonstrating great promise to be utilized in the safety of food and the environment. The Fe_3O_4@SiO_2@molecular imprinting polymer (MIP)

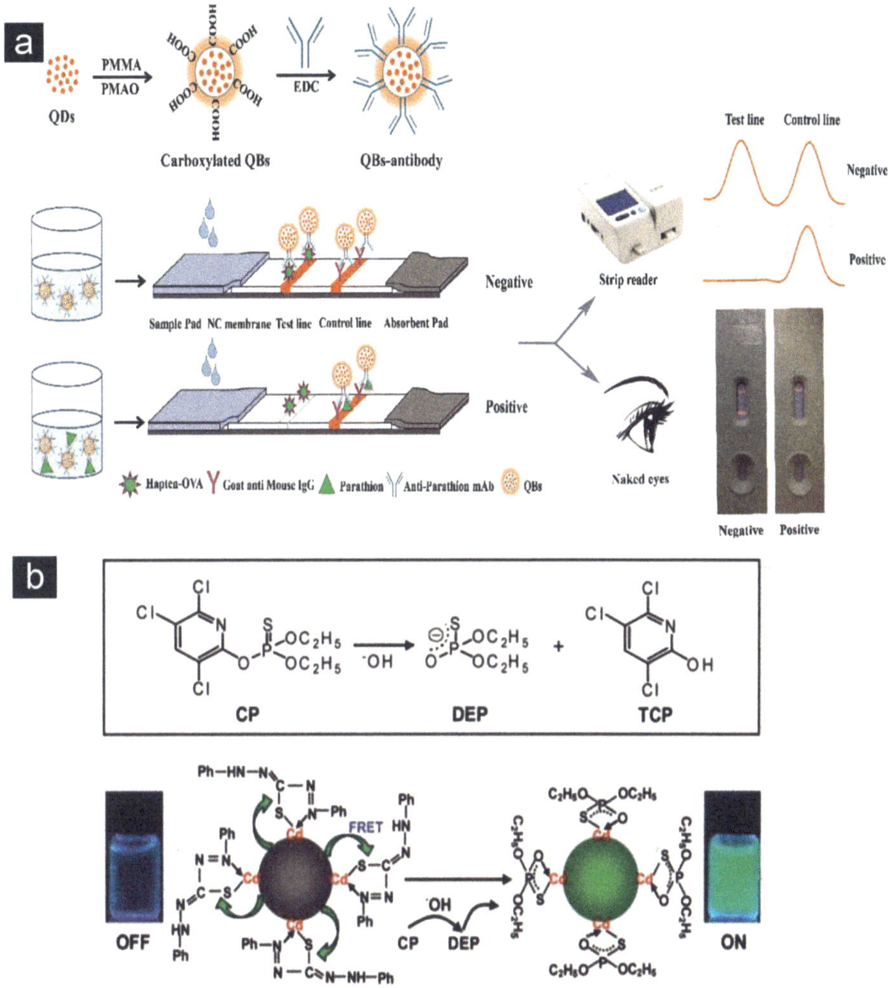

Fig. 10.8 Effects of QDs in the detection of pesticides in agricultural products
(A) immunochromatographic assay designed by antibody-tagged CdSe/ZnS QD submicrobeads for the detection of parathion; (B) fluorescence switch observed due to ligand replacement mechanism for detecting organophosphorothioate pesticides.
Panel (A) reprinted with permission from Liu, B., Li, P., Wang, Y., Guo, Y., Zhang, H., Dong, S., Xiong, Y. and Zhang, C., 2020. Quantum dot submicrobead-based immunochromatographic assay for the determination of parathion in agricultural products. Food Analyt. Methods 13(9), 1736–1745; (B) from Zhang, K., Mei, Q., Guan, G., Liu, B., Wang, S. and Zhang, Z., 2010. Ligand replacement-induced fluorescence switch of quantum dots for ultrasensitive detection of organophosphorothioate pesticides. Analyt. Chem. 82(22), 9579–9586.

was used for the development of biomimetic antibody which has characteristic features of multiple recognition sites and increased adsorption limit.

Besides these, researchers also used QDs as labels to develop the biomimetic fluorescence immunoassay method. This strategy can simultaneously determine various pesticide residues such as methyl parathion, chlorpyrifos, and trichlorfon. These pesticide haptens were labeled utilizing QDs with various emission wavelengths (Jiang et al., 2019). Some of the research groups synthesized CdTe QDs fluorescence probes, and the surface of the CdTe QDs was coated by dithizone through the coordination between surface cadmium ions and dithizone molecules for the detection of organophosphorothioate pesticides. Detection was based on fluorescence resonance energy transfer (FRET) mechanism. When organophosphorothioate pesticides bind with these QDs, the dithizone ligands at the CdTe QD surface are replaced by the hydrolyzate of the organophosphorothioate, and consequently, the fluorescence is generated as presented in Fig. 10.8B (Zhang et al., 2010). This QD-based sensor was able to detect pesticide residues even at very low concentrations.

10.4.5 Enhancing the sustainability of greenhouses by QDs

The growth of a plant can be controlled by optimizing various factors such as temperature, nutrients, light, and atmospheric humidity. Greenhouse, a structure developed using transparent materials, is an efficient way of increasing the growth of plants by regulating the climatic conditions and optimizing various parameters. Factors such as nutrients, moisture, and temperature are important for plant growth and photosynthesis but provisions for optimum light inside the greenhouse have proved to be difficult in the past years. Visible light of wavelength 380 to 780 nm and quantum energies between 1.6 and 3.2 eV are important entities for affecting the rate of photosynthesis in plants (Gudkov et al., 2020).

Solar radiation is the only source of visible light for autotrophic organisms. Unavailability of proper solar radiation and a poor level of light can decrease the rate of photosynthesis in plants. Artificial light sources are used to intensify the photosynthesis rate, and it is also found that by controlling the intensity of light using artificial sources can increase the yield by 25% to 30% (Ouzounis et al., 2015). An electric light source with spectral composition is the most widely used artificial light source in greenhouse cultivation. However, an electrical source has proved to be economically unattractive as 100 W to 1 kW of electricity is required per 1 square meter of the greenhouse area for about 10–12 h (Pavlov et al., 2017). Therefore, studies have been focused on various QDs to provide an efficient and cheap light source with proper spectral composition to intensify the photosynthetic rate of plants.

From the literature, it is known that red light of wavelength 600 nm to 750 nm is of primary importance for the process of photosynthesis in plants. Red and orange rays of light are also responsible for flowering in plants (Wolf and Blankenship, 2019). The wavelength of 380 to 490 nm (blue-violet spectra) is also directly associated with the photosynthesis process (Petroutsos et al., 2016). Blue-violet rays of light help in various synthesis processes and accelerate the flowering process during short-day conditions. Carotenoids and aquatic plants generally absorb the green (490–565 nm) and yellow (565–595 nm) radiation and are also considerably important for agricultural cultivation (Smith et al., 2017). In a study, effect of photoconversion QD-coated polymer films for the growth of *Solanum lycopersicum* was observed by converting UV light into blue and red emissions that are considered most important for plant growth (Fig. 10.9B). The conversion of light by QDs helped in increasing the plant biomass as compared to the plant grown in white light (Simakin et al., 2020). Some of the studies have found that providing monolithic radiation is not efficient enough to accelerate the photosynthetic rate, a combination of all the spectrum of light is necessary for plant growth as evidenced by the researchers shown in Fig. 10.9A. Therefore, providing a controlled wavelength of each required spectrum helps in achieving the maximum yield in plants (Parrish et al., 2021).

In recent years, researchers have shown interest in QDs because of their size and surface tunable optoelectronic properties. QDs can be used as an economical source for providing a composed spectrum of light in the greenhouse structure and can substitute the use of an electrical light source in the near future. Semiconductor QDs such as CdSe, CdTe, ZnSe, ZnTe, and ZnS have already attracted researchers as these QDs are known to emit radiation in the visible spectrum (Pavlov et al., 2017). However, the toxic property of heavy metals has limited the use of semiconductor QDs in the agricultural field. Silicon and graphene quantum dots are also being investigated to overcome the limitations of semiconductor quantum dots. The nontoxic and biodegradable nature of silicon and graphene quantum dots makes them suitable in various agricultural applications. The property of silicon and graphene quantum dots to emit radiation in between visible spectra and near-infrared spectra and their low cost of production make them the most suitable candidate as an artificial light source for greenhouse plantation (Maxwell, 2019).

10.5 Mechanism of action of QDs in enhancing crop productivity

10.5.1 QDs in enhancing photosynthesis

Sunlight is an inexhaustible energy source, but its intermittent availability requires storage of this energy. Such stored energy can be beneficial in artificial photosynthesis mediated by H_2O splitting. Semiconductor-type QDs are considered ideal material for efficient photogeneration of H_2 by H_2O splitting which gets involved in the photosystem-based photosynthesis. The quantum confinement of QDs can be advantageous in rendering broad solar spectrum range. The splitting and photogeneration of H_2 involve three steps (Li et al., 2018): first being the light absorption wherein the solar energy is absorbed due to quantum confinement of the QDs; second, charge separation, where the absorbed light is separated into an exciton; and finally, redox catalysis, where the exciton with high energy redox potential moves to the surface of QDs. From the exciton, the electrons are processed by the cathodic hydrogen evolution reaction cocatalyst, whereas the holes are processed through the anodic cocatalyst.

During photosynthesis, chloroplasts play an essential role in converting light energy to chemical energy. These plant organelles show emission in visible region

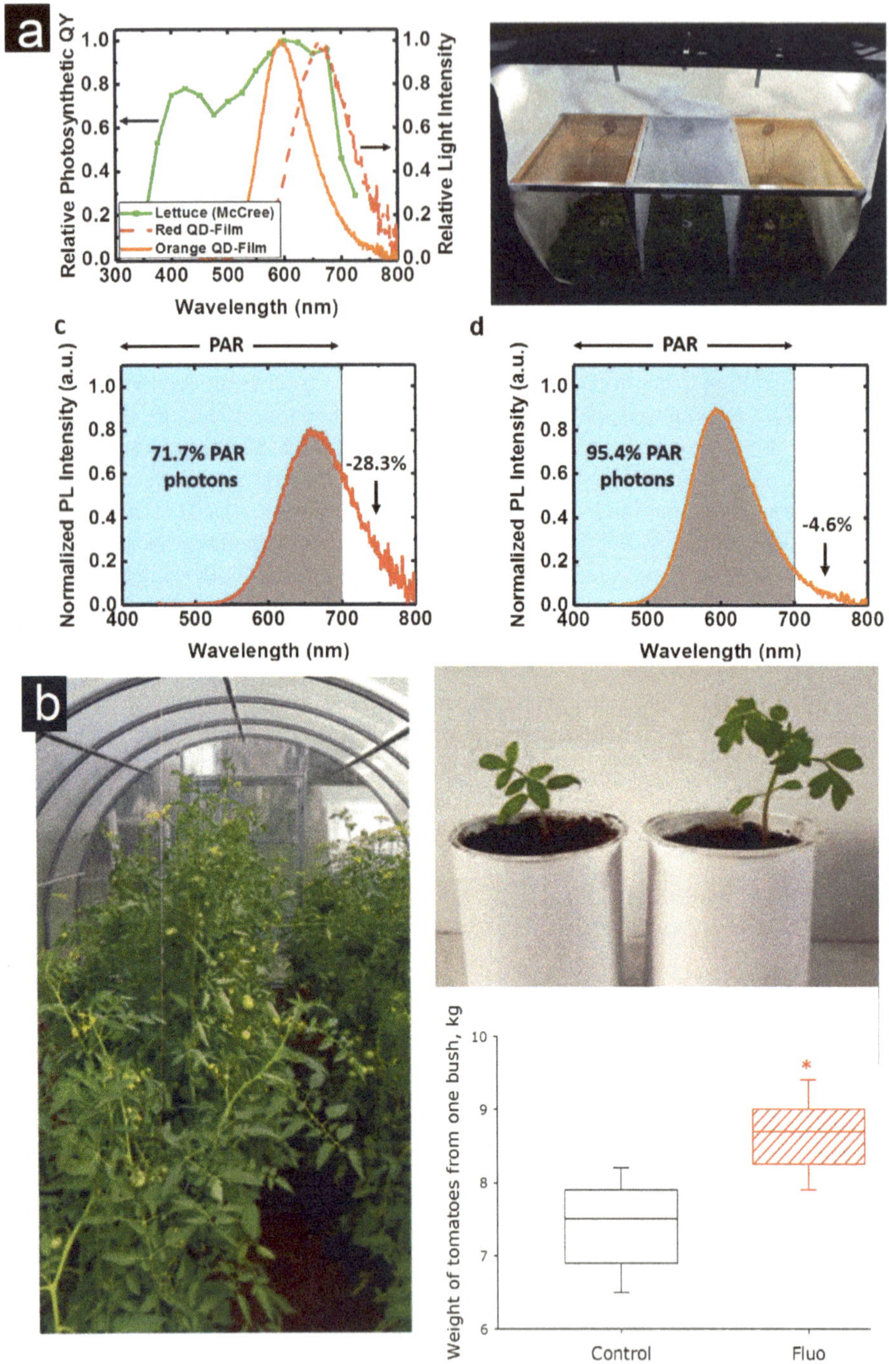

Fig. 10.9 See figure legend on opposite page

and not in UV region. To absorb maximum solar energy, it is prerequisite to harvest maximum UV light. In this regard, Li et al. (2020) synthesized SiQDs to convert UV absorption into blue emission. When these QDs were incubated with the chloroplasts, the complex showed strong absorption in UV light region. Simultaneously, photoluminescence of QDs decreased as compared to nontreated QD groups indicating the absorption of emitted blue light through chloroplast. This subsequent reabsorption of blue light made UV light harvesting by the chloroplast feasible. When CQDs were incubated with the chloroplasts, the dots seem to cover their entire surface. The blue light emission by the CQDs was absorbed by the chloroplasts with a simultaneous decrease in the photoluminescence of the dots in quantum yield-dependent manner (Li et al., 2021). Further surface passivation of CQDs can enhance their charge transfer potential. In a study, amine-functionalized CQDs conjugate strongly to chloroplast surface leading to enhanced electron transfer and accelerated pathway involved in light reaction (Chandra et al., 2014). Li et al. (2021) reported that CQDs are, however, involved in the conversion of harmful UV radiation to harmless photosynthetically active radiation without any significant transfer of energy suggesting safe use of UV light in photosynthesis with the help of QDs. Such conversion is achieved by QDs with moderate quantum yield only because the lower yielded QDs interfere with the light absorption, whereas the high-yielded QDs may lead to photoinhibition.

The conversion of UV light also influences the electron transport rates and actual photosynthetic efficiency of photosystem II. In a study, the Hill reaction was also evaluated in the presence of SiQDs to exhibit UV light absorption by chloroplasts. A decrease in absorption is usually observed due to the reduction of DCPIP (terminal electron acceptor) to DCPID during photoreaction in chloroplasts. DCPIP accepts electron from the water in photosystem II during photosynthesis. The decrease in absorption was intensified with time when chloroplasts were treated with the QDs. The results suggested that the photosynthetic activity of chloroplast is elevated by QDs by enhancing their light-harvesting capability mediated by the transfer of electron during the reaction (Li et al., 2020).

RuBisCo is a crucial enzyme in plants which influences the rates of photosynthesis and control carbon assimilation (Wang et al., 2018). In rice seedlings, CQDs increase the carboxylase activity of RuBisCo, thereby promoting carbon assimilation and hence the photosynthesis leading to better seedling growth. Several researches have

Fig. 10.9 Effects of QDs in enhancing the sustainability of greenhouses for better crop yield (A) growth of lettuce under different light exposures on $CuInS_2/ZnS$ QDs; (B) effect of photoconversion QD-coated polymer films for the growth of *Solanum lycopersicum* leading to increase in biomass by converting UV light into blue and red emissions which are considered most important for plant growth.
Panel (A) reprinted with permission from Parrish, C.H., Hebert, D., Jackson, A., Ramasamy, K., McDaniel, H., Giacomelli, G.A. and Bergren, M.R., 2021. Optimizing spectral quality with quantum dots to enhance crop yield in controlled environments. Commun. Biol. 4(1), 1–9; (B) from Simakin, A.V., Ivanyuk, V.V., Dorokhov, A.S. and Gudkov, S.V., 2020. Photoconversion Fluoropolymer Films for the Cultivation of Agricultural Plants Under Conditions of Insufficient Insolation. Appl. Sci. 10(22), 1–10.

reported enhanced RuBisCo activity in CQD-treated *Vigna radiata*, *Trifolium repens*, *Oryza sativa*, and *Arabidopsis thaliana* plants (Li et al., 2018, 2019; Wang et al., 2018). In plants, CQDs are also degraded by hydrogen peroxide and horseradish peroxidase to generate CO_2 and plant hormone analogues. The CO_2 can be converted to carbohydrates via Calvin cycle whereas the plant hormone analogues can help in plant growth. Huang et al. (2018) reported the degradation of GQDs leading to increase in plant yield by the generation of growth hormone analogues. The presence of moisture in seeds can significantly enhance their germination rate. CQDs are also known to enhance the moisture content of the plants by increasing their water uptake capability (Wang et al., 2018). The hydrophilic groups like the presence of hydroxyl and carboxyl moieties on QD surface allow enough binding sites for water molecules, allowing their movement along with the QDs. Such binding efficiencies can also allow adsorption of metal ions like calcium, potassium, magnesium, copper, zinc, iron, and manganese leading to effective uptake of nutritional content to the plants for better growth (Li et al., 2019). Majumdar et al. (2019) have reported the influence of CdS QDs in upregulating metabolic pathways like glycolysis, citric acid cycle, fatty acid oxidation, biosynthesis of secondary metabolites, and amino acids by upregulating the enzymes involved in the processes.

10.5.2 QDs in controlling stress conditions

The eradication of abiotic and biotic stress is highly warranted for higher crop productivity. QDs play a significant role in reducing such stress in plants through different mechanisms. The generation of ROS is primarily responsible for stress-mediated effects on plant growth. The stress in plants can lead to damage to enzymes, lipids, chlorophyll, stomatal closure, and photoinhibition and decrease photosynthetic rates. Proline works as an osmoprotectant, thereby preventing plants from oxidative stress. Proline-tagged CQDs showed better inhibition of stress in plants due to sustained release of proline for its action (Gohari et al., 2021). CQDs are reported to have radical scavenging properties mediated by the transfer of hydrogen atoms, acceptance of carboxyl and amine groups (Zhao et al., 2015). CQDs are also reported to increase the activities of superoxide dismutase, peroxidase, and catalase, thereby relieving the peanut plants from the drought stress (Su et al., 2018). Li et al. (2019) and Xiao et al. (2019) further demonstrated that CQDs prevented exposure of wheat and citrus seedlings with heavy metals like cadmium ions by adsorbing these ions onto the surface, thereby inhibiting their absorption in plants. CQDs also averted metal ion-mediated plant growth reduction by restoring the content of anthocyanin in plants, inhibiting cell membrane damage, and increasing antioxidant enzyme activities. CQDs also upregulate the expression of thionin gene responsible for regulating the plant-pathogen interaction, thereby increasing disease resistance in the crops (Li et al., 2018). Melatonin and serotonin localized in chloroplast are essential in reducing the oxidative stress in the plants. QDs have made the bioimaging of these natural antioxidants possible within the plants by their adsorption onto the surface of QDs (Erland et al., 2019). CdS QDs also regulate the oxidative stress in plants by enhancing the pathogenicity-related gene's expression leading to the activation of salicylic acid signaling pathway (Majumdar et al., 2019). The crop diseases can also be managed by the antibacterial activity of CQDs. Li et al. (2018) synthesized CQDs from vitamin C demonstrating broad-spectrum antibacterial and antifungal activity against gram-positive and gram-negative bacteria and pathogenic fungus. CQDs can render such activity by degrading the cell walls and nuclear contents of the microorganisms.

10.6 Toxicity profile of QD

Recently, QDs have been frequently used as a sensor for the detection of biological samples, and food safety. Enormous and important physiochemical properties of QDs have made them quite efficient for diverse applications. But negative effects of QDs have also been observed on environment and humans (Manshian et al., 2016, 2017; Liu et al., 2017). It is assumed that there are numerous factors that are responsible for the cytotoxicity of QDs like surface properties, size, concentration, exposure route, exposure time duration, and coating materials (Oh et al., 2016). Certain QDs which are prepared from heavy metal ions like mercury, cadmium, lead, and arsenic exhibited toxicity. It should be noted that even having such high potential, the use of Cd-based QDs is limited because of their toxicity that has proved hazardous for biological and environmental use. The toxic effects were investigated by researchers and they concluded that for the clinical research/biological implementation, CdSe QDs showed the toxic effect when exposed to UV because of photolysis features (Liu et al., 2015). However, some studies indicate the damage of cells and organs due to the release of Cd ion from Cd-QDs (Wang et al., 2016). This released ion can attach with mitochondrial proteins and results in cell lysis. Frequent use and very high surface charge on QDs easily interact with biomolecules rendering dose-dependent toxic effects, thereby still limiting their clinical use. Pristine CdSe QDs were observed to interact with the cell walls of *Picea omorika*, specifically cellulose and lignin via hydroxyl groups to induce structural changes and representative toxicity (Sun et al., 2021). Researchers also observed that the ratio of reduced to oxidized glutathione levels in plants significantly represents the oxidative stress generated by CdSe/ZnS QDs (Navarro et al., 2012). Many kinds of literature have suggested the release of Cd ions for eliciting the toxicological effects, but such release is completely dependent on the physiochemical stability of the dots, synthesis procedures, passivation, and environmental conditions (Pagano et al., 2018).

QDs' cytotoxicity can be generated either by degradation of QD core or by generation of ROS. Studies performed in different plants also highlight the genotoxic effects of QDs. DNA nicking was observed when both QD and DNA are incubated in UV light and dark conditions leading to the formation of photo- and surface-generated free radicals (Svitkova et al., 2017). In a study conducted on the undifferentiated cells of *Medicago sativa*, increased DNA breaks (single or double stranded) were observed upon exposure to mercaptopropanoic acid-functionalized CdSe/ZnS QDs (Santos et al., 2013). In some studies, researchers observed low toxicity against cells at higher concentrations of very small-sized QDs representing huge role of size onto the so far presented toxicity of QDs (Vandarkuzhali et al., 2017). In a study, it was emphasized that at a higher concentration, PEGylated CQDs cause less toxicity to cells and therefore can be used in clinical researches for bioimaging technology. These results suggest that carbon-based QDs are safer alternatives as compared to semiconductor QDs. Some researchers hypothesized that polymer and peptide coating on the surface can overcome the effects of toxicity of QDs (Havrdova et al., 2016). Due to many advantages of QDs, they are used in the agriculture sector for better growth and development of plants and subsequent increase in the productivity rate. But QDs have also some toxicities to the plants; however, the actual mechanism still needs to be explored. Recently, it was demonstrated that nitrogen-doped CQDs (N-CDs) inhibit the growth of primary root in *A. thaliana*, but they slow down the cell division process in the tip of the root due to the downregulation of the cell cycle and DNA repair genes. It was observed that auxin level got decreased in root tip because of the accumulation of N-CDs in the nucleus of meristem cells (Yan et al., 2021). Rispail et al. (2014) have reported that when MPA-QDs were used to eradicate fungal disease from the plants, low toxicity was observed due to the presence of QDs.

Increased utilization of QDs has raised health concerns highlighting the need for measuring the impact of these nanomaterials onto the ecosystem. Most of the QDs are exposed to the environment through indirect routes interacting with biota and in the food chain leading to biomagnification. Hence, studying the interactions of QDs in agri-food sector is highly desired. Reports have suggested that plant root microbiome plays a significant role in detoxification or stabilization of QDs before their uptake through roots, thereby causing detrimental effects on plants and soil (Sillen et al., 2015). Therefore, development of safer QDs is highly desired for the sustainable use in the ecosystem. The critical step that becomes essential to stabilize the QDs to reduce their toxicity by inhibiting the release of ions as in the Cd-based QDs (Pagano et al., 2018). Surface passivation plays a significant role in defining the chemical stability for QDs. Carbon-based QDs are considered more safer than the other semiconductor-based QDs; therefore, use of CQDs should be more preferred in agriculture compared to other QDs. However, Cd- or metal-based QDs can be fostered in detecting the pesticides, herbicides, stress conditions, and pathogens (Iavicoli et al., 2017; Kumar et al., 2015). For the past decades, researchers are more focused toward the development of core/shell-type QDs to eliminate ion release-based toxic effects. Research has also been shifted in using transition element-doped QDs in spite of heavy metals. Including paramagnetic ions in QD lattice can also prevent the leakage of ions into the environment and thereby will reduce toxic effects posed so far (Reshma and Mohanan, 2019).

10.7 Conclusions and future perspectives

QDs have been developed as powerful tools for biosensing, bioimaging, drug delivery, diagnosis, and therapeutics. The advancements in synthesis procedures and passivation strategies successfully controlled the size, shape, stability, surface charge, and quantum yield. Especially, conjugation of biologically derived molecules has opened great avenues in imaging cells and cellular processes. They have also proven to be efficient fluorescent probes but targeted delivery still requires more analysis making them smart multifunctional nanodots. However, with increasing complexity in applications, quantification, and control over the distribution of functional groups onto the QD surface requires more studies. Application of QDs in agriculture is the emerging theme and a lot of research in this regard are recently booming. Different QDs have significantly influenced the productivity of the crops by enhancing the growth, monitoring the presence of herbicides/pesticides, and controlling the plant-related diseases. QDs also protect crops from abiotic and biotic stresses by scavenging free radicals, expressing disease resistant genes, and rendering antibacterial effects. Although significant progress has been attained by QDs but concerns associated with the size-dependent toxicity have limited their clinical use. In plants, several biomarkers based on proteins and genes required for photosynthesis, cell viability, and/or DNA damage are already under practice for detecting exposure of engineered nanomaterials. Such biomarker technologies can be used for the risk management of QDs. Rapid development of nanotechnology has expanded the field of nanotoxicology which can be explored for determining the physiological measurements and simultaneous management of biological pathways. This will be helpful in characterizing the actual risk to ensure the use of QDs in sustainable manner across different applications. Future ecotoxicological analysis needs to be performed to determine QDs exposure to the environment. To reduce toxicity, use of heavy metal-based QDs should be replaced with transition element-based QDs. As of now, the research is mostly directed toward carbon and graphene QDs only. Establishing novel QDs in future will further enhance the efficiencies and will open new options for researchers toward determining potentials in different applications.

References

Abbasi, E., Kafshdooz, T., Bakhtiary, M., Nikzamir, N., Nikzamir, N., Nikzamir, M., Mohammadian, M., Akbarzadeh, A., 2016. Biomedical and biological applications of quantum dots. Artif. Cells Nanomed. Biotechnol. 44, 885–891.

AbouElhamd, A.R., Al-Sallal, K.A., Hassan, A., 2019. Review of core/shell quantum dots technology integrated into building's glazing. Energies 12, 1–22.

Ananthanarayanan, A., Wang, Y., Routh, P., Sk, M.A., Than, A., Lin, M., Zhang, J., Chen, J., Sun, H., Chen, P., 2015. Nitrogen and phosphorus co-doped graphene quantum dots: synthesis from adenosine triphosphate, optical properties, and cellular imaging. Nanoscale 7, 8159–8165.

Andra, S., Balu, S.K., Jeevanandham, J., Muthalagu, M., Vidyavathy, M., San Chan, Y., Danquah, M.K., 2019. Phytosynthesized metal oxide nanoparticles for pharmaceutical applications. Naunyn Schmiedebergs Arch. Pharmacol. 392, 755–771.

Arie, T., 2019. Fusarium diseases of cultivated plants, control, diagnosis, and molecular and genetic studies. J. Pestic. Sci. 44, 275–281.

Arul, V., Sethuraman, M.G., 2018. Facile green synthesis of fluorescent N-doped carbon dots from Actinidia deliciosa and their catalytic activity and cytotoxicity applications. Opt. Mater. 78, 181–190.

Athanassiou, C.G., Kavallieratos, N.G., Benelli, G., Losic, D., Rani, P.U., Desneux, N., 2018. Nanoparticles for pest control: current status and future perspectives. J. Pestic. Sci. 91, 1–15.

Bacon, M., Bradley, S.J., Nann, T., 2014. Graphene quantum dots. Part. Part. Syst. Charact. 31, 415–428.

Badıllı, U., Mollarasouli, F., Bakirhan, N.K., Ozkan, Y., Ozkan, S.A., 2020. Role of quantum dots in pharmaceutical and biomedical analysis, and its application in drug delivery. TrAC Trends Anal. Chem. 131, 1–12.

Bajwa, N., Mehra, N.K., Jain, K., Jain, N.K., 2016. Pharmaceutical and biomedical applications of quantum dots. Artif. Cells Nanomed. Biotechnol. 44, 758–768.

Baptista, A., Silva, F., Porteiro, J., Míguez, J., Pinto, G., 2018. Sputtering physical vapour deposition (PVD) coatings: a critical review on process improvement and market trend demands. Coat 8, 1–22.

Beloglazova, N.V., Goryacheva, I.Y., Shmelin, P.S., Kurbangaleev, V., De Saeger, S., 2015. Preparation and characterization of stable phospholipid-silica nanostructures loaded with quantum dots. J. Mater. Chem. B 3, 180–183.

Bi, C., Kershaw, S.V., Rogach, A.L., Tian, J., 2019. Improved stability and photodetector performance of CsPbI3 perovskite quantum dots by ligand exchange with aminoethanethiol. Adv. Funct. Mater. 29, 1–9.

Blanco-Canosa, J.B., Wu, M., Susumu, K., Petryayeva, E., Jennings, T.L., Dawson, P.E., Algar, W.R., Medintz, I.L., 2014. Recent progress in the bioconjugation of quantum dots. Coord. Chem. Rev. 263, 101–137.

Bonilla, C.M., Kouznetsov, V.V., 2016. "Green" quantum dots: basics, green synthesis, and nanotechnological applications. In: Green Nanotechnology-Overview and Further Prospects. Intechopen, Croatia, pp. 174–192.

Borovaya, M.N., Burlaka, O.M., Yemets, A.I., Blume, Y.B., 2015. Biosynthesis of quantum dots and their potential applications in biology and biomedicine. In: Fesenko, O., Yatsenko, L. (Eds.), Nanoplasmonics, Nano-Optics, Nanocomposites, and Surface Studies. Springer, Switzerland, pp. 339–362.

Borovaya, M., Pirko, Y., Krupodorova, T., Naumenko, A., Blume, Y., Yemets, A., 2015. Biosynthesis of cadmium sulphide quantum dots by using Pleurotus ostreatus (Jacq.) P. Kumm. Biotechnol. Biotechnol. Equip. 29, 1156–1163.

Brichkin, S.B., Razumov, V.F., 2016. Colloidal quantum dots: synthesis, properties and applications. Russ. Chem. Rev. 85, 1297–1312.

Brooks, J., Lefebvre, D.D., 2017. Optimization of conditions for cadmium selenide quantum dot biosynthesis in Saccharomyces cerevisiae. Appl. Microbiol. Biotechnol. 101, 2735–2745.

Bruna, N., Collao, B., Tello, A., Caravantes, P., Díaz-Silva, N., Monrás, J.P., Órdenes-Aenishanslins, N., Flores, M., Espinoza-Gonzalez, R., Bravo, D., Pérez-Donoso, J.M., 2019. Synthesis of salt-stable fluorescent nanoparticles (quantum dots) by polyextremophile halophilic bacteria. Sci. Rep. 9, 1–13.

Cassidy, J., Zamkov, M., 2020. Nanoshell quantum dots: quantum confinement beyond the exciton Bohr radius. J. Chem. Phys. 152, 1–10.

Cha, W., Kim, H.J., Lee, S., Kim, J., 2017. Size-controllable and stable organometallic halide perovskite quantum dots/polymer films. J. Mater. Chem. C 5, 6667–6671.

Chakravarty, D., Erande, M.B., Late, D.J., 2015. Graphene quantum dots as enhanced plant growth regulators: effects on coriander and garlic plants. J. Sci. Food Agric. 95, 2772–2778.

Chandra, S., Pradhan, S., Mitra, S., Patra, P., Bhattacharya, A., Pramanik, P., Goswami, A., 2014. High throughput electron transfer from carbon dots to chloroplast: a rationale of enhanced photosynthesis. Nanoscale 6, 3647–3655.

Chen, Y., Cordero, J.M., Wang, H., Franke, D., Achorn, O.B., Freyria, F.S., Coropceanu, I., Wei, H., Chen, O., Mooney, D.J., Bawendi, M.G., 2018. A ligand system for the flexible functionalization of quantum dots via click chemistry. Angew. Chem. 130 (17), 4742–4746.

Chen, R., Han, B., Yang, L., Yang, Y., Xu, Y., Mai, Y., 2016. Controllable synthesis and characterization of CdS quantum dots by a microemulsion-mediated hydrothermal method. JOL 172, 197–200.

Chen, X., Peng, L., Huang, K., Shi, Z., Xie, R., Yang, W., 2016. Non-injection gram-scale synthesis of cesium lead halide perovskite quantum dots with controllable size and composition. Nano Res. 9, 1994–2006.

Chen, J., Shu, J., Anqi, Z., Juyuan, H., Yan, Z., Chen, J., 2016. Synthesis of carbon quantum dots/TiO2 nanocomposite for photo-degradation of Rhodamine B and cefradine. Diamond Relat. Mater. 70, 137–144.

Chen, G., Wang, P., 2020. Electroanalytical methods for detecting pesticides in agricultural products: a review and recent developments. Int. J. Electrochem. Sci. 15, 2700–2712.

Chinnathambi, S., Chen, S., Ganesan, S., Hanagata, N., 2014. Silicon quantum dots for biological applications. Adv. Healthc. Mater. 3, 10–29.

Chiodi, C., Concheri, G., Squartini, A., Ravi, S., Broccanello, C., Moro, M., Stevanato, P., 2021. Quantification of rhizomania virus by automated RNA isolation and PCR based methods in sugar beet. Virus Dis. 32, 161–166.

Chistyakov, A.A., Zvaigzne, M.A., Nikitenko, V.R., Tameev, A.R., Martynov, I.L., Prezhdo, O. V., 2017. Optoelectronic properties of semiconductor quantum dot solids for photovoltaic applications. J. Phys. Chem. Lett. 8, 4129–4139.

Choi, S.H., 2017. Unique properties of graphene quantum dots and their applications in photonic/electronic devices. J. Phys. D Appl. Phys. 50, 1–10.

Dahi, A., Colson, P., Jamin, C., Cloots, R., Lismont, M., Dreesen, L., 2016. Radio-frequency magnetron sputtering: a versatile tool for CdSe quantum dots depositions with controlled properties. J. Mater. Environ. Sci. 7, 2277–2287.

Debruyne, D., Deschaume, O., Coutiño-Gonzalez, E., Locquet, J.P., Hofkens, J., Van Bael, M. J., Bartic, C., 2015. The pH-dependent photoluminescence of colloidal CdSe/ZnS quantum dots with different organic coatings. Nanotechnology 26, 1–9.

Dhawan, S., Dhawan, T., Vedeshwar, A.G., 2014. Growth of Nb2O5 quantum dots by physical vapor deposition. Mater. Lett. 126, 32–35.

Dimos, K., 2016. Carbon quantum dots: surface passivation and functionalization. Curr. Org. Chem. 20, 682–695.

Ding, X., Peng, F., Zhou, J., Gong, W., Slaven, G., Loh, K.P., Lim, C.T., Leong, D.T., 2019. Defect engineered bioactive transition metals dichalcogenides quantum dots. Nat. Commun. 10, 1–13.

Ding, H., Yu, S.B., Wei, J.S., Xiong, H.M., 2016. Full-color light-emitting carbon dots with a surface-state-controlled luminescence mechanism. ACS Nano 10, 484–491.

Dogan, Ü., Kasap, E., Cetin, D., Suludere, Z., Boyaci, I.H., Türkyılmaz, C., Ertas, N., Tamer, U., 2016. Rapid detection of bacteria based on homogenous immunoassay using chitosan modified quantum dots. Sens. Actuators B 233, 369–378.

Dohnalová, K., Gregorkiewicz, T., Kůsová, K., 2014. Silicon quantum dots: surface matters. J. Phys. Condens. Matter 26, 1–28.

Du, F., Cheng, Z., Lai, Z., Ruan, G., Zhao, C., 2019. Red-emissive nitrogen doped carbon quantum dots for highly selective and sensitive fluorescence detection of the alachlor herbicide in soil samples. New J. Chem. 43, 18695–18701.

Dutta, S., Chatterjee, S., Mallem, K., Cho, Y.H., Yi, J., 2019. Control of size and distribution of silicon quantum dots in silicon dielectrics for solar cell application: a review. Renew. Energy 144, 2–14.

Elshafey, R., Radi, A.E., 2018. Electrochemical impedance sensor for herbicide alachlor based on imprinted polymer receptor. J. Electroanal. Chem. 813, 171–177.

Epp, J.B., Schmitzer, P.R., Crouse, G.D., 2018. Fifty years of herbicide research: comparing the discovery of trifluralin and halauxifen-methyl. Pest Manag. Sci. 74, 9–16.

Erland, L.A., Yasunaga, A., Li, I.T., Murch, S.J., Saxena, P.K., 2019. Direct visualization of location and uptake of applied melatonin and serotonin in living tissues and their redistribution in plants in response to thermal stress. J. Pineal Res. 66, e12527.

Facure, M.H., Schneider, R., Mercante, L.A., Correa, D.S., 2020. A review on graphene quantum dots and their nanocomposites: from laboratory synthesis towards agricultural and environmental applications. Environ. Sci. Nano 7, 3710–3734.

Fan, L., Zhu, M., Lee, X., Zhang, R., Wang, K., Wei, J., Zhong, M., Wu, D., Zhu, H., 2013. Direct synthesis of graphene quantum dots by chemical vapor deposition. Part. Part. Syst. Charact. 30, 764–769.

Faucher, S., Charron, G., Lützen, E., Le Coustumer, P., Schaumlöffel, D., Sivry, Y., Lespes, G., 2018. Characterization of polymer-coated CdSe/ZnS quantum dots and investigation of their behaviour in soil solution at relevant concentration by asymmetric flow field-flow fractionation-multi angle light scattering-inductively coupled plasma-mass spectrometry. Anal. Chim. Acta 1028, 104–112.

Feng, P., Geng, B., Cheng, Z., Liao, X., Pan, D., Huang, J., 2019. Graphene quantum dots-induced physiological and biochemical responses in mung bean and tomato seedlings. Braz. J. Bot. 42, 29–41.

Foubert, A., Beloglazova, N.V., Rajkovic, A., Sas, B., Madder, A., Goryacheva, I.Y., De Saeger, S., 2016. Bioconjugation of quantum dots: review & impact on future application. TrAC Trends Anal. Chem. 83, 31–48.

Fu, C.C., Wu, C.Y., Chien, C.C., Hsu, T.H., Ou, S.F., Chen, S.T., Wu, C.H., Hsieh, C.T., Juang, R.S., Hsueh, Y.H., 2020. Polyethylene Glycol6000/carbon nanodots as fluorescent bio-imaging agents. Nanomaterials 10, 1–15.

Giraldo, J.P., Wu, H., Newkirk, G.M., Kruss, S., 2019. Nanobiotechnology approaches for engineering smart plant sensors. Nat. Nanotechnol. 14, 541–553.

Gohari, G., Panahirad, S., Sepehri, N., Akbari, A., Zahedi, S.M., Jafari, H., Dadpour, M.R., Fotopoulos, V., 2021. Enhanced tolerance to salinity stress in grapevine plants through application of carbon quantum dots functionalized by proline. Environ. Sci. Pollut. Res., 1–14.

Gomase, A., Sangale, S., Mundhe, A., Gadakh, P., Nikam, V., 2019. Quantum dots: method of preparation and biological application. J. Drug Deliv. Ther. 9, 670–672.

Grandhi, G.K., Viswanatha, R., 2016. Understanding the role of surface capping ligands in passivating the quantum dots using copper dopants as internal sensor. J. Phys. Chem. C 120, 19785–19795.

Gudkov, S.V., Simakin, A.V., Bunkin, N.F., Shafeev, G.A., Astashev, M.E., Glinushkin, A.P., Grinberg, M.A., Vodeneev, V.A., 2020. Development and application of photoconversion fluoropolymer films for greenhouses located at high or polar latitudes. J. Photochem. Photobiol. B 213, 1–11.

Guo, L.P., Zhang, Y., Li, W.C., 2017. Sustainable microalgae for the simultaneous synthesis of carbon quantum dots for cellular imaging and porous carbon for CO2 capture. J. Colloid Interface Sci. 493, 257–264.

Haro-González, P., Martínez-Maestro, L., Martín, I.R., García-Solé, J., Jaque, D., 2012. High-sensitivity fluorescence lifetime thermal sensing based on CdTe quantum dots. Small 8, 2652–2658.

Hatakeyama, H., Nakahata, Y., Yarimizu, H., Kanzaki, M., 2017. Live-cell single-molecule labeling and analysis of myosin motors with quantum dots. Mol. Biol. Cell 28, 173–181.
Havrdova, M., Hola, K., Skopalik, J., Tomankova, K., Petr, M., Cepe, K., Polakova, K., Tucek, J., Bourlinos, A.B., Zboril, R., 2016. Toxicity of carbon dots-effect of surface functionalization on the cell viability, reactive oxygen species generation and cell cycle. Carbon 99, 238–248.
Hu, Y., Li, J., Ma, L., Peng, Q., Feng, W., Zhang, L., He, S., Yang, F., Huang, J., Li, L., 2010. High efficiency transport of quantum dots into plant roots with the aid of silwet L-77. Plant Physiol. Biochem. 48, 703–709.
Hu, M.Z., Zhu, T., 2015. Semiconductor nanocrystal quantum dot synthesis approaches towards large-scale industrial production for energy applications. Nanoscale Res. Lett. 10, 1–15.
Huang, C., Dong, H., Su, Y., Wu, Y., Narron, R., Yong, Q., 2019. Synthesis of carbon quantum dot nanoparticles derived from byproducts in bio-refinery process for cell imaging and in vivo bioimaging. Nanomaterials 9, 1–11.
Huang, C., Xia, T., Niu, J., Yang, Y., Lin, S., Wang, X., Yang, G., Mao, L., Xing, B., 2018. Transformation of 14C-labeled graphene to 14CO2 in the shoots of a rice plant. Angew. Chem. 130, 9907–9911.
Iavicoli, I., Leso, V., Beezhold, D.H., Shvedova, A.A., 2017. Nanotechnology in agriculture: opportunities, toxicological implications, and occupational risks. Toxicol. Appl. Pharmacol. 329, 96–111.
Imran, M., Saif, M.J., Kuznetsov, A.E., Idrees, N., Iqbal, J., Tahir, A.A., 2019. Computational investigations into the structural and electronic properties of Cd n Te n (n = 1–17) quantum dots. RSC Adv. 9, 5091–5099.
Irandoust, M., Haghighi, M., Taherpour, A.A., Jafarzadeh, M., 2018. Electrochemical sensing of trifluralin in water by fluconazole-immobilized Fe3O4/SiO2 nanomagnetic core-shell linked to carbon nanotube modified glassy carbon electrode; an experimental and theoretical modeling. J. Iran. Chem. Soc. 15, 719–732.
Jafari, M., Hasanzadeh, M., Karimian, R., Shadjou, N., 2019. Sensitive detection of Triluralin in untreated human plasma samples using reduced graphene oxide modified by polyethylene imine and silver nanoparticles: a new platform on the analysis of peticides and chemical injuries. Microchem. J. 147, 741–748.
Jeevanandam, J., Balu, S.K., Andra, S., Danquah, M.K., Vidyavathi, M., Thalagu, M., 2021. Quantum dots synthesis and application. In: Mubarak, N.M., Khalid, M., Walvekar, R., Numan, A. (Eds.), Contemporary Nanomaterials in Material Engineering Applications. Springer, Cham, pp. 229–265.
Jeevanandam, J., Barhoum, A., Chan, Y.S., Dufresne, A., Danquah, M.K., 2018. Review on nanoparticles and nanostructured materials: history, sources, toxicity and regulations. Beilstein J. Nanotechnol. 9, 1050–1074.
Jeevanandam, J., Sundaramurthy, A., Sharma, V., Murugan, C., Pal, K., Kodous, M.H.A., Danquah, M.K., 2020. Sustainability of one-dimensional nanostructures: fabrication and industrial applications. In: Szekely, G., Livingston, A. (Eds.), Sustainable Nanoscale Engineering. Elsevier, pp. 83–113.
Jiang, M., He, J., Gong, J., Gao, H., Xu, Z., 2019. Development of a quantum dotlabelled biomimetic fluorescence immunoassay for the simultaneous determination of three organophosphorus pesticide residues in agricultural products. Food Agric. Immunol. 30, 248–261.
Jin, H., Jin, Q., Liang, Z., Liu, Y., Qu, X., Sun, Q., 2019. Quantum dot based fluorescent traffic light nanoprobe for specific imaging of avidin-type biotin receptor and differentiation of cancer cells. Anal. Chem. 91, 8958–8965.

Kalita, H., Palaparthy, V.S., Baghini, M.S., Aslam, M., 2020. Electrochemical synthesis of graphene quantum dots from graphene oxide at room temperature and its soil moisture sensing properties. Carbon 165, 9–17.

Karakoti, A.S., Shukla, R., Shanker, R., Singh, S., 2015. Surface functionalization of quantum dots for biological applications. Adv. Colloid Interface Sci. 215, 28–45.

Karasali, H., Pavlidis, G., Marousopoulou, A., Ambrus, A., 2017. Occurrence and distribution of trifluralin, ethalfluralin, and pendimethalin in soils used for long-term intensive cotton cultivation in central Greece. J. Environ. Sci. Health B 52, 719–728.

Karmakar, R., 2015. Quantum dots and it method of preparations-revisited. Prajnan O Sadhona A Sci. Annu. 2, 116–142.

Kasibabu, B.S.B., D'souza, S.L., Jha, S., Singhal, R.K., Basu, H., Kailasa, S.K., 2015. One-step synthesis of fluorescent carbon dots for imaging bacterial and fungal cells. Anal. Methods 7, 2373–2378.

Khanna-Chopra, R., Semwal, V.K., Lakra, N., Pareek, A., 2019. Proline—a key regulator conferring plant tolerance to salinity and drought. In: Hasanuzzaman, M., Fujita, M., Oku, H., Islam, M.T. (Eds.), Plant Tolerance to Environmental Stress: Role of Phytoprotectants. CRC Press, Boca Raton, pp. 59–80.

Kim, Y.T., Han, J.H., Hong, B.H., Kwon, Y.U., 2010. Electrochemical synthesis of CdSe quantum-dot arrays on a graphene basal plane using mesoporous silica thin-film templates. Adv. Mater. 22, 515–518.

Kim, M., Kim, B.G., Kim, J.Y., Jang, W., Wang, D.H., 2020. Enhanced colloidal stability of perovskite quantum dots via split-ligand re-precipitation for efficient bi-functional interlayer in photovoltaic application. J. Ind. Eng. Chem. 88, 137–147.

Kim, M., Osone, S., Kim, T., Higashi, H., Seto, T., 2017. Synthesis of nanoparticles by laser ablation: a review. KONA Powder Part. J., 1–11.

Kominkova, M., Milosavljevic, V., Vitek, P., Polanska, H., Cihalova, K., Dostalova, S., Hynstova, V., Guran, R., Kopel, P., Richtera, L., Masarik, M., 2017. Comparative study on toxicity of extracellularly biosynthesized and laboratory synthesized CdTe quantum dots. J. Biotechnol. 241, 193–200.

Korsch, A.R., Nguyen, G.N., Schmidt, M., Ebler, C., Valentin, S.R., Lochner, P., Rothfuchs, C., Wieck, A.D., Ludwig, A., 2019. Temperature and bias anomalies in the photoluminescence of InAs quantum dots coupled to a Fermi reservoir. Phys. Rev. B 99, 1–11.

Kou, E., Yao, Y., Yang, X., Song, S., Li, W., Kang, Y., Qu, S., Dong, R., Pan, X., Li, D., Zhang, H., 2021. Regulation mechanisms of carbon dots in the development of lettuce and tomato. ACS Sustain. Chem. Eng. 9, 944–953.

Kumar, S., Aziz, S.T., Girshevitz, O., Nessim, G.D., 2018. One-step synthesis of N-doped graphene quantum dots from chitosan as a sole precursor using chemical vapor deposition. J. Phys. Chem. C 122, 2343–2349.

Kumar, P., Kim, K.H., Deep, A., 2015. Recent advancements in sensing techniques based on functional materials for organophosphate pesticides. Biosens. Bioelectron. 70, 469–481.

Kumar, M.S., Yasoda, K.Y., Kumaresan, D., Kothurkar, N.K., Batabyal, S.K., 2018. TiO2-carbon quantum dots (CQD) nanohybrid: enhanced photocatalytic activity. Mater. Res. Express. 5, 1–11.

Lai, Z., Guo, X., Cheng, Z., Ruan, G., Du, F., 2020. Green synthesis of fluorescent carbon dots from cherry tomatoes for highly effective detection of trifluralin herbicide in soil samples. ChemistrySelect 5, 1956–1960.

Langemeier, C.B., Robertson, A.E., Wang, D., Jackson-Ziems, T.A., Kruger, G.R., 2017. Factors affecting the development and severity of Goss's bacterial wilt and leaf blight of corn, caused by Clavibacter michiganensis subsp. nebraskensis. Plant Dis. 101, 54–61.

Lee, J.S., Youn, Y.H., Kwon, I.K., Ko, N.R., 2018. Recent advances in quantum dots for biomedical applications. J. Pharm. Investig. 48, 209–214.
Li, Y., Dai, C., Wang, X., Lv, W., Zhou, H., Zhao, G., Li, L., Sun, Y., Wu, Y., Zhao, M., 2019. A novel strategy to create bifunctional silica-protected quantum dot nanoprobe for fluorescence imaging. Sens. Actuators B 282, 27–35.
Li, H., Huang, J., Liu, Y., Lu, F., Zhong, J., Wang, Y., Li, S., Lifshitz, Y., Lee, S.T., Kang, Z., 2019. Enhanced RuBisCO activity and promoted dicotyledons growth with degradable carbon dots. Nano Res. 12, 1585–1593.
Li, H., Huang, J., Lu, F., Liu, Y., Song, Y., Sun, Y., Zhong, J., Huang, H., Wang, Y., Li, S., Lifshitz, Y., 2018. Impacts of carbon dots on rice plants: boosting the growth and improving the disease resistance. ACS Appl. Bio Mater. 1, 663–672.
Li, H., Huang, J., Song, Y., Zhang, M., Wang, H., Lu, F., Huang, H., Liu, Y., Dai, X., Gu, Z., Yang, Z., 2018. Degradable carbon dots with broad-spectrum antibacterial activity. ACS Appl. Mater. Interfaces 10, 26936–26946.
Li, Q., Jin, X., Yang, Y., Wang, H., Xu, H., Cheng, Y., Wei, T., Qin, Y., Luo, X., Sun, W., Luo, S., 2016. Nd2 (S, Se, Te) 3 colloidal quantum dots: synthesis, energy level alignment, charge transfer dynamics, and their applications to solar cells. Adv. Funct. Mater. 26, 254–266.
Li, Y., Li, W., Zhang, H., Dong, R., Li, D., Liu, Y., Huang, L., Lei, B., 2019. Biomimetic preparation of silicon quantum dots and their phytophysiology effect on cucumber seedlings. J. Mater. Chem. B 7, 1107–1115.
Li, Y., Li, W., Zhang, H., Liu, Y., Ma, L., Lei, B., 2020. Amplified light harvesting for enhancing Italian lettuce photosynthesis using water soluble silicon quantum dots as artificial antennas. Nanoscale 12, 155–166.
Li, Y., Liu, Y., Yang, D., Jin, Q., Wu, C., Cui, J., 2020. Multifunctional molybdenum disulfide-copper nanocomposite that enhances the antibacterial activity, promotes rice growth and induces rice resistance. J. Hazard. Mater. 394, 1–11.
Li, Y., Pan, X., Xu, X., Wu, Y., Zhuang, J., Zhang, X., Zhang, H., Lei, B., Hu, C., Liu, Y., 2021. Carbon dots as light converter for plant photosynthesis: augmenting light coverage and quantum yield effect. J. Hazard. Mater. 410, 124534.
Li, X.B., Tung, C.H., Wu, L.Z., 2018. Semiconducting quantum dots for artificial photosynthesis. Nat. Rev. Chem. 2, 160–173.
Li, G., Wang, H., Zhang, T., Mi, L., Zhang, Y., Zhang, Z., Zhang, W., Jiang, Y., 2016. Solvent-polarity-engineered controllable synthesis of highly fluorescent cesium lead halide perovskite quantum dots and their use in white light-emitting diodes. Adv. Funct. Mater. 26, 8478–8486.
Li, J., Xiao, L., Cheng, Y., Cheng, Y., Wang, Y., Wang, X., Ding, L., 2019. Applications of carbon quantum dots to alleviate Cd2+ phytotoxicity in Citrus maxima seedlings. Chemosphere 236, 1–11.
Li, Y., Xu, J., Liu, H., Song, J., Li, Y., Cheng, B., Li, J., Li, X., 2017. A template/electrochemical deposition method for fabricating silver nanorod arrays based on porous anodic alumina. Nanomater. Nanotechnol. 7, 1847980417717543.
Li, Y., Xu, X., Wu, Y., Zhuang, J., Zhang, X., Zhang, H., Lei, B., Hu, C., Liu, Y., 2020. A review on the effects of carbon dots in plant systems. Mater. Chem. Front. 4, 437–448.
Li, W., Zheng, Y., Zhang, H., Liu, Z., Su, W., Chen, S., Liu, Y., Zhuang, J., Lei, B., 2016. Phytotoxicity, uptake, and translocation of fluorescent carbon dots in mung bean plants. ACS Appl. Mater. Interfaces 8, 19939–19945.
Liang, Y., Duan, Y., Fan, C., Dong, H., Yang, J., Tang, J., Tang, G., Wang, W., Jiang, N., Cao, Y., 2019. Preparation of kasugamycin conjugation based on ZnO quantum dots for improving its effective utilization. Chem. Eng. J. 361, 671–679.

Lim, S.Y., Shen, W., Gao, Z., 2015. Carbon quantum dots and their applications. Chem. Soc. Rev. 44, 362–381.

Liu, B., Li, P., Wang, Y., Guo, Y., Zhang, H., Dong, S., Xiong, Y., Zhang, C., 2020. Quantum dot submicrobead-based immunochromatographic assay for the determination of parathion in agricultural products. Food Anal. Methods 13, 1736–1745.

Liu, Q., Li, H., Xia, Q., Liu, Y., Xiao, K., 2015. Role of surface charge in determining the biological effects of CdSe/ZnS quantum dots. Int. J. Nanomedicine 10, 7073–7088.

Liu, J., Yang, C., Liu, J., Hu, R., Hu, Y., Chen, H., Law, W.C., Swihart, M.T., Ye, L., Wang, K., Yong, K.T., 2017. Effects of cd-based quantum dot exposure on the reproduction and offspring of Kunming mice over multiple generations. Nanotheranostics 1, 23–37.

Liu, A.A., Zhang, Z., Sun, E.Z., Zheng, Z., Zhang, Z.L., Hu, Q., Wang, H., Pang, D.W., 2016. Simultaneous visualization of parental and progeny viruses by a capsid-specific HaloTag labeling strategy. ACS Nano 10, 1147–1155.

Liu, Y., Zhao, Y., Zhang, T., Chang, Y., Wang, S., Zou, R., Zhu, G., Shen, L., Guo, Y., 2019. Quantum dots-based immunochromatographic strip for rapid and sensitive detection of acetamiprid in agricultural products. Front. Chem. 7, 1–10.

Liu, Z., Zou, H., Wang, N., Yang, T., Peng, Z., Wang, J., Li, N., Huang, C., 2018. Photoluminescence of carbon quantum dots: coarsely adjusted by quantum confinement effects and finely by surface trap states. Sci. China Chem. 61, 490–496.

Luo, J., Zhao, S., Wu, P., Zhang, K., Peng, C., Zheng, S., 2015. Synthesis and characterization of new Cd-doped ZnO/ZnS core-shell quantum dots with tunable and highly visible photoluminescence. J. Mater. Chem. C 3, 3391–3398.

Lv, W., Li, L., Xu, M., Hong, J., Tang, X., Xu, L., Wu, Y., Zhu, R., Chen, R., Huang, W., 2019. Improving the stability of metal halide perovskite quantum dots by encapsulation. Adv. Mater. 31, 1–28.

Ma, L., Tu, C., Le, P., Chitoor, S., Lim, S.J., Zahid, M.U., Teng, K.W., Ge, P., Selvin, P.R., Smith, A.M., 2016. Multidentate polymer coatings for compact and homogeneous quantum dots with efficient bioconjugation. J. Am. Chem. Soc. 138, 3382–3394.

Maestro, L.M., Jacinto, C., Silva, U.R., Vetrone, F., Capobianco, J.A., Jaque, D., Solé, J.G., 2011. CdTe quantum dots as nanothermometers: towards highly sensitive thermal imaging. Small 7, 1774–1778.

Mahajan, S., Rani, M., Dubey, R.B., Mahajan, J., 2013. Synthesis of CdSe crystal using hot injection method. Int. J. Latest Res. Sci. Technol. 2, 518–521.

Mai, B.T., Barthel, M.J., Lak, A., Avellini, T., Panaite, A.M., Rodrigues, E.M., Goldoni, L., Pellegrino, T., 2020. Photo-induced copper mediated copolymerization of activated-ester methacrylate polymers and their use as reactive precursors to prepare multi-dentate ligands for the water transfer of inorganic nanoparticles. Polym. Chem. 11, 2969–2985.

Majumdar, S., Pagano, L., Wohlschlegel, J.A., Villani, M., Zappettini, A., White, J.C., Keller, A.A., 2019. Proteomic, gene and metabolite characterization reveal the uptake and toxicity mechanisms of cadmium sulfide quantum dots in soybean plants. Environ. Sci. Nano 6, 3010–3026.

Mala, J.G.S., Rose, C., 2014. Facile production of ZnS quantum dot nanoparticles by Saccharomyces cerevisiae MTCC 2918. J. Biotechnol. 170, 73–78.

Manshian, B.B., Abdelmonem, A.M., Kantner, K., Pelaz, B., Klapper, M., Nardi Tironi, C., Parak, W.J., Himmelreich, U., Soenen, S.J., 2016. Evaluation of quantum dot cytotoxicity: interpretation of nanoparticle concentrations versus intracellular nanoparticle numbers. Nanotoxicology 10, 1318–1328.

Manshian, B.B., Martens, T.F., Kantner, K., Braeckmans, K., De Smedt, S.C., Demeester, J., Jenkins, G.J., Parak, W.J., Pelaz, B., Doak, S.H., Himmelreich, U., 2017. The role of intracellular trafficking of CdSe/ZnS QDs on their consequent toxicity profile. J. Nanobiotechnol. 15, 1–14.

Martynenko, I.V., Litvin, A.P., Purcell-Milton, F., Baranov, A.V., Fedorov, A.V., Gun'Ko, Y. K., 2017. Application of semiconductor quantum dots in bioimaging and biosensing. J. Mater. Chem. B 5, 6701–6727.

Maxwell, T.J., 2019. Coated Quantum Dots: Engineering of Surface Chemistry for Biomedical and Agricultural Applications. Doctoral dissertation, University of Central Florida).

Milhome, M.A., de Lima, L.K., de A. Nobre, C., de AF Lima, F., do Nascimento, R.F., 2018. Effect of ozonization in degradation of trifluralin residues in aqueous and food matrices. J. Environ. Sci. Health B 53, 786–792.

Moghaddam, E., Youzbashi, A.A., Kazemzadeh, A., Eshraghi, M.J., 2015. Preparation of surface-modified ZnO quantum dots through an ultrasound assisted sol-gel process. Appl. Surf. Sci. 346, 111–114.

Molaei, M.J., 2019. Carbon quantum dots and their biomedical and therapeutic applications: a review. RSC Adv. 9, 6460–6481.

Namdari, P., Negahdari, B., Eatemadi, A., 2017. Synthesis, properties and biomedical applications of carbon-based quantum dots: an updated review. Biomed. Pharmacother. 87, 209–222.

Nandan, Y., Mehata, M.S., 2019. Wavefunction engineering of type-I/type-II excitons of CdSe/ CdS core-shell quantum dots. Sci. Rep. 9, 1–11.

Navarro, D.A., Bisson, M.A., Aga, D.S., 2012. Investigating uptake of water-dispersible CdSe/ZnS quantum dot nanoparticles by Arabidopsis thaliana plants. J. Hazard. Mater. 211, 427–435.

Neo, D.C., Cheng, C., Stranks, S.D., Fairclough, S.M., Kim, J.S., Kirkland, A.I., Smith, J.M., Snaith, H.J., Assender, H.E., Watt, A.A., 2014. Influence of shell thickness and surface passivation on PbS/CdS core/shell colloidal quantum dot solar cells. Chem. Mater. 26, 4004–4013.

Nguyen, K.A., Förster, H., Adaskaveg, J.E., 2018. Efficacy of copper and new bactericides for managing olive knot in California. Plant Dis. 102, 892–898.

Nuruzzaman, M.D., Rahman, M.M., Liu, Y., Naidu, R., 2016. Nanoencapsulation, nano-guard for pesticides: a new window for safe application. J. Agric. Food Chem. 64, 1447–1483.

Nxele, S.R., Nyokong, T., 2016. Conjugation of azide-functionalisedCdSe/ZnS quantum dots with tetrakis (5-hexyn-oxy) Fe (II) phthalocyanine via click chemistry for electrocatalysis. Electrochim. Acta 194, 26–39.

Nykiel-Szymańska, J., Bernat, P., Słaba, M., 2018. Potential of *Trichoderma koningii* to eliminate alachlor in the presence of copper ions. Ecotoxicol. Environ. Saf. 162, 1–9.

Oh, E., Liu, R., Nel, A., Gemill, K.B., Bilal, M., Cohen, Y., Medintz, I.L., 2016. Meta-analysis of cellular toxicity for cadmium-containing quantum dots. Nat. Nanotechnol. 11, 479–486.

Omar, N.A.S., Fen, Y.W., Abdullah, J., Zaid, M.H.M., Daniyal, W.M.E.M.M., Mahdi, M.A., 2019. Sensitive surface plasmon resonance performance of cadmium sulfide quantum dots-amine functionalized graphene oxide based thin film towards dengue virus E-protein. Opt. Laser Technol. 114, 204–208.

Osman, H., Li, W., Zhang, X., Chun, F., Deng, W., Moatasim, M., Zheng, X., Deng, W., Zhang, H., Yang, W., 2019. One-step hot injection synthesis of gradient alloy CdxZn1-xSySe1-y quantum dots with large-span self-regulating ability. JOL 206, 565–570.

Ouzounis, T., Rosenqvist, E., Ottosen, C.O., 2015. Spectral effects of artificial light on plant physiology and secondary metabolism: a review. Hortic. Sci. 50, 1128–1135.

Pagano, L., Maestri, E., White, J.C., Marmiroli, N., Marmiroli, M., 2018. Quantum dots exposure in plants: minimizing the adverse response. Curr. Opin. Environ. Sci. Health 6, 71–76.

Panagiotopoulou, M., Salinas, Y., Beyazit, S., Kunath, S., Duma, L., Prost, E., Mayes, A.G., Resmini, M., Tse Sum Bui, B., Haupt, K., 2016. Molecularly imprinted polymer coated quantum dots for multiplexed cell targeting and imaging. Angew. Chem. 128, 8384–8388.
Parrish, C.H., Hebert, D., Jackson, A., Ramasamy, K., McDaniel, H., Giacomelli, G.A., Bergren, M.R., 2021. Optimizing spectral quality with quantum dots to enhance crop yield in controlled environments. Commun. Biol. 4, 1–9.
Pavlov, S.A., Sherstneva, N.E., Koryakin, S.L., Maksimova, E.Y., Makovetskiy, V.V., Krikushenko, V.V., Antipov, E.M., 2017. Features of light conversion process with covering materials containing quantum dots and their application in agriculture. In: Gusev, A., Dyatcheck, T., Godymchuk, A. (Eds.), Nano Hybrids and Composites. Trans Tech Publications Ltd, pp. 162–175.
Petroutsos, D., Tokutsu, R., Maruyama, S., Flori, S., Greiner, A., Magneschi, L., Cusant, L., Kottke, T., Mittag, M., Hegemann, P., Finazzi, G., 2016. A blue-light photoreceptor mediates the feedback regulation of photosynthesis. Nature 537, 563–566.
Pisanicli, T.R., Zhang, Y., Wang, T.H., 2014. Quantum dots in diagnostics and detection: principles and paradigms. Analyst 139, 2968–2981.
Prasad, R., Bhattacharyya, A., Nguyen, Q.D., 2017. Nanotechnology in sustainable agriculture: recent developments, challenges, and perspectives. Front. Microbiol. 8, 1–13.
Prusty, D., Paramanik, L., Parida, K., 2021. Recent advances on alloyed quantum dots for photocatalytic hydrogen evolution: a mini-review. Energy Fuel 35, 4670–4686.
Pu, Y., Cai, F., Wang, D., Wang, J.X., Chen, J.F., 2018. Colloidal synthesis of semiconductor quantum dots toward large-scale production: a review. Ind. Eng. Chem. Res. 57, 1790–1802.
Qi, J., Liu, W., Biswas, C., Zhang, G., Sun, L., Wang, Z., Hu, X., Zhang, Y., 2015. Enhanced power conversion efficiency of CdS quantum dot sensitized solar cells with ZnO nanowire arrays as the photoanodes. Opt. Commun. 349, 198–202.
Qi, B.P., Zhang, X., Shang, B.B., Xiang, D., Zhang, S., 2018. Solvothermal tuning of photoluminescent graphene quantum dots: from preparation to photoluminescence mechanism. J. Nanopart. Res. 20, 1–9.
Ramanan, V., Thiyagarajan, S.K., Raji, K., Suresh, R., Sekar, R., Ramamurthy, P., 2016. Outright green synthesis of fluorescent carbon dots from eutrophic algal blooms for in vitro imaging. ACS Sustain. Chem. Eng. 4, 4724–4731.
Redfern, J., Geerts, L., Seo, J.W., Verran, J., Tosheva, L., Wee, L.H., 2018. Toxicity and antimicrobial properties of ZnO@ ZIF-8 embedded silicone against planktonic and biofilm catheter-associated pathogens. ACS Appl. Nano Mater. 1, 1657–1665.
Reshak, A.H., 2017. Quantum dots in photocatalytic applications: efficiently enhancing visible light photocatalytic activity by integrating CdO quantum dots as sensitizers. Phys. Chem. Chem. Phys. 19, 24915–24927.
Reshma, V.G., Mohanan, P.V., 2019. Quantum dots: applications and safety consequences. JOL 205, 287–298.
Rispail, N., De Matteis, L., Santos, R., Miguel, A.S., Custardoy, L., Testillano, P.S., Risueno, M. C., Pérez-de-Luque, A., Maycock, C., Fevereiro, P., Oliva, A., 2014. Quantum dot and superparamagnetic nanoparticle interaction with pathogenic fungi: internalization and toxicity profile. ACS Appl. Mater. Interfaces 6, 9100–9110.
Rose, M.T., Cavagnaro, T.R., Scanlan, C.A., Rose, T.J., Vancov, T., Kimber, S., Kennedy, I.R., Kookana, R.S., Van Zwieten, L., 2016. Impact of herbicides on soil biology and function. Adv. Agron. 136, 133–220.
Rossi, D., Wang, H., Dong, Y., Qiao, T., Qian, X., Son, D.H., 2018. Light-induced activation of forbidden exciton transition in strongly confined perovskite quantum dots. ACS Nano 12, 12436–12443.

Ruiyi, L., Sili, Q., Zhangyi, L., Ling, L., Zaijun, L., 2017. Histidine-functionalized graphene quantum dot-graphene micro-aerogel based voltammetric sensing of dopamine. Sens. Actuators B 250, 372–382.

Safarpour, H., Safarnejad, M.R., Tabatabaei, M., Mohsenifar, A., Rad, F., Basirat, M., Shahryari, F., Hasanzadeh, F., 2012. Development of a quantum dots FRET-based biosensor for efficient detection of Polymyxabetae. Can. J. Plant Pathol. 34, 507–515.

Salaheldin, A.M., Walter, J., Herre, P., Levchuk, I., Jabbari, Y., Kolle, J.M., Brabec, C.J., Peukert, W., Segets, D., 2017. Automated synthesis of quantum dot nanocrystals by hot injection: mixing induced self-focusing. Chem. Eng. J. 320, 232–243.

Santana, C.P., Mansur, A.A., Carvalho, S.M., da Silva-Cunha Jr, A., Mansur, H.S., 2019. Bi-functional quantum dot-polysaccharide-antibody immunoconjugates for bioimaging and killing brain cancer cells in vitro. Mater. Lett. 252, 333–337.

Santos, A.R., Miguel, A.S., Macovei, A., Maycock, C., Balestrazzi, A., Oliva, A., Fevereiro, P., 2013. CdSe/ZnS Quantum Dots trigger DNA repair and antioxidant enzyme systems in *Medicago sativa* cells in suspension culture. BMC Biotechnol. 13, 1–11.

Schiffman, J.D., Balakrishna, R.G., 2018. Quantum dots as fluorescent probes: synthesis, surface chemistry, energy transfer mechanisms, and applications. Sens. Actuators B 258, 1191–1214.

Sedighi, A., Krull, U.J., 2016. Rapid immobilization of oligonucleotides at high density on semiconductor quantum dots and gold nanoparticles. Langmuir 32, 13500–13509.

Segal, E., Haleva, E., Salomon, A., 2019. Ultrasensitive plasmonic sensor for detecting sub-PPB levels of alachlor. ACS Appl. Nano Mater. 2, 1285–1293.

Sillen, W.M., Thijs, S., Abbamondi, G.R., Janssen, J., Weyens, N., White, J.C., Vangronsveld, J., 2015. Effects of silver nanoparticles on soil microorganisms and maize biomass are linked in the rhizosphere. Soil Biol. Biochem. 91, 14–22.

Simakin, A.V., Ivanyuk, V.V., Dorokhov, A.S., Gudkov, S.V., 2020. Photoconversion fluoropolymer films for the cultivation of agricultural plants under conditions of insufficient insolation. Appl. Sci. 10, 1–10.

Singh, M., Goyal, M., Devlal, K., 2018. Size and shape effects on the band gap of semiconductor compound nanomaterials. J. Taibah Univ. Sci. 12, 470–475.

Singh, R.K., Kumar, R., Singh, D.P., Savu, R., Moshkalev, S.A., 2019. Progress in microwave-assisted synthesis of quantum dots (graphene/carbon/semiconducting) for bioapplications: a review. Mater. Today Chem. 12, 282–314.

Singh, A.K., Pal, P., Gupta, V., Yadav, T.P., Gupta, V., Singh, S.P., 2018. Green synthesis, characterization and antimicrobial activity of zinc oxide quantum dots using Eclipta alba. Mater. Chem. Phys. 203, 40–48.

Smith, H.L., McAusland, L., Murchie, E.H., 2017. Don't ignore the green light: exploring diverse roles in plant processes. J. Exp. Bot. 68, 2099–2110.

Su, L.X., Ma, X.L., Zhao, K.K., Shen, C.L., Lou, Q., Yin, D.M., Shan, C.X., 2018. Carbon nanodots for enhancing the stress resistance of peanut plants. ACS Omega 3, 17770–17777.

Su, S.L., Singh, D.N., Baghini, M.S., 2014. A critical review of soil moisture measurement. Measurement 54, 92–105.

Sun, Y., Chen, G., Deng, H., Yang, L., Liu, C., Sui, Y., Lv, S., Wei, M., Yang, J., 2017. Enhanced photovoltaic performance of CdS quantum dot-sensitized solar cells using 4-tertbutylbenzoic acid as self-assembled monolayer on ZnO photoanode. Phys. Status Solidi A 214, 1–7.

Sun, H., Wang, M., Lei, C., Li, R., 2021. Cell wall: an important medium regulating the aggregation of quantum dots in maize (*Zea mays* L.) seedlings. J. Hazard. Mater. 403, 1–9.

Sundin, G.W., Wang, N., 2018. Antibiotic resistance in plant-pathogenic bacteria. Annu. Rev. Phytopathol. 56, 161–180.

Svitkova, V., Blaskovicova, J., Tekelova, M., Kallai, B.M., Ignat, T., Horackova, V., Skladal, P., Kopel, P., Adam, V., Farkasova, D., Labuda, J., 2017. Assessment of CdS quantum dots effect on UV damage to DNA using a DNA/quantum dots structured electrochemical biosensor and DNA biosensing in solution. Sens. Actuators B 243, 435–444.

Tan, T.L., Zulkifli, N.A., Zaman, A.S.K., Binti Jusoh, M., Yaapar, M.N., Rashid, S.A., 2021. Impact of photoluminescent carbon quantum dots on photosynthesis efficiency of rice and corn crops. Plant Physiol. Biochem. 162, 737–751.

Tayyebi, A., Akhavan, O., Lee, B.K., Outokesh, M., 2018. Supercritical water in top-down formation of tunable-sized graphene quantum dots applicable in effective photothermal treatments of tissues. Carbon 130, 267–272.

Thambiraj, S., Shankaran, R., 2016. Green synthesis of highly fluorescent carbon quantum dots from sugarcane bagasse pulp. Appl. Surf. Sci. 390, 435–443.

Tian, P., Tang, L., Teng, K.S., Lau, S.P., 2018. Graphene quantum dots from chemistry to applications. Mater. Today Chem. 10, 221–258.

Ung, T.D.T., Tran, T.K.C., Pham, T.N., Nguyen, D.N., Dinh, D.K., Nguyen, Q.L., 2012. CdTe and CdSe quantum dots: synthesis, characterizations and applications in agriculture. Adv. Nat. Sci. Nanosci. Nanotechnol. 3, 1–11.

Valizadeh, A., Mikaeili, H., Samiei, M., Farkhani, S.M., Zarghami, N., Akbarzadeh, A., Davaran, S., 2012. Quantum dots: synthesis, bioapplications, and toxicity. Nanoscale Res. Lett. 7, 1–14.

Vandarkuzhali, S.A.A., Jeyalakshmi, V., Sivaraman, G., Singaravadivel, S., Krishnamurthy, K. R., Viswanathan, B., 2017. Highly fluorescent carbon dots from pseudo-stem of banana plant: applications as nanosensor and bio-imaging agents. Sens. Actuators B 252, 894–900.

Vardar, D.O., Aydin, S., Hocaoglu, I., Acar, F.H.Y., Basaran, N., 2018. Effects of silver sulfide quantum dots coated with 2-mercaptopropionic acid on genotoxic and apoptotic pathways in vitro. Chem. Biol. Interact. 291, 212–219.

Vasudevan, D., Gaddam, R.R., Trinchi, A., Cole, I., 2015. Core-shell quantum dots: properties and applications. J. Alloys Compd. 636, 395–404.

Vo, N.T., Ngo, H.D., Vu, D.L., Duong, A.P., Lam, Q.V., 2015. Conjugation of E. coli O157: H7 antibody to CdSe/ZnS quantum dots. J. Nanomater. 2015, 1–7.

Vu, H.T., Chiang, R.K., Huang, C.Y., Chen, C.J., Yu, H.C., Lien, J.Y., Su, Y.K., 2016. Enhanced thermal stability of green-emission quantum-dot light-emitting diodes via composition-gradient thick-shell quantum dots. Appl. Phys. Express 9, 1–5.

Wagner, A.M., Knipe, J.M., Orive, G., Peppas, N.A., 2019. Quantum dots in biomedical applications. Acta Biomater. 94, 44–63.

Wang, X., Feng, Y., Dong, P., Huang, J., 2019. A mini review on carbon quantum dots: preparation, properties, and electrocatalytic application. Front. Chem. 7, 1–9.

Wang, T., Hsieh, H., Hsieh, Y., Chiang, C., Sun, Y., Wang, C., 2012. The in vivo biodistribution and fate of CdSe quantum dots in the murine model: a laser ablation inductively coupled plasma mass spectrometry study. Anal. Bioanal. Chem. 404, 3025–3036.

Wang, M., Wang, J., Sun, H., Han, S., Feng, S., Shi, L., Meng, P., Li, J., Huang, P., Sun, Z., 2016. Time-dependent toxicity of cadmium telluride quantum dots on liver and kidneys in mice: histopathological changes with elevated free cadmium ions and hydroxyl radicals. Int. J. Nanomedicine 11, 2319–2328.

Wang, B., Wang, Y., Wu, H., Song, X., Guo, X., Zhang, D., Ma, X., Tan, M., 2014. A mitochondria-targeted fluorescent probe based on TPP-conjugated carbon dots for both one-and two-photon fluorescence cell imaging. RSC Adv. 4, 49960–49963.

Wang, L., Wu, B., Li, W., Wang, S., Li, Z., Li, M., Pan, D., Wu, M., 2018. Amphiphilic graphene quantum dots as self-targeted fluorescence probes for cell nucleus imaging. Adv. Biosyst. 2, 1–8.

Wang, H., Zhang, M., Song, Y., Li, H., Huang, H., Shao, M., Liu, Y., Kang, Z., 2018. Carbon dots promote the growth and photosynthesis of mung bean sprouts. Carbon 136, 94–102.

Wani, S.H., Kumar, V., Khare, T., Guddimalli, R., Parveda, M., Solymosi, K., Suprasanna, P., Kishor, P.K., 2020. Engineering salinity tolerance in plants: progress and prospects. Planta 251, 1–29.

Wolf, B.M., Blankenship, R.E., 2019. Far-red light acclimation in diverse oxygenic photosynthetic organisms. Photosynth. Res. 142, 349–359.

Wu, X., Liu, H., Liu, J., Haley, K.N., Treadway, J.A., Larson, J.P., Ge, N., Peale, F., Bruchez, M. P., 2003. Immunofluorescent labeling of cancer marker Her2 and other cellular targets with semiconductor quantum dots. Nat. Biotechnol. 21, 41–46.

Wu, H., Tito, N., Giraldo, J.P., 2017. Anionic cerium oxide nanoparticles protect plant photosynthesis from abiotic stress by scavenging reactive oxygen species. ACS Nano 11, 11283–11297.

Xiao, L., Guo, H., Wang, S., Li, J., Wang, Y., Xing, B., 2019. Carbon dots alleviate the toxicity of cadmium ions (Cd 2+) toward wheat seedlings. Environ. Sci. Nano 6, 1493–1506.

Xie, H., Chen, E., Ye, Y., Xu, S., Guo, T., 2020. Highly stabilized gradient alloy quantum dots and silica hybrid nanospheres by core double shells for photoluminescence devices. J. Phys. Chem. Lett. 11, 1428–1434.

Xie, Q., Wu, D., Wang, X., Li, Y., Fang, F., Wang, Z., Ma, Y., Su, M., Peng, S., Liu, H., Wang, K., 2019. Branched capping ligands improve the stability of cesium lead halide (CsPbBr 3) perovskite quantum dots. J. Mater. Chem. C 7, 11251–11257.

Xu, Y., Wang, Z., Guo, Z., Huang, H., Xiao, Q., Zhang, H., Yu, X.F., 2016. Solvothermal synthesis and ultrafast photonics of black phosphorus quantum dots. Adv. Opt. Mater. 4, 1223–1229.

Yan, L., Hu, B., Chen, G., Zhao, M., Walcott, R.R., 2017. Further evidence of cucurbit host specificity among *Acidovoraxcitrulli* groups based on a detached melon fruit pathogenicity assay. Phytopathology 107, 1305–1311.

Yan, X., Wang, J., Li, D., Feng, J., Xu, Q., Chen, H., Han, R., 2021. Response of primary root to nitrogen-doped carbon dots in Arabidopsis thaliana: alterations in auxin level and cell division activity. Environ. Sci. Nano 8, 1352–1363.

Yuan, T., Meng, T., He, P., Shi, Y., Li, Y., Li, X., Fan, L., Yang, S., 2019. Carbon quantum dots: an emerging material for optoelectronic applications. J. Mater. Chem. C 7, 6820–6835.

Zaytseva, O., Neumann, G., 2016. Carbon nanomaterials: production, impact on plant development, agricultural and environmental applications. Chem. Biol. Technol. Agric. 3 (1), 1–26.

Zeng, Z., Xiao, F.X., Phan, H., Chen, S., Yu, Z., Wang, R., Nguyen, T.Q., Tan, T.T.Y., 2018. Unraveling the cooperative synergy of zero-dimensional graphene quantum dots and metal nanocrystals enabled by layer-by-layer assembly. J. Mater. Chem. A 6, 1700–1713.

Zhan, N., Palui, G., Mattoussi, H., 2015. Preparation of compact biocompatible quantum dots using multicoordinating molecular-scale ligands based on a zwitterionic hydrophilic motif and lipoic acid anchors. Nat. Protoc. 10, 859–874.

Zhang, J., Crisp, R.W., Gao, J., Kroupa, D.M., Beard, M.C., Luther, J.M., 2015. Synthetic conditions for high-accuracy size control of PbS quantum dots. J. Phys. Chem. Lett. 6, 1830–1833.

Zhang, C., Lian, L., Yang, Z., Zhang, J., Zhu, H., 2019. Quantum confinement-tunable ultrafast charge transfer in a PbS quantum dots/WSe2 0D–2D hybrid structure: transition from the weak to strong coupling regime. J. Phys. Chem. Lett. 10, 7665–7671.

Zhang, K., Mei, Q., Guan, G., Liu, B., Wang, S., Zhang, Z., 2010. Ligand replacement-induced fluorescence switch of quantum dots for ultrasensitive detection of organophosphorothioate pesticides. Anal. Chem. 82, 9579–9586.

Zhang, T., Qu, Z., Li, B., Yang, Z., 2019. Simultaneous determination of atrazine, pendimethalin, and trifluralin in fish samples by QuEChERS extraction coupled with gas chromatography-electron capture detection. Food Anal. Methods 12, 1179–1186.

Zhao, S., Lan, M., Zhu, X., Xue, H., Ng, T.W., Meng, X., Lee, C.S., Wang, P., Zhang, W., 2015. Green synthesis of bifunctional fluorescent carbon dots from garlic for cellular imaging and free radical scavenging. ACS Appl. Mater. Interfaces 7, 17054–17060.

Zhao, P., Li, X., Baryshnikov, G., Wu, B., Ågren, H., Zhang, J., Zhu, L., 2018. One-step solvothermal synthesis of high-emissive amphiphilic carbon dots via rigidity derivation. Chem. Sci. 9, 1323–1329.

Zhao, C., Liu, B., Piao, S., Wang, X., Lobell, D.B., Huang, Y., Huang, M., Yao, Y., Bassu, S., Ciais, P., Durand, J.L., 2017. Temperature increase reduces global yields of major crops in four independent estimates. Proc. Natl. Acad. Sci. 114, 9326–9331.

Zhao, M.X., Zhu, B.J., Yao, W.J., Chen, D.F., Wang, C., 2018. The delivery of doxorubicin of multifunctional β-cyclodextrin-modified CdSe/ZnS quantum dots for bioactivity and nano-probing. Chem. Biol. Drug Des. 91, 285–293.

Zheng, K., Lu, M., Rutkowski, B., Dai, X., Yang, Y., Taccardi, N., Stachewicz, U., Czyrska-Filemonowicz, A., Hüser, N., Boccaccini, A.R., 2016. ZnO quantum dots modified bioactive glass nanoparticles with pH-sensitive release of Zn ions, fluorescence, antibacterial and osteogenic properties. J. Mater. Chem. B 4, 7936–7949.

Micro- and nanoencapsulation techniques in agriculture

11

Snehal Desai, Manish Singh, Anamika Chavan, Nilesh S. Wagh, and Jaya Lakkakula
Amity Institute of Biotechnology, Amity University Maharashtra, Mumbai, Maharashtra, India

11.1 Introduction

Agricultural chemicals play a very important role in keeping the growing demand for food. However, the harmful effects of agrochemicals on the environment and health are a major problem in their use. The application of agrochemicals has notable effects on the environment and human health because most of them are tenacious and persistent natural pollutants. The global goal of reducing the use of agrochemicals is focused on doing agricultural practices with diminished harmful effects environmentally. Agrochemicals include pesticides, fertilizers, and other growth enhancements (Sinha et al., 2019).

The usage of the nanotechnological advances in the agricultural field is taking a leap so that the value of food is enhanced, lesser input of the agricultural products is done, and most importantly the nutritional quality and contents along with maximized shelf life are increased. It also assists in reducing the toxic chemical contents. Numerous nanomaterials considering nanoinsecticides, nanopesticides, nanofertilizers, and different packaging materials are adopted to enhance the agricultural outcomes. Different varieties of materials are utilized to prepare coat nanomaterials, e.g., plant extracts, metal oxides, emulsion, lipids, polymers. Surface coating with such materials to the fertilizers assists in holding the material more tightly to the plant because of higher surface tension. The shell consists of the active compound which can be a chemical or biological agent which provides protection against various diseases as well as pests (Chand Mali et al., 2020).

Conventional pesticides used with particular formulations are associated with certain disadvantages such as very poor dispersibility, nonbiodegradability, higher content of the organic solvents, and leaching problem. Such issues have motivated the development of new research in recent years. To mention, some of the research's encapsulation technologies have a great impact in reducing the problems faced due to the conventional pesticide or agrochemical usage by its application. Minimal amount or quantity of the pesticides or agrochemicals is to be used which can be achieved by improving the retention time period of the pesticides used with improved precised efficiency (Wani et al., 2019).

Agricultural Nanobiotechnology. https://doi.org/10.1016/B978-0-323-91908-1.00009-2

Utilization of the nanotechnological processes in the agricultural application is still in its developing state and in its infancy. Overall, by the application of the nanotechnological processes and mechanisms to provide the appropriate safety, more interaction with other particular compounds, reduction in toxicity is yet to be improved. As such, much of the investigation is to be done in the field of nanotechnology for its application in the agricultural field so as to deduce in exactly which form the agrochemicals should be utilized (Wani et al., 2019).

The trend of utilization of the fertilizers with slow-release fertilizers so as to reduce the agrochemical consumption and minimize the detrimental environmental effects motivated the development of encapsulated fertilizers wherein the NPK fertilizers can be enclosed and entrapped within the nanoparticles for its better survival and achieve the property of controlled release of the fertilizers (Corradini et al., 2010).

Although the wide known use of these techniques is found to be very beneficial in the agricultural and food sector, the limited use of this encapsulation technique is done due to the costing for the procedure. Development of such methodologies for yield improvement and reducing the environmental hazards is required so as to increase the health benefits and marketability.

11.2 Application of nanoencapsulation and microencapsulation techniques in agriculture

Numerous techniques and methods have been developed for the encapsulation at the micro- and nanoscale. Broadly, those approaches are classified into various categories, i.e., physiomechanical methods that consist of spray-drying, freeze-drying, etc. Physicochemical method includes emulsification, coacervation, and chemical methods such as in situ polymerization and molecular inclusion. The selection of the appropriate method to be adopted for the encapsulation will be conditional on the wall material and the properties of the core, the release rate, processing steps, size of particles, and also on the eventual application of the particle encapsulated.

Microencapsulation and nanoencapsulation are the techniques that basically include a combination of solid, liquid, or gaseous substances into some other materials that eventually increase and proffer useful properties to the product. Microencapsulation involves conversion of the surface properties of the coating material in such a manner that it encapsulates an active compound. Nanoencapsulation is similar to microencapsulation but what differs is it utilizes particles of size in the range of nano.

11.2.1 Nanofertilizers

Nutrient fertilization serves a key function in sustaining soil fertility and enhancing the crop production and quality. The precise nutrient management relies on chemical fertilizers. These chemical or traditional fertilizers are cost-expensive and deliver environmental risks. With the help of nanotechnology, nanofertilizers are emerging and promising alternative to these chemical fertilizers in the agricultural sector. Nanofertilizers offered a unique outreach in the development of nutrient uptake in plants, greater efficacy, increased yield of the grain, biocontrol agent, and increased rate of germination and in the precision farming. Several instances of metal oxides, polymers, and organic and inorganic nanoencapsulated nanofertilizers researches have been conducted with respect to their applications in the agricultural fields. Similar example has been portrayed later which highlights the incorporation of the manganese zinc ferrite nanoparticles at different concentrations as a foliar nanofertilizer for *Cucurbita pepo* L.

In a study, Shebl et al. utilized template-free microwave-assisted hydrothermal green synthesis technique in the production of manganese-zinc ferrite nanoparticles at different concentrations (0, 10, 20, and 30 ppm) as foliar nanofertilizer for *C. pepo* L. The method was facile and acted as an effective way to control the particle size distribution and provide macroscopic morphology. The average particle size estimated from the TEM (transmission electron microscopy) graphs ranged from 10 to 12.5 nm. These manganese-zinc ferrite nanoparticles exhibited cubical shape, the regularity of which increased as the holding synthesis temperature (T-100, T-120, T-140, T-160, and T-180) rose with respect to the particle size.

The application of $Mn_{0.5}Zn_{0.5}Fe_2O_4$ NPs, a foliar nanofertilizer, showed impact on the growth and characteristics of the fruit of *C. pepo* L. plant during two successive seasons, 2017 and 2018. The highest fruit yield of squash *C. pepo* L. per hectare (54.8 and 55.2 t/ha) with the nanofertilizer preparations was at T-160 or 160°C at the concentrations of 10 ppm. Thus, finally the overall studies demonstrated that the influence of the synthesis of the temperature on the physicochemical properties of the prepared manganese zinc ferrite nanofertilizer that includes surface, pore structure, particle size and shape, growth criteria of the plant, and yield of *C. pepo* L. when being applied at different concentrations (Shebl et al., 2020).

Individual encapsulation has been a widely used approach but this example represents a unique method of coencapsulation (encapsulating two or more compounds in a single-carrier system) which is being used for the first time agricultural purpose wherein the chitosan, being the biodegradable biopolymer, makes the nanofertilizer comprised of copper (Cu) and salicylic acid (SA) upregulate the source activity promoting higher crop productivity.

The study of Sharma et al. highlights the chitosan nanofertilizer which comprised of copper (Cu) and salicylic acid (SA). The synthesis of chitosan nanofertilizer was performed using the ionic gelation method coupled with the coencapsulation. This was the first time that the usage of Cu and SA coencapsulated chitosan nanofertilizer was applied on the maize (*Zea mays* L.) via seed treatment and its foliar application. The size of chitosan nanofertilizer was recorded to be approximately 540 nm. Fourier transform infrared spectroscopy (FTIR) results presented the hypothetical model depicting the interactions of Cu and SA. The symmetric porous, spherical structure of the chitosan nanofertilizer was analyzed using scanning electron microscopy (SEM). Further, scanning electron microscopy-energy-dispersive X-ray spectroscopy (SEM-EDX) studies gave a confirmation of the presence of Cu into the chitosan nanofertilizer.

The study further revealed the effects attributed to the enhanced activities of alpha-amylase (1.7-fold) and protease enzymes (threefold) in germinating seeds. Chitosan nanofertilizer's foliar application (0.01%–0.16%) showed an increase in antioxidant enzymes (1.06–1.91-fold), reduction in malondialdehyde, increased chlorophyll contents (twofold) in leaves, and remarkable induction of sucrose translocated in the internodes. Thus, in this overall study, primary aim of the research was to develop strong biodegradable chitosan-based nanofertilizer which can be used to boost and upregulate source activity, supporting the sink strength for the increased crop production of maize (Sharma et al., 2020).

Another example that deciphers the enormous potential of chitosan-silicon nanofertilizer (Cs-Si NF) as a next-generation nanofertilizer with the vision of sustainable farming. Kumaraswamy et al. studied the novel chitosan-silicon nanofertilizer (Cs-Si NF) wherein the encapsulation of (Si) silicon was performed using chitosan-tripolyphosphate (TPP) nanomatrix. The size of synthesized nanofertilizer was recorded to be around 360.5 nm in diameter. Ionic gelation method was adopted in synthesis of the nanofertilizer.

The characterization was followed by dynamic light scattering (DLS), which resulted in understanding of the hydrodynamic diameter, polydispersity index (PDI), and zetapotential. FTIR analysis was performed to understand the interactions of the chitosan, TPP, and Si. Further, TEM and SEM observations revealed the porous network structure and spherical shape of Cs-Si NF. On the exhibition of Cs-Si NF, slow release of Si and the high growth and yield in the crop were observed. Different concentrations of Cs-Si NF (0.04%–0.12%, *w*/v) seed primed depicted 3.7-fold increase in seedling vigor index (SVI) than SiO_2.

The foliar application was significant enough to induce antioxidant defense enzyme activities. It balanced O_2 and H_2O_2 contents in leaves by equilibrated cellular redox homeostasis. Overall, in the field experiments, the 0.08% Cs-Si NF gave 43.4% high yield/plot and also increased plant height, cob length, and test weight in comparison with SiO_2. The results deciphered that the highest potential of Cs-Si-Ni nanofertilizer is promising to be used as a next-generation nanofertilizer on the account of sustainable agriculture (Kumaraswamy et al., 2021).

With the goal of sustainability in agricultural sector, the comparative study of using nanofertilizer and non-nanofertilizer is essential to prove that the use of nanofertilizers being beneficial with respect to yield growth quality parameters and crop productivity. A similar study was carried out in the following example. Abd El-Azeim et al. gave an insight into the effects of nano- and non-nanofertilizers on potato quality and its production. The average particle size was observed (NPK nanofertilizer) for nitrogen—5.78 to 9.33 nm, phosphorus—4.98 to 7.66 nm, and potassium—5.6 to 9.45 nm.

In comparison with the control treatments, plots that received foliar application of 25% or 50% recommended NPK nanofertilizers were found to be at an absolute economical treatment (23.59-ton ha), rates of starch (79.62%), NPK nutrient used efficiencies (67.74, 278.92, and 118.54 kg potato/kg nutrient), and index of harvest (59.24%) which indicated higher potato yields. Economically, a high profit:cost ratio of the productivity of the potato (*Solanum tuberosum* L.) was observed. The productivity increased due to the foliar application of NPK nanofertilizers which depicted a positive effect on harvest indexes. Also, the nano-NPK boosted plants in absorption of water and nutrients, improving photosynthesis; hence, they are considered to be the biological pumps in increasing the biomass and yield of the crop. Besides, foliar application proved beneficial compared to soil addition technique as its impact of nano- and non-nanofertilizer was observed on the potato yield and its growth quality metrics.

Thus, the overall focus in this research here lies on the impacts of the NPK chemical fertilizers which acted as control treatments in comparison with NPK nanofertilizer in its equivalent or lower rates (100%, 50%, 25%) by the technique of foliar application and soil addition on the production and quality of the potato (Abd El-Azeim et al., 2020).

In an interesting study by Yusefi-Tanha et al., the zinc oxide nanoparticles (ZnONPs) served as a novel nanofertilizer for the enrichment of soil-deficient Zn with Zn. Its influence on the yield of the seed and the antioxidant defense system in the (*Glycine* max cv. Kowsar) soil-grown soybean was observed. The samples of ZnONPs were prepared by sol-gel method of encapsulation. The strategy was designed to enable synthesizing three types of different morphological ZnONPs: spherical 38 nm, floral-like 59 nm, and rod-like >500 nm. The result obtained was that the spherical ZnONPs were most protective compared to the other two morphological characteristics.

The findings also noted that a significance of particle size, its morphological characteristics, and concentration-dependent influence of foliar application of encapsulated ZnONPs were observed on seed yield, increased chlorophyll content, shoot height, grain yield, lipid peroxidations, and varied biomarkers of antioxidants in soybean. These studies summarized that the ZnONPs are the novel nanofertilizers for the crops grown in soil deficient of Zn and could eradicate malnutrition globally (Yusefi-Tanha et al., 2020).

Biologically synthesized NPs have the immense potential to enhance the plant growth and the yield of the bioactive compounds. For this, the biosynthesis of nanoperlite/MnO_2 using the medicinal plants and understanding the mechanism for the generation of new nanofertilizer is vital.

Thus, Jafarirad et al. reported a new green route for the preparation of manganese oxide (MnO_2) or perlite nanocomposites (NCs) by the extract of leaf of *Hypericum perforatum,* a medicinal plant. According to the field emission scanning electron microscopy (FE-SEM) images, the formation of the sheets and spherical nanoparticles for the nanoperlite and MnO_2 were depicted through FE-SEM. The placing of the MnO_2 NPs on the surface of nanoperlite sheets for the production of MnO_2/perlite-NCs is depicted in Fig. 11.1A–D. Besides, the different sizes of the nanoparticles ranged from 20 to 50 nm are observed in Fig. 11.1D. The average size of the nanoparticles was found around 50.5 nm which was quantified using Debye-Scherrer equation. In this study, the production of hypericin and pseudohypericin was raised in treated shoot cultures with lower concentrations of nanoperlites. MnO_2 was depicted suitable in plant fertilization and producing qualified medicinal plants when used at specific concentrations. Metabolomics thus resulted that both the nanoperlite and MnO_2 can be used in the generation of new nanofertilizers and also for producing the secondary metabolites (Jafarirad et al., 2021).

Studies on overcoming the deficiencies of the certain metal oxides in the agricultural plants through the inclusion of nanoparticles have been significantly increased. Comparison of chemical synthesis and biological synthesis of metal oxide nanorods

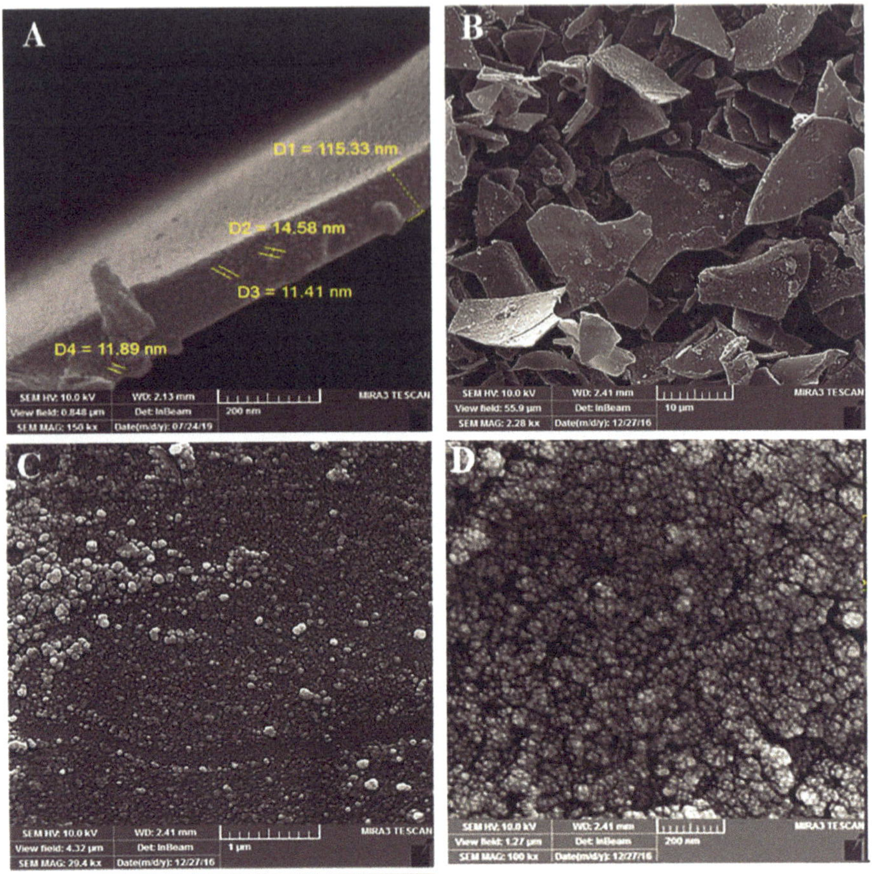

Fig. 11.1 FE-SEM images of nanoperlite (A) and MnO_2/perlite NCs, MnO_2 nanoparticles immobilized on surface of perlite in different magnifications (B–D).
Reprinted with the permission from Jafarirad, S., Kosari-Nasab, M., Mohammadpour Tavana, R., Mahjouri, S., & Ebadollahi, R. (2021). Impacts of manganese bio-based nanocomposites on phytochemical classification, growth and physiological responses of *Hypericum perforatum* L. shoot cultures. Ecotoxicol. Environ. Safety, 209(September 2020), 111841.

will prove a wider perspective on its field application. Hasan et al. investigated the comparative response of chemically synthesized and biologically synthesized iron oxide nanorods. These nanorods used *Moringa oleifera* along with bulk ferric chloride ($FeCl_3$) on *Zea mays*.

The study was designed to investigate the impact of iron oxide nanorods (FeO NRs) synthesized by biological and chemical methods through the coprecipitation on *Z. mays* with aspects related to growth, physiology, and antioxidative and biochemical traits. The released iron helped in the formation of biologically synthesized NRs which resulted in an improved leaf area 13% leaves per plant 26%, total chlorophyll content 80%, and nitrate content 6%. These synthesized FeO NRs were eco-friendly as well as effective for the growth of the plant. This study encourages the usage of green-synthesized FeO NRs for *Z. mays* to manage the iron nutrition in crops and also for multiple uses in agriculture in future (Hasan et al., 2020).

A similar study on metal oxide nanoparticles by Younes et al. studied the impact of the synthesized metal oxide nanoparticles zinc oxide and titanium dioxide (ZnO and TiO_2) on the production of seedlings of *Solanum melongena* L., *Capsicum annum* L., *Lycopersicon esculentum* L., three Solanaceae crops. TEM confirmed the size of the nanoparticles to be around 43 nm. The gel-coating encapsulation method was used to encapsulate the two metal oxide nanoparticles mixed with the seeds of eggplant, pepper, and tomato crops at four different concentrations (0, 50, 100, and 150 mg/L). This resulted in the effect on the germination characters and the growth of the seedlings.

Besides, the performance of growing transplants has been validated with the nanoparticles—gel-coated seedlings than the control treatments, as it could be an efficient and innovative way for the production of transplants which could withstand harsh environmental conditions (Younes et al., 2020).

The necessity for the usage of nanoparticles with green route modification for the sustainable agriculture is of paramount importance. Thus, the following example similarly cites the study using metal oxide NPs. Cyriac et al. investigated that zinc oxide nanoparticles (ZnONPs) using green route that employed bioderived structure-directing agents in an aqueous medium over the neutral conditions of pH. 1 M citric acid was capped on ZnONPs. These hybrid-natured citric acid ZnONPs were successfully characterized by the use of FT-IR spectroscopy and UV-Vis spectroscopy. These citric acid-capped ZnONPs had flower-like structures with an average particle size ranging from 20 to 30 nm when revealed through HR-TEM images.

This novel hybrid approach of citric acid being used as a template and using NaOH flakes for the adjustment of pH to the circumneutral pH conditions, and distinct morphologies could be obtained with regard to ZnO nanostructures. ZnONPs capped with 1 M citric acid reported a higher seed germination rate and the root growth. This impact was observed using the millimolar concentrations of these colloidal nanoparticles on *Oryza sativa* L. (rice) and *Vigna unguiculata* L. (pea) plants. Thus, the study summarized that the effect of ZnONPs with 1 M citric acid and ZnONPs ensures the effect of bioavailability crop improvement than compared to macromolecular zinc oxide. This approach can be used as a sustainable version in the crop production (Cyriac et al., 2019).

An another instance of utilization of ZnO nanoparticles and its impact on the activity of photosynthesis of *Vicia faba* was explored by Pedruzzi et al. (2020). The particle size played a key role in this photosynthetic functioning of 25 nm diameter compared to 70 nm. With the help of foliar spray application, ZnONPs have great potential as a foliar nanofertilizer to increase the growth and yield of food crops. Their capacity for the generation of reactive oxygen species (ROS) and releasing Zn ions was analyzed by fluorescence and dialysis experiments, respectively. This is also the mechanism of how ZnONPs affect the leaves of *V. faba*, in overall impacting its photosynthetic performance. Thus, the study reveals that ZnONPs were phytotoxic at higher concentrations over 200 mg/L. Despite the phytotoxic effects, ZnONPs may be applied as a nanofertilizer in agricultural aspects as Zn being the fundamental micronutrient and acts as a regulatory cofactor of important enzymes (Pedruzzi et al., 2020).

Nanopriming using different NPs has given impressive results with regard to germination of seeds and growth of crops. Among those, silver nanoparticles have a more feasible entry into the soil through several ways. Less known impacts of aged priming of seeds made the following study inquisitive about this concept.

Abdel-Aziz et al. demonstrated the effect of priming of 7-year-old seed consisting of varied concentrations of silver nanoparticles (AgNPs) and its effect on the growth of broad bean seedlings. Trisodium method was used in the generation of AgNPs. The characterization of AgNPs was performed using UV-Vis spectrophotometry at 250 nm. The AgNPs appeared spherical in shape and had an average mean size of 42 nm.

With the use of seed priming technique for AgNPs, it increased shoot and root growth parameters in comparison with the control. While the lower concentrations of AgNPs are in use to improve the growth of the seeds, starch, total carbohydrate, and increased chlorophyll content, for different concentrations of AgNPs, the effects of the priming on the growth of the seedlings of the aged broad bean seedlings could be observed. Seed priming with AgNPs resulted in the increased shoot and root lengths in comparison with the control (Fig. 11.2). Further studies are in the need to have a glance on the effects of AgNPs on the productivity of the crop and also among various crop species (Abdel-Aziz and Rizwan, 2019).

Nanofertilizers in agriculture play a pivotal role to enrich the productivity of the crops while minimizing environmental pollution problems. A research studied by Abdel-Aziz et al. investigated the effects of two engineered nanomaterials chitosan nanoparticles or carbon nanotubes (CNTs) loaded with NPK acting as fertilizer upon French bean plants. This method was approached for the comparison between the two methodologies: seed priming and foliar application. The obtained Cs and Cs-NPK

Fig. 11.2 The effects of priming with different concentrations of AgNPs on seedling growth of aged broad bean seedlings. Bar: cm.
Reprinted with permission from Abdel-Aziz, H. M. M., & Rizwan, M. (2019). Chemically synthesized silver nanoparticles induced physio-chemical and chloroplast ultrastructural changes in broad bean seedlings.

nanoparticles had an average size of 20–22 nm and 95–97 nm, whereas the cylindrical shape was observed with the help of TEM.

Seed priming caused a significant decrease in parameters of growth lower than the control wherein the foliar application improved the growth parameters higher in comparison with control when French bean plants were grown in clay-sandy soil. Overall resultant study conveys the use of chitosan (Cs) nanoparticles or CNTs alone or with NPK as foliar application fertilizer for French bean plants. Better growth and yield were found with the use of nanochitosan rather than carbon nanotubes (Abdel-Aziz and Rizwan, 2019).

In an another study on determining the efficiency of ZnONPs, Awan et al. focused on ZnONPs on the growth of *Brassica oleracea* var. italic. The study deals with the synthesis of ZnONPs using green biosynthesis method from the $ZnSO_4$ solution and the extract of the blacks seeds of *Nigella sativa* L. which acted as a reducing agent. These newly formed nanoparticles were characterized using UV-Vis spectrophotometer, SEM. The average particle size observed analyzing was about 24 nm. ZnONP application to the broccoli resulted in 37.5% seed germination, 56.6% root length, 16.6% shoot length, 41% seedling's weight, 11.5% in a number of leaves, 17.1% height of the plant, and 24.4% leaf area.

As compared to distilled water treatment with the absence of nanoparticles, the presence of nanoparticles gave rise to chlorophyll 50%, phenolic content 67.4%, proline 14.6%, and sugar content 36.2%. This study has been the first time reported the use of ZnONPs as a nanofertilizer for the broccoli crop. Results also concluded that ZnONPs are the efficient form of nanofertilizers with the positive impacts on the morphological and biochemical content of the plant (Awan et al., 2021).

Basil (*Ocimum basilicum* L.), being the most important vegetable and aromatic plant, was chosen for the following study to address the impacts of Zn and Cu nanoparticles (NPs).

In the present study, Abbasifar et al. researched upon that the green synthesis method of Zn and Cu nanoparticles (NPs) was performed by reducing Zn and Cu ions and exposing them to basil (*O. basilicum* L.) extract. When observed through electron microscopy, Zn and Cu appeared quasi-spherical and diameters were less than 100 nm.

An experiment was conducted to investigate the effect of Zn and CuNPs on the basil plant, wherein four levels of zinc nanoparticles (ZnNPs) (0, 1000, 2000, and 4000 ppm) and four levels of copper nanoparticles (CuNPs) (0, 1000, 2000, and 4000 ppm). The results occurred were 4000 ppm ZnNPs+2000 ppm CuNPs significantly affected concentrations of chlorophyll *a*, chlorophyll *b*, total chlorophyll content, highest anti-oxidant activity (plants treated with 4000 ppm ZnNPs+0 ppm), and the content of carotenoid in the basil leaves. This foliar application method of Zn and Cu nanofertilizers was successful and an essential for better quantity and quality in the production of basil plant (Abbasifar et al., 2020).

Introduction of iron smart fertilizer for the agricultural application is a novel strategy to induce environment-friendly conditions and raw materials as well as nontoxic to human fibroblast cells. Thus, the following research demonstrates a similar approach. The aim of Jahangirian et al. was to prepare zeolite/iron (III) oxide nanocomposites (zeolite/Fe_2O_3-NCs) as a smart fertilizer in improving the yield of the crop and productivity of the soil. Zeolite/Fe_2O_3-NC was produced by loading of Fe_2O_3-NPs onto the zeolite by the method of green precipitation.

The product formed was characterized using near and far Fourier-transform infrared spectroscopy (FT-IR), powder X-ray diffraction, energy-dispersive X-ray spectroscopy (EDX), SEM, and TEM. On characterization, results occurred were mean particle sizes of 1.45, 2.19, and 2.20 nm on the zeolite's surface with respective amounts of 4, 7, and 12 wt% Fe_2O_3-NPs. Three types of nontoxic nanocomposites were tested in order to prove the iron smart fertilizer beneficial and slow release of iron ions. The study overall comprised of producing nontoxic nanocomposites as a smart fertilizer in developing the crops, usage of low-cost raw materials with lower amounts of energy, environment-friendly, and successful zeolite/iron (III) oxide nanoparticles as iron smart nanofertilizer for agricultural applications (Jahangirian et al., 2020).

11.2.2 Nanofungicides

Nanofungicides are used as an eco-friendly alternative that plays a crucial role in plant and crop disease management by serving as an alternative to conventional synthetic fungicides for the agricultural practices. Because of the vast functionalization and physiochemical properties, nanofungicides can be utilized for the plant disease management.

According to the study conducted by Weisany et al., the encapsulation of Iranian thyme (*Thymus daenensis* L.) and dill (*Anethum graveolens* L.) essential oils is carried out in the copper nanoparticles. The encapsulation of these essential oils into the copper nanoparticles was assisted in the evaluation of their antifungal activity against the phytopathogen *Colletotrichum nymphaeae*. This phytopathogen is the causative agent of anthracnose that leads to sunken necrotic lesions on the leaves, stem, and fruits on various commercially important plants such as strawberry, beans, and also other crop plants.

Encapsulation of the essential oil in the copper nanoparticles effectively reduced the mycelium growth by about 90% after the 9 days of incubation treatment. The SEM analysis indicated that the copper nanoparticles used for the encapsulation were of size about 100–125 nm (Fig. 11.3A). Encapsulation has led to the inhibition of conidia germination of *C. nymphaeae*. The SEM study conducted also suggested that the encapsulation has resulted in the deformation of the mycelial growth and also led to the hypha twisting. The prominent peaks were observed for the FT-IR spectra of copper nanoparticles formed confirming the formation (Fig. 11.3B).

Gas chromatography-mass spectrometry (GC-MS) analysis for the essential oil composition revealed that the thyme and dill essential oil composition is influenced by the encapsulation with the Cu nanoparticles which is attributed to the susceptibility of the core material for the degradation process. Thus, the overall study conducted indicated that the copper-based nanoencapsulation technique can be developed as a useful delivery system for the fungicides for the purpose of plant protection (Weisany et al., 2019b).

Due to the tremendous losses observed in the crop cultivation worldwide because of the invasion of various plant pathogens, usage of fungicides was adopted so as to

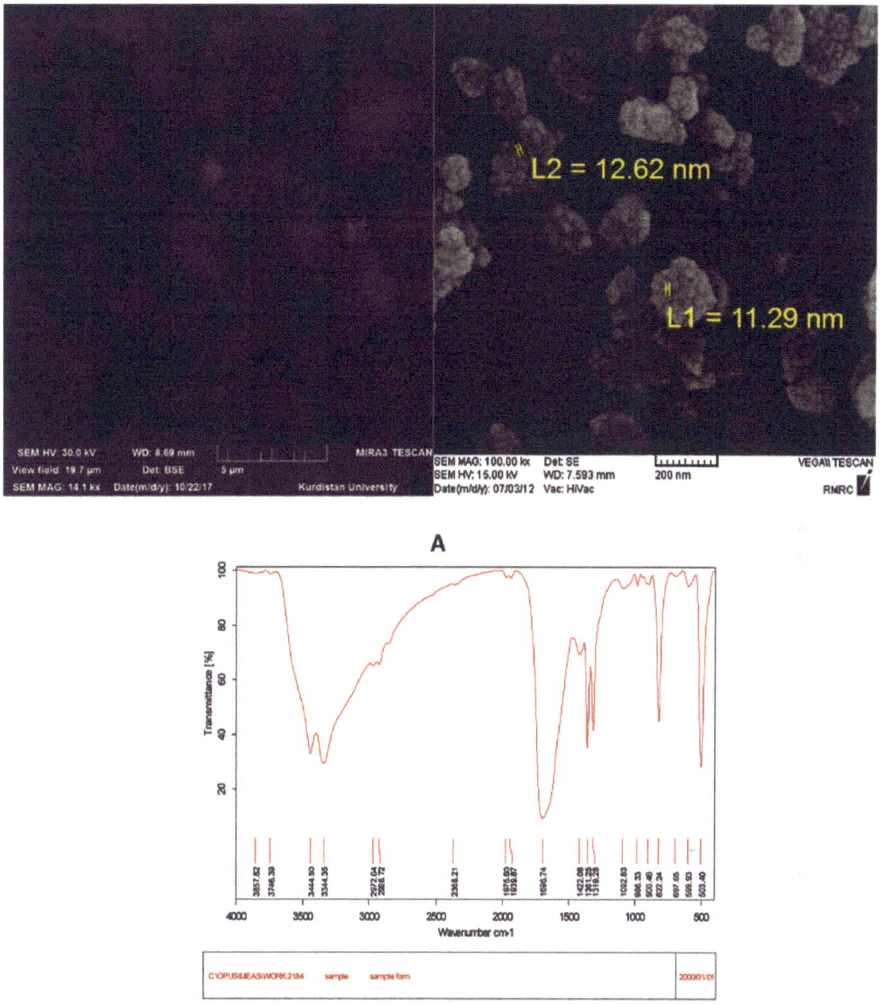

Fig. 11.3 (A) Scanning electron micrographs of copper nanoparticles. (B) FT-IR spectra of copper nanoparticles.

control the postharvest decay during cultivation. Although the use of synthetic fungicides is found to be effective, but due to the significant drawbacks, adoption of the eco-friendlier and more biodegradable alternative using nanofungicides was done. For example, according to the study conducted by Mohammadi et al., the nanoencapsulation of *Zataria multiflora* essential oil (ZEO) is synthesized using CsNPs so as to increase and enhance the antifungal activity. It also aids in enhancing the stability of the ZEO against the isolate of pathogen *Botrytis cinerea* Pers which is a causative agent of gray mold disease in important crops like tomato, grapevine, strawberry, and cucumber. ZEO is considered a good antimicrobial agent that reduces postharvesting decay.

The encapsulation of this essential oil in the chitosan nanoparticles is done by the process of ion gelation technique, with a view to enhance the bioactivity of the essential oil and lower their diffusion rate which eventually maintains the high concentration of the active compound on the surface. Characterization of ZEO-loaded CsNPs is carried out using high-resolution TEM and the average particle size obtained was 125–175 nm. Thus, this study indicates that the CsNPs can be used as a controlled release system for the ZEOs so as to enhance the antifungal activity for crop protection (Mohammadi et al., 2015).

Similarly, the use of chitosan nanoparticles can also be done so as to use as a promising natural fungicide due to its superior performance in the agricultural and food industry.

According to the study made by Khodaiyan et al., in order to enhance the antifungal efficiency, the encapsulation of the clove essential oil in the chitosan nanoparticles is carried out by following an emulsion gelation technique. The antimicrobial property of clove essential oil (CEO) and its main constituent eugenol is limited due to water-soluble constituents and high volatility. Thus, encapsulation helps in the prolonged activity of the EOs through controlled release of it and enhancing its bioavailability against *A. Niger* which is isolated from a spoilt pomegranate. FE-SEM results indicated spherical and size of around 40 and 100 nm for the most optimum obtained clove essential oil chitosan nanoparticles (CEO-CsNPs).

The DLS technique result showed that through the loading of the CEO into the CsNPs, the size increased considerably. The characterization was carried out by using the CEO-CsNPs of the lowest size and having a high surface-to-volume ratio. The study revealed that the encapsulation efficiency, average size, and loading capacity can be effectively controlled by the mass-to-charge ratio of chitosan NPs to CEO. It indicated that the encapsulation of EOs leads to enhanced performance against the *A. Niger* under the in vitro conditions and that is considered to be due to the controlled and preserved release of the encapsulated EOs thus leading to enhanced and better antifungal activity. Thus, according to this study, the CEO-ChNPs can be effectively considered and used as a natural fungicide for agricultural practices (Hasheminejad et al., 2019).

Various insects and pests are controlled by the usage of various agrochemicals but excessive use of such agrochemicals leads to reduced germination and also seed damage. The formulations are such modified that the excessive use of such agrochemicals is avoided by more eco-friendly alternatives. According to the study made by El-Naggar et al., cryogel materials are porous and lightweight structures that are used for the controlled release of tebuconazole fungicide on the seeds for controlling the pests and diseases during the agricultural practices, so the cryogel was prepared by blending and cross-linking glyoxal with the natural polymers such as hydroxyethyl cellulose and bacterial cellulose using the freeze-drying technique.

Different concentration of silver@titanium oxide nanoparticles (Ag@TiO_2NPs) was inserted within the cryogel formed with a view to increase the antimicrobial efficiency and to obtain the controlled release of the tebuconazole fungicide by its encapsulation within the cryogel. For determining the crystallinity of the structures formed, X-ray diffraction analysis was carried out depending upon the diffraction peaks observed (Fig. 11.4A). By performing TEM, the spherical shape and small size of the NPs were determined (Fig. 11.4B).

The elemental analysis through the energy-dispersive X-ray spectroscopy (EDX) revealed the presence of C and O elements along with Ag and Ti elements confirming the presence of the nanoparticles within the cryogel. It also indicated that the porosity of the cryogel is released due to the presence of the nanoparticles in it and the presence of Ag@TiO_2NPs improves the mechanical properties of the cryogel as well. Encapsulation efficiency is calculated by the use of UV-Vis spectroscopy, and it was calculated around 91.4% to 97.3% and core material is efficiently released at a reduced rate.

According to the MIC calculations, the antimicrobial activity against the tested organism is significantly increased due to the presence of Ag@TiO_2NPs even at the low concentrations. Considering all the excellent properties of the cryogel when doped with Ag@TiO_2NPs, it can be significantly used for agricultural applications (El-Naggar et al., 2020).

Infection to the various plants during cultivation and also the outbreak of diseases is preventing by using natural agrochemicals. Nowadays, the use of copper-based antimicrobials against plant pathogens is reviewed. For instance, through the study made by Fuchs et al., it is observed that the microencapsulation of the copper sulfate/copper hydroxide mixture was done into a fat matrix that led to the formation of CuCaps that were utilized for its better adhesion with the grapevines. Grapevine downy mildew is a disease observed due to biotrophic oomycete *Plasmopara viticola*. Thus, copper-containing fungicides can be effectively used due to their antifungal activities against a broad range of bacteria, fungi, and oomycetes. The CuCaps were formed by spray congealing technique and resulted in a wettable and dispersible powder with the good susceptibility results.

The leaf disk assay revealed that the CuCaps with lower amounts of the copper used have similar activity as that of the commercially available copper fungicide. The particle size analysis revealed the size of particles to around 75 μm. The dose-response

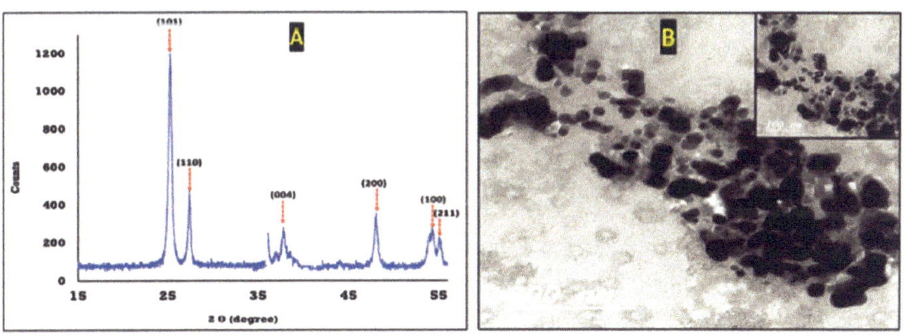

Fig. 11.4 (A) X-ray diffraction and (B) transmission electron microscope of Ag@TiO_2NPs. Reprinted with permission from El-Naggar, M. E., Hasanin, M., Youssef, A. M., Aldalbahi, A., Newehy, M., & Abdelhameed, R. M. (2020). Hydroxyethyl cellulose/bacterial cellulose cryogel dopped silver@titanium oxide nanoparticles: Antimicrobial activity and controlled release of Tebuconazole fungicide. Int. J. Biol. Macromol.

curve obtained by the experimental results showed a continuous increase in the efficacy. Thus, through study it can be concluded that the microencapsulation using copper nanoparticles can be used effectively for agricultural practices (Weitbrecht et al., 2021).

Wheat is found to be the most widely cultivated crop all over the world and is the second most crucial food crop, but due to various plant pathogens leading to diseases in the agricultural product it has led to a great concern. Thus, the eco-friendly alternative has been developed to control the fungal pathogens and it involves the usage of secondary metabolites. According to the study made by Amini et al., the nanoencapsulation of lemongrass oil (LGO) and clove essential oil (CEO) was done into the mesoporous silica nanoparticles (MSNPs) to use it against *Gaeumannomyces graminis* var. *tritici (Ggt),* which is a causative agent of take-all disease in the wheat crop which is a root fungal disease.

The nanoparticles are formed by following sol-gel technique and their characterization was further carried out by using SEM and TEM. The spherical nanoparticles of an average size of around 50–70 nm were obtained with hexagonal pores. By using UV-Vis spectroscopy, the encapsulation efficiency was calculated and it was found to be around 84.24% and 80.69% for LGO and CO, respectively. Then through the GC-MS analysis, it was revealed that the chemical compositions of both the EOs did not change after loading it into the MSNPs.

The in vitro analysis of the antifungal activities through minimum inhibitory concentration (MIC) and minimum fungicidal concentration (MFC) showed that these EOs have good antifungal effectivity against *Gaeumannomyces graminis* var. *tritici (Ggt)*. The release study using UV-Vis absorption indicated its effectivity and controlled release of the essential oils was confirmed. Thus, through the results mentioned earlier, it was concluded that MSNP is a good carrier system for the encapsulation of LGO and CEO for its enhanced antifungal activity for the wheat crop protection (Sattary et al., 2020).

Essential oils are the secondary metabolites that are unstable and are naturally occurring in the aromatic plants. These essential oils executed biological activities that led to the inhibition of the bacterial and fungal growth and, hence, act as an alternative for the disease management in various plants. For example, according to the report made by Weisany et al., the encapsulation of the thyme (*Thymus daenensis* L.) and dill (*Anethum graveolens* L.) essential oils was synthesized using silver nanoparticles to improve their antifungal activities against the *Colletotrichum nymphaeae*, a causative agent of anthracnose in many of the horticultural plants. Anthracnose is considered to be a serious disease that is mostly found in strawberry cultures worldwide.

SEM result analysis indicated that the average size of silver nanoparticles used for the purpose of encapsulation was found to be about 6–90 nm. GC-MS analysis reported the composition of the thyme and dill essential oils. Essential oil release study indicated that the release was comparatively slower for the encapsulated EOs than compared to that of the nonencapsulated EOs. According to the assay for the germination of the conidia, it was significantly reduced by 82% and 71% for thyme and dill, respectively, after the encapsulation of the essential oils into the AgNPs. It was observed that when these EOs were encapsulated in the AgNPs, mycelium inhibition by about 80% was observed for *C. nymphaeae*. Thus, in accordance with the results obtained, nanoencapsulation of the EOs in the silver nanoparticles has led to a controlled release of the active compound and enhanced antifungal activity which eventually results in an efficient agent for its application in the agricultural practices (Weisany et al., 2019a).

It is found that only 10% of the fungicides applied for the crop cultivation reach the targeted crop and that is due to the poor uptake, degradation, and also the runoff of the agrochemicals. To overcome this problem the usage of nanocarriers is done as an environment-friendly alternative which triggers the release of the actives at the particular site only by a controlled mechanism. Through the study made by Machdo et al., in order to find the alternative method to the currently used spraying agrochemicals method was done since it led to the usage of a high amount of agrochemicals that endangered the environment.

Cellulose-based nanocarriers (NCs) can be used as tools for the sustainable agricultural practices by having the controlled release of the agrochemicals used. Herein cellulose nanocarrier with agrochemicals which is selectively cleaved by the cellulase through the plant pathogens like *Neonectria ditissima* (*syn. Nectria galligena*) which is a causative agent of apple cankers is developed. The nanocarriers were developed by miniemulsion technique followed by the solvent evaporation technique which is used for the encapsulation of the agrochemicals for its controlled release.

The characterization of the nanocarriers with the help of SEM depicted the spherical structure and size of about 150–250 nm with encapsulation efficiency of 74 + 29%. Through the evaluation of the antifungal activities, which was followed by calculating the MIC values, strong inhibition of about 70% was observed for *N. ditissima*. Thus, according to the results obtained, the conclusion of usage of cellulose nanocarriers as an effective alternative to the conventional agrochemical delivery system due to its controlled release system, encapsulation efficiency, and inhibition can be made (Machado et al., 2021).

The more potent nanopesticides are developed than the conventional formulations that lead to the less contamination of the soil as well as of the environment. For example, Gopal et al. studied the nanoform of the preexisting fungicide hexaconazole to improve the plant or crop protection which is caused due to the fungal pathogens. Herein the encapsulation of the active agent is done into the nontoxic polymer like PEG-400(Polyethylene glycol). Through DLS and SEM results' analysis, the average size of nanohexaconazole was obtained to be around 100 nm. FTIR results indicated no change in the chemical structure due to encapsulation.

The in vitro fungicidal assay that conducted against *Rhizoctonia solani* represented the inhibition in the fungal growth at 0.01–0.1 ppm of about 18.8–71.1% and 9.3–41.8% for the nanohexaconazole and for the commercially available hexaconazole, respectively. But it was also understood that the effect of these nanoagrochemicals was not effective against the nitrate- and ammonia-oxidizing organisms. As the formulated agrochemical delivery system used is found to be more effective and less toxic as compared to that of the conventionally followed agrochemical delivery system, it acts as a convenient method for its utilization in the agricultural practices (Kumar et al., 2015).

The usage of the essential oil as a free compound is found to be hindered due to the issues of volatility, poor miscibility, and premature degradation of the active compounds. This can be avoided by the application of appropriate encapsulation technique for the essential oils which prevents the volatilization and leads to the increase of the stability of the active compounds.

According to Cadena et al., the encapsulation of the cinnamaldehyde (CNAD) into mesoporous silica nanoparticles (MSNPs) would give enhanced and prolonged efficacy of the bioactive agent which has antimicrobial capacity to protect from seed-borne disease better than cinnamon (*Cinnamomum zeylanicum*) bark essential oil. Encapsulation provided reduced volatility of the CNAD and also provides stability and immiscibility in the aqueous solutions. Synthesized MSNPs were found to be around 100 nm as observed by TEM.

CNAD which is a bioactive compound of cinnamon essential oil is encapsulated into the MSNPs and was inserted in the sodium alginate seed coating and his coating did not affect the shape and size of the seeds but promoted the germination of the seeds. The MSNPs treated with alginate seeds germinate faster than that of control plants. Moreover, the study revealed that the enhancement of the antimicrobial activity can be done by the MSNP encapsulation that allows less volatilization of the essential oils, thus requiring less concentration of the bioactive compound and giving the same treatment for the microbial disease during a crop cultivation (Bravo Cadena et al., 2018).

Over the last decade, mycobiocides that are originated from the plants have received a tremendous amount of attention due to their nontoxic characteristics that protect humans and animals from the deleterious health issues faced.

According to the studies made by Kalagatur et al., the development of the encapsulation technique for *Cymbopogon martinii* essential oil (CMEO) inside the chitosan nanoparticles is performed so that it can be effectively used against plant pathogen fungi *Fusarium graminearum* which is a causative agent for Fusarium head blight (FHB). Encapsulation was carried out by following the emulsification procedure. GC-MS technique was adopted for the determination of the components CMEO which were around 46 compounds and the most abundantly found was geraniol which was nearly 19.06%.

The characterization of chitosan-encapsulated *C. martini* essential oil nanoparticles (Ce-CMEO-NPs) was conducted, and through the FTIR analysis, positive quality identification and successful encapsulation of the NPs were determined. Through SEM, the spherical morphology for Ce-CMEO-NPs was revealed to be around 100 nm. The zetapotential ranging from 39.3 to 37.2 mV was determined which signified that NPs are not greatly affected by CMEO loading into it. Antifungal activities conducted on maize grains revealed that Ce-CMEO-NPs have potent fungicidal activity. The in vitro essential oil release assay also suggested a controlled release of encapsulated essential oil. Thus, referring to the results analyzed, Ce-CMEO-NPs can be efficiently applicable for safeguarding agricultural crops from the mycotoxigenic fungi (Kalagatur et al., 2018).

11.2.3 Nanoherbicides

By the application of nanotechnology techniques and formulation of herbicides into nanoherbicides, the herbicides overuse can be precluded and thus gives an eco-friendly alternative for the agricultural application. Herbicide's formulation provides an safer and more effectual delivery system by providing a controlled release of agrochemicals, and thus, it substantially reduces the herbicide consumption rate and also increases the crop productivity.

According to the study conducted by Taban et al., the investigation of the phytotoxicity effect by the nanoencapsulated savory essential oil was synthesized. The encapsulation of the savory EOs is loaded into the protein natural polymers such as arabic gum gelatin, gelatin, and apple pectin. The herbicide test was conducted on tomato and amaranth. Metribuzin is a fertilizer used in the tomato field, but over the time, biotypes of amaranth and *Chenopodium album* have become resistant to this herbicide. By following the GC-MS analysis, the components of the EOs were determined.

The encapsulation efficiency was conducted by spectrophotometer and the highest of the encapsulation efficiency was found for the Arabic gum-gelatin AGG (e) 3 that was 91.3% and lowest was for G(e)3 and that was 76.6%. The ability to suppress Arabic gum-gelatin (AGG) and apple pectin (AP) when cross-linked with citric acid was found to be comparable with that of the metribuzin. Lower germination and fresh weight percent revealed the phytotoxic effect on the amaranth plant and have a negligible effect on the tomato plant and seedling. Thus, this encapsulation of the savory essential oil formulation as an alternative to metribuzin herbicide for tomato cultivation can be done effectively (Taban et al., 2021).

Bioaccumulation of the agrochemicals into the soil can lead to nontargeted plants cultivation during the crop production. Thus, the controlled release system has been developed reducing the quantities of the active compound required and the increased bioavailability. In accordance with the study conducted by Pereira et al., encapsulation of atrazine herbicide inside the nanoparticles of poly(epsilon-caprolactone) is done so as to evaluate the herbicidal activity and genotoxicity and for the control of weeds during cultivation. Bioaccumulation of agrochemical atrazine can lead to adverse environmental hazards, and thus, encapsulation is a more secure alternative for its application.

The nanoparticles of poly(epsilon-caprolactone) are prepared by an oil-in-water method, and after the encapsulation, the decrease in the nanocapsules size was found to be around 483.1 ± 2.5 nm. The encapsulation efficiency for nanocapsules was obtained to be $93.4 \pm 1.2\%$. The controlled release of the active ingredients was carried out for the same, and on testing on various nontargeted and targeted organisms, it revealed that encapsulation caused no harm to the nontargeted organism (*Zea mays*) and was found to be more effective for the targeted organism (*Brassica sp.*). Various soil column experiments conducted indicated that the bioavailability of the active component was enhanced due to the encapsulation. The experimental results also indicated that the encapsulated herbicide has a less genotoxic effect than that of the free herbicide used, making it more preferable for agricultural practices and applications (Pereira et al., 2014).

Due to the increased solubility of the herbicides used in crop cultivation used to avoid the off-target, plant cultivation can lead to the contamination of the river and groundwater thus very less amount of the herbicides applied to the crops reaches to the actual target. In order to develop a method which releases the active compound in a controlled manner and only at the targeted site, encapsulation technique has found to be the best alternative.

For example, according to the study conducted by Grillo et al., the investigation of less toxic and efficient herbicidal formulation by following encapsulation technique is done as an alternative to the usage of paraquat herbicide that deals with the issue of nontargeted toxicity while application. The encapsulation of the paraquat herbicide is made into the chitosan and tripolyphosphate (TPP) herbicide by following the ion gelation technique. The herbicidal activity test was done on maize (*Z. mays*) and mustard (*Brassica sp*) cultivation. The stability test for CS/TPP nanoparticles was evaluated over a period of 60 days. The atomic force microscopy (AFM) analysis conducted revealed the particle size of the nanoparticles to be approximately 420 nm when encapsulated with paraquat and nanoparticles were spherical in shape.

The encapsulation efficiency for the nanoparticles was found to be around 62%. By experimental analysis, the reduced soil sorption was observed with a modulated diffusion rate. *A.cepa* assay results and signifies less chromosome damage by the usage of nanoparticle bound paraquat compared to that of free herbicide. And, the MTT test conducted revealed less toxicity by the application of encapsulation techniques for herbicide. Thus, in accordance with the findings discussed previously, the conclusion for the encapsulation-based herbicide formulation for a more stable and less toxic version of herbicide was found to be much more effective and efficient for the agricultural application (Grillo et al., 2014).

In order to mitigate the problem of toxicity and to increase the efficiency liposomes and biodegradable nanoparticles with greater biocompatibility were used for the study. Through the study conducted by Grillo et al., in order to reduce the toxicity due to the herbicides used while agricultural practices, as an alternative to it the formulation for making a carrier system of poly(caprolactone) (PCL) nanocapsules which is a biodegradable polymer, and encapsulating three triazine herbicides, namely atrazine, simazine, and ametryn, in it can be utilized. Triazines can be utilized for the control of weed development in sugarcane cultivation, sorghum, etc.

The analysis of microcapsules morphology was done by AFM and TEM, and the average size of the nanocapsules was observed to be around 160–250 nm. Through high-performance liquid chromatography (HPLC) analysis, the results interpreted that simazine-encapsulated nanocapsules were released at a slower rate than that of the other two mentioned herbicides. The zetapotential was -30 mV while the polydispersity value was below 0.2. The formulation was stable for around 250 days without major changes in the size of the particles.

The genotoxicity test conducted on *Allium cepa* cells and human lymphocytes revealed that the nanoencapsulated herbicide has fewer toxic effects when compared with herbicide in their free form. Thus, considering all the results evaluated, this encapsulated formulation for the herbicide can be considered as an environmentally friendly alternative for the agricultural practices (Grillo et al., 2012).

Because of the low biodegradability and half-life in water of the herbicides which is used for the weed control, it is found to be the threat to the public health, and thus, the urge for the development of the safer and eco-friendlier alternative was found in order to reduce their environmental toxicity. Polymer-based controlled release system for the agrochemical release for crop cultivation is utilized which potentially reduces the toxicity of the agrochemicals.

For example, based on the studies made by Taverna et al., the formulation for the controlled release of atrazine herbicide through the organosolv and ionic lignin microparticles was made so as to provide a less toxic alternative during the agricultural practices. The microspheres were made by solvent extraction/evaporation techniques. The mean particle size for the formulated spherical microcapsules was found to be 25.3 ± 6.7 µm. The organosolv lignin microcapsules show a higher amount of encapsulation efficiency. The release profile for atrazine indicated that approximately 98% of atrazine herbicide is released within 2 days of application.

The leaching experiment demonstrated that the microencapsulated atrazine was found less in the leachates than that of the free herbicide which eventually increased the dosing time. Thus, this microcapsule-based formulation can be effectively used as herbicide for obtaining high-valued products and having less environmental effects (Taverna et al., 2018).

11.2.4 Nanopesticides

Nanopesticides are an emerging node of technological advancements on a nanoscale that will minimize the adverse effects of conventional pesticides used in agriculture. Nanopesticides have a wider range of benefits in accordance with the increased efficacy, durability, and being eco-friendly.

Numerous nanoformulations have been used such as nanoemulsions, nanocapsules, metal oxides, and engineered nanoparticles, and their applications on enhancing crop yield, pest management control, economical, lower toxicity content, biocompatibility, and a better stability seemed promising than the usage of the conventional pesticides in the agricultural sector. The following study of the synthesis of CNT-*graft*-poly (citric acid)-encapsulated pesticides (CNT-g-PCA-EP hybrid materials), resulting in novel nanocomposite for plant disease treatment of *Alternaria alternata fungi.*, was acknowledged with its efficacy than the normal bulk pesticide.

In a study conducted by Sarlak et al., nanopesticides were synthesized, by trapping pesticides (zineb and mancozeb) in aqueous solution of multiwall carbon nanotubes-graft-poly (citric acid) to get MWCNT-g-PCA hybrid materials. The use of functionalized carbon nanotubes and the development of a novel nanocomposite for plant disease treatment (*A. alternata fungi.*) seemed promising. The conjugated citric acid branches present in polycitric acid shell made the synthesized MWCNT-g-PCA hybrid material soluble in water as well as trap water-soluble chemical species and metal ions.

This encapsulation of pesticides on CNT-g-PCA hybrid material gained confirmation via TEM analysis which clearly indicates polycitric acid shells around CNTs and encapsulated pesticide nanoparticles. The influence of the encapsulated pesticide was studied onto malady *A. alternata*, supplying Potato Dextrose Agar as a growth medium which in resultant showed that it is more stable and effective than bulk pesticide. Zineb and mancozeb being encapsulated into MWCNT-g-PCA hybrid material led to the conversion of the bulk pesticide to the pesticide nanoparticles. The UV-Vis spectroscopy gave a wider perspective on optimum pH, time of stirring, and temperature encapsulation of the zineb and mancozeb by polymeric shell of CNT-g-PCA.

This study has been the first report of usage of water-soluble polymerized carbon nanotubes for the encapsulation method of pesticides and also the application of new nanoparticles in malady agent treatment. Thus, the overall study emphasized the new CNT-g-PCA-EP hybrid material in comparison with bulk pesticide had an extraordinary influence on *A. alternata fungi* (Sarlak et al., 2014).

With regard to nanopesticides, the encapsulation of active ingredients of pesticides within the nanocarriers enhancing their deliverability to the targeted pests helped in enabling control over their release kinetics. The nanocarriers have controlled released mechanisms which yield a better efficacy when encapsulated pesticides are used compared to nonencapsulated formulations.

For an instance, Peters et al. focused their study based on recombinant catalytic modular protein D4S2, the creation of silicon nanocapsules containing encapsulated fipronil (SNC), and their modified version, poly(ethyleneimine)-coated SNC (SNC-PEI), and tested them in the field against termite colonies *Coptotermes lacteus*. The high encapsulation efficiency of 97% approximately was achieved with the help of an integrated biomolecular bioprocess developed to produce D4S2 for manufacturing the SNC consisting of fipronil. Formulations of fipronil containing the bait toxicants were exposed to nine termite mounds. This was in regard to prolong fipronil which in turn allowed enough time for the termites in order to relocate the baits, while distributing throughout the colony of *C. lacteus*, resulting in colony elimination.

As the overall studies concentrate upon, the application of silica nanocapsules can be observed here wherein the producing D4S2 protein was performed in the microbial cell factories further synthesis of nanocapsules of silica and SNC-PEI-coated nanocapsules. The D4S2 protein-stabilized nanoemulsions were used as a template. Worker termites hunting for bait toxicants spread the bait to other termite castes (e.g., queen, king, and soldiers) inside of the mound. The continuous release properties of silica nanocapsules would allow enough time for the dispersion of the baits, resulting in the eradication of the colony. In this trial study, despite the relocation of baits into the mounds, the colony eradication was not sufficient enough to prove the study, and thus, further studies are vital for understanding the behavior and their mechanism of resistance (Peters et al., 2019).

An another example demonstrates ability in creating a method for controlling pests and also inducing higher crop yields with the influence of metal oxides such as TiO_2 in the form of nanoparticles. In this work by Shaker et al., an economically important pest, cotton leafworm, which has a broader host plant range, was able to create a novel method for controlling *S. littoralis* by the successful use of TiO_2 nanoparticles. The average size of the crystal was detected around 95 nm, in which the titanium oxide grains were detected with the homogenous distribution at lesser degree of coalescence.

Experimentations were carried out using two treatments: application of TiO_2 nanoparticles and a negative control (distilled water) against the second and fourth in star larvae *S. littoralis*. The LC 50 value of second instar larvae was 62.5 ppm, whereas for fourth instar larvae, it was 125 ppm. Six varied concentrations (1000, 500, 250, 125, 62.5, and 31.25 ppm) were used in treatments wherein they helped in further evaluations of mortality %, the LC50 values, and other biological parameters of the insect.

The TiO_2 nanoparticle treatment on the cotton leafworm observed effective toxic activity in all concentrations in second instar than the fourth instar larvae of *S. littoralis*. The titanium oxide nanoparticles minimized the problems which were caused by *S. littoralis* to the host crops. Apart from pesticidal activity, TiO_2 was reported to promote the growth of the plant and higher photosynthetic activity, to reduce the diseases, and to enhance the crop yield by 30% and grain weight by 20% (Shaker et al., 2017).

With the incorporation of nanotechnology into an agricultural sector, its sustainability should also be taken into account. To this, the nanocarrier systems consisted of zein (corn protein) is an added example that is capable of improving the physicochemical characteristics of biopesticides and also reducing the toxicity apart from that it can be used in the protection of the crops.

The following study of Pascoli et al. revolves around the zein nanoparticles encapsulating neem oil-based nanopesticides. The eco-friendly version and its formulations were used in the application of sustainable agriculture. Neem oil extracted from *Azadirachta indica* consists of active ingredient azadirachtin which is a commercial biopesticide. Evaluation of the neem oil-loaded zein nanoparticles and their toxicity effect was carried out using organisms such as *A. cepa* and *Caenorhabditis elegans*.

The result of toxicity assays with *A. cepa* showed that the nanoparticles favored increased DNA-relative damage index with the usage of neem oil, whereas in *C. elegans,* the nanoparticles showed the absence of toxic effect, but the interference of neem oil observed a decrease in GST-4 gene expression by 66% where it showed a decrease in the fluorescence intensity emitted by the worms treated with neem oil. The nanoencapsulation of neem almost nullified the toxic effects in *C. elegans*. Thus, the study acknowledges the perspectives of the neem oil-loaded zein nanoparticles in toxicity studies and their usage in protecting the crop in organic farming (Pascoli et al., 2019).

In the upcoming years, essential oils have proved to be efficient against numerous stored grain pests. Moreover, the medicinal plant essential oils have proven to be a promising substitute to these synthetic insecticides. Moreover, nanopesticides established on the essential oils and also polymer-based encapsulation is an evolutionary technique to overcome many limitations such as high volatility, rapid oxidations, and environmental conditions. Nanoformulations of the essential oils could resolve issues regarding solubility and bioavailability. A similar study of polymer-based nanoencapsulations and usage of medicinal plant as EOs was observed in this example. Rajkumar et al. developed the peppermint essential oil encapsulation in chitosan nanoparticles which had a biological effectivity against the stored grain in the action of pest control.

This study encompasses polymeric-based encapsulation of EOs. By ionic gelation method, the insecticidal activity helped in the improvement of two stored grain pests. The nanoencapsulation of chitosan/peppermint oil nanoparticles (CS/PO NPs) was assured by UV-Vis spectrophotometry and XRD technique wherein the <563.3 nm size was detected after characterization of nanoparticles and encapsulation of efficiency $>64\%$. AChE activity in vivo deciphered 52.43%–37.71% inhibition in *S. oryzae* and inhibitory activity for *T. castaneum* was 37.80%–31.29%. The study further demonstrated that CS/PO NPs have a significant effect against *S. oryzae* and

T. castaneum, thus is the promising design that could be used in the stored food pest management in agriculture (Rajkumar et al., 2020).

Another study with the usage of biodegradable nanospheres containing EOs in controlling the pest is a remarkable example. Christofoli et al. highlight the insecticidal effect of *Zanthoxylum rhoifolium* (Rutaceae) leaves essential oil in *Bemisia tabaci*. The essential oil from *Z. rhoifolium* has an average yield of about 0.03%, including the components β-elemene, D-germacrene, β-caryophyllene, δ-elemene, β-cedrene, bicyclogermacrene, and *E*-caryophyllene. These poly-€-(caprolactone) (PCL) nanospheres had an encapsulation efficiency of 96%.

In comparison with the control, the results signified that *in natura* and nanoencapsulated essential oils reduce 95% of eggs and nymphs. Collectively, this signified that both the *in natura* and in nanoencapsulated essential oils from *Z. rhoifolium* leaves can be a potential agent to be used in *B. tabaci* control assisting in integration with pest management control in the agricultural sector (Christofoli et al., 2015).

In recent years, excessive use of synthetic pesticides had led to serious issues such as contamination of groundwater, spread of the toxic content through leaching, rainfall, and volatilization in the environment. To investigate new means for the usage of pesticides is absolutely essential. Nanoencapsulation of an acaricide using polyethylene glycol (PEG) has shown improved persistence and solubility. PEG has received wide attention for pesticide nanoformulations over the years.

One such example to note is the following study by Ahmadi et al. They studied the nanoencapsulation of *Acaricide and clofentenzine*, using a polyethylene glycol by method of simple and green route against the two-spotted spider mites—*Tetranychus urticae*. This *Thrips urticae* is commonly controlled by synthetic acaricides, giving rise to detrimental side effects on the environment. The necessity of new means of pesticide in minimizing the environmental side effects was inevitable. This was depicted by the usage of clofentenzine nanocapsules, with regard to its persistence and improved water solubility, which maintains effective bioactivity than the normally formulated clofentenzine. This acaricide delivery system had the capacity to continuously release clofentenzine over a period of 80 h. Hence, these spherical nanoparticles of around approximately 250 nm showed that the acaricidal efficacy of the nanoencapsulated clofentenzine improved.

The characterization of acaricide-loaded PEG NPs was performed using FTIR and TEM. FTIR confirmed the nature of acaricide in the clofentenzine nanoformulations. The findings also demonstrated the clofentenzine nanoparticles in controlling *T. urticae* at much lower concentrations than the required normal formulations of the clofentenzine. Thus, this study highlights the potential of clofentenzine nanoparticles as a selective acaricide, economical and environment-friendly, and its importance in the agricultural work (Ahmadi et al., 2020).

Nanoencapsulation of the pesticides has been advantageous in terms of efficient use of pesticides, minimizing the environmental risks, and improvement in the crop yield and guarantees eco-protection. The investigation presented later highlights the similar study of the delivery of pesticide in the form of nanoformulations in plants and observing its ultimate effects. Kumar et al. researched on the synthesis and characterization of the properties of pesticide loaded with sodium alginate nanoparticles.

Pesticide loading in the nanoparticles was determined with the help of UV-Vis spectrophotometer while its characterization was followed with the usage of TEM, zetasizer, and FTIR which is reflected in Fig. 11.5.

As depicted in Fig. 11.5A and B, it represents the low-magnification and high-magnification images. With the help of TEM, the particle size was found to be around 50–100 nm. The zetasizer was used to analyze the particle size which worked on the principle of DLS. The particle size obtained using zetasizer was 150 nm which is shown in Fig. 11.5C. Because of the aggregation of nanoparticles, a wide range of sizes was appeared. Fig. 11.5D describes the FTIR spectrum of pesticide, sodium alginate, and pesticide-loaded sodium alginate nanoformulation. FTIR spectrum depicted that the imidacloprid band at 1560.41 cm^{-1} is because of the =CH— stretch of the pyridine group, the $—NO_2$ symmetric stretch of the band was located at 1234.4 cm^{-1}.

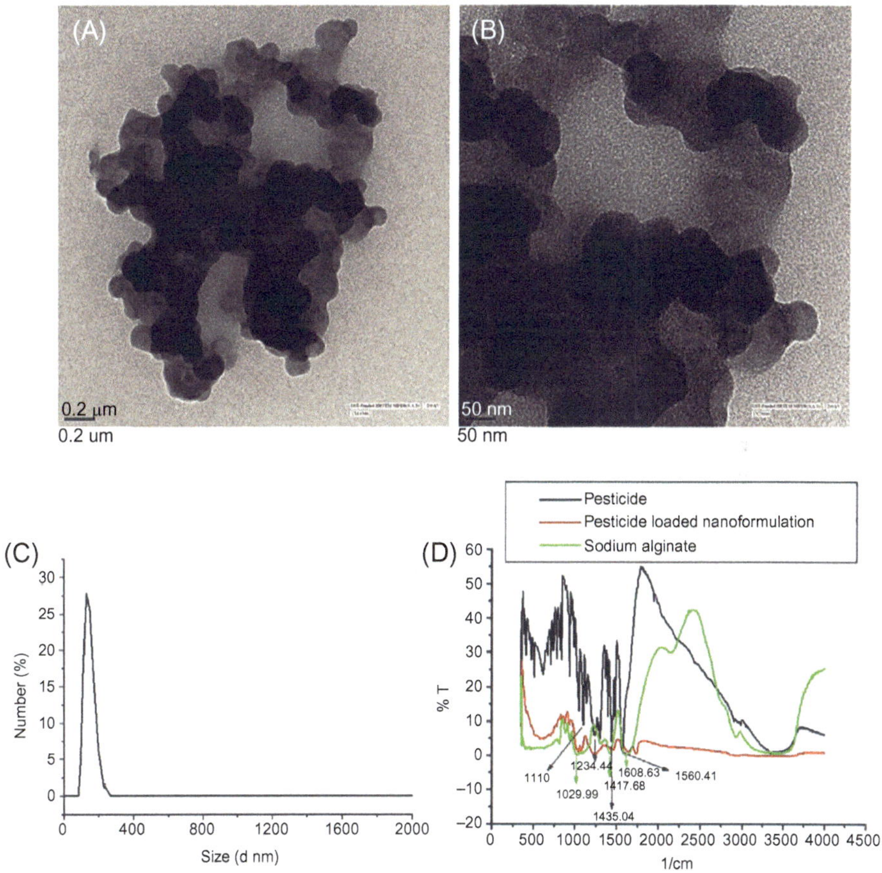

Fig. 11.5 (A and B) TEM micrographs of pesticide-loaded sodium alginate nanoparticles, (C) particle size analysis, and (D) FTIR of pesticide-loaded sodium alginate nanoparticles. Reprinted with the permission from Kumar, S., Bhanjana, G., Sharma, A., Sidhu, M.C., Dilbaghi, N., 2014. Synthesis, characterization and on field evaluation of pesticide loaded sodium alginate nanoparticles. Carbohydrate Polymers 101, 1061–1067.

The aromatic =CH— was at the location of 1435.04 cm^{-1} and aryl-Cl stretching was at 1110 cm^{-1}. FTIR gave the confirmation of the presence of pesticides in the pesticide-loaded sodium alginate nanoformulations. Also, there were no chemical interactions observed between sodium alginate and pesticide (Kumar et al., 2014).

This cohesive study on sodium alginate nanoformulations showcases that the encapsulated pesticide-loaded nanoformulations are more effective, have lower toxicity content, and show biocompatibility and better stability than the normal pesticide. The on-field evaluation on crop pests such as jassids/leafhoppers is performed by spraying nanoencapsulated pesticides that resulted in increased efficacy of the crops. Thus, it can be used commercially as it is advantageous for the environment in a decreased amount of residue of pesticides in foods and increased safety for the farmer and agricultural products (Kumar et al., 2014).

11.3 Conclusions and future perspectives

An extensive variety of encapsulation techniques have been developed and are used for various applications. Encapsulation technique enables controlled release of the sensitive molecules in the nanocarrier system. The sophisticated encapsulation techniques such as microencapsulation and nanoencapsulation are designed with such modifications that they enable triggering procedural advancement with maximum efficiency outputs by their utilization. Selection of the appropriate encapsulation technology is contingent on several factors to mention such as various physicochemical features, appropriate particle size, various release properties, nanocarrier used, delivery method utilized, and procedural cost.

The advantages for adopting the encapsulation technique serve as a well-structured option to provide efficient controlled release of the agrochemicals and also for the protection of crops. It equips a more convenient method to increase the bioavailability for the agrochemicals or for the targeted molecules. This nanotechnology-based process of encapsulation lacks much of its implication for commercial agricultural production. Despite the enormous advantages of the encapsulation technique, it faces many challenges and theoretical problems such as disbalance during the compatibility of the wall material, the stability issue which might be faced while storing at the different environmental conditions, lack of appropriate and much accurate characterization technique, etc. In fact, much of the focus and attention is expected to be drawn at developing the encapsulation technique, with more efficient stability under a wide range of conditions, and toward the procedure with more performance uniformity.

References

Abbasifar, A., Shahrabadi, F., Valizadeh Kaji, B., 2020. Effects of green synthesized zinc and copper nano-fertilizers on the morphological and biochemical attributes of basil plant. J. Plant Nutr. 43 (8), 1104–1118.

Abd El-Azeim, M.M., Sherif, M.A., Hussien, M.S., Tantawy, I.A.A., Bashandy, S.O., 2020. Impacts of nano- and non-nanofertilizers on potato quality and productivity. Acta Ecol. Sin. 40 (5), 388–397.

Abdel-Aziz, H.M.M., Rizwan, M., 2019. Chemically synthesized silver nanoparticles induced physio-chemical and chloroplast ultrastructural changes in broad bean seedlings. Chemosphere 235, 1066–1072.

Ahmadi, Z., Saber, M., Mahdavinia, G.R., 2020. Nanoencapsulation of clofentezine with enhanced acaricidal activity against the two spotted mite, *Tetranychus urticae* Koch (Acari: Tetranychidae). Toxin Rev. 40, 1–9.

Awan, S., Shahzadi, K., Javad, S., Tariq, A., Ahmad, A., Ilyas, S., 2021. A preliminary study of influence of zinc oxide nanoparticles on growth parameters of *Brassica oleracea* var italic. J. Saudi Soc. Agric. Sci. 20 (1), 18–24.

Bravo Cadena, M., Preston, G.M., van der Hoorn, R.A.L., Flanagan, N.A., Townley, H.E., Thompson, I.P., 2018. Enhancing cinnamon essential oil activity by nanoparticle encapsulation to control seed pathogens. Ind. Crop. Prod. 124, 755–764.

Chand Mali, S., Raj, S., Trivedi, R., 2020. Nanotechnology a novel approach to enhance crop productivity. Biochem. Biophys. Rep. 24, 100821.

Christofoli, M., Costa, E.C.C., Bicalho, K.U., de Cássia Domingues, V., Peixoto, M.F., Alves, C.C.F., Araújo, W.L., de Melo Cazal, C., 2015. Insecticidal effect of nanoencapsulated essential oils from *Zanthoxylum rhoifolium* (Rutaceae) in *Bemisia tabaci* populations. Ind. Crop Prod. 70, 301–308.

Corradini, E., de Moura, M.R., Mattoso, L.H.C., 2010. A preliminary study of the incorparation of NPK fertilizer into chitosan nanoparticles. Express Polym Lett 4, 509–515.

Cyriac, J., Melethil, K., Thomas, B., Sreejit, M., Varghese, T., 2019. Synthesis of biogenic ZnO nanoparticles and its impact on seed germination and root growth of *Oryza sativa* L. And *Vigna unguiculata* L. Mater. Today: Proc. 25, 224–229.

El-Naggar, M.E., Hasanin, M., Youssef, A.M., Aldalbahi, A., El-Newehy, M.H., Abdelhameed, R.M., 2020. Hydroxyethyl cellulose/bacterial cellulose cryogel dopped silver@titanium oxide nanoparticles: antimicrobial activity and controlled release of Tebuconazole fungicide. Int. J. Biol. Macromol. 165, 1010–1021.

Grillo, R., dos Santos, N.Z.P., Maruyama, C.R., Rosa, A.H., de Lima, R., Fraceto, L.F., 2012. Poly(e{open}-caprolactone)nanocapsules as carrier systems for herbicides: physico-chemical characterization and genotoxicity evaluation. J. Hazard. Mater. 231–232, 1–9.

Grillo, R., Pereira, A.E.S., Nishisaka, C.S., de Lima, R., Oehlke, K., Greiner, R., Fraceto, L.F., 2014. Chitosan/tripolyphosphate nanoparticles loaded with paraquat herbicide: an environmentally safer alternative for weed control. J. Hazard. Mater. 278, 163–171.

Hasan, M., Rafique, S., Zafar, A., Loomba, S., Khan, R., Hassan, S.G., Khan, M.W., Zahra, S., Zia, M., Mustafa, G., Shu, X., Ihsan, Z., Mahmood, N., 2020. Physiological and anti-oxidative response of biologically and chemically synthesized iron oxide: *Zea mays* a case study. Heliyon 6 (8), e04595.

Hasheminejad, N., Khodaiyan, F., Safari, M., 2019. Improving the antifungal activity of clove essential oil encapsulated by chitosan nanoparticles. Food Chem. 275, 113–122.

Jafarirad, S., Kosari-Nasab, M., Mohammadpour Tavana, R., Mahjouri, S., Ebadollahi, R., 2021. Impacts of manganese bio-based nanocomposites on phytochemical classification, growth and physiological responses of *Hypericum perforatum* L. shoot cultures. Ecotoxicol. Environ. Saf. 209, 111841.

Jahangirian, H., Rafiee-Moghaddam, R., Jahangirian, N., Nikpey, B., Jahangirian, S., Bassous, N., Saleh, B., Kalantari, K., Webster, T.J., 2020. Green synthesis of zeolite/Fe_2O_3 nanocomposites: toxicity & cell proliferation assays and application as a smart iron nanofertilizer. Int. J. Nanomedicine 15, 1005–1020.

Kalagatur, N.K., Nirmal Ghosh, O.S., Sundararaj, N., Mudili, V., 2018. Antifungal activity of chitosan nanoparticles encapsulated with *Cymbopogon martinii* essential oil on plant pathogenic fungi *Fusarium graminearum*. Front. Pharmacol. 9, 1–13.

Kumar, S., Bhanjana, G., Sharma, A., Sidhu, M.C., Dilbaghi, N., 2014. Synthesis, characterization and on field evaluation of pesticide loaded sodium alginate nanoparticles. Carbohydr. Polym. 101, 1061–1067.

Kumar, R., Nair, K.K., Alam, M.I., Gogoi, R., Singh, P.K., Srivastava, C., Gopal, M., Goswami, A., 2015. Development and quality control of nanohexaconazole as an effective fungicide and its biosafety studies on soil nitifiers. J. Nanosci. Nanotechnol. 15, 1350–1356.

Kumaraswamy, R.V., Saharan, V., Kumari, S., Chandra Choudhary, R., Pal, A., Sharma, S.S., Rakshit, S., Raliya, R., Biswas, P., 2021. Chitosan-silicon nanofertilizer to enhance plant growth and yield in maize (*Zea mays* L.). Plant Physiol. Biochem. 159, 53–66.

Machado, T.O., Beckers, S.J., Fischer, J., Sayer, C., de Araújo, P.H.H., Landfester, K., Wurm, F. R., 2021. Cellulose nanocarriers via miniemulsion allow Pathogen-Specific agrochemical delivery. J. Colloid Interface Sci. 601, 678–688.

Mohammadi, A., Hashemi, M., Hosseini, S.M., 2015. Nanoencapsulation of Zataria multiflora essential oil preparation and characterization with enhanced antifungal activity for controlling Botrytis cinerea, the causal agent of gray mould disease. Innov. Food Sci. Emerg. Technol. 28, 73–80.

Pascoli, M., Jacques, M.T., Agarrayua, D.A., Avila, D.S., Lima, R., Fraceto, L.F., 2019. Neem oil based nanopesticide as an environmentally-friendly formulation for applications in sustainable agriculture: an ecotoxicological perspective. Sci. Total Environ. 677, 57–67.

Pedruzzi, D.P., Araujo, L.O., Falco, W.F., Machado, G., Casagrande, G.A., Colbeck, I., Lawson, T., Oliveira, S.L., Caires, A.R.L., 2020. ZnO nanoparticles impact on the photosynthetic activity of *Vicia faba*: effect of particle size and concentration. NanoImpact 19, 100246.

Pereira, A.E.S., Grillo, R., Mello, N.F.S., Rosa, A.H., Fraceto, L.F., 2014. Application of poly (epsilon-caprolactone) nanoparticles containing atrazine herbicide as an alternative technique to control weeds and reduce damage to the environment. J. Hazard. Mater. 268, 207–215.

Peters, B.C., Wibowo, D., Yang, G.Z., Hui, Y., Middelberg, A.P.J., Zhao, C.X., 2019. Evaluation of baiting fipronil-loaded silica nanocapsules against termite colonies in fields. Heliyon 5, e02277.

Rajkumar, V., Gunasekaran, C., Paul, C.A., Dharmaraj, J., 2020. Development of encapsulated peppermint essential oil in chitosan nanoparticles: characterization and biological efficacy against stored-grain pest control. Pestic. Biochem. Physiol. 170, 104679.

Sarlak, N., Taherifar, A., Salehi, F., 2014. Synthesis of nanopesticides by encapsulating pesticide nanoparticles using functionalized carbon nanotubes and application of new nanocomposite for plant disease treatment. J. Agric. Food Chem. 62, 4833–4838.

Sattary, M., Amini, J., Hallaj, R., 2020. Antifungal activity of the lemongrass and clove oil encapsulated in mesoporous silica nanoparticles against wheat's take-all disease. Pestic. Biochem. Physiol. 170, 104696.

Shaker, A.M., Zaki, A.H., Abdel-Rahim, E.F.M., Khedr, M.H., 2017. TiO_2 nanoparticles as an effective nanopesticide for cotton leaf worm. Int. J. Agric. Eng. 2017, 61–68.

Sharma, G., Kumar, A., Devi, K.A., Prajapati, D., Bhagat, D., Pal, A., Raliya, R., Biswas, P., Saharan, V., 2020. Chitosan nanofertilizer to foster source activity in maize. Int. J. Biol. Macromol. 145, 226–234.

Shebl, A., Hassan, A.A., Salama, D.M., Abd El-Aziz, M.E., Abd Elwahed, M.S.A., 2020. Template-free microwave-assisted hydrothermal synthesis of manganese zinc ferrite as a nanofertilizer for squash plant (*Cucurbita pepo* L). Heliyon 6, e03596.

Sinha, T., Bhagwatwar, P., Krishnamoorthy, C., 2019. Polymers for Agri-Food Applications. pp. 5–28.

Taban, A., Saharkhiz, M.J., Kavoosi, G., 2021. Development of pre-emergence herbicide based on Arabic gum-gelatin, apple pectin and savory essential oil nano-particles: a potential green alternative to metribuzin. Int. J. Biol. Macromol. 167, 756–765.

Taverna, M.E., Busatto, C.A., Lescano, M.R., Nicolau, V.V., Zalazar, C.S., Meira, G.R., Estenoz, D.A., 2018. Microparticles based on ionic and organosolv lignins for the controlled release of atrazine. J. Hazard. Mater. 359, 139–147.

Wani, T.A., Masoodi, F.A., Baba, W.N., Ahmad, M., Rahmanian, N., Jafari, S.M., 2019. Nanoencapsulation of agrochemicals, fertilizers, and pesticides for improved plant production. In: Advances in Phytonanotechnology, pp. 279–298.

Weisany, W., Amini, J., Samadi, S., Hossaini, S., Yousefi, S., Struik, P.C., 2019a. Nano silver-encapsulation of *Thymus daenensis* and *Anethum graveolens* essential oils enhances antifungal potential against strawberry anthracnose. Ind. Crop Prod. 141, 111808.

Weisany, W., Samadi, S., Amini, J., Hossaini, S., Yousefi, S., Maggi, F., 2019b. Enhancement of the antifungal activity of thyme and dill essential oils against *Colletotrichum nymphaeae* by nano-encapsulation with copper NPs. Ind. Crop Prod. 132, 213–225.

Weitbrecht, K., Schwab, S., Rupp, C., Bieler, E., Dürrenberger, M., Bleyer, G., Schumacher, S., Kassemeyer, H.H., Fuchs, R., Schlücker, E., 2021. Microencapsulation—an innovative technique to improve the fungicide efficacy of copper against grapevine downy mildew. J. Crop. Prod. 139, 105382.

Younes, N.A., Hassan, H.S., Elkady, M.F., Hamed, A.M., Dawood, M.F.A., 2020. Impact of synthesized metal oxide nanomaterials on seedlings production of three Solanaceae crops. Heliyon 6 (1), e03188.

Yusefi-Tanha, E., Fallah, S., Rostamnejadi, A., Pokhrel, L.R., 2020. Zinc oxide nanoparticles (ZnONPs) as a novel nanofertilizer: influence on seed yield and antioxidant defense system in soil grown soybean (Glycine max cv. Kowsar). Sci. Total Environ. 738, 140240.

Plant essential oil-based nanoemulsions: A novel asset in the crop protection arsenal

12

Prithusayak Mondal[a], *Rosalin Laishram*[b,c], *Prahlad Sarkar*[d], *Rajesh Kumar*[e], *Rajib Karmakar*[f], *Dipak Kumar Hazra*[f], *Kaushik Banerjee*[g], *Kumaresh Pal*[c], and *Ashok Choudhury*[h]
[a]Regional Research Station (Terai Zone), Uttar Banga Krishi Viswavidyalaya, Cooch Behar, West Bengal, India, [b]Division of Biochemistry, ICAR-Indian Agricultural Research Institute, New Delhi, India, [c]Department of Biochemistry, Faculty of Agriculture, Uttar Banga Krishi Viswavidyalaya, Cooch Behar, West Bengal, India, [d]All India Network Project on Jute and Allied Fibres, Uttar Banga Krishi Viswavidyalaya, Cooch Behar, West Bengal, India, [e]Division of Agricultural Chemicals, ICAR-Indian Agricultural Research Institute, New Delhi, India, [f]Department of Agricultural Chemicals, Bidhan Chandra Krishi Viswavidyalaya, Mohanpur, West Bengal, India, [g]National Reference Laboratory, ICAR-National Research Centre for Grapes, Pune, Maharashtra, India, [h]Directorate of Research, Uttar Banga Krishi Viswavidyalaya, Cooch Behar, West Bengal, India

12.1 Introduction

Agriculture is reckoned as the stanchion of economic and other developments of any nation. The entire world is striving to find ways and means to keep its escalating population amply fed. On the contrary, it is facing challenges of slumping productivity due to a variety of pests like insects, mites, nematodes, bacteria, fungi, viruses, and even weeds. Due to the urgency to improve food production and to control pest populations, synthetic crop protection chemicals, known as pesticides, are extensively applied (Tudi et al., 2021). The synthetic approaches of fabricating chemicals for pest management keep on generating considerable threats to the environmental components (Mondal et al., 2021). Researchers worldwide are spending hours in the laboratories to look for green approaches to pesticide development using natural products that may emerge as viable alternatives with respect to environmental sustainability. Plant-derived substances or botanicals have immense potential to curb pest population without hampering the ecosystem balance (Mondal et al., 2018). The progressive need for developing novel and safe pesticides with unique modes of action requires discovery and commercialization of natural products with meager environmental load through the intervention of green chemistry.

Plant essential oil (EO) is one of the botanicals that have long chronicles of human exercise as flavoring and fragrance agents, and more recently, these have been used in aromatherapy, nutraceuticals, and other foods and beverages (Isman, 2020). It has

Agricultural Nanobiotechnology. https://doi.org/10.1016/B978-0-323-91908-1.00007-9

become a trending research interest nowadays in integrated pest management programs because of their safe toxicological profiles. EOs are generally scented and volatile secondary metabolites secreted from distinct plant parts and have been reported to exhibit a vast range of biological activities including antioxidant (Bouzabata et al., 2015), insect repellent (Lalthazuali and Mathew, 2017), insecticidal (Moura et al., 2019), antibacterial (Avetisyan et al., 2017), antifungal (Swamy et al., 2016), antiviral (Brand et al., 2016), nematicidal (Eloh et al., 2020), and antiprotozoal (Sauter et al., 2012) activities. The phytoconstituents of EOs are actually intricate mixtures of a wide diversity of volatile compounds, especially terpenoids and phenolics, having numerous undefined modes of action that may enhance their performance because of the synergistic effect between constituents. These appear to be biocompatible, are fast penetrant, and leave low to almost nil toxic residues in treated products. However, their volatility, water insolubility, and chemical instability in presence of varied environmental factors are the prime concerns regarding their effective utilization that ultimately leads to the generation of new formulation technologies.

Nanotechnology has frequently been used in industrial, medicinal, and energy sectors and recently has appeared as a boon in agriculture and allied sectors. Nanotech-mediated strategies deal with particles at nanoscale level and can bring eco-friendly alternatives in terms of reduced toxicity, enhanced shelf-life, decent solubility for hydrophobic (water insoluble or poorly soluble) agrochemicals, minimal use with utmost effect, controlled release with prolonged effect (Mondal et al., 2017; Mondal and Kumar, 2016). Biopesticide formulations containing EOs tailored via nanotechnological interventions can play a pivotal role in minimizing these problems associated with their applications. Nanoemulsion (NE) is one of such nanoformulations which is considered as a metastable nanocarrier system containing a blend of immiscible liquids, in which the dispersed fine droplets are of nanometer size (~ 20–500 nm) (Gupta et al., 2016). The composition, especially the ratio between oil, water and surfactant and fabrication technique is essential to maintain a long-term stable formulation. These nanoformulations have been proved to be more potent than their bulk (free) counterparts in the form of uniform dispersion, decreased volatility, increased water miscibility, resistance to hydrolysis and enzymatic breakdown, and improved stability (Mondal and Kumar, 2016; Pavoni et al., 2020; Mondal et al., 2017).

This chapter overviews the various aspects regarding the characteristics and development of plant EO-based NEs and their effectiveness in managing different pests of agricultural importance.

12.2 Constituents of essential oil

EOs are basically odoriferous secondary metabolites reported to exhibit a wide range of biological activities and can be extracted from numerous plant parts, including leaves (*Eucalyptus* sps.), flowers (*Rosa* sps.), fruits (*Citrus* sps.), roots (*Chrysopogon* sps.), rhizomes (*Curcuma* sps.), woods (*Cedrus* sps.), barks (*Cinnamomum* sps.), seeds (*Foeniculum* sps.), grasses (*Cymbopogon* sps.), gum

(*Boswellia* sps.), tree blossoms (*Cananga* sps.), bulbs (*Allium* sps.), and dried flower buds (*Syzygium* sps.) (Tisserand and Young, 2013). These appear to be biocompatible, fast penetrant and leave low to almost nil toxic residues in treated products. Majority of the EOs are colorless or pale in color, lighter than water, having high refractive index and optimal rotation (Reddy, 2019). EOs are basically hydrophobic and largely soluble in lipids, alcohols, ethers, and most organic solvents. Moreover, they are susceptible to oxidization and may generate resinous products via polymerization (Li et al., 2014). In general, the weather and climatic situations, geographic locale, and numerous other conditions directly or indirectly determine the relative composition of EOs (Si et al., 2012).

The phytoconstituents of EOs are complex blends of numerous aromatic compounds (Zuzarte and Salgueiro, 2015). However, these are chiefly categorized into two major groups consisting of phenolic compounds—terpenoids (mainly monoterpene and sesquiterpene hydrocarbons, and their oxygenated derivatives) and phenylpropanoids (phenols, phenolic ethers, and aldehydes) (Table 12.1).

12.3 Characteristics of nanoemulsions

12.3.1 Types of emulsions

Chemically, emulsion is a colloidal system where both phases are liquids. The International Union of Pure and Applied Chemistry (IUPAC) defined emulsion as "*A fluid colloidal system in which liquid droplets and/or liquid crystals are dispersed in a liquid. The droplets often exceed the usual limits for colloids in size. An emulsion is denoted by the symbol O/W if the continuous phase is an aqueous solution and by W/O if the continuous phase is an organic liquid (an 'oil'). More complicated emulsions such as O/W/O (i.e., oil droplets contained within aqueous droplets dispersed in a continuous oil phase) are also possible*" (IUPAC, 2014). In addition to aqueous and oil phases, an emulsion formulation typically requires another chemical known as emulsifier.

Emulsions are generally classified based on the system configuration, particularly the droplet size and the thermodynamic and associated physical properties of emulsions. The size range of droplets for each kind of emulsion is sometimes found arbitrary; therefore, the stability of the entire emulsion system including the dispersion of droplets must be considered while defining any type of emulsions (Sheth et al., 2020). The conventional emulsion, known as macroemulsion with mean droplet diameter in the micrometer range, is the most common type of emulsions found in nature. However, these emulsions are vulnerable to physical instability (e.g., flocculation and coalescence, and gravitational segregation), especially when subjected to environmental stresses (Peter, 2009). Nanoemulsion and microemulsion formulations are typically composed of three components—oil, water, and emulsifier (McClements, 2010). Although these two formulations are structurally similar in terms of dispersions of nanoscale droplets, however, the principal difference between these lies in their thermodynamics (Gupta et al., 2016). The comparison among different types of emulsions is explained in Table 12.2.

Table 12.1 Phytoconstituents of essential oils.

Major class	Subtypes	Structural type	Functional group	Examples
Terpenoids	Monoterpenes	Acyclic	Hydrocarbons	Myrcene, ocimene
			Alcohols	Geraniol, nerol, linalool, citronellol, lavandulol
			Aldehydes	Citronellal, geranial, neral
			Ketones	Tagetone,
			Esters	Linalyl acetate or propionate, citronellyl acetate
		Monocyclic	Hydrocarbons	p-Cymene, α-phellandrene
			Alcohols	Thymol, carvacrol, menthol, α-terpineol
			Ketones	Menthones, carvone, pulegone, piperitone
			Keto-alcohols	Diosphenol
			Esters	menthyl or α-terpinyl acetate
		Bicyclic	Hydrocarbons	Limonene
			Alcohols	Borneol
			Ketones	Camphor, fenchone, thujone, pinocamphone, pinocarvone
			Ethers	1,8-Cineole, menthofuran
			Peroxides	Ascaridole
			Esters	Isobornyl acetate
		Tricyclic	Hydrocarbons	Tricyclene
	Sesquiterpenes	Acyclic	Hydrocarbons	Farnesenes
			Alcohols	β-Nerolidol, farnesol
		Monocyclic	Hydrocarbons	β-Bisabolene, elemenes, curcumenes, zingiberene
			Alcohols	Bisabolol
			Ketones	Germacrone, cis-longipinan-2,7-dione, turmerones
		Bicyclic	Hydrocarbons	Azulene, cadinenes, b-caryophyllene
			Alcohols	Carotol, β-santalol
			Ketones	Nootkatone, β-vetinone
			Epoxides	Humulene epoxides
		Tricyclic	Hydrocarbons	Longifolene
			Alcohols	Cedrol, patchoulol, viridiflorol
			Epoxides	Caryophyllene oxide
Phenylpropanoids	Cinnamic acid derivatives	Phenols	Alcohols	Eugenol, chavicol
		Phenolic ethers	Ethers	Myristicin, safrole, anethole, estragole
		Phenolic aldehydes	Aldehydes	Cinnamaldehyde, vanillin

Table 12.2 Comparison among various types of emulsions.

Emulsion type	Droplet size and shape	Preparation techniques	Polydispersity index	Stability
Macroemulsion	Large (> 1 μm); spherical	High- and low-energy methods	Often high (>40%); droplets are chiefly polydisperse due to high surface tension	Thermodynamically unstable, weakly kinetically stable; broadening of droplets size due to coalescence resulting in phase separation rapidly (in minutes or hours)
Nanoemulsion	Small (~ 100 nm); spherical	High- and low-energy methods	Typically low (<10%–20%); droplets are not monodisperse, but polydispersity is significantly less than macroemulsion	Thermodynamically unstable, kinetically stable; phase separation relatively slower (over weeks or months)
Microemulsion	Very small (~ 10 nm); spherical, lamellar	Spontaneous low-energy methods	Significantly low (<10%); droplets tend to be monodisperse from thermodynamic self-assembly	Thermodynamically stable phase; micelles remain dispersed infinitely if state variables are kept unchanged

12.3.2 Composition

In some kinds of literature, NE has been termed as miniemulsion or submicron emulsion or ultrafine emulsion. Like other emulsions, NE also has three distinct phases—oil or oleaginous phase, water or aqueous phase, and emulsifying agent(s).

12.3.2.1 Oleaginous phase

In typical emulsion systems, oil phase refers to any carrier oil, wherein lipophilic active ingredient(s) is/are dissolved or oil itself as bioactive component (such as essential oil in this chapter). A variety of nonpolar molecules like free fatty acids, mono-, di-, and triacylglycerols, waxes, mineral oils, and other lipophilic may be utilized to formulate emulsion (McClements and Rao, 2011). Nanoemulsion is a colloidal system whose formation, stability, and other properties are governed by the physicochemical peculiarities of the oil chosen, viz. density, viscosity, water solubility, interfacial tension, polarity, and refractive index (McClements and Rao, 2011; de Oca-Ávalos et al., 2017; Gupta et al., 2016). Oils with lower viscosity and lesser liquid (water)-liquid (oil) interfacial tension usually tend to form smaller droplets as it requires shorter period for external energy inputs to disrupt low viscosity oils and lesser interfacial tension can facilitate easy size reduction with minimal energy (Jafari et al., 2008). Furthermore, highly lipophilic oils impede the formation of nanoemulsion via the phase inversion method (Witthayapanyanon et al., 2006). From that point of view, EOs of highly polar nature having low viscosity and interfacial tension may become an excellent choice for formulating nanoemulsion (Saxena et al., 2017).

Despite being a good option, NEs are vulnerable to destabilization phenomena like coalescence or Ostwald ripening that happens due to the difference in chemical potential generated between multisized droplets (McClements and Rao, 2011). Therefore, in some cases, incorporation of highly nonpolar components with high molarity, such as medium-chain triglycerides (MCTs) or long-chain triglycerides (LCTs) or ester gums or vegetable oils (sesame, corn, sunflower oil, etc.), is encouraged as "ripening inhibitors" to slow down the occurrence of Ostwald ripening (Pavoni et al., 2019; Donsì et al., 2012). However, cases of bioactivity reduction were reported in some literature due to the addition of ripening inhibitors (Wan et al., 2018; Chang et al., 2012). Some researchers suggested applying carrier oils to prevent easy evaporation of EOs and to decelerate the rate of absorption of EOs that resulted in prolonged effect (Viljoen et al., 2015).

12.3.2.2 Aqueous phase

The selection of appropriate aqueous phase is vital for the formulation as it can influence the critical micelle concentration (CMC) and structural properties of the emulsifiers, thereby improving the overall physicochemical properties of the developed NEs (Yuliani and Noveriza, 2019). The aqueous phase may be comprised of only water or other polar solvents or solvents containing polar compounds, carbohydrates, proteins, acids, minerals, or alcoholic cosolvents (McClements and Rao, 2011). In recent years, attention has been imposed to replace traditional organic solvents with

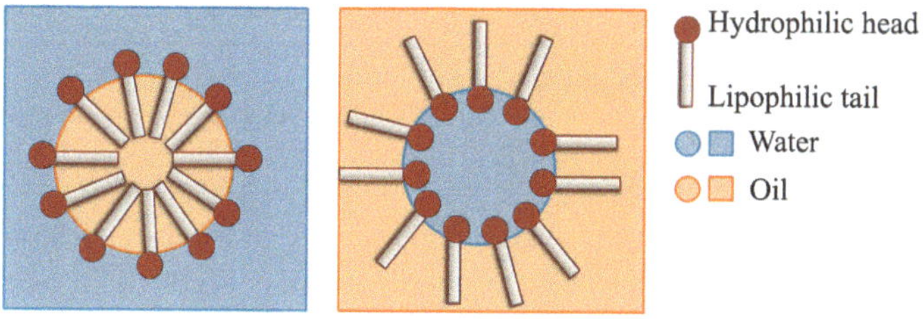

A B

Fig. 12.1 Schematic representation of alignment of a typical emulsifier in (A) oil-in-water (O/W) and (B) water-in-oil (W/O) emulsions.

green solvents to accomplish more eco-friendly pesticide nanoformulations. Some solvents such as *N*, *N*-methyl oleate (Feng et al., 2016), dimethyldecanamide (AMD-10), and D-limonene (Santos et al., 2016) are considered as safe biosolvents for agrochemical industries posing low risk to farmers and customers.

12.3.2.3 Emulsifying agent(s)

Emulsifiers are surface-active agents that lower the surface and interfacial tension among fluids (IUPAC, 2014). For these reasons, these compounds are also called as surfactants that are characterized by contrasting features between two parts within the molecule—one part showing lipophilic nature, whereas the other part displaying hydrophilic attributes (Fig. 12.1) (Salvia-Trujillo et al., 2017). The amphiphilic nature allows the surfactants to be absorbed spontaneously at the oil–water interface and prevents the aggregation of droplets resulting in the formation of sound dispersion of two immiscible phases (Muzaffar et al., 2013; Kralova and Sjöblom, 2009). The selection of an ideal emulsifier depends on various factors, such as the characteristics of the desired product, the methodology of emulsion preparation, the amount of the emulsifier added, the physical and chemical characteristics of every phase involved, and the presence of other functional components (if any) in the emulsion.

Emulsifiers are one type of stabilizers that prevent liquids that ordinarily do not mix from separating. The electrical properties of the surfactants significantly influence the development and stability of any formulation. Therefore, based on electrical charges, surfactants can be categorized into four types—anionic, cationic, nonionic, and amphoteric or zwitterionic (Table 12.3) (McClements, 2015). Generally, electrostatic interactions act as stabilizers for the ionic surfactants, whereas the nonionic surfactants are stabilized by hydrogen bond, dipole interactions, and steric repulsions (Muzaffar et al., 2013). Out of these, nonionic surfactants are usually preferred because of their safer toxicological profile, rapid adsorption kinetics, and capability to induce a steric hindrance through the bulky molecular functional groups projected toward the continuous medium (Salvia-Trujillo et al., 2017).

Most frequently used surfactants include sorbitan fatty acid esters, sucrose esters, glycerol fatty acid esters (polyglycerols), polyoxyethylene sorbitan fatty acid esters (polysorbates), and polyoxyethylene ethers (Pavoni et al., 2020). For EO-based

Table 12.3 Classification of surfactants.

Types	Subtypes	Features	Examples
Ionic	Anionic	Carries negative charge	Sodium lauryl sulfate and lignosulfonate
	Cationic	Carries positive charge	Cetrimonium bromide and benzalkonium chloride
	Amphoteric	Both cationic and anionic centers attached to the same molecule	Cocamidopropyl betaine, and lauryl hydroxysultaine
Nonionic	Ester	Chargeless	Sorbitan alkyl esters (Span series)
	Ether	Chargeless	Polyoxyethylene lauryl ether
	Amine	Chargeless	Polyethoxylated tallow amine
	Alcohol	Chargeless	Cetostearyl alcohol ethoxylate
Mixed	Nonionic-cationic	Mixture of chargeless and positively charged surfactants	Cetyltrimethylammonium bromide (CTAB) + polysorbate 60 (Tween 60)
	Nonionic-anionic	Mixture of chargeless and negatively charged surfactants	Sodium dodecyl sulfate (SDS) + octyl phenol ethoxylate (Triton X-100)

colloidal systems, polysorbates (Tween series, especially Tween 80 and Tween 20) are well utilized as they can make stable formulations without use of any cosurfactants (Balasubramani et al., 2017; Abd-Elsalam and Khokhlov, 2015). Both Tween surfactants maintain safe toxicological profile; however, in most of the cases, Tween 80 appeared to be better performing than Tween 20 in terms of droplet size and NE stability, which may be due to the structural difference in the nonpolar tail of the molecules (Ghosh et al., 2014; Dai et al., 1997). Natural surfactant molecules like proteins and polysaccharides are also ventured in developing EO-based colloidal dispersions. Although some of them can make stable NEs, yet they respond only under low-energy method and their efficiency is reduced due to their tardy absorption during homogenization and limited potency to bring down interfacial tension (McClements and Rao, 2011). However, opposite image came to the fore when chemically modified starch was used as emulsifying agents for the formulation of peppermint and clove EO-based NEs (Liang et al., 2012; Majeed et al., 2016).

Surfactants are also classified by an empirical number, known as hydrophilic–lipophilic balance (HLB) value that varies from 0 to 20 considering the both hydrophilic and lipophilic segments of the molecules. For An HLB, a value of 0 (zero) signifies a completely lipophilic moiety, whereas a value of 20 corresponds to an entirely hydrophilic entity (Griffin, 1949, 1954). Generally, for O/W emulsions generally surfactants are chosen with higher HLB values (e.g., Tween series), and lower HLB values (e.g., Span series) are preferred for W/O emulsions. Surfactants with intermediate to high HLB values, usually between 11 and 16, are reportedly preferred for the preparations of emulsions from EOs (Chang and McClements, 2014). The HLB value of surfactants may influence the physicochemical properties of NE system as a function of the oil phase chosen for the formulation. If the HLB value of the surfactant approaches nearer to the value needed by the oil phase, monodispersed droplets with the finest sizes can be attained (Fernandes et al., 2013).

Sometimes, another compound, known as cosurfactant, is used in addition to surfactant to help NE formation, which works synergistically with the surfactant and can further lessen the interfacial tension while augmenting the fluidity of the hydrocarbon region at the interface, which ultimately increases the entropy of the system (Pavoni et al., 2020). Cosurfactants are typically surface-active amphiphilic molecules that, because of their small polar heads, cannot stabilize an emulsion themselves; however, they allow a higher penetration of the oil between the surfactant tails favoring the optimal curvature of the interfacial film. Mostly, the NEs prepared via energy-saving (low-energy) methods require the contribution of cosurfactants to get the desired effect (Shafiq-un-Nabi et al., 2007). The most commonly used cosurfactants are short-chain/medium-chain alcohols that readily get access to the oil–water interface because of rapid diffusion between the participating phases. These cosurfactants intercalate between surfactant molecules undermining both the polar heads and the hydrocarbon tail interactions leading to the formation of a more flexible interfacial film that undergoes facile deformation around droplets. The ratio between surfactant and cosurfactant is a crucial factor controlling the phase characteristics, which relies on differences in the packing parameters of surfactant and cosurfactant at the O/W interface. Thence, it is difficult to establish a unanimous ratio, as they are dependent on all the components of the system. To figure out the optimum quali-quantitative composition, it is preferable to conduct a formulation study with the help of pseudoternary phase diagram (Pavoni et al., 2020).

12.4 Preparation of nanoemulsions

NE formation is not spontaneous and it is expressed by the following equation:

$$\Delta G = \Delta A\gamma - T\Delta S;$$

where ΔG = the total free energy required for stable emulsion formation, ΔA = variation in the interfacial area, γ = tension at the O/W interface, T = temperature, and ΔS = variation of the entropy of the dispersion system (Lawrence and Rees, 2012).

From the equation, it is clear that the formation of NEs requires external or internal energy as it is a thermodynamically unstable system. The NE fabrication methods can be classified into two types—high-energy (energy-intensive) and low-energy (energy-saving) methods.

NEs prepared from EOs are obviously O/W formulations that are stabilized to suit target applications. Generally, NEs require lower concentrations (~5%–10%) of surfactants as compared to other emulsion formulations (Tadros et al., 2004). The composition and techniques through which NEs are developed are vital as they determine the droplet size, stability, and other attributes of the entire emulsion system. Low-energy methods are generally preferred for producing NEs with small droplet sizes; however, their applications are limited as it requires a higher amount of surfactants and only a few selective oils and emulsifiers can be used for this. On the contrary, high-energy methods are considered more versatile in terms of a wide selection of oils and emulsifiers (Ghosh et al., 2013). The time (duration) and force (intensity) of the applied energy as well as the characteristics of the surfactant and oil phase determine the reduction in droplet size. As the EOs have low viscosity and interfacial tension as compared to the conventional oils, NEs developed from EOs can easily achieve small droplet size (McClements and Rao, 2011). High energy is obvious to generate a pressure of 10–100 atm surpassing the Laplace pressure for stabilization of NEs by shearing the droplets. The input energy can be generated via high-energy and/or low-energy methods, as depicted in Fig. 12.2 (Mustafa and Hussein, 2020). High-energy methods are usually top-down approaches, where NEs are developed from macroemulsion using energy-intensive devices without atomic-level control, whereas in bottom-up methods, NEs are fabricated using the energy generated through atomic and molecular level interaction between the components during the emulsification process.

12.4.1 High-energy techniques

Top-down fabrication techniques require mechanical appliances to produce intense energy that can blend and rupture the dispersion medium to form tiny oil droplets (Gutiérrez et al., 2008; Leong et al., 2009). The droplet size is inversely proportional

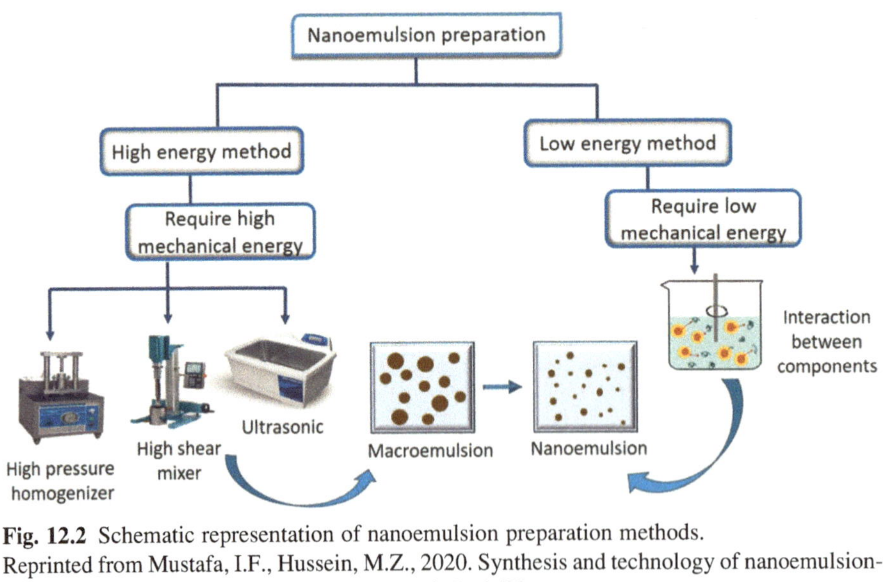

Fig. 12.2 Schematic representation of nanoemulsion preparation methods.
Reprinted from Mustafa, I.F., Hussein, M.Z., 2020. Synthesis and technology of nanoemulsion-based pesticide formulation. Nanomaterials 10(8), 1608.

to the intensity and duration of the energy inputs. The energy-intensive top-down technology includes various techniques, such as high-pressure homogenization (HPH), microfluidization, and ultrasonication (Gupta et al., 2016).

HPH approach generally involves two steps. In the first step, the primary emulsion is formed through mechanical means to produce oil droplets whose size lies in micron level (Rao and McClements, 2011). The second step aids in the conversion of the primary emulsion into a NE through the disruption of oil droplets into finer ones employing energy-intensive mechanical appliances, i.e., high-pressure homogenizers. The applied energy forces by high-pressure valve homogenizers charge the fluid through a specified valve under elevated pressure that creates a tremendous impact resulting in disruption of the droplets and formation of fine colloidal dispersion. The microfluidization technique uses "microfluidizer" devices that force the sample to pass through microchannels utilizing a high-pressure positive displacement pump leading to the formation of tiny droplets of the submicron range. Microfluidization is a direct emulsification technique (Azmi et al., 2019). Unlike HPH, this method lacks preemulsification as the dispersed phase can be directly introduced into the continuous phase (Abdelrasoul et al., 2018; Hassanin et al., 2017a, b, 2019). Among all the energy-consuming techniques, ultrasonication appears to be the simplest technique in formulating NEs. The strong agitative force generated by the ultrasonic processors brings about extreme shear that disintegrates droplets to create NEs. Prolonged sonication time produces more kinetic energy capable to reduce droplet size; however, it needs to be standardized for optimum results (Kumari et al., 2018).

The sequence of the steps taken during initial preparation may influence the stability of NEs. Numerous researches have suggested that altering the addition sequence of organic and aqueous phases sometimes accomplishes maximal stability (Feng et al., 2016). This theory is in line with the result of another study, where the addition of the aqueous phase into the organic phase brought off the maximum stability to the emulsions (Feng et al., 2020). Lastly, the emulsification treatment at the final step also performs convincingly in transformation of macroemulsion into NE.

12.4.2 Low-energy techniques

The bottom-up methods depend on the internal chemical energy of the system (Sole et al., 2006). These techniques involve instantaneous formation of tiny oil droplets in the oil–water-emulsifier-mixed systems, while changing either solution or composition and environmental factors such as temperature (Freitas et al., 2005; Bouchemal et al., 2004). Internal interaction of the components in the system facilitates the emulsification process, where surfactants' behavior plays a significant role. These methods include spontaneous emulsification, membrane emulsification, phase inversion temperature (PIT), emulsion inversion point (EIP) or phase inversion composition (PIC), and solvent displacement.

Spontaneous emulsification, also termed as self-emulsification or emulsification by solvent diffusion (ESD), is steered by diverse ways such as diffusion of solutes between two phases without changing the surfactant curvature in the system, turbulence at the interfaces, surface tension gradient, condensation, or dispersion mechanism (Salem and Ezzat, 2018). It can be achieved by the aqueous phase or oil phase titration process. The preparation method consists of varying oil and surfactant ratios, and based on the values of all the three components (oil, water, and surfactant), pseudoternary diagrams are generated (Anwer et al., 2014). The isotropic region depicted by the triangular phase diagram reveals several combinations of formulations. All NE regions in the phase diagram are deemed to be the optimum formulations with the lowest concentration of surfactant selected.

During the emulsifying process, if changes are made in surfactant curvature, the process is known as the phase inversion method. The transition between two phases is brought about by two factors—temperature or composition, hence called as phase inversion temperature (PIT) and phase inversion composition (PIC), respectively. All surfactants are welcomed in the PIC method; however, the PIT method allows only temperature-sensitive surfactants (e.g., polyoxyethylene-type surfactants (Ren et al., 2019). PIT method is considered as more healthful than the self-emulsifying method as the former method can be totally free from the use of organic solvents (Jintapattanakit, 2018). Another modified method, known as in situ phase inversion, operated without any strong force or equipment, relies on adding extra emulsifier to obtain nanodroplets (Qin et al., 2019). Green approach is also gaining popularity nowadays due to its reduced cost and less environmental impacts. One example of this approach is solvent-free low-energy method (Jesus et al., 2017). The nonheating strategies have been suggested where EOs are used in NE formulation to retain the quality of the thermolabile aromatic compounds (Da Botas et al., 2017).

12.5 Characterizations of nanoemulsions

The basic physicochemical properties of NE include particle size and shape, size distribution, zeta potential, pH, and viscosity. Therefore, the morphological and rheological properties and stabilities need to be characterized before application in crop protection as these may determine the efficacy of the formulation.

12.5.1 Particle size and distribution

Dynamic light scattering (DLS) is the most prevalent technique for the determination of micelle size and distribution of NE droplets (Fig. 12.3A–C). Before analysis, samples are diluted with deionized water to thwart the multiple scattering caused by the aggregation phenomenon happened through electrostatic interaction. However, the true droplet size can be determined by electron microscopic techniques. The polydispersity index (PDI) is also a vital parameter to indicate the system stability. Any PDI value of less than 0.5 is considered acceptable for agricultural applications; however, attempts are made to bring down the PDI value to less than 0.3, which is desirable. Smaller PDI value suggests monodisperse property of the particles. Samples with higher PDI values are generally rejected due to the higher degree of nonuniformity of size distribution of particles (Díaz-Blancas et al., 2016).

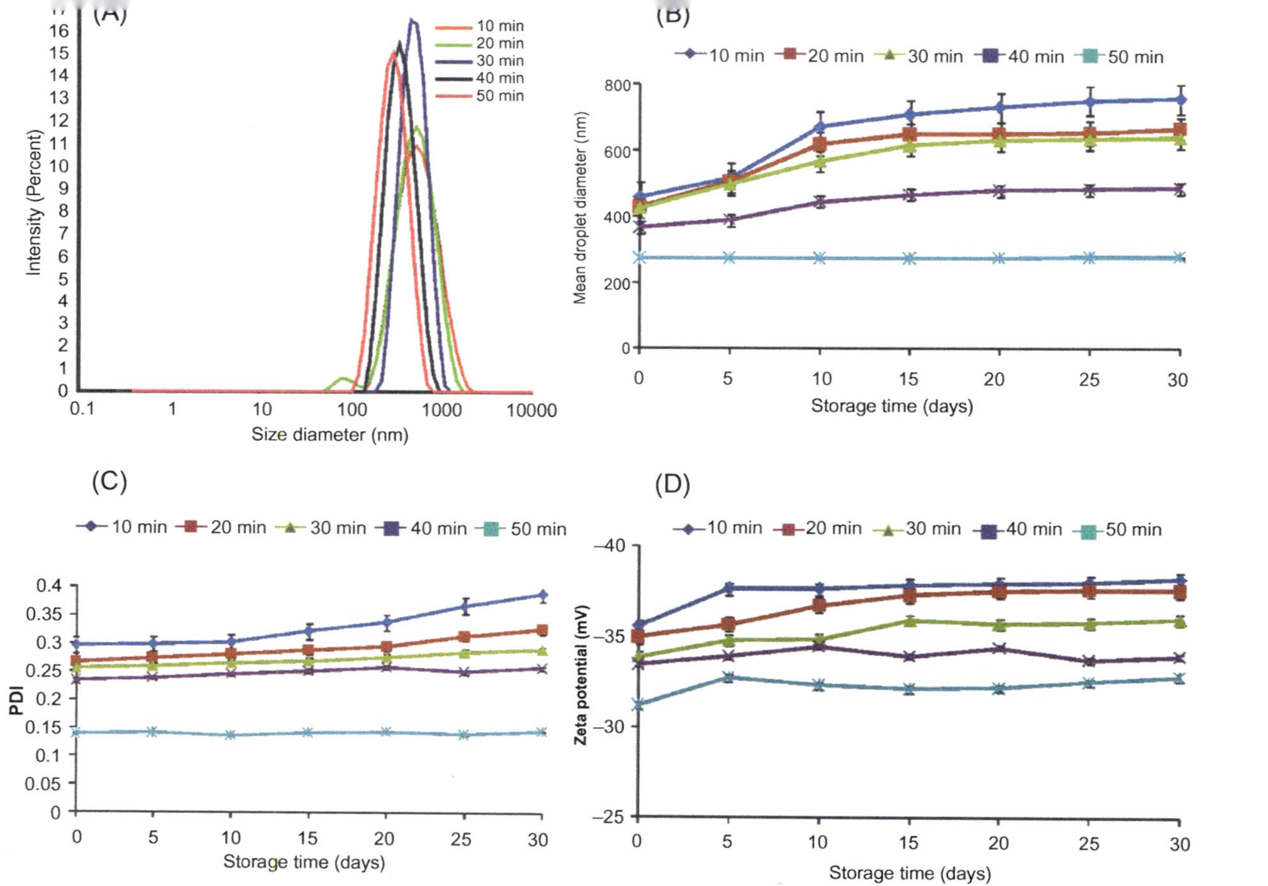

Fig. 12.3 DLS study of thymol nanoemulsions (A) size distribution of 10–50 min sonicated nanoemulsions at 0 days (B) a comparative graphical representation of mean droplet diameter (*Z*-averages), (C) polydispersity index, and (D) zeta potential of nanoemulsions (from 0 to 30 days) developed by different sonication times (Error bars represent ± standard error).

Reprinted from Kumari, S., Kumaraswamy, R.V., Choudhary, R.C., Sharma, S.S., Pal, A., Raliya, R., Biswas, P., Saharan, V., 2018. Thymol nanoemulsion exhibits potential antibacterial activity against bacterial pustule disease and growth promotory effect on soybean. Sci. Rep. 8, 6650.

The size of NEs is reported to be affected by numerous factors, but majorly dominated by the surfactant concentration and its packing parameters. The surfactant packing parameters are vital as they may cause alteration of surfactant curvature, which ultimately offers smaller droplets (Zeng et al., 2017). It has been found that increasing the ratio of surfactant to oil can make the droplet size fall up to a certain level (Goodarzi and Zendehboudi, 2019).

12.5.2 Viscosity, zeta potential, and pH measurement

Zeta potential, which measures the surface potential of the suspended droplets, is considered as one of the vital properties for stability assessment, which are measured by Zetasizer equipment (Fig. 12.3D). Zeta value may be positive or negative depending on the NE composition. Generally, a zeta potential greater than ± 30 mV is considered adequate to ensure the physical stability of NE for at least two years (Gurpret and Singh, 2018). High zeta potential value can prevent the aggregation of the droplets due to electrostatic repulsion between the droplets and thus, assures stability. As the biological cells are negatively charged, therefore, NEs, if positively charged, can easily penetrate the pathogen cells due to electrostatic attraction and enhance pesticidal effects (Mondal et al., 2017). The pH of the developed formulations also determines the stability of the dispersed system and it is generally measured by a pH meter. Viscosity, determined by an Ostwald viscometer, is dependent on components of organic phase, nature of the surfactants, and viscosity of the oil(s). EO NEs due to high water loading exhibit low viscosity. However, any change in the oil: water: surfactant ratio can significantly change the properties of the NEs (Heydari et al., 2020; Al-Assiuty et al., 2019).

12.5.3 Morphology and stability test

The shape and morphology of nanoemulsions can be determined employing various electron microscopic techniques like atomic force microscope (AFM), transmission electron microscopy (TEM), and cryogenic-field emission scanning electron microscopy (Cryo-FESEM) (Fig. 12.4). The EO-based NEs are generally spherical under microscope (Ali et al., 2017; Badawy et al., 2018) or these show core shell-like structures (Surendhiran et al., 2019) because of some cluster of nanomicelles developed during the synthesis process.

Shelf life of NEs can be estimated with variable storage time and temperatures. The stability tests are generally performed for a few days, but can go over twelve months depending upon the type and purpose of developed NEs. Three different storage temperatures are usually fixed—low temperature (0–5°C), standard laboratory temperature (~25°C), and high temperature (50 ± 5°C). During the storage period, the study focuses on observing the external appearance of NEs and computing their physicochemical parameters such as micelle or particle size and zeta potential at predefined time intervals. The samples without any abnormalities in their appearances like phase separation, creaming, flocculation, coalescence, and sedimentation are generally assumed stable formulations. To ensure stability, zeta potential and droplet size of

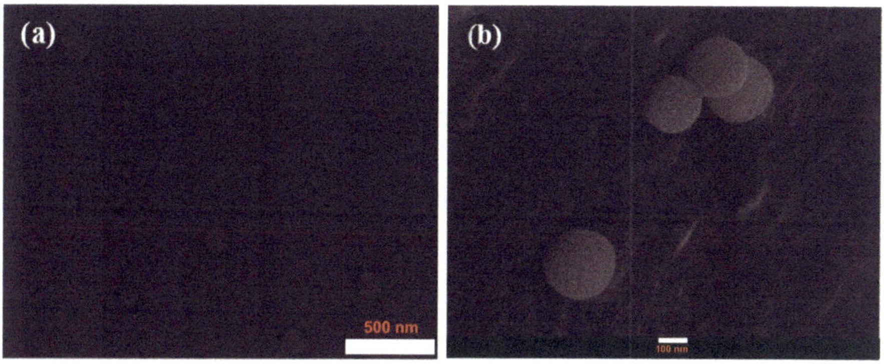

Fig. 12.4 Morphological characterization of thymol nanoemulsion: (A) TEM image at 45 kx magnification (scale bar = 500 nm); (B) Cryo-FESEM image at 80 kx magnification (scale bar = 100 nm).
Partially reprinted from Kumari, S., Kumaraswamy, R.V., Choudhary, R.C., Sharma, S.S., Pal, A., Raliya, R., Biswas, P., Saharan, V., 2018. Thymol nanoemulsion exhibits potential antibacterial activity against bacterial pustule disease and growth promotory effect on soybean. Sci. Rep. 8, 6650.

NEs are determined several times during and after storage and compared with initial values before storage. The samples maintaining optimum values are considered stable for future applications.

High temperature has been reported to cause NE instability due to the particle movement and dissolution of emulsifier into water leading to the conglomeration of particles in the emulsion system (Guan et al., 2018). Another major concern regarding NE instability is Ostwald ripening which generally starts taking place a few days after the preparation, and then flocculation and coalescence will occur at a later stage, as shown in Fig. 12.5. The oil phase fraction in NE system may control the Ostwald ripening. The coarsening phenomenon is prominent in NEs with low oil phase fraction (up to 5%) that can be stabilized with suitable emulsifier or emulsifier mixtures (Koroleva et al., 2018). Oswald ripening, a phenomenon that is concerned with the growth of smaller particles (droplets) into larger ones with low curvature radius, is really a major concern for long-duration storage. However, this event can be prevented through the enhancement of droplet elasticity and the addition of surfactant, which might lower interfacial free energy creating a mechanical barrier against coalescence (Patel and Joshi, 2012).

12.5.4 Retention and contact angle measurement

Agrochemicals, for successful pest control, must adhere to the plant surface properly. Retention and contact angle of leaves are needed to be measured for checking the affinity of NE toward leaf surfaces. Attempts are made to increase the adhesion property of the formulation, which ultimately increases NE efficiency. Retention can be estimated by dipping and microweighing method, while contact angle of leaves is

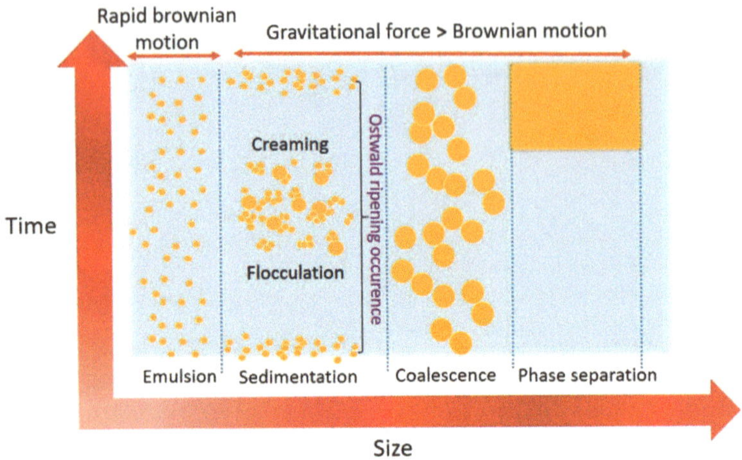

Fig. 12.5 Increase in droplet size of nanoemulsion owing to instability phenomenon as a function of storage time.
Reprinted from Mustafa, I.F., Hussein, M.Z., 2020. Synthesis and technology of nanoemulsion-based pesticide formulation. Nanomaterials 10(8), 1608.

determined using a precise measuring device equipped with a CCD (charge-coupled device) camera. EOs have low interfacial tension; therefore, NEs show a decline in contact angle with increasing oil content, which ultimately facilitates pesticide diffusion in the plant surface (Qian et al., 2018).

12.6 Application of nanoemulsions in crop protection

Researchers worldwide have come up with various approaches to utilize NEs in pest control programs. The developed NEs based on total essential oil or from any specific component of EO have been reported to be effective in eco-friendly pest management. In almost all cases, it has been found that the optimization of various parameters of NE formulation, such as surfactant concentration, oil-surfactant mixing ratio, and emulsification time, is crucial to obtain stable dispersion with strong and favorable pesticidal outcomes.

12.6.1 Activities against macropests

Macropests are pests that are bigger in size and easily visible by naked eye. These include insects, mites, and weeds.

EO-based NEs have been found effective against a vast range of insects and act as phagodeterrents, toxicants, repellents, or insect growth regulators (IGRs), affecting both larvae and adults of different insect orders. A study highlighted the development of NEs containing Mint (*Mentha longifolia*) essential oil, water, and two emulsifiers (Tween 80 and glyceryl monostearate) with the help of high-speed homogenizer, where mean droplet diameter varied from 14 to 36 nm. The developed formulation showed good insecticidal efficacy against Mediterranean flour moth (*Ephestia kuehniella*) with LC_{50} value of 14,068 ppm (Louni et al., 2018). A very promising insecticidal activity was found against red flour beetle (*Tribolium castaneum*) using NEs developed from aniseed (*Pimpinella anisum*) EO (Hashem et al., 2018). The formulation was prepared using Tween 80 surfactant and ethanol cosurfactant by centrifugation method and characterized by photon correlation spectroscopy (mean droplet size 198.9 nm; polydispersity index 0.303; zeta potential -25.4 ± 4.47 mV; viscosity 0.8872 cP.). The NE was found toxic against *T. castaneum* ($LC_{50} = 9.3\%$ *v*/v) with considerable destruction of its progeny. Morphological and histological impairments triggered by feeding and exposure to the aniseed NE were also analyzed by microscopic techniques. Fig. 12.6 is a clear rendition of destructions caused to the insect cuticle by the nanoformulation.

In another study conducted by Massoud et al. (2018), NE was developed by blending peppermint EO, deionized water, and Tween 80 surfactant followed by ultrasonication. The droplets looked spherical in shape and the average size of droplets recorded was 43.55 nm with PDI of 0.787. The nanoformulation proved to be a potent nano-

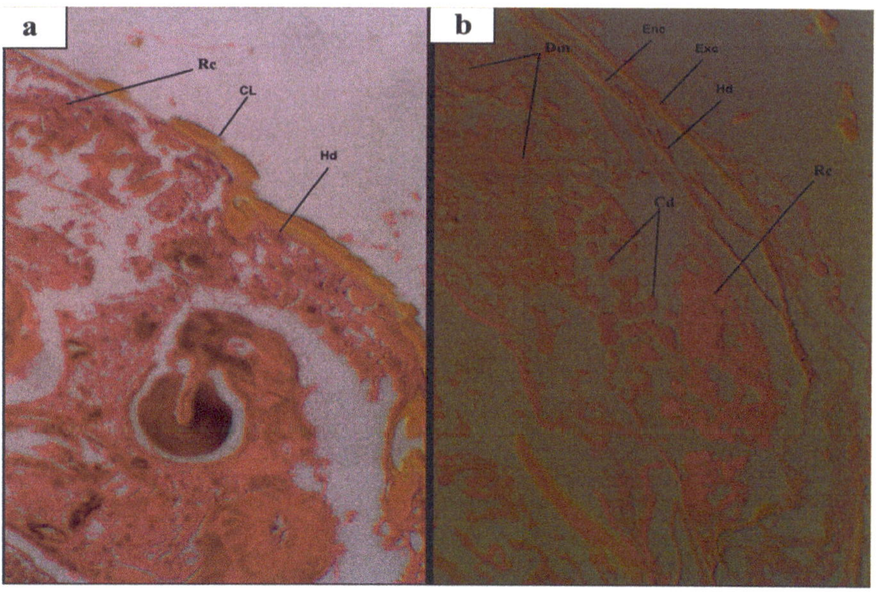

Fig. 12.6 Light microscopy images of *Tribolium castaneum* cuticle: (A) control insects; (B) beetles exposed to the nanoemulsion (Abbreviations: CL = cuticular layer, Hd = hypodermis, Rc = regenerative cells, Dm = destroyed muscular layer, Enc = endocuticle, Exc = exocuticle, Cd = cellular debris).
Reprinted with permission from Hashem, A.S., Awadalla, S.S., Zayed, G.M., Maggi, F., Benelli, G., 2018. Pimpinella anisum essential oil nanoemulsions against Tribolium castaneum—insecticidal activity and mode of action. Environ. Sci. Pollut. Res. 25(19), 18802–18812. Copyright© 2018, Springer-Verlag GmbH Germany, part of Springer Nature.

bioinsecticide against storage pest rice weevil (*Sitophilus oryzae*) with LC_{50} values of 0.127, 0.106, and 0.095 μL/cm^2 after 24, 48, and 72 h of exposure, respectively.

EOs are safe and influential substances to check the population of phytophagous mites. Therefore, attempts are being made recently to fabricate NEs based on EOs and evaluate those as miticidal agents. In a research project handled by Mossa et al. (2019), rosemary EO, deionized water, and Tween 20 were mixed and subjected to sonication by ultrasound to create NEs. The formulation presented a good dispersion with spherical droplets. The lowest droplet size recorded was 139.9 nm under particle size analyzer (DLS) and 39 nm under TEM. Spray application of the fabricated nanoformulation showed high acaricidal activity against immature and adult two-spotted spider mites (*Tetranychus urticae*) with LC_{50} 723.71 and 865.68 μg ml^{-1}, respectively.

Some NEs were also explored for managing the unwanted plants such as weeds and grasses, as those plants usually compete with crops for space and essential nutrients. Hazrati et al. (2017) studied NE containing summer savory (*Satureja hortensis*) EO and evaluated the herbicidal potential of the nanoformulation against worldwide weeds *Amaranthus retroflexus* and *Chenopodium album*. The NE showed an initial droplet size of 92.7 ± 2.6 nm and a polydispersity index of 0.29 ± 0.01. Under laboratory conditions, the maximum inhibition of germination was observed at the concentration of 800 μL/L. This NE exhibited strong herbicidal activity owing to the presence of phenolic monoterpene carvacrol known for strong phytotoxic effects.

12.6.2 Activities against meso- and micropests

Micropests are very small in size and can be seen only with the help of a microscope. These cover pathogens like bacteria, fungi, protozoa, and viruses. Mesopests generally include nematodes due to their characteristic size ranging from microscopic to overt small entities.

NEs containing EOs have been found to generate significant negative impacts on nematode survival, development, reproduction, behavior, and metabolic pathways. In a recent study, NEs co-encapsulating chili oil, cinnamon oil, and neem oil were fabricated through a composite approach of homogenization and sonication (Nguyen et al., 2020). The mean particle size, PDI, and zeta potential of the prepared nanoformulation were 156.7 nm, 0.193, and −36.8 mV, respectively. This novel formulation showed promising nematicidal effects both in vitro (82.3%) and ex vitro (75% and 91.3% for sprinkler and drip irrigation, respectively).

The growth of fungus has been reported to be suppressed substantially by the plant EO-based NEs. Garlic oil NE was developed using oil, water, and surfactant mixture (Tween 80 and Span 80) by ultrasonication method and characterized by DLS (micelle size around 60 nm and PDI around 0.3) and Raman spectroscopic technique (Long et al., 2020). The fungicidal activity against *Penicillium italicum* was evaluated by broth microdilution assay, which was reported to destroy fungal protein with a low minimum inhibitory concentration (MIC) value of 0.23% as compared to 3.7% in the case of raw garlic oil. The SEM image confirmed the devastation in spore and mycelia of fungal pathogen by garlic oil NE (Fig. 12.7).

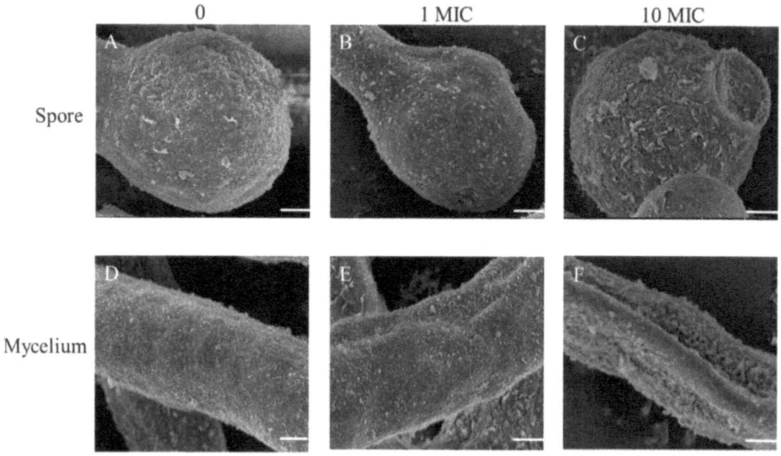

Fig. 12.7 SEM images of spores and mycelia after the treatment with NEs of various concentrations. (A) Spore as control; (B) spore treated with NE of 1 MIC; (C) spore treated with NE of 10 MIC; (D) mycelium as control; (E) mycelium treated with NE of 1 MIC; and (F) mycelium treated with NE of 10 MIC. (Scale bars = 1 μm).
Reprinted from Long, Y., Huang, W., Wang, Q., Yang, G., 2020. Green synthesis of garlic oil nanoemulsion using ultrasonication technique and its mechanism of antifungal action against Penicillium italicum. Ultrason. Sonochem. 64, 104970. Copyright© 2020, Elsevier B.V.

Hassanin et al. (2017a, b) prepared thyme EO-based NEs using nonionic surfactant (Tween 80) by ultrasonic emulsification method. The DLS study revealed mono- or di-dispersed system of NEs, whereas the TEM micrograph revealed the spherical shape of droplets with a diameter ranging from 26.6 to 45.3 nm. The antifungal assay was conducted using the paper disk method against *Sclerotinia sclerotiorum*, which demonstrated that NEs drastically affected mycelial growth of the pathogen. Abd-Elsalam and Khokhlov (2015) developed eugenol oil NE by mixing oil, water, and Tween 20 surfactant followed by ultrasonication. The fabricated NEs appeared as fairly monodispersed colloidal system under Zetasizer (DLS) with a mean droplet diameter of 50–110 nm (TEM). In vitro antifungal activities of prepared NEs were determined through agar well diffusion assay that showed significant growth inhibition of *Fusarium oxysporum* f. sp. *vasinfectum*.

Antibacterial potential of nanoemulsions also was tested against plant pathogenic bacteria. Using saponin from *Quillaja saponaria* as a natural surfactant, Kumari et al. (2018) prepared and characterized thymol NE. The most stable formulation developed with 50 min of ultrasonication showed values of mean droplet diameter (274 ± 2 nm), polydispersity index (0.1), and zeta potential (−31 mV). Thymol NE showed promising antibacterial activity against *Xanthomonas axonopodis* pv. *glycine*, causal organism for bacterial pustule disease in soybean, and no bacterial colony was found in the concentration range of 0.02–0.06 (*v*/v) of NE as compared to control and thymol and saponin.

A list of EO-based NEs used for crop protection purposes is presented in Table 12.4.

Table 12.4 Nanoemulsions based on essential oil and their applications in agricultural pest management.

Sl. No.	Components	Fabrication technique	Physicochemical properties	Activity	Target species	References
1	Clove (*Syzygium aromaticum*) EO and lemongrass (*Cymbopogon citratus*) EO, DW, Tween 20, COE-40, and PG	LEE method	MDD 76.73 ± 4.8; PDI 0.207 ± 0.02; V 26.9 ± 1.9 cP.	Fungicidal; IC_{50} 1410.08 mg/L	*Fusarium oxysporum* f. sp. *lycopersici*	Sharma et al. (2018)
2	Peppermint (*Mentha piperita*) EO, DW, and Tween 80	U/S method	MDD 20–40 nm; ZP −16.2 ± 0.20	Fungicidal; 87.5% inhibition of disease severity at 21st day	*Alternaria solani*	Pandey et al. (2020)
3	Cinnamon (*Cinnamomum zeylanicum*) EO, DW, and Tween 80	U/S method	MDD 65.98 nm; PDI 0.15; V 1.67 mPa.s.	Fungicidal; diameter of inhibition zones varying from 16.34 to 23.74 mm	*Aspergillus niger*, *Rhizopus arrhizus*, *Penicillium sp.*, and *Colletotrichum gloeosporioides*	Pongsumpun et al. (2020)
4	Garlic (*Allium sativum*) EO, DW, and Tween 20	U/S method	MDD 93.4 nm ± 0.532; ZP −31.67 ± 2.4 mV	Acaricidal; LC_{50} 298.23 and 309.63 μg/mL	Eriophyid olive mites (*Aceria oleae* and *Tegolophus hassani*)	Mossa et al. (2018)
5	Eucalyptus (*Eucalyptus globulus*) EO, aqueous filtrate of *Pongamia glabra* and *Jatropha curcas*, Suprol T-60, and PG	LEE method	MDD 77 nm; PDI 0.23	Insecticidal; LC_{50} 5.49 mg/L	Red flour beetle (*Tribolium castaneum*)	Pant et al. (2014)

Table 12.4 Continued

Sl. No.	Components	Fabrication technique	Physicochemical properties	Activity	Target species	References
6	Camphor (*Eucalyptus globulus*) EO, DW, and Tween 20	U/S method	MDD 99.0 ± 0.605 nm	Insecticidal; LC_{50} 181.49 μg/g; 100% mortality after 72 h	Wheat weevil (*Sitophilus granarius*)	Mossa et al. (2017)

N.B. *In every case, the best NE formulation was selected having the best outcomes.*
Abbreviations: CFG = centrifugation, COE-40 = Castor oil ethoxylate-40, DW = Deionized water, EO = Essential oil, IC_{50} = 50% Inhibitory concentration, LC_{50} = 50% Lethal concentration, LEE = Low-energy emulsification, MDD = Mean droplet diameter, PDI = Polydispersity index, PG = Propylene glycol, U/S = Ultrasonication, V = Viscosity, ZP = Zeta potential.
Units: cP = centipoise, μg/g = microgram per gram, μg/ml = microgram per milliliter, mg/l = milligram per liter, mPa s = millipascal-second, mV = millivolts, nm = nanometer.

12.7 Current and future perspectives

Nanoemulsification is an easy and convenient approach to protect and enhance the bioavailability of volatile oleaginous constituents. EOs have been extensively applied in traditional medicine, food preservation, and pest control for decades and categorized under Generally Recognized as Safe (GRAS) category by both the United States Food and Drug Administration (US FDA) and the Environmental Protection Agency (EPA). However, their mode of action is still obscure. Even though nanotechnology keeps offering favorable exploitation of EOs, the ambiguity and inadequacy of trustworthy scientific assessment of the potential risks of nanomaterials on the environmental entities, including humans, limit their boundless applications. Proper toxicological/safety profiling of the emulsifiers needs to be evaluated as most of the surfactants and cosurfactants are synthetic or semisynthetic and reported to be potentially hazardous. The dynamics of interactions among plants, pests, and nanomaterials are not clearly comprehensible, and target-specific controlled delivery of botanical nanopesticides for agricultural applications is still wanting. There also exists a lack of perception of the possible industrial applications of many of the fabrication strategies as the physics of NE generation is semiempirical till now. The immediate intervention of the rational scale-up procedure of NEs for commercial utilizations is the need of the hour. The exquisite optimization of the synthetic protocols to obtain formulations with robust stability, increased shelf life, satisfying physicochemical attributes, and invention of unique delivery systems may make NEs reach the apex of customer preference and market trends.

12.8 Conclusion

The arena of NE technology is still evolving. NEs based on EO components have shown promising results in foods and medicines owing to their unusual characteristics, such as large surface area, transparent or translucent appearance, sound stability, adjustable rheology, and promising functionality; however, their applications against phytopests are scarce. Majority of the EO-based NEs applied in crop protection systems are generated via either ultrasonication (high-energy approach) or spontaneous emulsification (low-energy approach). The interests of NEs development and their subsequent applications in crop defense are growing; however, the reports of their commercial applications are missing. Pesticide industries and government organizations must come forward in upscaling research efforts to standardize and validate protocols of NE fabrication incorporating EOs and assessment of crop protection as well as environmental safety in experimental setups to make EO-based NE as a full-fledged technology in the coming days.

Acknowledgments

Dr. Prithusayak Mondal acknowledges the Department of Science & Technology and Biotechnology, Government of West Bengal, for furnishing monetary support under the project entitled

"*Development of a novel pesticide formulation from a new turmeric variety in Terai region of West Bengal*" (Ref. No. 174(Sanc.)/ST/P/S&T/1G-36/2018, Dt. 14-02-2019).

References

Abdelrasoul, M., Ahmed, A., Badawy, M., 2018. Formulation, characterizations and antibacterial activity of some nanoemulsions incorporating monoterpenes. J. Plant Prot. Pathol. 9, 697–705.

Abd-Elsalam, K.A., Khokhlov, A.R., 2015. Eugenol oil nanoemulsion: antifungal activity against fusarium oxysporum f. sp. Vasinfectum and phytotoxicity on cottonseeds. Appl. Nanosci. 5 (2), 255–265.

Al-Assiuty, B.A., Nenaah, G.E., Ageba, M.E., 2019. Chemical profile, characterization and acaricidal activity of essential oils of three plant species and their nanoemulsions against Tyrophagus putrescentiae, a stored-food mite. Exp. Appl. Acarol. 79, 359–376.

Ali, E.O.M., Shakil, N.A., Rana, V.S., Sarkar, D.J., Majumder, S., Kaushik, P., Singh, B.B., Kumar, J., 2017. Antifungal activity of nano emulsions of neem and citronella oils against phytopathogenic fungi, *Rhizoctonia solani* and *Sclerotium rolfsii*. Ind. Crop Prod. 108, 379–387.

Anwer, M.K., Jamil, S., Ibnouf, E.O., Shakeel, F., 2014. Enhanced antibacterial effects of clove essential oil by nanoemulsion. J. Oleo Sci. 63, 347–354.

Avetisyan, A., Markosian, A., Petrosyan, M., Sahakyan, N., Babayan, A., Aloyan, S., Trchounian, A., 2017. Chemical composition and some biological activities of the essential oils from basil Ocimum different cultivars. BMC Complement. Altern. Med. 17 (1), 60. https://doi.org/10.1186/s12906-017-1587-5.

Azmi, N.A.N., Elgharbawy, A.A.M., Motlagh, S.R., Samsudin, N., Salleh, H.M., 2019. Nanoemulsions: factory for food, pharmaceutical and cosmetics. Processes 7, 617.

Badawy, M.E.I., Abdelgaleil, S.A.M., Mahmoud, N.F., Marei, A.E.-S.M., 2018. Preparation and characterizations of essential oil and monoterpene nanoemulsions and acaricidal activity against two-spotted spider mite (Tetranychus urticae Koch). Int. J. Acarol. 44, 330–340.

Balasubramani, S., Rajendhiran, T., Moola, A.K., Kumari, R., Diana, B., 2017. Development of nanoemulsion from *Vitex negundo* L. essential oil and their efficacy of antioxidant, antimicrobial and larvicidal activities (*Aedes aegypti* L.). Environ. Sci. Pollut. Res. 24, 15125–15133.

Bouchemal, K., Briançon, S., Perrier, E., Fessi, H., 2004. Nano-emulsion formulation using spontaneous emulsification: solvent, oil and surfactant optimisation. Int. J. Pharm. 280 (1–2), 241–251.

Bouzabata, A., Cabral, C., Gonçalves, M.J., Cruz, M.T., Bighelli, A., Cavaleiro, C., Casanova, J., Tomi, F., Salgueiro, L., et al., 2015. *Myrtus communis* L. as source of a bioactive and safe essential oil. Food. Chem. Toxicol. 75, 166–172. https://doi.org/10.1016/j.fct.2014.11.009.

Brand, Y.M., Roa-Linares, V.C., Betancur-Galvis, L.A., Durán-García, D.C., Stashenko, E., 2016. Antiviral activity of Colombian Labiatae and Verbenaceae family essential oils and monoterpenes on Human Herpes viruses. J. Essent. Oil Res. 28 (2), 130–137. https://doi.org/10.1080/10412905.2015.1093556.

Chang, Y., McClements, D.J., 2014. Optimization of orange oil nanoemulsion formation by isothermal low-energy methods: influence of the oil phase, surfactant, and temperature. J. Agric. Food Chem. 62, 2306–2312.

Chang, Y., McLandsborough, L., McClements, D.J., 2012. Physical properties and antimicrobial efficacy of thyme oil nanoemulsions: influence of ripening inhibitors. J. Agric. Food Chem. 60, 12056–12063.

Da Botas, G.S., Cruz, R.A.S., De Almeida, F.B., Duarte, J.L., Araújo, R.S., Souto, R.N.P., Ferreira, R., Carvalho, J.C.T., Santos, M.G., Rocha, L., Pereira, V.L.P., Fernandes, C. P., 2017. Baccharis reticularia DC. and limonene nanoemulsions: promising larvicidal agents for *Aedes aegypti* (Diptera: Culicidae) control. Molecules 22, 1990.

Dai, L., Li, W., Hou, X., 1997. Effect of the molecular structure of mixed non-ionic surfactants on the temperature of miniemulsion formation. Colloids Surf. A Physicochem. Eng. Asp. 125, 27–32.

de Oca-Ávalos, J.M.M., Candal, R.J., Herrera, M.L., 2017. Nanoemulsions: stability and physical properties. Curr. Opin. Food Sci. 16, 1–6.

Díaz-Blancas, V., Medina, D.I., Padilla-Ortega, E., Bortolini-Zavala, R., Olvera-Romero, M., Luna-Bárcenas, G., 2016. Nanoemulsion formulations of fungicide tebuconazole for agricultural applications. Molecules 21, 1271.

Donsì, F., Annunziata, M., Vincensi, M., Ferrari, G., 2012. Design of nanoemulsion-based delivery systems of natural antimicrobials: effect of the emulsifier. J. Biotechnol. 159, 342–350.

Eloh, K., Kpegba, K., Sasanelli, N., Koumaglo, H.K., Caboni, P., 2020. Nematicidal activity of some essential plant oils from tropical West Africa. Int. J. Pest Manag. 66 (2), 131–141.

Feng, J., Shi, Y., Yu, Q., Sun, C., Yang, G., 2016. Effect of emulsifying process on stability of pesticide nanoemulsions. Colloids Surf. A Physicochem. Eng. Asp. 497, 286–292.

Feng, J., Wang, R., Chen, Z., Zhang, S., Yuan, S., Cao, H., Jafari, S.M., Yang, W., 2020. Formulation optimization of D-limonene-loaded nanoemulsions as a natural and efficient biopesticide. Colloids Surf. A Physicochem. Eng. Asp. 596, 124746.

Fernandes, C.P., Mascarenhas, M.P., Zibetti, F.M., Lima, B.G., Oliveira, R.P.R.F., Rocha, L., Falcao, D.Q., 2013. HLB value, an important parameter for the development of essential oil phytopharmaceuticals. Rev. Bras. Farm. 23, 108–114.

Freitas, S., Merkle, H.P., Gander, B., 2005. Microencapsulation by solvent extraction/evaporation: reviewing the state of the art of microsphere preparation process technology. J. Control. Release 102 (2), 313–332.

Ghosh, V., Saranya, S., Mukherjee, A., Chandrasekaran, N., 2013. Antibacterial microemulsion prevents sepsis and triggers healing of wound in wistar rats. Colloids Surf. B Biointerfaces 105, 152–157.

Ghosh, V., Mukherjee, A., Chandrasekaran, N., 2014. Eugenol-loaded antimicrobial nanoemulsion preserves fruit juice against, microbial spoilage. Colloids Surf. B Biointerfaces 114, 392–397.

Goodarzi, F., Zendehboudi, S., 2019. A comprehensive review on emulsions and emulsion stability in chemical and energy industries. Can. J. Chem. Eng. 97 (1), 281–309.

Griffin, W.C., 1949. Classification of surface-active agents by 'HLB'. J. Soc. Cosmet. Chem. 1, 311–326.

Griffin, W.C., 1954. Calculation of HLB values of non-ionic surfactants. J. Soc. Cosmet. Chem. 5, 249–256.

Guan, W., Tang, L., Wang, Y., Cui, H., 2018. Fabrication of an effective avermectin nanoemulsion using a cleavable succinic ester emulsifier. J. Agric. Food Chem. 66, 7568–7576.

Gupta, A., Eral, H.B., Hatton, T.A., Doyle, P.S., 2016. Nanoemulsions: formation, properties and applications. Soft Matter 12, 2826–2841.

Gurpret, K., Singh, S.K., 2018. Review of nanoemulsion formulation and characterization techniques. Indian J. Pharm. Sci. 80 (5), 781–789.

Gutiérrez, J.M., González, C., Maestro, A., Solè, I., Pey, C.M., Nolla, J., 2008. Nano-emulsions: new applications and optimization of their preparation. Curr. Opin. Colloid Interface Sci. 13 (4), 245–251.

Hashem, A.S., Awadalla, S.S., Zayed, G.M., Maggi, F., Benelli, G., 2018. *Pimpinella anisum* essential oil nanoemulsions against *Tribolium castaneum*—insecticidal activity and mode of action. Environ. Sci. Pollut. Res. 25 (19), 18802–18812.

Hassanin, M., Abd-El-Sayed, M., Abdallah, M., 2017a. Antifungal activity of some essential oil emulsions and nanoemulsions against *Fusarium oxysporum* pathogen affecting cumin and geranium plants. Sci. J. Flowers Ornam. Plants 4, 245–258.

Hassanin, M.M.H., Halawa, A.E.A., Ali, A.A., 2017b. Evaluation of the activity of thyme essential oil nanoemulsion against Sclerotinia rot of fennel. Egypt. J. Agric. Res. 95 (3), 1037–1050.

Hassanin, M.M.H., Mohamed, N.T., Abd-El-Sayed, M.A., 2019. Fungicidal activity of nanoemulsified essential oils against Botrytis leaf blight of Poinsettia (Euphorbia pulcherrima) in Egypt. Egypt. J. Agric. Res. 96, 1259–1273.

Hazrati, H., Saharkhiz, M.J., Niakousari, M., Moein, M., 2017. Natural herbicide activity of *Satureja hortensis* L. essential oil nanoemulsion on the seed germination and morphophysiological features of two important weed species. Ecotoxicol. Environ. Saf. 142, 423–430.

Heydari, M., Amirjani, A., Bagheri, M., Sharifian, I., Sabahi, Q., 2020. Eco-friendly pesticide based on peppermint oil nanoemulsion: preparation, physicochemical properties, and its aphicidal activity against cotton aphid. Environ. Sci. Pollut. Res. 27, 6667–6679.

Isman, M., 2020. Bioinsecticides based on plant essential oils: a short overview. Z. Naturforsch. C J. Biosci. 75 (7–8), 179–182.

IUPAC, International Union of Pure and Applied Chemistry, 2014. Compendium of chemical terminology. In: McNaught, A.D., Wilkinson, A. (Eds.), The "Gold Book", second ed. Blackwell Scientific Publications, Oxford, ISBN: 0-9678550-9-8, https://doi.org/10.1351/goldbook. Online version (3.0.1) created by S. J. Chalk. Retrieved from: https://goldbook.iupac.org/files/pdf/goldbook.pdf.

Jafari, S.M., Assadpoor, E., He, Y., Bhandari, B., 2008. Re-coalescence of emulsion droplets during high-energy emulsification. Food Hydrocoll. 22, 1191–1202.

Jesus, F.L.M., De Almeida, F.B., Duarte, J.L., Oliveira, A.E., Cruz, R.A.S., Souto, R.N.P., Ferreira, R.M.A., Kelmann, R.G., Carvalho, J.C.T., Lira-guedes, A.C., 2017. Preparation of a Nanoemulsion with *Carapa guianensis* aublet (Meliaceae) oil by a low-energy/solvent-free method and evaluation of its preliminary residual larvicidal activity. Evid. Based Complement. Alternat. Med. 2017, 6756793.

Jintapattanakit, A., 2018. Preparation of nanoemulsions by phase inversion temperature (PIT) method. Pharm. Sci. Asia 45, 1–12.

Koroleva, M., Nagovitsina, T., Yurtov, E., 2018. Nanoemulsions stabilized by non-ionic surfactants: stability and degradation mechanisms. Phys. Chem. Chem. Phys. 20, 10369–10377.

Kralova, I., Sjöblom, J., 2009. Surfactants used in food industry: a review. J. Dispers. Sci. Technol. 30, 1363–1383.

Kumari, S., Kumaraswamy, R.V., Choudhary, R.C., Sharma, S.S., Pal, A., Raliya, R., Biswas, P., Saharan, V., 2018. Thymol nanoemulsion exhibits potential antibacterial activity against bacterial pustule disease and growth promotory effect on soybean. Sci. Rep. 8, 6650.

Lalthazuali, Mathew, N., 2017. Mosquito repellent activity of volatile oils from selected aromatic plants. Parasitol. Res. 116 (2), 821–825. https://doi.org/10.1007/s00436-016-5351-4.

Lawrence, M.J., Rees, G.D., 2012. Microemulsion-based media as novel drug delivery systems. Adv. Drug Deliv. Rev. 64, 175–193.

Leong, T., Wooster, T., Kentish, S., Ashokkumar, M., 2009. Minimising oil droplet size using ultrasonic emulsification. Ultrason. Sonochem. 16 (6), 721–727.

Li, Y., Fabiano-Tixier, A.S., Chemat, F., 2014. Essential oils: from conventional to green extraction. In: Li, Y., et al. (Eds.), Essential Oils as Reagents in Green Chemistry. Springer Briefs in Molecular Science. Springer, Cham.

Liang, R., Xu, S., Shoemaker, C.F., Li, Y., Zhong, F., Huang, Q., 2012. Physical and antimicrobial properties of peppermint oil nanoemulsions. J. Agric. Food Chem. 60, 7548–7555.

Long, Y., Huang, W., Wang, Q., Yang, G., 2020. Green synthesis of garlic oil nanoemulsion using ultrasonication technique and its mechanism of antifungal action against Penicillium italicum. Ultrason. Sonochem. 64, 104970.

Louni, M., Shakarami, J., Negahban, M., 2018. Insecticidal efficacy of nanoemulsion containing mentha longifolia essential oil against ephestia kuehniella (Lepidoptera: Pyralidae). J. Crop Prot. 7, 171–182.

Majeed, H., Liu, F., Hategekimana, J., Sharif, H.R., Qi, J., Ali, B., Bian, Y.-Y., Ma, J., Yokoyama, W., Zhong, F., 2016. Bactericidal action mechanism of negatively charged food grade clove oil nanoemulsions. Food Chem. 197, 75–83.

Massoud, M.A., Adel, M.M., Zaghloul, O.A., Mohamed, M.I.E., Abdel-Rheim, K.H., 2018. Eco-friendly nano-emulsion formulation of *Mentha piperita* against stored product pest *Sitophilus oryzae*. Adv. Crop Sci. Technol. 6 (6), 404.

McClements, D.J., 2010. Emulsion design to improve the delivery of functional lipophilic components. Annu. Rev. Food Sci. Technol. 1, 241–269.

McClements, D.J., 2015. Food Emulsions: Principles, Practices, and Techniques. CRC Press, Boca Raton, FL, USA, ISBN: 1498726690.

McClements, D.J., Rao, J., 2011. Food-grade nanoemulsions: formulation, fabrication, properties, performance, biological fate, and potential toxicity. Crit. Rev. Food Sci. Nutr. 51 (4), 285–330.

Mondal, P., Kumar, R., 2016. Preparation of azomethine based nano-chemicals and antibacterial activity against nitrifying bacteria. Pestic. Res. J. 28 (2), 194–200.

Mondal, P., Kumar, R., Gogoi, R., 2017. Azomethine based nano-chemicals: development, in vitro and in vivo fungicidal evaluation against *Sclerotium rolfsii*, *Rhizoctonia bataticola* and *Rhizoctonia solani*. Bioorg. Chem. 70, 153–162. https://doi.org/10.1016/j.bioorg.2016.12.006.

Mondal, P., Biswas, S., Pal, K., Ray, D.P., 2018. Annona squamosa as a potential botanical insecticide for agricultural domains: a review. Int. J. Bioresour. Sci. 5 (1), 81–89.

Mondal, P., Handore, A.V., Mukherjee, K., Ghosh, S.K., Karmakar, R., 2021. Pesticide residues: analysis, impact and mitigation. In: Rakshit, A., Singh, H.B., Singh, A.K., Abhilash, P.C., Sarma, B.K., Tripathi, V.K., Parewa, H.P. (Eds.), Handbook of Agriculture, Environment and Biotechnology. Association of Agriculture, Environment and Biotechnology, pp. 425–449.

Mossa, A.T.H., Abdelfatta, N.A.H., Mohafrash, S.M.M., 2017. Nanoemulsion of camphor (*Eucalyptus globulus*) essential oil, formulation, characterization and insecticidal activity against wheat weevil, *Sitophilus granarius*. Asian J. Crop Sci. 9, 50–62.

Mossa, A.T.H., Afia, S.I., Mohafrash, S.M.M., Abou-Awad, B.A., 2018. Formulation and characterization of garlic (*Allium sativum* L.) essential oil nanoemulsion and its acaricidal activity on eriophyid olive mites (Acari: Eriophyidae). Environ. Sci. Pollut. Res. 25, 10526–10537.

Mossa, A.H., Afia, S.I., Mohafrash, S.M.M., Abou-Awad, B.A., 2019. Rosemary essential oil nanoemulsion, formulation, characterization and acaricidal activity against the two-spotted spider mite Tetranychus urticae Koch (Acari: Tetranychidae). J. Plant Prot. Res. 59 (1), 102–112.

Moura, E.d.S., Faroni, L.R.D.'.A., Zanuncio, J.C., Helono, F.F., Prates, L.H.F., 2019. Insecticidal activity of *Vanillosmopsis arborea* essential oil and of its major constituent α-bisabolol against *Callosobruchus maculatus* (Coleoptera: Chrysomelidae). Sci. Rep. 9, 3723.

Mustafa, I.F., Hussein, M.Z., 2020. Synthesis and technology of nanoemulsion-based pesticide formulation. Nanomaterials 10 (8), 1608. https://doi.org/10.3390/nano10081608.

Muzaffar, F., Singh, U.K., Chauhan, L., 2013. Review on microemulsion as futuristic drug delivery. Int. J. Pharm. Pharm. Sci. 5 (3), 39–53.

Nguyen, M.-H., Vu, N.-B.-D., Nguyen, T.-H.-N., Tran, T.-N.-M., Le, H.-S., Tran, T.-T., Le, X.-C., Le, V.-T., Nguyen, N.-T.-T., Trinh, N.-A., 2020. Effective biocontrol of nematodes using lipid nanoemulsions co-encapsulating chili oil, cinnamon oil and neem oil. Int. J. Pest Manag. https://doi.org/10.1080/09670874.2020.1861361.

Pandey, S., Giri, V.P., Tripathi, A., Kumari, M., Narayan, S., Bhattacharya, A., Srivastava, S., Mishra, A., 2020. Early blight disease management by herbal nanoemulsion in Solanum lycopersicum with bio-protective manner. Ind. Crop Prod. 150, 112421.

Pant, M., Dubey, S., Patanjali, P.K., Naik, S.N., Sharma, S., 2014. Insecticidal activity of eucalyptus oil nanoemulsion with karanja and jatropha aqueous filtrates. Int. Biodeter. Biodegr. 91, 119–127.

Patel, P.R., Joshi, J.R., 2012. An overview on nanoemulsion: a novel approach. Int. J. Pharm. Sci. Res. 3, 4640–4650.

Pavoni, L., Benelli, G., Maggi, F., Bonacucina, G., 2019. Green nanoemulsion interventions for biopesticide formulations. In: Koul, O. (Ed.), Nano-Biopesticides Today and Future Perspectives. Academic Press, Cambridge, MA, USA, pp. 133–160.

Pavoni, L., Perinelli, D.R., Bonacucina, G., Cespi, M., Palmieri, G.F., 2020. An overview of micro- and nanoemulsions as vehicles for essential oils: formulation, preparation and stability. Nanomaterials 10 (1), 135.

Peter, G.J., 2009. Encapsulation of flavors in emulsions for beverages. Curr. Opin. Colloid Interface Sci. 14, 43–47.

Pongsumpun, P., Iwamoto, S., Siripatrawan, U., 2020. Response surface methodology for optimization of cinnamon essential oil nanoemulsion with improved stability and antifungal activity. Ultrason. Sonochem. 60, 104604.

Qian, K., Zhao, H., He, L., 2018. Preparation of fenpropathrin nanoemulsion by phase inversion temperature method and its study on performance. Int. J. Nanosci. 17, 1–7.

Qin, H., Zhou, X., Gu, D., Li, L., Kan, C., 2019. Preparation and characterization of a novel waterborne lambda-cyhalothrin/alkyd nanoemulsion. J. Agric. Food Chem. 67 (38), 10587–10594.

Rao, J., McClements, D.J., 2011. Formation of flavor oil microemulsions, nanoemulsions and emulsions: influence of composition and preparation method. J. Agric. Food Chem. 59, 5026–5035.

Reddy, D.N., 2019. Essential oils extracted from medicinal plants and their applications. In: Akhtar, M., Swamy, M., Sinniah, U. (Eds.), Natural Bio-active Compounds. Springer, Singapore, pp. 237–283.

Ren, G., Sun, Z., Wang, Z., Zheng, X., Xu, Z., Sun, D., 2019. Nanoemulsion formation by the phase inversion temperature method using polyoxypropylene surfactants. J. Colloid Interface Sci. 540, 177–184.

Salem, M.A., Ezzat, S.M., 2018. Nanoemulsions in food industry. In: Milani, J.M. (Ed.), Some New Aspects of Colloidal Systems in Foods. IntechOpen, pp. 31–51.

Salvia-Trujillo, L., Soliva-Fortuny, R., Rojas-Graü, M.A., McClements, D.J., Martin-Belloso, O., 2017. Edible nanoemulsions as carriers of active ingredients: a review. Annu. Rev. Food Sci. Technol. 8, 439–466.

Santos, J., Trujillo-Cayado, L.A., Calero, N., Alfaro, M.C., Muñoz, J., 2016. Development of eco-friendly emulsions produced by microfluidization technique. J. Ind. Eng. Chem. 36, 90–95.

Sauter, I.P., Rossa, G.E., Lucas, A.M., Cibulski, S.P., Roehe, P.M., Alves da Silva, L.A., Rott, M.B., Vargas, R.M.F., Cassel, E., von Poser, G.L., 2012. Chemical composition and amoebicidal activity of *Piper hispidinervum* (Piperaceae) essential oil. Ind. Crops Prod. 40, 292–295. https://doi.org/10.1016/j.indcrop.2012.03.025.

Saxena, A., Maity, T., Paliwal, A., Wadhwa, S., 2017. Technological aspects of nanoemulsions and their applications in the food sector. In: Oprea, A.E., Grumezescu, A.M. (Eds.), Nanotechnology Applications in Food. Elsevier, Amsterdam, The Netherlands, pp. 129–152.

Shafiq-un-Nabi, S., Shakeel, F., Talegaonkar, S., Ali, J., Baboota, S., Ahuja, A., Khar, R.K., Ali, M., 2007. Formulation development and optimization using nanoemulsion technique: a technical note. AAPS PharmSciTech 8, E12–E17.

Sharma, A., Sharma, N.K., Srivastava, A., Kataria, A., Dubey, S., Sharma, S., Kundu, B., 2018. Clove and lemongrass oil based non-ionic nanoemulsion for suppressing the growth of plant pathogenic fusarium oxysporum f. sp. lycopersici. Ind. Crop Prod. 123, 353–362.

Sheth, T., Seshadri, S., Prileszky, T., Helgeson, M.E., 2020. Multiple nanoemulsions. Nat. Rev. Mater. 5, 214–228.

Si, L., Chen, Y., Han, X., Zhan, Z., Tian, S., Cui, Q., Wang, Y., 2012. Chemical composition of essential oils of *Litsea cubeba* harvested from its distribution areas in China. Molecules 17, 7057–7066.

Sole, I., Maestro, A., Pey, C., Gonzalez, C., Solans, C., Gutierrez, J., 2006. Nano-emulsions preparation by low energy methods in an ionic surfactant system. Colloids Surf. A Physicochem. Eng. Asp. 288 (1–3), 138–143.

Surendhiran, M., Raja, K., Jerlin, R., Marimuthu, S., Srivignesh, S., 2019. Nano emulsion seed invigouration for improved germination and seedling vigour in maize. Int. J. Agric. Sci. Res. 9, 333–340.

Swamy, M.K., Akhtar, M.S., Sinniah, U.R., 2016. Antimicrobial properties of plant essential oils against human pathogens and their mode of action: an updated review. Evid. Based Compl. Altern. Med. 2016, 3012462. https://doi.org/10.1155/2016/3012462.

Tadros, T., Izquierdo, P., Esquena, J., Solans, C., 2004. Formation and stability of nano-emulsions. Adv. Colloid Interface Sci. 108–109, 303–318.

Tisserand, R., Young, R., 2013. Essential Oil Safety: A Guide for Health Care Professionals. Elsevier Health Sciences, United Kingdom, ISBN: 9780443062414.

Tudi, M., Ruan, H.D., Wang, L., Lyu, J., Sadler, R., Connell, D., Chu, C., Phung, D.T., 2021. Agriculture development, pesticide application and its impact on the environment. Int. J. Environ. Res. Public Health 18, 1112.

Viljoen, J.M., Cowley, A., Du Preez, J., Gerber, M., Du Plessis, J., 2015. Penetration enhancing effects of selected natural oils utilized in topical dosage forms. Drug Dev. Ind. Pharm. 41, 2045–2054.

Wan, J., Zhong, S., Schwarz, P., Chen, B., Rao, J., 2018. Influence of oil phase composition on the antifungal and mycotoxin inhibitory activity of clove oil nanoemulsions. Food Funct. 9, 2872–2882.

Witthayapanyanon, A., Acosta, E.J., Harwell, J.H., Sabatini, D.A., 2006. Formulation of ultralow interfacial tension systems using extended surfactants. J. Surfactant Deterg. 9, 331–339.

Yuliani, S., Noveriza, R., 2019. Effect of carrier oil and co-solvent on the formation of clove oil nanoemulsion by phase inversion technique. IOP Conf. Ser. Earth Environ. Sci. 309, 012036.

Zeng, L., Xin, X., Zhang, Y., 2017. Development and characterization of promising Cremophor EL-stabilized o/w nanoemulsions containing short-chain alcohols as a cosurfactant. RSC Adv. 7, 19815–19827.

Zuzarte, M., Salgueiro, L., 2015. Essential oils chemistry. In: de Sousa, D.P. (Ed.), Bioactive Essential Oils and Cancer. Springer International Publishing AG, Cham, Switzerland, pp. 19–62.

Nanofertilizers for sustainable agriculture

13

Saurav Das[a] *and Sahila Beegum*[b]
[a]Department of Agronomy and Horticulture, University of Nebraska–Lincoln, Lincoln, NE, United States, [b]Nebraska Water Center, Part of the Robert B. Daugherty Water for Food Global Institute, University of Nebraska, Lincoln, NE, United States

13.1 Introduction

Exponential increase in the world population, which is estimated to exceed 9.7 billion by 2050, has also anticipated current crop production to increase by 70%–100% to fulfill the demand. However, the greatest challenge is to sustainably use the fertilizer to increase the crop production. A large portion of the conventional fertilizer ends up as environmental loss either in the form of leachate or runoff. Nutrient use efficiency has hardly increased for conventional fertilizers with nitrogen (N) 30%–35%, phosphorus (P) 18%–20%, and potassium (K) 35%–40%, and maximum amount of applied fertilizer faces environmental loss (Subramanian et al., 2015). In particular, fertilizers, like N, which are very prone to environmental losses, create major concern due to their large application in crop production and associated water and air pollution. Despite the agronomic improvement, over the years, nitrogen use efficiency remained significantly low and on an average, more than 50% of the applied N is potentially lost to the environment (Mejias et al., 2021). Additionally, the nutrient use efficiency of conventional fertilizers also depends on the final concentration of the fertilizers which reach to the target. This can be largely varied and often much below the minimum desired concentration due to leaching loss, drift, runoff, hydrolysis, evaporation, photolytic, or microbial degradation. Repeated use of conventional fertilizer adversely affects the nutrient equilibrium of soil (Shang et al., 2019). Therefore, the major goal of modern agriculture is to deliver the nutrient more efficiently without causing environmental pollution and deterioration of soil ecosystem.

Nanotechnology is the study, design, synthesis, manipulation, and application of nano-sized (<100 nm) materials. Nanomaterials differ from their original compound, as the decrease in size provides a high surface area to volume ratio for higher chemical reactivity. A wide range of materials were used for the creation of nanoparticles like metal oxides, ceramics, semiconductors, quantum dots, lipids, polymers, dendrimers, and emulsions (Benelmekki, 2015; Ruiz-Cañas et al., 2020). Nanotechnological application can also be widely found in the agricultural field as nanosized fungicides, bactericides, herbicides, growth-regulator, and fertilizers. Nanofertilizer can be defined as nano-sized (~30–40 nm) macro and micronutrients' carriers for plants, which can be delivered in a controlled manner (Shang et al., 2019).

Agricultural Nanobiotechnology. https://doi.org/10.1016/B978-0-323-91908-1.00005-5

Application of nanofertilizers and the ability to control their release rate have provided the potential to minimize the environmental loss and increase crop production in a sustainable way. For example, application of porous nanomaterials, zeolite, clay, or chitosan can significantly reduce nitrogen loss by regulating plant demand-based release. Ammonium-charged zeolite can increase the solubility of phosphate mineral and thus improve the phosphorus availability and uptake (Dwivedi et al., 2016). Graphene oxide nanofilms can prolong the potassium nitrate release and thus reduce the leaching loss (Shalaby et al., 2016). Application of nano-calcite ($CaCO_3$–40%) with nano SiO_2 (4%), MgO (1%), Fe_2O_3 (1%) has been found to improve the uptake of Ca, Mg, Fe, Zn, Mn, and P (Sabir et al., 2014).

There is ample evidence describing improved crop production, soil quality, and environmental quality post-application of nanofertilizers. However, the number of their research studies and their on-farm application are still limited. This chapter will discuss about different types of nanofertilizers, their synthesis method, their effect on crop production, fate in the environment, and their effect on human health. This chapter will also discuss the future prospect of nanofertilizers for developing a sustainable agricultural production system.

13.2 Nanofertilizer formulation

13.2.1 Types of nano-fertilizers and release mechanism

Based on the type of formulations, nanofertilizers can be classified into three categories viz. a. nanoscale fertilizer, which is a nanoscale conventional fertilizer, b. nanoscale additive fertilizers, which are conventional fertilizers with nanoscale additive or supplemental materials like dicyandiamide (DCD) and nitrapyrin, and c. nanoscale coating fertilizers, which mostly refer to nutrients encapsulated by nanofilm of chemical or biological compounds. Release of nutrient immobilized or encapsulated into nano-carrier is activated by bio-physicochemical nature of the environment. Biological release factors are mostly extracellular enzymes present in the soil released by different bacteria and fungi. Chemical triggered release mostly depends on the moisture, pH variation, and ion-exchange reaction, whereas physical factors are mostly ultrasound, magnetic field, heat, and diffusion parameters which control the release of nutrients from nano-carriers (Ribeiro and Carmo, 2019; Weeks and Hettiarachchi, 2019). Based on the functional properties and mode of action, nanofertilizers can be further classified as control or slow-release fertilizers, magnetic fertilizers, and nano-composite fertilizers containing a wide range of micro and macronutrients (Shang et al., 2019).

13.2.2 Methods of synthesis

Nanofertilizer can be synthesized using top-down (physical), bottom-up (chemical), or biological (plant, bacteria, and fungi) approaches.

13.2.2.1 Physical (top-down)

The most common method of nanofertilizer synthesis is top-down. In the top-down method, the substrate or absorbent (a bulk-sized) for nano-fertilizers is fragmented for several hours using different methods to achieve the nano-dimension. Some of the common physical methods applied for the synthesis of nanoparticles are gas condensation, aerosol synthesis method, high energy ball mill, mechanical attrition, molecular beam epitaxy, thermolysis, vacuum arc deposition, and expansion-cooling by temperature (reference). A study by Manikandan and Subramanian (2014) reported ball milling of zeolite at 0, 1, 2, 4, and 6 h, which had reduced the dimension in the order of 1078, 475, 398, 357, and 203 nm, respectively. The size reduction also increased the corresponding surface area as 41, 55, 72, 83, and 110 $m^2 g^{-1}$ respectively. Though physical method is simple in approach, the structural homogeneity of the synthesized product is hard to maintain. Structural heterogeneity and agglomeration are major drawbacks in the top-down approach. Nanofertilizers can readily lose their nano-properties due to quick aggregation when applied to an aqueous solution or when put in the presence of soil moisture (Weeks and Hettiarachchi, 2019).

13.2.2.2 Chemical (bottoms-up)

In bottoms up, the nanofertilizers are synthesized from atoms, molecules, and monomers. Most common chemical methods are sol–gel (polyol), microemulsion, thermolysis, and electrochemical synthesis. There are several studies where sub-micrometer metal oxides viz. Y_2O_3, V_xO_y, Mn_3O_4, ZnO, $CoTiO_3$, SnO_2, PbO, and TiO_2 were created using the sol–gel method (Cele, 2020). Ramírez-Rodríguez et al. (2020a, 2020b) prepared biomimetic calcium phosphate (CaP) nanoparticles doped with K and N (as nitrate and urea) using the co-precipitation method. Khan et al. (2021), in a study, also used co-precipitation to synthesize nano-zeolite using sodium silicate and ammonium sulfate solution in the presence of ethylene glycol and sodium hydroxide solution. Kubavat et al. (2020) prepared chitosan-based nanoparticles by polymerization of chitosan with methacrylic acid and later incorporated with potassium to produce chitosan-potassium nanofertilizers. Chemical methods are reliable for producing homogenous and stable nanoscale particles. However, metal precursors, toxic reducing agents, and stability agents are responsible for creating secondary environmental issues.

13.2.2.3 Biological (bottoms-up)

The biological method mostly utilizes the function of microbial (bacteria and fungi) bio-mineralization property and plant extract as a reducing agent. Biosynthesis is an environmentally safe procedure, and it does not produce secondary toxic compounds as part of the synthesis process. Conventional chemical reduction methods are also expensive and require organic solvents as reducing agents which are toxic to human health and the environment. Microorganisms like bacteria and fungi can control the structure, phase, orientation, and nano-structural topography of inorganic crystals. Bioactive compounds of plants and microorganisms provide natural capping agents without environmental pollution. There are several reports on the biosynthesis of nanoparticles especially those of silver, gold, and titania (Bansal et al., 2005) (Table 13.1). A study by Priyam et al. (2019) reported the biosynthesis of nano-hydroxyapatite (30 ± 5 nm) using *Bacillus licheniformis* with different phosphate concentrations (2%–20% *w/v*) of $K_2H_2PO_4$. Similarly, synthesis of ZnO nanoparticles ranging from 20 to 57 nm was achieved using bacterial culture viz. *Aeromonas hydrophila*, and plant extract viz. *Elaeagnus angustifolia*, *Coriandrum sativum*, *Calotropis gigantea*, and *Acalypha indica* (Singh et al., 2018). Ponmurugan et al. (2016) synthesized copper nanofertilizers using *Streptomyces griseus*. There are several studies which reported the use of bacteria for the synthesis of iron and manganese NF (de França Bettencourt et al., 2020). Table 13.1 represents the studies which reported the application of plants, fungi, and bacteria for the synthesis of nanosized materials.

13.3 Characterization of nanofertilizers

Synthesized nanofertilizers can be characterized using the particle size analyzer (PSA), zeta analyzer, Fourier transform infrared spectroscopy (FTIR), Raman spectroscopy, X-ray diffraction (XRD), scanning electron microscope (SEM), energy-dispersive X-ray spectroscopy (EDAX), transmission electron microscope (TEM), and atomic force microscope (AFM) to determine the size, shape, charge distribution, functional groups, and elemental composition. Shebl et al. (2019) used XRD, SEM, and TEM to characterize nano-oxide particles (ferric, manganese, and zinc). Kubavat et al. (2020) used FTIR spectroscopy to analyze the formation of chitosan nanofertilizers and used SEM, TEM, and AFM for size distribution, and elemental composition. Extensive research has been carried out to characterize nanofertilizers based on nitrogen (Manikandan and Subramanian, 2014; Ramírez-Rodríguez et al., 2020b), phosphorus (Ramírez-Rodríguez et al., 2020a), potassium (Nido et al., 2019), sulfatic (Li and Zhang, 2010), and zinc fertilizers (Yusefi-Tanha et al., 2020).

13.4 Basic forms of nanofertilizers

Most applied nanofertilizers in agricultural systems are zeolites, nanocomposites, super adsorbent fertilizers, and carbon nanotubes. Zeolites are microscopic crystalline forms of aluminosilicate minerals. There are different types of zeolites viz. clinoptilolite, stilbite, chabazite, analcime, and natrolite. Nanocomposites are multiphase matrix of silicate. The incorporation of nanocomposite provides unique characteristics to nanofertilizers. Application of ammonium-loaded clinoptilolite with a combination of phosphorite was found to facilitate the slow release of N and P (Barbarick et al., 1990). Super adsorbents are polymers that are cross-linked with the nanofertilizer for improved water holding and nutrient use efficiency (Butt and Naseer, 2020).

Table 13.1 Biosynthesis of nanoparticles using various plants, bacteria, and fungi.a

Biological entity	Type	Nanoparticles	Size (nm)	Extracellular/ intracellular	Reference
Plant					
Avena sativa	Plant	Au	5–85	Intracellular	Armendariz et al. (2004)
Azadirachta indica	Plant	Ag, Au, and Ag/Au bimetallic	50–100	Extracellular	Shankar et al. (2004)
Aloe vera	Plant	Ag	15.2 ± 4.2	–	Chandran et al. (2006)
Alfalfa	Plant	Ti/Ni bimetallic	1–4	–	Schabes-Retchkiman et al. (2006)
Emblica Officinalis	Plant	Ag and Au	10–25	Extracellular	Ankamwar et al. (2005)
Fungi					
Aspergillus fumigatus	Fungi	Ag	5–25	Extracellular	Bhainsa and D'Souza (2006)
Colletotrichum sp.	Fungi	Au	20–40	Extracellular	Shankar et al. (2003)
Fusarium oxysporium	Fungi	Au	20–40	Extracellular	Mukherjee et al. (2002)
Fusarium oxysporum	Fungi	Zirconia	3–11	Extracellular	Bansal et al. (2004)
Fusarium oxysporum	Fungi	CdS	5–20	Extracellular	Ahmad et al. (2002)
Bacteria					
Magnetotactic bacteria	Bacteria	Magnetic (Fe_3O_4), greigite (Fe_3S_4)	–	–	Lefèvre et al. (2011)
Pseudomonas aeruginosa	Bacteria	Au	15–30	Extracellular	Husseiny et al. (2007)
Desulfovibrio desulfuricans	Bacteria	Palladium	–	–	Yong et al. (2002)
Pseudomonas stutzeri	Bacteria	Ag	> = 200		Joerger et al. (2000)

[a] Note: "–" indicates the missing/no data.

13.5 Method of application

13.5.1 Soil

A most common method of fertilizer application is direct broadcast to soil. Direct application to the rootzone, especially near the root tip, provides easy access for plant to nanofertilizers. However, the complex dynamic physicochemical nature (pH, CEC, different nutrients, root exudates, organic matter) of soil along with its biotic community (microorganisms) complicates the nanofertilizer uptake mechanisms (Chugh et al., 2021). Localized application of large concentrations of nanoparticles can influence the soil or rhizospheric microbial communities (Raliya et al., 2018). Furthermore, the physicochemical properties of the soil can induce agglomeration of the particle. A comparative study between the soil application and foliar spray showed a significant advantage of foliar spray for nutrient uptake (Raliya et al., 2018).

13.5.2 Foliar

Soil application of fertilizers is the most preferred and convenient method of application. However, soil application of nutrients, especially phosphorus, iron, often gets fixed to the soil as insoluble forms and half of the nutrient leaches or runs off during rainfall or irrigation events before a crop can uptake. Besides, root restricting conditions like critical compaction can also restrict crops from soil nutrient uptake (Alshaal and El-Ramady, 2017). Foliar application provides a better way of nutrient application and minimizes environmental pollution by improving nutrient use efficiency. However, the foliar application also comes with its problems like it needs an excessive amount of fertilizer, then there is a very short penetration time, and the rest of the fertilizers wash off to the ground. Penetration is the major structural barrier for foliar application due to the nano-scale pore size (5–20 nm) of the cell wall (Fleischer et al., 1999; Schwab et al., 2016). According to the polar pore model, for penetration of polar and ionic solutes to the cuticle, the exclusion limit of pore radius has been estimated as 2–2.4 nm and stomal diffusion pore ration >20 nm. Application of nanofertilizers has been found promising and efficient in many studies. Engineered nanoparticles with size <50 nm were reported for efficient stomatal penetration in *Vicia fabal* L. (Eichert et al., 2008). In a study by Wang et al. (2013), a nanofertilizer with size 27.3–46.7 nm was found to be better for stomatal penetration in watermelon.

13.6 Effect of nanofertilizer on plant growth

The concept of nanofertilizer emerges to overcome the environmental issues associated with conventional fertilizers. Synthesized nanofertilizers have functional properties like high solubility, stability, effectiveness, time-controlled release, and enhanced target activity for high nutrient use efficiency. They are also easy to deliver and dispose of to minimize ecotoxicity. Nutrients are often absorbed, ligand-bound, entrapped, or encapsulated in the polymeric shell of nanoparticles or directly synthesized constituting the nutrient itself. For example, nanosized urea fertilizers modified with hydroxyapatite showed a gradual release of nutrients relating to crop growth. These nanofertilizers showed a slow release of nitrogen up to 60 days of plant growth compared to the 30 days of conventional urea fertilizer (Kottegoda et al., 2011). In one study, mesoporous silica nanoparticle encapsulated urea showed a five-fold improvement in the release curve (Wanyika et al., 2012). The slow–release nature of the nanofertilizer prevents environmental loss and improves crop productivity (Fig. 13.1). The application of nanoparticles like Al_2O_3, TiO_2, CeO_2, FeO, and ZnO has improved agricultural productivity in different crops (Table 13.2). The application of nanofertilizers also has a significant effect on seed germination and seedling growth (Fig. 13.1). The seed treated with nanoparticles viz. Au, CuO, and TiO_2 has shown increased seed germination rates (Dhir, 2021). The application of calcium

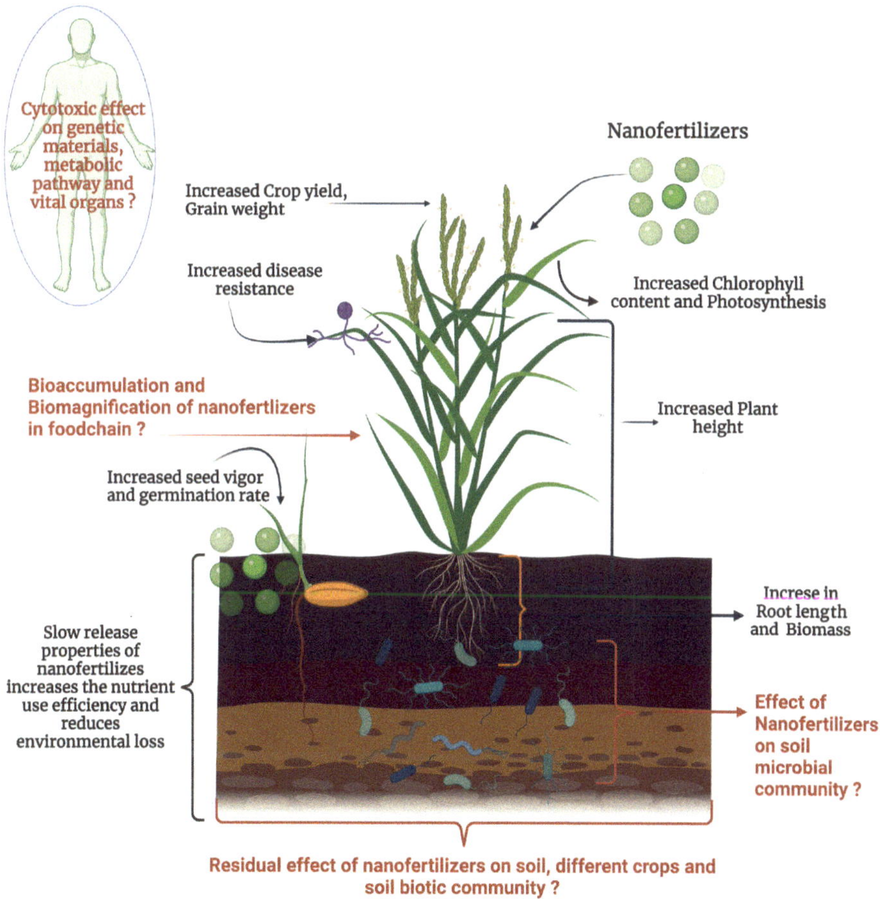

Fig. 13.1 Illustration of the positive impacts of nanofertilizers is represented in black color in the figure and the effect which needs evaluation before successful commercial and field-scale applications is represented in the red color fonts.

Table 13.2 Effect of different types of nanoparticles and their size on crop growth.

Nanoparticle	Size (nm)	Crop	Effect	Reference
Titanium dioxide	–	Spinach (*Spinacia oleracea*)	• Photosynthesis rate increased ~3 times, • chlorophyll-a (~45%) and chlorophyll-b (~28%) were increased	Zheng et al. (2005)
Iron and copper	7.5–20.5	Potato (*Solanum tuberosum*)	The weight of sprouts was increased by 50% with Fe and it was not significant with Cu	Chalenko et al. (2010)
Multi-walled carbon nanotubes	30	Mustard (*Brassica juncea*)	Seed germination was increased by 99%	Mondal et al. (2011)
Carbon nanotubes	10–30	Gram (*Cicer arietinum*)	Water absorption was increased by 50%	Tripathi et al. (2011)
Zinc oxide	25	Peanut (*Arachis hypogaea*)	Crop yield was increased by 25%–30%	Prasad et al. (2012)
Zinc oxide, Iron oxide and Zinc/iron/copper oxide	20, 100, 40	Mung (*Vigna radiata*)	Root and shoot biomass ware increased by 40% and 44% with ZnO, 68%, and 48% with FeO, 42%, and 84% with Zn-Fe-Cu, respectively	Dhoke et al. (2013)
Hydroxyapatite	<200	Rice (*Oryza Sativa*)	The sugar level in rice straw and rice husk was reduced by 21%–41%	Dutta et al. (2014)
Iron oxide nanoparticles	20	Peanut (*A. hypogaea*)	Increased root length, plant height, biomass, and chlorophyll content of peanut plants	Rui et al. (2016)

phosphate nanogel fertilizer has improved the germination in *Oryza Sativa*, *Arachis hypogea*, and *Amaranthus spinosus* (Dhir, 2021). Nanofertilizer increases the water permeability of seed and vigor of seed. High germination of peanut seeds and root growth have been observed after the application of nano-ZnO (Prasad et al., 2012). Application of nanoparticles also improves the plant growth and yield in different crops including watermelon (Acharya et al., 2020), mustard (Arora et al., 2012), maize (Choudhary et al., 2019; Van Nguyen et al., 2021), wheat (Razzaq et al., 2015), millet (Sathiyabama and Manikandan, 2021), fenugreek (Sadak, 2019), etc.

13.7 Summary of few recent studies

- Van Nguyen et al. (2021) reported the application of Cu nanoparticles to improve the anthocyanin, chlorophyll, and carotenoid content in maize under drought conditions.
- Kumaraswamy et al. (2021) reported that seeds of maize primed with chitosan-silicon nanofertilizer (Cu-Si-NF) at 0.04%–0.12% improve the seedling vigor by 3.7-fold, and application of Cu-Si-NF at a rate of 0.08% increased the yield/plot by 43.40%.
- In a study, Ramírez-Rodríguez et al. (2020a, 2020b) showed comparable growth rate of durum wheat (kernel weight) with 50% of nano-U (nano-urea) (75 kg of N ha^{-1}) fertilizer as compared to 100% of conventional fertilizer (150 N ha^{-1}) in a controlled chamber.
- Naseem et al. (2020) demonstrated that nanocomposites of zinc aluminosilicate ($ZnAl_2Si_{10}O_{24}$) loaded with urea provide delayed release of nutrients and also showed a significantly higher yield in *Oryza sativa* L. compared to commercial urea.
- In a study by Al-juthery et al. (2019) in Husseiniya Province of Babylon, found that foliar application of nano-chelate silicon fertilizers, nano-chelated potato specific fertilizers, and nano-chelated complete micro fertilizers at a rate of 2 kg ha^{-1} increases starch content (control = 12.22%, NF = 17.10%), crude protein (C = 7.79%, NF = 9.08%), and ascorbic acid concentration (C = 136.33 mg kg^{-1} NF = 185.33 mg kg^{-1}) in potatoes. They also found that combination effect of the three nanofertilizer increased the vegetative yield (control = 2.24, NF = 2.90 Mg ha^{-1}).
- Choudhary et al. (2017) showed that the application of Cu-chitosan nanoparticles can control Curvularia leaf spot disease in maize, in a controlled environment at 0.04%–0.16% concentration and in the field condition at 0.12%–0.16% concentration. The study also reported that the application of Cu-chitosan improved plant height, diameter, root length, root number, and chlorophyll content in the pot experiment. In the field experiment, Cu-chitosan improved the plant height, ear length, ear weight, grain yield, and 100-grain weight.
- Tarafdar et al. (2014) reported that application of ZnO of size 15–25 nm in pearl millet has increased the shoot length (15.1%), root length (4.2%), root area (24.2%), chlorophyll content (24.4%), total soluble leaf protein (38.7%), plant dry biomass (12.5%), and enzyme activity of acid phosphatase (76.9%), alkaline phosphatase (61.7%), phytase (322.2%), and dehydrogenase (21%).

13.8 Environmental and human health impacts

The application of nanofertilizers in agriculture has significantly improved the production, for their efficient nutrient release mechanism and easy penetration to soil profile and root system. However, several studies indicate the agglomerating nature of nanoparticles, which eventually reduces their efficacy and promotes the formation of reactive oxygen species. To avoid agglomeration and provide stabilization, chemical compounds such as chitosan, oleyl-amine, gluconic acid, cellulose, polyethylene glycol, poly-N-vinyl-2-pyrrolidone (PVP) are used. Studies reported that continuous use of these compounds can have major impacts on human health and the environment. Depending on the size and surface reactivity, nanofertilizers can be toxic for human body for their penetration efficiency through different efflux pumps. Nanofertilizers can penetrate through the skin and via the respiratory, gastrointestinal, and genital routes. A study by Larese Filon et al. (2015) showed that nanoparticles <4 nm can penetrate and permeate intact skin, and particles in size 4–20 nm can potentially damage the skin. Agglomeration of the nanoparticle is considered as one of the toxic phenomena. A study by Braydich-Stolle et al. (2005) showed the cytotoxic nature of silver nanoparticles in mammalian germline stem cells compared to the no-significant effect of corresponding salts. Although there is no direct report on human disease, which has been linked to the nanoparticle. However, the risk associated with the application of nanoparticles or nano-sized fertilizers must be evaluated for safe application. There are several lab-scale studies which indicate the potential effect of nano-sized particles on their cellular penetration ability and interaction with biomolecular structures like DNA, protein, ribosomes. A multidisciplinary human and environmental risk assessment study is needed for ethical, social, and political consideration of safe application. A systematic quantitative evaluation of ecosystem effect including the change in macro-microflora soil and water community after the application of nanofertilizers and their residual effects must be evaluated before commercialization of nanofertilizers.

13.9 Conclusion and future perspectives

Nanofertilizers are much more beneficial compared to traditional fertilizers for their unique functional and structural characteristics like slow-release nutrient curve, precise delivery method, and less environmental loss. However, in the future, research must be carried out to understand the safety and toxicity of the application of nanofertilizers in the agricultural system. There is so much uncertainty in the interaction of nanofertilizers with the environment and the impact on human beings. Understanding the applicability of a certain nanofertilizer for a particular crop and soil is important. A lifecycle assessment is necessary to evaluate the impact of nanofertilizers on human health and the environment. Some of the major concerns regarding the application, which needs to be addressed before field-scale applications are:

- **(I)** Field-scale investigation into the impact of nanofertilizers on the environment in addition to laboratory-scale experiments.
- **(II)** Scientific safety assessment and metabolic fates of nanofertilizer application.
- **(III)** Understanding the transport, translocation, and fate of nanofertilizers in the food chain including residual effects on plant, soil, and environment.
- **(IV)** Standardize the dose–response curve specific to specific crops and agroecosystem.

(V) Understanding the effect of different doses of nanofertilizers on biotic communities and bioaccumulation, biomagnification properties.
(VI) Cost–benefit study should be carried out to analyze the real economic improvements.
(VII) Understanding the potential impact of nanofertilizers and development of mitigation methods to tackle the accidental release of nanofertilizers during manufacturing, use, and disposal.
(VIII) Establishing regulatory laws for field-scale application for different nanofertilizers.
(IX) Development of educational and outreach program along with the research program to bridge the gap between science and farm.

References

Acharya, P., Jayaprakasha, G.K., Crosby, K.M., Jifon, J.L., Patil, B.S., 2020. Nanoparticle-mediated seed priming improves germination, growth, yield, and quality of watermelons (*Citrullus lanatus*) at multi-locations in Texas. Sci. Rep. 10, 5037. https://doi.org/10.1038/s41598-020-61696-7.

Ahmad, A., Mukherjee, P., Mandal, D., Senapati, S., Khan, M.I., Kumar, R., Sastry, M., 2002. Enzyme mediated extracellular synthesis of CdS nanoparticles by the fungus, Fusarium oxysporum. J. Am. Chem. Soc. 124, 12108–12109. https://doi.org/10.1021/ja027296o.

Al-juthery, H.W.A., Al-taee, R.A.H.G., Al-Obaidi, Z.H.H., Ali, E.A.H.M., Nal-Shami, Q.M., 2019. Influence of foliar application of some nano-fertilizers in growth and yield of potato under drip irrigation. J. Phys. Conf. Ser. 1294. https://doi.org/10.1088/1742-6596/1294/9/092024, 092024.

Alshaal, T., El-Ramady, H., 2017. Foliar application: from plant nutrition to biofortification. Env. Biodivers. Soil Secur. 1, 71–83. https://doi.org/10.21608/jenvbs.2017.1089.1006.

Ankamwar, B., Damle, C., Ahmad, A., Sastry, M., 2005. Biosynthesis of gold and silver nanoparticles using Emblica officinalis fruit extract, their phase transfer and transmetallation in an organic solution. J. Nanosci. Nanotechnol. 5, 1665–1671. https://doi.org/10.1166/jnn.2005.184.

Armendariz, V., Herrera, I., Peralta-videa, J.R., Jose-yacaman, M., Troiani, H., Santiago, P., Gardea-Torresdey, J.L., 2004. Size controlled gold nanoparticle formation by *Avena sativa* biomass: use of plants in nanobiotechnology. J. Nanopart. Res. 6, 377–382. https://doi.org/10.1007/s11051-004-0741-4.

Arora, S., Sharma, P., Kumar, S., Nayan, R., Khanna, P.K., Zaidi, M.G.H., 2012. Gold-nanoparticle induced enhancement in growth and seed yield of *Brassica juncea*. Plant Growth Regul. 66, 303–310. https://doi.org/10.1007/s10725-011-9649-z.

Bansal, V., Rautaray, D., Ahmad, A., Sastry, M., 2004. Biosynthesis of zirconia nanoparticles using the fungus Fusarium oxysporum. J. Mater. Chem. 14, 3303. https://doi.org/10.1039/b407904c.

Bansal, V., Rautaray, D., Bharde, A., Ahire, K., Sanyal, A., Ahmad, A., Sastry, M., 2005. Fungus-mediated biosynthesis of silica and titania particles. J. Mater. Chem. 15, 2583–2589. https://doi.org/10.1039/B503008K.

Barbarick, K.A., Lai, T.M., Eberl, D.D., 1990. Exchange fertilizer (phosphate rock plus ammonium-zeolite) effects on Sorghum-Sudangrass. Soil Sci. Soc. Am. J. 54, 911–916. https://doi.org/10.2136/sssaj1990.03615995005400030050x.

Benelmekki, M., 2015. Designing Hybrid Nanoparticles. Morgan & Claypool Publishers.

Bhainsa, K.C., D'Souza, S.F., 2006. Extracellular biosynthesis of silver nanoparticles using the fungus aspergillus fumigatus. Colloids Surf. B Biointerfaces 47, 160–164. https://doi.org/10.1016/j.colsurfb.2005.11.026.

Braydich-Stolle, L., Hussain, S., Schlager, J.J., Hofmann, M.-C., 2005. In vitro cytotoxicity of nanoparticles in mammalian germline stem cells. Toxicol. Sci. 88, 412–419. https://doi.org/10.1093/toxsci/kfi256.

Butt, B.Z., Naseer, I., 2020. Nanofertilizers. In: Javad, S. (Ed.), Nanoagronomy. Springer International Publishing, Cham, pp. 125–152, https://doi.org/10.1007/978-3-030-41275-3_8.

Cele, T., 2020. Preparation of nanoparticles. In: Engineered Nanomaterials—Health and Safety. IntechOpen, https://doi.org/10.5772/intechopen.90771.

Chalenko, G.I., Gerasimova, N.G., Vasyukova, N.I., Ozeretskovskaya, O.L., Kovalenko, L. V., Folmanis, G.E., Tananaev, I.G., 2010. Metal nanoparticles induce germination and wound healing in potato tubers. Dokl. Biol. Sci. 435, 428–430. https://doi.org/10.1134/s0012496610060165.

Chandran, S.P., Chaudhary, M., Pasricha, R., Ahmad, A., Sastry, M., 2006. Synthesis of gold nanotriangles and silver nanoparticles using aloevera plant extract. Biotechnol. Prog. 22, 577–583. https://doi.org/10.1021/bp0501423.

Choudhary, R.C., Kumaraswamy, R.V., Kumari, S., Sharma, S.S., Pal, A., Raliya, R., Biswas, P., Saharan, V., 2017. Cu-chitosan nanoparticle boost defense responses and plant growth in maize (*Zea mays* L.). Sci. Rep. 7, 9754. https://doi.org/10.1038/s41598-017-08571-0.

Choudhary, R.C., Kumaraswamy, R.V., Kumari, S., Sharma, S.S., Pal, A., Raliya, R., Biswas, P., Saharan, V., 2019. Zinc encapsulated chitosan nanoparticle to promote maize crop yield. Int. J. Biol. Macromol. 127, 126–135. https://doi.org/10.1016/j.ijbiomac.2018.12.274.

Chugh, G., Siddique, K.H.M., Solaiman, Z.M., 2021. Nanobiotechnology for agriculture: smart technology for combating nutrient deficiencies with nanotoxicity challenges. Sustainability 13, 1781. https://doi.org/10.3390/su13041781.

de França Bettencourt, G.M., Degenhardt, J., Zevallos Torres, L.A., de Andrade Tanobe, V.O., Soccol, C.R., 2020. Green biosynthesis of single and bimetallic nanoparticles of iron and manganese using bacterial auxin complex to act as plant bio-fertilizer. Biocatal. Agric. Biotechnol. 30, 101822. https://doi.org/10.1016/j.bcab.2020.101822.

Dhir, B., 2021. Nanofertilizers and their applications. In: Kumar, R., Kumar, R., Kaur, G. (Eds.), New Frontiers of Nanomaterials in Environmental Science. Springer, Singapore, pp. 229–241, https://doi.org/10.1007/978-981-15-9239-3_10.

Dhoke, S.K., Mahajan, P., Kamble, R., Khanna, A., 2013. Effect of nanoparticles suspension on the growth of mung (*Vigna radiata*) seedlings by foliar spray method. Nanotechnol. Dev. 3, e1. https://doi.org/10.4081/nd.2013.e1.

Dutta, N., Mukhopadhyay, A., Dasgupta, A.K., Chakrabarti, K., 2014. Improved production of reducing sugars from rice husk and rice straw using bacterial cellulase and xylanase activated with hydroxyapatite nanoparticles. Bioresour. Technol. 153, 269–277. https://doi.org/10.1016/j.biortech.2013.12.016.

Dwivedi, S., Saquib, Q., Al-Khedhairy, A.A., Musarrat, J., 2016. Understanding the role of nanomaterials in agriculture. In: Singh, D.P., Singh, H.B., Prabha, R. (Eds.), Microbial Inoculants in Sustainable Agricultural Productivity. Functional Applications, vol. 2. Springer India, New Delhi, pp. 271–288, https://doi.org/10.1007/978-81-322-2644-4_17.

Eichert, T., Kurtz, A., Steiner, U., Goldbach, H.E., 2008. Size exclusion limits and lateral heterogeneity of the stomatal foliar uptake pathway for aqueous solutes and water-suspended nanoparticles. Physiol. Plant. 134, 151–160. https://doi.org/10.1111/j.1399-3054.2008.01135.x.

Fleischer, A., O'Neill, M.A., Ehwald, R., 1999. The pore size of non-graminaceous plant cell walls is rapidly decreased by borate ester cross-linking of the pectic polysac charide rhamnogalacturonan II. Plant Physiol. 121, 829–838. https://doi.org/10.1104/pp.121.3.829.

Husseiny, M.I., El-Aziz, M.A., Badr, Y., Mahmoud, M.A., 2007. Biosynthesis of gold nanoparticles using Pseudomonas aeruginosa. Spectrochim. Acta A Mol. Biomol. Spectrosc. 67, 1003–1006. https://doi.org/10.1016/j.saa.2006.09.028.

Joerger, R., Klaus, T., Granqvist, C.G., 2000. Biologically produced silver–carbon composite materials for optically functional thin-film coatings. Adv. Mater. 12, 407–409. https://doi.org/10.1002/(SICI)1521-4095(200003)12:6<407::AID-ADMA407>3.0.CO;2-O.

Khan, M.Z.H., Islam, M.R., Nahar, N., Al-Mamun, M.R., Khan, M.A.S., Matin, M.A., 2021. Synthesis and characterization of nanozeolite based composite fertilizer for sustainable release and use efficiency of nutrients. Heliyon 7. https://doi.org/10.1016/j.heliyon.2021.e06091, e06091.

Kottegoda, N., Munaweera, I., Madusanka, N., Karunaratne, V., 2011. A green slow-release fertilizer composition based on urea-modified hydroxyapatite nanoparticles encapsulated wood. Curr. Sci. 101, 73–78.

Kubavat, D., Trivedi, K., Vaghela, P., Prasad, K., Anand, G.K.V., Trivedi, H., Patidar, R., Chaudhari, J., Andhariya, B., Ghosh, A., 2020. Characterization of a chitosan-based sustained release nanofertilizer formulation used as a soil conditioner while simultaneously improving biomass production of *Zea mays* L. Land Degrad. Dev. 31, 2734–2746. https://doi.org/10.1002/ldr.3629.

Kumaraswamy, R.V., Saharan, V., Kumari, S., Chandra Choudhary, R., Pal, A., Sharma, S.S., Rakshit, S., Raliya, R., Biswas, P., 2021. Chitosan-silicon nanofertilizer to enhance plant growth and yield in maize (*Zea mays* L.). Plant Physiol. Biochem. 159, 53–66. https://doi.org/10.1016/j.plaphy.2020.11.054.

Larese Filon, F., Mauro, M., Adami, G., Bovenzi, M., Crosera, M., 2015. Nanoparticles skin absorption: new aspects for a safety profile evaluation. Regul. Toxicol. Pharmacol. 72, 310–322. https://doi.org/10.1016/j.yrtph.2015.05.005.

Lefèvre, C.T., Menguy, N., Abreu, F., Lins, U., Pósfai, M., Prozorov, T., Pignol, D., Frankel, R. B., Bazylinski, D.A., 2011. A cultured Greigite-producing magnetotactic bacterium in a novel group of sulfate-reducing bacteria. Science 334, 1720–1723. https://doi.org/10.1126/science.1212596.

Li, Z., Zhang, Y., 2010. Use of surfactant-modified zeolite to carry and slowly release sulfate. Desalin. Water Treat. 21, 73–78. https://doi.org/10.5004/dwt.2010.1226.

Manikandan, A., Subramanian, K.S., 2014. Fabrication and characterisation of nanoporous zeolite based N fertilizer. Afr. J. Agric. Res. 9, 276–284. https://doi.org/10.5897/AJAR2013.8236.

Mejias, J.H., Salazar, F., Pérez Amaro, L., Hube, S., Rodriguez, M., Alfaro, M., 2021. Nanofertilizers: a cutting-edge approach to increase nitrogen use efficiency in grasslands. Front. Environ. Sci. 9. https://doi.org/10.3389/fenvs.2021.635114.

Mondal, A., Basu, R., Das, S., Nandy, P., 2011. Beneficial role of carbon nanotubes on mustard plant growth: an agricultural prospect. J. Nanopart. Res. 13, 4519. https://doi.org/10.1007/s11051-011-0406-z.

Mukherjee, P., Senapati, S., Mandal, D., Ahmad, A., Khan, M.I., Kumar, R., Sastry, M., 2002. Extracellular synthesis of gold nanoparticles by the fungus Fusarium oxysporum. ChemBioChem 3, 461–463. https://doi.org/10.1002/1439-7633(20020503)3:5<461::AID-CBIC461>3.0.CO;2-X.

Naseem, F., Zhi, Y., Farrukh, M.A., Hussain, F., Yin, Z., 2020. Mesoporous ZnAl 2 Si 10 O 24 nanofertilizers enable high yield of Oryza sativa L. Sci. Rep. 10, 10841. https://doi.org/10.1038/s41598-020-67611-4.

Nido, P.J., Migo, V., Maguyon-Detras, M.C., Alfafara, C., 2019. Process optimization potassium nanofertilizer production via ionotropic pre-gelation using alginate-chitosan carrier. MATEC Web Conf. 268. https://doi.org/10.1051/matecconf/201926805001, 05001.

Ponmurugan, P., Manjukarunambika, K., Elango, V., Gnanamangai, B.M., 2016. Antifungal activity of biosynthesised copper nanoparticles evaluated against red root-rot disease in tea plants. J. Exp. Nanosci. 11, 1019–1031. https://doi.org/10.1080/17458080.2016.1184766.

Prasad, T.N.V.K.V., Sudhakar, P., Sreenivasulu, Y., Latha, P., Munaswamy, V., Reddy, K.R., Sreeprasad, T.S., Sajanlal, P.R., Pradeep, T., 2012. Effect of nanoscale zinc oxide particles on the germination, growth and yield of peanut. J. Plant Nutr. 35, 905–927. https://doi.org/10.1080/01904167.2012.663443.

Priyam, A., Das, R.K., Schultz, A., Singh, P.P., 2019. A new method for biological synthesis of agriculturally relevant nanohydroxyapatite with elucidated effects on soil bacteria. Sci. Rep. 9, 15083. https://doi.org/10.1038/s41598-019-51514-0.

Raliya, R., Saharan, V., Dimkpa, C., Biswas, P., 2018. Nanofertilizer for precision and sustainable agriculture: current state and future perspectives. J. Agric. Food Chem. 66, 6487–6503. https://doi.org/10.1021/acs.jafc.7b02178.

Ramírez-Rodríguez, G.B., Dal Sasso, G., Carmona, F.J., Miguel-Rojas, C., Pérez-de-Luque, A., Masciocchi, N., Guagliardi, A., Delgado-López, J.M., 2020a. Engineering biomimetic calcium phosphate nanoparticles: a green synthesis of slow-release multinutrient (NPK) nanofertilizers. ACS Appl. Bio Mater. 3, 1344–1353. https://doi.org/10.1021/acsabm.9b00937.

Ramírez-Rodríguez, G.B., Miguel-Rojas, C., Montanha, G.S., Carmona, F.J., Dal Sasso, G., Sillero, J.C., Skov Pedersen, J., Masciocchi, N., Guagliardi, A., Pérez-de-Luque, A., Delgado-López, J.M., 2020b. Reducing nitrogen dosage in *Triticum durum* plants with urea-doped nanofertilizers. Nanomaterials 10, 1043. https://doi.org/10.3390/nano10061043.

Razzaq, A., Ammara, R., Jhanzab, H.M., Mahmood, T., Hafeez, A., Hussain, S., 2015. A novel nanomaterial to enhance growth and yield of wheat. J. Nanosci. Nanotechnol. 2, 55–58.

Ribeiro, C., Carmo, M., 2019. Why nonconventional materials are answers for sustainable agriculture. MRS Energy Sustain. 6. https://doi.org/10.1557/mre.2019.7.

Rui, M., Ma, C., Hao, Y., Guo, J., Rui, Y., Tang, X., Zhao, Q., Fan, X., Zhang, Z., Hou, T., Zhu, S., 2016. Iron oxide nanoparticles as a potential iron fertilizer for peanut (Arachis hypogaea). Front. Plant Sci. 7. https://doi.org/10.3389/fpls.2016.00815.

Ruiz-Cañas, M.C., Quintero, H.I., Corredor, L.M., Manrique, E., Romero Bohórquez, A.R., 2020. New nanohybrid based on hydrolyzed polyacrylamide and silica nanoparticles: morphological, structural and thermal properties. Polymers 12, 1152. https://doi.org/10.3390/polym12051152.

Sabir, A., Yazar, K., Sabir, F., Kara, Z., Yazici, M.A., Goksu, N., 2014. Vine growth, yield, berry quality attributes and leaf nutrient content of grapevines as influenced by seaweed extract (Ascophyllum nodosum) and nanosize fertilizer pulverizations. Sci. Hortic. 175, 1–8. https://doi.org/10.1016/j.scienta.2014.05.021.

Sadak, M.S., 2019. Impact of silver nanoparticles on plant growth, some biochemical aspects, and yield of fenugreek plant (*Trigonella foenum-graecum*). Bull. Natl. Res. Cent. 43, 38. https://doi.org/10.1186/s42269-019-0077-y.

Sathiyabama, M., Manikandan, A., 2021. Foliar application of chitosan nanoparticle improves yield, mineral content and boost innate immunity in finger millet plants. Carbohydr. Polym. 258, 117691. https://doi.org/10.1016/j.carbpol.2021.117691.

Schabes-Retchkiman, P.S., Canizal, G., Herrera-Becerra, R., Zorrilla, C., Liu, H.B., Ascencio, J.A., 2006. Biosynthesis and characterization of Ti/Ni bimetallic nanoparticles. Opt. Mater. 29, 95–99. https://doi.org/10.1016/j.optmat.2006.03.014. Selected papers from the Second Topical Meeting on Nanostructured Materials and Nanotechnology NANOTECH-2005.

Schwab, F., Zhai, G., Kern, M., Turner, A., Schnoor, J.L., Wiesner, M.R., 2016. Barriers, pathways and processes for uptake, translocation and accumulation of nanomaterials in plants—critical review. Nanotoxicology 10, 257–278. https://doi.org/10.3109/17435390.2015.1048326.

Shalaby, T.A., Bayoumi, Y., Abdalla, N., Taha, H., Alshaal, T., Shehata, S., Amer, M., Domokos-Szabolcsy, É., El-Ramady, H., 2016. Nanoparticles, soils, plants and sustainable agriculture. In: Ranjan, S., Dasgupta, N., Lichtfouse, E. (Eds.), Nanoscience in Food and Agriculture 1, Sustainable Agriculture Reviews. Springer International Publishing, Cham, pp. 283–312, https://doi.org/10.1007/978-3-319-39303-2_10.

Shang, Y., Hasan, M.K., Ahammed, G.J., Li, M., Yin, H., Zhou, J., 2019. Applications of nanotechnology in plant growth and crop protection: a review. Molecules 24. https://doi.org/10.3390/molecules24142558.

Shankar, S.S., Ahmad, A., Pasricha, R., Sastry, M., 2003. Bioreduction of chloroaurate ions by geranium leaves and its endophytic fungus yields gold nanoparticles of different shapes. J. Mater. Chem. 13, 1822–1826. https://doi.org/10.1039/B303808B.

Shankar, S.S., Rai, A., Ahmad, A., Sastry, M., 2004. Rapid synthesis of Au, Ag, and bimetallic Au core–Ag shell nanoparticles using neem (Azadirachta indica) leaf broth. J. Colloid Interface Sci. 275, 496–502. https://doi.org/10.1016/j.jcis.2004.03.003.

Shebl, A., Hassan, A.A., Salama, D.M., Abd El-Aziz, M.E., Abd Elwahed, M.S.A., 2019. Green synthesis of nanofertilizers and their application as a foliar for Cucurbita pepo L. J. Nanomater. 2019. https://doi.org/10.1155/2019/3476347, e3476347.

Singh, A., Singh, N.B., Afzal, S., Singh, T., Hussain, I., 2018. Zinc oxide nanoparticles: a review of their biological synthesis, antimicrobial activity, uptake, translocation and biotransformation in plants. J. Mater. Sci. 53, 185–201. https://doi.org/10.1007/s10853-017-1544-1.

Subramanian, K.S., Manikandan, A., Thirunavukkarasu, M., Rahale, C.S., 2015. Nanofertilizers for balanced crop nutrition. In: Rai, M., Ribeiro, C., Mattoso, L., Duran, N. (Eds.), Nanotechnologies in Food and Agriculture. Springer International Publishing, Cham, pp. 69–80, https://doi.org/10.1007/978-3-319-14024-7_3.

Tarafdar, J.C., Raliya, R., Mahawar, H., Rathore, I., 2014. Development of zinc nanofertilizer to enhance crop production in pearl millet (Pennisetum americanum). Agric. Res. 3, 257–262. https://doi.org/10.1007/s40003-014-0113-y.

Tripathi, S., Kumar Sonkar, S., Sarkar, S., 2011. Growth stimulation of gram (Cicer arietinum) plant by water soluble carbon nanotubes. Nanoscale 3, 1176–1181. https://doi.org/10.1039/C0NR00722F.

Van Nguyen, D., Nguyen, H.M., Le, N.T., Nguyen, K.H., Nguyen, H.T., Le, H.M., Nguyen, A. T., Dinh, N.T.T., Hoang, S.A., Van Ha, C., 2021. Copper nanoparticle application enhances plant growth and grain yield in maize under drought stress conditions. J. Plant Growth Regul. https://doi.org/10.1007/s00344-021-10301-w.

Wang, W.-N., Tarafdar, J., Biswas, P., 2013. Nanoparticle synthesis and delivery by an aerosol route for watermelon plant foliar uptake. J. Nanopart. Res. 15. https://doi.org/10.1007/s11051-013-1417-8.

Wanyika, H., Gatebe, E., Kioni, P., Tang, Z., Gao, Y., 2012. Mesoporous silica nanoparticles carrier for urea: potential applications in agrochemical delivery systems. J. Nanosci. Nanotechnol. 12, 2221–2228. https://doi.org/10.1166/jnn.2012.5801.

Weeks, J.J., Hettiarachchi, G.M., 2019. A review of the latest in phosphorus fertilizer technology: possibilities and pragmatism. J. Environ. Qual. 48, 1300–1313. https://doi.org/10.2134/jeq2019.02.0067.

Yong, P., Rowson, N.A., Farr, J.P.G., Harris, I.R., Macaskie, L.E., 2002. Bioreduction and biocrystallization of palladium by Desulfovibrio desulfuricans NCIMB 8307. Biotechnol. Bioeng. 80, 369–379. https://doi.org/10.1002/bit.10369.

Yusefi-Tanha, E., Fallah, S., Rostamnejadi, A., Pokhrel, L.R., 2020. Zinc oxide nanoparticles (ZnONPs) as a novel nanofertilizer: influence on seed yield and antioxidant defense system in soil grown soybean (*Glycine max* cv. Kowsar). Sci. Total Environ. 738, 140240. https://doi.org/10.1016/j.scitotenv.2020.140240.

Zheng, L., Hong, F., Lu, S., Liu, C., 2005. Effect of nano-TiO2 on strength of naturally aged seeds and growth of spinach. Biol. Trace Elem. Res. 104, 83–91. https://doi.org/10.1385/BTER:104:1:083.

Role of engineered nanomaterials in sustainable agriculture and crop production

14

Sougata Ghosh[a,c] *and Bhavtosh Kikani*[b]
[a]Department of Microbiology, School of Science, RK University, Rajkot, Gujarat, India, [b]Department of Biological Sciences, P.D. Patel Institute of Applied Sciences, Charotar University of Science and Technology, Changa, Gujarat, India, [c]Department of Physics, Faculty of Science, Kasetsart University, Bangkok, Thailand

14.1 Introduction

Survival of humankind is largely dependent on the productive and sustainable agriculture. No doubt, the current practices ensure production of sufficient food with improved quality (Lowry et al., 2019), and this may need intervention of technological advancements to fulfill future needs (Willett et al., 2019). Inadequate agricultural resources with limited technological knowledge may further make the transformation difficult. Of course, the green revolution has significantly improved the scenario year-by-year with visible gains. For instance, the green revolution has provided 2884 kcal per capita per day (about 8×10^{15} kcal per year) in 2013 (Lowry et al., 2019). However, current agricultural practices are inherently inefficient and unsustainable and have led to unacceptable environmental degradation (Willett et al., 2019). Moreover, the degraded soil (Pimentel and Burgess, 2013) and climate change have flattened the yields of several crops. Together, enormous and nonscientific applications of pesticides and herbicides disturb the biological diversity and ecological balance in any agricultural land (Beketov et al., 2013; Goulson, 2013). All these factors may contribute adversely to the current agricultural system. To achieve an objective of sustainability with provision of quality food in sufficient quantity, a new agricultural revolution is emerging. Newer techniques, tools, and methodologies are being employed with serious directed efforts. The gene editing tool, CRISPR-Cas9 (Jaganathan et al., 2018) and applications of plant-growth-promoting bacteria (Glick, 2012) are the examples.

Engineered nanomaterials (ENMs) have the potential to lead this change. Uniquely, smaller size of ENMs (<100 nm) facilitates to cross the biological barriers easily with efficient diffusion within plant tissues. This, in turn, aids a way to deliver nutrients and pesticides ably. Several ENMs offer a regulated release of micronutrients which help in plant growth (Kah et al., 2019). Furthermore, the structure and surface attributes of the ENMs can be modified easily, which supports proper targeted delivery within plants and soils (Lowry et al., 2019).

Agricultural Nanobiotechnology. https://doi.org/10.1016/B978-0-323-91908-1.00015-8

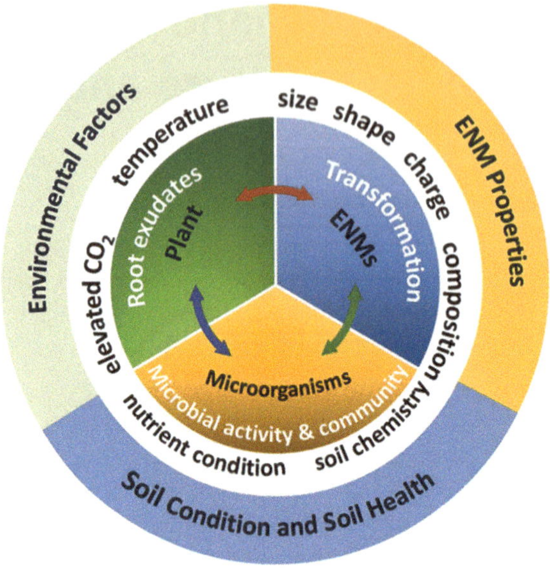

Fig. 14.1 Relationship between the plant growth and microbial diversity along with the transformation of engineered nanoparticles in the soil.
Reprinted with permission from Zhang, P., Guo, Z., Zhang, Z., Fu, H., White, J.C. and Lynch, I., 2020. Nanomaterial transformation in the soil–plant system: implications for food safety and application in agriculture. *Small*, *16*(21), p.2000705. Copyright © 2020. Wiley.

In the present chapter, we try to highlight the interactions of engineered nanoparticles with plants and soil microbiota. Furthermore, the adverse impacts of excessive use of ENMs are also discussed with an emphasis over the regulatory policies to gain maximum advantages in the agricultural sector as illustrated in Fig. 14.1 (Zhang et al., 2020).

14.2 Engineered nanomaterials (ENMs)

Nanomaterials are the one having very small size (around 100 nm) and may be naturally occurring or engineered. The ENMs are dimensionally tuned in a range to make them suitable for the commercial purposes (Berube et al., 2010). Nanomaterials are also classified based on their inherent attributes, mainly chemical nature, dimensionality, shape, and size. They are classified as organic (example, lipid micelles) or inorganic (silver or magnetite nanoparticles) based on chemical nature. They may consist of a single repeating constituent or multiple repeating constituents, also referred to as a nanocomposite. Further to add, the nanomaterials can be categorized as either 0D, 1D, 2D, or 3D, as per their dimensions. For example, quantum dots and nanodispersions have all dimensions in nanoscale and so are known as 0D nanomaterials. In comparison, nanotubes and nanowires are 1D nanomaterials. On the same note, nanofilms are referred to as 2D nanomaterials, whereas mesoporous structures and nanoclusters are classified as 3D nanomaterials (Sun et al., 2014). Based on shape, an large variety of nanomaterials are synthesized, such as nanoparticles, nanotubes, nanocubes, nanowires, nanosheets, and nanospheres (Khan, 2020).

Certain unique properties of ENMs, such as very small size and high volume to specific surface area ratio encouraged the scientific fraternity to study them in depth. Due to very small size, the nanomaterials can penetrate through the tissues quite easily. Furthermore, greater volume to surface area makes them lighter and also provides more surface for the interactions. In addition, the structures and surface properties can be easily modified as per the application requirements. The increased stabilities of ENMs against physical, chemical, and biological stress also add a value to their potential applications. In the following section, ENMs with various agricultural applications are discussed in detail.

14.3 Soil quality improvement

The soil inherently harbors a range of natural and ENMs. Natural nanomaterials include clay, iron oxides, and dead-decaying organic materials (Klaine et al., 2008). In contrast, ENMs include several candidates such as TiO_2, ZnO, SiO_2, FeOx, AlOx, CeOx, CNTs fullerenes, Ag and quantum dots, which are released into the soils during their manufacturing or applications or during disposal process (Piccinno et al., 2012). According to Keller et al. (2013), around 0.1%–2% of ENMs are released at the time of the production process. At the same time, it was reported that the ENMs used in the electronics, paper, plastic, or packaging industries are introduced to the soils during the landfilling process.

It was recently reported that approximately 63%–91% of ENMs are disposed via landfills (Keller et al., 2013). In addition, many of the wastes including biomedical wastes are treated via incineration process. The ENMs released in the air via multiple means may eventually reach the soil. Alternatively, wastewater may contain ample amount of ENMs which are localized in the soil during irrigation.

Once deposited in the soil, the fate of ENMs depends on the type of the soil and properties of nanomaterials too. For instance, the ENMs either get dissolved or may interact with the charged soil particles. Sometimes, they may get accumulated within different soil microorganisms, commonly referred to as bioaccumulation (Klaine et al., 2008). Besides, the properties of the ENMs, such as size, shape, chemical attributes, and surface properties also determine their fate in the soil. Generally, the nanomaterials undergo weathering process with some chemical transformations, aggregations, and sometimes disaggregation as well (Bundschuh et al., 2018). Furthermore, the chemical transformation process includes dissolution, sulfidation, adsorption, and desorption. The production, application, and disposal process of the ENMs are summarized in Fig. 14.2 (Ahmed et al., 2018).

ENMs can effectively be used to improve soil quality and thereby agricultural production. It is strongly advised that the sustainable agriculture should prioritize increased and cost effective food production with no or least disturbance to the natural ecosystem. Albeit healthy soil is the key component to be emphasized to achieve the objectives, in reality, the scenario is completely reverse. The current agricultural

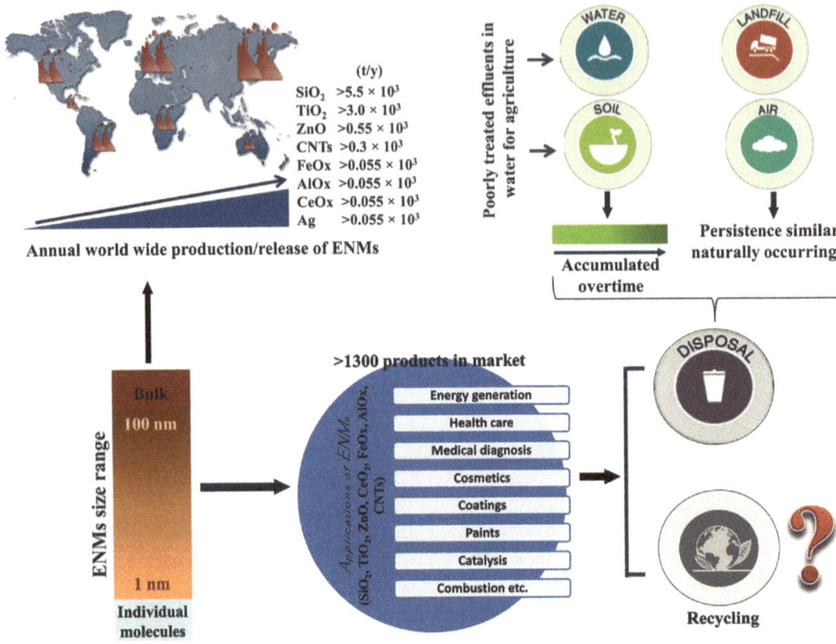

Fig. 14.2 Production, application, and disposal of ENMs—a global scenario. Reprinted with permission from Ahmed, B., Khan, M.S., Saquib, Q., Al-Shaeri, M. and Musarrat, J., 2018. Interplay between ENMs and edible plants: a current perspective. In Phytotoxicity of nanoparticles (pp. 63–102). Springer, Cham. Copyright © 2018. Springer.

practices are not completely scientific and systematic, which leads to soil deterioration and loss of productivity and damage to the ecosystem (Bandala and Berli, 2019).

ENMs are now gradually being considered as one of the key players to improve soil health in order to enhance sustainable agriculture. The properties of ENMs can efficiently be explored to achieve the goals. The properties include smaller size, larger surface area, better cation exchange capacities, prolonged nutrient release, and efficient nutrient delivery. Thul et al. (2013) reported successful use of ENMs to restore soil properties. Different ENMs are reported to exhibit direct impact on plant growth and productivity (Rai and Ingle, 2012; Mukhopadhyay, 2014; Liu and Lal, 2015). Yet, a serious concern has emerged about their relative toxicity on plants, soil microorganisms, mainly plant growth promoting rhizobacteria and human beings as well. Thus, judicious use of ENMs is strongly advocated with stringent regulatory policies.

14.4 Soil restoration

The soil may get contaminated with different pollutants frequently due to several reasons. Therefore, it is especially emphasized to restore soil properties for its extended use without disturbing its fertility. Varieties of microbial and phytoremediation approaches are discussed in past. Among, the use of ENMs has emerged as an alternate treatment in place of the conventional approaches, as sometimes these conventional approaches prove impractical and costly (Rajapaksha et al., 2015).

Zero valent ions are successfully used to achieve significant removal of metal and metalloid pollutants from the soil. They are used to achieve restoration of ground water too (Yan et al., 2013). The in situ experiments revealed their effective use to treat soil samples contaminated by various contaminants (Cecchin et al., 2017). In United States and Europe, the zero valent ions are used to remove chlorinated solvents in soil and ground water samples (Virkutyte and Varma, 2014).

Another candidate, the self-assembled monolayers on mesoporous supports (SAMMS) are also of great interest. It can remove organic pollutants from soil habitats by sequestration process and can extract heavy metals and polyaromatic hydrocarbons from liquid wastes and soil (Brandl et al., 2015).

Other ENMs are also frequently used, such as silver oxide, iron oxide, and manganese oxide, which harbor great capacity to remove metals. The properties of manganese oxides, such as better specific surface and acidic pH, make them interesting to explore further. They can be used efficiently to treat the samples containing the pollutants in the oxidized state. However, their applications become limited when the pollutants are of reduced chemical species (Watanabe et al., 2013; Villalobos et al., 2014). Furthermore, the metal-based nano-oxides are used extensively in the biogenic applications too (Zhou et al., 2015). Moreover, the concept is widely being explored at lab scale as it accounts for less concern about environmental hazards and pollutions.

Interestingly, nanoscale biopolymers are also given great attention as they are reportedly used as efficient soil conditioners (Mohammadi and Khalafi-Nezhad, 2012). On a similar note, the soil health of arid soils can be improved using a nanosized sulfonated polyaniline, collectively referred to as nSPANI. Particularly, nSPANI delays crust formation on the top soil layer. This, in turn, decreases runoff and subsequent erosion. It has exhibited moderate toxicity too (El-Din et al., 2016). Similarly, nanoclays are the soil additives, which can be used as growth enhancers for pot-based vegetation (Boroghani et al., 2011). Nanoclays are also used to stabilize sandy soil, which thereby improve its water holding capacities and reduced soil erosion (Öztaş et al., 2002). Noticeably, water saturated soil was predominantly used to study such nanotechnology-based soil restoration approaches at small scale and field level. In contrast, unsaturated soil samples should also be studied with the same research objectives (Cecchin et al., 2017; Bandala and Berli, 2019).

14.5 Suppression of plant diseases

The ENMs exhibit beneficial and harmful effects on agricultural activities (Gardea-Torresdey et al., 2014). They can control plant diseases, improve plant growth, and development (Baker et al., 2005; Kim et al., 2008; Jo et al., 2009; Nowack et al., 2011; Srilatha, 2011; Gogos et al., 2012; Mishra and Singh, 2015). On the similar line, it was also advocated that the beneficial effects largely depend on soil properties, types of the crops, and their physiology and extent of photosynthesis (Zheng et al., 2005; Gao et al., 2008). On the other hand, there are numerous reports where the ENMs inhibit plant growth (Frenk et al., 2013; Dimkpa, 2014). They are reported to interact with the soil microorganisms and interfere with the plant-pathogen interactions. For example, photocatalytic TiO_2/Ag nanomaterials decorated with sodium dodecyl sulfate were conjugated with dimethomorph as an efficient pesticide since several years (Khot et al., 2012). On another note, antagonistic effect of ENMs on plant pathogens is also widely studied. The mechanism of action is immensely dynamic. The ENMs may directly inhibit growth of the plant pathogens (Khan et al., 2018). Sometimes, they may release metal ions which in turn inhibit the growth of pathogens (Elmer and White, 2018). In addition, the ENMs are also reported to enhance growth and activities of soil microorganisms which inhibit the growth of phytopathogens (Servin et al., 2015).

Silver nanoparticles (AgNPs) have been extensively studied. Incidences of mycosis to ryegrasses were remarkably reduced with the use of AgNPs (Jo et al., 2009). Use of AgNPs inhibited fungal growth and reduced the disease in the plants as well. Recently, plant viral infections were reduced with the help of different nanomaterials, such as magnetite nanoparticles and TiO_2 by hydrolyzing the coat proteins (Hao et al., 2018). Interestingly, a mixture of ENMs including Ag, SiO_2, TiO_2, and ZnO induced anticandidal activities of *Pseudomonas* species (Khan et al., 2018). In addition, *Fusarium* species have raised some serious issues in the present day agriculture. ZnONPs could effectively control the growth of *F. graminearum* in lab scale studies, followed by field-based experiments (Dimkpa et al., 2013). The study was further supported by Savi et al. (2015), where they reported that the infection of *F. graminearum* to the wheat plants could be significantly controlled by ZnONPs without any notable phytotoxicity. In another study, a set of carbon nanomaterials was used to inhibit the growth of *F. graminearum* and *F. poae*. The mechanisms of inhibition were surface deposition over the surface of fungal spores, reduced water uptake, and cell lysis (Wang et al., 2014).

14.6 Interaction with plants and plant growth promoting bacteria

It is well evident that the nanomaterials are not extensively used in the agriculture. However, certain products are available commercially (Sekhon, 2014; Servin et al., 2015). This, in turn, indicates that the ENMs are appended in the soil to gain better yield. The nanomaterials, therefore, interfere with the plant-microbe interactions in the soil. It is advocated that the plants secrete certain chemicals or metabolites which normally interact with the soil microbiome. Hence, the presence of ENMs would affect microbial growth. Sometimes, these secreted metabolites of plants and soil microorganisms may alter chemical properties of nanomaterials and thereby affect the growth of plants and soil microbiome.

14.6.1 Direct impact on plant growth

The ENMs were reported to get absorbed from soil via plant roots. It was also proposed that they get transported to various plant parts later and thus affect plant growth immensely (Lin et al., 2009). On the same note, plant growth or inhibition depends on chemical composition, size, surface charges, and relative concentrations of ENMs in the agricultural soil (Husen and Siddiqi, 2014; Faizan et al., 2018). Plant growth is equally affected by the extent of the transportation of ENMs within different plant tissues via either apoplastic or symplastic routes (Pérez-de-Luque, 2017).

Apparently, the TiO_2NPs and multiwalled carbon nanotubes (MWCNTs) did not transport much from soil to wheat plant (Gogos et al., 2016). Numerous means have been reported via which the toxicity of nanomaterials may affect plants (Khodakovskaya et al., 2011; Servin et al., 2015). The commonly found is the oxidative stress, which was determined by increased expressions of superoxide dismutase and peroxide dismutase. Moreover, it has been reported that a mixture of carbon nanomaterials induced expression of the phytohormones. Along with this, increased activities of superoxide dismutase and peroxide dismutase were documented (Hao et al., 2018). On a similar note, the AgNPs inhibited the germination of ryegrass and flax (El-Temsah and Joner, 2012). In another study, CeO_2 was found accumulated in the rhizospheric region and was translocated in the plant tissues as well (Hernandez-Viezcas et al., 2013). It was also reported that CeO_2NPs could enhance plant growth when applied in trace concentrations (Medina-Velo et al., 2017). The CeO_2NPs were taken up by the roots which then entered into the xylem and later got translocated in the aerial parts. This was confirmed by fluorescent isothiocyanate (FITC) labeled CeO_2NPs and confocal microscopy as evident from Fig. 14.3.

On a similar note, ZnONPs decreased chlorophyll content of plants, which proved toxic to the plant growth (Wang et al., 2016). Poly(acrylic acid) nanoceria increased carbon assimilation rates by 67% in *Arabidopsis thaliana* plants. Actually, nanoceria served to protect plants against abiotic stress by scavenging free radicals (Wu et al., 2017). Evidently, the CuONPs, at very high concentrations, inhibited denitrification, nitrification, and soil respiration processes. In contrast, certain growth enhancing impacts after nanomaterial applications are also well-documented (Faizan et al., 2018; Zhu et al., 2019). Application of TiO_2NPs promoted plant growth by increasing chlorophyll and carotene contents and thereby increased rate of photosynthesis (Singh and Lee, 2016). Similarly, graphene oxide (GO) nanomaterials increased water retention ability, which, in turn, promoted seed germination (He et al., 2018). ZnONPs promoted the growth of tomato plants by providing Zn as a micronutrient (Faizan et al., 2018). A study reported improved growth of *Cicer arietinum* upon treatment with the water soluble wood pyrolysis wastes (Sonkar et al., 2012). Sometimes, ENMs serve as micronutrients and carriers for other nutrients as well (Jampílek and Král'ová, 2017). Smaller size and efficient penetration power make them quite important nutrient carriers (Pérez-de-Luque, 2017). The plant biomass was reportedly increased with higher numbers of the leaves in *A. thaliana* after application of nano-cerium (Tumburu et al., 2017).

The exposure of carbon nanotubes (CNTs) in the tobacco field altered gene expression pattern. The genes regulating cell division or water transport were upregulated in trace concentrations of CNTs. This alterations improved cell growth (Khodakovskaya et al., 2012). Even, magnetite nanoparticles also promoted plant growth and increased activities of antioxidants (Iannone et al., 2016). Albeit several such reports are available, some systematic explorations are still needed with reference to the optimum dose of ENMs used, so that they cannot prove toxic to the plant health.

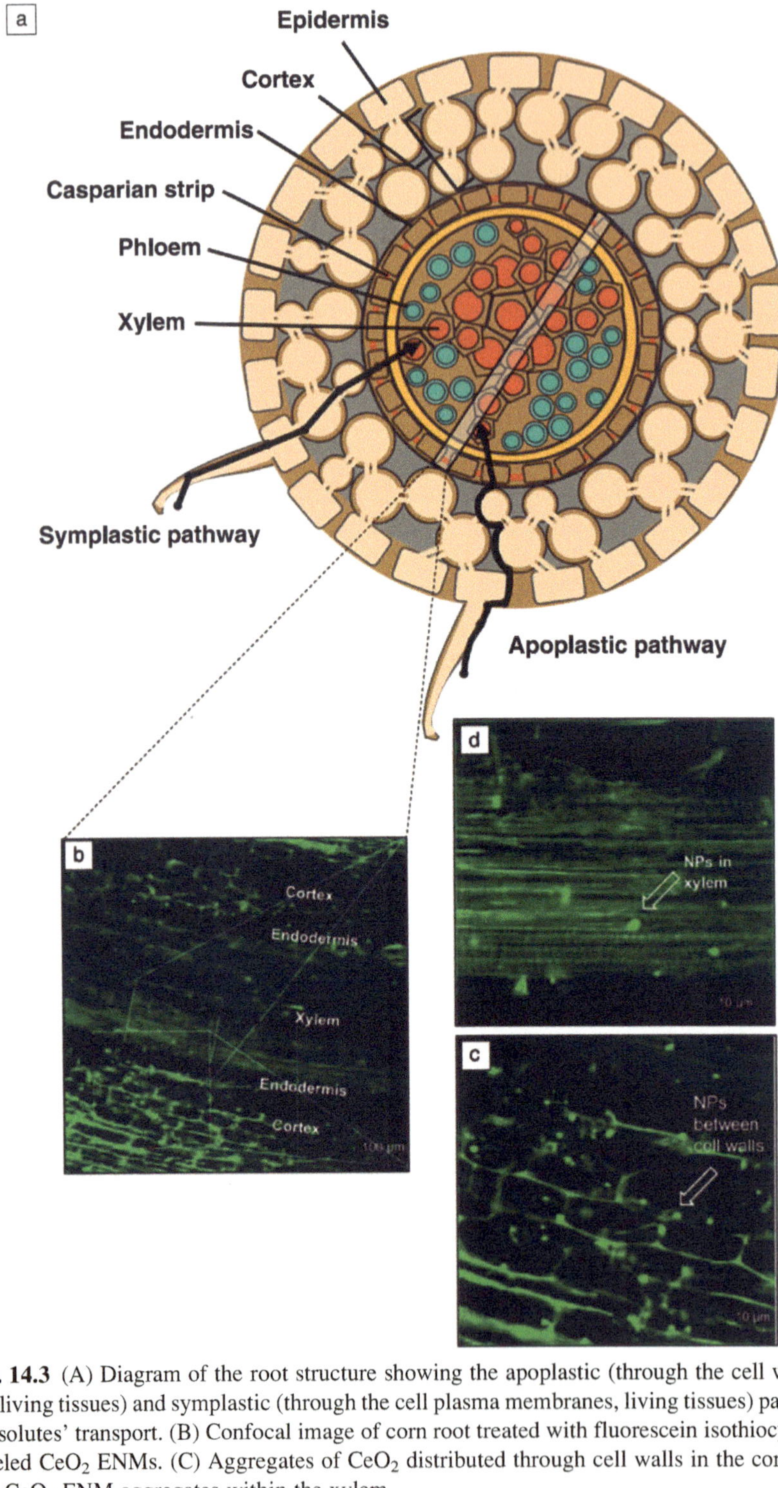

Fig. 14.3 (A) Diagram of the root structure showing the apoplastic (through the cell walls, nonliving tissues) and symplastic (through the cell plasma membranes, living tissues) pathways for solutes' transport. (B) Confocal image of corn root treated with fluorescein isothiocyanate-labeled CeO_2 ENMs. (C) Aggregates of CeO_2 distributed through cell walls in the cortex. (D) CeO_2 ENM aggregates within the xylem.

Reprinted with permission from Medina-Velo, I.A., Peralta-Videa, J.R., Gardea-Torresdey, J. L., 2017. Assessing plant uptake and transport mechanisms of ENMs from soil. MRS Bulletin, 42(5), pp. 379–384. Copyright © 2017. MRS Bulletin.

14.6.2 Effect on soil and plant microbiome

Apart from direct impact on the plant growth, the ENMs also influence it indirectly by influencing growth of soil microorganisms dwelling in the rhizosphere and even plant pathogens. Several ENMs inherently possess antimicrobial properties, which may be due to interference in their proliferation or essential metabolic pathways. This, in turn, impacts different biogeochemical cycles within the soil habitat. For instance, reduced GO inhibited activity of PS II and therefore, overall photosynthesis was adversely affected (Chen et al., 2019). Furthermore, soil microorganisms play a vital role in maintenance of different biogeochemical cycles. Thus, the use of ENMs may affect the metabolic status of these soil microorganisms. For instance, the ENMs affect the nitrogen fixation process by both free-living and symbiotic microorganisms. It was observed that TiO_2 nanoparticles inhibited growth, nitrogen fixation rate, and nitrogen assimilation properties of *Anabaena variabilis* (Cherchi and Gu, 2010).

On the same line, high concentrations of CuNPs also diminished carbon and nitrogen cycles. However, lower concentrations did not inhibit significantly (Simonin et al., 2018). Silver is known to have antimicrobial activity. The silver nanoparticles (AgNPs) inhibited growth of *Azotobacter vinelandii*, which, in turn, decreased the rate of nitrogen fixation. The modes of action include damage to the cell structure and integrity, generation of oxidative stress, and inhibition of nitrogenase activities (Zhang et al., 2018). The ENMs also influenced metabolism of PGPR and thereby have significant impact on the plant growth (Radniecki et al., 2011; Karunakaran et al., 2014; Chai et al., 2015). One of the reports compared relative toxicities of TiO_2 and ZnONPs and concluded that ZnONPs were more toxic than the former. It was also reported that the growth of nitrogen fixers and methane oxidizers reduced significantly upon the treatment with ZnONPs (Ge et al., 2011). In contrast, bacterial communities mediating decomposition of the organic pollutants were enriched in the soil. In another similar report, impact of TiO_2NPs on the microbial communities of wheat rhizosphere was evaluated. Although some of the bacterial populations altered, extent of root colonization and overall growth of the plant did not alter. However, the dynamics in the alterations of the bacterial communities were proposed to be used to determine the presence of the said nanomaterials in the soil (Moll et al., 2017). Furthermore, AgNPs exhibited inhibitory effects on ammonia oxidizing bacteria, whereas these enhanced growth of actinomycetes and bacteroides remarkably. The treatment with AgNPs reduced microbial biomass, leucine aminopeptidase activity, and nitrogen-fixing microorganisms, as well (Grün et al., 2018). Interestingly, the mixture of TiO_2 and polystyrene nanomaterials inhibited the growth of the rhizospheric soil bacteria, which further resulted in reduction of root and shoot growth (Kibbey and Strevett, 2019).

One of the recent studies reported that the soil irrigated by water containing a range of ENMs supported the growth of cyanobacteria and an unidentified group belonging to archaea (Liu et al., 2018). Same ENMs may elicit different responses in different soils or vegetation. For instance, the CNTs did not alter species' diversity of rhizosphere of the tomato cultivation (Khodakovskaya et al., 2013). On the other hand, carbon nanomaterials inhibited the rhizosphere microbiome of paddy and thereby influenced rice cultivation (Hao et al., 2018).

In contrast, it has also been reported that such ENMs in limited concentrations may promote the growth of plants and PGPR (Yuan et al., 2017; Khan et al., 2018). To support, soil was treated with MWCNTs in very low concentrations, which in turn resulted in enhanced nitrogen fixation and biomass production (Yuan et al., 2017). Similarly, treatment of the legume-*Rhizobium* symbiosis with a mixture containing single-walled carbon nanotubes (SWCNTs), multiwalled carbon nanotubes (MWCNTs), and graphene oxide (GO) increased the rate of nitrogen fixation and plant biomass (Yuan et al., 2017).

The antimicrobial activity of the ENMs is very widely studied. The mechanisms to control microbial growth include mainly cell membrane disruption, interference in the metabolic functions, protein denaturation, damage to the nucleic acids, impairment of DNA replication, formation of reactive oxygen species (ROS), and subsequent oxidative stress induction, and even lethal mutations (Khan et al., 2016). Conclusively, a serious concern has emerged upon interactions of ENMs with the soil microbiomes and plants. Many studies reported that ENMs may incur an adverse impact on microbial communities and their metabolic potentials, which has strongly advocated that further depth investigations are essentially required to understand how ENMs affect microbiomes of different types of soils. Such understanding may lead to value addition to our existing knowledge and may help in meaningful use. On the same note, a group of microbes may show higher sensitivity compared to the others. The concept can be explored to formulate ENMs to achieve better inhibition of plant pathogens with no or insignificant impact on plant growth promoting bacteria.

14.7 Regulatory policies

Many countries do not practice strict and defined regulations to use and market nanotechnology-based products. Yet, the European policy is considered as stringent and well-documented. In general, it includes a few regulations explicating particular provisions of nanomaterials. However, the concerns over the regulations on nanomaterials in the EU countries have significantly drawn attentions during recent past. This has resulted in the formulation of the EU soft laws relating to the nanoregulatory framework (Jain et al., 2018).

It is evident that insufficient knowledge about the physicochemical properties of various ENMs and few techniques for their qualitative and quantitative determination makes execution of the risk assessment approaches very difficult (Scott-Fordsmand et al., 2017; López-Moreno et al., 2018). Yet, certain quantitative techniques are frequently adopted to achieve the same objectives. Such techniques would include in vitro high-throughput screening (Collins et al., 2017), high content analysis (HCA), and proteomics analysis (Driessen et al., 2015). Additionally, computational toxicology and statistical approaches have also emerged as promising strategies to predict the toxicity. These tools can be used to foresee relevant environmental hazards as well (Chen et al., 2017). To control the toxicity level and reduce environmental hazards, the NANoREG project has been proposed to formulate the stringent regulatory policies associated with the impacts of ENMs on human health. It is a collaborative effort between the European Chemical Agency and the Organization for European Economic Cooperation and Development. The main emphasis was given to Registration, Evaluation, Authorization, and Restriction of Chemicals, shortly referred to as REACH. Furthermore, the objectives of the NAnoREG include analysis, practical guidance, and judicious applications of the REACH regulatory policies and also defining various strategies to facilitate its efficient implementation. It also includes environmentally safe synthesis of ENMs along with their fate and risk assessment in the natural environment. However, the European Joint Research Centre (JRC) published a report on the NANoREG initiative in 2017 (Gottardo et al., 2017), where it was clearly mentioned that the REACH protocol could not be completely applied to all ENMs. The reasons include inadequate detection methods, little available information about their entire fate, and inability to distinguish between the ENMs and other nanoparticles present in the soil. Therefore, more generous efforts in the qualitative and quantitative testing are needed to achieve effective outcomes.

Similarly, the United States Environmental Protection Agency (USEPA) has also been exploring potential usage of the nanomaterials and associated risk managements. According to its report in 2007, it was mentioned that the use of NMs in society increases, and it is reasonable to assume that their presence in environmental media will increase proportionately, with consequences on human and environmental exposure. Committees on the toxicity, carcinogenicity, and mutagenicity of chemicals in food, consumer products, and the environment have also identified the risk assessment of ENMs as an area of interest in their "Joint Statement on Nanomaterials Toxicology" (López-Moreno et al., 2018). The positive and negative impacts of ENMs can be summarized in Fig. 14.4.

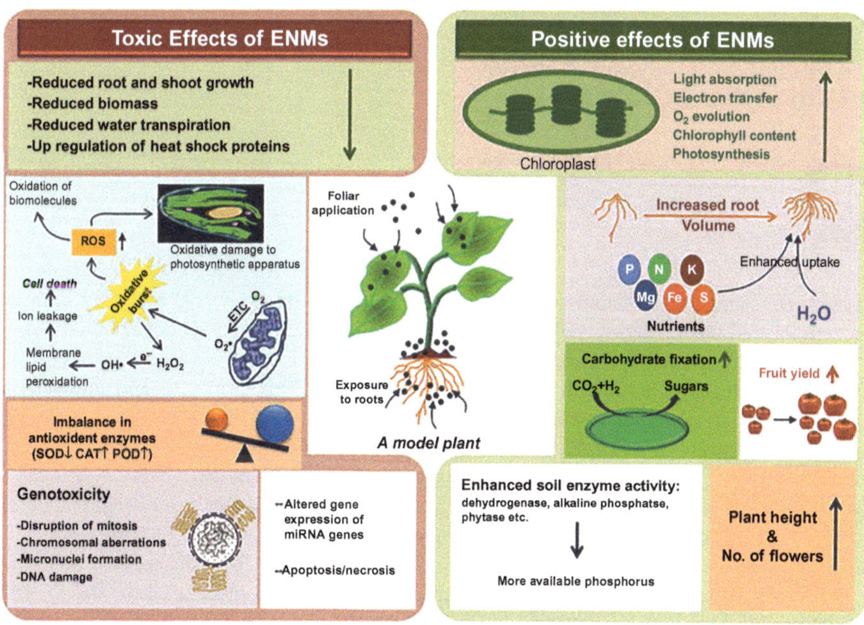

Fig. 14.4 The overall impact of ENMs on the plant growth.
Reprinted with permission from Ahmed, B., Khan, M.S., Saquib, Q., Al-Shaeri, M. and Musarrat, J., 2018. Interplay between ENMs and edible plants: a current perspective. In *Phytotoxicity of nanoparticles* (pp. 63–102). Springer, Cham. Copyright © 2018. Springer.

14.8 Conclusion and future perspectives

Current agricultural practices need goal oriented, dedicated, and systemic efforts to satisfy future demands. Along with other advanced tools and technologies, such as CRISPR Cas9, bioinformatics, and omics approaches, nanotechnology may serve a very good alternative in revolutionizing agricultural practices. The ENMs may transform the ongoing agricultural processes to more efficient and sustainable processes. The ENMs harbor various advantages inherently, mainly smaller size, higher mass to surface ratio, tunable surface properties, and morphology. The ENMs are reported to help in the soil restoration process and may also interact with the plants and soil microbiota. Albeit, such interaction may be positive and negative too. On the same line, interactions of ENMs with the plant pathogens inhibit their growth significantly. Therefore, further in-depth investigations may aid our knowledge in the same research area. The ENMs are also reported to influence the biogeochemical cycles by interfering with the microbial metabolism. Thus, certain strict government policies with their regulated optimum use may improve agricultural yields qualitatively and quantitatively.

Acknowledgment

The authors acknowledge their host institutes, School of Science, RK University, Rajkot 360020, Gujarat, India and P.D. Patel Institute of Applied Sciences, Charotar University of Science and Technology, Changa for necessary support.

Declaration of interest

Authors hereby declare no conflict of interest.

References

Ahmed, B., Khan, M.S., Saquib, Q., Al-Shaeri, M., Musarrat, J., 2018. Interplay between engineered nanomaterials (ENMs) and edible plants: a current perspective. In: Phytotoxicity of Nanoparticles. Springer, Cham, pp. 63–102.

Baker, C., Pradhan, A., Pakstis, L., Pochan, D.J., Shah, S.I., 2005. Synthesis and antibacterial properties of silver nanoparticles. J. Nanosci. Nanotechnol. 5 (2), 244–249.

Bandala, E.R., Berli, M., 2019. Engineered nanomaterials (ENMs) and their role at the nexus of food, energy, and water. Mater. Sci. Energy Technol. 2 (1), 29–40.

Beketov, M.A., Kefford, B.J., Schäfer, R.B., Liess, M., 2013. Pesticides reduce regional biodiversity of stream invertebrates. Proceedings Nat Acad. Sci. 110 (27), 11039–11043.

Berube, D.M., Searson, E.M., Morton, T.S., Cummings, C.L., 2010. Project on emerging nanotechnologies-consumer product inventory evaluated. Nanotechnol. Law Bus. 7, 152.

Boroghani, M., Mirnia, S.K., Vahhabi, J., Ahmadi, S.J., Charkhi, A., 2011. Nanozeolite synthesis and the effect of on the runoff and erosion control under rainfall simulator. Aust. J. Basic Appl. Sci. 5 (12), 1156–1164.

Brandl, F., Bertrand, N., Lima, E.M., Langer, R., 2015. Nanoparticles with photoinduced precipitation for the extraction of pollutants from water and soil. Nat. Commun. 6 (1), 1–10.

Bundschuh, M., Filser, J., Lüderwald, S., McKee, M.S., Metreveli, G., Schaumann, G.E., Schulz, R., Wagner, S., 2018. Nanoparticles in the environment: where do we come from, where do we go to? Environ. Sci. Eur. 30 (1), 1–17.

Cecchin, I., Reddy, K.R., Thomé, A., Tessaro, E.F., Schnaid, F., 2017. Nanobioremediation: integration of nanoparticles and bioremediation for sustainable remediation of chlorinated organic contaminants in soils. Intern. Biodet. Biodeg. 119, 419–428.

Chai, H., Yao, J., Sun, J., Zhang, C., Liu, W., Zhu, M., Ceccanti, B., 2015. The effect of metal oxide nanoparticles on functional bacteria and metabolic profiles in agricultural soil. Bull. Environ. Contam. Toxicol. 94 (4), 490–495.

Chen, G., Peijnenburg, W., Xiao, Y., Vijver, M.G., 2017. Current knowledge on the use of computational toxicology in hazard assessment of metallic engineered nanomaterials. Int. J. Mol. Sci. 18 (7), 1504.

Chen, L., Wang, C., Yang, S., Guan, X., Zhang, Q., Shi, M., Yang, S.T., Chen, C., Chang, X.L., 2019. Chemical reduction of graphene enhances in vivo translocation and photosynthetic inhibition in pea plants. Environ. Sci. Nano 6 (4), 1077–1088.

Cherchi, C., Gu, A.Z., 2010. Impact of titanium dioxide nanomaterials on nitrogen fixation rate and intracellular nitrogen storage in *Anabaena variabilis*. Environ. Sci. Technol. 44 (21), 8302–8307.

Collins, A.R., Annangi, B., Rubio, L., Marcos, R., Dorn, M., Merker, C., Estrela-Lopis, I., Cimpan, M.R., Ibrahim, M., Cimpan, E., Ostermann, M., 2017. High throughput toxicity screening and intracellular detection of nanomaterials. Wiley Interdiscip. Rev. Nanomed. Nanobiotechnol. 9 (1), e1413.

Dimkpa, C.O., 2014. Can nanotechnology deliver the promised benefits without negatively impacting soil microbial life? J. Basic Microbiol. 54 (9), 889–904.

Dimkpa, C.O., McLean, J.E., Britt, D.W., Anderson, A.J., 2013. Antifungal activity of ZnO nanoparticles and their interactive effect with a biocontrol bacterium on growth antagonism of the plant pathogen *Fusarium graminearum*. Biometals 26 (6), 913–924.

Driessen, M.D., Mues, S., Vennemann, A., Hellack, B., Bannuscher, A., Vimalakanthan, V., Riebeling, C., Ossig, R., Wiemann, M., Schnekenburger, J., Kuhlbusch, T.A., 2015. Proteomic analysis of protein carbonylation: a useful tool to unravel nanoparticle toxicity mechanisms. Part. Fibre Toxicol. 12 (1), 1–18.

El-Din, A.Z., Hassan, H.H.A., El-Kheir, M.M.A., Youssef, R.M., 2016. Controlling soil surface crust formation using nanosized sulfonated polyaniline. J. Soil Water Conserv. 1 (1), 1–9.

Elmer, W., White, J.C., 2018. The future of nanotechnology in plant pathology. Annu. Rev. Phytopathol. 56, 111–133.

El-Temsah, Y.S., Joner, E.J., 2012. Impact of Fe and Ag nanoparticles on seed germination and differences in bioavailability during exposure in aqueous suspension and soil. Environ. Toxicol. 27 (1), 42–49.

Faizan, M., Faraz, A., Yusuf, M., Khan, S.T., Hayat, S., 2018. Zinc oxide nanoparticle-mediated changes in photosynthetic efficiency and antioxidant system of tomato plants. Photosynthetica 56 (2), 678–686.

Frenk, S., Ben-Moshe, T., Dror, I., Berkowitz, B., Minz, D., 2013. Effect of metal oxide nanoparticles on microbial community structure and function in two different soil types. PLoS One 8 (12), e84441.

Gao, F., Liu, C., Qu, C., Zheng, L., Yang, F., Su, M., Hong, F., 2008. Was improvement of spinach growth by nano-TiO_2 treatment related to the changes of Rubisco activase? Biometals 21 (2), 211–217.

Gardea-Torresdey, J.L., Rico, C.M., White, J.C., 2014. Trophic transfer, transformation, and impact of engineered nanomaterials in terrestrial environments. Environ. Sci. Technol. 48 (5), 2526–2540.

Ge, Y., Schimel, J.P., Holden, P.A., 2011. Evidence for negative effects of TiO_2 and ZnO nanoparticles on soil bacterial communities. Environ. Sci. Technol. 45 (4), 1659–1664.

Glick, B.R., 2012. Plant growth-promoting bacteria: mechanisms and applications. Scientifica 2012. https://doi.org/10.6064/2012/963401. Article ID: 963401.

Gogos, A., Knauer, K., Bucheli, T.D., 2012. Nanomaterials in plant protection and fertilization: current state, foreseen applications, and research priorities. J. Agric. Food Chem. 60 (39), 9781–9792.

Gogos, A., Moll, J., Klingenfuss, F., van der Heijden, M., Irin, F., Green, M.J., Zenobi, R., Bucheli, T.D., 2016. Vertical transport and plant uptake of nanoparticles in a soil mesocosm experiment. J. Nanobiotechnol. 14 (1), 1–11.

Gottardo, S., Alessandrelli, M., Amenta, V., Atluri, R., Barberio, G., Bekker, C., Bergonzo, P., Bleeker, E., Booth, A., Borges, T., Buttol, P., 2017. NANoREG Framework for the Safety Assessment of Nanomaterials. European Commission Joint Research Centre.

Goulson, D., 2013. An overview of the environmental risks posed by neonicotinoid insecticides. J. Appl. Ecol. 50 (4), 977–987.

Grün, A.L., Straskraba, S., Schulz, S., Schloter, M., Emmerling, C., 2018. Long-term effects of environmentally relevant concentrations of silver nanoparticles on microbial biomass, enzyme activity, and functional genes involved in the nitrogen cycle of loamy soil. J. Environ. Sci. 69, 12–22.

Hao, Y., Ma, C., Zhang, Z., Song, Y., Cao, W., Guo, J., Zhou, G., Rui, Y., Liu, L., Xing, B., 2018. Carbon nanomaterials alter plant physiology and soil bacterial community composition in a rice-soil-bacterial ecosystem. Environ. Pollut. 232, 123–136.

He, Y., Hu, R., Zhong, Y., Zhao, X., Chen, Q., Zhu, H., 2018. Graphene oxide as a water transporter promoting germination of plants in soil. Nano Res. 11 (4), 1928–1937.

Hernandez-Viezcas, J.A., Castillo-Michel, H., Andrews, J.C., Cotte, M., Rico, C., Peralta-Videa, J.R., Ge, Y., Priester, J.H., Holden, P.A., Gardea-Torresdey, J.L., 2013. In situ synchrotron X-ray fluorescence mapping and speciation of CeO_2 and ZnO nanoparticles in soil cultivated soybean (*Glycine max*). ACS Nano 7 (2), 1415–1423.

Husen, A., Siddiqi, K.S., 2014. Carbon and fullerene nanomaterials in plant system. J. Nanobiotechnol. 12 (1), 1–10.

Iannone, M.F., Groppa, M.D., de Sousa, M.E., van Raap, M.B.F., Benavides, M.P., 2016. Impact of magnetite iron oxide nanoparticles on wheat (*Triticumaestivum* L.) development: evaluation of oxidative damage. Environ. Exp. Bot. 131, 77–88.

Jaganathan, D., Ramasamy, K., Sellamuthu, G., Jayabalan, S., Venkataraman, G., 2018. CRISPR for crop improvement: an update review. Front. Plant Sci. 9, 985.

Jain, A., Ranjan, S., Dasgupta, N., Ramalingam, C., 2018. Nanomaterials in food and agriculture: an overview on their safety concerns and regulatory issues. Crit. Rev. Food Sci. Nutr. 58 (2), 297–317.

Jampílek, J., Kráľová, K., 2017. Nanomaterials for delivery of nutrients and growth-promoting compounds to plants. In: Prasad, R., Kumar, M., Kumar, V. (Eds.), Nanotechnology. Springer, Singapore, pp. 177–226, https://doi.org/10.1007/978-981-10-4573-8_9.

Jo, Y.K., Kim, B.H., Jung, G., 2009. Antifungal activity of silver ions and nanoparticles on phytopathogenic fungi. Plant Dis. 93 (10), 1037–1043.

Kah, M., Tufenkji, N., White, J.C., 2019. Nano-enabled strategies to enhance crop nutrition and protection. Nat. Nanotechnol. 14 (6), 532–540.

Karunakaran, G., Suriyaprabha, R., Manivasakan, P., Rajendran, V., Kannan, N., 2014. Influence of nano and bulk SiO_2 and Al_2O_3 particles on PGPR and soil nutrient contents. Curr. Nanosci. 10 (4), 604–612.

Keller, A.A., McFerran, S., Lazareva, A., Suh, S., 2013. Global life cycle releases of engineered nanomaterials. J. Nanopart. Res. 15 (6), 1–17.

Khan, S.T., 2020. Interaction of engineered nanomaterials with soil microbiome and plants: their impact on plant and soil health. In: Sustainable Agriculture Reviews. vol. 41. Springer, Cham, pp. 181–199.

Khan, S.T., Ahmad, J., Ahamed, M., Jousset, A., 2018. Sub-lethal doses of widespread nanoparticles promote antifungal activity in Pseudomonas protegens CHA0. Sci. Total Environ. 627, 658–662.

Khan, S.T., Musarrat, J., Al-Khedhairy, A.A., 2016. Countering drug resistance, infectious diseases, and sepsis using metal and metal oxides nanoparticles: current status. Colloids Surf. B Biointerfaces 146, 70–83.

Khodakovskaya, M.V., De Silva, K., Biris, A.S., Dervishi, E., Villagarcia, H., 2012. Carbon nanotubes induce growth enhancement of tobacco cells. ACS Nano 6 (3), 2128–2135.

Khodakovskaya, M.V., de Silva, K., Nedosekin, D.A., Dervishi, E., Biris, A.S., Shashkov, E.V., Galanzha, E.I., Zharov, V.P., 2011. Complex genetic, photothermal, and photoacoustic analysis of nanoparticle-plant interactions. Proc. Natl. Acad. Sci. U. S. A. 108 (3), 1028–1033.

Khodakovskaya, M.V., Kim, B.S., Kim, J.N., Alimohammadi, M., Dervishi, E., Mustafa, T., Cernigla, C.E., 2013. Carbon nanotubes as plant growth regulators: effects on tomato growth, reproductive system, and soil microbial community. Small 9 (1), 115–123.

Khot, L.R., Sankaran, S., Maja, J.M., Ehsani, R., Schuster, E.W., 2012. Applications of nanomaterials in agricultural production and crop protection: a review. Crop Prot. 35, 64–70.

Kibbey, T.C., Strevett, K.A., 2019. The effect of nanoparticles on soil and rhizosphere bacteria and plant growth in lettuce seedlings. Chemosphere 221, 703–707.

Kim, H.S., Kang, H.S., Chu, G.J., Byun, H.S., 2008. Antifungal effectiveness of nanosilver colloid against rose powdery mildew in greenhouses. In: Solid State Phenomena. vol. 135. Trans Tech Publications Ltd, pp. 15–18.

Klaine, S.J., Alvarez, P.J., Batley, G.E., Fernandes, T.F., Handy, R.D., Lyon, D.Y., Mahendra, S., McLaughlin, M.J., Lead, J.R., 2008. Nanomaterials in the environment: behavior, fate, bioavailability, and effects. Environ. Toxicol. Chem. 27 (9), 1825–1851.

Lin, S., Reppert, J., Hu, Q., Hudson, J.S., Reid, M.L., Ratnikova, T.A., Rao, A.M., Luo, H., Ke, P.C., 2009. Uptake, translocation, and transmission of carbon nanomaterials in rice plants. Small 5 (10), 1128–1132.

Liu, J., Williams, P.C., Geisler-Lee, J., Goodson, B.M., Fakharifar, M., Peiravi, M., Chen, D., Lightfoot, D.A., Gemeinhardt, M.E., 2018. Impact of wastewater effluent containing aged nanoparticles and other components on biological activities of the soil microbiome, Arabidopsis plants, and earthworms. Environ. Res. 164, 197–203.

Liu, R., Lal, R., 2015. Potentials of engineered nanoparticles as fertilizers for increasing agronomic productions. Sci. Total Environ. 514, 131–139.

López-Moreno, M.L., Cassé, C., Correa-Torres, S.N., 2018. Engineered nanomaterials interactions with living plants: benefits, hazards and regulatory policies. Curr. Opin. Environ. Sci. Health 6, 36–41.

Lowry, G.V., Avellan, A., Gilbertson, L.M., 2019. Opportunities and challenges for nanotechnology in the Agri-tech revolution. Nat. Nanotechnol. 14 (6), 517–522.

Medina-Velo, I.A., Peralta-Videa, J.R., Gardea-Torresdey, J.L., 2017. Assessing plant uptake and transport mechanisms of engineered nanomaterials from soil. MRS Bull. 42 (5), 379–384.

Mishra, S., Singh, H.B., 2015. Biosynthesized silver nanoparticles as a nano-weapon against phytopathogens: exploring their scope and potential in agriculture. Appl. Microbiol. Biotechnol. 99 (3), 1097–1107.
Mohammadi, S., Khalafi-Nezhad, A., 2012, March. Chitosan supported hexa-sulfooxymethyl melamine nanoparticles: a green and recyclable biopolymer catalyst for multicomponent reaction. In: Proceedings of the Fourth International Conference on Nanostructures (ICNS4), pp. 12–14.
Moll, J., Klingenfuss, F., Widmer, F., Gogos, A., Bucheli, T.D., Hartmann, M., van der Heijden, M.G., 2017. Effects of titanium dioxide nanoparticles on soil microbial communities and wheat biomass. Soil Biol. Biochem. 111, 85–93.
Mukhopadhyay, S.S., 2014. Nanotechnology in agriculture: prospects and constraints. Nanotechnol. Sci. Appl. 7, 63.
Nowack, B., Krug, H.F., Height, M., 2011. 120 Years of nanosilver history: implications for policy makers. Environ. Sci. Technol. 1, 1177–1183.
Öztaş, T., Özbek, A.K., Aksakal, E.L., 2002, June. Structural developments in soil treated with polyvinylalcohol. In: International Conference on Sustainable Land Use and Management. Soil Sci. Soc. of Turkey Int. Symp, pp. 143–148.
Pérez-de-Luque, A., 2017. Interaction of nanomaterials with plants: what do we need for real applications in agriculture? Front. Environ. Sci. 5, 12.
Piccinno, F., Gottschalk, F., Seeger, S., Nowack, B., 2012. Industrial production quantities and uses of ten engineered nanomaterials in Europe and the world. J. Nanopart. Res. 14 (9), 1–11.
Pimentel, D., Burgess, M., 2013. Soil erosion threatens food production. Agriculture 3 (3), 443–463.
Radniecki, T.S., Stankus, D.P., Neigh, A., Nason, J.A., Semprini, L., 2011. Influence of liberated silver from silver nanoparticles on nitrification inhibition of *Nitrosomonas europaea*. Chemosphere 85 (1), 43–49.
Rai, M., Ingle, A., 2012. Role of nanotechnology in agriculture with special reference to management of insect pests. Appl. Microbiol. Biotechnol. 94 (2), 287–293.
Rajapaksha, A.U., Ahmad, M., Vithanage, M., Kim, K.R., Chang, J.Y., Lee, S.S., Ok, Y.S., 2015. The role of biochar, natural iron oxides, and nanomaterials as soil amendments for immobilizing metals in shooting range soil. Environ. Geochem. Health 37 (6), 931–942.
Savi, G.D., Piacentini, K.C., Souza, S.R., Costa, M.E.B., Santos, C.M.R., Scussel, V.M., 2015. Efficacy of zinc compounds in controlling Fusarium head blight and deoxynivalenol formation in wheat (Triticum aestivum L.). Int. J. Food Microbiol. 205, 98–104.
Scott-Fordsmand, J.J., Peijnenburg, W.J., Semenzin, E., Nowack, B., Hunt, N., Hristozov, D., Marcomini, A., Irfan, M.A., Jiménez, A.S., Landsiedel, R., Tran, L., 2017. Environmental risk assessment strategy for nanomaterials. Int. J. Environ. Res. Public Health 14 (10), 1251.
Sekhon, B.S., 2014. Nanotechnology in agri-food production: an overview. Nanotechnol. Sci. Appl. 7, 31.
Servin, A., Elmer, W., Mukherjee, A., De la Torre-Roche, R., Hamdi, H., White, J.C., Bindraban, P., Dimkpa, C., 2015. A review of the use of engineered nanomaterials to suppress plant disease and enhance crop yield. J. Nanopart. Res. 17 (2), 1–21.
Simonin, M., Cantarel, A.A., Crouzet, A., Gervaix, J., Martins, J.M., Richaume, A., 2018. Negative effects of copper oxide nanoparticles on carbon and nitrogen cycle microbial activities in contrasting agricultural soils and in presence of plants. Front. Microbiol. 9, 3102.
Singh, J., Lee, B.K., 2016. Influence of nano-TiO2 particles on the bioaccumulation of Cd in soybean plants (Glycine max): a possible mechanism for the removal of Cd from the contaminated soil. J. Environ. Manage. 170, 88–96.
Sonkar, S.K., Roy, M., Babar, D.G., Sarkar, S., 2012. Water soluble carbon nano-onions from wood wool as growth promoters for gram plants. Nanoscale 4 (24), 7670–7675.

Srilatha, B., 2011. Nanotechnology in agriculture. J. Nanosci. Nanotechnol. 2 (7), 5.
Sun, Z., Liao, T., Dou, Y., Hwang, S.M., Park, M.S., Jiang, L., Kim, J.H., Dou, S.X., 2014. Generalized self-assembly of scalable two-dimensional transition metal oxide nanosheets. Nat. Commun. 5 (1), 1–9.
Thul, S.T., Sarangi, B.K., Pandey, R.A., 2013. Nanotechnology in agroecosystem: implications on plant productivity and its soil environment. Expert Opin. Environ. Biol. 2, 7.
Tumburu, L., Andersen, C.P., Rygiewicz, P.T., Reichman, J.R., 2017. Molecular and physiological responses to titanium dioxide and cerium oxide nanoparticles in Arabidopsis. Environ. Toxicol. Chem. 36 (1), 71–82.
Villalobos, M., Escobar-Quiroz, I.N., Salazar-Camacho, C., 2014. The influence of particle size and structure on the sorption and oxidation behavior of birnessite: I. adsorption of As (V) and oxidation of As (III). Geochim. Cosmochim. Acta 125, 564–581.
Virkutyte, J., Varma, R.S., 2014. Greener and sustainable remediation using iron nanomaterials. In: Green Catalysts for Energy Transformation and Emission Control. American Chemical Society, pp. 1–21.
Wang, X., Liu, X., Chen, J., Han, H., Yuan, Z., 2014. Evaluation and mechanism of antifungal effects of carbon nanomaterials in controlling plant fungal pathogen. Carbon 68, 798–806.
Wang, X., Yang, X., Chen, S., Li, Q., Wang, W., Hou, C., Gao, X., Wang, L., Wang, S., 2016. Zinc oxide nanoparticles affect biomass accumulation and photosynthesis in Arabidopsis. Front. Plant Sci. 6, 1243.
Watanabe, J.I., Tani, Y., Chang, J., Miyata, N., Naitou, H., Seyama, H., 2013. As (III) oxidation kinetics of biogenic manganese oxides formed by *Acremonium strictum* strain KR21-2. Chem. Geol. 347, 227–232.
Willett, W., Rockström, J., Loken, B., Springmann, M., Lang, T., Vermeulen, S., Garnett, T., Tilman, D., DeClerck, F., Wood, A., Jonell, M., 2019. Food in the Anthropocene: the EAT–Lancet Commission on healthy diets from sustainable food systems. Lancet 393 (10170), 447–492.
Wu, H., Tito, N., Giraldo, J.P., 2017. Anionic cerium oxide nanoparticles protect plant photosynthesis from abiotic stress by scavenging reactive oxygen species. ACS Nano 11 (11), 11283–11297.
Yan, W., Lien, H.L., Koel, B.E., Zhang, W.X., 2013. Iron nanoparticles for environmental clean-up: recent developments and future outlook. Environ. Sci.: Processes Impacts 15 (1), 63–77.
Yuan, Z., Zhang, Z., Wang, X., Li, L., Cai, K., Han, H., 2017. Novel impacts of functionalized multi-walled carbon nanotubes in plants: promotion of nodulation and nitrogenase activity in the rhizobium-legume system. Nanoscale 9 (28), 9921–9937.
Zhang, P., Guo, Z., Zhang, Z., Fu, H., White, J.C., Lynch, I., 2020. Nanomaterial transformation in the soil–plant system: implications for food safety and application in agriculture. Small 16 (21), 2000705.
Zhang, L., Wu, L., Si, Y., Shu, K., 2018. Size-dependent cytotoxicity of silver nanoparticles to *Azotobacter vinelandii*: growth inhibition, cell injury, oxidative stress and internalization. PLoS One 13 (12), e0209020.
Zheng, L., Hong, F., Lu, S., Liu, C., 2005. Effect of nano-TiO_2 on strength of naturally aged seeds and growth of spinach. Biol. Trace Elem. Res. 104 (1), 83–91.
Zhou, D., Kim, D.G., Ko, S.O., 2015. Heavy metal adsorption with biogenic manganese oxides generated by *Pseudomonas putida* strain MnB1. J. Ind. Eng. Chem. 24, 132–139.
Zhu, Y., Xu, F., Liu, Q., Chen, M., Liu, X., Wang, Y., Sun, Y., Zhang, L., 2019. Nanomaterials and plants: positive effects, toxicity and the remediation of metal and metalloid pollution in soil. Sci. Total Environ. 662, 414–421.

Nanopesticides for crop protection

15

Sanjoy Kumar Paul, Hossain Sohrawardy, Nur Uddin Mahmud, Paritosh Chandra Roy, and Tofazzal Islam
Institute of Biotechnology and Genetics Engineering (IBGE), Bangabandhu Sheikh Mujibur Rahman Agricultural University, Gazipur, Bangladesh

15.1 Introduction

Since the starting of agriculture about 10,000 years ago, the growers have had to fight against various enemies (pests) of crop plants including animals, insects, and microorganisms, such as fungi, bacteria, viruses, nematodes, and mycoplasma (Oerke, 2006). Crop pests are still a vital challenge in crop production globally (Singh et al., 2015; Bhattacharyya et al., 2016). Crop damages due to various pests can be varied from 16% to 80% (Oerke, 2006). As a plant protection input, about 4.6 million tons of synthetic chemical pesticides are used every year in the world. Most of these applied chemical pesticides go into the environment and as a residue in agricultural products and reallocate in the ecological cycle (Malaj et al., 2014; Köhler and Triebskorn, 2013). Conventional agricultural chemicals increase pest resistance, water and soil pollution, incorporation in the food chain, and accumulation in biodiversity (Khan et al., 2017; Zhao et al., 2017a). Because of foliar application, the rate of pesticide loss and breakdown is typically more than 99%, which are caused by runoff, spray drift, and rolling down at the time of field application (Nuruzzaman et al., 2016; Song et al., 2017; Zhao et al., 2017a). Solving these emerging issues in crop cultivation, nanoparticle-based pesticides could be used as efficient approaches for the management of pests. Nanopesticides are found effective in controlling plant pathogens (Prasad et al., 2017). Nanomaterials (1–100 nm in size) have a more surface area, high penetration capacity, and effectiveness that keep away them from accumulating as residues in the environment (Qureshi et al., 2018). Herein, nanopesticides represent a progressive and substitute method to overcome the limitations of currently used pesticide formulations. Nanopesticides can be made by the active ingredients of nanosized particles having pesticidal properties or by nanoscaled particles having pesticide characteristics. They have the capacity to enhance wettability as well as dispersion of different formulations used in agriculture such as lessening the runoff of the organic solvent and movement of the pesticide (Bhattacharyya et al., 2016). Nanopesticides carry wonderful pesticidal properties. They are required in small quantities for efficient pest management, which saves the environment from the harmful effects of chemical pesticides (Kumar et al., 2015; Saini et al., 2014). Nanoformulations behave the same as conventional pesticides but they possess slow

Agricultural Nanobiotechnology. https://doi.org/10.1016/B978-0-323-91908-1.00014-6

and targeted exposure of the active ingredient as well as enhance the seeming solubility of a poorly soluble active ingredient, which can protect against premature degradation (Athanassiou et al., 2018). Thus, nanoscience has been emerged as a promising and highly efficient area of research for the development of nanopesticides to manage plant pests in crop production (Parisi et al., 2015).

Nanotechnology deals with materials having nanosized (1–100 nm) characteristics. These materials have some properties that are different from bulk materials, enabling unique applications (Li et al., 2017; Jiang et al., 2017; Iqbal et al., 2017; Elizabath et al., 2019). However, the instantaneous colloidal particles at nanoscale dimensions of 10–1000 nm can be used as nanoparticles in agriculture and related fields (Nakache et al., 2000; US Department of Agriculture, 2002; Islam, 2019). Nanotechnology links with many interdisciplinary sciences, such as nanomaterials, physics, engineering, chemistry, and biology, improved a lot in the few decades as a growing field (Prasad et al., 2014; Jiang et al., 2017; Khan et al., 2018). Nanoparticles containing nanosized insecticides, herbicides, fungicides, etc., targeting the specific plant cellular organelles or macromolecules can discharge their components as "magic bullets" (Siddiqui et al., 2015; Islam et al., 2018; Hossain et al., 2020). Presently, designed nanoparticles are considered emerging technologies that act as powerful tools for promoting a new generation of sustainable agriculture. Nanopesticides show higher efficiency against plant fungi and are found potentially effective to control other pathogens like fish pathogen, pathogenic bacteria, and oomycetes. Hoque et al. (2021) fabricated two nanocomposites that are poorly oxidized silver oxide/silver/tin (IV) oxide and found that these materials are antipathogenic against four harmful pathogens, viz., a pathogenic bacterium of fish (*Stenotrophomonas maltophilia*), oomycete (*Phytophthora cactorum* strain), and another two strains of strawberry fungus (*Colletotrichum siamense*). Moreover, nanotechnology can be used in the agri-food sector to mitigate many environmental issues arisen from agricultural practices and ensure food security (Eltarahony et al., 2018). Nanotechnology will conserve agriculture and reduce environmental challenges by producing nanoscale pesticides and other pest control agents. Lower doses of nanoparticles can effectively control pests and they are considered environment-friendly (Singh et al., 2015; Bhattacharyya et al., 2016). However, in terms of the application of nanotechnology, agriculture ranked the second-highest as some of the the nanotools are increasing the fertility of the soil and crop yields. Micron-level chemical pesticides are used in agriculture for a long time but nowadays, scientists shift them to nanoscale pesticides. The efficiency of material used in the nanovesicle form can be improved, thus reducing the quantities used with greater effectiveness (Singh et al., 2015). Therefore, applications of nanobiotechnology are currently under intense research. To tackle the global challenges such as burgeoning population, climatic change, and reduction of soil fertility and nutrients as well as the emerge of the new pests, technological innovation in agriculture is important. Currently, nanobiotechnology is thought to be one of the viable solutions to face the obstacles in agriculture (Maurya et al., 2018). Various aspects of the usefulness of nanopesticides in agriculture have recently been reviewed (Deka et al., 2021; Raj et al., 2021). This chapter aims to review the current knowledge of nanopesticides and update our understanding of the potential of these new technologies in crop protection. Furthermore, the research gap and the future way forward in the use of nanotechniques in plant protection have also been discussed.

15.2 Nanopesticide and disease management of crops

Plant pathology is highly influenced by the progress of science and technology. Due to disease infestation, approximately 14%–25% of crop yield is lost which causes approximately $2 trillion financial loss worldwide. The use of different chemical formulations as pest control agents is a traditional practice to minimize crop damage. At present, almost 4.6 million tons of agrochemicals are consumed annually and more than 90% of them run off into the environment or exist as a residue in agricultural products. Most of the active ingredients (AIs) of the pesticides are insoluble in water. Thus, to make a successful spray formulation, use of the carrier, dispersant, emulsifier, solvent, etc., are essential. The actual usefulness of a pesticide on the target is <0.1% due to rolling down, spray drift, rainwater leaching at the time of field application, etc., making the utilization of the pesticide extremely inefficient (Deshpande, 2019). In recent years, use of nanomaterials (NMs) to generate novel formulations has proved their great potentiality in improving the efficacy and safety of pesticides in plant protection through a controlled release mechanism (Zhao et al., 2017a). Nanopesticides are used in various ways on seeds, leaves, and soils for plant disease management (Khan and Rizvi, 2014). Herein, the application of NP is thought to be an effective, alternative, and environmentally safe method for crop pest management (Fig. 15.1). In the case of nanoparticles, a huge variation was observed in the nanoform from their macroforms in physiochemical studies. The impact of nanoparticles on pathogens and microorganisms can be tested. The technology is used to control plant pathogens, especially pathogenic bacteria and fungi. Because of their ultrasmall size, NPs have high reactivity against microorganisms (Ismail et al., 2017). However, nanopesticides are usually developing in two ways: (1) directly processing into nanosized pesticides (nanoparticles) and (2) loading pesticide particles with nanocarriers in delivery systems (Ghormade et al., 2011).

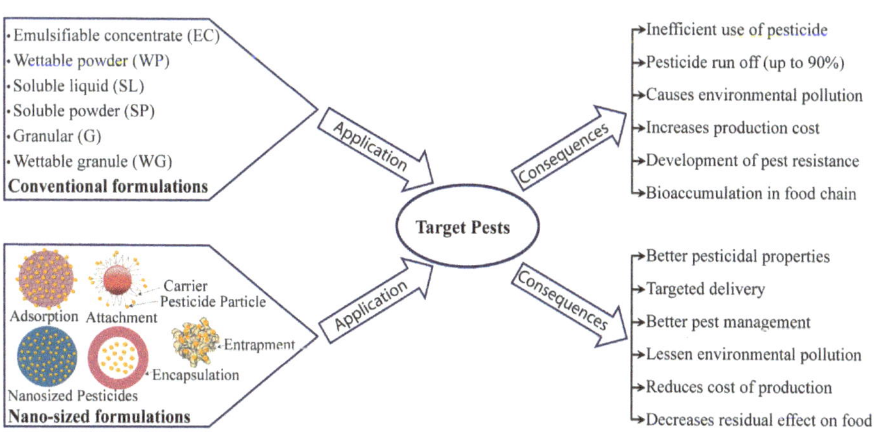

Fig. 15.1 Pesticide formulations (conventional and nanosized) and their consequences.

15.2.1 Directly processing nanosized fungicides

The NMs are used in plant protection for their antimicrobial activity against pathogens. The widely discovered nanosized materials are NPs of metals and carbon polymer-based NMs. Among these, metal NPs such as copper (Cu), silver (Ag), and zinc (Zn) are widely examined for their antimicrobial characteristics (Baker et al., 2017). However, copper nanoparticles (CuNPs) have antifungal activity against selected crop fungal pathogen and act as a disinfectant agent in the poultry industry as well as in animal husbandry-related activities (Kanhed et al., 2014). Similarly, the synthesized spherical $Cu_2O@OAm$ (oleylamine, 30 nm) NPs showed great antifungal activity against the yeast, *Saccharomyces cerevisiae* (Giannousi et al., 2014).

In the recent years, nanotechnology is obtaining a remarkable impulse for its ability to modify metals into their nanosize, which significantly changes the chemical, physical, biological, and optical properties of metals. Currently, many NPs are used to control plant pathogens, and the rate is increasing. For instance, sheath blight disease of rice by *Rhizoctonia solani* Kühn and false smut by *Ustilaginoidea virens* are becoming the serious diseases that cause devastating yield losses worldwide. Both the diseases are usually controlled by using different pesticides. However, silver nanoparticles (AgNPs) showed a noticeable inhibitory effects on the lesion development in leaves by both *R. solani* (Nejad et al., 2017) and *U. virens* (Bhargava et al., 2018). Islam et al. (2018) showed that Ag-Dopa-CP inhibited the growth of wheat blast fungi and has no visible phytotoxic effects on the wheat plant. Additionally, the antibacterial activity of Ag-Dopa-CP may be used for the control of bacterial diseases, however, a further study is necessary to confirm their usefulness as antifungal and antibacterial agent. The efficacy of AgNPs at different concentrations was evaluated against four soil-borne plant pathogens namely, *Sclerotinia sclerotiorum*, *Sclerotium rolfsii*, *R. solani*, and *Fusarium oxysporum*. The results exhibited that at 100-ppm concentration, AgNPs significantly interrupt the growth of mycelia of the tested pathogens (Kaman and Dutta, 2017). Moreover, the antifungal property of nitrogen-doped carbon nanohorn (NCNH) in inhibiting a plant pathogenic fungus *R. solani* was investigated by Dharni et al. (2016).

In the apple plant, *Fusarium* wilt and apple scab diseases are caused by *Fusarium solani* and *Venturia inaequalis*. An excellent inhibition of these fungi was observed by Ag-doped (hollow) TiO_2 NPs (Boxi et al., 2016). Another wilt disease in chili (*Capsicum annuum* L.) caused by a soil-borne fungus, *F. oxysporum* threatens chili production. Commercially available traditional fungicides are being used to control the outbreaks of this disease which is hazardous to the environment. The NPs like cobalt and nickel nanoferrites ($CoFe_2O_4$ and $NiFe_2O_4$) illustrated in the Fig. 15.2 showed significant inhibitory activity against three plant pathogens viz., *F. oxysporum*, *Colletotrichum gloeosporioides*, and *Dematophora necatrix*, which infect capsicum roots (Sharma et al., 2017).

All over the world, citrus fruit crop plays a major role in economic, social, and health influences on the overall human lifestyle. From the bloom to harvesting stage, several pathogens attack citrus fruits. Besides, postharvest pathogens also affect the yield of fruits and significantly decline their quality. In citrus, the most plentiful fungal

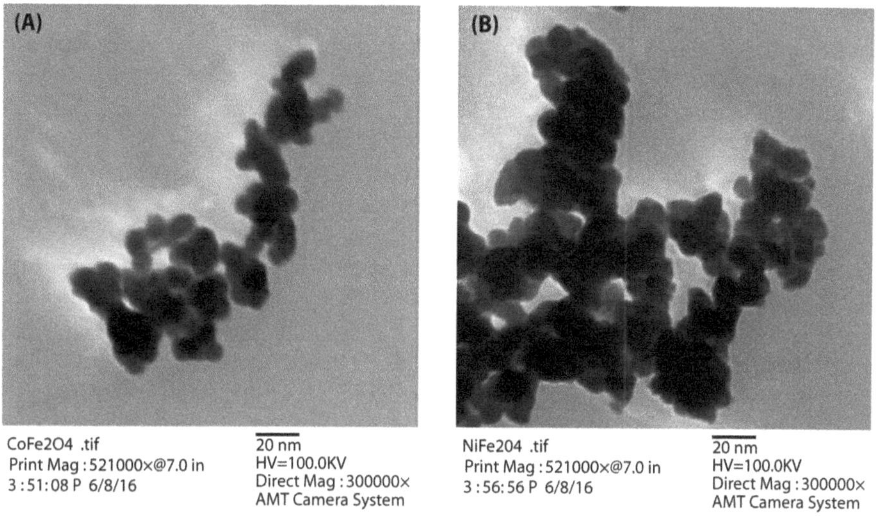

Fig. 15.2 The formation of ferrites showed spherical and uniform with an average size of 25 nm. TEM images of (A) $CoFe_2O_4$ and (B) $NiFe_2O_4$ nanoferrites synthesized at 800°C. These exhibited potential inhibitor properties against three phytopathogens, such as *F. oxysporum*, *Colletotrichum gloeosporioides*, and *Dematophora necatrix*, which infect capsicum (*Capsicum annuum* L.) roots.
Reproduced with permission from Sharma, P., Sharma, A., Sharma, M., Bhalla, N., Estrela, P., Jain, A., Thakur, P., Thakur, A., 2017. Nanomaterial fungicides: in vitro and in vivo antimycotic activity of cobalt and nickel nanoferrites on phytopathogenic fungi. Global Chall. 1, 1700041 under a Creative Commons license.

pathogen is *Alternaria alternata* followed by *Alternaria citri* and *Penicillium digitatum*. The AgNPs displayed effective antifungal properties against these fungi that have opened the gateway of new generation nano-based fungicides. The nanoformulations are less harmful that could replace the cytotoxic effect of the conventional fungicides (Abdelmalek and Salaheldin, 2016).

In agriculture, *Phytophthora* and *Alternaria* blights are marked as the most injurious plant diseases. The influence of the treatment of tomato (*Solanum lycopersicum*) with an aqueous dispersion of sodium tallow amphopolycarboxyglycinate-stabilized AgNPs was examined on the dispersion's protective properties against the aforesaid blight pathogens in the field conditions. The estimated yield of tomatoes treated with AgNPs was higher than the yield of the untreated control. The productivity of tomato was increased under the stimulation of nanosilver. There was a decrease in the number of stricken plants with late blight (*P. infestans*) and early blight (*A. solani*) diseases. These findings could be applied for the fabrication of plant protection agents using colloidal silver (Zakharova et al., 2017). The antifungal potential of AgNPs was associated with the morphological modifications of fungal hyphae. In addition, the synthesized AgNPs could penetrate and disperse throughout the fungal cell as well as interact with the components of the cell (Abdel-Hafez et al., 2016). On the other hand, nanostructured silver fungicides were applied as exogenous nontraditional compounds to control *Pyrenophora teres* L. which causes net blotch disease in a highly susceptible Egyptian barley variety. The efficacy of disease control by nanosilver fungicide could be helpful to understand the basic mechanism for creating the new resistant barley cultivars for decreasing the dependency of environmentally hazardous chemical fungicide application (Abdelaal et al., 2020). However, Aleksandrowicz-Trzcińska describes that CuNPs as metal nanoparticles play an influential role in controlling damping-off and wood-decay fungi (*P. cactorum*, *Fusarium redolens*, *R. solani*, and *F. oxysporum*) under in vitro conditions. The NPs show selective antifungal activity, which specified the considerable toxicity of AgNPs and CuNPs to specific pathogens (Aleksandrowicz-Trzcińska et al., 2018). Moreover, $Cu_3(PO_4)_2{\cdot}3H_2O$ nanosheets were synthesized and examined the stimulation of coordinating anion, dissolution profile, as well as particle morphology on *F. oxysporum* f. sp. *niveum* which is pathogenic to watermelon. In a greenhouse study, $Cu_3(PO4)_2{\cdot}3H_2O$ nanosheets significantly suppressed (58%) fungal disease as calculated by the yield (Borgatta et al., 2018). Similarly, Bramhanwade et al. (2015) investigated the CuNPs which were used as novel antifungals against crop pathogenic fungi. As shown in Fig. 15.3, CuNPs inhibited 93.98% of disease severity in tomato and dragon fruit caused by *Fusarium* sp. (Viet et al., 2016).

Tomato, potato, apple, and grape are economically valuable crops in agriculture. Sulfur nanoparticles (SNPs) were applied to two phytopathogens in organic agriculture. *F. solani* isolates from an infected tomato leaf which is a causal organism of the early blight and *Fusarium* wilt diseases. Besides, *V. inaequalis* causes apple scab disease. Application of the SNPs were found effective in inhibiting these fungal growth

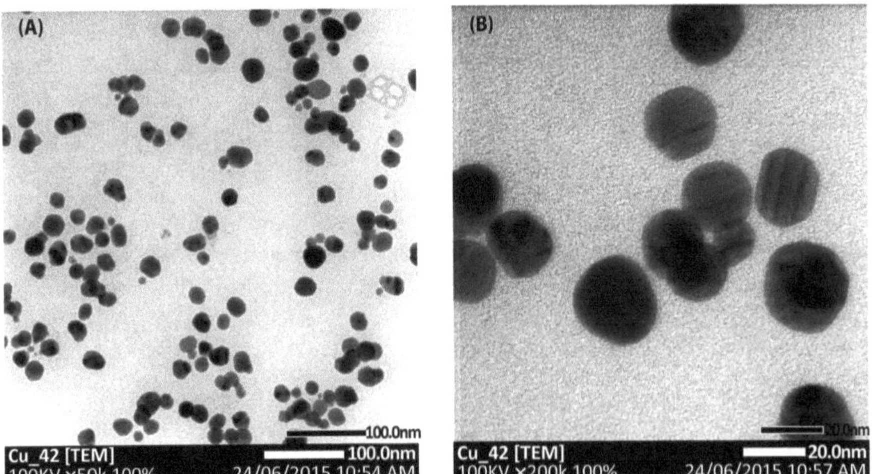

Fig. 15.3 Transmission electron microscopic images of uniform-sized spherical CuNPs for various scales. TEM image shows (A) 100 nm and (B) 20 nm in size. The nanomaterials effectively inhibited disease severity of tomato and dragon fruit caused by *Fusarium* sp. Reproduced with permission from Viet, P.V., Nguyen, H.T., Cao, T.M., Hieu, L.V., 2016. Fusarium antifungal activities of copper nanoparticles synthesized by a chemical reduction method. J. Nanomater. 2016 under a Creative Commons license.

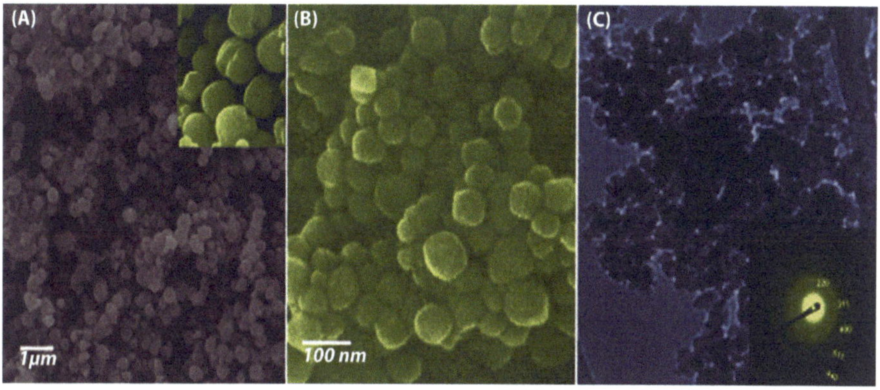

Fig. 15.4 SEM images of mesoporous alumina NPs. (A) and (B) present the stable particle morphologies of calcined material. The fabricated nanoparticles showed spherical morphologies with uniform sizes from 100 to 250 nm. Moreover, (C) the TEM image discovered a spherical-shaped morphology and worm-like mesopore structures connected to the large regions of alumina NP domains. NPs performed as an effective alternative to conventional pesticide formulations to control *Fusarium* rot root in tomatoes.
Under a Creative Commons license, the images were reproduced with permission from Creative Commons license from Shenashen, M., Derbalah, A., Hamza, A., Mohamed, A., El Safty, S., 2017. Antifungal activity of fabricated mesoporous alumina nanoparticles against root rot disease of tomato caused by *Fusarium oxysporum*. Pest Manag. Sci. 73, 1121–1126.

(Rao and Paria, 2013). Mesoporous alumina sphere (MAS) NPs synthesized were assessed for their biological activity on tomato root rot caused by *F. oxysporum* over the conventional fungicide, tolclofos-methyl. The application of cylindrically cubic MAS (illustrated in Fig. 15.4) was demonstrated as an effective nanopesticide in controlling *Fusarium* rot root in tomatoes (Shenashen et al., 2017). Another study examined the inhibitory activity of some nanoparticles of magnesium oxide, titanium dioxide, and zinc oxide (MgO, TiO_2, and ZnO NPs) in sugar beet against *S. rolfsii* and *R. solani* to control root rot diseases. These NPs showed a fantastic disease controlling activity against damping-off and root rot diseases in an environmentally safe manner (El-Argawy et al., 2017). Researchers looked at the effects of engineered nanomaterials (ENMs) Ag_2S, polyvinylpyrrolidone (PVP)-coated Ag ENMs (PVP-Ag), and Ag^+ on arbuscular mycorrhizal fungi (AMF) on tomato plants. The results of that study indicated that Ag_2S ENMs were less poisonous to plants, less troublesome to plant-mycorrhizal symbiosis, as well as less inhibitory to the soil microbes compared to PVPAg ENMs or Ag^+ (Judy et al., 2015).

Soybean (*Glycine max* L.) is frequently infected by the plant pathogenic fungi belonging to the genera of *Rhizoctonia*, *Fusarium*, *Sclerotium*, *Colletotrichum*, *Alternaria*, and *Macrophomina* spp. Zinc (ZnNPs) and manganese (MnNPs) nanoparticles were tested against these six different soybean plant pathogenic fungi in potato dextrose agar (PDA) media. The results proved that Zn- and MnNPs are inhibitory against the test fungi in vitro (Dhoble and Kulkarni, 2016). Moreover, ZnONPs (shown in Fig. 15.5) were proved as potential antifungal agent on the mycelial development of *Erythricium salmonicolor* (Arciniegas-Grijalba et al., 2017).

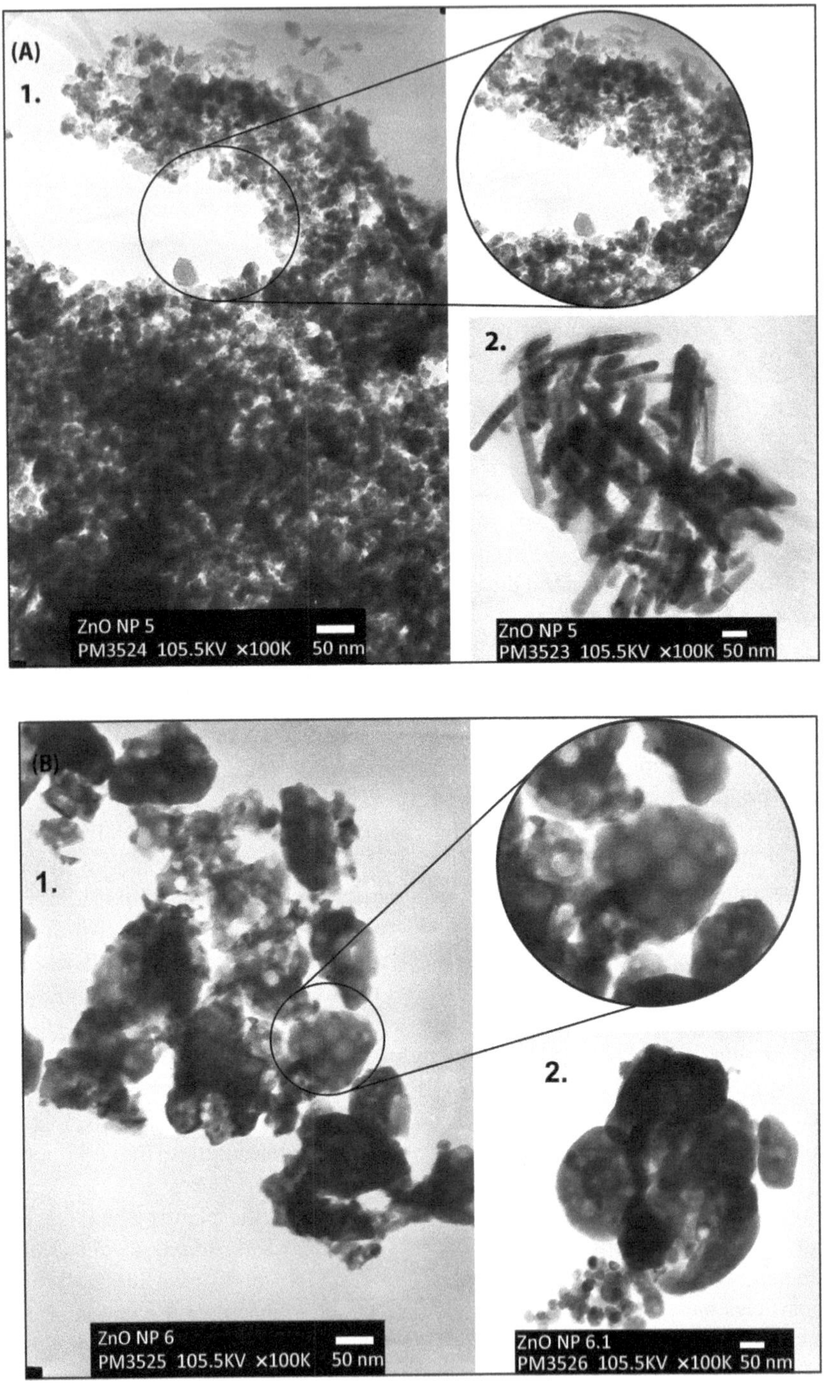

Fig. 15.5 See Figure legends on opposite page.

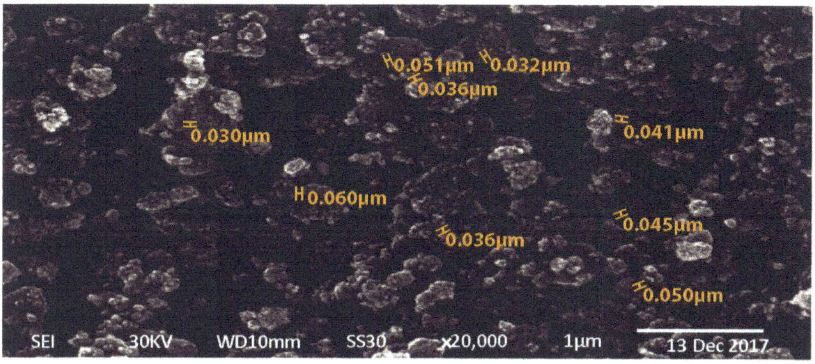

Fig. 15.6 Scanning electron microscopy (SEM) images of NSi showed the regular spherical shape with size ranging from 30 to 60 nm. NSi was found as an alternative in controlling the chocolate spot disease compared to traditional harmful fungicides.
Reproduced with permission from Hasan, K.A., Soliman, H., Baka, Z., Shabana, Y.M., 2020. Efficacy of nano-silicon in the control of chocolate spot disease of *Vicia faba* L. caused by *Botrytis fabae*. Egypt. J. Basic Appl. Sci. 7, 53–66 under the license of Creative Commons.

Several pathogens are causing a significant reduction in the yield of cucumber production. Cucumber powdery mildew disease is one of them, which is caused by *Podosphaera xanthii*. It is a widespread and destructive disease mainly in semiarid areas where low temperature and high humidity are predominant (Tanaka et al., 2017). Hafez et al. (2020) described that the bioagent *Bacillus subtilis* + nano-ZnO effectively control powdery mildew disease in cucumber plants. They showed that the combination of the bioagent and nanomaterial generated an inhibitory effect by decreasing electrolyte leakage. Similarly, in the case of *Vicia faba* L., nano-silicon (NSi) (shown in Fig. 15.6) was found to be an alternative to controlling the chocolate spot disease compared to traditional harmful fungicides. The application of NSi boosted the resistance of *V. faba* plants against infection by the pathogen *Botrytis fabae* by enhancing the production of protective compounds like proline and phenols and improving the action of defense-related enzymes (Hasan et al., 2020).

Fig. 15.5 TEM images of synthesized ZnONPs. Here, A1, ZnONP 0.15/400 showed spherical morphology and size range between 20 and 35 nm, while A2—the needles-like growth mechanism—was shaped by the ordered aggregation of ZnO nanospheres, and B1 ZnONP 0.1/600 observed with acicular morphology and size range between 30 and 45 nm, while B2 grouped themselves by creating "clusters," containing a limited number of NPs. NPs proved their potential as an antifungal agent on the mycelial development of *E. salmonicolor*.
Under a Creative Commons license, the images were reproduced with permission from Arciniegas-Grijalba, P.A., Patiño-Portela, M.C., Mosquera-Sánchez, L.P., Guerrero-Vargas, J. A., Rodríguez-Páez, J.E., 2017. ZnO nanoparticles (ZnO-NPs) and their antifungal activity against coffee fungus *Erythricium salmonicolor*. Appl. Nanosci. 7, 225–241.

15.2.1.1 Nanofungicides extracted from botanicals

The emergence of resistant fungal pathogens occurs for the indiscriminate use of conventional synthetic fungicides. The green synthesis process for metal NP fabrication played a very potential role that allows eco-friendly and nonhazardous production of nanopesticides. Bio-efficacy of silver nanoparticles (AgNPs) produced from botanicals were found excellent in controlling *Alternaria zinniae*, which causes leaf spot disease of marigold. It is effective for mycelial growth reduction. The AgNPs biosynthesized by plants have been accepted as potential tools to battle against parasitic fungi (Negi, 2016). The extract-mediated synthesis of AgNPs from leaf biomass of *Saccharum officinarum* exhibits moderate-to-good inhibition of *Cladosporium fulvum*, *Colletotrichum acutatum*, and *Phytophthora capsici* (Velu et al., 2017). *Amaranthus retroflexus* is used in producing AgNPs as green products, which also inhibit many plant pathogens, such as *A. alternata*, *Macrophomina phaseolina*, and *F. oxysporum* (Bahrami-Teimoori et al., 2017). *Tagetes patula* leaf extract was used to synthesize AgNPs that act as reducing mediators. It is also highly toxic to phytopathogenic fungi *Colletotrichum chlorophyti*. Green synthesis of these NPs are considered as safer alternatives to the synthetic chemical fungicides (Sukhwal et al., 2017). Moreover, AgNPs synthesized with an aqueous extract of *Artemisia absinthium* L. inhibited *Phytophthora* infection, which played a vital role in plant survival (Ali et al., 2015a). Furthermore, AgNPs synthesized using *Aspergillus versicolor* showed an antifungal action against two diseases of strawberry (*Fragaria × ananassa*) such as white mold caused by *S. sclerotiorum* and gray mold caused by *Botrytis cinerea* (Elgorban et al., 2016). Commercially available Zinc NPs (Zn- and ZnONPs) show inhibition of spore germination and infectivity of Oomycete pathogen *Peronospora tabacina* that causes blue mold on tobacco leaves which is highly destructive toward the seed plants (Wagner et al., 2016). Synthesized AgNPs show antifungal activities in vitro on cultures of *Neofusicoccum parvum* (Khatami et al., 2017). Moreover, platinum (Pt) NPs fabricated through green synthesis from *Prunus × yedoensis* tree gum extract displayed inhibitory effects against two phytopathogens viz., *C. acutatum* and *C. fulvum* (Velmurugan et al., 2016). Additionally, the disease symptoms caused by *Dothiorella sarmentorum* include bud necrosis, wood cankers, branch and shoot dieback in diseased trees. To control the mycelial growth of *D. sarmentorum*, AgNPs were biosynthesized using *Lathyrus* sp. and *Stachys lavandulifolia* that show tremendous antifungal effects on PDA medium (Azizi et al., 2016). *Descurainia sophia*-mediated synthesis of AgNPs was proved as a potential biological resource, which inhibited the growth of *R. solani* in vitro (Khatami et al., 2016).

15.2.1.2 Nanofungicides extracted from the microbes

Microbial nanotechnology serves as an excellent biosynthesis process for rapid mass-scale production of nanoparticles. Synthesis of biological AgNPs extracellularly from the fungus *Trichoderma longibrachiatum* stabilizes the synthesis process and act as a reducing agent. The application of AgNPs reduces the number of colonies for numerous plant pathogenic fungi up to 90% of efficiencies against *Fusarium verticillioides*, *Pyricularia grisea*, *Penicillium brevicompactum*, *Helminthosporium oryzae*, and

Fusarium moniliforme (Elamawi et al., 2018). Similarly, AgNPs were prepared biogenetically by using the fungus *Trichoderma harzianum*. These NPs are potential inhibitors against the sclerotia production by white mold fungus. These nanoparticles were also found inhibitory against *S. sclerotiorum* fungus. It inhibits germination of sclerotia and mycelial growth (Guilger et al., 2017). Mycosynthesized AgNPs (using an endophytic fungi, *Taxus baccata* L. Iranian yew) inhibits the growth and proliferation of *Salmonella typhi*, *B. subtilis*, *Staphylococcus* sp., *Escherichia coli*, and *Pseudomonas aeruginosa*. Antibacterial properties showed by a maximum inhibitory zone against *B. subtilis*. The endophytic fungus, *Nemania* sp., could also be useful fungal system for AgNP production through the extracellular mycosynthesis process, which is considered as an eco-friendly method for nanoparticle synthesis (Farsi and Farokhi, 2018). Another study revealed that AgNPs developed through mycosynthesis display an effective inhibitory activity against diverse fungal pathogens, such as *A. solani*, which caused early blight disease of tomato. The AgNPs also act as an effective fungicide in the postharvest stage of pome fruits to control many fungal pathogens. The nanoparticles synthesized from two yeasts viz., *Cryptococcus laurentii* and *Rhodotorula glutinis*, had comparable efficacy to a fungicide namely iprodione which is usually sprayed for controlling postharvest diseases (Fernández et al., 2016). On the other hand, *Bipolaris maydis* was used to synthesize AgNPs as the produced NPs are antifungal against *Exserohilum turcicum*, which causes northern corn leaf blight in maize. These results clearly indicate that mycosynthesis of NPs are a promising process for developing new environment-friendly antimicrobial agents for controlling fungal, oomyceteous, and bacterial phytopathogens (Huang et al., 2018a,b).

Bird's eye leaf spot disease of tea incurs a large amount of economic loss every year in tea trading activities of many tea-producing countries. Nanoparticles like AgNPs and CuNPs are biosynthesized because they have antagonistic properties. Many biological control agents like *Pseudomonas fluorescens*, *Trichoderma atroviride*, and *Streptomyces sannanensis* are used in the biosynthesis of useful nanoparticles. Biosynthesized CuNPs from *S. sannanensis* exhibited their antagonistic property against *Cercospora theae* during irregular disease occurrence (Gnanamangai et al., 2017). Table 15.1 presents some nanoparticles that show fungicidal effects useful for plant protection.

15.2.2 Nanoinsecticides

Nanoparticles are also becoming promising tools for insect control in agriculture. Several nano-based insecticides are used for plant protection, e.g., imidacloprid (Chhipa, 2017), thiamethoxam (Wibowo et al., 2014), and thiram (Fraceto et al., 2016), which are now available in the market. Actually, the potential application of nanopesticides is the reduction of the toxic impact of conventional chemical pesticides by decreasing the rate of pesticide use (Nuruzzaman et al., 2016). Shifa et al. (2019) evaluated the bioefficacy of nanoscale formulation of deltamethrin and conventional commercial analog against greenhouse whitefly, *Trialeurodes vaporariorum* under in vitro conditions through contact or residual bioassay. They found that the recommended strength of commercial formulation decreased by any factor ranging 5–10 yielded using the concentration of nano deltamethrin. Besides, nano-calcium carbonate was used for the protection of plants as well as the supply of nutrients. These NMs showed better

Table 15.1 Some nanoparticles that show fungicidal effects for crop protection.

Nanoparticles	Dose	Action/outcome	References
AgNPs	2.6 µg/mL	Ag-Dopa-CP exhibited antifungal activity against wheat blast fungus, *Magnaporthe oryzae Triticum* pathotype	Islam et al. (2018)
AgNPs	10 µg/mL	*Artemisia absinthium*-mediated AgNPs inhibit *Phytophthora* spp. without affecting normal plant physiology	Ali et al. (2015a, b)
AgNPs	150 ppm	Antifungal activity against *Sclerotinia sclerotiorum* and *Botrytis cinerea* that cause white and gray molds of strawberry	Elgorban et al. (2016)
AgNPs	50 ppm	Effectively suppress the lesion development of rice due to the rice sheath blight disease (*Rhizoctonia solani* Kühn).	Nejad et al. (2017)
AgNPs	40 µg/mL	Inhibit mycelia growth of *Dothiorella sarmentorum* (dieback).	Azizi et al. (2016)
AgNPs	10 ppm	Decrease up to 73.9% mycelial growth of *Alternaria zinnia* which caused leaf spot disease in marigold	Negi (2016)
AgNPs	10 ppm	Inhibit *A. solani* that causes early blight disease of tomato	Abdel-Hafez et al. (2016)
AgNPs	100 ppm	Inhibit phytopathogenic fungi *Colletotrichum chlorophyti*	Sukhwal et al. (2017)
AgNPs	25 µg/mL	*Descurainia sophia*-mediated AgNPs exhibited the highest inhibition against *R. solani*	Khatami et al. (2016)
AgNPs	150 ppm	Show inhibitory activity against *A. alternata*, *A. citri*, and *Penicillium digitatum*	Abdelmalek and Salaheldin (2016)
AgNPs	3 ppm	Exhibit antifungal activity against postharvest diseases in pome fruits	Fernández et al. (2016)
AgNPs	250 µg/mL	Display moderate-to-good antifungal action against *Phytophthora capsici*, *Cladosporium fulvum*, and *Colletotrichum acutatum*	Velu et al. (2017)

AgNPs	3.16×10^{12} NPs/mL	Cytotoxic and genotoxic effects against germination of white mold disease caused by *Sclerotinia sclerotiorum* fungus without any negative effect on the growth of soybeans	Guilger et al. (2017)
AgNPs	100 ppm	Compare protection level of nanosilver from *A. solani* (73.1%) and *Phytophthora infestans* (59.1%)	Zakharova et al. (2017)
AgNPs	400 µg/mL	Show potent antifungal activities against plant pathogenic fungi *Macrophomina phaseolina* (charcoal rot) and *Fusarium oxysporum* (Fusarium wilt). *A. alternata* (leaf spot)	Bahrami-Teimoori et al. (2017)
AgNPs	100 ppm	Exhibit significant inhibition of mycelial growth of pathogens viz., *Sclerotinia sclerotiorum*, *Fusarium oxysporum*, *Sclerotium rolfsii*, and *R. solani*	Kaman and Dutta (2017)
AgNPs	30 ppm	Exhibit strong impact on quantitative inhibition and disruption of antifungal growth of *Neofusicoccum parvum*, a causal agent of decline and dieback symptoms in grapevine	Khatami et al. (2017)
AgNPs	100 µM	ECAgNPs coated with epicatechin plus gentamicin confer inhibitory activity against the fungus *Aspergillus niger*	Ikram et al. (2017, 2018)
AgNPs	10 ppm	Synthesized NPs from *Phomopsis liquidambaris* strain SA1 show effective antimicrobial and mosquitocidal activity.	Seetharaman et al. (2018)
AgNPs	1 mM	Exhibit inhibitory activity against *Bacillus subtilis*	Farsi and Farokhi (2018)
AgNPs	–	Reduce many pathogenic fungal colonies to great extent (up to 90%)	Elamawi et al. (2018)
AgNPs	200 µg/mL	Complete inhibition of conidia germination was achieved against northern corn leaf blight causing fungus *Exserohilum turcicum*	Huang et al. (2018a,b)

Continued

Table 15.1 Continued

Nanoparticles	Dose	Action/outcome	References
AgNPs	–	Control effectively barley net blotch disease of a susceptible Egyptian barley "Giza 2000" caused by *Pyrenophora teres* L.	Abdelaal et al. (2020)
AgNPs	100 ppm	Show the best performance (67.11%) followed by silicon carbide NPs (60.17%) to reduce false smut disease of rice under in vitro conditions which causal agent is *Ustilaginoidea virens*	Bhargava et al. (2018)
AgNPs	20 ppm	Exhibit potential effectiveness to control *Colletotrichum gloeosporioides*, the causal agent of anthracnose disease in many plants	Shanmugam et al. (2015)
AgNPs	15 mM	Nanomaterials containing tulsi extract shows the maximum zone of inhibition against *Xanthomonas axonopodis* pv. *punicae*	Sherkhane et al. (2018)
AgNPs	–	Daphina exposed to a mixture of fullerenes and an organophosphate (tribufos) conferred antagonistic interaction	Louder (2015)
AgNPs	1 mM	Act as an antiviral agent against the viral disease of faba bean (bean yellow mosaic virus)	Elbeshehy et al. (2015)
AgNPs	30 ppm	Control many plant pathogens effectively	Ali et al. (2015b)
CuNPs	450 ppm	Inhibit *Fusarium* sp. effectively (93.98%) which is the causal agent of disease in tomato and dragon fruit	Viet et al. (2016)
CuNPs	10 mg/L	$Cu_3(PO_4)_2{\cdot}3H_2O$ nanosheets decrease disease progression (58%) and suppress fungal diseases significantly.	Borgatta et al. (2018)
AgNPs and CuNPs	1.5 ppm	Freshly prepared extracellular AgNPs and CuNPs inhibit (59.42%–79.76%) bird's eye spot disease of tea caused by *Cercospora theae*.	Gnanamangai et al. (2017)

CuNPs and AgNPs	50 ppm	Prove their antifungal activity against wood decay fungi viz., *Fusarium oxysporum*, *F. redolens*, *R. solani*, *Sparassis crispa*, *Grifola frondosa*, *Fistulina hepatica*, *Meripilus giganteus*, and *Phytophthora cactorum*	Aleksandrowicz-Trzcińska et al. (2018)
Ag_2S ENMs	1 mg/kg Ag_2S ENMs	Exhibit less toxicity to plants and less disruptive effect to plant-mycorrhizal symbiosis and reduce inhibitory result to the soil microbes compared to PVP-Ag ENMs and Ag^+	Judy et al. (2015)
Graphene oxide (GO)	500 µg/mL	Have a potential effect on the reproduction of some selected pathogens, which kill approximately 90% of the bacteria and suppress 80% of germination of conidia.	Chen et al. (2014)
CuNPs	0.003 M	Show fungicidal activity against *Fusarium* causing crop diseases	Bramhanwade et al. (2015)
	20 µg/6 mm disc	Control plant pathogens better than chemical fungicide bavistin and showed inhibitory activity against *Phoma destructiva*, *Curvularia lunata*, and *Alternaria alternata*. Also used in poultry and animal husbandry as a disinfectant	Kanhed et al. (2014)
CuNPs		Show effectiveness to control *Xanthomonas oryzae* pv. *oryzae*, which caused bacterial leaf blight disease of rice	Kala et al. (2016)
CuONPs	250 mg/mL	Act as a biocompatible alternative for soil-borne disease management	Chen et al. (2019)
CuO, Cu_2O, Cu/ Cu_2O@OAmNPs	3.73 µg/mL	Produce reactive oxygen species and control pathogenic *Saccharomyces cerevisiae*	Giannousi et al. (2014)
ZnO/nanocopper composite (ZnO-nCuSi)	0.22 kg/ha	ZnO-nCuSi showed significant antimicrobial properties against canker disease of citrus under in vitro conditions	Young et al. (2017)
ZnONPs	0.1 and 0.15 M	Inhibit pathogenic *Erythricium salmonicolor* significantly	Arciniegas-Grijalba et al. (2017)

Continued

Table 15.1 Continued

Nanoparticles	Dose	Action/outcome	References
Zinc oxide (ZnO) combined with *Bacillus subtilis*	250 ppm/L	ZnO along with *Bacillus subtilis* control *Podosphaera xanthii*, the causal agent of powdery mildew in cucumber	Hafez et al. (2020)
ZnO-nanocopper-loaded silica gel	–	Effective in controlling citrus canker disease in "Ruby" grapefruit trees	Young et al. (2017)
ZnONPs	8 mg/L	Successfully inhibit leaf infection of *Peronospora tabacina*, which are the blue mold in tobacco plants	Wagner et al. (2016)
ZnONPs and MnNPs	100 ppm	Significant inhibition was observed on some selected plant pathogenic fungi viz., *R. solani*, *Fusarium oxysporum*, *Sclerotium rolfsii*, *A. alternata*, *Colletotrichum truncatum*, and *Macrophomina phaseolina* of soybean (*Glycine max*)	Dhoble and Kulkarni (2016)
AgNPs and ZnONPs	ZnONPs (≥200 μg/mL) and AgNPs (≥25 μg/mL)	Antifungal activities of these NPs are useful in developing efficient management strategies to control economically important fungal diseases	Li et al. (2017)
Chitosan NPs	3 mg/mL	Exhibit excellent inhibitory activity to control wilt disease of tomato in vitro and results showed up to 81% of protection.	Sathiyabama and Charles (2015)
Chitosan NPs	1000 ppm	Inhibit *F. graminearum* in vitro, which causes head blight (FHB) disease of wheat	Kheiri et al. (2016)
Chitosan NPs	2 mg/mL	Oleoyl-chitosan significantly inhibit the growth of plant pathogens, i.e., *Botryosphaeria dothidea*, *Nigrospora oryzae*, and *A. tenuissima*	Xing et al. (2016)
Chitosan NPs	1000 ppm	Have significant antifungal and insecticidal activities against fungi and insects	Sahab et al. (2015)
Quaternized chitosan-capped mesoporous silica (MSNs)	–	In the case of nanoformulation, a similar result shows against *Phomopsis asparagi* (Sacc.) using the half dose	Cao et al. (2016)

Chitosan-encapsulated hexaconazole and/or dazomet NPs	–	Able to boost defense mechanisms to the disease incidence and able to eradicate the pathogenic *Ganoderma* fungus infection on the oil palm seedlings	Maluin et al. (2020)
Alginate-chitson	0.05%	Alginate-chitson-encapsulated acetamiprid has good encapsulation efficiency up to 62% and the highest release performance at pH 10 over pH 7 and pH 4	Kumar et al. (2015)
Cu-chitosan NPs	0.1%	Impressive growth inhibition of *A. alternata*, *Macrophomina phaseolina*, and *R. solani*.	Saharan et al. (2013)
Cu-chitosan NPs	0.12%	Show maximum effectiveness in disease control against *Fusarium* wilt in tomato plants	Saharan et al. (2015)
Ag-doped hollow TiO_2	0.75 mg/plate for *V. inaequalis* and 0.43 mg/plate against *F. solani*	Enhance fungicidal activity against *Fusarium solani* and *Venturia inaequalis* under visible light exposure	Boxi et al. (2016)
Sulfur NPs	22 mM	Excellent effectiveness against infected tomato leaves and apple scab disease, caused by *Fusarium solani* and *Venturia inaequalis*, respectively	Rao and Paria (2013)
Nano-silicon (NSi)	1.5 and 3 mM	Improve the resistance of *Vicia faba* L. plants against *Botrytis fabae* disease	Hasan et al. (2020)
Mesoporous silica NPs	–	Show low risk of drug accumulation in cucumber plants	Zhao et al. (2017b)
MgO-, TiO_2-, and ZnONPs	100 ppm	Exhibit effective and environmentally friendly alternative of fungicides to control damping-off and root rot disease of sugar beet	El-Argawy et al. (2017)
MgONPs	0.05% and 0.1%	Induce systemic resistance in the tomato plants against pathogen *Ralstonia solanacearum*	Imada et al. (2016)
Titanium oxide (TiO_2NPs)	0.43 mg/plate for *Fusarium solani* and 0.75 mg/plate for *Venturia inaequalis*	Display enhanced activities against *Fusarium solani* as well as *Venturia inaequalis* when exposed to light	Boxi et al. (2016)
Platinum (Pt) NP	4 and 8 μg/mL	Show zone inhibition against in vitro culture of *Colletotrichum acutatum* and *Cladosporium fulvum*	Velmurugan et al. (2016)

Continued

Table 15.1 Continued

Nanoparticles	Dose	Action/outcome	References
Mesoporous alumina NPs	400 mg/L	About 78% growth inhibition of *Fusarium oxysporum* fungus in tomato plants by mesoporous nanoparticles	Shenashen et al., 2017
Nitrogen-doped carbon nanohorn (NCNH)	100 µg/mL	The growth of the pathogen *R. solani* decreases considerably	Dharni et al. (2016)
Cobalt and nickel nanoferrites ($CoFe_2O_4$ and $NiFe_2O_4$)	500 ppm	Show significant antifungal activity against *Fusarium oxysporum*, *Colletotrichum gloeosporioides*, and *Dematophora necatrix* in vitro conditions and reduce the in vivo incidence of *Fusarium* wilt in capsicum	Sharma et al. (2017)
Nanomicelle	0.40–0.74 mg/L	Increase the bioefficacy of fungicides against *R. solani*	Koli et al. (2015)
Amphiphilic polymer	50 mg/L and 100 mg/L	Nanocarriers can maintain a sufficient amount of active ingredients available for a considerable period after application, which reduces the amount of pesticides	Majumder et al. (2020)
Solid lipid NPs (SLN)		Have no phytotoxic effects on the germination of *Zea mays*, *Brassica rapa*, and *Pisum sativum*	Nakasato et al. (2017)
Solid lipid nanoparticle	1% (m/v)	Exhibit reduced toxicity and controlled release properties of fungicides	Campos et al. (2015)
Lipid-based nanoemulsions, SLN, and NLC	–	Show faster penetration into the plant roots	Nguyen et al. (2016)
CS-Cu, MV-Cu, FQ-Cu nanocomposite	100 µg/mL	Under in vitro conditions, copper composites potentially lessen bacterial spot disease severity of tomato, which is caused by *Xanthomonas perforans* strain	Strayer-Scherer et al. (2018)
Silver (Ag) in a matrix of bovine submaxillary mucin (BSM) BSM-Ag	400 mg/L	Have antimicrobial activity against devastating bacterial plant diseases of cucurbit crops	Makarovsky et al. (2018)
Chitosan and laminarian	–	Reduce the concentration of numerous amino acids, preventing downy mildew of grapevine caused by *Plasmopara viticola*	Garde-Cerdán et al. (2017)

Pectin-conjugated hollow silica microcapsules	–	The microcapsules of kasugamycin pose good pectinase as well as glutathione-responsive characteristics, high drug-loading capacity, sustained antimicrobial efficacy, and good photothermal stability against *Erwinia carotovora*	Fan et al. (2017)
Silica NP	500 ppm	Nanosilica can successfully suppress *Spodoptera littoralis* on squash	El-Helaly et al. (2016)
Solid lipid NPs	1.25% and 0.63%	Increase pesticidal efficacy against *Phthorimaea operculella* in the field	Adel et al. (2015)
Nanomicelle	0.052 mg/L	Increase stability and bioefficacy of poorly soluble diethyl phenylacetamide	Balaji et al. (2015)
Core-shell PDA@PNIPAm	60 μg/mL	The system has the key potential in controlled drug release as well as agriculture-related fields as a delivery system for pesticides with a photothermal-responsive behavior	Xu et al. (2017)
Nanospheres (polyvinyl alcohol) PVA	–	Emamectin-benzoate (EMB) slow-release microspheres provide attractive properties for pesticidal efficacy improvement and prolong the impact of EMB in the atmosphere	Wang et al. (2017)
NeemAzal/neem oil (azadirachtin and nimbidin)	–	Encapsulated PVA/Alg-MMT nanocomposite capsule beads have the potential for controlled release of pesticides	Rashidzadeh et al. (2014)

CA, citric acid; *CMC*, carboxymethyl cellulose; *EOs*, essential oils; *ICZ*, ZVI NPS-coated zeolite; *IMZ*, ZVI NPS and zeolite mixture; *NE*, lipid-based nanoemulsions; *NLC*, nanostructured lipid carriers; *OAm*, oleylamine; *OC*, organic carbon content; *OLNs*, oil-loaded nanogels; *OW*, organic waste; *PDA@PNIPAm*, polydopamine@poly (*N*-isopropyl acrylamide) polymer shell; *Pt*, platinum; *PVP-Ag ENMs*, polyvinylpyrrolidone (PVP)-coated Ag ENMs (engineered nanomaterials); *SLN*, solid lipid nanoparticles; *TS*, total solids; *TSS*, total suspension solids; *VS*, volatile solids; *VSS*, volatile-suspended solids.

pest control against California red scale (*Aonidiella aurantii*), as well as oriental fruit flies (*Bactrocera dorsalis*) compared to colloidal calcium (Hua et al., 2015).

The nanoinsecticidal power of alumina dust is more than efficient than conventional dust formulation. Pesticide-derived powders show better results for controlling various stored pests such as *Sitophilus oryzae* and *Rhyzopertha dominica* compared to available non-nanoinsecticides (Buteler et al., 2015). Besides, Balaji et al. (2015) created an active nanometric composition of a water-soluble drug, diethyl phenylacetamide (DEPA). By examining the comparative effectiveness of DEPA and nano-DEPA compounds, it was demonstrated that nano-DEPA has better efficacy compared to its standard form against *Culex tritaeniorhynchus* encephalitis vector. Therefore, nano-DEPA could be an effective means of controlling vector distribution at a reduced volume (Balaji et al., 2015).

Hydrophobic silica nanoparticles (SNPs) are used for powdery mildew treatment in various crop plants, such as *Macrotyloma uniflorum*, *Vigna mungo*, *Cicer arietinum*, *V. radiata*, *V. unguiculata*, and *Cajanus cajan* and also to treat beetle infection in the storage area. *Callosobruchus maculatus* shows a decrease in oviposition, adult emergence, and seed hurt by the effect of SNPs. Therefore, the optimal state of SNPs was shown as a seed protection agent against *C. maculatus* (Arumugam et al., 2016). In addition, in an attempt to test the effect of nano-silica compared to Silica and Diazinon as a recommended insecticide, it was applied as a foliar spray to squash plants in the control environment. That study recommended that the foliar use of nano-silica at 500-ppm concentration significantly control *Spodoptera littoralis* pest (El-Helaly et al., 2016).

One of the key benefits of nanosized formulations is the requirement of lesser volumes for the same effectiveness as conventional counterparts. For example, a biocontrol agent, *Photorhabdus luminescens*, was applied to control a diverse pest population, including mealybugs, aphids, and mites (Kulkarni et al., 2016). On the other hand, the synergistic consequence of epicatechin-capped silver nanoparticles (ECAgNPs) with gentamicin was studied in vitro. The ECAgNPs as a novel type of potential antimicrobial agent with gentamicin showed significant antifungal properties compared to the activities of epicatechin, gentamicin, and silver nitrate (Ikram et al., 2017). Furthermore, when epicatechin was used as a coating agent to synthesize epicatechin-coated silver nanoparticles (ECAgNPs), the produced NPs exhibited remarkable selectivity for gentamicin in tap water, blood plasma, and serum (Ikram et al., 2018). Table 15.2 presents some nanoparticles that show promising insecticidal effects against some important insect pests.

15.2.2.1 Plant-derived NPs as insecticides

The lethal and sublethal effect of citrus peel Eos-like emulsions containing polyethylene glycol (PEG) nanoparticles (EONPs) (Fig. 15.7) are effective against the tomato parasite, *Tuta absoluta*. Increased pesticidal activity with reduced toxicity of EONPs is seen on plants (Campolo et al., 2017). Similarly, camphor EO nanoemulsion has shown potential grain protection from *Sitophilus granarius*, and other pests (Mossa et al., 2017). In addition, decaying nanospheres containing natural nanoencapsulation

Table 15.2 Some nanoparticles that show insecticidal and miticidal effects for crop protection.

Nanomaterials	Dose	Action/outcome	References
Silica-epichlorohydrin-carboxymethyl cellulose	3–12 mg/L	Encapsulated emamectin benzoate shows cellulose enzyme-responsive properties which helps to control *Myzus persicae*	Guo et al. (2015)
Ag NP dust	1000 mg/L	Proves its potent efficiency for the control of castor semilooper, *Achaea janata* (L.), and the oriental leafworm moth, *Spodoptera litura* (F.) on *Ricinus communis* L. leaves	Yasur and Rani (2015)
PCL nanospheres-*Zanthoxylum rhoifolium*	5%	Optimize the control system against *Bemisia tabaci* and offer improved protection against degradation and oxidation of NPs from sunlight and heat	Christofoli et al. (2015)
Calcium carbonte NPs	26% for the field test and 19.5% for the laboratory test	Provide advanced performance in respect of protective efficacy against punctures of oviposition from oriental fruit flies, an insect of Indian jujube	Hua et al. (2015)
Nano-Diatomaceous earth	–	Acts as a valuable tool in pest management programs of *Tribolium confusum* and *T. castaneum*	Sabbour et al. (2015)
Myristic acid-chitosan nanogel loaded with *Cuminum cyminum* essential oil (EOs)	16 μL/L	It is more toxic than *Cuminum cyminum* oil against beetle pests *Sitophilus granarius* and *T. confusum*	Ziaee et al. (2014)
Cypermethrin nanofibers as pheromone nanogels	0.87 mg/L	Show controlled release of pheromone and insecticide to attract and kill *Grapholita molesta*	Czarnobai De Jorge et al. (2017)
Nano-Al_2O_3 dust	250 ppm	Exhibits superior mortality rates and show more effectiveness against *S. oryzae* and *Rhyzopertha dominica*	Buteler et al. (2015)
Alginate-grafted anisotropic silica NPs (Alg-SiO_2)	–	pH-responsive Alg-SiO_2 NPs and the model drug λ-cyhalothrin were embedded in the emulsions. These results provide a controlled pesticide delivery system	Chen et al. (2017)

Continued

Table 15.2 Continued

Nanomaterials	Dose	Action/outcome	References
Alginate sodium-encapsulated copper(II) Schiff base modified mesoporous silica NPs	–	Show sustainable and slow release of chlorpyrifos	Huayao et al. (2016)
Silica nanoparticles (SNPs)	500 and 1000 ppm	Provide reduced infestation against pulse beetle in stored conditions	Arumugam et al. (2016)
Amino-grafted silica nanoparticles	–	Deliver a higher rate of enzyme immobilization and improve nematicidal offshoot against *Caenorhabditis elegans*	Qin et al. (2015)
Green silver nanoparticles (GSN)	17 mg/mL	Show effectiveness for root-knot nematode (*Ulva lactuca* and *Turbinaria turbinata*) and it is effective for eggplants	Abdellatif et al. (2016)
Hollow mesoporous silica nanoparticles	–	High *Plutella xylostella* larval mortality for enzymatic response of chlorantraniliprole	Kaziem et al. (2017)
CA-CMC-bentonite hydrogel composites	–	Exhibit the beneficial effect for effective control of insects that are alkaline	Sarkar and Singh (2017)
Citrus peel essential oil nanoformulations	23.10 mg/mL	Successfully applied against *Tuta absoluta* under integrated pest management (IPM) programs	Campolo et al. (2017)
Camphor essential oil nanoemulsion	181.49 μg/g	Shows potentials against *S. granarius* and other insects to protect grains	Mossa et al. (2017)

Al_2O_3, aluminum oxide; *CA*, citric acid; *CMC*, carboxymethyl cellulose; *EOs*, essential oils; *NPs*, nanoparticles; *PCL*, poly (epsilon-caprolactone).

Fig. 15.7 The SEM analysis established that the size distribution of all EO-NPs was within approximately 50 nm. Additionally, all EO-NPs (A) Bare PEG; (B) Lemon EO-NP; (C) Mandarin EO-NP; and (D) Sweet orange was consisting of clusters of spherical features within the submicrometer range. These EO-NPs increased pesticidal activity with reduced toxicity on plants as well as effective against tomato parasite.
Reproduced with permission from Campolo, O., Cherif, A., Ricupero, M., Siscaro, G., Grissa-Lebdi, K., Russo, A., Cucci, L.M., Di Pietro, P., Satriano, C., Desneux, N., Biondi, A., 2017. Citrus peel essential oil nanoformulations to control the tomato borer, *Tuta absoluta*: chemical properties and biological activity. Sci. Rep. 7, 1–10 under a Creative Commons license.

oils from *Zanthoxylum rhoifolium* have antimicrobial effects on *Bemisia tabaci* in the IPM system (Christofoli et al., 2015). NeemAzal was impregnated with polyvinyl alcohol/alginate-montmorillonite (PVA/Alg-MMT) to get a modified nanocomposite. This nanocomposite had a significant effect on preventing the release of the powerful NeemAzal-regulating pesticide (Rashidzadeh et al., 2014). Similarly, the release of a well-managed light herbicide particle (LCHP) with a shell structure was prepared based on biochar, ATP, Gly, AZO, and ASO. Therefore, this technology provides an easy way to control the release of pesticides and reduce the risk of environmental pollution (Chen et al., 2018). In addition, as bionanomaterial, nano-Diatomaceous earth (nano-DE) was compared to natural Diatomaceous earth (DE) nanoparticles as important tools in pest control programs for flour, *Tribolium confusum*, and red beetle flour *Tribolium castaneum* (Sabbour et al., 2015). When geranium EO is loaded with strong lipid nanoparticles (SLNs), it provided a high percentage of *Phthorimaea operculella* deaths in the field conditions (Adel et al., 2015).

15.2.2.2 Nanoinsecticides extracted from microbes

Nanoinsecticides fabricated from microbial origin are considered as a potential alternative to the synthetic chemical pesticides. Natural sources are usually environment-friendly and free from adverse effects. Silver nanoparticles synthesized by an endophytic fungus, *Phomopsis liquidambaris* strain SA1 (SAAgNPs) shows high efficacy as an antimicrobial and mosquitocidal agent (Seetharaman et al., 2018). The synthesis of nanoinsecticides using various microorganisms is an interesting area of research which has high potential for practical implications.

15.2.3 Chitosan nanomaterials as pesticides

Nanofungicides are expected to have higher uptake by the plant by minimizing the run-off, volatilization, and leaching losses. The chitosan is a biopolymer derived from the exoskeleton of crustaceans and cell wall of fungi. This natural substance has growth promoting and immunity inducing effects on plants. The chitosan NMs (CNMs) show high promise as effective antimicrobial pesticides toward pathogenic plant microbes, such as bacteria, fungi, and viruses (Kumaraswamy et al., 2018). Chitosan-based nanoparticles were tested to control the *R. solani*, *A. alternata*, and *M. phaseolina*. The obtained results are evident that chitosan and Cu-chitosan NPs have incredible potentials in crop protection (Saharan et al., 2013). Similarly, Cu-chitosan NPs were evaluated for their antifungal and growth promotion efficacy in tomato (*S. lycopersicum*) against *F. oxysporum* and *A. solani* in in vitro conditions. The results of that investigation established significant growth promotion and antifungal competencies of these NPs (Saharan et al., 2015). Sathiyabama and Charles (2015) reported that cell wall polymer (chitosan) (CWP-NP) isolated from *F. oxysporum* f. sp. *lycopersici* exhibited antifungal activity under in vitro conditions. When CWP-NPs are applied as a foliar application to tomato, it reduced the severity of wilt disease and gave the higher yield. Besides, microencapsulation techniques were used to enhance insecticidal efficacy of the chemical pesticides. In this connection, the nanogels of myristic acid-chitosan loaded by EO (essential oil) extracted from cumin, *Cuminum cyminum* L., were more toxic than *C. cyminum* oil alone against two beetle pests, *S. granarius* and *T. confusum* (Ziaee et al., 2014). Chitosan (CS)-*g*-poly (acrylic acid) (PAA) (CS-*g*-PAA) nanoparticles were well dispersed and stable in an aqueous solution. These NPs show a significant antifungal effect against different fungi isolated from soybean seeds *Aspergillus niger*, *F. solani*, and *F. oxysporum* (Sahab et al., 2015). Additionally, mycelium growth experiments were conducted using oleoyl-chitosan (O-chitosan) NPs against *Alternaria tenuissima*, *Nigrospora sphaerica*, *N. oryzae*, and *Botryosphaeria dothidea*. The findings suggest that O-chitosan NPs could be a convenient alternative to synthetic chemical pesticides for controlling pathogenic fungi (Xing et al., 2016).

In the vineyard, downy mildew disease caused by *Plasmopara viticola* is a destructive obligate biotrophic pathogen which significantly reduce the yield of quality of grapevine fruits. Copper-based pesticide formulations are considered detrimental to the quality of wine from the fruits. To control grapevine downy mildew (GDM) disease, LamE (laminarian) and chitosan application individually or alternately with copper hydroxide were found as a suitable alternative that would not affect the grape quality (Garde-Cerdán et al., 2017). Another nano-chitosan (CS-*g*-PAA) (chitosan-*g*-polyacrylic acid) shows the highest effect against the three insects of soybean. The findings of that experiment indicated that the mean value of eggs deposited/female significantly decreased by the application of CS-*g*-PAA in laboratory and semifield conditions. Similar trends of efficacy of this nanochitosan were also reported in controlling the *Callosobruchus maculatus* and *C. maculatus* insects (Sahab et al., 2015).

Recently, fungicides of hexaconazole and/or dazomet were encapsulated into chitosan NPs to combat basal stem rot (BSR) disease, which is caused by *Ganoderma boninense*. The results indicate that chitosan NPs act as a biocide and/or as a nanocarrier for the conventional fungicide. In addition, the particle size of the chitosan-based nanoformulations plays an important role in suppressing the disease. The synergistic outcome of chitosan-hexaconazole-dazomet NPs displays higher disease reduction than the untreated infected oil palm seedlings (Maluin et al., 2020).

15.2.4 Nanoparticles as a nematicide

Root-knot nematodes are recognized as significant disease-causing agents to the eggplants. Novel approaches of green silver nanoparticles (GSN) are applied for controlling plant-parasitic nematodes to avoid synthetic nematicide exposures. The GSN has been synthesized using the algal extract solution prepared from two algal genera (*Ulva lactuca* and *Turbinaria turbinata*). Nematicidal effects of GSN were assessed under the controlled condition on eggplants (*Solanum melongena* cv. Login). The GSN was found more active in reducing the second juvenile (J2s) stage of *Meloidogyne javanica* population in the soil. On the other hand, GSN is proven effective for controlling the root-knot nematode (both *U. lactuca* algae and *T. turbinata*) that is similar to the chemical control method without causing any phytotoxic effects in eggplants under treatment (Abdellatif et al., 2016).

15.2.5 Nanobactericides

Copper (Cu) bactericides are widely used as crop protection agents in agriculture. However, Cu leaching into the surrounding ecosystem, Cu accumulation in soil, and Cu resistance development in phytopathogenic bacteria are the major concerns associated with the Cu-based bactericides. The Cu application could be reduced by adding Zn and increasing Cu bioavailability via nanotechnology. The prepared nano-copper-loaded silica gel (ZnO-nCuSi) demonstrates strong in vitro antimicrobial characteristics against various bacterial species. Young et al. (2017) proved that the ZnO-nCuSi is a successful citrus canker disease-controlling element at less than half the metallic rate of the conventional cuprous oxide/zinc oxide formulations in the field conditions.

Bacterial blight is caused by *Xanthomonas axonopodis* pv. *punicae* in pomegranate. The silver nanoparticles can reduce silver ions in aqueous solutions with the help of green synthesis method. The leaf extracts of tulsi, neem, tridax, and drumstick were used in AgNP synthesis. Thus, synthesized AgNPs potentially control the growth of *X. axonopodis* pv. *punicae* (Sherkhane et al., 2018). Besides, synthesized AgNPs in bovine submaxillary mucin (BSM) have potential antimicrobial importance on plant pathogenic bacterial strains (Gram-negative and Gram-positive). The results showed that melon seeds treated with BSM-AgNPs successfully stop seed-to-seedling transmission of *Acidovorax citrulli*, one of the most aggressive pathogens of cucurbit crop grown worldwide (Makarovsky et al., 2018).

Bacterial wilt is a serious problem that affects many important food crops like tomato and tobacco due to infection by *Ralstonia solanacearum*. The substantial antimicrobial efficacy of NPs against phytopathogens is widely studied in agriculture. However, few reports have concentrated on soilborne plant pathogenic bacteria. Imada et al. (2016) showed that disease incidence significantly reduces after drenching tomato roots with MgONP suspension earlier to inoculation. Fig. 15.8 illustrates the size dimension of MgO (magnesium oxide) NP, which induces a systemic resistance in tomato plants. Besides, copper oxide NPs (CuONPs) synthesized using papaya leaf extracts have the bactericidal action against *R. solanacearum*, the causal agent of bacterial wilt disease in tobacco plants under controlled conditions using laboratory and greenhouse. The results suggest that CuONPs have an antibacterial property in soilborne disease management (Chen et al., 2019). Previously, greenhouse investigations validate that all copper composites potentially reduced the severity of bacterial spot disease (Strayer-Scherer et al., 2018).

15.2.6 Nanomaterials as antiviral agents

The ability of the virus to infect, transport nucleic acid genome to a definite site in the host cell, self-replicating system, and package of the nucleic acid, and finally come out of the host cell accurately in a systematic manner has fulfilled the requirement to be

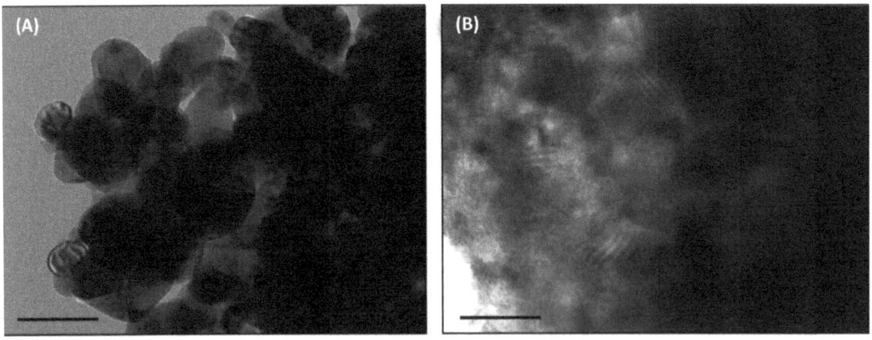

Fig. 15.8 TEM micrographs of MgO (magnesium oxide) NPs. *Bars* indicate (A) 200nm and (B) 20nm. The NP induces systemic resistance in tomato plants.
Reproduced with permission from Imada, K., Sakai, S., Kajihara, H., Tanaka, S., Ito, S., 2016. Magnesium oxide nanoparticles induce systemic resistance in tomato against bacterial wilt disease. Plant Pathol. 65, 551–560 under a Creative Commons license.

used in nanotechnology for controlling virus diseases in plants (Singh et al., 2015). The AgNPs were synthesized using some newly isolated bacterial strains (*Bacillus pumilus*, *Bacillus persicus*, and *Bacillus licheniformis*). These AgNPs are capable of controlling bean yellow mosaic virus infections of faba beans (Elbeshehy et al., 2015). Antiviral nanomaterials should be an interesting area of research for the management of destructive viral diseases in the crop plants.

15.2.7 Nanomolluscicides

The land snail (*Eobania vermiculata*) causes considerable damage to agriculture (Ali et al., 2015b). Metaldehyde is being used to show defense against invasion and damage caused by snails, slugs, and molluscs (Abubakar et al., 2017). But, this commercially available traditional molluscicide has limitations like costly, nonspecific, and injurious effects on nontargeted organisms (Coelho and Caldeira, 2016). The AgNPs biologically synthesized using *Raphanus sativus* var. *aegyptiacus* was effectively used to control the land snail (Ali et al., 2015b). Moreover, Omobhude et al. (2017) prepared a molluscicidal polymeric NP of polylactic acid by encapsulating curcumin and nisin and evaluated its efficacy against *Biomphalaria pfeifferi*. This nanoformulation showed a high efficacy against *B. pfeifferi*.

15.3 Weed management in the crop field using nanoherbicides

Weeds are undesirable plants that reduce crop yield, thus needing timely eradication of them from the crop fields. The application of conventional herbicides destroy weeds from the aboveground parts of a plant but does not eradicate tubers or roots that sprout new weed plants under suitable conditions (Ali et al., 2014). Furthermore, the uncontrolled and repeated use of the same conventional herbicide may lead to herbicide resistance in the target weed species (Seehra, 2017). In these circumstances, the use of nanoherbicides could be an alternative and effective method to eliminate weeds from the agricultural field (Hossain et al., 2020; Thangadurai et al., 2020). Nanoparticles can serve as a good carrier as well astern into nanoformulation when added with conventional herbicides. Nano-based formulations of herbicides are prepared by exploiting the nanoscience potentials for the effective delivery of biological or chemical pesticides. Weed control using nano-based herbicide could reduce the unexpected herbicide resistance, preserve the activity of the active ingredient (AI), and extend their release capacity over a longer period of time (Manjunatha et al., 2016). The nanoparticle-based herbicides have several advantages like improved solubility in water, lower weed resistance, and decreased toxicity in comparison with conventional herbicides (Pereira et al., 2014). The development of particular herbicide particles encapsulated with nanomaterial targets at definite receptors exists at the root system of the targeted species of weed (Seehra, 2017). Commercial herbicide formulations can be embedded in specific engineered nanomaterials (ENMs) to achieve targeted delivery into the root system of weeds. These engineered herbicides can be taken up by the roots of weeds and transported to target cells due to the high penetrability of ENMs. After the uptake of nanoherbicides, significant hindrance of metabolic pathways like glycolysis takes place, which causes phytoxicity and finally death of respective weed species (Duhan et al., 2017). Moreover, these nanoscale herbicides could also be applied in rain-fed areas to reduce vapourization loss of herbicides due to inadequate soil moisture. Based on the controlled release properties of herbicides via encapsulation, the weeds can be significantly destroyed. Currently, nanomaterials are also used as adjuvants to enhance the herbicidal activity of conventional formulations. These nanoformulations help in overcoming the prime drawback of the traditional herbicide industry, as there is an evolution of herbicide resilient plants (Seehra, 2017). Table 15.3 presents some nanoparticles that show herbicidal effects on the management of weeds.

15.3.1 Polymeric nanoparticles

A variety of nanoparticles were applied for nanoherbicide formulation; among them polymeric nanoparticles are the smartest form that are prepared either in the form of nanocapsules or nanospheres. As a polymer, epsilon-caprolactone is repeatedly used for the encapsulation of atrazine herbicide because it has good physicochemical properties along with enhanced bioavailability and biocompatibility (Seehra, 2017). For instance, the effectiveness of nanoatrazine was assessed in the case of postemergent control of the tolerant weed like *Digitaria insularis* at the initial phase of development. These results suggest that nanoencapsulation of atrazine shows its postemergent herbicidal potentiality against *D. insularis* (sourgrass) plants at the time of initial development (Sousa et al., 2020). Previously, the application of atrazine-bearing polycaprolactone nanocapsules (PCL) exhibited their potential for the postemergence control of *Amaranthus viridis* and *Bidens pilosa* (Sousa et al., 2018).

The nanoparticle-based systems of herbicide delivery are generally composed of biodegradable polymeric substances that are nontoxic metabolites (Seehra, 2017). Nano-enabled herbicides play a significant role in the sustainable elimination of undesirable weeds (Kah and Hofmann, 2014; Iavicoli et al., 2017). For instance, poly(epsilon-caprolactone) nanocapsules are considered as an efficient carrier for atrazine and are prepared as an alternative system to cut the harmful environmental impacts of the herbicide. Their results proved that encapsulated atrazine into poly(epsilon-caprolactone) nanocapsules could be used for effective control of weeds without any phytotoxicity to susceptible crop species compared with nanoscale atrazine (Preisler et al., 2020). Similar results were found using PCL comprising atrazine against maize plants (Oliveira et al., 2015) and nanoformulation of metribuzin (Fig. 15.9) against soybean (Diyanat and Saeidian, 2019). Schnoor et al. (2018) evaluated the herbicidal efficacy of synthesized poly(lactic-*co*-glycolic-acid) (PLGA) loaded with atrazine, using potato as the test plant. In an in vitro condition, the potato root length is significantly inhibited by nanoformulation. Based on their findings, this nanoformulation can be treated as a nanoherbicide for controlling weed growth. Similarly, herbicide (metsulfuron-methyl) containing pectin (polysaccharide) nanoparticles showed improved efficacy and environmental safety (Kumar et al.,

Table 15.3 Some nanoparticles that show herbicidal effects for crop protection.

Nanomaterials	Dose	Action/outcome	References
Poly(epsilon-caprolactone)	1 mg/mL	Effective against a target organism compared to the conventional one	Pereira et al. (2014)
Poly(epsilon-caprolactone)	1 µg/mL	Decreases growth properties of purslane without any side effect on soybean	Diyanat and Saeidian (2019)
Poly(epsilon-caprolactone)	200 g/ha	Controls weed efficiently without showing any phytotoxicity to susceptible crops compared with its nonnanoformulations	Preisler et al. (2020)
Poly(epsilon-caprolactone)	0.1 mg/mL	Harmless and effective for weed management without affecting crop growth	Oliveira et al. (2015)
Poly(epsilon-caprolactone) (PCL) nanocapsules	200 g/ha	Proves their significant potentiality for postemergence controlling of annual herbs *Bidens pilosa* and *Amaranthus viridis*	Sousa et al. (2018)
Poly(epsilon-caprolactone) nanocapsules	1000 or 2000 g/ha	Shows postemergent herbicidal activity against *Digitaria insularis* plants at initial developmental stages, favoring the control of this atrazine-tolerant weed	Sousa et al. (2020)
Pectin (polysaccharide) nanoparticles	–	Shows improved effectiveness and better environmental protection compared to nonnanoformulations	Kumar et al. (2017)
Poly(lactic-*co*-glycolic-acid) (PLGA)	54.0 µg/mL	Expresses a major reduction in vegetative growth where the root length is significantly affected in vitro	Schnoor et al. (2018)
Poly(epsilon-caprolactone)	–	Nanocapsulated triazine herbicide formulation reduces the quantity of herbicides used for weed control	Clemente et al. (2014)
Poly(epsilon-caprolactone)	0.1 mg/mL	Shows potential herbicidal properties at lower concentrations compared to conventional formulations and proved a substantial increase in respect of herbicidal efficacy	Bombo et al. (2019)
Solid lipid nanoparticles (SLN)	3 kg/ha	SLN reduces the toxicity of herbicides, which is considered an environment-friendly method of controlling agricultural weeds	de Oliveira et al. (2015)
Nanomicelle	1.0×10^{-6} mol/L	It is used in the controlled release of 2,4-D.	Ding et al. (2016)
Chitosan NPs	10 mg/L	Encapsulation of herbicides like paraquat reduces their toxicity	Grillo et al. (2015)
Alginate microparticles	6 g/L	Acts as a carrier system for herbicides, which has a suitable controlled release capacity for effective weed control	Faria et al. (2017)
Alginate/chitosan NPs Chitosan/tripolyphosphate NPs	400 g/ha 400 g/ha	Have reduced toxicity toward nontarget organisms but maintain satisfactory herbicidal activity against the target weed species, *Bidens pilosa*	Maruyama et al. (2016)

Continued

Table 15.3 Continued

Nanomaterials	Dose	Action/outcome	References
Chitosan/tripolyphosphate (CH/TPP) NPs	10^{10} particles/mL	Exhibit enhanced efficacy and provide significant molecular interactions between chitosan, tripolyphosphate, and paraquat	Moreno et al. (2019)
Chitosan/tripolyphosphate nanoparticles	0.48 mg/mL	Reduce health hazards and adverse environmental effects	Grillo et al. (2014)
Glutathione-responsive carboxymethyl chitosan NPs	3 kg/ha	Diuron-loaded NPs prove effectiveness against *Echinochloa crusgalli* and do not affect plant growth of nontarget species (*Zea mays*). It poses considerable controlled release properties of herbicides	Yu et al. (2015)
Mesoporous silica nanoparticles	–	Show higher herbicide (2,4-dichlorophenoxyacetic acid) loading capacity and significant bioactivity on target plant as well as do not have any harmful effects on nontarget species	Cao et al. (2017)
Rice husk nanosorbents	2.5 kg/ha	Present enhanced herbicidal action against the target plant like *Brassica* sp. compared with nonnano 2,4-D	Abigail et al. (2016)
Nanoemulsion of *Thymus capitatus* L. (wild thyme) *Thymus capitatus* L. (cultivated thyme) *Majorana hortensis* L. (marjoram)	10 ppm	Results prove that nanoformulations could be contributed to perennial weed control	Balah and Abd El Azim (2016)
LCHP with nanocomposite	–	Performs a significant adhesion ability on the leaf surface of weeds. The characteristics improve the control efficacy of weed species	Chen et al. (2018)
Sepiolite	5 g/L	Reduces leaching of herbicides and exhibits anticipated bioactivity for weed control	del Carmen Galán-Jiménez et al. (2020)
Isocyanate-functionalized silica cross-linked PEI	–	Shows considerable efficiency of drug loading, protects drugs from photo- and thermal degradation as well as exhibit a controlled release profile of the formulations	Liang et al. (2017)

NPs, nanoparticles; CH/TPP, chitosan/tripolyphosphate; LCHP, light responsively controlled release herbicide particle; SNPs, silica nanoparticles; PCL, poly(epsilon-caprolactone); PEI, polyethylenimine; PLGA, poly(lactic-*co*-glycolic-acid).

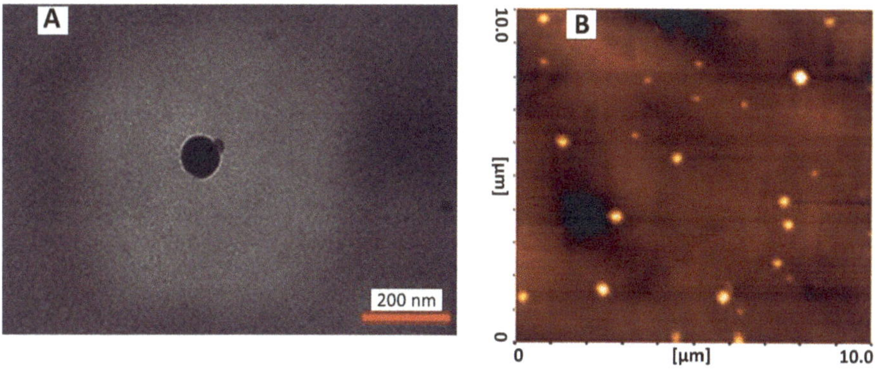

Fig. 15.9 Transmission electron microscopy images are showing nanoformulation of metribuzin through poly(epsilon-caprolactone) nanocapsules (A) Polycaprolactone (PCL) nanocapsules containing metribuzin (MET) have regularly spherical morphology with a size range of 150–250 nm and (B) Photographs of PCL/MET nanocapsules (7 days after preparation). The images exhibited regularly spherical features without any aggregation. The nanocapsules could be used for effective control of weed without any phytotoxicity to susceptible crop species.
Reproduced with permission from Diyanat, M., Saeidian, H., 2019. The metribuzin herbicide in polycaprolactone nanocapsules shows less plant chromosome aberration than non-encapsulated metribuzin. Environ. Chem. Lett. 17, 1881–1888 with kind permission from Springer Nature.

2017). In addition, improvement of the interactions among the nanomaterial PLGA [poly(lactic-*co*-glycolic-acid)], herbicide atrazine, and solvents are depend on a better understanding of the toxicity mechanism of nanoformulations against weeds. For instance, a nanocapsule like poly(epsilon-caprolactone)-loaded ametryn and atrazine is more effective against algae (*Pseudokirchneriella subcapitata*) as compared to their conventional formulation (Clemente et al., 2014). Furthermore, Bombo et al. (2019) stated similar findings of good herbicidal activity and efficacy of nanoencapsulated atrazine pesticide at a low concentration.

Grillo et al. (2015) reported numerous applications of the polymeric NPs. These NPs act as a carrier for the pesticide delivery. As nanocarriers, alginate/chitosan (Ag/Cs) NPs were preferred to prepare an encapsulated paraquat herbicide. Moreno et al. (2019) provided insights regarding molecular interactions among chitosan, tripolyphosphate, and paraquat. In another study, polymeric microparticles containing tebuthiuron were found as an appropriate carrier for the controlled release herbicides, which could control weeds effectively (Faria et al., 2017). Moreover, the encapsulation showed a considerable reduction in herbicidal toxicity on nontarget organisms but maintained a satisfactory herbicidal activity against targeted weed species like *B. pilosa* (Maruyama et al., 2016) as well as reduced adverse impacts on human health and the environment (Grillo et al., 2014). Furthermore, the herbicidal activity of the encapsulated herbicide was also assayed against a weed species, *Echinochloa crusgalli*. It revealed that the diuron-loaded NPs, which has controlled release properties during the delivery of herbicides are effective in controlling *E. crusgalli* (Yu et al., 2015).

A novel amphiphilic polymer-based nanopesticide poses self-assemble capacity into photoresponsive micelles in liquid solutions for controlled discharge of herbicide particles. It revealed that in the absence of light no release was sensed while the rate of cumulative release from the micelles enhanced gradually based on solar-simulated radiation (Ding et al., 2016). Besides, solid-lipid NPs (SLN) loaded with herbicides atrazine and simazine displayed significant physicochemical stability as well as higher encapsulation efficiencies. Cytotoxicity assays displayed that the presence of SLN NPs reduces the toxicity of herbicides, which are considered an environment-friendly technique for weed management in agriculture (de Oliveira et al., 2015).

15.3.2 Inorganic nanoparticles

In the recent past, silica dioxide NPs (SiO_2NPs) were reported as a carrier of inorganic herbicides for pH-sensitive active substances. These SiO_2NPs preserve the ideal concentration of herbicides with a related reduction in the frequency of the rate of herbicide consumption. This system can protect and stabilize the herbicides as well as decrease their wastage through an easy deposit on leaves of the plant. Another novel nanoformulation such as functionalized positively charged mesoporous silica NPs (MSNPs), which has the potential bioactivity against the target plant species without showing any antagonistic effects on the nontarget plant. This strategy could widely be used for agrochemicals, which are carrying charges and carriers having opposite charges to reduce the harmful effects on the environment (Cao et al., 2017).

15.3.3 Agro-industrial waste-based nanoparticles

Abigail et al. (2016) prepared nanosized rice husk waste and applied it to test the capacity as a nanocarrier for herbicide 2,4-dichlorophenoxyacetic acid (2,4-D). The developed nanocarrier from rice husk exhibited improved and reversible sorption of herbicide 2,4-D, revealing its exceptionality to serve as a good carrier for encapsulation of herbicides. Adsorption of herbicide 2,4-D onto the nanosized rice husk did not reduce the herbicidal action when verified against the weed viz., *Brassica* sp., in comparison with conventional 2,4-D. The nanosized rice husk-based delivery technique of herbicide could be amazing because of its multiple benefits of constructive utilization of the waste apart from efficiently using the herbicide with environmental pollution. Thus, the evolution of herbicide resistance due to the indiscriminate use of conventional herbicides could successfully be eradicated by using these nanostructure-based herbicide carriers.

15.3.4 Other nanomaterials

The herbicidal activity of EOs (essential oils), as well as their Mac-E (macroemulsion) and Nano-E (nanoemulsion) formulations of *Thymus capitatus* L. (cultivated and wild thyme) and *Majorana hortensis* L. (marjoram) was examined against seed and seedling growth of *Convolvulus arvensis* and *Setaria viridis*. The herbicidal activity depends on the types, concentration, and formulation of oils as well as the growth stage of the weeds. These experimental findings indicate that nanoformulations could be used in perennial type weed control programs (Balah and Abd El Azim, 2016).

Recently, glyphosate has been declared as an apparent carcinogen by the International Agency for Research on Cancer, an organization under WHO (Guyton et al., 2015). Therefore, nano-enabled herbicide technique was tested as an alternative to minimize glyphosate resistance and development of glyphosate-tolerant GM crops. In relation to this, Chen and coworkers in 2018 investigated the herbicidal potentials of nanocomposite combining azobenzene (AZO), amino silicone oil (ASO), attapulgite (ATP), glyphosate, and biochar. Photoresponsive controlled release performance was significantly stimulated by the nanocomposite when tested against Bermuda weeds. The robust adhesion property of the nanocomposite on the leaf surface was proposed by the authors as a mechanism that enhances the efficacy of nanomaterials, which reduces run-off loss and unwanted environmental hazards(Chen et al., 2018).

The mesotrione (MS) [2-(4-mesyl-2-nitrobenzoyl) cyclohexane-1,3-dione] herbicide was reported as SRF (slow-release formulation) for prolonged weed control compared to conventional counterpart. In the case of maize, the use of SRF and conventional Callisto as postemergence indicated that the quantity of herbicide used in the soil from SRF was almost two times higher compared to Callisto. For longer times, no significant dissimilarities were observed in soil accumulation as well as in the crop yield. Consequently, mesotrione (MS) leaching was significantly reduced due to the utilization of the prepared sepiolite-based formulation over the traditional formulation while continuing the anticipated bioactivity for weed control (del Carmen Galán-Jiménez et al., 2020). Additionally, microcapsules are greatly desirable for accomplishing the most effective application of the pesticide as well as reducing pollution of the environment. The microcapsule formulations of silica-IPTS-PEI had long-term durability and more herbicidal action against specific weeds compared with pendimethalin emulsifiable concentration. It has less genotoxicity than conventional pendimethalin as well as is proved as an eco-friendly nanoherbicide preparation (Liang et al., 2017).

15.4 Pesticides with nanocarriers in delivery systems

Conventional pesticides are encapsulated in nanoscale carriers that are trapped by the stem or root system of plants. These nanocarriers would curtail the loss of pesticides, resulting in higher commercial profit and improved environmental protection. The pesticide penetration capacity through the nanocarrier system into plant roots and their upward transportation from the root to the stem were examined by many researchers. For this purpose, solid lipid nanoparticles (SLN), nanostructured lipid carriers (NLC), and lipid-based nanoemulsions (NE) were successfully prepared that provide a significant basic background essential to produce a new cohort of pesticides (Nguyen et al., 2016).

15.4.1 Controlled release systems (CRS)

The controlled release property of active ingredients (a.i.) is a very good characteristic of any pesticides. Nanotechnology has recently allowed the production of formulations for the controlled release of active ingredients. The nanoencapsulation technique of pesticides maintains slow/controlled release during delivery into the plant, which can be termed as a controlled release system (CRS). As a controlled release (CR) formulation, nano-Mancozeb was prepared using polyethylene glycol (PEG)-based functionalized amphiphilic-type copolymers to control early blight in tomatoes. The findings of that experiment showed optimized release characteristics of Mancozeb to achieve disease control up to the desired period after application that reduced the number of applications as well as the amount of pesticide use (Majumder et al., 2020). Additionally, carbendazim (MBC) (methyl-2-benzimidazole carbamate) and tebuconazole (TBZ) [(*RS*)-1-(4-chlorophenyl)-4,4-dimethyl-3-(1*H*-1,2,4-triazol-1-ylmethyl)pentan-3-ol] are widely used in agriculture for the prevention and control of fungal diseases (Fig. 15.10). Moreover, SLNPs and polymeric nanocapsules are considered carrier systems that offer benefits including changes in the release profiles of bioactive compounds and their transfer methods to the active site, which reduced losses because of leaching or degradation and decreased toxicity in the environment and humans. The conventional pesticide release profiles were altered when the non-nanocompounds such as MBC and TBZ were loaded in the nanoscale particles. Furthermore, the cytotoxicity analysis indicated that encapsulation of conventional fungicides reduced their toxicity level. These new systems offered novel alternatives for the management of plant fungal diseases (Campos et al., 2015). Moreover, novel core-shell PDA@PNIPAm (polydopamine@poly(*N*-isopropylacrylamide)) nanocomposites are used as near infrared (NIR), which have light controlled release targeted system for pesticide delivery. The core PDA microspheres offer additional active sites for pesticides and act as an alternative to conventional gold-based nanoparticles. These NIR-sensitive, cost-effective, and photothermal-responsive nanomaterials are widely applicable as agricultural pesticide or pharmaceutical drug delivery systems (Xu et al., 2017). Similarly, various fungal diseases caused by plant pathogen *R. solani* can effectively be controlled by a single application of nano-carbendazim for the entire lifespan of crops (Koli et al., 2015).

The controlled release systems (CRS) of pesticide formulation are considered the latest and most innovative techniques. For example, the incorporation of cypermethrin in nanofibers for *Grapholita molesta* (Busck) suggested a controlled release of pheromone and insecticide (Czarnobai De Jorge et al., 2017). Moreover, Yasur and Rani (2015) demonstrated that $AgNO_3$ has a significant negative impact on responses of growth and feeding of two lepidopteran pests, namely Asian armyworm (*Spodoptera litura* F.) and castor semilooper (*Achaea janata* L.) which attacks *Ricinus communis* L. commonly known as a castor plant. In another study, enzymatic-responsive controlled release formulations (CRFs) were fabricated by anchoring mechanically interlocked molecules using α-cyclodextrin onto the surface pore rims of hollow mesoporous silica (HMS) to control the larval infestation of *Plutella xylostella* (diamondback moth) which is among the most injurious insect pests of

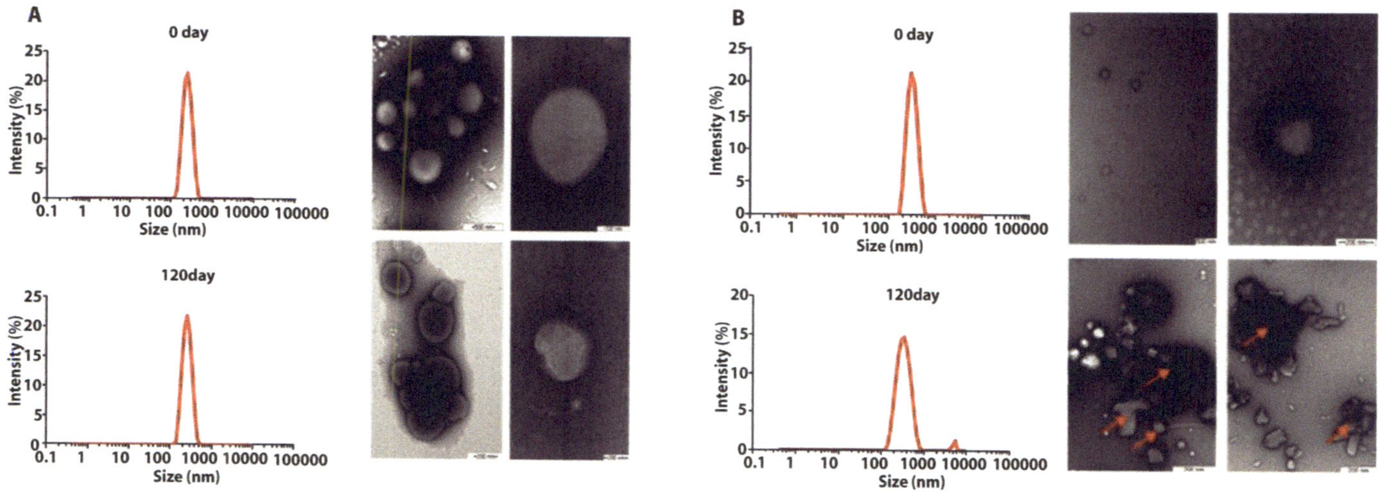

Fig. 15.10 TEM images and morphological analysis of NPs containing MBC (methyl-2-benzimidazole carbamate) and tebuconazole (TBZ). The nanocapsule polycaprolactone (NC-PCL) and solid lipid nanoparticles (SLNs) were spherical. The size distribution of NCs was 490 and SLNs was 191.6 nm. Herein, (A) the NCs remained unchanged after 120 days of storage at ambient temperature but (B) the SLNs revealed the presence of several nanocrystals *(arrow marked)*, which were possibly composed of fungicides. Furthermore, the cytotoxicity analysis proved the encapsulation of conventional fungicides. These new systems offered novel alternatives for the management of plant fungal diseases due to their reduced toxicity level. Reproduced with permission from Campos, E.V.R., De Oliveira, J.L., Da Silva, C.M.G., Pascoli, M., Pasquoto, T., Lima, R., Abhilash, P.C., Fraceto, L.F., 2015. Polymeric and solid lipid nanoparticles for sustained release of carbendazim and tebuconazole in agricultural applications. Sci. Rep. 5, 1–14 under a Creative Commons license.

cabbage and other crucifer crops throughout the world (Kaziem et al., 2017). Additionally, pH-sensitive citric acid crosslinked carboxymethyl cellulose hydrogel composites were prepared and encapsulation of thiamethoxam in hydrogel composites was achieved which will be useful for better control of insect pests through selectively triggering the release of insecticide in their alkaline guts (Sarkar and Singh, 2017).

Alginate-chitosan-encapsulated acetamiprid has good encapsulation efficiency and maximum release behavior at pH 10. The nanoformulation was proved superior compared to commercial formulation in terms of controlled release capacity. Therefore, this nanoformulation would be helpful in lessening pesticide consumption (Kumar et al., 2015). Additionally, Guo et al. (2015) showed that silica-epichlorohydrin-carboxymethyl cellulose-encapsulated emamectin benzoate has excellent cellulose enzyme-responsive properties and a sustained pesticide efficacy against *Myzus persicae*. Moreover, the optimal emamectin benzoate (EMB) pesticide slow-release microspheres had outstanding antiphotolysis, good leaf distribution, stability, and controlled release features. Therefore, EMB slow-release microspheres were an excellent candidate for upgrading pesticidal efficacy and extending the control effect of EMB in the atmosphere (Wang et al., 2017). Moreover, a novel system of pesticide delivery was prepared by combining *N*-(2-hydroxyl) propyl-3-trimethylammonium CS chloride (HTCC)-coated mesoporous silica NPs (MSNPs). HTCC-coated MSNPs displayed superior performance compared to bare MSNPs. It showed significant performance in fungicidal actions on *Phomopsis asparagi* and reduced the pesticide amounts (Cao et al., 2016).

A series of amino, carboxylic, and aldehydic surface-grafted silica NPs (SNPs) were prepared based on SiO_2NYSi40 NPs to develop a delivery system, which will be efficient, biocompatible as well as cost-effective. Experimental results showed that SNPs posed significant delivery capacity to carry a large protein into *Caenorhabditis elegans*. Therefore, prepared SNPs exhibit functions as an effective nanocarrier for the delivery of macromolecular biopesticides or drugs, which are proteic in nature and have important applications in agriculture as well as biotechnological field (Qin et al., 2015). Moreover, mesoporous silica NPs are utilized as pesticide nanocarriers in plants, which have been considered a novel method to reduce the injudicious use of conventional pesticides. Mesoporous silica nanoparticles were prepared to achieve pyrimethanil-loaded NPs, which had nearly no harmful effect on the rate of circulation and dissipation in cucumber plants. The application of pesticide-loaded NPs on leaves of plants poses a minor risk of pyrimethanil accumulation in the edible plant part (Zhao et al., 2017b).

The research on ecologically responsive nanopesticides is in an emerging phase. Application of nano-based formulations could be a recognized strategy for the effective and precise release of chemical ingredients on the particular living being. Moreover, it checks the accumulation of agrochemical such as pesticides in nature. Therefore, it is very challenging to prepare nanobiopesticides that will be stable and active under diverse climatic conditions. Huang et al. (2018a,b) emphasized that if ecological conditions such as humidity, changes of enzyme, light, temperature, and pH are considered during pesticide application, nanoformulations can boost up the efficiency of pesticides and decrease environmental pollution. For example, coumarin copolymer was used to fabricate a photo-responsive 2,4-D that acts as a nano controlled release preparation (Atta et al., 2015). Compared to the conventional counterpart, the controlled release pesticides displayed improved thermal stability and lesser dissolvability.

Nanoencapsulation is another method that can protect active substances from adverse atmospheres while also serving controlled release and target-oriented delivery. Nanoencapsulated products are made to defend active ingredient of agrochemicals from degradation as well as for enhanced stability, a slow-release property, solubility, and specificity over an extended period (Prasad et al., 2017). Furthermore, nanoencapsulated products are utilized at lower concentrations than non-nano ones. These products will be eco-friendly and less detrimental to human beings (Nuruzzaman et al., 2016). Efforts are being made to encapsulate commercial pesticides with nanomaterials. Encapsulation of conventional pesticides is done using solid lipid NPs, polymer-based NMs, inorganic porous NMs, layered double hydroxides (LDHs), and nanoclays (Nuruzzaman et al., 2016). Moreover, pH-responsive alginate-grafted anisotropic silica (Alg-SiO_2) nanoparticles and the model drug λ-cyhalothrin were embedded in the emulsions. These results offer a new method for pH-responsive emulsion application as a controlled delivery system of pesticides (Chen et al., 2017). Besides, pH- and ion-responsive system of alginate sodium modified chlorpyrifos/copper(II) Schiff base mesoporous silica was prepared, and after encapsulation by ALG, it showed sustainable decreased release of chlorpyrifos with increasing pH (Huayao et al., 2016). On the other hand, kasugamycin is considered an aminoglycoside antibiotic widely used for plant disease management that is isolated from *Streptomyces kasugaensis*. But, kasugamycin has photothermal instability and low efficiency that bordered its utilization. For this reason, a unique formulation of kasugamycin was prepared via silica microcapsules, which respond to different biological stimuli produced by various pests. The prepared silica-SS-pectin microcapsules have higher loading efficiency, improved thermal properties, and light stability of nano-kasugamycin compared to their conventional counterparts. Thus, microcapsules are considered improved and sustainable antimicrobial practices that pose controlled release characteristics against *Erwinia carotovora* (Fan et al., 2017).

15.5 Research gap and the future way forward

Overall, the chapter includes different roles of nanopesticides for crop protection in agriculture. The different properties of nanopesticides have elevated the standards and overcome the limitations posed by the conventional agroparticle formulations, which have deleterious effects on the surroundings as well as living beings (Baker et al., 2019). However, there are noteworthy limitations and knowledge gaps that remain regarding the frontier nanotechniques. The limitations and knowledge gaps are summerized below.

1. As plant disease-suppressing agents, more field studies are necessary to establish the real efficacy because the reproducibility of ENM is affected under field conditions. Because most of the studies were conducted under controlled conditions, uses of commercial potting

soil and studies under field conditions are needed. Recent data on nanopesticides do not guarantee us to expect the similar results would be obtained in the natural soil conditions. Most of the nanomaterials are patented and highly expensive in preparation. These issues are limiting the practical applications of most of the nanopesticides discovered so far. Therefore, to confirm the efficacies of any nanopesticide formulations, experiments under natural field conditions are strongly recommended.

2. Considering the application of nanopesticides, there are no experimental findings on possible development of resistant pest species. There is no clear understanding regarding doses, specificity of plants, and possible collateral influences of nanoherbicides that are leading to the development of tolerant species. It is necessary to confirm that nanopesticides have no effect on nontargeted plant growth-promoting microbes or other symbiotic microbial communities living in the plant or soil. Furthermore, critical assessment for the development of resistance upon repeated application of ENM-based nanopesticides should be accomplished (Adisa et al., 2019).
3. The majority of ENMs are metallic. Some of them have already been proven to be very stable under specific ecological conditions. As such, large-scale utilization of those ENMs could possibly lead to metal contamination of soil that can create an unfamiliar environmental concern. Thus, appropriate dose and establishment of an efficient targeted delivery strategy are required before operation (Gao and Lowry, 2018; Kumar et al., 2019; Worrall et al., 2018; Gomez et al., 2020). Although the potential toxicity of many metallic and carbon NPs to crop plants has already been reported, very few studies have estimated the phytotoxic effects of polymers and SNPs. To mitigate the environmental damage, further study is required for the future use of nanoparticles in agriculture (Nakasato et al., 2017). More studies have to be conducted for a clear understanding of this issue.
4. Regarding human health, there are no findings on specific impacts due to the consumption of food obtained from crops exposed to nanosized agrochemicals (Gomez et al., 2020). However, day by day, human exposure to NPs is increasing, and its influence on human health and surroundings has become a matter of public concern. Although enormous potentials are being presented by nanotechnology, the accumulation of nanosized materials in human bodies, and the environment has triggered different health hazard and safety-related issues. Thus, a uniform and integrated international regulatory framework will require resolving the issues regarding nanomaterial use (Ghorbanpour et al., 2020).
5. Numerous studies have shown that the use of nanotechnology in agriculture improves efficiency and yield. However, the interactions among agrochemicals are comparatively unknown. Therefore, the study of pesticide toxicity and toxicokinetics of nanomaterials is necessary. With the increasing use of nanomaterials as antimicrobials, studies are desirable to inspect possible interactions and revision of risk (Louder, 2015).
6. The study of the benefit/cost ratio highlighting the real economic improvement of the application of nanoenabled pesticides is required. Finally, as long as these issues remain untouched, the possibility of large-scale application of nanoenabled strategies will be low in crop protection in agriculture.

15.6 Concluding remarks

The introduction of nanotechnology has provided breakthroughs in the field of agriculture as well as in the agrochemical industry. Nanopesticides provide the higher efficiency and target specificity in pest control compared to traditional chemical pesticides. As losses of applied nanopesticides are low, the implementation of these novel technologies significantly reduce environmental pollution and increase farm profitability. Due to unique properties, the impacts of nanopesticides on nontarget organisms, human health, biodiversity, and the environment would be less compared with its conventional ones. A plethora of nanopesticides has already been tested in the field conditions and showed their higher efficacies compared to synthetic chemical pesticides. Besides, the carrier material can destroy the pesticide active ingredient after a significant pest control stage. Considering the simple preparation technique, brilliant antipest action, and free from toxic organic solvent, the additive-based delivery system of pesticides has a bright future in plant protection and sustainable agriculture in the changing climate. However, further research is needed to confirm the safety of human health and environment before recommending any nanopesticides for promoting agricultural productivity and food security.

References

Abdelaal, K.A., EL-Shawy, E.A., Hafez, Y.M., Abdel-Dayem, S.M., Chidya, R.C.G., Saneoka, H., El Sabagh, A., 2020. Nano-silver and non-traditional compounds mitigate the adverse effects of net blotch disease of barley in correlation with up-regulation of antioxidant enzymes. Pak. J. Bot. 52, 1065–1072.

Abdel-Hafez, S.I., Nafady, N.A., Abdel-Rahim, I.R., Shaltout, A.M., Daròs, J.A., Mohamed, M. A., 2016. Assessment of protein silver nanoparticles toxicity against pathogenic *Alternaria solani*. 3 Biotech 6, 1–12.

Abdellatif, K.F., Abdelfattah, R.H., El-Ansary, M.S.M., 2016. Green nanoparticles engineering on root-knot nematode infecting eggplants and their effect on plant DNA modification. Iran. J. Biotechnol. 14, 250.

Abdelmalek, G.A.M., Salaheldin, T.A., 2016. Silver nanoparticles as a potent fungicide for citrus phytopathogenic fungi. J. Nanomed. Res. 3 (5), 00065.

Abigail, M.E.A., Samuel, S.M., Chidambaram, R., 2016. Application of rice husk nanosorbents containing 2,4-dichlorophenoxyacetic acid herbicide to control weeds and reduce leaching from soil. J. Taiwan Inst. Chem. Eng. 63, 318–326.

Abubakar, A., Bala, A.Y., Singh, K., 2017. Plant molluscicides and their modes of action: a review. Int. J. Sci. Res. Technol. 2, 37–49.

Adel, M.M., Atwa, W.A., Hassan, M.L., Salem, N.Y., Farghaly, D.S., Ibrahim, S.S., 2015. Biological activity and field persistence of *Pelargonium graveolens* (Geraniales: Geraniaceae) loaded solid lipid nanoparticles (SLNs) on *Phthorimaea operculella* (Zeller)(PTM)(Lepidoptera: Gelechiidae). Int. J. Sci. Res. 4, 514–520.

Adisa, I.O., Pullagurala, V.L.R., Peralta-Videa, J.R., Dimkpa, C.O., Elmer, W.H., Gardea-Torresdey, J.L., White, J.C., 2019. Recent advances in nano-enabled fertilizers and pesticides: a critical review of mechanisms of action. Environ. Sci. 6, 2002–2030.

Aleksandrowicz-Trzcińska, M., Szaniawski, A., Olchowik, J., Drozdowski, S., 2018. Effects of copper and silver nanoparticles on growth of selected species of pathogenic and wood-decay fungi in vitro. For. Chron. 94, 109–116.

Ali, M.A., Rehman, I., Iqbal, A., Din, S., Rao, A.Q., Latif, A., Samiullah, T.R., Azam, S., Husnain, T., 2014. Nanotechnology, a new frontier in agriculture. Adv. Life Sci. 1, 129–138.

Ali, M., Kim, B., Belfield, K.D., Norman, D., Brennan, M., Ali, G.S., 2015a. Inhibition of Phytophthora parasitica and *P. capsici* by silver nanoparticles synthesized using aqueous extract of *Artemisia absinthium*. Phytopathology 105, 1183–1190.

Ali, M.S., Yousef, M.H.N., Nafady, N.A., 2015b. The land snail *E. vermiculata* is an important crop pest causing considerable damage in agriculture and the snail used as bioindicator organism of environmental pollution. J. Nanomater. 2015. Article ID: 218904.

Arciniegas-Grijalba, P.A., Patiño-Portela, M.C., Mosquera-Sánchez, L.P., Guerrero-Vargas, J. A., Rodríguez-Páez, J.E., 2017. ZnO nanoparticles (ZnO-NPs) and their antifungal activity against coffee fungus *Erythricium salmonicolor*. Appl. Nanosci. 7, 225–241.

Arumugam, G., Velayutham, V., Shanmugavel, S., Sundaram, J., 2016. Efficacy of nanostructured silica as a stored pulse protector against the infestation of bruchid beetle, *Callosobruchus maculatus* (Coleoptera: Bruchidae). Appl. Nanosci. 6, 445–450.

Athanassiou, C.G., Kavallieratos, N.G., Benelli, G., Losic, D., Rani, P.U., Desneux, N., 2018. Nanoparticles for pest control: current status and future perspectives. J. Pestic. Sci. 91, 1–15.

Atta, S., Paul, A., Banerjee, R., Bera, M., Ikbal, M., Dhara, D., Singh, N.P., 2015. Photoresponsive polymers based on a coumarin moiety for the controlled release of pesticide 2, 4-D. RSC Adv. 5, 99968–99975.

Azizi, Z., Pourseyedi, S., Khatami, M., Mohammadi, H., 2016. *Stachys lavandulifolia* and *Lathyrus* sp. mediated for green synthesis of silver nanoparticles and evaluation its antifungal activity against *Dothiorella sarmentorum*. J. Clust. Sci. 27, 1613–1628.

Bahrami-Teimoori, B., Nikparast, Y., Hojatianfar, M., Akhlaghi, M., Ghorbani, R., Pourianfar, H.R., 2017. Characterisation and antifungal activity of silver nanoparticles biologically synthesised by *Amaranthus retroflexus* leaf extract. J. Exp. Nanosci. 12, 129–139.

Baker, S., Volova, T., Prudnikova, S.V., Satish, S., Prasad, N., 2017. Nanoagroparticles emerging trends and future prospect in modern agriculture system. Environ. Toxicol. Pharmacol. 53, 10–17.

Baker, S., Satish, S., Prasad, N., Chouhan, R.S., 2019. Nano-agromaterials: influence on plant growth and crop protection. In: Industrial Applications of Nanomaterials. Elsevier, pp. 341–363.

Balah, M.A., Abd El Azim, W.M., 2016. Emulsions and nanoemulsions formation from wild and cultivated thyme and marjoram essential oils for weeds control. J. Plant Prot. Pathol. 7, 641–648.

Balaji, A.P.B., Mishra, P., Kumar, R.S., Mukherjee, A., Chandrasekaran, N., 2015. Nanoformulation of poly (ethylene glycol) polymerized organic insect repellent by PIT emulsification method and its application for Japanese encephalitis vector control. Colloids Surf. B Biointerfaces 128, 370–378.

Bhargava, P., Kumar, A., Kumar, S., Azad, C.S., 2018. Impact of fungicides and nanoparticles on *Ustilaginoidea virens* causing false smut disease of rice. J. Pharmacogn. Phytochem. 7, 1541–1544.

Bhattacharyya, A., Duraisamy, P., Govindarajan, M., Buhroo, A.A., Prasad, R., 2016. Nanobiofungicides: emerging trend in insect pest control. In: Advances and Applications Through Fungal Nanobiotechnology. Springer, Cham, pp. 307–319.

Bombo, A.B., Pereira, A.E.S., Lusa, M.G., de Medeiros Oliveira, E., de Oliveira, J.L., Campos, E.V.R., de Jesus, M.B., Oliveira, H.C., Fraceto, L.F., Mayer, J.L.S., 2019. A mechanistic view of interactions of a nanoherbicide with target organism. J. Agric. Food Chem. 67, 4453–4462.

Borgatta, J., Ma, C., Hudson-Smith, N., Elmer, W., Plaza Pérez, C.D., De La Torre-Roche, R., Zuverza-Mena, N., Haynes, C.L., White, J.C., Hamers, R.J., 2018. Copper based nanomaterials suppress root fungal disease in watermelon (*Citrullus lanatus*): role of particle morphology, composition and dissolution behavior. ACS Sustain. Chem. Eng. 6, 14847–14856.

Boxi, S.S., Mukherjee, K., Paria, S., 2016. Ag doped hollow TiO_2 nanoparticles as an effective green fungicide against *Fusarium solani* and *Venturia inaequalis* phytopathogens. Nanotechnology 27, 085103.

Bramhanwade, K., Shende, S., Bonde, S., Gade, A., Rai, M., 2015. Fungicidal activity of cu nanoparticles against *Fusarium* causing crop disease. Environ. Chem. Lett. 14, 229–235.

Buteler, M., Sofie, S.W., Weaver, D.K., Driscoll, D., Muretta, J., Stadler, T., 2015. Development of nanoalumina dust as insecticide against *Sitophilus oryzae* and *Rhyzopertha dominica*. Int. J. Pest Manag. 61, 80–89.

Campolo, O., Cherif, A., Ricupero, M., Siscaro, G., Grissa-Lebdi, K., Russo, A., Cucci, L.M., Di Pietro, P., Satriano, C., Desneux, N., Biondi, A., 2017. Citrus peel essential oil nanoformulations to control the tomato borer, *Tuta absoluta*: chemical properties and biological activity. Sci. Rep. 7, 1–10.

Campos, E.V.R., De Oliveira, J.L., Da Silva, C.M.G., Pascoli, M., Pasquoto, T., Lima, R., Abhilash, P.C., Fraceto, L.F., 2015. Polymeric and solid lipid nanoparticles for sustained release of carbendazim and tebuconazole in agricultural applications. Sci. Rep. 5, 1–14.

Cao, L., Zhang, H., Cao, C., Zhang, J., Li, F., Huang, Q., 2016. Quaternized chitosan-capped mesoporous silica nanoparticles as nanocarriers for controlled pesticide release. Nanomaterials 6, 126.

Cao, L., Zhou, Z., Niu, S., Cao, C., Li, X., Shan, Y., Huang, Q., 2017. Positive-charge functionalized mesoporous silica nanoparticles as nanocarriers for controlled 2, 4-dichlorophenoxy acetic acid sodium salt release. J. Agric. Food Chem. 66, 6594–6603.

Chen, J., Peng, H., Wang, X., Shao, F., Yuan, Z., Han, H., 2014. Graphene oxide exhibits broad-spectrum antimicrobial activity against bacterial phytopathogens and fungal conidia by intertwining and membrane perturbation. Nanoscale 6, 1879–1889.

Chen, K., Yu, G., He, F., Zhou, Q., Xiao, D., Li, J., Feng, Y., 2017. A pH-responsive emulsion stabilized by alginate-grafted anisotropic silica and its application in the controlled release of λ-cyhalothrin. Carbohydr. Polym. 176, 203–213.

Chen, C., Zhang, G., Dai, Z., Xiang, Y., Liu, B., Bian, P., Zheng, K., Wu, Z., Cai, D., 2018. Fabrication of light-responsively controlled-release herbicide using a nanocomposite. Chem. Eng. Sci. 349, 101–110.

Chen, J., Mao, S., Xu, Z., Ding, W., 2019. Various antibacterial mechanisms of biosynthesized copper oxide nanoparticles against soilborne *Ralstonia solanacearum*. RSC Adv. 9, 3788–3799.

Chhipa, H., 2017. Nanofertilizers and nanopesticides for agriculture. Environ. Chem. Lett. 15, 15–22.

Christofoli, M., Costa, E.C.C., Bicalho, K.U., de Cássia Domingues, V., Peixoto, M.F., Alves, C.C.F., Araújo, W.L., de Melo Cazal, C., 2015. Insecticidal effect of nanoencapsulated essential oils from *Zanthoxylum rhoifolium* (Rutaceae) in *Bemisia tabaci* populations. Ind. Crop Prod. 70, 301–308.

Clemente, Z., Grillo, R., Jonsson, M., Santos, N.Z.P., Feitosa, L.O., Lima, R., Fraceto, L.F., 2014. Ecotoxicological evaluation of poly (ε-caprolactone) nanocapsules containing triazine herbicides. J. Nanosci. Nanotechnol. 14, 4911–4917.

Coelho, P.M.Z., Caldeira, R.L., 2016. Critical analysis of molluscicide application in schistosomiasis control programs in Brazil. Infect. Dis. Poverty 5, 1–6.

Czarnobai De Jorge, B., Bisotto-de-Oliveira, R., Pereira, C.N., Sant'Ana, J., 2017. Novel nanoscale pheromone dispenser for more accurate evaluation of *Grapholita molesta* (Lepidoptera: Tortricidae) attract-and-kill strategies in the laboratory. Pest Manag. Sci. 73, 1921–1926.

de Oliveira, J.L., Campos, E.V.R., Goncalves da Silva, C.M., Pasquoto, T., Lima, R., Fraceto, L. F., 2015. Solid lipid nanoparticles co-loaded with simazine and atrazine: preparation, characterization, and evaluation of herbicidal activity. J. Agric. Food Chem. 63, 422–432.

Deka, B., Babu, A., Baruah, C., Barthakur, M., 2021. Nanopesticides: a systematic review of their prospects with special reference to tea pest management. Front. Nutr. 8, 686131. https://doi.org/10.3389/fnut.2021.686131.

del Carmen Galán-Jiménez, M., Morillo, E., Bonnemoy, F., Mallet, C., Undabeytia, T., 2020. A sepiolite-based formulation for slow release of the herbicide mesotrione. Appl. Clay Sci. 189, 105503.

Deshpande, M.V., 2019. Nanobiopesticide perspectives for protection and nutrition of plants. In: Nano-Biopesticides Today and Future Perspectives. Academic Press, pp. 47–68.

Dharni, S., Sanchita, Unni, S.M., Kurungot, S., Samad, A., Sharma, A., Patra, D.D., 2016. In vitro and in silico antifungal efficacy of nitrogen-doped carbon nanohorn (NCNH) against *Rhizoctonia solani*. J. Biomol. Struct. Dyn. 34, 152–162.

Dhoble, S.M., Kulkarni, N.S., 2016. Antimycotic activity of zinc and manganese nanoparticles on commercially important phytopathogens of soybean (*Glycine max* (L.) Merril). Sch. Acad. J. Biosci. 4, 1032–1037.

Ding, K., Shi, L., Zhang, L., Zeng, T., Yin, Y., Yi, Y., 2016. Synthesis of photoresponsive polymeric propesticide micelles based on PEG for the controlled release of a herbicide. Polym. Chem. 7, 899–904.

Diyanat, M., Saeidian, H., 2019. The metribuzin herbicide in polycaprolactone nanocapsules shows less plant chromosome aberration than non-encapsulated metribuzin. Environ. Chem. Lett. 17, 1881–1888.

Duhan, J.S., Kumar, R., Kumar, N., Kaur, P., Nehra, K., Duhan, S., 2017. Nanotechnology: the new perspective in precision agriculture. Biotechnol. Rep. 15, 11–23.

Elamawi, R.M., Al-Harbi, R.E., Hendi, A.A., 2018. Biosynthesis and characterization of silver nanoparticles using *Trichoderma longibrachiatum* and their effect on phytopathogenic fungi. Egypt. J. Biol. Pest Control 28, 1–11.

El-Argawy, E., Rahhal, M.M.H., El-Korany, A., Elshabrawy, E.M., Eltahan, R.M., 2017. Efficacy of some nanoparticles to control damping-off and root rot of sugar beet in El-Behiera Governorate. Asian J. Plant Pathol. 11, 35–47.

Elbeshehy, E.K., Elazzazy, A.M., Aggelis, G., 2015. Silver nanoparticles synthesis mediated by new isolates of *Bacillus spp.*, nanoparticle characterization and their activity against bean yellow mosaic virus and human pathogens. Front. Microbiol. 6, 453.

Elgorban, A.M., Aref, S.M., Seham, S.M., Elhindi, K.M., Bahkali, A.H., Sayed, S.R., Manal, M. A., 2016. Extracellular synthesis of silver nanoparticles using *Aspergillus versicolor* and evaluation of their activity on plant pathogenic fungi. Mycosphere 7, 844–852.

El-Helaly, A.A., El-Bendary, H.M., Abdel-Wahab, A.S., El-Sheikh, M.A.K., Elnagar, S., 2016. The silica-nano particles treatment of squash foliage and survival and development of *Spodoptera littoralis* (Bosid.) larvae. Pest Control 5, 6.

Elizabath, A., Babychan, M., Mathew, A.M., Syriac, G.M., 2019. Application of nanotechnology in agriculture. Int. J. Pure Appl. Biosci. 7, 131–139.

Eltarahony, M., Zaki, S., ElKady, M., Abd-El-Haleem, D., 2018. Biosynthesis, characterization of some combined nanoparticles, and its biocide potency against a broad spectrum of pathogens. J. Nanomater. 2018, 1–16.

Fan, C., Guo, M., Liang, Y., Dong, H., Ding, G., Zhang, W., Tang, G., Yang, J., Kong, D., Cao, Y., 2017. Pectin-conjugated silica microcapsules as dual-responsive carriers for increasing the stability and antimicrobial efficacy of kasugamycin. Carbohydr. Polym. 172, 322–331.

Faria, D.M., Dourado Júnior, S.M., Nascimento, J.P.L.D., Nunes, E.D.S., Marques, R.P., Rossino, L.S., Moreto, J.A., 2017. Development and evaluation of a controlled release system of TBH herbicide using alginate microparticles. Mater. Res. 20, 225–235.

Farsi, M., Farokhi, S., 2018. Biosynthesis of antibacterial silver nanoparticles by endophytic fungus *Nemania sp.* isolated from *Taxus baccata* L. (Iranian yew). Zahedan J. Res. Med. Sci. 20.

Fernández, J.G., Fernández-Baldo, M.A., Berni, E., Camí, G., Durán, N., Raba, J., Sanz, M.I., 2016. Production of silver nanoparticles using yeasts and evaluation of their antifungal activity against phytopathogenic fungi. Process Biochem. 51, 1306–1313.

Fraceto, L.F., Grillo, R., de Medeiros, G.A., Scognamiglio, V., Rea, G., Bartolucci, C., 2016. Nanotechnology in agriculture: which innovation potential does it have? Front. Environ. Sci. Eng. 4, 20.

Gao, X., Lowry, G.V., 2018. Progress towards standardized and validated characterizations for measuring physicochemical properties of manufactured nanomaterials relevant to nano health and safety risks. NanoImpact 9, 14–30.

Garde-Cerdán, T., Mancini, V., Carrasco-Quiroz, M., Servili, A., Gutiérrez-Gamboa, G., Foglia, R., Pérez-Álvarez, E.P., Romanazzi, G., 2017. Chitosan and Laminarin as alternatives to copper for *Plasmopara viticola* control: effect on grape amino acid. J. Agric. Food Chem. 65, 7379–7386.

Ghorbanpour, M., Bhargava, P., Varma, A., Choudhary, D.K. (Eds.), 2020. Biogenic Nano-Particles and Their Use in Agro-Ecosystems. Springer.

Ghormade, V., Deshpande, M.V., Paknikar, K.M., 2011. Perspectives for nano-biotechnology enabled protection and nutrition of plants. Biotechnol. Adv. 29, 792–803.

Giannousi, K., Sarafidis, G., Mourdikoudis, S., Pantazaki, A., Dendrinou-Samara, C., 2014. Selective synthesis of Cu_2O and Cu/Cu_2O NPs: antifungal activity to yeast *Saccharomyces cerevisiae* and DNA interaction. Inorg. Chem. 53, 9657–9666.

Gnanamangai, B.M., Ponmurugan, P., Jeeva, S.E., Manjukarunambika, K., Elango, V., Hemalatha, K., Kakati, J.P., Mohanraj, R., Prathap, S., 2017. Biosynthesised silver and copper nanoformulation as foliar spray to control bird's eye spot disease in tea plantations. IET Nanobiotechnol. 11, 917–928.

Gomez, A., Narayan, M., Zhao, L., Jia, X., Bernal, R.A., Lopez-Moreno, M.L., Peralta-Videa, J. R., 2020. Effects of nano-enabled agricultural strategies on food quality: current knowledge and future research needs. J. Hazard. Mater., 123385.

Grillo, R., Pereira, A.E., Nishisaka, C.S., De Lima, R., Oehlke, K., Greiner, R., Fraceto, L.F., 2014. Chitosan/tripolyphosphate nanoparticles loaded with paraquat herbicide: an environmentally safer alternative for weed control. J. Hazard. Mater. 278, 163–171.

Grillo, R., Clemente, Z., de Oliveira, J.L., Campos, E.V.R., Chalupe, V.C., Jonsson, C.M., de Lima, R., Sanches, G., Nishisaka, C.S., Rosa, A.H., Oehlke, K., 2015. Chitosan nanoparticles loaded the herbicide paraquat: the influence of the aquatic humic substances on the colloidal stability and toxicity. J. Hazard. Mater. 286, 562–572.

Guilger, M., Pasquoto-Stigliani, T., Bilesky-Jose, N., Grillo, R., Abhilash, P.C., Fraceto, L.F., De Lima, R., 2017. Biogenic silver nanoparticles based on *Trichoderma harzianum*: synthesis, characterization, toxicity evaluation and biological activity. Sci. Rep. 7, 1–13.

Guo, M., Zhang, W., Ding, G., Guo, D., Zhu, J., Wang, B., Punyapitak, D., Cao, Y., 2015. Preparation and characterization of enzyme-responsive emamectin benzoate microcapsules based on a copolymer matrix of silica-epichlorohydrin-carboxymethylcellulose. RSC Adv. 5, 93170–93179.

Guyton, K.Z., Loomis, D., Grosse, Y., El Ghissassi, F., Benbrahim-Tallaa, L., Guha, N., Scoccianti, C., Mattock, H., Straif, K., International Agency for Research on Cancer Monograph Working Group ILF, 2015. Carcinogenicity of tetrachlorvinphos, parathion, malathion, diazinon, and glyphosate. Lancet Oncol. 16, 490–491.

Hafez, Y.M., Attia, K.A., Kamel, S., Alamery, S.F., El-Gendy, S., Al-Doss, A.A., Mehiar, F., Ghazy, A.I., Ibrahim, E.I., Abdelaal, K.A., 2020. *Bacillus subtilis* as a bio-agent combined with nano molecules can control powdery mildew disease through histochemical and physiobiochemical changes in cucumber plants. Physiol. Mol. Plant Pathol. 111, 101489.

Hasan, K.A., Soliman, H., Baka, Z., Shabana, Y.M., 2020. Efficacy of nano-silicon in the control of chocolate spot disease of *Vicia faba* L. caused by *Botrytis fabae*. Egypt. J. Basic Appl. Sci. 7, 53–66.

Hoque, M.I.U., Chowdhury, A.N., Islam, M.T., Firoz, S.H., Luba, U., Alowasheeir, A., Rahman, M.M., Rehman, A.U., Ahmad, S.H.A., Holze, R., Hossain, M.S.A., 2021. Fabrication of highly and poorly oxidized silver oxide/silver/tin (IV) oxide nanocomposites and their comparative anti-pathogenic properties towards hazardous food pathogens. J. Hazard. Mater. 408, 124896.

Hossain, A., Kerry, R.G., Farooq, M., Abdullah, N., Islam, M.T., 2020. Application of nanotechnology for sustainable crop production systems. In: Nanotechnology for Food, Agriculture, and Environment. Springer, Cham, pp. 135–159.

Hua, K.H., Wang, H.C., Chung, R.S., Hsu, J.C., 2015. Calcium carbonate nanoparticles can enhance plant nutrition and insect pest tolerance. Pestic. Sci. 40, 208–213.

Huang, B., Chen, F., Shen, Y., Qian, K., Wang, Y., Sun, C., Zhao, X., Cui, B., Gao, F., Zeng, Z., Cui, H., 2018a. Advances in targeted pesticides with environmentally responsive controlled release by nanotechnology. Nanomaterials 8, 102.

Huang, W., Fang, X., Wang, H., Chen, F., Duan, H., Bi, Y., Yu, H., 2018b. Biosynthesis of AgNPs by *B. maydis* and its antifungal effect against *Exserohilum turcicum*. IET Nanobiotechnol. 12, 585–590.

Huayao, C., Yueshun, L., Hongjun, Z., Xinhua, Z., Sheng, G., Hua, X., 2016. Highly efficient alginate sodium encapsulated chlorpyrifos/copper (II) Schiff base mesoporous silica sustained release system with pH and ion response for pesticide delivery. RSC Adv. 6, 114714–114721.

Iavicoli, I., Leso, V., Beezhold, D.H., Shvedova, A.A., 2017. Nanotechnology in agriculture: opportunities, toxicological implications, and occupational risks. Toxicol. Appl. Pharmacol. 329, 96–111.

Ikram, F., Qayoom, A., Ikram, N., Shah, M.R., 2017. Synergistic effect of epicatechin coated silver nanoparticles on antimicrobial activity of gentamicin against *Aspergillus niger*. J. New Technol. Mater. 277, 1–6.

Ikram, F., Qayoom, A., Shah, M.R., 2018. Synthesis of epicatechin coated silver nanoparticles for selective recognition of gentamicin. Sens. Actuators B 257, 897–905.

Imada, K., Sakai, S., Kajihara, H., Tanaka, S., Ito, S., 2016. Magnesium oxide nanoparticles induce systemic resistance in tomato against bacterial wilt disease. Plant Pathol. 65, 551–560.

Iqbal, S.M., Raj, C.S., Michael, J.J., Irfan, A.M., 2017. A comparative investigation of Al_2O_3/H_2O, SiO_2/H_2O and ZrO_2/H_2O nanofluid for heat transfer applications. Dig. J. Nanomater. Biostruct. 12.

Islam, M.T., 2019. Applications of nanomaterials for future food security: challenges and prospects. Malays. J. Halal Res. 2, 6–9.

Islam, M.S., Akter, N., Rahman, M.M., Shi, C., Islam, M.T., Zeng, H., Azam, M.S., 2018. Mussel-inspired immobilization of silver nanoparticles toward antimicrobial cellulose paper. ACS Sustain. Chem. Eng. 6, 9178–9188.

Ismail, M., Prasad, R., Ibrahim, A.I., Ahmed, A.I., 2017. Modern prospects of nanotechnology in plant pathology. In: Nanotechnology. Springer, Singapore, pp. 305–317.

Jiang, B., Li, C., Dag, Ö., Abe, H., Takei, T., Imai, T., Hossain, S.A., Islam, T., Wood, K., Henzie, J., Yamauchi, Y., 2017. Mesoporous metallic rhodium nanoparticles. Nat. Commun. 8, 15581.

Judy, J.D., Kirby, J.K., Creamer, C., McLaughlin, M.J., Fiebiger, C., Wright, C., Cavagnaro, T. R., Bertsch, P.M., 2015. Effects of silver sulfide nanomaterials on mycorrhizal colonization of tomato plants and soil microbial communities in biosolid-amended soil. Environ. Pollut. 206, 256–263.

Kah, M., Hofmann, T., 2014. Nanopesticide research: current trends and future priorities. Environ. Int. 63, 224–235.

Kala, A., Soosairaj, S., Mathiyazhagan, S., Raja, P., 2016. Green synthesis of copper bionanoparticles to control the bacterial leaf blight disease of rice. Curr. Sci., 2011–2014.

Kaman, P.K., Dutta, P., 2017. In vitro evaluation of biosynthesized silver nanoparticles (Ag NPs) against soil borne plant pathogens. Int. J. Nanotechnol. Appl. 11, 261–264.

Kanhed, P., Birla, S., Gaikwad, S., Gade, A., Seabra, A.B., Rubilar, O., Duran, N., Rai, M., 2014. In vitro antifungal efficacy of copper nanoparticles against selected crop pathogenic fungi. Mater. Lett. 115, 13–17.

Kaziem, A.E., Gao, Y., He, S., Li, J., 2017. Synthesis and insecticidal activity of enzyme-triggered functionalized hollow mesoporous silica for controlled release. J. Agric. Food Chem. 65, 7854–7864.

Khan, M.S., Qureshi, N.A., Jabeen, F., 2017. Assessment of toxicity in fresh water fish Labeo rohita treated with silver nanoparticles. Appl. Nanosci. 7, 167–179.

Khan, M.R., Rizvi, T.F., 2014. Nanotechnology: scope and application in plant disease management. Plant Pathol. J. 13, 214–231.

Khan, J.H., Lin, J., Young, C., Matsagar, B.M., Wu, K.C., Dhepe, P.L., Islam, M.T., Rahman, M., Shrestha, L.K., Alshehri, S.M., Ahamad, T., 2018. High surface area nanoporous carbon derived from Bangladeshi jute. Mater. Chem. Phys. 216, 491–495.

Khatami, M., Mehnipor, R., Poor, M.H.S., Jouzani, G.S., 2016. Facile biosynthesis of silver nanoparticles using *Descurainia sophia* and evaluation of their antibacterial and antifungal properties. J. Clust. Sci. 27, 1601–1612.

Khatami, M., Mortazavi, S.M., Kishani-Farahani, Z., Amini, A., Amini, E., Heli, H., 2017. Biosynthesis of silver nanoparticles using pine pollen and evaluation of the antifungal efficiency. Iran. J. Biotechnol. 15, 95.

Kheiri, A., Jorf, S.M., Malihipour, A., Saremi, H., Nikkhah, M., 2016. Application of chitosan and chitosan nanoparticles for the control of fusarium head blight of wheat (*Fusarium graminearum*) in vitro and greenhouse. Int. J. Biol. Macromol. 93, 1261–1272.

Köhler, H.R., Triebskorn, R., 2013. Wildlife ecotoxicology of pesticides: can we track effects to the population level and beyond? Science 341, 759–765.

Koli, P., Singh, B.B., Shakil, N.A., Kumar, J., Kamil, D., 2015. Development of controlled release nanoformulations of carbendazim employing amphiphilic polymers and their bioefficacy evaluation against *Rhizoctonia solani*. J. Environ. Sci. Health B 50, 674–681.

Kulkarni, R.A., Prabhuraj, A., Ashoka, J., Hanchinal, S.G., Hiregoudar, S., 2016. Generation and evaluation of nanoparticles of supernatant of *Photorhabdus luminescens* (Thomas and Poinar) against mite and aphid pests of cotton for enhanced efficacy. Curr. Sci., 2312–2316.

Kumar, S., Chauhan, N., Gopal, M., Kumar, R., Dilbaghi, N., 2015. Development and evaluation of alginate-chitosan nanocapsules for controlled release of acetamiprid. Int. J. Biol. Macromol. 81, 631–637.

Kumar, S., Bhanjana, G., Sharma, A., Dilbaghi, N., Sidhu, M.C., Kim, K.H., 2017. Development of nanoformulation approaches for the control of weeds. Sci. Total Environ. 586, 1272–1278.

Kumar, S., Nehra, M., Dilbaghi, N., Marrazza, G., Hassan, A.A., Kim, K.H., 2019. Nano-based smart pesticide formulations: emerging opportunities for agriculture. J. Control. Release 294, 131–153.

Kumaraswamy, R.V., Kumari, S., Choudhary, R.C., Pal, A., Raliya, R., Biswas, P., Saharan, V., 2018. Engineered chitosan based nanomaterials: bioactivities, mechanisms and perspectives in plant protection and growth. Int. J. Biol. Macromol. 113, 494–506.

Li, J., Sang, H., Guo, H., Popko, J.T., He, L., White, J.C., Parkash, D.O., Jung, G., Xing, B., 2017. Antifungal mechanisms of ZnO and Ag nanoparticles to *Sclerotinia homoeocarpa*. Nanotechnology 28, 155101.

Liang, Y., Guo, M., Fan, C., Dong, H., Ding, G., Zhang, W., Tang, G., Yang, J., Kong, D., Cao, Y., 2017. Development of novel urease-responsive pendimethalin microcapsules using silica-IPTS-PEI as controlled release carrier materials. ACS Sustain. Chem. Eng. 5, 4802–4810.

Louder, J.K., 2015. Nanotechnology in Agriculture: Interactions between Nanomaterials and Cotton Agrochemicals. (Doctoral dissertation).

Majumder, S., Kaushik, P., Rana, V.S., Sinha, P., Shakil, N.A., 2020. Amphiphilic polymer based nanoformulations of mancozeb for management of early blight in tomato. J. Environ. Sci. Health B 55, 501–507.

Makarovsky, D., Fadeev, L., Salam, B.B., Zelinger, E., Matan, O., Inbar, J., Jurkevitch, E., Gozin, M., Burdman, S., 2018. Silver nanoparticles complexed with bovine submaxillary mucin possess strong antibacterial activity and protect against seedling infection. Appl. Environ. Microbiol. 84.

Malaj, E., Peter, C., Grote, M., Kühne, R., Mondy, C.P., Usseglio-Polatera, P., Brack, W., Schäfer, R.B., 2014. Organic chemicals jeopardize the health of freshwater ecosystems on the continental scale. Proc. Natl. Acad. Sci. U. S. A. 111, 9549–9554.

Maluin, F.N., Hussein, M.Z., Azah Yusof, N., Fakurazi, S., Idris, A.S., Zainol Hilmi, N.H., Jeffery Daim, L.D., 2020. Chitosan-based agronanofungicides as a sustainable alternative in the basal stem rot disease management. J. Agric. Food Chem. 68, 4305–4314.

Manjunatha, S.B., Biradar, D.P., Aladakatti, Y.R., 2016. Nanotechnology and its applications in agriculture: a review. J. Farm Sci. 29, 1–13.

Maruyama, C.R., Guilger, M., Pascoli, M., Bileshy-José, N., Abhilash, P.C., Fraceto, L.F., De Lima, R., 2016. Nanoparticles based on chitosan as carriers for the combined herbicides imazapic and imazapyr. Sci. Rep. 6, 1–15.

Maurya, S., Pal, D., Jaiswal, G., Philip, S., 2018. Smart agriculture: role of nanotechnology in agriculture. Int. J. Adv. Manag. Technol. Eng. Sci. 8, 1648–1654.

Moreno, A., Jordana, A., Grillo, R., Fraceto, L.F., Jaime, C., 2019. A study on the molecular existing interactions in nanoherbicides: a chitooligosaccharide/tripolyphosphate loaded with paraquat case. Colloids Surf. A Physicochem. Eng. Asp. 562, 220–228.

Mossa, A.T.H., Abdelfattah, N.A.H., Mohafrash, S.M.M., 2017. Nanoemulsion of camphor (*Eucalyptus globulus*) essential oil, formulation, characterization and insecticidal activity against wheat weevil, *Sitophilus granarius*. Asian J. Crop Sci. 9, 50–62.

Nakache, E., Poulain, N., Candau, F., Orecchioni, A.M., Irache, J.M., 2000. Biopolymer and polymer nanoparticles and their biomedical applications. In: Handbook of Nanostructured Materials and Nanotechnology. Academic Press, pp. 577–635.

Nakasato, D.Y., Pereira, A.E., Oliveira, J.L., Oliveira, H.C., Fraceto, L.F., 2017. Evaluation of the effects of polymeric chitosan/tripolyphosphate and solid lipid nanoparticles on germination of Zea mays, *Brassica rapa* and *Pisum sativum*. Ecotoxicol. Environ. Saf. 142, 369–374.

Negi, M., 2016. Bio-Efficacy of Silver Nanoparticles of Botanicals Against *Alternaria zinniae* Causing Leaf Spot Disease in Marigold. (Doctoral dissertation).

Nejad, S.M., Bonjar, G.H.S., Khatami, M., Amini, A., Aghighi, S., 2017. In vitro and in vivo antifungal properties of silver nanoparticles against *Rhizoctonia solani*, a common agent of rice sheath blight disease. IET Nanobiotechnol. 11, 236–240.

Nguyen, M.H., Hwang, I.C., Bui, C.B., Park, H.J., 2016. Effects of the physical state of nanocarriers on their penetration into the root and upward transportation to the stem of soybean plants using confocal laser scanning microscopy. J. Crop Prot. 87, 25–30.

Nuruzzaman, M.D., Rahman, M.M., Liu, Y., Naidu, R., 2016. Nanoencapsulation, nano-guard for pesticides: a new window for safe application. J. Agric. Food Chem. 64, 1447–1483.

Oerke, E.-C., 2006. Crop losses to pest. J. Agric. Sci. 144, 31–43.

Oliveira, H.C., Stolf-Moreira, R., Martinez, C.B., Sousa, G.F., Grillo, R., de Jesus, M.B., Fraceto, L.F., 2015. Evaluation of the side effects of poly (epsilon-caprolactone) nanocapsules containing atrazine toward maize plants. Front. Chem. 3, 61.

Omobhude, M.E., Morenikeji, O.A., Oyeyemi, O.T., 2017. Molluscicidal activities of curcumin-nisin polylactic acid nanoparticle on *Biomphalaria pfeifferi*. PLoS Negl. Trop. Dis. 11, e0005855.

Parisi, C., Vigani, M., Rodríguez-Cerezo, E., 2015. Agricultural nanotechnologies: what are the current possibilities? Nano Today 10, 124–127.

Pereira, A.E.S., Grillo, R., Mello, N.F.S., Rosa, A.H., Fraceto, L.F., 2014. Application of poly(epsilon-caprolactone) nanoparticles containing atrazine herbicide as an alternative technique to control weeds and reduce damage to the environment. J. Hazard. Mater. 268, 207–215.

Prasad, R., Kumar, V., Prasad, K.S., 2014. Nanotechnology in sustainable agriculture: present concerns and future aspects. Afr. J. Biotechnol. 13, 705–713.

Prasad, R., Bhattacharyya, A., Nguyen, Q.D., 2017. Nanotechnology in sustainable agriculture: recent developments, challenges, and perspectives. Front. Microbiol. 8, 1014.

Preisler, A.C., Pereira, A.E., Campos, E.V., Dalazen, G., Fraceto, L.F., Oliveira, H.C., 2020. Atrazine nanoencapsulation improves pre-emergence herbicidal activity against *Bidens pilosa* without enhancing long-term residual effect on *Glycine max*. Pest Manag. Sci. 76, 141–149.

Qin, X., Xiang, X., Sun, X., Ni, H., Li, L., 2015. Preparation of nanoscale *Bacillus thuringiensis* chitinases using silica nanoparticles for nematicide delivery. Int. J. Biol. Macromol. 82, 13–21.

Qureshi, A., Singh, D.K., Dwivedi, S., 2018. Nano-fertilizers: a novel way for enhancing nutrient use efficiency and crop productivity. Int. J. Curr. Microbiol. App. Sci., 3325–3335.

Raj, S.N., Anooj, E.S., Rajendran, K., Vallinayagam, S., 2021. A comprehensive review on regulatory invention of nano pesticides in agricultural nano formulation and food system. J. Mol. Struct. 1239, 130517.

Rao, K.J., Paria, S., 2013. Use of sulfur nanoparticles as a green pesticide on *Fusarium solani* and *Venturia inaequalis* phytopathogens. RSC Adv. 3, 10471–10478.

Rashidzadeh, A., Olad, A., Salari, D., Jalil Hejazi, M., 2014. On the encapsulation of natural pesticide using polyvinyl alcohol/alginate–montmorillonite nanocomposite for controlled release application. Polym. Eng. Sci. 54, 2707–2714.

Sabbour, M.M., Abd El-Aziz, S.E., Shadia, E., 2015. Efficacy of nano-diatomaceous earth against red flour beetle, *Tribolium castaneum* and confused flour beetle, *Tribolium confusum* (Coleoptera: Tenebrionidae) under laboratory and storage conditions. Bull. Environ. Pharmacol. Life Sci. 4, 54–59.

Sahab, A.F., Waly, A.I., Sabbour, M.M., Nawar, L.S., 2015. Synthesis, antifungal and insecticidal potential of chitosan (CS)-g-poly (acrylic acid)(PAA) nanoparticles against some seed borne fungi and insects of soybean. Int. J. ChemTech Res. 8, 589–598.

Saharan, V., Mehrotra, A., Khatik, R., Rawal, P., Sharma, S.S., Pal, A., 2013. Synthesis of chitosan based nanoparticles and their in vitro evaluation against phytopathogenic fungi. Int. J. Biol. Macromol. 62, 677–683.

Saharan, V., Sharma, G., Yadav, M., Choudhary, M.K., Sharma, S.S., Pal, A., Raliya, R., Biswas, P., 2015. Synthesis and in vitro antifungal efficacy of Cu–chitosan nanoparticles against pathogenic fungi of tomato. Int. J. Biol. Macromol. 75, 346–353.

Saini, P., Gopal, M., Kumar, R., Srivastava, C., 2014. Development of pyridalyl nanocapsule suspension for efficient management of tomato fruit and shoot borer (*Helicoverpa armigera*). J. Environ. Sci. Health B 49, 344–351.

Sarkar, D.J., Singh, A., 2017. Base triggered release of insecticide from bentonite reinforced citric acid crosslinked carboxymethyl cellulose hydrogel composites. Carbohydr. Polym. 156, 303–311.

Sathiyabama, M., Charles, R.E., 2015. Fungal cell wall polymer based nanoparticles in protection of tomato plants from wilt disease caused by *Fusarium oxysporum* f. sp. *lycopersici*. Carbohydr. Polym. 133, 400–407.

Schnoor, B., Elhendawy, A., Joseph, S., Putman, M., Chacón-Cerdas, R., Flores-Mora, D., Bravo-Moraga, F., Gonzalez-Nilo, F., Salvador-Morales, C., 2018. Engineering atrazine loaded poly (lactic-co-glycolic acid) nanoparticles to ameliorate environmental challenges. J. Agric. Food Chem. 66, 7889–7898.

Seehra, M. (Ed.), 2017. Nanostructured Materials: Fabrication to Applications. BoD–Books on Demand.

Seetharaman, P.K., Chandrasekaran, R., Gnanasekar, S., Chandrakasan, G., Gupta, M., Manikandan, D.B., Sivaperumal, S., 2018. Antimicrobial and larvicidal activity of eco-friendly silver nanoparticles synthesized from endophytic fungi *Phomopsis liquidambaris*. Biocatal. Agric. Biotechnol. 16, 22–30.

Shanmugam, C., Gunasekaran, D., Duraisamy, N., Nagappan, R., Krishnan, K., 2015. Bioactive bile salt-capped silver nanoparticles activity against destructive plant pathogenic fungi through in vitro system. RSC Adv. 5, 71174–71182.

Sharma, P., Sharma, A., Sharma, M., Bhalla, N., Estrela, P., Jain, A., Thakur, P., Thakur, A., 2017. Nanomaterial fungicides: in vitro and in vivo antimycotic activity of cobalt and nickel nanoferrites on phytopathogenic fungi. Global Chall. 1, 1700041.

Shenashen, M., Derbalah, A., Hamza, A., Mohamed, A., El Safty, S., 2017. Antifungal activity of fabricated mesoporous alumina nanoparticles against root rot disease of tomato caused by *Fusarium oxysporum*. Pest Manag. Sci. 73, 1121–1126.

Sherkhane, A.S., Suryawanshi, H.H., Mundada, P.S., Shinde, B.P., 2018. Control of bacterial blight disease of pomegranate using silver nanoparticles. J. Nanomed. Nanotechnol. 9, 500.

Shifa, F.A., Mukhtar, M., Pandit, A., Murtaza, I., Nazir, N., Hakeem, K.A., 2019. A critical study of reduced pesticide application rates of nano-deltamethrin in comparison to its conventional analogue against *Trialeurodes vaporariorum*. J. Entomol. Zool. Stud. 7, 969–974.

Siddiqui, M.H., Al-Whaibi, M.H., Firoz, M., Al-Khaishany, M.Y., 2015. Role of nanoparticles in plants. Nanotechnol. Plant Sci., 19–35.

Singh, S., Singh, B.K., Yadav, S.M., Gupta, A.K., 2015. Applications of nanotechnology in agricultural and their role in disease management. J. Nanosci. Nanotechnol. 5, 1–5.

Song, M., Ju, J., Luo, S., Han, Y., Dong, Z., Wang, Y., Gu, Z., Zhang, L., Hao, R., Jiang, L., 2017. Controlling liquid splash on superhydrophobic surfaces by a vesicle surfactant. Sci. Adv. 3, e1602188.

Sousa, G.F., Gomes, D.G., Campos, E.V., Oliveira, J.L., Fraceto, L.F., Stolf-Moreira, R., Oliveira, H.C., 2018. Post-emergence herbicidal activity of nanoatrazine against susceptible weeds. Front. Environ. Sci. 6, 12.

Sousa, B.T., Pereira, A.D.E.S., Fraceto, L.F., Caixeta de Oliveira, H., Dalazen, G., 2020. Effectiveness of nanoatrazine in post-emergent control of the tolerant weed *Digitaria insularis*. J. Plant Prot. Res., 185–192.

Strayer-Scherer, A., Liao, Y.Y., Young, M., Ritchie, L., Vallad, G.E., Santra, S., Freeman, J.H., Clark, D., Jones, J.B., Paret, M.L., 2018. Advanced copper composites against copper-tolerant *Xanthomonas perforans* and tomato bacterial spot. Phytopathology 108, 196–205.
Sukhwal, A., Jain, D., Joshi, A., Rawal, P., Kushwaha, H.S., 2017. Biosynthesised silver nanoparticles using aqueous leaf extract of *Tagetes patula* L. and evaluation of their antifungal activity against phytopathogenic fungi. IET Nanobiotechnol. 11, 531–537.
Tanaka, K., Fukuda, M., Amaki, Y., Sakaguchi, T., Inai, K., Ishihara, A., Nakajima, H., 2017. Importance of prumycin produced by *Bacillus amyloliquefaciens* SD-32 in biocontrol against cucumber powdery mildew disease. Pest Manag. Sci. 73, 2419–2428.
Thangadurai, D., Sangeetha, J., Prasad, R. (Eds.), 2020. Nanotechnology for Food, Agriculture, and Environment. Springer.
US Department of Agriculture, 2002. Nanoscale Science and Engineering for Agriculture and Food Systems. United States Department of Agriculture, National Planning Workshop, Washington, DC.
Velmurugan, P., Shim, J., Kim, K., Oh, B.T., 2016. *Prunus × yedoensis* tree gum mediated synthesis of platinum nanoparticles with antifungal activity against phytopathogens. Mater. Lett. 174, 61–65.
Velu, M., Lee, J.H., Chang, W.S., Lovanh, N., Park, Y.J., Jayanthi, P., Palanivel, V., Oh, B. T., 2017. Fabrication, optimization, and characterization of noble silver nanoparticles from sugarcane leaf (*Saccharum officinarum*) extract for antifungal application. 3 Biotech 7, 1–9.
Viet, P.V., Nguyen, H.T., Cao, T.M., Hieu, L.V., 2016. Fusarium antifungal activities of copper nanoparticles synthesized by a chemical reduction method. J. Nanomater. 2016.
Wagner, G., Korenkov, V., Judy, J.D., Bertsch, P.M., 2016. Nanoparticles composed of Zn and ZnO inhibit *Peronospora tabacina* spore germination in vitro and *P. tabacina* infectivity on tobacco leaves. Nanomaterials 6, 50.
Wang, Y., Wang, A., Wang, C., Cui, B., Sun, C., Zhao, X., Zeng, Z., Shen, Y., Gao, F., Liu, G., Cui, H., 2017. Synthesis and characterization of emamectin-benzoate slow-release microspheres with different surfactants. Sci. Rep. 7, 1–9.
Wibowo, D., Zhao, C.X., Peters, B.C., Middelberg, A.P., 2014. Sustained release of fipronil insecticide in vitro and in vivo from biocompatible silica nanocapsules. J. Agric. Food Chem. 62, 12504–12511.
Worrall, E.A., Hamid, A., Mody, K.T., Mitter, N., Pappu, H.R., 2018. Nanotechnology for plant disease management. J. Agron. 8, 285.
Xing, K., Shen, X., Zhu, X., Ju, X., Miao, X., Tian, J., Feng, Z., Peng, X., Jiang, J., Qin, S., 2016. Synthesis and in vitro antifungal efficacy of oleoyl-chitosan nanoparticles against plant pathogenic fungi. Int. J. Biol. Macromol. 82, 830–836.
Xu, X., Bai, B., Wang, H., Suo, Y., 2017. A near-infrared and temperature-responsive pesticide release platform through core–shell polydopamine@ PNIPAm nanocomposites. ACS Appl. Mater. Interfaces 9, 6424–6432.
Yasur, J., Rani, P.U., 2015. Lepidopteran insect susceptibility to silver nanoparticles and measurement of changes in their growth, development and physiology. Chemosphere 124, 92–102.
Young, M., Ozcan, A., Myers, M.E., Johnson, E.G., Graham, J.H., Santra, S., 2017. Multimodal generally recognized as safe ZnO/nanocopper composite: a novel antimicrobial material for the management of citrus phytopathogens. J. Agric. Food Chem. 66, 6604–6608.
Yu, Z., Sun, X., Song, H., Wang, W., Ye, Z., Shi, L., Ding, K., 2015. Glutathione-responsive carboxymethyl chitosan nanoparticles for controlled release of herbicides. Int. J. Mater. Sci. Appl. 6, 591.

Zakharova, O.V., Gusev, A.A., Zherebin, P.M., Skripnikova, E.V., Skripnikova, M.K., Ryzhikh, V.E., Lisichkin, G.V., Shapoval, O.A., Bukovskii, M.E., Krutyakov, Y.A., 2017. Sodium tallow amphopolycarboxyglycinate-stabilized silver nanoparticles suppress early and late blight of *Solanum lycopersicum* and stimulate the growth of tomato plants. J. Bionanosci. 7, 692–702.

Zhao, X., Cui, H., Wang, Y., Sun, C., Cui, B., Zeng, Z., 2017a. Development strategies and prospects of nano-based smart pesticide formulation. J. Agric. Food Chem. 66, 6504–6512.

Zhao, P., Cao, L., Ma, D., Zhou, Z., Huang, Q., Pan, C., 2017b. Synthesis of pyrimethanil-loaded mesoporous silica nanoparticles and its distribution and dissipation in cucumber plants. Molecules 22, 817.

Ziaee, M., Moharramipour, S., Mohsenifar, A., 2014. MA-chitosan nanogel loaded with *Cuminum cyminum* essential oil for efficient management of two stored product beetle pests. J. Pestic. Sci. 87, 691–699.

Nanoherbicides for field applications

16

Sougata Ghosh[a,c], *Bishwarup Sarkar*[b], *and Sirikanjana Thongmee*[c]
[a]Department of Microbiology, School of Science, RK University, Rajkot, Gujarat, India, [b]College of Science, Northeastern University, Boston, MA, United States, [c]Department of Physics, Faculty of Science, Kasetsart University, Bangkok, Thailand

16.1 Introduction

Consistent rise in global population urges a concomitant increase in agriculture productivity in order to meet the growing demand for food supply, and thus, extensive amounts of pesticides such as herbicides, fungicides, and insecticides are applied in agricultural fields. However, excessive application of such pesticides has resulted in contamination of land, water, and air (Craig, 2018). Such overuse of pesticides has caused harmful effects on nontarget organisms as well as adversely affected human health (Liu et al., 2019).

Herbicides are an important class of agrochemicals that control weed growth in agricultural fields in order to increase crop yield. Herbicides are classified into 28 groups by the Herbicide Resistance Action Committee (HRAC) that was based on their site of action that was suggested to be used for selecting proper herbicide against certain herbicide-resistant weeds (Retzinger and Mallory-Smith, 1997). Herbicides have several target sites for exhibiting their inhibitory actions. Glyphosate herbicide, for example, inhibits enzymatic activity of 3-enolpyruvylshikimate 5-phosphate (EPSP) synthase that is involved in amino acid biosynthesis (Jablonkai, 2011). Likewise, commercial herbicides haloxyfop and diclofop are known to be acetyl-CoA carboxylase (ACC) inhibitors. Glutamine synthetase (GS) has been reported to be suppressed by phosphinothricin. Isoxazole herbicides target 4-hydroxyphenylpyruvate dioxygenase (HPPD) enzyme that catalyzes conversion of 4-hydroxyphenyl-pyruvate (HPP) into homogentisate (HGA) (Jablonkai, 2011).

Chronic toxicity caused by prolonged exposure to herbicides can cause a variety of health complications such as liver deformations, different types of cancer, as well as a reduction in body weight (Kniss, 2017). In addition, microbial population of soil is dependent on nature and dose of herbicide used (Lone et al., 2014). Herbicides such as metribuzin negatively alter the soil microbiota, whereas some herbicides such as sulfonylurea do not create a significant alteration in microbial populations (Lone et al., 2014).

Nanoencapsulation of herbicides into polymeric materials can therefore become an efficient strategy for sustained release of herbicides with increased specificity toward target organism along with a reduced sorption that increases bioavailability of

Agricultural Nanobiotechnology. https://doi.org/10.1016/B978-0-323-91908-1.00010-9

herbicides (Pascoli et al., 2018). In view of the background, this chapter explains in detail synthesis, characterization, and application of several nanoherbicides for increased efficacy as listed in Table 16.1. Later, future outlooks are also discussed in view of implementation of such nanoherbicides as efficient agrochemical derivatives for weed management.

16.2 Polymeric nanomaterials

Several herbicides are encapsulated into polymeric nanomaterials for their application against weed species. Alginate/chitosan (AG/CS) polymers were used by dos Santos Silva et al. (2011) to prepare nanoparticles that were used as a carrier system for paraquat herbicide. Nanoparticles were prepared by mechanically mixing 0.063% *w/v* of sodium alginate and 18 mM of $CaCl_2$ for 60 min followed by the addition of chitosan. After nanoparticle formation, 200 μg/L of paraquat was incorporated into the mixture. Hydrodynamic diameter of the AG/CS nanoparticles containing paraquat was 635 ± 12 nm, whereas zeta potential and polydispersion values were estimated to be -22.8 ± 2.3 mV and 0.518, respectively. Hence, the system was suggested to have considerable colloidal stability; however, particles were nonhomogeneous in nature. Transmission electron microscopy (TEM) showed spherical, dense, and solid particles with a size range of 197 to 305 nm. Differential scanning calorimetry (DSC) results indicated paraquat fusion with a narrow endothermic peak at 70°C and broader peak at 62°C that suggested the presence of physical mixture of AG/CS, whereas physical mixture of AG/CS nanoparticles and paraquat was confirmed by the presence of an intense broad peak at 68°C. Later on, peaks corresponding to only AG/CS nanoparticles were observed indicating proper dispersion of paraquat in the matrix of nanoparticles due to electrostatic interaction between negatively charged carboxylic groups of alginate and positively charged paraquat as well as van der Waals interaction of chitosan polymer and herbicide. An amount of organic matter present in soil samples was observed to affect the paraquat sorption in both free state and associated with nanoparticles while sustained release of herbicide was observed only when nanoherbicides were applied in soil samples, thus showing effective strategy of reducing harmful impacts of paraquat.

In another study, Namasivayam et al. (2014) prepared silver nanoparticle-chitosan-encapsulated synthetic herbicide paraquat (AgNP-CS-PQ) and evaluated its herbicidal activity against *Eichhornia crassipes*. Ionic gelation method was carried out for the synthesis of AgNP-CS-PQ wherein 1 mM $AgNO_3$ was reduced using 1 mM $NaBH_4$ to obtain a brown-colored nanoparticle precipitate after which chitosan solution dissolved in 1% acetic acid and 0.0001% paraquat solution was added under stirring conditions for 2 h. Ultraviolet–visible (UV–Vis) spectroscopy results showed a peak at 420 nm indicating surface plasmon resonance (SPR) of silver atoms. Scanning electron microscopy (SEM) analysis showed uniform and spherical morphology of nanoformulation with a size of 100 nm. Energy-dispersive X-ray (EDX) spectroscopy also confirmed elemental composition of the nanoformulation that are silver nanoparticles (AgNPs), chitosan, and paraquat herbicide. Further, spectrophotometric analysis revealed 90% paraquat

Table 16.1 Nanoherbicides used for weed management.

Nanocarrier	Herbicide	Encapsulation efficiency	Activity	References
Polymeric nanomaterials				
Alginate/chitosan (AG/CS) nanoparticles	Paraquat	–	–	dos Santos Silva et al. (2011)
Silver nanoparticles with chitosan encapsulation (AgNP-CS)	Paraquat	90%	Necrotic lesions in *Eichhornia crassipes*	Namasivayam et al. (2014)
Chitosan/alginate (CS/ALG) nanoparticles and chitosan/tripolyphosphate (CS/TPP) nanoparticles	Imazapic and imazapyr	Imazapic $58.6 \pm 4.5\%$ (CS/TPP) and $62.3 \pm 3.2\%$ (CS/ALG) Imazapyr $69.6 \pm 5.5\%$ (CS/TPP) and $71.3 \pm 2.8\%$ (CS/ALG)	Reduced growth in roots and aerial parts of *Bidens pilosa*	Maruyama et al. (2016)
CS/TPP nanoparticles	Paraquat	$62.66 \pm 0.77\%$	Limp leaf necrotic activity against *Zea mays*	Grillo et al. (2014)
CS/TPP nanoparticles	Paraquat	$54.34 \pm 0.29\%$	Lipid peroxidation, decrease in chlorophyll content and glutathione peroxidase activity, increase in superoxide dismutase activity in spinach leaf tissues	Pontes et al. (2021)
Glutathione-responsive carboxymethyl chitosan nanoparticles	Diuron	$78.28 \pm 2.36\%$	Reduction in dry weight biomass of *Echinochloa crus-galli*	Yu et al. (2015)
Pectin nanoparticles	Metsulfuron methyl	$63 \pm 2\%$	Decrease in biomass of *Chenopodium album*	Kumar et al. (2017)

Continued

Table 16.1 Continued

Nanocarrier	Herbicide	Encapsulation efficiency	Activity	References
Pectin/chitosan/tripolyphosphate (PEC/CS/TPP) nanoparticles	Paraquat	89.41 ± 5.7%	Leaf necrosis and reduction in dry biomass in *Z. mays* and *Brassica* sp.	Rashidipour et al. (2019)
Poly (lactic-*co*-glycolic acid) (PLGA) nanoherbicides	Atrazine	50%	Decrease in stem and root length in potato plant	Schnoor et al. (2018)
Poly(ε-caprolactone) (PCL) nanocapsules	Pretilachlor	99.5 ± 1.3%	Decrease in *E. crus-galli* shoot length	Diyanat et al. (2019)
PCL nanocapsules	Ametryn, atrazine, and simazine	85%	Reduced mitotic and chromosomal aberration indices against *Allium cepa*	Grillo et al. (2012)
PCL nanocapsules	Atrazine	–	Leaf necrosis and chlorotic petiole in *Brassica juncea*	Oliveira et al. (2015)
PCL nanoparticles and PCL nanospheres	Atrazine	93.4 ± 1.2% (PCL nanoparticles) and 92.7 ± 1.2% (PCL nanospheres)	Phytotoxic effect against *Brassica* sp.	Pereira et al. (2014)
PCL nanocapsules	Atrazine	–	Root and shoot growth inhibition, biomass reduction in *Amaranthus viridis* and *B. pilosa*	Sousa et al. (2018)
Poly(methylmethacrylate) (PMMA) nanocapsules	Haloxyfop-R-methyl	95%	–	Mahmoudian et al. (2020)
Solid lipid nanoparticles (SLNs)	Simazine and atrazine	89.7 ± 0.02% (atrazine) and 97.3 ± 0.05% (simazine)	Inhibitory effect in roots and aerial parts of *Raphanus raphanistrum*	de Oliveira et al. (2015)

Table 16.1 Continued

Nanocarrier	Herbicide	Encapsulation efficiency	Activity	References
Other nanomaterials				
Layered double hydroxide (LDH) anionic nanoclay and Cloisite 10A (Clo10A) cationic nanoclay	Imazamox	–	–	Khatem et al. (2019)
Mesoporous silica nanoparticles (MSNs)	Sodium salt of 2,4-D	$27.7 \pm 0.4\%$	Inhibition of *Cucumis sativus* L.	Cao et al. (2017)
Rice husk biochar-based nanocarriers	2,4-D	–	Necrosis in *Brassica* sp.	Abigail (2019)
Rice husk nanosorbents	2,4-D	–	100% Mortality of *Brassica* sp.	Abigail et al. (2016)

entrapment efficiency with a total release of around 90% at 24 h. No significant change in physicochemical parameters, enzyme activity, as well as microbial population of soil was observed after treatment with nanoherbicides, whereas free herbicide treatment resulted in a reduction of soil microflora. Treatment of *E. crassipes* leaves with 25 μg/mL of AgNP-CS-PQ resulted in necrotic lesions of 1.5 cm highlighting herbicidal activity. Moreover, treatment of *Vigna mongo* seeds with nanoherbicide showed 90.1% germination as compared to 23.5% seed germination after free herbicide treatment. Parameters of nanoherbicides-treated seedlings such as total chlorophyll content, shoot length, surface area of leaf, and number of new branches that emerged were observed to increase at a growth rate similar to that of control, whereas free herbicide-treated seedlings showed reduced chlorophyll content and stunted growth with respect to control plants.

Maruyama et al. (2016) prepared nanomaterials based on chitosan that is used as carrier systems for imazapic and imazapyr herbicides. Two different chitosan-based nanoparticles were prepared, namely chitosan/alginate (CS/ALG) nanoparticles and chitosan/tripolyphosphate (CS/TPP) nanoparticles. Ionotropic gelification method was employed for the synthesis of CS/ALG nanoparticles wherein imazapic and imazapyr herbicides were added into sodium alginate solution. On the contrary, CS/TPP nanoparticles were prepared using 0.1% chitosan dissolved in 0.3% acetic acid followed by the addition of herbicides and tripolyphosphate (TPP). Dynamic light scattering (DLS) results revealed an average size of CS/TPP and CS/ALG nanoparticles to be 478.6 ± 52.3 nm and 377.7 ± 9.7 nm, respectively, while nanoparticle tracking analysis (NTA) method gave an average size distribution of 233.5 ± 10.7 nm and 246.8 ± 2.6 nm for CS/TPP and CS/ALG nanoparticles, respectively. In addition, concentrations of CS/TPP and CS/ALG nanoparticles were 1.20×10^9 particles/mL and 1.83×10^9 particles/mL, respectively. CS/ALG nanoparticles showed no change in stability in both the presence and absence of nanoparticles with a constant zeta potential value of approximately −30 mV that prevents particle aggregation. On the contrary, CS/TPP nanoparticles without any herbicides showed a variation in stability having a zeta potential value of +15 mV indicative of particle aggregation that was increased to +26 mV after encapsulation of imazapic and imazapyr herbicides. Imazapic herbicide encapsulation efficiencies into CS/TPP and CS/ALG nanoparticles were $58.6 \pm 4.5\%$ and $62.3 \pm 3.2\%$, respectively, whereas imazapyr encapsulation efficiencies into CS/TPP and CS/ALG nanoparticles were $69.6 \pm 5.5\%$ and $71.3 \pm 2.8\%$, respectively.

It was further speculated that the presence of -CH_3 group in imazapic may increase its hydrophobicity that could lead to a decrease in interactions with nanoparticle polymers and thus reduce its encapsulation efficiency. Release profile analysis revealed that 55% of free imazapic and 97% of free imazapyr were released after 300 min. However, after nanoparticle encapsulation, release of imazapic and imazapyr was decreased to 30% and 20%, respectively, after 300 min. The mechanism of imazapic release was suggested to involve diffusion of the herbicide while imazapyr was thought to be transported using Case II transport system.

Toxicity of the obtained nanoherbicides was evaluated against *Allium cepa* that showed a decrease in genotoxicity after encapsulation of herbicides as compared to free commercial herbicide formulations. Relative alteration index values of CS/ALG nanoparticles were lowest as compared to other nanoformulations that indicated slower release of herbicides that in turn caused lesser cellular damage within 24 h of treatment. Likewise, comet assay results highlighted higher damage index values after free herbicide treatment as compared to nanoherbicides. Furthermore, efficacy of nanoherbicides was investigated against *Bidens pilosa* that showed reduced growth in roots as well as aerial parts of the weed as compared to negative control.

CS/TPP nanoparticles were also prepared by Grillo et al. (2014) for encapsulating paraquat herbicide and evaluated its use as a safer method of weed control. CS/TPP nanoparticles were prepared using ionic gelification method with 0.1% chitosan dissolved in 0.2% acetic acid and 0.1% TPP. Chitosan solution was mixed with 12 mg of paraquat before the formation of nanoparticles such that final paraquat concentration would be 0.48 mg/mL. Efficiency of paraquat encapsulation was 62.66 ± 0.77% that showed that a considerable affinity of components of nanoformulation that could be a result of electrostatic interaction is between positively charged herbicide and negatively charged TPP anion. Physicochemical properties were observed to remain stable for a period of 60 days at 25°C with hydrodynamic diameter of around 300 nm and polydispersion index values were observed in the range of 0.20 to 0.35 that highlighted homogeneity of particles that were monodispersed in medium. The presence of cationic chitosan polymer resulted in a positive zeta potential value of around 40 mV that was also speculated to avoid coalescence and aggregation between particles. Atomic force microscopy (AFM) images showed spherical morphology of nanoformulations containing paraquat with an average size of around 420 nm. Release kinetics of paraquat showed that 92% of free herbicide that was accessible out of which 70% was released after 350 min. Further, soil sorption kinetics showed that equilibrium was attained within 30 min wherein less sorption of herbicide was seen when associated with a nanoparticle that was proposed to be advantageous as more herbicide would be available for plants.

Investigation of cytotoxicity and genotoxicity of CS/TPP nanoparticles containing paraquat was carried out with Chinese hamster ovary (CHO) cells and seeds of *A. cepa*, respectively. Cytotoxicity assay revealed inhibitory concentration-50 (IC_{50}) value of free paraquat herbicide to be around 0.12 mg/mL while paraquat encapsulated in CS/TPP nanoparticles virtually nontoxic for a time period of 24 h. Likewise, cytotoxicity assays showed only a small decrease in cellular viability when herbicide associated with nanoparticles was applied as compared to unloaded nanoparticles. Genotoxicity assay showed greater damage caused by free paraquat as compared to nanoparticles containing paraquat indicating protective ability of CS/TPP nanoparticles against toxicity of herbicide. In addition, herbicidal activity of prepared nanoformulation was evaluated against *Zea mays* and *Brassica* sp. wherein limp leaf necrosis was observed after 48 h with greater necrosis in the case of *Z. mays* when nanoherbicide was used as compared to a free herbicide that was suggested to be a result of better nanoparticle adhesion on the leaves as well as reduced decomposition of herbicide.

Evaluation of paraquat-containing CS/TPP nanoparticles (CS/TPP: PQ) on photosystem I (PSI) was investigated by Pontes et al. (2021). Concentration of synthesized CS/TPP: PQ was 5×10^6 nanoparticles/mL with a size distribution of 191 ± 82 nm. SEM results revealed that nanoherbicides were spherical in shape with homogeneous particles present within the sample. Encapsulation efficiency of 54.34 ± 0.29% was observed for paraquat herbicide. Release kinetics showed 50% release of paraquat from CS/TPP nanoparticles within an approximate 320 min via non-Fickian mechanism of transport. Square-wave voltammetry experiments revealed that empty CS/TPP nanoparticles, chitosan, and tripolyphosphate were non-electroactive materials while paraquat showed two-step electrochemical reduction. In addition, interaction of paraquat with chitosan component of nanoparticles did not significantly affect this redox reaction while the introduction of tripolyphosphate component in the nanoherbicide complex presented only one peak that suggested the function of tripolyphosphate to act as a glue between glucosamine residues of chitosan and molecules of paraquat.

Also, it was implicated that tripolyphosphate may decrease redox reaction feasibility on the surface of working electrode that can alter the electroactivity of nanoherbicides. Treatment of spinach leaves with analytical grade paraquat as well as CS/TPP: PQ nanoherbicides showed significant membrane damage as compared to unloaded CS/TPP nanoparticles that were attributed to damage caused by oxidative stress. Maximum levels of thiobarbituric acid reactive substances (TBARS) were produced by leaf tissues of spinach plants that were an indicator of lipid peroxidation. Chlorophyll content of spinach leaves was also reduced after treatment with nanoherbicides. Activity of superoxide dismutase (SOD) enzyme was also observed to be significantly increased in the presence of CS/TPP: PQ nanoparticles that showed the induction of increased $O_2^{\bullet -}$ levels. On the contrary, glutathione peroxidase (GPx) activity decreased after treatment with nanoherbicides, whereas catalase activity was least affected. Hence, it was assumed that an increase in SOD activity along with a decrease in GPx activity would result in more production of reactive oxygen species (ROS) and less H_2O_2 scavenging, respectively, that would cumulatively lead to increased tissue damage in target plant. Moreover, relative amounts of $P700^+$ were more prominently decreased after treatment with nanoherbicide as compared to free paraquat treatment. This decrease in $P700^+$ was speculated to be a consequence of a reduction in NAPH/$NADP^+$ ratio after nanoherbicide exposure that may lead to deactivation of electron transport chain (ETC).

Yu et al. (2015) prepared glutathione-responsive carboxymethyl chitosan nanoparticles having disulfide cross-linkages for its use in the controlled release of diuron herbicide. Synthesis of nanoparticles involved dissolution of carboxymethyl chitosan-11-mercapto undecanoic acid conjugates (CMCS-MUA) in deionized water under dark and shaking conditions at 25°C followed by ultrasonication. TEM images showed spherical morphology of CMCS-MUA nanoaggregates having outer core-inner shell structure along with a size of approximately 250 nm in diameter. Amphiphilic properties of obtained nanoformulations were suggested to involve self-aggregation of undecane chains present on CMCS-MUA that led to the formation of hydrophobic cores, whereas carboxylic groups were assumed to stretch out to form hydrophilic shells. Further application of ultrasonication was suggested to oxidize neighboring thiol groups present in the hydrophobic cores in order to form disulfide cross-linkages. No significant change in size, zeta potential value, and polydispersity index was observed that indicated high stability of prepared nanoparticles; however, in the presence of 2 mM of glutathione (GSH), an increase in size as well as polydispersity index values was observed due to cleavage of disulfide linkages.

Later on, diuron was used as a model herbicide that showed a loading and encapsulation efficiency of 7.26 ± 1.53% and 78.28 ± 2.36%, respectively, when 1:2 M ratio of MUA:CMCS precursor units was taken for the synthesis of nanoparticles. Further, in vitro release of diuron was investigated in the presence of phosphate buffer solution (PBS) at pH 7.0 and varying concentrations of GSH wherein 2 mM of GSH showed 58.18% release after 24 as compared to 13.02% release in the absence of GSH. Hence, it was proposed that cleavage of disulfide bonds facilitated the release of diuron herbicide that could be utilized for controlled release of herbicide in agriculture. Herbicidal activity of prepared CMCS-MUA nanoherbicides containing diuron was investigated against *Echinochloa crus-galli* as target species along with *Z. mays* as nontarget species. A significant change in dry weight of *E. crusgalli* was seen when nanoherbicide treatment was performed as compared to free diuron while no dry weight change was observed in *Z. mays* as it is resistant to diuron because of the presence of cytochrome P450 monooxygenase enzyme.

Fabrication of pectin nanoparticles for loading of metsulfuron methyl herbicide was performed by Kumar et al. (2017), and then, its herbicidal activity against weeds was evaluated. Nanocapsules were prepared using the emulsification of two solutions made up of pectin and dioctyl sodium sulfosuccinate (AOT) dissolved in dichloromethane. Later on, polyvinyl alcohol (PVA) was added onto this primary emulsion followed by the addition of $CaCl_2$ under stirring conditions for 24 h at room temperature. Size of pectin nanoparticles and their encapsulation efficiency were observed to be dependent upon concentrations of pectin and AOT. Thus, optimized concentrations of 1% pectin and 10% AOT were used for nanoformulation preparation for further characterization. Encapsulation efficiency of optimized nanoparticles was 63 ± 2% having a particle size of 164 nm. Loading efficiency of metsulfuron methyl herbicide into pectin nanoparticles was 6.3% with no change in size of nanoparticles after encapsulation. Zeta potential value of obtained nanoherbicide was −35.9 mV that indicated the stability of nanoformulation. Fourier transform infrared (FTIR) spectra of metsulfuron methyl-loaded pectin nanoparticles confirmed the presence of pectin as well as metsulfuron methyl herbicide. Herbicidal action of metsulfuron methyl-loaded pectin nanoparticles was then analyzed against *Chenopodium album* weed species that were grown in *Triticum aestivum* crop plant. Complete control of weed plants was observed within three to five weeks after nanoherbicide treatment without any inhibitory effects against wheat crops. Quantitative estimation of biomass of weed plants showed a decrease from 48 g/m^2 in control plant to 5 g/m^2 after nanoherbicide treatment, whereas free herbicide treatment decreased biomass to 19 g/m^2 which highlighted increased specificity and controlled release after nanoencapsulation.

Novel pectin/chitosan/tripolyphosphate (PEC/CS/TPP) nanoparticles were prepared by Rashidipour et al. (2019) for investigating its ability to act as a carrier system for paraquat herbicide. Optimum concentrations of 0.04% chitosan, 0.01% tripolyphosphate, 0.08% pectin, and 5 mg of paraquat herbicide were used for the synthesis of nanoparticles that provided 89.41 ± 5.7% efficiency of encapsulation.

Average size distribution of nanoparticle size increased from 249.85 ± 2.43 nm to 520.00 ± 4.94 nm after loading of paraquat herbicide that was further increased to 763.10 ± 2.32 nm after 60 days. Polydispersity index of both empty and loaded nanoparticles was around 0.40 that indicated monodispersed nature of particles in the medium with no significant change observed even after 60 days. SEM micrographs and AFM results showed semispherical structure of particles that were homogeneously distributed throughout the medium. DSC thermograms showed stabilization of paraquat after encapsulation as evidenced from removal of paraquat endothermal peak in thermal spectrum of paraquat-loaded nanoparticles that were attributed to electrostatic interaction between positively charged paraquat and negatively charged polymer of nanoparticles. Release kinetics demonstrated 74.36% of paraquat release after 360 min with a release constant of 0.60 min^{-1} and release exponent of 0.68 that stipulated diffusion and swelling controlled release of paraquat. Majority of paraquat was retained in 0–10 cm soil column depth when soil sample was treated with nanoherbicide. Herbicidal activity of prepared nanoformulations was observed against *Z. mays* and *Brassica* sp. within 48 h of treatment wherein leaf necrosis and a significant reduction in dry biomass were observed for both target plants.

Schnoor et al. (2018) engineered poly (lactic-*co*-glycolic acid) (PLGA) nanoherbicides loaded with atrazine. Synthesis of nanoherbicides was carried out using modified nanoprecipitation method wherein 1 mg/mL of 1,2-distearoyl-*sn*-glycero3-phosphoethanolamine-N-[amino(polyethylene glycol)] (DSPE-PEG-NH_2) and lecithin solution were mixed followed by heating the reaction mixture at 68°C for 5 min under light conditions. Later on, 2 mg/mL of PLGA solution containing atrazine was added into the reaction mixture to prepare nanoherbicides. The structure of nanoherbicide obtained exhibited a core composed of PLGA while DSPE-PEG-NH_2 formed the shell and atrazine was embedded in the polymer matrix composed of PLGA. A copolymer ratio of 75:25, that is, 75% of lactide and 25% of glycolide monomers were used for the preparation of nanoherbicide that makes PLGA matrix slightly more hydrophobic in nature as a result of steric effects caused by methyl groups of lactide monomer while on the contrary, sufficient hydrophilicity is also maintained for sustained release of herbicides. TEM images showed spherical morphology of PLGA nanoherbicides with an average size range of 110 ± 10 nm prior to lyophilization. Efficiency of encapsulation was 50% with infrared (IR) spectra of PLGA nanoherbicides demonstrating a band at 3389 cm^{-1}, 1760 cm^{-1}, and 1069 cm^{-1} indicating the presence of PLGA component, whereas additional peaks at 1623 cm^{-1} and 1555 cm^{-1} confirmed the presence of atrazine. The affinity of acetonitrile when was used as a solvent was highest toward atrazine with interaction energy calculated to be −4.5 kcal/mol while interaction energy of atrazine with acetone and PLGA were −4 kcal/mol and −3.9 kcal/mol, respectively. Hence, it was assumed that if PLGA and atrazine are dissolved in acetonitrile, there will be favorable interaction between solvent and atrazine that was observed in a reduction of encapsulation efficiency to 8% inside PLGA matrix. However, acetone was observed to have an equivalent affinity toward PLGA and atrazine; hence, 50% encapsulation efficiency was achieved using acetone as the solvent for dissolution of PLGA and atrazine herbicide. PLGA nanoherbicides were observed to release 50% of atrazine in

72 h. DLS data revealed the size of prelyophilized nanoherbicides to be around 110 ± 10 nm. Further, in vitro potato plant model was used for the evaluation of PLGA nanoherbicide that showed a significant decrease in stem and root length, number of leaves as well as fresh and dry weight. Similar results were obtained upon the treatment of plant model with free atrazine, whereas PLGA nanoparticles showed negligible effect on the growth parameters of model. Microscopic observations of plants' roots after nanoherbicide treatment showed cellular detriment along with decrease in radical length of plant as shown in Fig. 16.1.

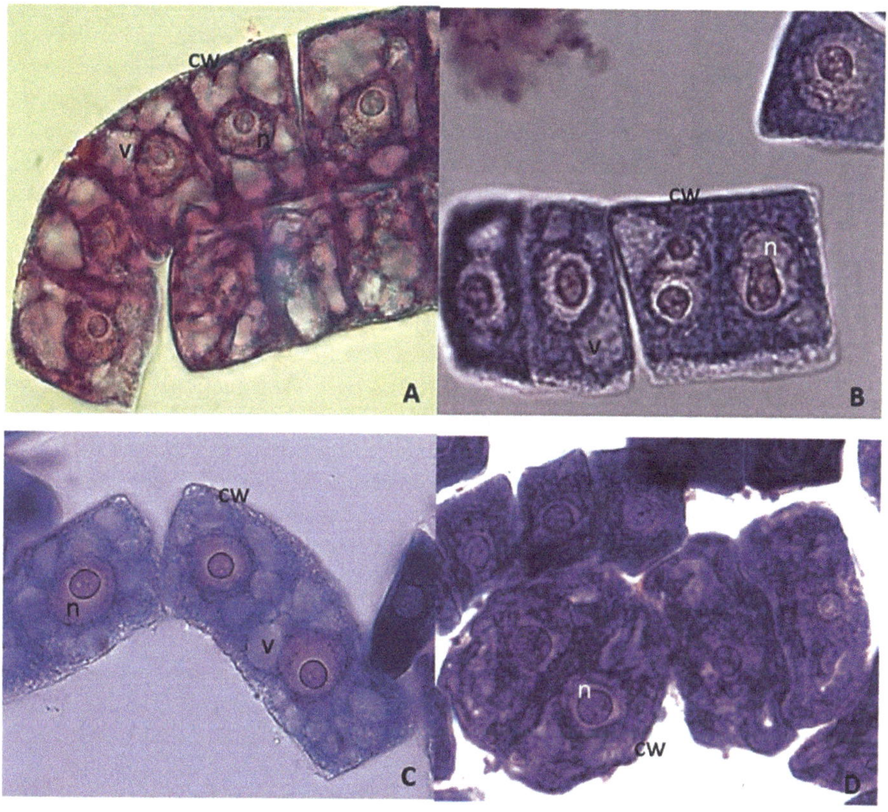

Fig. 16.1 Cellular structure of the control and samples treated with three different nanoherbicides' doses. (A) Control, (B) 0.7 μg/mL, (C) 6.3 μg/mL, and (D) 54 μg/mL; n = nucleus, cw = cell wall, and v = vesicles. In A–C, we observe evidence that indicates that the plant cells are about to undergo cellular division because of the great number of vesicles (v) derived from diverse organelles and cytoplasm. It is important to note that in D, we can see cells with damage in the ultracellular structure. It is evident that the cell wall (cw) has lost rigidity. All the micrographics were obtained at 100×.
Reprinted with permission from Schnoor, B., Elhendawy, A., Joseph, S., Putman, M., Chacón-Cerdas, R., Flores-Mora, D., Bravo-Moraga, F., Gonzalez-Nilo, F., Salvador-Morales, C., 2018. Engineering atrazine loaded poly (lactic-*co*-glycolic acid) nanoparticles to ameliorate environmental challenges. J. Agric. Food Chem. 66(30), 7889–7898. Copyright © 2018 American Chemical Society.

Diyanat et al. (2019) prepared poly(ε-caprolactone) PCL nanocapsules that were used as a carrier system for pretilachlor herbicide. After preparation of nanoherbicide, 0.5% of free pretilachlor was found in the solution, indicating $99.5 \pm 1.3\%$ of herbicide was encapsulated. Significant colloidal stability was seen after 60 days with a negligible change in concentration of free herbicide in the mixture. Further, hydrodynamic size of empty PCL nanocapsules was 220 ± 20 nm, whereas PCL nanocapsules enveloping pretilachlor had a similar hydrodynamic size of 199 ± 9 nm. Likewise, polydispersion index of empty PCL nanocapsules and pretilachlor-containing nanocapsules was observed to be 0.27 ± 0.01 and 0.24 ± 0.02, respectively, that indicated that nanoformulations were monodispersed in medium. Zeta potential values of nanocapsules in the presence and absence of herbicides were -23 ± 2.0 and -25.8 ± 1.9 mV, respectively, that highlighted negative charge on the surface of nanocapsules because of carboxylic group present in PCL polymer. FTIR spectroscopy results of free herbicide showed peak at 1671 cm^{-1} that corresponded to amide group along with other bands at 1457 and 2966 cm^{-1} that indicated C—H bond, whereas peak at 1117 cm^{-1} showed the presence of C—O ether bonds. FTIR spectra of PCL nanocapsules showed peaks at 3295, 2921, and 1736 cm^{-1} that suggested the presence of -OH groups, C—H bonds and ester groups, respectively, whereas nanoherbicide spectrum was observed to be similar to that of empty PCL nanocapsules. TEM images of nanoherbicide showed irregular morphology wherein loading of pretilachlor into PCL nanocapsules was further observed as white spots with a particle size range of 70–200 nm. Herbicidal activity of prepared nanoformulations was evaluated against seeds of *Oryza sativa* and *E. crus-galli*. Maximum decrease in shoot length was observed in *E. crus-galli* after treatment with PCL nanocapsules containing pretilachlor, whereas empty PCL nanocapsules showed no growth inhibitory effects. Concentration of pretilachlor was maximum at soil depth of 5–10 m after the application of both free herbicide and nanoherbicide. Moreover, nanoherbicide was speculated to prevent absorption of pretilachlor by soil particles that increase its availability to act against target plant. In addition, genotoxicity of prepared nanoformulations was investigated using onion cells that revealed a significant reduction in mitotic index after treatment with commercial pretilachlor, whereas chromosomal aberration index was lower in the case of treatment with PCL nanocapsules containing pretilachlor. Hence, it was assumed that pretilachlor encapsulation may reduce genotoxic effects of herbicide that can lead to decreased DNA damage.

Grillo et al. (2012) reported the development of PCL nanocapsules to be used as a carrier system for three different kinds of herbicides, namely ametryn, atrazine, and simazine. Preparation of nanocapsules was based on interfacial deposition of preformed polymer wherein organic phase composed of 100 mg PCL, 30 mL acetone, 200 mg Miglyol 810 oil, 40 mg Span 60 surfactant, and 10 mg herbicide was mixed with aqueous phase that was made up of Tween 80 solution. The mixture was agitated for 10 min followed by the evaporation of organic solvent. Maximum release rates of PCL nanoherbicides containing atrazine and ametryn within 5 days were 72% and 61%, respectively, while encapsulation efficiency for both these herbicides were 85% highlighting difference in interaction affinities of herbicides with the polymeric phases. On the contrary, release rate of simazine reached only 25% after 5 days that was speculated to be a result of stronger interaction with PCL as well as hydrophobic nature of simazine herbicide. Efficiencies of atrazine, ametryn, and simazine herbicides release were 56.5%, 47.2%, and 18.7%, respectively. Later on, herbicide release profile data were interpreted using Korsmeyer-Peppas model that showed that release exponent of all three herbicides was in the range of 0.87–0.99 that suggested Type II transport involving swelling of polymer and relaxation of polymer matrix. Likewise, release constant of atrazine, ametryn, and simazine was calculated to be $6.8 \times 10^{-4}\,min^{-1}$, $3.2 \times 10^{-4}\,min^{-1}$, and $2.4 \times 10^{-4}\,min^{-1}$, respectively. Size of empty nanocapsule and PCL nanocapsule containing herbicides was similar in size of around 260 nm. Stability of nanoherbicides was observed to be intact throughout 270 days when stored at room temperature. Polydispersity index values of all three nanoherbicides were below 0.2 that indicating significant homogeneity of particles. Further, zeta potential values of nanoherbicides were approximately −30 mV during 270 days of the experiment that showed that no aggregation of nanoformulations and thus remained stable throughout. AFM images of PCL nanocapsules containing ametryn showed spherical morphology of nanoherbicide that were uniformly distributed having an average size range of 200–300 nm. IR spectra of empty PCL nanocapsules revealed four peaks at $3340\,cm^{-1}$, $2960\,cm^{-1}$, $1740\,cm^{-1}$, and $1190\,cm^{-1}$ that indicated stretching of O—H bond, C—H bond, C=O bond, and angular deformation of C=O bond, respectively. PCL nanocapsules containing ametryn, atrazine, and simazine showed a similar banding pattern as that of empty nanocapsules that suggested association between polymers of nanocapsule and herbicide. Comet test showed greater genotoxicity of free herbicides as compared to nanoherbicides encapsulated in PCL nanocapsules. Damage index value was decreased to around 50% in the case of encapsulation of ametryn highlighting 50% reduction in cellular DNA damage. Moreover, *A. cepa* analyses were performed to evaluate mitotic and chromosomal aberration indices that showed greater values when *A. cepa* was treated with herbicide alone when compared to nanoherbicides that indicated a reduction in cytotoxicity of herbicides after encapsulation.

In another experiment, PCL nanocapsules were developed by Oliveira et al. (2015) to enhance postemergence herbicidal activity of atrazine. NTA showed Brownian motion of nanocapsules with an approximate concentration of 7.5×10^{12} particles/mL. Further, average distribution of empty PCL nanocapsules was observed to be 221.0 ± 2.5 nm, whereas PCL nanocapsules containing atrazine herbicide had an average size of 240.7 ± 2.9 nm. Herbicidal activity assays were performed on *Brassica juncea* plant with toxicity of nanocapsules containing atrazine showing toxicity after 3 days of treatment as shown in Fig. 16.2, while after 7 days of treatment, majority of leaves were necrotic and wilted having a chlorotic petiole. Similar results were observed when *B. juncea* plants were treated with commercially available atrazine and ten-fold dilution of nanoherbicide. Dilution of commercial atrazine resulted in approximately 50% increase in dry weight of shoot indicating reduced effectiveness of herbicidal activity that was not observed in treatment with same dilution of nanoherbicides. Chlorophyll *a* fluorescence measurement over 72 h after treatment with diluted nanoherbicide showed a reduction in maximum quantum yield of photosystem II (PSII) that was intensified within 24 and 48 h of treatments with undiluted

Fig. 16.2 Symptom evolution in *B. juncea* leaves treated with the formulations. Symptoms were recorded (A) 3 and (B) 7 days after the plants were sprayed with 3.1 mL of water, empty PCL nanocapsules (NC), commercial atrazine (ATZ), or PCL nanocapsules containing atrazine (NC + ATZ). The formulations containing atrazine at 1 mg/mL were used undiluted or after 10-fold dilution in water (1/10), resulting in atrazine dosages equivalent to 2000 or 200 g ha^{-1}, respectively.
Reprinted from Oliveira, H.C., Stolf-Moreira, R., Martinez, C.B.R., Grillo, R., de Jesus, M.B., Fraceto, L.F., 2015. Nanoencapsulation enhances the post-emergence herbicidal activity of atrazine against mustard plants. PLoS One 10(7), e0132971. (Open access).

nanoherbicides. Net photosynthesis rate and stomatal conductance were also decreased after 24 h of treatment with nanocapsules containing atrazine while intracellular CO_2 concentrations in leaves were increased. Likewise, transpiration rate was decreased as well after 48 h of nanoherbicide treatment. Higher lipid peroxidation in leaves was observed after 72 h of treatment with nanoherbicide as compared to the application of commercial atrazine.

Pereira et al. (2014) also prepared PCL nanoparticles encapsulating atrazine for weed management. Oil-in-water method was used for the formation of nanocapsules wherein 200 mg of myritol was added in the organic phase that was made up of PCL dissolved in dichloromethane and atrazine dissolved in acetone. Similar protocol was followed for the preparation of PCL nanospheres as well; however, only the solution containing polymer was used. Sizes of unloaded nanocapsules and nanospheres were 513 ± 7.5 nm and 365.1 ± 0.16 nm, respectively, that were decreased to 483.1 ± 2.5 nm for nanocapsules and 408.5 ± 2.5 nm for nanospheres after herbicide encapsulation. Nanospheres were more stable than nanocapsules after encapsulation with polydispersion index value for nanospheres remaining near 0.2 even after 90 days while it increased in the case of nanocapsules indicating particle aggregation that may lead to decrease in stability. It was proposed that such decrease in stability of nanocapsules may be due to oxidation of oil resulting in the release of internal components and thus forming an emulsion with surfactant. Encapsulation efficiencies of nanocapsules and nanospheres were 93.4 ± 1.2% and 92.7 ± 1.2%, respectively. Korsmeyer-Peppas model showed release coefficient (*n*) values of nanocapsules and nanospheres to be 0.70 and 0.65, respectively, which suggested the release of atrazine from nanoformulations is a result of anomalous transport that may involve diffusion of herbicide along with the relaxation of polymer matrix. Release constant values (*k*) of nanocapsules and nanospheres were 0.015 and 0.019 min^{-1}, respectively, indicating faster release by nanospheres. Herbicidal activity of nanoparticles containing atrazine herbicide was more prominent in *Brassica* sp. that was considered as a target plant as compared to *Z. mays* that was taken as a nontarget plant as shown Fig. 16.3. In the case of *Brassica* sp., phytotoxic effect was exhibited by empty nanocapsules, whereas nanospheres without atrazine showed no effect. Similar herbicidal activity was observed with nanoherbicide and free atrazine indicating no change alteration in mechanism of action. Quantification of free as well as encapsulated atrazine at different soil depths showed a concentration gradient wherein maximum herbicide was found in the upper ring in the case of free atrazine while nanoparticles containing atrazine were evenly distributed in between 0 and 4-cm and 4–8-cm rings that were attributed to the application of surfactant and decreased soil sorption due to encapsulation. Hence, nanoherbicide showed greater distribution of atrazine throughout the soil column.

Germination index of *Brassica* sp. for all treatments was 90% when compared to control as shown in Fig. 16.3B. Treatment of soil from 8 to 12 and 12–16 cm depths with nanoherbicide showed less efficiency as compared to free herbicide treatment. Thus, it was concluded that the encapsulation of herbicide may lead to decrease in mobility. On the contrary, treatment of soil from 0 to 4 cm soil depth with nanoformulations showed efficient growth restriction as compared to free atrazine. Nanospheres showed better performance when treated on soil layer from 4 to 8 cm. Genotoxicity studies were performed on *A. cepa* that showed fewer chromosomal aberrations by low concentration of nanoherbicides when compared to low amount of free herbicide suggesting a reduction in DNA damaging ability of herbicide due to association with nanoparticles. Furthermore, herbicidal activity assay using nanocapsules containing atrazine showed alteration in seed germination.

Postemergence herbicidal activity of PCL nanocapsules containing atrazine was investigated by Sousa et al. (2018) against two susceptible weeds, namely *Amaranthus viridis* and *Bidens pilosa*. Interfacial deposition method was carried out using preformed polymer for the preparation of nanocapsules. A significant reduction in maximum quantum yield (Fv/Fm) and relative electron transport rate of PSII (rETR) were observed in the postemergence treatment of *A. viridis* and *B. pilosa*. Moreover, treatment with nanocapsules devoid of herbicide decreased rETR of *B. pilosa* by 6.9%, whereas Fv/Fm remained constant when compared to control. Similarly, empty nanocapsules decreased rETR and Fv/Fm values in *A. viridis* by 13.6% and 6.9%, respectively. Furthermore, the application of nanoherbicide showed a significant decrease of Fv/Fm in *B. pilosa* and *A. viridis* by 54% and 53%, respectively, as compared to the control treatment. Likewise, rETR was decreased upon nanoherbicide treatment by 97% and 98% in *A. viridis* and *B. pilosa*, respectively. Also, *B. pilosa*

Fig. 16.3 Preemergent atrazine treatment of the nontarget species (*Zea mays*) and the target species (*Brassica* sp.). (A) *Z. mays* plants before collection; (B) *Brassica* sp. plants before collection.

Reprinted with permission from Pereira, A.E., Grillo, R., Mello, N.F., Rosa, A.H., Fraceto, L.F., 2014. Application of poly (epsilon-caprolactone) nanoparticles containing atrazine herbicide as an alternative technique to control weeds and reduce damage to the environment. J. Hazard. Mater. 268, 207–215. Copyright © 2014 Elsevier B.V.

plants treated with atrazine nanoherbicide showed a decrease in relative growth rate of roots and shoots by −17.88 and −3.70 mg/g/d, respectively, that highlighted a reduction in biomass, whereas *A. viridis* plants treated with free atrazine and nanoherbicide showed similar inhibition in root and shoot growth.

Mahmoudian et al. (2020) prepared poly(methylmethacrylate) (PMMA) nanocapsules loaded with haloxyfop-R-methyl herbicide using emulsion polymerization method and studied its behavior of release. *N,N′*-methylenebis(acrylamide) was used as a cross-linking agent during polymerization that did not allow the formation of secondary particles and facilitated formation of nanocapsules with uniform shape and size. FTIR spectral analysis of PMMA nanoherbicide showed the absence of any peak in between 2261 and 2135 cm^{-1}, suggesting proper encapsulation of haloxyfop-R-methyl herbicide into PMMA shell. Field emission-scanning electron microscopy (FE-SEM) results displayed uniform and smooth-surfaced particles that were spherical in shape with a size range of 110–150 nm. Moreover, an increase in the concentration of herbicide resulted in a reduction in surface smoothness of particles, and hence, such nanoformulations were shown to have a raspberry morphology. Maximum loading and encapsulation efficiency of 45% and 95%, respectively, were observed for prepared nanoherbicides with varying parameters such as ratio of monomers and herbicides, type of surfactant used, and concentration of cross-linking agent. Release of herbicides was mediated through fine pores of PMMA polymer with highest amount of haloxyfop-R-methyl herbicide released from PMMA nanocapsules was found to be 4.4 ppm after 6 days.

Simazine and atrazine herbicides were encapsulated into solid lipid nanoparticles (SLNs) by de Oliveira et al. (2015) and investigated their herbicidal activity. Emulsification method was carried out for the synthesis of SLN followed by evaporation of solvent. The hydrodynamic diameter of obtained SLN in the absence of herbicides was 272.2 ± 1.35 nm that was reduced to 255.43 ± 3.01 nm after encapsulation of herbicides. Increase in hydrodynamic diameters was observed after 15 days of storage after which nanoherbicides remained stable for 120 days. Polydispersity index of nanoherbicides was observed to have values below 0.15 after synthesis that increased slightly over a period of time. However, value of polydispersity index remained below 0.2, highlighting significant stability. Zeta potential measurements revealed the surface charge of particles with an average value of −15 mV that remained stable throughout the study. Encapsulation efficiencies of atrazine and simazine were 89.7 ± 0.02% and 97.3 ± 0.05%, respectively. TEM images revealed a spherical structure of empty SLN having average sizes of 111 ± 2.3 nm and 178 ± 3.1 nm in the presence and absence of herbicides, respectively. Encapsulation of herbicides into SLN was shown to cause deformation in the shape of nanoparticles. DSC analysis of herbicides showed two peaks at 179°C and 232°C that corresponded to atrazine and simazine fusion points, respectively, whereas SLN containing herbicides showed characteristic peak corresponding to fusion point of α and β forms of tripalmitin lipids but no symmetrical peaks of herbicides were observed thus, indicating dispersion of herbicides throughout the matrix of SLN. In addition, IR spectra of nanoherbicides containing both atrazine and simazine showed characteristic peaks corresponding to poly(vinyl) alcohol and tripalmitin along with low intense peaks indicating deformation of C=C and C=N bonds of herbicides. NTA technique was carried out for further characterization of SLN suspensions that showed particle number concentration and average diameter to be $(9.09 \pm 0.070) \times 10^{12}$ particles/mL and 240.37 ± 95.45 nm, respectively, in the absence of herbicides. On the contrary, encapsulation of herbicides in SLN resulted in particle number concentration of $(11.75 \pm 0.64) \times 10^{12}$ particles/mL along with an average hydrodynamic diameter of 252.18 ± 91.56 nm. Release kinetic assays revealed slower release profiles of nanoherbicides through 1 kDa pores of a cellulose membrane as compared to free herbicides wherein 50% of free atrazine and simazine was release within 2.5 h and 5.3 h, respectively, while after encapsulation in SLN, same amount of atrazine and simazine were released in 52.9 h and 51.1 h, respectively. In vitro cytotoxicity of nanoherbicides was evaluated using 3-(4,5-dimethylthiazol-2-yl)-2,5-diphenyl tetrazolium bromide (MTT) reduction test using 3T3 fibroblast cells that showed 100% cellular viability in the presence of empty SLN which was reduced to 80% in the presence of nanoherbicide formulation. Further reduction in cellular viability to 64% was seen after the application of commercial-free herbicide formulations. Herbicidal activity against *Raphanus raphanistrum* as the target species showed inhibitory effects on aerial as well as root parts of the plant, whereas no activity was observed against *Z. mays* that was considered as a nontarget species.

16.3 Other nanomaterials

Apart from polymeric nanocarriers, several other nanomaterials are also reported that can act as efficient carrier system for herbicides. In one such study, Khatem et al. (2019) synthesized nanoformulations of anionic clay made of layered double hydroxide (LDH) to be used as host nanocarriers for imazamox herbicide. Direct synthesis of LDH nanoherbicides involving conventional coprecipitation method was followed as well as method of regeneration of calcined LDH at 500°C was also carried out for obtaining desired nanoherbicides. A commercial organic clay termed Cloisite 10A (Clo10A) that was cationic in nature was also assayed for synthesis of nanoherbicides. X-ray diffraction (XRD) patterns of formulated nanoherbicides showed a proper intercalation of imazamox using both methods of synthesis. Nanoherbicides prepared using regeneration method was proposed to have 115 cmol/kg of imazamox with an anion exchange capacity (AEC) of 30% as compared to direct synthesis nanoformulations having 55 cmol/kg of herbicide content and 14% AEC. Hence, such higher content of imazamox in nanoherbicides produced by regeneration was suggested to conform an organic bilayer and a basal spacing value of 21.44 Å, whereas direct synthesis-mediated nanoformulations conformed to a monolayer having 8.70 Å of basal spacing. In the case of Clo10A nanoherbicides, imazamox was present in the interlayer space as well with increased basal spacing reflection to 22.68 Å. Verification of the presence of imazamox in nanoherbicide complex was further confirmed through FTIR spectroscopy wherein band at 1575 cm^{-1} was seen in LDH nanoherbicides prepared through regeneration that corresponded to anionic form of imazamox. Average particle size of LDH nanoherbicides prepared through regeneration and direct synthesis methods was 306 nm and 225 nm, respectively, whereas unloaded LDH nanoformulations had a particle size of 429 nm. Such a decrease in average particle size was hypothesized to be a result of herbicide intercalation that produced crystal tension and thus produced fragmentation leading to formation of smaller sized particles. SEM micrographs displayed typical hexagonal shape of unloaded LDH nanoclays, whereas imazamox-containing LDH nanoformulations were irregular agglomerates and nonuniformly distributed. Similarly, SEM images of Clo10A nanoherbicides showed aggregation where irregular stacks were formed due to lipophilic forces. All release profiles of prepared nanoformulations showed slower release of herbicide as compared to free imazamox herbicide formulation.

Application of mesoporous silica nanoparticles (MSNs) in controlled release of sodium salt of 2,4-D herbicide was reported by Cao et al. (2017). Pristine MSNs (P-MSNs) were synthesized through mechanism involving liquid crystal templating wherein cetyl trimethylammonium bromide (CTAB) was utilized as the structure-directing agent along with tetraethyl orthosilicate (TEOS) as the source of silica. Thereafter, P-MSNs that were synthesized were positive-charge functionalized using grafting method in order to incorporate trimethylammonium (TA) groups that led to the formation of MSN-TA. SEM and TEM micrographs displayed spherical morphology of MSN-TA having an average particle size of around 423 nm as shown in Fig. 16.4. Loading of 2,4-D sodium salt at herbicide to carrier ratio of 1:1 along with pure water as solvent and room temperature conditions was most efficient with maximum loading and encapsulation efficiency of $21.7 \pm 0.3\%$ and $27.7 \pm 0.4\%$, respectively. Release kinetics of herbicides from prepared nanoherbicides was dependent on pH with 43% release observed at a pH value of 10.0, and hence, a controlled-release mechanism was proposed based on electrostatic interactions. Under acidic conditions, MSN-TA nanoherbicides were speculated to have an overall positive charge that may facilitate strong electrostatic attraction with negatively charged 2,4-D sodium salt, whereas at alkaline conditions, silanol groups of MSN-TA nanoherbicides are deprotonated that may create significant electrostatic repulsion which would increase the release rate of herbicide. Herbicide release was also dependent on the temperature wherein 96% of 2,4-D sodium salt was released after 900 min at 40°C. Bioactivity of MSN-TA nanoherbicides was assessed against *Cucumis sativus* L. and *Triticum aestivum* L. wherein selective inhibition was observed against *C. sativus* L. target plant with similar effectivity as that of free herbicide while no effect in *T. aestivum* L. was seen after application of MSN-TA nanoherbicide containing sodium salt of 2,4-D.

In another study, Abigail (2019) fabricated biochar-based nanocarriers for sustained release of 2,4-dichlorophenoxyacetic acid (2,4-D) herbicide. Nanoparticles from rice husk biochar (rBC) were prepared via pyrolysis at 350°C and were loaded with 0.25 wt% ratio of 2,4-D herbicide. Mechanism of adsorption of 2,4-D onto rBC nanoparticles was found to follow Freundlich isotherm model that inferred attachment of 2,4-D molecules onto rBC nanoparticles through a heterogeneous reversible multilayer mode of adsorption. In addition, sorption kinetic results revealed a pseudosecond-order model that depicted the process of strong reversible

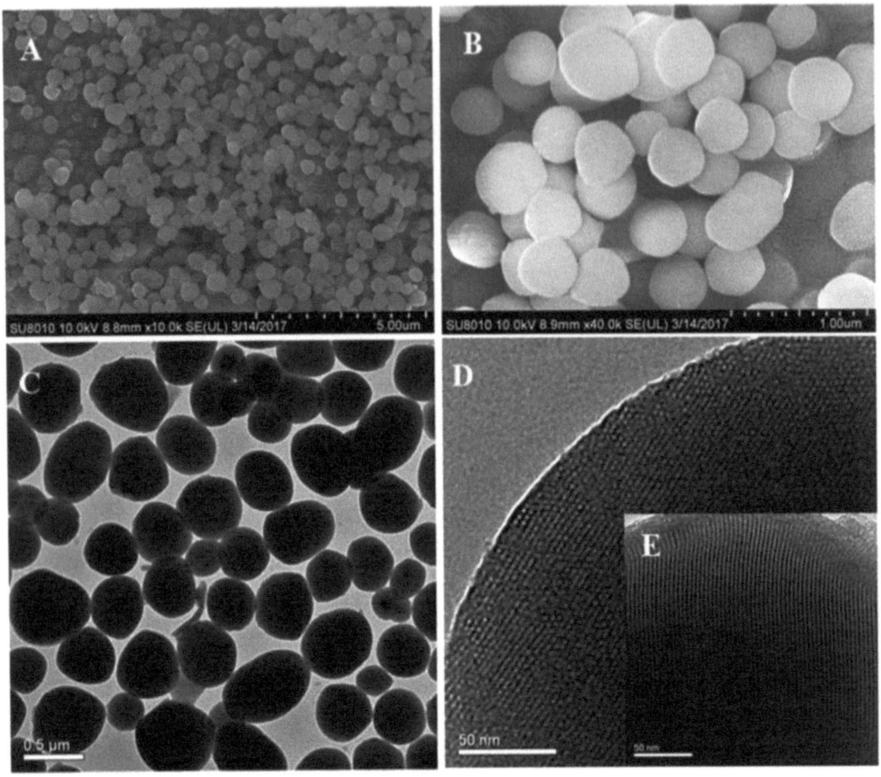

Fig. 16.4 SEM (A, B) and TEM (C, D, E) images of the TA groups functionalized MSN (MSN-TA). TEM images (D) and (E) were viewed along and perpendicular to the channel direction. Scale bars: (A) 5.0 µm, (B) 1.0 µm, (C) 0.5 µm, and (D, E) 50 nm.
Reprinted with permission from Cao, L., Zhou, Z., Niu, S., Cao, C., Li, X., Shan, Y. and Huang, Q., 2017. Positive-charge functionalized mesoporous silica nanoparticles as nanocarriers for controlled 2, 4-dichlorophenoxy acetic acid sodium salt release. J. Agric. Food Chem. 66(26), 6594–6603. Copyright © 2018 American Chemical Society.

chemisorption that facilitated higher loading of herbicide. Hydrodynamic particle size of fabricated rBC nanoherbicides containing 2,4-D was 256.5 nm with a negative zeta potential value of −28.8 mV, highlighting high stability of nanoformulation. FTIR results showed the presence of carboxyl group, aromatic carbon, and siloxane that facilitated 2,4-D adsorption onto surface of rBC nanoparticles. EDX spectrum also displayed the presence of carbon, oxygen, and silicon elements in rBC nanoherbicides along with an additional peak of chlorine that was hypothesized to be a result of sorption process. Sustained release of 2,4-D from rBC nanoherbicides in water was demonstrated for approximately 55 days at a pH value of 7.0 and at room temperature. Pattern of 2,4-D release was analyzed using Korsmeyer-Peppas model that exhibited a controlled diffusion process in spherical monolithic matrices. Likewise, 2,4-D release in soil was slower as compared to free 2,4-D control experiment. Bioactivity of rBC nanoherbicides was observed against *Brassica* sp. as the target plant with faster necrosis effect using 2,4-D nanoformulations as compared to the treatment with free 2,4-D, whereas no herbicidal activity of 2,4-D nanoherbicides was seen when used against *Z. mays* that indicated specific activity.

Abigail et al. (2016) also prepared rice husk nanosorbents for its application as a carrier of 2,4-D herbicide. Rapid and efficient adsorption of herbicide was observed at 1:0.10 (%) ratio of rice husk nanosorbents to 2,4-D where equilibrium was attained within-contact time of 90 min. Langmuir isotherm model was best fit with highest regression coefficient value of 0.998 while D-R isotherm showed mean sorption energy of 15.81 kJ/mol. Likewise, a pseudosecond-order reaction kinetics model was considered to be best fit which suggested that sorption process may be a combination of diffusion of film as well as intraparticle. DLS results of rice husk nanosorbents showed an average hydrodynamic particle size of 42.1 nm while zeta potential measurements revealed a negative value of 28.8 mV. Furthermore, SEM micrographs displayed rice husk nanosorbents were irregular squares in shape that had a smooth surface. Adsorption of 2,4-D onto nanosorbents resulted in the formation of aggregates that were suggested to affect surface properties as well. Sustained release of herbicide was obtained for about 6 days with a release mechanism that involved diffusion. Bioactivity of rice husk nanoherbicides was analyzed against *Brassica* sp. wherein 100% mortality was attained with enhanced activity of nanoformulation as compared to technical grade 2,4-D. On the contrary, germination index of 92% was maintained in the case of *Z. mays* after nanoherbicide application that depicted enhanced specificity and increased bioavailability of 2,4-D after nanoadsorption. Therefore, utilization of waste materials like rice husk for preparation of nanomaterials could be used as potential nanocarriers for herbicide.

16.4 Conclusion and future prospects

Herbicides are widely used in agriculture for effective control of weeds. These chemicals can alter various physiological processes, such as inhibition of photosynthesis, regulating plant growth, blocking biosynthesis of amino acid, inhibition of cell elongation, and cell division (Kniss, 2017). Hence, exploring the mechanism of action of herbicides is most significant for developing effective weed control strategies. Nanotechnology-driven solutions have led to the fabrication of nanoherbicides with enhanced weed inhibitory activity. Various polymeric and inorganic nanocarriers have been used for the preparation of nanoherbicides. However, the physical and chemical approaches for generation of nanoparticles involve corrosive chemicals and hazardous reaction conditions (Ghosh et al., 2016a,b,c,d). Hence, there is a chance that the nanocarriers developed by these routes may be toxic for the crops and environment.

Biological route for the generation of nanoparticles can be a promising alternative for delivery of herbicides. Bacteria, algae, fungi, and medicinal plants are reported for the synthesis of metal and metal oxide nanoparticles with exotic shapes and attractive physicochemical properties (Rokade et al., 2018, 2017; Adersh et al., 2015; Jamdade et al., 2019; Shende et al., 2017; Ghosh, 2018). Further, biogenic nanoparticles have antibacterial, antifungal, and insecticidal properties (Bloch et al., 2021; Ranpariya et al., 2021). Hence, using biogenic nanoparticles as carriers for herbicides may help in enhanced crop protection due to their broad-spectrum bactericidal, fungicidal, and larvicidal properties (Shende et al., 2018; Bhagwat et al., 2018; Shinde et al., 2018).

The nanoparticles can be surface functionalized with classical photosystem-II (PSII) inhibitors, quinone-binding protein, enzyme inhibitors that may help in targeted delivery and sustained release resulting in more effective weed control (Ghosh et al., 2015; Kitture et al., 2015). Similarly, herbicides can be encapsulated in liposomes or nanocomposites for targeting enzymes, such as 5-enolpyruvylshikimate 3-phosphate (EPSP) synthase, acetohydroxyacid synthase (AHAS), protoporphyrinogen oxidase (PPO), 4-hydroxyphenylpyruvate dioxygenase (HPPD), acetyl-CoA carboxylase (ACC), or glutamine synthase (GS), thereby altering the metabolic pathways eventually resulting in inhibition of weeds. However, before using the nanoherbicides, their toxicity profiles should be thoroughly studied in order to ensure the safety of the farmers, soil health, and crops. In view of the background, it is evident that nanoherbicides are the most promising agents that can revolutionize the face of future agriculture.

Acknowledgment

Dr. Sougata Ghosh acknowledges Kasetsart University, Bangkok, Thailand for Post Doctoral Fellowship and funding under Reinventing University Program (Ref. No. 6501.0207/10870 dated 9th November, 2021).

References

Abigail, M.E.A., 2019. Biochar-based nanocarriers: fabrication, characterization, and application as 2,4-dichlorophenoxyacetic acid nanoformulation for sustained release. 3 Biotech 9 (8), 317.

Abigail, M.E.A., Samuel, S.M., Chidambaram, R., 2016. Application of rice husk nanosorbents containing 2,4-dichlorophenoxyacetic acid herbicide to control weeds and reduce leaching from soil. J. Taiwan Inst. Chem. Eng. 63, 318–326.

Adersh, A., Kulkarni, A.R., Ghosh, S., More, P., Chopade, B.A., Gandhi, M.N., 2015. Surface defect rich ZnO quantum dots as antioxidant inhibiting α-amylase and α-glucosidase: a potential anti-diabetic nanomedicine. J. Mater. Chem. B 3, 4597–4606.

Bhagwat, T.R., Joshi, K.A., Parihar, V.S., Asok, A., Bellare, J., Ghosh, S., 2018. Biogenic copper nanoparticles from medicinal plants as novel antidiabetic nanomedicine. World J. Pharm. Res. 7 (4), 183–196.

Bloch, K., Pardesi, K., Satriano, C., Ghosh, S., 2021. Bacteriogenic platinum nanoparticles for application in nanomedicine. Front. Chem. 9, 624344.

Cao, L., Zhou, Z., Niu, S., Cao, C., Li, X., Shan, Y., Huang, Q., 2017. Positive-charge functionalized mesoporous silica nanoparticles as nanocarriers for controlled 2,4-dichlorophenoxy acetic acid sodium salt release. J. Agric. Food Chem. 66 (26), 6594–6603.

Craig, K., 2018. A review of the chemistry, pesticide use, and environmental fate of sulfur dioxide, as used in California. Rev. Environ. Contam. Toxicol. 246, 33–64.

de Oliveira, J.L., Campos, E.V.R., Goncalves da Silva, C.M., Pasquoto, T., Lima, R., Fraceto, L. F., 2015. Solid lipid nanoparticles co-loaded with simazine and atrazine: preparation, characterization, and evaluation of herbicidal activity. J. Agric. Food Chem. 63 (2), 422–432.

Diyanat, M., Saeidian, H., Baziar, S., Mirjafary, Z., 2019. Preparation and characterization of polycaprolactone nanocapsules containing pretilachlor as a herbicide nanocarrier. Environ. Sci. Pollut. Res. Int. 26 (21), 21579–21588.

dos Santos Silva, M., Cocenza, D.S., Grillo, R., de Melo, N.F.S., Tonello, P.S., de Oliveira, L.C., Cassimiro, D.L., Rosa, A.H., Fraceto, L.F., 2011. Paraquat-loaded alginate/chitosan nanoparticles: preparation, characterization and soil sorption studies. J. Hazard. Mater. 190 (1-3), 366–374.

Ghosh, S., 2018. Copper and palladium nanostructures: a bacteriogenic approach. Appl. Microbiol. Biotechnol. 101 (18), 7693–7701.

Ghosh, S., More, P., Derle, A., Kitture, R., Kale, T., Gorain, M., Avasthi, A., Markad, P., Kundu, G.C., Kale, S., Dhavale, D.D., Bellare, J., Chopade, B.A., 2015. Diosgenin functionalized iron oxide nanoparticles as novel nanomaterial against breast cancer. J. Nanosci. Nanotechnol. 15 (12), 9464–9472.

Ghosh, S., Patil, S., Chopade, N.B., Luikham, S., Kitture, R., Gurav, D.D., Patil, A.B., Phadatare, S.D., Sontakke, V., Kale, S., Shinde, V., Bellare, J., Chopade, B.A., 2016a. *Gnidia glauca* leaf and stem extract mediated synthesis of gold nanocatalysts with free radical scavenging potential. J. Nanomed. Nanotechnol. 7, 358.

Ghosh, S., Harke, A.N., Chacko, M.J., Gurav, S.P., Joshi, K.A., Dhepe, A., Dewle, A., Tomar, G.B., Kitture, R., Parihar, V.S., Banerjee, K., Kamble, N., Bellare, J., Chopade, B.A., 2016b. *Gloriosa superba* mediated synthesis of silver and gold nanoparticles for anticancer applications. J. Nanomed. Nanotechnol. 7, 4.

Ghosh, S., Chacko, M.J., Harke, A.N., Gurav, S.P., Joshi, K.A., Dhepe, A., Kulkarni, A.S., Shinde, V.S., Parihar, V.S., Asok, A., Banerjee, K., Kamble, N., Bellare, J., Chopade, B.A., 2016c. *Barleria prionitis* leaf mediated synthesis of silver and gold nanocatalysts. J. Nanomed. Nanotechnol. 7, 4.

Ghosh, S., Gurav, S.P., Harke, A.N., Chacko, M.J., Joshi, K.A., Dhepe, A., Charolkar, C., Shinde, V.S., Kitture, R., Parihar, V.S., Banerjee, K., Kamble, N., Bellare, J., Chopade, B.A., 2016d. *Dioscorea oppositifolia* mediated synthesis of gold and silver nanoparticles with catalytic activity. J. Nanomed. Nanotechnol. 7, 5.

Grillo, R., dos Santos, N.Z.P., Maruyama, C.R., Rosa, A.H., de Lima, R., Fraceto, L.F., 2012. Poly (ε-caprolactone) nanocapsules as carrier systems for herbicides: Physico-chemical characterization and genotoxicity evaluation. J. Hazard. Mater. 231–232, 1–9.

Grillo, R., Pereira, A.E., Nishisaka, C.S., De Lima, R., Oehlke, K., Greiner, R., Fraceto, L.F., 2014. Chitosan/tripolyphosphate nanoparticles loaded with paraquat herbicide: an environmentally safer alternative for weed control. J. Hazard. Mater. 278, 163–171.

Jablonkai, I., 2011. Molecular mechanism of action of herbicides. In: Herbicides–Mechanisms and Mode of Action. Intech, Croatia, pp. 3–24.

Jamdade, D.A., Rajpali, D., Joshi, K.A., Kitture, R., Kulkarni, A.S., Shinde, V.S., Bellare, J., Babiya, K.R., Ghosh, S., 2019. *Gnidia glauca* and *Plumbago zeylanica* mediated synthesis of novel copper nanoparticles as promising antidiabetic agents. Adv. Pharmacol. Sci. 2019, 9080279.

Khatem, R., Celis, R., Hermosín, M.C., 2019. Cationic and anionic clay nanoformulations of imazamox for minimizing environmental risk. Appl. Clay Sci. 168, 106–115.

Kitture, R., Chordiya, K., Gaware, S., Ghosh, S., More, P.A., Kulkarni, P., Chopade, B.A., Kale, S.N., 2015. ZnO nanoparticles-red sandalwood conjugate: a promising anti-diabetic agent. J. Nanosci. Nanotechnol. 15, 4046–4051.

Kniss, A.R., 2017. Long-term trends in the intensity and relative toxicity of herbicide use. Nat. Commun. 8, 14865.

Kumar, S., Bhanjana, G., Sharma, A., Dilbaghi, N., Sidhu, M.C., Kim, K.H., 2017. Development of nanoformulation approaches for the control of weeds. Sci. Total Environ. 586, 1272–1278.

Liu, L., Bilal, M., Duan, X., Iqbal, H.M., 2019. Mitigation of environmental pollution by genetically engineered bacteria- current challenges and future perspectives. Sci. Total Environ. 667, 444–454.

Lone, A.H., Raverkar, K.P., Pareek, N., 2014. In-vitro effects of herbicides on soil microbial communities. The Bioscan 9 (1), 11–16.

Mahmoudian, M., Torbati, S., AliMirzayi, N., Nozad, E., Kochameshki, M.G., Shokri, A., 2020. Preparation and investigation of poly (methylmethacrylate) nano-capsules containing haloxyfop-R-methyl and their release behavior. J. Environ. Sci. Health B 55 (4), 301–309.

Maruyama, C.R., Guilger, M., Pascoli, M., Bileshy-José, N., Abhilash, P.C., Fraceto, L.F., De Lima, R., 2016. Nanoparticles based on chitosan as carriers for the combined herbicides imazapic and imazapyr. Sci. Rep. 6, 19768.

Namasivayam, K.R.S., Aruna, A., Gokila, 2014. Evaluation of silver nanoparticles-chitosan encapsulated synthetic herbicide paraquate (AgNp-CS-PQ) preparation for the controlled release and improved herbicidal activity against *Eichhornia crassipes*. Res. J. Biotechnol. 9 (9), 19–27.

Oliveira, H.C., Stolf-Moreira, R., Martinez, C.B.R., Grillo, R., de Jesus, M.B., Fraceto, L.F., 2015. Nanoencapsulation enhances the post-emergence herbicidal activity of atrazine against mustard plants. PLoS One 10 (7), e0132971.

Pascoli, M., Lopes-Oliveira, P.J., Fraceto, L.F., Seabra, A.B., Oliveira, H.C., 2018. State of the art of polymeric nanoparticles as carrier systems with agricultural applications: a mini-review. Energy Ecol. Environ. 3, 137–148.

Pereira, A.E., Grillo, R., Mello, N.F., Rosa, A.H., Fraceto, L.F., 2014. Application of poly (epsilon-caprolactone) nanoparticles containing atrazine herbicide as an alternative technique to control weeds and reduce damage to the environment. J. Hazard. Mater. 268, 207–215.

Pontes, M.S., Antunes, D.R., Oliveira, I.P., Forini, M.M., Santos, J.S., Arruda, G.J., Caires, A. R., Santiago, E.F., Grillo, R., 2021. Chitosan/tripolyphosphate nanoformulation carrying paraquat: insights on its enhanced herbicidal activity. Environ. Sci.: Nano 8, 1336–1351.

Ranpariya, B., Salunke, G., Karmakar, S., Babiya, K., Sutar, S., Kadoo, N., Kumbhakar, P., Ghosh, S., 2021. Antimicrobial synergy of silver-platinum nanohybrids with antibiotics. Front. Microbiol. 11, 610968.

Rashidipour, M., Maleki, A., Kordi, S., Birjandi, M., Pajouhi, N., Mohammadi, E., Heydari, R., Rezaee, R., Rasoulian, B., Davari, B., 2019. Pectin/chitosan/tripolyphosphate nanoparticles: efficient carriers for reducing soil sorption, cytotoxicity, and mutagenicity of paraquat and enhancing its herbicide activity. J. Agric. Food Chem. 67 (20), 5736–5745.

Retzinger, E.J., Mallory-Smith, C., 1997. Classification of herbicides by site of action for weed resistance management strategies. Weed Technol. 11 (2), 384–393.

Rokade, S.S., Joshi, K.A., Mahajan, K., Tomar, G., Dubal, D.S., Parihar, V.S., Kitture, R., Bellare, J., Ghosh, S., 2017. Novel anticancer platinum and palladium nanoparticles from *Barleria prionitis*. Glob. J. Nanomed. 2 (5), 555600.

Rokade, S., Joshi, K., Mahajan, K., Patil, S., Tomar, G., Dubal, D., Parihar, V.S., Kitture, R., Bellare, J.R., Ghosh, S., 2018. *Gloriosa superba* mediated synthesis of platinum and palladium nanoparticles for induction of apoptosis in breast cancer. Bioinorg. Chem. Appl. 2018, 4924186.

Schnoor, B., Elhendawy, A., Joseph, S., Putman, M., Chacon-Cerdas, R., Flores-Mora, D., Bravo-Moraga, F., Gonzalez-Nilo, F., Salvador-Morales, C., 2018. Engineering atrazine loaded poly (lactic-co-glycolic acid) nanoparticles to ameliorate environmental challenges. J. Agric. Food Chem. 66 (30), 7889–7898.

Shende, S., Joshi, K.A., Kulkarni, A.S., Shinde, V.S., Parihar, V.S., Kitture, R., Banerjee, K., Kamble, N., Bellare, J., Ghosh, S., 2017. *Litchi chinensis* peel: a novel source for synthesis of gold and silver nanocatalysts. Glob. J. Nanomed. 3 (1), 555603.

Shende, S., Joshi, K.A., Kulkarni, A.S., Charolkar, C., Shinde, V.S., Parihar, V.S., Kitture, R., Banerjee, K., Kamble, N., Bellare, J., Ghosh, S., 2018. *Platanus orientalis* leaf mediated rapid synthesis of catalytic gold and silver nanoparticles. J. Nanomed. Nanotechnol. 9 (2). https://doi.org/10.4172/2157-7439.1000494.

Shinde, S.S., Joshi, K.A., Patil, S., Singh, S., Kitture, R., Bellare, J., Ghosh, S., 2018. Green synthesis of silver nanoparticles using *Gnidia glauca* and computational evaluation of synergistic potential with antimicrobial drugs. World J. Pharm. Res. 7 (4), 156–171.

Sousa, G.F., Gomes, D.G., Campos, E.V., Oliveira, J.L., Fraceto, L.F., Stolf-Moreira, R., Oliveira, H.C., 2018. Post-emergence herbicidal activity of nanoatrazine against susceptible weeds. Front. Environ. Sci. 6, 12.

Yu, Z., Sun, X., Song, H., Wang, W., Ye, Z., Shi, L., Ding, K., 2015. Glutathione-responsive carboxymethyl chitosan nanoparticles for controlled release of herbicides. Mater. Sci. Appl. 6, 591–604.

Nanosensors in agriculture

17

Sahila Beegum[a] *and Saurav Das*[b]
[a]Nebraska Water Center, Part of the Robert B. Daugherty Water for Food Global Institute, University of Nebraska, Lincoln, NE, United States, [b]Department of Agronomy and Horticulture, University of Nebraska–Lincoln, Lincoln, NE, United States

17.1 Introduction

The agricultural production system has to improve to feed 9.7 billion people by 2050, and this should be attained through sustainable farm intensification without resulting in adverse environmental impacts (Shang et al., 2019). Over the decades, the agricultural sector has made a significant increase in production, 200% in Asia, 200% in Latin America and United States, and 68% in western Europe. This is majorly achieved by infusing technology in crop variety, chemical fertilizers, pesticides, precision agriculture, automatic and in situ monitoring of the plant, soil, atmosphere, and interactions (van Dijk and Meijerink, 2014). At the same time, there have been several reports on the detrimental impacts of the agricultural practices that have impacted the environment and ecology. The excessive and uncontrolled application of fertilizers and chemicals has resulted in soil pollution, groundwater, and surface water bodies, which has become a serious threat to human health (Schwarzenbach et al., 2006). Analysis of the global database of pesticide application and pollution risk shows that 64% of global agricultural land (~24.5 million km^2) is at risk of pesticide pollution by more than one ingredient, and 31% is at high risk (Tang et al., 2021). This has also reduced the beneficial soil bacterial community and insects in the soil (Zhang et al., 2021). Long-term irrigation has resulted in the deterioration of the soil structure and its ability to perform ecosystem functions. The main soil degradation processes are reduction in the soil organic carbon, acidification, nutrient imbalance, and loss of soil fertility (Mihelič et al., 2021).

Among several possible approaches to improve agricultural productivity and food security and to address the limitations of the current agricultural practices in a sustainable manner, nanotechnology has gained good momentum in the last decade. The US Environmental Protection Agency defines nanotechnology as the science of understanding and control of matter at dimensions of roughly 1–100nm, where unique physical properties make novel applications possible (EPAUS, 2007). Some of the applications of nanotechnology in agriculture are (1) nanoformulations of pesticides and fertilizers for crop improvement; (2) nanosensors in crop protection for the identification of diseases and agrochemical residues; (3) nanodevices for the genetic engineering of plants; (4) plant disease diagnostics; (5) animal health, animal breeding, and poultry production; and (6) postharvest management (Ghidan and Al Antary, 2020).

Agricultural Nanobiotechnology. https://doi.org/10.1016/B978-0-323-91908-1.00012-2

Nanosensors are sensor devices that have at least one element of their sensing components smaller than 100 nm. The smaller size of the sensors assists interaction at a single-molecule scale and improves the sensitivity and detection level down to 10 parts per million (Bhushan, 2017). Nanosensing technology utilizes the novel physics at the nanoscale using nanomaterials that can deliver unique chemical and physical properties and a larger surface area for sensing biochemical interactions (Yang and Timothy, 2021). The nanomaterials such as metal (gold, silver, cobalt, etc.) nanoparticles, carbon nanotubes (CNTs), magnetic nanoparticle-based systems, nanoprobes, nanowires/nanofibers, nanosystems (cantilevers/nanofabrication's), nanoelectromechanical systems, and quantum dots (QD) have been actively investigated for their applications as sensors. The spatial and temporal resolution and the reliability of the sensors have been improved by integrating nanoscience, electronics, and computers.

A basic nanosensor system has the following components: (a) sample which can be in any state of matter; (b) analyte, which is the substance the nanosensors are designed to detect; (c) recognition element that discriminates analytes from other substances in the sample; (d) transducer that converts recognition events into a measurable signal; and (e) signal processor that converts transducer signals into actionable data. Sensitivity of the sensors, rapid detection, cost-effectiveness, and practicality in different mediums are some of the major advantages of nanosensors. Nanosensors enabled real-time monitoring of the crop condition, pathogen, pesticide, moisture content, and environmental leaching is helping traditional agriculture to perform in a precise way. This chapter will discuss different types of nanosensors and their applications in the modern agricultural system to sense the variable parameter of the soil, water, and air. This chapter will also discuss the future applicability of the technique.

17.2 Nanosensors in agriculture

The application of autonomous nanosensors in agriculture has enabled comprehensive surveying of crop progress in real time with precise spatiotemporal positioning. Real-time monitoring of crop progress has improved agronomic management decisions like planting and harvesting based on detection of soil nutrients, moisture, pH, temperature, and plant's morphogenetic changes. The structural change in the plant can be analyzed using piezoelectric nanosensors. These sensors can generate electric signals in response to changes in the mechanical environment of the plants (Dimri et al., 2020). Nanosensors can be used for sensing a wide variety of plant pathogens. It has wide applications in the detection of soil physicochemical properties. There are sensors for detecting the soil nutrients, soil pH, soil moisture, gases, temperature, etc. (Kaushal and Wani, 2017). Nanosensors are also used for monitoring the plant signaling pathways and metabolism (Kwak et al., 2017). Detection of soil nutrients has helped in reducing the environmental contamination. Low limit of detection and high sensitivity of nanosensors have transformed conventional agriculture into smart agriculture.

17.2.1 Major types of nanosensors

Fig. 17.1 shows major nanostructures and their applications in the agricultural sector (Srivastava et al., 2018).

Following are the major types of nanosensors used in agriculture.

17.2.1.1 Optical nanosensors

Optical nanosensors are devices that transduce chemical or biological molecules using an optical signal, having dimensions less than 1000nm. It has an inert, biofriendly matrix in which a sensing component and an optical reporter are entrapped (Aylott, 2003). In most cases, fluorescence is the transduction method used because of its high sensitivity and relative ease of measurement. Fluorescence-entrapped polyacrylamide nanoparticle-based sensors for pH measurements and graphene-based optical nanosensors for detecting heavy metal ions are examples of optical nanosensors (Keiji et al., 2006; Zhang et al., 2018). Another application of optical nanosensors is for the detection of arsenic taken up by the plants through roots (Lew et al., 2021).

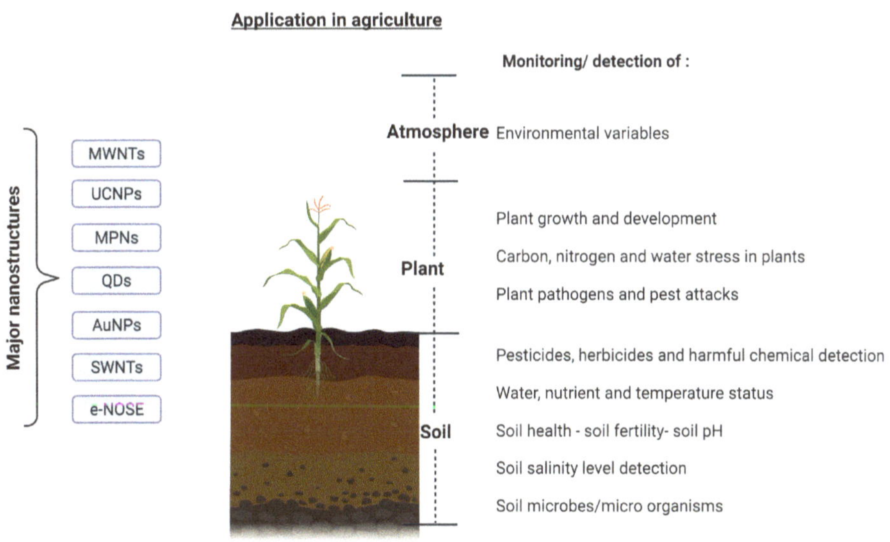

Fig. 17.1 Schematic representation of major nanostructures and their applications in agriculture. Major nanostructures are MWCNTs (multiwalled carbon nanotubes), UCNPs (upconversion nanoparticles), MPNs (magnetic nanoparticles), QDs (quantum dots), AuNPs (gold nanoparticles), and SWCNTs (single-walled carbon nanotubes).
Reproduced with permission from Srivastava, A.K., Dev, A., Karmakar, S., 2018. Nanosensors and nanobiosensors in food and agriculture. Environ. Chem. Lett. 16, 161–182. https://doi.org/10.1007/s10311-017-0674-7, Copyright © 2017, Springer International Publishing AG.

17.2.1.2 Electrochemical nanosensors

Electrochemical nanosensors function based on the principle of electrochemistry. An electrochemical sensor device detects an electrochemical signal and transforms it into signals that are analyzed based on their nature. Electrochemical sensors are made up of two main components. They are (a) the chemical recognition system (the essential component of the sensor) and (b) the physicochemical transducer device (that converts chemical responses into signals that can easily be detected through any electrical instrumentation) (Thangadurai et al., 2021). Some of the electrochemical methods used in the nanosensors are potentiometry (based on the potential difference), conductometry (based on resistance), amperometry, voltammetry (based on potential), etc. Minimal power requirements, high sensitivity of electrochemical transducers, compatibility with modern miniaturization technologies, low maintenance and cost, and rapid and low detection limits are the advantages of electrochemical nanosensors.

17.2.1.3 Nanobiosensors

A biosensor is a type of sensor system that detects analytes based upon a biological reaction with the detector components. These are used to detect biological products by combining physical and chemical transducers with a biological recognition element. Nanobiosensors are a form of biosensors with the analytical unit as a biologically derived sensitizing element (molecular imprints, lectins, antibodies, enzymes, nucleic acids, etc.), physiochemical transducer, and a reader attached with it. These sensors can sense the minor changes in bioactive compounds, metabolism transformation, heat, and pH concentration changes. Based on the type of transducers, the biosensors are classified into various types (e.g., optical, thermal, piezoelectric, electrochemical, and electrochemical luminescence biosensors (Plata et al., 2010; Rai et al., 2012). Nanobiosensors are also used to detect soil diseases caused by microorganisms, bacteria, viruses, and fungi. Several nanobiosensors are designed to detect contaminants; pests; nutrient content; and plant stress due to drought, temperature, or pressure. Nanobiosensors are also used for sensing the physiological changes in plants. Plant wearables, metalloid nanoparticles, quantum dots, and array-based are some of the nanobiosensors used for monitoring plant growth and development. Nanobiosensors are also used for soil quality assessments.

17.3 Detection of soil quality

Nanosensors can measure the soil pH, nutrients, residual particles, temperature, and moisture content. It can sense urea, glucose, and pesticides. Precise irrigation can be given based on the moisture content in the soil. Diagnose soil diseases caused by infecting soil microorganisms, such as viruses, bacteria, and fungi. Nanofibers in one dimension made up of potassium niobate are used for determining the soil humidity (Ganeshkumar et al., 2016). Nanosensors can be used for the detection of plant-microbiome interactions. Successful detection and modulation of the plant microbiome bacteria enable scientists to not only influence plant beneficial activities but also favorably influence soil health by reducing the use of chemical fertilizers. Interactions between plants, fungi, and microbes in the soil are of great importance for sustainable agriculture as these influence plant beneficial activities and soil health. Microbial diversity in the soil can be detected using nanosensors. Some of the nanomaterials used for the detection of microbes include carbon nanotubes (CNTs), metal nanoparticles (mNPs), quantum dots (QDs), silica nanoparticles, and upconversion nanoparticles (UCNPs) (Sashidhar et al., 2019). Some of the nanosensors detect the desired and undesired bacteria in the soil by comparing the amount of oxygen utilized by them during respiration (Kaushal and Wani, 2017). Nanotechnology-based microelectromechanical system sensors are used for sensing moisture and temperature of the soil. Some of the moisture-level-detecting nanosensors have microcantilever beams coated with water-sensitive nanopolymer and combined with the Wheatstone bridge piezoresistor circuit. The changes on the surface of the microcantilever based on the moisture in the soil are detected using the nanoresistor (Dimri et al., 2020). Nanosensors can be distributed throughout the field to identify the soil conditions and group and improve management decisions.

The presence of heavy metals like Pb^{2+}, Hg^{2+}, Ag^{+}, Cd^{2+}, and Cu^{2+} in the soil and water above the permissible limits is unsafe for humans and the environment. Reusing the wastewater from industries can result in the accumulation of heavy metals in the agricultural soils. ICTS nanosensors, optical nanosensors, and electrochemical nanosensors are used for detecting cadmium. Colorimetric nanosensors, multimodal nanosensors, and magnetic fluorescent nanosensors are used for detecting mercury. ICTS nanosensors, surface plasmon resonance nanosensors, and colorimetric nanosensors are used for lead detection (Sharma et al., 2021).

17.4 Detection of water quality

Water quality monitoring is challenging due to the low concentration of pollutants along with their complexity and redox variability. Promising results from nanosensors, especially nanowires (Cui et al., 2001) and quantum dots (Han et al., 2001), showed potential applicability in water quality measurements. Rapid analysis time and ultralow detection limit have increased the attention in nanosensor research. Environmental losses, especially nitrate (NO_3^-), due to the large application of nitrogen in agriculture have emerged as one of the major water pollutants. On a global scale, nitrate concentration can exceed values as high as $110\,mg\,L^{-1}$ in surface water and $440\,mg\,L^{-1}$ in groundwater. High concentration of nitrate poses a serious threat to human health, especially infants, pregnant women, and people with low stomach acid. Recommended level of nitrate in drinking water is $10\,mg\,L^{-1}$. Colorimetric detection of nitrate using gold nanorods (AuNRs) has been reported by Akbari et al. (2018). Gold surface-modified plasmon resonance sensor showed a nitrate detection limit of up to $3.0\,\mu g\,L^{-1}$(Miao et al., 2019). Fiber-optic surface plasmon resonance sensor showed high reliability for the low-cost probe design and ultratrace detection limit (Parveen et al., 2017).

Heavy metals that pose a major threat to human health for their carcinogenic effects have been efficiently detected with different nanocomposite sensors. Nanosensors with gold (Au)/polyaniline (PANI) were used to detect Hg^{2+} ions as low as 10^{-11} M in water samples based on Raman intensity response (RIS) (Wang et al., 2011). Biosensor using nonpathogenic *Escherichia coli* with luciferase gene has been used for the detection of arsenic in range of 5–100 $\mu g L^{-1}$(Stocker et al., 2003). Level of aluminum (Al) ions in the environment has significantly increased, especially in food and water samples. Colorimetric and fluorometric-based nanosensors functionalized with gold nanoparticles showed a promising application for the detection of Al^{3+} (Abubaker et al., 2019).

Optical sensor with silver nanoparticle stabilized with sodium sulfonate 3-mercaptopropane (SNP-MPS) showed sensitivity in the detection of Ni_2^+ as low as 0.3 ppm (Prosposito et al., 2016). The recent development of Internet of Things (IoT) has also promised the development of sensors networks for real-time data acquisition for water distribution chain. Deployment of sensor network will enable the detection of dead-end pipes, connection status between service mains, service lines, and premise plumbing. It can also check the spatiotemporal release of contaminants. If contamination is detected beyond the permissible limits, the service mains can be stopped to prevent the spread through the distribution system.

17.5 Detection of pesticide residue

Pesticides are an important ingredient for crop production; at the same time, they can also cause health risks if present in the soil above the permissible limits. Some of the pesticides are carcinogenic, mutagenic, and cytotoxic. Most of the conventional pesticide detection methods (high-pressure liquid chromatography, mass spectrometry, gas chromatography, flow, and antibody-based immunoassays) are time-consuming and impractical for on-site pesticide detection. Rapid, real-time, sensitive, inexpensive sensors have been developed to detect pesticide residue (Verdian, 2018). Some of the recent sensors had a biological recognition mechanism with a physical transduction technique. Nanosensors composed of TiO_2 nanotubes were used for atrazine detection in the soil (Yu et al., 2010). Phosphorus-containing amino acid-type herbicide in the soil is detected using nanosensors with a nanofilm-modified pencil graphite electrode. Nanofabrication of biosensors has further improved the functionality of nanosensors in terms of sensitivity, selectivity, and multiplexing capacity. Electrochemical nanobiosensors based on cholinesterase enzyme have shown promising results in the detection of organophosphate pesticides and carbamate residues for their high sensitivity (Zhou et al., 2017). Tyrosine-based biosensors are sensitive in the detection of carbamate pesticides and atrazine. Alkaline phosphate-based biosensors have been successfully used in the detection of organochlorine pesticides such as carbamate and fenitrothion (García Sánchez et al., 2003). Organophosphorus hydrolase and glutathione S-transferase-based biosensors have been used in the detection of parathion and methyl parathion (Du et al., 2010), paraoxon (Lee et al., 2010). Enzyme complex of acetylcholinesterase (AchE) and choline oxidase with CdTe quantum dots showed a detection limit of 4.49 nM for the organophosphate pesticides. Acetylcholinesterase biosensors stabilized by poly(*N*-vinyl-2-pyrrolidone) (PVP)-capped Cd quantum dots showed promising results in the detection of trichlorfon (Li et al., 2006). A list of nanoparticles and target compounds used for detecting soil nutrient availability, pesticide and herbicide residues are presented in Table 17.1.

Table 17.1 List of nanoparticles, target compounds, and the detection limit.

Nanoparticle	Target compound/ species	Detection limit
Titanium oxide (TiO_2)	Atrazine	0.1 ppt
Carbon	Herbicide detection	
Gold (Au)	Acetamiprid	5 nM
Gold (Au)	Urea and urease activity	5 μM and 1.8 U L^{-1}
Gold (Au)	*Pantoea stewartii* subsp.	7.8×10^3 cfu mL^{-1}
Gold (Au)	Herbicide detection	
Gold (Au)	Organophosphate's detection	
Gold nanorods	Cymbidium mosaic virus and Odontoglossum ringspot virus	48 and 42 pg mL^{-1}
Graphene	Herbicide detection	
Graphene oxide	Nitrate	10^{-5} M
Multiwalled carbon nanotubes (MWCNTs)	Glyphosate and glufosinate	0.35 and 0.19 ng mL^{-1}
MWC-chitosan nanocomposites	Methyl parathion	7.5×10^{-13} M
PEDOT nanofibers—graphene oxide nanosheet composites	Nitrate	0.68 mg L^{-1}
Quantum dots	DNA	3.55×10^{-9} M
Silver (Ag)	Herbicide detection	
ZnO-chitosan nanocomposite membranes	*Trichoderma harzianum*	1.0×10^{-19} mol L^{-1}
$TiO_2$and PbO_2	Trichlorfon	10^{-10}
MOF-5	Organophosphate	5×10^{-9} M
Aptamer-polymeric microsphere AuNPs	Malathion	3.3 μg m L^{-1}
AChE/L-cys/HGNs/Chits/GCE	Chlorpyrifos and carbofuran	0.06 and 0.08 μg dm^{-3}
Lateral flow immunochromatographic	Carbofuran and triazophos	32 and 4 μg L^{-1}
Eu(III)-AuNPs	Pinacolyl methylphosphonate, methylphosphonic acid, glyphosate	1 μmol dm^{-3}
Multiwalled carbon nanotubes, graphene sheets, ethyleneimine polymer, Au nanocomposites	Carbofuran	0.03 ng mL^{-1}

Continued

Table 17.1 Continued

Nanoparticle	Target compound/ species	Detection limit
Bis-aniline-bridged AuNPs composites	Mg^{2+}, Ca^{2+}, Sr^{2+}, Ba^{2+}	Femtomolar
Aptamer-AuNPs	Acetamiprid	0.1 ppm
AuNPs-multiwalled carbon nanotube-reduced graphene oxide nanoribbon composites	Acetamiprid	1.7×10^{-14} mol dm^{-3}
AChE/Chit-PB-MWCNTs-HGNs/ Au	Malathion, chlorpyrifos, monocrotophos, and carbofuran	Nanomolar range
MWCNTs-chitosan-GCE	Br^-	9.6×10^{-8} μg mL^{-1}

Reproduced with permission from Srivastava, A.K., Dev, A., Karmakar, S., 2018. Nanosensors and nanobiosensors in food and agriculture. Environ. Chem. Lett. 16, 161–182. https://doi.org/10.1007/s10311-017-0674-7, Copyright © 2017, Springer International Publishing AG and from Chhipa, H., 2019. Applications of nanotechnology in agriculture. In: Methods in Microbiology. Academic Press Inc., pp. 115–142. https://doi.org/10.1016/bs.mim.2019.01.002, Elsevier.

17.6 Detection of nutrient availability

Precise nutrient availability and fertilizer management are essential for crop productivity. Accurate estimation of the nutrient availability in the soil will help in increasing productivity and reducing the leaching of excess nutrients. Nanosensors can be used to detect the real-time nutrition status of the soil to calculate fertilizer requirements. Spectroscopic, amperometric, and impedance sensors are widely used in the detection of topsoil nutrients. The nanosensor detectors generally used for nitrate detection in the soil are graphene oxide-based nanosensors and nanofibers made of graphite oxide sheet and compound poly (3,4-ethylene dioxythiophene) (Ali et al., 2017). A study by Xu et al. (2021) showed the applicability of screen-printed electrodes of Co-nanoparticle for detecting phosphate from hydroponic solution.

17.7 Detection of plant stresses

H_2O_2, Ca^{2+}, sugars, nitric oxide, carbon monoxide, plant hormones (abscisic acid, jasmonic acid, etc.), and volatile organic compounds are major plant stress signaling molecules. The profiles of these signals get altered under different types of stresses in the plant. Nanosensors have been developed to sense the status of these signals, which can be correlated to the plant conditions (Wu and Li, 2021). Ratiometric quantum dot sensor and fluorescence tracking of QDs are generally used for detecting glucose. Hemin-complexed DNA aptamer-coated SWCNTs and nanoneedle transistor-based sensors are used for detecting NO and Ca^{2+}, respectively. Optical nanosensors and wearable sensors interfaced with plants can translate plant chemical signals to electric signals, radio waves, and optical waves, which can be monitored by electronic devices

(Giraldo et al., 2019). Stomatal opening controls the temperature, transpiration, photosynthesis, and leaf water potential in a plant. Real-time monitoring of the stomatal pore status is essential for assessing the plant growth status. Electrical conductometric nanosensors actuated by the stomatal pore have been implemented in the real-time tracking of the stomatal opening or closure (Wu and Li, 2021).

Nanosensors are also used for understanding the plant biology by studying the cellular concentration of small molecules, protein dynamics, biophysical parameters, nutrients, etc. Genetically encoded fluorescence resonance energy transfer (FRET) nanosensors are used for detecting DNA, metal ions, and organic compounds. Some of the biomolecules are detected using nanoparticle-based FRET sensors (Kwak et al., 2017).

17.8 Detection of plant pathogens

One of the major reasons for production and economic losses in the agricultural sector is caused by plant diseases. Plant pathogens cause a $21 billion loss in the agricultural sector each year in the United States (Carroll et al., 2017). In 2019, the yield loss estimates at the global level, majorly due to pathogens and pests for wheat, rice, maize, potato, and soybean, were 21.5%, 30.0%, 22.5%, 17.2%, and 21.4%, respectively (Savary et al., 2019). Early detection of plant pathogens can significantly help in preventing crop damage. Nanosensors can be used to study the cellular functions in the plant and the spatiotemporal dynamics of analytes. Plants emit chemical compounds as a means of communicating vital information about the surrounding environment (Kwak et al., 2017). The current nanoscience and nanotechnology have helped in sensing and detecting plant pathogens and biomolecules with the help of metabolite-dependent nanobiosensor. Some of the most common nanotechnological methods used in the detection of plant pathogens are as follows.

17.8.1 *Microneedle patches*

Microneedle patches are generally made of polyvinyl alcohol and fabricated through a simple vacuum base micromolding procedure. These can break the cell wall of the plant leaf tissue and can release the plant or pathogenic DNA. The DNA extraction method requires a sample preparation time of less than 1 min and the isolated DNA is ready for amplification without the need for additional sample purification steps (Li et al., 2020). Some of the recent researches focus on the integration of effective microneedle patch techniques with miniaturized DNA amplification assays. Paul et al. (2019) showcased the application of microneedle patches based on quick plant DNA extraction methods for detecting the *Phytophthora infestans* in the tomato leaves. The study also concluded that this is a minimally invasive, lysis-free, purification-free method and thus has a great potential for application toward a rapid molecular diagnosis of plant diseases on the field.

17.8.2 Nanopore sequencing platform

The nanopore sequencing platform uses motor protein to transport the single-stranded DNA or RNA through a protein- or solid-state nanopore to generate the nucleotide-dependent electronic current signals. This enables facile analysis of pathogenic polynucleotides. This sequencing platform has the advantage of real-time and long reads of sequence data, fast running times, portable, low cost, scalable genome mapping, and small sample loading.

17.8.3 Plant wearables

These are the ultrathin and ultralightweight nanosensor designs using the technology of microelectromechanical systems that can attach the flexible sensor devices directly to the plant leaves (Buja et al., 2021).

17.8.4 Quantum dots

These are the semiconductor nanocrystals comprised of elements from the periodic group II–VI, III–V, or IV–VI. They usually have a size of 10–15 nm. These have shown tremendous potential as optical nanosensors. Owing to their unique and advantageous photophysical properties such as narrow and symmetric emission spectra, broad absorption spectra that enable the simultaneous excitation of multiple fluorescent colors, and size-tunable light emissions, QDs have found successful applications as biosensors for plant imaging and disease identification (Wang et al., 2016).

17.9 Conclusions and future perspectives

Increasing population and intensification of agricultural production have detrimentally affected the environment and ecosystem. Real-time and reliable measurement of environmental changes can help in agronomic decision-making. Precision in fertilizer application by measuring the nutritional status of soil can reduce the environmental loss, especially of N fertilizer. Detection of real-time soil moisture levels can help in better management of irrigation regime and to save water. Detection of plants' morphogenetic changes in response to climate change (elevated temperature and CO_2) and pathogenic invasion can help develop a better agricultural production system. Measuring the nutrient and pesticide residuals in the environment can protect the ecosystem and humans from unprecedented risk events. Though there is ample research showing the potential applicability of nanotechnology and nanosensors in the agricultural system, to date, field-level implementation is nearly negligible. Federal regulatory bodies, policymakers, and environmental researchers need to evaluate the risk assessment and management before on-farm implementation. The scientific community needs to communicate the benefits of real-time data and precision farming for a sustainable environment and agriculture to the producers, the end-users of this technology.

References

Abubaker, M., Ngah, C.W.Z.C.W., Ahmad, M., Kuswandi, B., 2019. Chapter 19—Functionalized gold nanoparticles as optical nanosensors for determination of aluminum (III) ions in water samples. In: Shukla, A.K., Iravani, S. (Eds.), Green Synthesis, Characterization and Applications of Nanoparticles, Micro and Nano Technologies. Elsevier, pp. 459–484, https://doi.org/10.1016/B978-0-08-102579-6.00020-4.

Akbari, A., Derakhshesh, M., Abrishami, M.E., 2018. Nitrate ion effect on gold nanorods plasmon resonance behavior: a potential chemical sensor. Inorg. Chem. Commun. 98, 159–164. https://doi.org/10.1016/j.inoche.2018.10.023.

Ali, M.A., Jiang, H., Mahal, N.K., Weber, R.J., Kumar, R., Castellano, M.J., Dong, L., 2017. Microfluidic impedimetric sensor for soil nitrate detection using graphene oxide and conductive nanofibers enabled sensing interface. Sensors Actuators B Chem. 239, 1289–1299. https://doi.org/10.1016/j.snb.2016.09.101.

Aylott, J.W., 2003. Optical nanosensors—an enabling technology for intracellular measurements. Analyst 128, 309–312. https://doi.org/10.1039/B302174M.

Bhushan, B., 2017. Introduction to Nanotechnology. Springer Handbooks, https://doi.org/10.1007/978-3-662-54357-3_1.

Buja, I., Sabella, E., Monteduro, A.G., Chiriacò, M.S., De Bellis, L., Luvisi, A., Maruccio, G., 2021. Advances in plant disease detection and monitoring: from traditional assays to in-field diagnostics. Sensors (Basel) 21, 2129. https://doi.org/10.3390/s21062129.

Carroll, C.L., Carter, C.A., Goodhue, R.E., Lawell, C.C.L., Subbarao, K.V., Alston, J., Boik, A., Cameron, C., Muehlegger, E., Novan, K., Orazem, P., Rust, J., Schlenker, W., Scott, P., Sumner, D., 2017. Crop Disease and Agricultural Productivity. National Bureau of Economic Research, https://doi.org/10.3386/w23513.

Cui, Y., Wei, Q., Park, H., Lieber, C.M., 2001. Nanowire nanosensors for highly sensitive and selective detection of biological and chemical species. Science 293, 1289–1292. https://doi.org/10.1126/science.1062711.

Dimri, A., Pathak, N., Sharma, S., 2020. Nanosensors for root zone parameters influencing plant growth. In: Nanomaterials for Agriculture and Forestry Applications. Elsevier, pp. 387–406, https://doi.org/10.1016/b978-0-12-817852-2.00015-9.

Du, D., Chen, W., Zhang, W., Liu, D., Li, H., Lin, Y., 2010. Covalent coupling of organophosphorus hydrolase loaded quantum dots to carbon nanotube/Au nanocomposite for enhanced detection of methyl parathion. Biosens. Bioelectron. 25, 1370–1375. https://doi.org/10.1016/j.bios.2009.10.032.

Ganeshkumar, R., Sopiha, K.V., Wu, P., Cheah, C.W., Zhao, R., 2016. Ferroelectric $KNbO_3$ nanofibers: synthesis, characterization and their application as a humidity nanosensor. Nanotechnology 27. https://doi.org/10.1088/0957-4484/27/39/395607, 395607.

García Sánchez, F., Navas Díaz, A., Ramos Peinado, M.C., Belledone, C., 2003. Free and sol–gel immobilized alkaline phosphatase-based biosensor for the determination of pesticides and inorganic compounds. Anal. Chim. Acta 484, 45–51. https://doi.org/10.1016/S0003-2670(03)00310-6.

Ghidan, A.Y., Al Antary, T.M., 2020. Applications of nanotechnology in agriculture. In: Applications of Nanobiotechnology. IntechOpen, https://doi.org/10.5772/intechopen.88390.

Giraldo, J.P., Wu, H., Newkirk, G.M., Kruss, S., 2019. Nanobiotechnology approaches for engineering smart plant sensors. Nat. Nanotechnol. 14, 541–553. https://doi.org/10.1038/s41565-019-0470-6.

Han, M.-Y., Gao, X., Su, J., Nie, S., 2001. Quantum-dot-tagged microbeads for multiplexed optical coding of biomolecules. Nat. Biotechnol. 19, 631–635. https://doi.org/10.1038/90228.

Kaushal, M., Wani, S.P., 2017. Nanosensors: frontiers in precision agriculture. In: Nanotechnology: An Agricultural Paradigm. Springer, Singapore, pp. 279–291, https://doi.org/10.1007/978-981-10-4573-8_13.

Keiji, S., Zhong-You, S., Raoul, K., Hiroshi, M., 2006. Three-Dimensional pH Microprobing with an Optically-Manipulated Fluorescent Particle. pp. 141–142, https://doi.org/10.1246/cl.1996.141.

Kwak, S.Y., Wong, M.H., Lew, T.T.S., Bisker, G., Lee, M.A., Kaplan, A., Dong, J., Liu, A.T., Koman, V.B., Sinclair, R., Hamann, C., Strano, M.S., 2017. Nanosensor technology applied to living plant systems. Annu. Rev. Anal. Chem. 10, 113–140. https://doi.org/10.1146/annurev-anchem-061516-045310.

Lee, J.H., Park, J.Y., Min, K., Cha, H.J., Choi, S.S., Yoo, Y.J., 2010. A novel organophosphorus hydrolase-based biosensor using mesoporous carbons and carbon black for the detection of organophosphate nerve agents. Biosens. Bioelectron. 25, 1566–1570. https://doi.org/10.1016/j.bios.2009.10.013.

Lew, T.T.S., Park, M., Cui, J., Strano, M.S., 2021. Plant nanobionic sensors for arsenic detection. Adv. Mater. 33. https://doi.org/10.1002/ADMA.202005683.

Li, X.-H., Xie, Z., Min, H., Li, C., Liu, M., Xian, Y., Jin, L., 2006. Development of quantum dots modified acetylcholinesterase biosensor for the detection of trichlorfon. Electroanalysis 18, 2163–2167. https://doi.org/10.1002/elan.200603615.

Li, Z., Yu, T., Paul, R., Fan, J., Yang, Y., Wei, Q., 2020. Agricultural nanodiagnostics for plant diseases: recent advances and challenges. Nanoscale Adv. 2, 3083–3094. https://doi.org/10.1039/c9na00724e.

Miao, P., Liu, Z., Guo, J., Yuan, M., Zhong, R., Wang, L., Zhang, F., 2019. A novel ultrasensitive surface plasmon resonance-based nanosensor for nitrite detection. RSC Adv. 9, 17698–17705. https://doi.org/10.1039/C9RA02460C.

Mihelič, R., Pečnik, J., Glavan, M., Pintar, M., 2021. Impact of sustainable land management practices on soil properties: example of organic and integrated agricultural management. Land 10, 1–17. https://doi.org/10.3390/land10010008.

Parveen, S., Pathak, A., Gupta, B.D., 2017. Fiber optic SPR nanosensor based on synergistic effects of CNT/Cu-nanoparticles composite for ultratrace sensing of nitrate. Sensors Actuators B Chem. 246, 910–919. https://doi.org/10.1016/j.snb.2017.02.170.

Paul, R., Saville, A.C., Hansel, J.C., Ye, Y., Ball, C., Williams, A., Chang, X., Chen, G., Gu, Z., Ristaino, J.B., Wei, Q., 2019. Extraction of plant DNA by microneedle patch for rapid detection of plant diseases. ACS Nano. https://doi.org/10.1021/acsnano.9b00193.

Plata, M.R., Contento, A.M., Ríos, A., 2010. State-of-the-art of (bio) chemical sensor developments in analytical Spanish groups. Sensors 10, 2511–2576. https://doi.org/10.3390/S100402511.

Prosposito, P., Mochi, F., Ciotta, E., Casalboni, M., Matteis, F.D., Venditti, I., Fontana, L., Testa, G., Fratoddi, I., 2016. Hydrophilic silver nanoparticles with tunable optical properties: application for the detection of heavy metals in water. Beilstein J. Nanotechnol. 7, 1654–1661. https://doi.org/10.3762/bjnano.7.157.

Rai, V., Acharya, S., Dey, N., 2012. Implications of nanobiosensors in agriculture. J. Biomater. Nanobiotechnol. 3, 315–324. https://doi.org/10.4236/jbnb.2012.322039.

Sashidhar, P., Dubey, M.K., Kochar, M., 2019. Sensing soil microbes and interactions: how can nanomaterials help? In: Nanotechnology in the Life Sciences. Springer Science and Business Media B.V, pp. 213–236, https://doi.org/10.1007/978-3-030-16534-5_11.

Savary, S., Willocquet, L., Pethybridge, S.J., Esker, P., McRoberts, N., Nelson, A., 2019. The global burden of pathogens and pests on major food crops. Nat. Ecol. Evol. 3, 430–439. https://doi.org/10.1038/s41559-018-0793-y.

Schwarzenbach, R.P., Escher, B.I., Fenner, K., Hofstetter, T.B., Johnson, C.A., Von Gunten, U., Wehrli, B., 2006. The challenge of micropollutants in aquatic systems. Science. https://doi.org/10.1126/science.1127291.

Shang, Y., Kamrul Hasan, M., Ahammed, G.J., Li, M., Yin, H., Zhou, J., 2019. Applications of nanotechnology in plant growth and crop protection: a review. Molecules. https://doi.org/10.3390/molecules24142558.

Sharma, P., Pandey, V., Sharma, M.M.M., Patra, A., Singh, B., Mehta, S., Husen, A., 2021. A review on biosensors and nanosensors application in agroecosystems. Nanoscale Res. Lett. 16, 136. https://doi.org/10.1186/s11671-021-03593-0.

Srivastava, A.K., Dev, A., Karmakar, S., 2018. Nanosensors and nanobiosensors in food and agriculture. Environ. Chem. Lett. 16, 161–182. https://doi.org/10.1007/s10311-017-0674-7.

Stocker, J., Balluch, D., Gsell, M., Harms, H., Feliciano, J., Daunert, S., Malik, K.A., van der Meer, J.R., 2003. Development of a set of simple bacterial biosensors for quantitative and rapid measurements of arsenite and arsenate in potable water. Environ. Sci. Technol. 37, 4743–4750. https://doi.org/10.1021/es034258b.

Tang, F.H.M., Lenzen, M., McBratney, A., Maggi, F., 2021. Risk of pesticide pollution at the global scale. Nat. Geosci. 14, 206–210. https://doi.org/10.1038/s41561-021-00712-5.

Thangadurai, D., Shettar, A.K., Sangeetha, J., Adetunji, C.O., Islam, S., Al-Tawaha, A.R.M.S., 2021. Nanosensors for detection and evaluation of organic compounds in soil. In: Nanomaterials for Soil Remediation. Elsevier, pp. 205–219, https://doi.org/10.1016/b978-0-12-822891-3.00010-4.

USEPA, 2007. Environmental Protection Agency Nanotechnology White Paper. EPA, Ed.

van Dijk, M., Meijerink, 2014. A review of food security scenario studies: gaps and ways forward.

Wang, X., Shen, Y., Xie, A., Li, S., Cai, Y., Wang, Y., Shu, H., 2011. Assembly of dandelion-like Au/PANI nanocomposites and their application as SERS nanosensors. Biosens. Bioelectron. 26, 3063–3067. https://doi.org/10.1016/j.bios.2010.11.044.

Verdian, A., 2018. Apta-nanosensors for detection and quantitative determination of acetamiprid—a pesticide residue in food and environment. Talanta 176, 456–464. https://doi.org/10.1016/j.talanta.2017.08.070.

Wang, P., Lombi, E., Zhao, F.-J., Kopittke, P.M., 2016. Nanotechnology: a new opportunity in plant sciences. Trends Plant Sci. 21, 699–712. https://doi.org/10.1016/j.tplants.2016.04.005.

Wu, H., Li, Z., 2021. Recent advances in nano-enabled agriculture for improving plant performance. Crop J. https://doi.org/10.1016/j.cj.2021.06.002.

Xu, F., Wang, P., Bian, S., Wei, Y., Kong, D., Wang, H., 2021. A Co-nanoparticles modified electrode for on-site and rapid phosphate detection in hydroponic solutions. Sensors (Basel) 21, 299. https://doi.org/10.3390/s21010299.

Yang, T., Timothy, V.D., 2021. Challenges and potential solutions for nanosensors intended for use with foods. Nat. Nanotechnol. 16, 251–265. https://doi.org/10.1038/s41565-021-00867-7.

Yu, Z., Zhao, G., Liu, M., Lei, Y., Li, M., 2010. Fabrication of a novel atrazine biosensor and its subpart-per-trillion levels sensitive performance. Environ. Sci. Technol. 44, 7878–7883. https://doi.org/10.1021/es101573s.

Zhang, L., Peng, D., Liang, R.P., Qiu, J.D., 2018. Graphene-based optical nanosensors for detection of heavy metal ions. TrAC Trends Anal. Chem. 102, 280–289. https://doi.org/10.1016/J.TRAC.2018.02.010.

Zhang, M., Zhang, X., Zhang, L., Zeng, L., Liu, Y., Wang, X., He, P., Li, S., Liang, G., Zhou, W., Ai, C., 2021. The stronger impact of inorganic nitrogen fertilization on soil bacterial community than organic fertilization in short-term condition. Geoderma 382. https://doi.org/10.1016/j.geoderma.2020.114752, 114752.

Zhou, L., Zhang, X., Ma, L., Gao, J., Jiang, Y., 2017. Acetylcholinesterase/chitosan-transition metal carbides nanocomposites-based biosensor for the organophosphate pesticides detection. Biochem. Eng. J. 128, 243–249. https://doi.org/10.1016/j.bej.2017.10.008.

Regulatory affairs, commercialization, and economic aspects of nanomaterials used for agriculture

18

Sougata Ghosh[a,d], *Bishwarup Sarkar*[b], *Ajay Kumar*[c], *and Sirikanjana Thongmee*[d]
[a]Department of Microbiology, School of Science, RK University, Rajkot, Gujarat, India, [b]College of Science, Northeastern University, Boston, MA, United States, [c]Department of Postharvest Science, Agricultural Research Organization (ARO), Volcani Center, Rishon LeZion, Israel, [d]Department of Physics, Faculty of Science, Kasetsart University, Bangkok, Thailand

18.1 Introduction

Agriculture has been considered as one of the prime sectors at a global level for contribution to food security along with promotion of social and economic equity (Fraceto et al., 2016). Furthermore, with a constant increase in world population, change in climatic conditions, environmental pollution, and increased requirements of water and energy resources, new advancements and improved technologies are needed to be implemented in agricultural sectors in order to meet the growing demand for food supply (Raj et al., 2021). Hence, nanotechnology is considered to be one of the potent solutions for significantly increasing food production using lower cost, energy, and waste (Servin and White, 2016). The main concern for a decrease in plant productivity is maintenance of plant health (Mittal et al., 2020). Several disease-causing organisms termed pests often induce damage to plants that may lead to decrease in crop yield. Therefore, pesticides are applied in order to protect crops from such disease-inflicting factors (Jampílek and Králová, 2017). However, pesticides impose toxicity and majority of them are lethal to animal as well as human health. Nanotechnology has thus been reported to be useful in preparation of nanopesticides in the form of nanoherbicides, nanoinsecticides, and nanofungicides that ameliorate the drawbacks of conventional formulations of pesticides for treatment of plant diseases. In one such study, Ocsoy et al. (2013) prepared DNA-directed silver nanoparticles (AgNPs) on graphene oxide (GO) that showed antibacterial activity against *Xanthomonas perforans* that is known to cause bacterial spot disease in tomatoes. Application of Ag@dsDNA@GO composites significantly decreased severity of bacterial spot disease in tomato transplants when compared to untreated plants. Similarly, $Cu_3(PO4)_2{\cdot}3H_2O$ nanosheets were fabricated by Borgatta et al. (2018) that repressed growth of *Fusarium oxysporum* f. sp. *niveum* that is the causative agent of root fungal

Agricultural Nanobiotechnology. https://doi.org/10.1016/B978-0-323-91908-1.00008-0

disease in *Citrullus lanatus*. Rapid dissolution of copper was seen in organic acid solutions with a significant decrease in disease progress when dip method was performed for the application of $Cu_3(PO4)_2 \cdot 3H_2O$ nanosheets which was proposed to effectively increase homogeneity of coverage during treatment. AgNPs were also synthesized by Ali et al. (2015) using aqueous extracts of *Artemisia absinthium* that was able to inhibit mycelial growth, zoospore germination, germ tube elongation, and zoospore production of several *Phytophthora* spp. such as *P. parasitica*, *P. infestans*, *P. palmivora*, *P. cinnamomi*, *P. tropicalis*, *P. capsici*, and *P. katsurae* that causes several economically important diseases that are related to crops. In another study, Huang et al. (2018) reported antimicrobial properties of magnesium hydroxide nanoparticles that had low cytotoxicity with no significant tissue injury observed against tomato plants. Growth of *Xanthomonas alfalfa*, *Pseudomonas syringae*, and *Escherichia coli* was inhibited significantly in the presence of nanoparticles. Further addition of water-soluble capping agents such as trisodium citrate and betaine further increased the bioavailability of magnesium hydroxide nanoparticles.

Nanotechnology has also been used in the formulation of nanoinsecticides as well. One such example is nanostructured alumina (NSA) that can act as an insecticide (Stadler et al., 2018). These NSA particles are negatively charged that strongly bind to insects that are positively charged that are generated by triboelectrification. Sorption of NSA particles on the insect then causes dehydration that ultimately leads to death. Likewise, in another study, Ayoub et al. (2018) reported the synthesis of copper oxide (CuO) and calcium oxide (CaO) nanoparticles that exhibited entomotoxic effects against *Spodoptera littoralis*. Polymer-based nanocarriers that are easily biodegradable and biocompatible have also been utilized for encapsulation of pesticides. Carbofuran pesticide was encapsulated by Shakil et al. (2010) in nanospheric micelles that were prepared by enzymatic copolymerization of polyethylene glycols (PEGs) and dimethyl 5-hydroxyisophthalate. Herbicides are also encapsulated in carrier systems to attain a better and eco-friendly approach for selective targeting of weed species along with a reduction in toxic effects of herbicides against nontarget crops. Oliveira et al. (2015) reported the fabrication of such nanoherbicides wherein atrazine was encapsulated in poly(epsilon-caprolactone) (PCL) nanocapsules that provided effective postemergence activity of atrazine herbicide. Amphiphilic carboxymethyl chitosan derivatives (CMCS-MUA) were also used in a study by Yu et al. (2015) wherein self-assembly of CMCS-MUA followed by cross-linking with disulfide bonds showed a controlled release of diuron herbicides loaded that were loaded inside the nanoformulation. The prepared nanoherbicides were responsive to glutathione and effective herbicidal activity against *Echinochloa crusgalli* that was considered as the target weed species.

Fungal pathogens have also been targeted by nanoparticles, for example, Jo et al. (2009) showed a significant antifungal activity of AgNPs against *Bipolaris sorokiniana* and *Magnaporthe grisea* that cause seedling blight and rice blast diseases on crops, respectively. Other metallic nanoparticles such as zinc oxide (ZnO) and magnesium oxide (MgO) nanoparticles were also reported by Wani and Shah

(2012) to act as nanofungicides against several pathogenic fungi such as *Alternaria alternate*, *F. oxysporum*, *Rhizopus stolonifer*, and *Mucor plumbeus* with significant inhibition in sporulation of these fungal strains. Moreover, nanotechnology has also been applied in the production of eco-friendly fertilizers that reduce nutrient loss and provides controlled nutrient release at target sites, thus minimizing contamination and leaching of fertilizers in the environment.

In a similar study, Li et al. (2018) applied molybdenum sulfide (MoS_2) that exhibited a considerable increase of root and shoot lengths, chlorophyll content index (CCI) as well as upregulated aquaporin gene expression in rice plants that promoted its growth and development. Likewise, Liu et al. (2016) showed a significant increase in *Lactuca sativa* seedling germination after application of iron oxide nanoparticles (FeOx NPs) that showed the potential of such metallic oxide nanoparticles for its application as effective nanofertilizers.

18.2 Toxicological impact of using nanotechnology in agriculture

Although a significant research has been reported with respect to the application of distinct nanomaterials for increasing agronomic productivity, toxicity assessment of prepared nanoformulations with regard to environmental and human health is required. Nanoagrochemicals can be exposed to humans through various modes as depicted in Fig. 18.1. For instance, growing demand for food crops and population growth can increase exposure to nanoparticles applied in plants as well as animal-based food products, food waste released from affluent industries along with pre- and postharvesting of food crops may also facilitate exposure of nanoformulations applied in agricultural fields. Size, concentration, and other features of nanoparticles play an important role in imparting toxic effects. Chances of generating hazardous effects are also increased with decreased size of nanoparticles that may lead to a subsequent increase in surface area resulting in higher uptake and interaction of these nanomaterials with the plants (Nel et al., 2006).

Concentration of nanomaterials also plays an important factor in imparting toxic effects, for instance, Zhao et al. (2013) showed higher concentrations (800 mg/kg) of zinc oxide (ZnO) and cerium oxide (CeO_2) nanoparticles showed decrease in yield of cucumber plants as compared to lower doses (400 mg/kg). The most common mechanism of imparting toxicity by nanoparticles involves formation of reactive oxygen species (ROS), especially by removing ions from the biological samples (Yan and Chen, 2019). In addition, phytohormones are also altered after nanoparticle exposure since it causes decrease in levels of auxins, cytokinins and salicylic acid that could, in turn, affect overall metabolism of plants (O'Brien and Benková, 2013; Vankova et al., 2017). Several other studies highlighting the negative impact in growth and development of plants after nanoparticle exposure are summarized in Table 18.1.

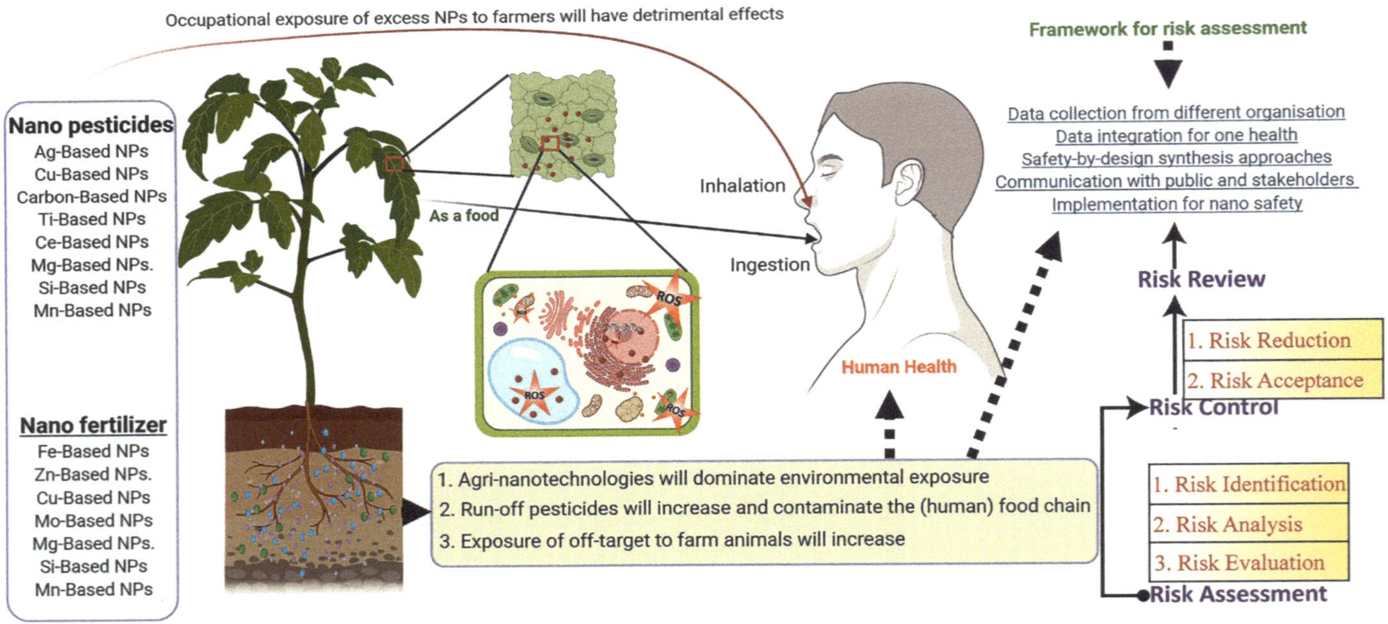

Fig. 18.1 Framework for the risk assessment of nanotechnology in the agrifood sector: The direct delivery of nanopesticides and nanofertilizers to plants can induce ROS, leading to intracellular organelles and function disruptions. Indirect release (i.e., released from agricultural products as well as food processing and packaging) results in the occupational exposure (i.e., from industrial processes, direct contact with nanopesticides and nanofertilizers) to humans. Risk framework for risk assessment, risk management, policymaking, and to support risk communication among the different stakeholders to improve agriculture nanotechnology sustainability.

Reprinted from Mittal, D., Kaur, G., Singh, P., Yadav, K., Ali, S.A., 2020. Nanoparticle-based sustainable agriculture and food science: recent advances and future outlook. Front. Nanotechnol. 2, 579,954.

Table 18.1 Detailed information on the toxicity of nanoparticle-based agrochemicals with various concentrations, sizes, and plant species tested.

Nanoparticles (NPs)	Concentration	Size	Plant species	Effects
Toxicity of metal and metal-oxide NP-based agrochemicals				
Fe_3O_4	30, 100, and 500 mg/L	–	*Lolium perenne* L. and *Cucurbita mixta* cv. *white cushaw*	Oxidative stress, increased superoxide dismutase, and catalase enzyme activities, lipid peroxidation
Citrate-coated Ag and ZnO	200 and 1000 µg/mL, respectively	56.1 and 17.4 nm, respectively	*Zea mays* L. and *Brassica oleracea* var. *capitata* L.	Alterations in cellular structures, tunneling effect with nZnO, metaxylem count changes by citrate-coated Ag
Ag	100 mg/L	20, 51, and 73 nm	*Vicia faba*	Altered PSII activity, stomatal conductance, and CO_2 assimilation
ZnO	5, 10, 25, 50, 75, 100, 125, 250, and 500 mg/L	<50 nm	*Brassica napus*	Inhibits root elongation at highest concentration (250 and 500 mg/L), effects on shoot elongation were less severe
ZnO	300, 600, and 1000 mg/kg	<100 nm	*Solanum lycopersicum*	Increased APX and SOD activity levels, and lipid peroxidation in roots through generation of H_2O_2, altered contents of nonenzymatic antioxidants, total phenols, flavonoids, −carotene, and lycopene in fruits were significantly reduced

Continued

Table 18.1 Continued

Nanoparticles (NPs)	Concentration	Size	Plant species	Effects
TiO_2 and CeO	0, 250, 500, and 1000 μg/mL	130–295 nm TiO_2, 230–290 nm CeO	*Lactuca sativa*, *Lycopersicon lycopersicum*, *Brassica oleracea*, *Glycine* max, *Daucus carota*, *L. perenne*, *Z. mays*, *Cucumis sativus*, *Avena sativa*, and *Allium cepa*	Plant species-dependent effects were observed, differential phytotoxicity depending on the nanoparticle type, significant changes in seed germination but no acute toxicity, lettuce, onion, ryegrass, tomato were shown effective for both the nanoparticles
CuO and Al_2O_3	0, 20, 200, and 2000 μg/mL	18 nm CuO and 21 nm Al_2O_3	*Solanum lycopersicon*	ROS-mediated membrane damage and membrane lipid peroxidation are observed along with the activity of antioxidant enzymes, DNA conformation by NPs is affected by both intercalative and nonintercalative modes, Al_2O_3 NPs induced lesser change in T_m DNA conformation. Macromolecular change in amide-I and II of proteins and carbohydrates, CuO with more toxicity than Al_2O_3, concentration-dependent decrease in fresh and dry biomass
CuO	0, 50, 100, 200, and 500 mg/kg	25, 50, and 250 nm	*G.* max cv. Kowsar	Particle size and concentration-dependent toxicity, altered, antioxidant biomarkers with CuO NPs of 25 nm in soil

TiO_2 and ZnO	0, 5, 10, 20, 40, and 80 mg/kg	~30 nm	*Hordeum vulgare* L.	The antioxidant enzyme activities are diversely affected but no changes in seed germination and root elongation have been observed, ZnO was more toxic than TiO_2
CuO	10, 50, 100, and 500 mg/L	20–40 nm	*Brassica oleracea* var. *botrytis* and *S. lycopersicum*	100 and 500 mg/L of CuO NPs have resulted in a reduction of total chlorophyll and sugar content, augmented lipid peroxidation, electrolyte leakage, and antioxidant enzyme activity, concentration-dependent increase in superoxide and hydrogen peroxide formation in leaves, deposition of lignin in roots seen at higher concentration
CuO	10 g/L	30–50 nm	*Hordeum sativum distichum*	Ultrastructural changes in chloroplast, vacuoles, mitochondria, and stomata, impaired germination rate, root morphology, cell wall, epidermis, cortical layers, and vascular bundles, inhibits root and shoot length
FeO	30, 40, or 50 mg/L	12 nm	*Lemna minor*	The chlorophyll content decreased at high iron oxide NP concentrations, which disrupted the light absorption mechanism, ROS production, diminishes photosynthetic pigments or chla/chlb content ratio
TiO_2, ZnO, Al_2O_3, and CuO	0.05, 0.5, 2, and 5 mg/mL	148–248 nm	*Raphanus sativus*, *C. sativus*, *Solanum lycopersicon*, and *Medicago sativa*	Stimulates metallothionein production, inhibited relative seed germination, root/shoot tolerance index however differential effects were observed based upon the plant species and nanoparticles type

Continued

Table 18.1 Continued

Nanoparticles (NPs)	Concentration	Size	Plant species	Effects
CuO, TiO_2, and Fe_2O_3	–	–	*Triticum aestivum*	Decreased contents of Fe and Zn by CuO treatment and reduced human essential amino acids, overall increased amino acid concentration by TiO_2
CuO, ZnO, and NiO	100, 250, 500, and 1000 mg/L	25–55 nm CuO, 18 nm ZnO, and 10–20 nm NiO	*Abelmoschus esculentus*	Suppressed plant growth in a concentration-dependent manner, reduced shoot and root length, decreased chlorophyll content, significantly altered total phenolic and total flavonoid contents, enhanced ROS, and malondialdehyde production
Se	0, 0.5, 1, 10, and 30 mg/L	20 nm	*Capsicum annuum*	Impose epigenetic changes, upregulated bZIP1 and WRKY1transcription factors, altered growth metabolism and anatomy of plants, caused DNA hypermethylations
Se	0, 1, 4, 10, 30, and 50 mg/L	10–30 nm	*Momordica charantia*	More concentrations cause stem bending, impaired root meristem, and severe toxicity, cause variation in DNA cytosine methylation, reflecting the epigenetic modification, also stimulates activities of catalase and peroxidases, disrupts xylem conducting tissues

CeO_2	0, 50, 100, and 1000 mg	<25 nm	*L. sativa*	Higher concentration treatment significantly deteriorates plant growth and biomass production, Superoxide dismutase, Peroxidase, Malondialdehyde activity was disrupted
Toxicities of nanotubes (NTs)				
MWCNTs and SWCNTs	1 and 10 mg/kg	1–4-nm-outer-diameter and 5–30-μm-length SWCNT-OH 30–50-nm-outer-diameter and 0.5–2.0-μm-length MWCNT-COOH	*S. lycopersicum*	Delay in early growth and flowering, MWCNT effects were seen on first fruit, SWCNT has effects on physiology, such as increased salicylic acid
Hollow MWCNT, Fe-filled carbon, and Fe-Co-filled carbon NTs	0, 10, 50, and 300 mg/L	5 nm thickness	*Oryza sativa* L.	Inhibited rice growth, decreased concentrations of endogenous plant hormones, and decreased nitrogen assimilation
MWCNTs	0, 125, 250, 500, and 1000 μg/mL	Outer diameter 5–15 nm and length 50 μm	*Cucurbita pepo* L.	Toxicity effects on germination rates or slow germination of seeds, enhanced cellular injuries, and inactivated antioxidant enzymes
MWCNTs-COOH	0, 25, 50, and 100 mg/L	–	*Ocimum basilicum*	High dosage (100 mg L^{-1}) of MWCNT-COOH leads to toxicity effects

Reprinted from Mittal, D., Kaur, G., Singh, P., Yadav, K., Ali, S.A., 2020. Nanoparticle-based sustainable agriculture and food science: recent advances and future outlook. Front. Nanotechnol. 2, 579,954.

18.3 Regulation for using nanotechnology in agriculture

At present, the regulation of nanomaterials is not governed by any unique legislation in the European Union (EU) as well as in other countries; however, efforts are being made with regard to addressing and regulating proper production, handling, and use of nanomaterials by either legislation or through recommendations and guidance (Amenta et al., 2015; Arts et al., 2014). However, the need for potential risk assessments of the application of nanomaterials in agriculture has been proposed by European Parliament as well as other nongovernmental organizations (NGOs), and thus, several expert bodies are active in this field as listed in Table 18.2.

Safety of substances and products present on the market in the EU is ensured by a binding framework provided by horizontal and sector-specific legislation that includes nanomaterials as well as nanoenabled products (Amenta et al., 2015). At present, nanomaterials are governed under several sections of legislation, for instance, the Regulation on the Provision of Food Information to Consumers (1169/20119), the Regulation on Plastic Food Contact Materials and Articles (10/2011), the Regulation on Active and Intelligent Materials and Articles (450/2009), the Biocidal Products Regulation (528/2012), and the Cosmetic Products Regulation(1223/2009), whereas some legislation such as Novel Food Regulation (258/97) or the Annexes to the Registration, Evaluation, Authorisation and Restriction of Chemicals (REACH) Regulation (1907/2006) are under revision (Amenta et al., 2015). Moreover, in order to maintain consistency in definition of "nanomaterial" for its use in legislation and policies, the European Commission (EC) published a "recommendation on the definition of nanomaterial" that acts as a reference and can be altered with respect to legislation of a specific product. This definition of nanomaterial provided by the EC only considers the size range of 1–100 nm as the defining property of the nanomaterial. Although methods of existing risk assessment and other regulatory aspects can be considered to be applicable for nanomaterials, still several key

Table 18.2 Organizations that are currently active in implementation and assessment of regulatory aspects, potential risk, and safe use of nanomaterials.

Name of country	Organization
The United States of America (USA)	Food and Drug Administration (FDA) and US Environmental Protection Agency (US-EPA)
Canada	Canadian Food Inspection Agency (CFIA) and Public Health Agency of Canada (PHAC)
Switzerland	Federal Office of Public Health (FOPH)
Turkey	Ministry of Environment and Urban Planning
Russia	Russian Corporation of Nanotechnologies
China	National Centre for Nanoscience and Technology (NCNST) and Commission on Nanotechnology Standardization
Iran	Iran Nanotechnology Initiative Council (INIC)
Taiwan	Taiwan Nanotechnology Industry Development Association (TANIDA)

aspects such as sample preparation, characterization, dosimetry, effect endpoints, exposure data, and models are under development wherein the EU Scientific Committee on Emerging and Newly Identified Health Risks (SCENIHR) has recommended a case-by-case approach in order to assess the risk factors involved in the application of nanomaterials (SCENIHR, 2007). Furthermore, the European Food Safety Authority (EFSA) has published a "guidance on the risk assessment of the application of nanoscience and nanotechnologies in the food and feed chain" that issued potential risk assessment measures of nanomaterials and nanoenabled products used in food additives, enzymes, flavorings, food contact materials, novel foods, feed additives, and pesticides (Amenta et al., 2015). Switzerland also ensures safe use of nanomaterials in pesticides by proper regulations and procedures wherein specific information on the composition, shape, particle size, surface area, state of aggregation, and presence of any functionalized groups have to be reported for registration, whereas in other countries such as the USA, the FDA has not yet adopted any regulatory definition for nanomaterial, but it considers a comprehensive approach when inspecting nanotechnology-based products (Hamburg, 2012). The FDA also issued certain guidances for the application of nanotechnology in food-related products wherein a preliminary safety assessment of "food ingredients and food-contact substances produced at nanoscale" is recommended that should be relevant to a food substance in its nanometer version (Tyler, 2012). Still, no food products that are specifically engineered at the nanometer scale are generally recognized as safe (GRAS) by FDA (Amenta et al., 2015). Similarly, no specific provisions for nanomaterials are present in the US Environmental Protection Agency (US-EPA) that regulates pesticides. Several other countries such as Australia, New Zealand, Canada, Malaysia, Taiwan, Japan, China, India, and South Africa are establishing committees and setting up organizations that can provide legislation with respect to the use of nanotechnology in agriculture that can be incorporated for safety regulations.

Also, in recent years, several important government, scientific, and institutional reports have stated that the application of nanotechnology could significantly aid in meeting the increased demands for safe water, reliable sources of energy, health care, education, and other basic human development needs as proposed by the United Nations Millennium Development Goals (Mukhopadhyay, 2014). In addition, a public opinion survey revealed that majority of the USA respondents considered nanotechnology to be comparatively neutral, safe, and more beneficial than other advanced technologies such as genetic modification in organisms, chemical pesticides, and disinfectants, whereas many respondents considered higher risk and less efficiency of nanotechnology when compared to solar power, vaccination, hydroelectric power, and computer display screens (Currall et al., 2006).

18.4 Commercial nanoproducts for agricultural applications

Nanotechnology has proven to be beneficial in agricultural sector but still, only a handful of nanoproducts are currently available on the market as a majority of agrochemical companies are interested in exploring the potential of nanomaterials that can provide increased crop yield along with reduced environmental effects in order to alleviate drawbacks of the application of conventional agricultural products (Campos, 2021). Around 46% of commercial nanobased products intended for agricultural applications are nanofertilizers that enhance uptake and utilization of nutrients by target plants, whereas 26% of commercial nanoformulations comprise nanopesticides and other nanoenabled products that improve survival of plants against diseases and harsh climatic conditions and 4% of nanoproducts aid in enhancing soil properties (Campos, 2021). Table 18.3 further summarizes commercially available nanoproducts in the market for agricultural purposes. According to StatNano database, India produces the most varied nanoproducts that are used in agriculture followed by other countries such as Germany, the United Kingdom (UK), the USA, and Vietnam (Campos, 2021). However, due to high development costs and lack of specific regulations in most countries, the commercialization of nanoformulations is limited (Campos, 2021; Lade et al., 2019). Therefore, five major factors are needed to be addressed in order to improve the commercialization of nanobased agricultural products: (i) drafting of proper legislation at an international level; (ii) proper communication of public and private organizations in order to create a network of nanobased product developers; (iii) standardization of nanomaterial definition; (iv) proper risk-benefit establishment with respect to nanomaterials; and (v) proper evaluation of nanomaterial impact on environment, human health, and ecosystem (Campos, 2021).

18.5 Economical aspects

As nanotechnology has made a broad and rudimentary impact on major sectors of the global economy such as medicine, energy, electronics, cosmetics, defense, environment, automotive along with others (Stirling, 2018), it becomes difficult to analyze the statistical significance of nanotechnology on the global economy. However, a significant rise in nanotechnology investments by both governmental and private organizations is evident with an estimated market of approximately $290.93 billion by 2028 as published by a report from the Emergen Research that stated a compound annual growth rate of around 18.3% for the upcoming decades (Emergen Research, 2021). In another report by BCC Research, the global market by 2025 for nanotechnology in food packaging is estimated to reach $29.40 billion (BCC Research, 2020). According to a report by Transparency Market Research, agriculture is widely integrated with nanotechnology in the USA, Western European nations, and Japan, whereas countries like Brazil, India, China, South Korea, South Africa, and Thailand are showing considerable growth in nanotechnology (Transparency Market Research, 2017). The major companies using nanotechnology in the field of agriculture which are currently active in the market are Nanosys Inc., ASML Holding, Zyvex Labs, Oxford Instruments plc, Nanoco Group plc, and ThalesNano Inc. Thus, varied potential applications of nanotechnology have generated a worldwide interest in nanotechnology research programs that have resulted in several interesting discoveries.

Due to such huge scope of nanotechnology in agriculture, all these discoveries have significant potential in developing agricultural practices as well as enhancing food security under adverse environmental conditions (Parisi et al., 2015). In addition,

Table 18.3 Nanotechnology-based commercial products for agriculture applications.

Commercial name	Manufacturer	Product type	Type of nanomaterial	Applications
NANOK	Aqua-YieldHuB (USA)	Fertilizer	Unspecified nanocontents	Sustained release of potassium to crops
NanoZn	Aqua-YieldHuB (USA)	Fertilizer	Unspecified nanocontents	Plant growth regulation, through the delivery of zinc to target crop tissues
NANO FERTILIZERS	SilvertechKimyaSanayaveTicaret Ltd. (Turkey)	Fertilizer	ZnO and TiO_2 nanoparticles/ nanopowders	Plant growth acceleration and crop yield enhancement through increased chlorophyll formation
NovaLand-F	Land Green & Technology Co (Taiwan)	Fertilizer	Unspecified nanomagnets	Plant growth acceleration and regulation through plant metabolism acceleration
NovaLand Nano-Mn, Cu, Fe, Zn, Mo, N	Land Green & Technology Co (Taiwan)	Fertilizer	Unspecified nanoparticles loading microelements (Mn, Cu, Fe, Zn, Mo, N)	Plant growth acceleration and regulation through increased chlorophyll formation and increased plant resistance against adverse conditions
Hyper Feed	Bio Nano Technology (Egypt)	Fertilizer	Unspecified nanocontents	Nutritional formulation with balanced major and microelements
Nano-fertilizer	NANOPAC (M) Sdn Bhd (Malaysia)	Fertilizer	Unspecified nanocontents	Plant growth acceleration and increases resistance against bacteria, fungus, and diseases
Groagro 4: Super Kalium Catalyst + T. E	Microwell Bio Solutions Sdn Bhd (Malaysia)	Fertilizer	Unspecified nanocomposite containing potassium	Plant growth acceleration and nutritional value improvement
Groagro 1: Super Nano-Catalyst Kalium + T. E	Microwell Bio Solutions Sdn Bhd (Malaysia)	Fertilizer	Unspecified nanocomposite containing potassium	Plant growth acceleration and nutritional value improvement. Recommended for peat soil

Continued

Table 18.3 Continued

Commercial name	Manufacturer	Product type	Type of nanomaterial	Applications
PADI 3 (17:3:25:2) + Groagro 4	Malaysian NPK Fertilizer Sdn Bhd (Malaysia)	Fertilizer	Unspecified nanocomposites	Unspecified properties
PADI 1 (17.5:15.5:10) + Groagro 4	Malaysian NPK Fertilizer SdnBhd (Malaysia)	Fertilizer	Unspecified nanocomposite containing potassium	Unspecified properties
NUBIOTEK Hyper Fe + Mg	Bioteska (Mexico)	Fertilizer	Fe nanoparticles/ nanopowders and M nanocapsules	Plant growth acceleration/regulation and plant resistance enhancement
NUBIOTEC ULTRA Ca	Bioteska (Mexico)	Fertilizer	Ca nanoparticles/ nanopowders	Plant growth acceleration/regulation and increases plant resistance against adverse conditions
FERTIL CALMAG	HPL Agronegocios (Brazil)	Fertilizer	Mg nanocapsules	Plant growth regulation and increases plant resistance against fungi and bacteria and diseases due to malnutrition of Ca and Mg
Nanovec TSS 80	Laboratorios Bio-Médicin (Brazil)	Fertilizer	Manganese, molybdenum, and zinc nanoencapsulated	Improves seed germination, root system, the absorbent radicals, and plant growth acceleration
FERTILE CALCIUM 25	HPL Agronegocios (Brazil)	Fertilizer	Calcium nanoparticles/ nanopowders	Increases plant productivity through potentialization of fertilizer effects
Lithocal	Litho Plant (Brazil)	Fertilizer	Ca and Mn nanoparticles/ nanopowders	Increases absorption of water and nutrients from the soil and increases plant resistance against drought
Liquid NPK	Nutri Solo Fertile (Brazil)	Fertilizer	Nanoparticulate fertilize derived from phytohormonic plant extract	Acts directly on plant nutrition through metabolism improvement and activation of nutritional balance by supplying any lacking nutrients

FOSVIT K30	Kimitec Group (Spain)	Fertilizer	P and K nanoparticles/nanopowders	Induces plant defense through the induction of the formation of phytoalexins that are toxic to pathogenic organisms
AZteroid Fungicide	Vive Crop Protection (Canada)	Fertilizer	Allosperse (polymer nanoparticles shuttles)	Seed germination promotion and provides broad-spectrum control for a variety of seed and seedling diseases
Nano Nutrients for crop	NanoLand Baltic (Lithuania)	Fertilizer	Unspecified nanosized micronutrients	Directly delivers the nutrients to chlorophyll when sprayed to leaves
Nano Chelated Complete Micro Fertilizer	KHAZRA (Iran)	Fertilizer	Unspecified nanocontents	Slow release of chelated fertilizers for optimal use of nutrients by plants and resistant and stable in both alkaline and acidic soils
Nano Chelated NPK 20-20-20	KHAZRA (Iran)	Fertilizer	Unspecified nanocontents	Slow release of chelated fertilizers for optimal use of nutrients by plants and resistant and stable in both alkaline and acidic soils
Nano Chelated Zinc	KHAZRA (Iran)	Fertilizer	Unspecified nanocontents	Slow release of chelated fertilizers for optimal use of nutrients by plants and resistant and stable in both alkaline and acidic soils
Nano Chelated Magnesium	KHAZRA (Iran)	Fertilizer	Unspecified nanocontents	Slow release of chelated fertilizers for optimal use of nutrients by plants and resistant and stable in both alkaline and acidic soils
Nano Chelated NPK 12-12-36	KHAZRA (Iran)	Fertilizer	Unspecified nanocontents	Slow release of chelated fertilizers for optimal use of nutrients by plants and resistant and stable in both alkaline and acidic soils

Continued

Table 18.3 Continued

Commercial name	Manufacturer	Product type	Type of nanomaterial	Applications
Nano Calcium	KHAZRA (Iran)	Fertilizer	Unspecified nanocontents	Slow release of chelated fertilizers for optimal use of nutrients by plants and resistant and stable in both alkaline and acidic soils
SomGuard SilCop	AgriLife (Greece)	Plant nutrient	Silver nanoparticles/ nanopowders	Nontoxic disinfectant highly toxic to microorganisms
HydroBloom	Blue Planet Environmental (USA)	Plant nutrient	Oxygen-rich nanobubble technology (NBT)	Increases nutrient uptake by plant root system in hydroponic system, resulting in plant growth regulation and acceleration
Teaklik/Hortiklik/ Florklik and Agriklik	Nanosolutions (India)	Plant nutrient	Unspecified nanocontents	Biologically activated product from plants' extract, which enhances soil quality and plant nutrition
Nano Zinc Chelate Fertilizer	AFME TRADING GROUP (UK)	Fertilizer	Zn nanoparticles/ nanopowders	Chelated fertilizer that provides to plants' essential micronutrients (Zn), resulting in plant growth acceleration/ regulation
Nano Iron Chelate Fertilizer	AFME TRADING GROUP (UK)	Fertilizer	Fe nanoparticles/ nanopowders	Chelated fertilizer that provides to plants' essential micronutrients (Fe), resulting in plant growth acceleration/ regulation
Nano Calcium Chelate Fertilizer	AFME TRADING GROUP (UK)	Fertilizer	Ca nanoparticles/ nanopowders	Chelated fertilizer that provides to plants' essential micronutrients (Ca), resulting in plant growth acceleration/ regulation

Nano Potassium Chelate Fertilizer	AFME TRADING GROUP (UK)	Fertilizer	K nanoparticles/ nanopowders	Chelated fertilizer that provides to plants' essential micronutrients (K), resulting in plant growth acceleration/ regulation
Water Soluble Nano-SiO_2 Powders used for fertilizer	HongwuInternation Group Ltd. (China)	Fertilizer	Silicon dioxide nanoparticles	Improves photosynthesis, root activity, and pest resistance of treated plants
Nano fertilizer	Lazuriton Nano Biotechnology Co., Ltd. (Taiwan)	Fertilizer	Unspecified nanocontents	Increases plant protection due to a protective layer on the leaf surface, preventing infection
Rich Herba Green	Richfield Fertilizers Pvt. Ltd. (India)	Plant nutrient	Unspecified nanocontents	Enhances the natural resistance of a plant and crop yield, increased plant resistance, and protect the crop against pests
Nano Green	Nano Green Sciences, Inc. (India)	Fertilizer	Unspecified nanocontents	Increases plants' photosynthesis with a corresponding increase in carbon dioxide consumption
The Nano-Ag Answer	Urth Agriculture (USA)	Fertilizer	Ag nanoparticles/ nanopowders	Creates a balance and growth of beneficial organisms for fertilization in soil
Nano-Gro	Agro Nanotechnology Crop (USA)	Fertilizer	Zeolite	Increases growth and yield uniformly. Increases resistance to severe weather and extreme atmospheric conditions
Nanosept/Nanosept Aqua/Nanosept Aqua Super	Nanosept (Hungary)	Plant-disinfecting agent	Silver nanoparticles	Bactericide, fungicide, and virucide
TAG NANO ZINC	Tropical Agrosystem India (P) Ltd. (India)	Fertilizer	"4G" Nanonutrient technologies	Improves growth and yield, increases photosynthetic activity, and reduces chlorosis
TAG NANO CAL	Tropical Agrosystem India (P) Ltd. (India)	Fertilizer	"4G" Nanonutrient technologies	Improves growth and yield and improves physiological activities of the plants

Continued

Table 18.3 Continued

Commercial name	Manufacturer	Product type	Type of nanomaterial	Applications
TAG NANO POTASH	Tropical Agrosystem India (P) Ltd. (India)	Fertilizer	"4G" Nanonutrient technologies	Acceleration of photosynthesis and consequently the production of carbohydrates, fats, and proteins
TAG NANO PHOS	Tropical Agrosystem India (P) Ltd. (India)	Fertilizer	"4G" Nanonutrient technologies	Improves growth and yield and reduces utilization of chemical phosphorus fertilizer
TAG NANO NPK	Tropical Agrosystem India (P) Ltd. (India)	Fertilizer	"4G" Nanonutrient technologies	Improves growth and yield, increases photosynthetic activity, and reduces chlorosis
Nano Bor 20%	Alert Biotech (India)	Fertilizer	Boron nanoparticles/ nanopowders	Allow crop development in regions with extreme rainfall, with acid and sandy soil, and with soil with low organic matter
Nano Zinc (Chelated) and Nano Zinc (Soil Application 21%)	Alert Biotech (India)	Fertilizer	Zinc nanoparticles/ nanopowders	Provides zinc to plants, which is very important for plants' growth, protein synthesis, and energy production
Plant Growth Promoter	Nanjing High Technology Nano Material (China)	Plant growth regulator	Unspecified nanooxides	Promotes plant growth, improves plant production and pests and disease resistance
Nano Silicate+	Plant Vitality Ltd. (UK)	Plant growth regulator	Unspecified nanocontents	Improves plant seed germination and helps to unlock dormant seeds. Increases resistance to drought and diseases
Hydro:Grow	Plant Vitality Ltd. (UK)	Seed speed	Unspecified nanocontents	Increases significantly initial growth and root development
Root Zone Mass	Plant Vitality Ltd. (UK)	Seed speed	Unspecified nanocontents	Promotes faster hatching of seeds and helps unlock dormant seeds
NANO-BORON	Kanak Biotech (India)	Plant growth regulator	Boron nanoparticles/ nanopowders	Helps in plant development through maintaining the balance between sugar and starch

NANO-MOLYBDENUM	Kanak Biotech (India)	Plant growth regulator	Molybdenum nanoparticles/ nanopowders	Provides better absorption of molybdenum by plants, which is an important component of enzymes
NANO-SILVER	Kanak Biotech (India)	Plant growth regulator	Silver nanoparticles/ nanopowders	Protect plants against microorganism infections
NANO-MAGENSIUM	Kanak Biotech (India)	Plant growth regulator	Magnesium nanocapsules	Helps to activate many enzymes important for plants' growth and induces chlorophyll formation
NANO-IRON	Kanak Biotech (India)	Plant growth regulator	Iron nanoparticles/ nanopowders	Increases the availability of iron to plants and helps in respiration and photosynthesis process
NANO-PHOSPHORUS	Kanak Biotech (India)	Plant Growth Regulator	Phosphorus nanoparticles/ nanopowders	Increases growth and reproduction
NANO-ZINC	Kanak Biotech (India)	Plant growth regulator	Zinc oxide nanoparticles/ nanopowders	Provides health development of plants since this compound is an important structural compound or regulatory cofactor of a wide range of enzymes
NeuCytokin	Neufarm GmbH (Germany)	Plant growth regulator	Unspecified nanocontents	Natural potassium and seaweed-based foliar fertilizer, which helps growth and development of crops
NANOCU	Bio Nano Technology (Egypt)	Pesticide	Copper nanoparticles	Plant growth acceleration and fungicide and bactericide properties
Pesticide Synergistic	Nanjing High Technology Nano Material (China)	Pesticide	Unspecified nanomaterials	Catalytic nanoscale materials, which potentiates biological activity and reduces toxicity
NANOSEPT AQUA SUPER/NANOSEPT AQUA/NANOSEPT	Nanosept (Hungary)	Plants-disinfecting agent	Silver nanoparticles/ nanopowders	Long-lasting antiseptic formulation with a broad-spectrum antimicrobial effect
Hydro: Flower/Coco: Flower/Coco: grow/Soil: Flower/Soil: grow	Plant Vitality Ltd. (UK)	Soil enhancer	Unspecified nanocontents	Mix of nanoparticles that improve plant growth, seed germination, and root activity

Reprinted with permission from Campos, E.V.R. 2021. Commercial nanoproducts available in world market and its economic viability. In: Jogaiah, S., Singh, H.B., Fraceto, L.F., de Lima, R., (Eds.) Advances in Nano-Fertilizers and Nano-Pesticides in Agriculture - A smart delivery system for crop Improvement. pp. 561–593. Copyright © 2021 Elsevier Inc.

the compound annual growth rate for nanotechnology particularly in agribusiness is around 28% which is greater than most other areas of application of nanotechnology (Campos, 2021).

18.6 Conclusion and future perspectives

Nanotechnology has proven to be a powerful tool when integrated with the field of agriculture. Biologically synthesized metal nanostructures like gold, silver, copper, platinum, and palladium nanoparticles may be used for developing nanopesticides, nanofertilizers, and other nanoformulations (Ghosh et al., 2016a, 2016b, 2016c, 2016d; Shende et al., 2017; Shende et al., 2018; Salunke et al., 2014; Sant et al., 2013; Shinde et al., 2018; Jamdade et al., 2019; Ghosh, 2018; Rokade et al., 2018; Rokade et al., 2017; Bhagwat et al., 2018; Ghosh et al., 2015a, 2015b; Bloch et al., 2021; Ranpariya et al., 2021). Similarly, zinc oxide nanoparticles with superior enzyme inhibitory, antimicrobial, antioxidant activity, and biocompatibility may be used for developing nanofertilizer with additional pesticidal activity (Asok et al., 2015; Robkhob et al., 2020; Karmakar et al., 2020; Kitture et al., 2015). Likewise, magnetic nanoparticles can be multifunctionalized with pesticides, herbicides, and growth hormones for targeted delivery and sustained release (Ghosh et al., 2015c). However, evaluation of safety and risk assessment of such nanoformulations are compulsory prior to the commercialization of these products in the agriculture market. At present, very limited information is known with regard to a widespread application of nanoproducts in agriculture sector. Therefore, the formation of systematic legislation is required for boosting the market of these nanoenabled agricultural products. Beginning with standardization of the definition of nanomaterials, systematic regulations are needed that can provide overall updates in global acceptance and development of nanoformulations intended to be applicable in agriculture. National initiative programs in order to increase public awareness, proper handling, and reuse of agriculture nanoproducts are required as well. Comprehensive risk assessment of these nanoproducts could also be carried out through the development of smart tools that can investigate the impact of nanoproducts on environment and its microflora and fauna as well as effect on human, animal, and plant health. Furthermore, systematic nano-biointerface-specific guidelines for generating toxicity studies can also be implemented prior to endorsement of policies in order to ideally classify the nanoproducts into specific groups such as toxic, nontoxic, or chemically hazardous after a thorough assessment. Finally, public acceptance of such nanobased products should be addressed for the successful utilization of these products. Hence, appropriate standardization and laboratories for testing, validation, and certification of agricultural nanomaterials would play a vital role in eliminating the constrictions of the commercialization of nanobased agricultural products that could in turn ameliorate the problems faced by the agricultural sector.

Acknowledgment

Dr. Sougata Ghosh acknowledges Kasetsart University, Bangkok, Thailand for Post Doctoral Fellowship and funding under Reinventing University Program (Ref. No. 6501.0207/10870 dated 9th November, 2021).

References

Ali, M., Kim, B., Belfield, K.D., Norman, D., Brennan, M., Ali, G.S., 2015. Inhibition of *Phytophthora parasitica* and *P. capsici* by silver nanoparticles synthesized using aqueous extract of *Artemisia absinthium*. Phytopathology 105 (9), 1183–1190.

Amenta, V., Aschberger, K., Arena, M., Bouwmeester, H., Moniz, F.B., Brandhoff, P., Gottardo, S., Marvin, H.J., Mech, A., Pesudo, L.Q., Rauscher, H., Schoonjans, R., Vettori, M.V., Weigel, S., Peters, R.J., 2015. Regulatory aspects of nanotechnology in the Agri/feed/food sector in EU and non-EU countries. Regul. Toxicol. Pharmacol. 73 (1), 463–476.

Arts, J.H.E., Hadi, M., Keene, A.M., Kreiling, R., Lyon, D., Maier, M., Michel, K., Petry, T., Sauer, U.G., Warheit, D., Wiench, K., Landsiedel, R., 2014. A critical appraisal of existing concepts for the grouping of nanomaterials. Regul. Toxicol. Pharmacol. 70 (2), 492–506.

Asok, A., Ghosh, S., More, P.A., Chopade, B.A., Gandhi, M.N., Kulkarni, A.R., 2015. Surface defect rich ZnO quantum dots as antioxidants inhibiting α-amylase and α-glucosidase: a potential anti-diabetic nanomedicine. J. Mater. Chem. B 3, 4597–4606.

Ayoub, H.A., Khairy, M., Elsaid, S., Rashwan, F.A., Abdel-Hafez, H.F., 2018. Pesticidal activity of nanostructured metal oxides for generation of alternative pesticide formulations. J. Agric. Food Chem. 66 (22), 5491–5498.

BCC Research, 2020. Nanotechnology for Food Packaging Market—A Global Market and Regional Analysis: Focus on (Product, Application, Industry Outlook, Region and Country Analysis)—Analysis and Forecast, 2019–2025. https://www.bccresearch.com/partners/bis-market-research/nanotechnology-for-food-packaging-market.html. (Accessed on 13/08/2021).

Bhagwat, T.R., Joshi, K.A., Parihar, V.S., Asok, A., Bellare, J., Ghosh, S., 2018. Biogenic copper nanoparticles from medicinal plants as novel antidiabetic nanomedicine. World J. Pharm. Res. 7 (4), 183–196.

Bloch, K., Pardesi, K., Satriano, C., Ghosh, S., 2021. Bacteriogenic platinum nanoparticles for application in nanomedicine. Front. Chem. 9, 624344.

Borgatta, J., Ma, C., Hudson-Smith, N., Elmer, W., Plaza Perez, C.D., De La Torre-Roche, R., Zuverza-Mena, N., Haynes, C.L., White, J.C., Hamers, R.J., 2018. Copper based nanomaterials suppress root fungal disease in watermelon (*Citrullus lanatus*): role of particle morphology, composition and dissolution behavior. ACS Sustain. Chem. Eng. 6, 14847–14856.

Campos, E.V.R., 2021. Commercial nanoproducts available in world market and its economic viability. In: Jogaiah, S., Singh, H.B., Fraceto, L.F., de Lima, R. (Eds.), Advances in Nano-Fertilizers and Nano-Pesticides in Agriculture—A Smart Delivery System for Crop Improvement. Elsevier, pp. 561–593.

Currall, S.C., King, E.B., Lane, N., Madera, J., Turner, S., 2006. What drives public acceptance of nanotechnology? Nat. Nanotechnol. 1 (3), 153–155.

Emergen Research, 2021. Nanotechnology Market Size, Share, Trends, By Type (Nanomaterials, Nanocomposites, Nano devices, Nano tools), By Industry (Food and

Agriculture, Healthcare, Information and Technology, Environment, Energy, Cosmetics), and By Region, Forecast to 2028. https://www.emergenresearch.com/industry-report/nanotechnology-market. Accessed on 13/08/2021.

Fraceto, L.F., Grillo, R., de Medeiros, G.A., Scognamiglio, V., Rea, G., Bartolucci, C., 2016. Nanotechnology in agriculture: which innovation potential does it have? Front. Environ. Sci. 4, 20.

Ghosh, S., 2018. Copper and palladium nanostructures: a bacteriogenic approach. Appl. Microbiol. Biotechnol. 102 (18), 7693–7701.

Ghosh, S., More, P., Nitnavare, R., Jagtap, S., Chippalkatti, R., Derle, A., Kitture, R., Asok, A., Kale, S., Singh, S., Shaikh, M.L., Ramanamurthy, B., Bellare, J., Chopade, B.A., 2015a. Antidiabetic and antioxidant properties of copper nanoparticles synthesized by medicinal plant *Dioscorea bulbifera*. J. Nanomed. Nanotechnol. S6, 007.

Ghosh, S., Nitnavare, R., Dewle, A., Tomar, G.B., Chippalkatti, R., More, P., Kitture, R., Kale, S., Bellare, J., Chopade, B.A., 2015b. Novel platinum-palladium bimetallic nanoparticles synthesized by *Dioscorea bulbifera*: anticancer and antioxidant activities. Int. J. Nanomedicine 10 (1), 7477–7490.

Ghosh, S., More, P., Derle, A., Kitture, R., Kale, T., Gorain, M., Avasthi, A., Markad, P., Kundu, G.C., Kale, S., Dhavale, D.D., Bellare, J., Chopade, B.A., 2015c. Diosgenin functionalized iron oxide nanoparticles as novel nanomaterial against breast cancer. J. Nanosci. Nanotechnol. 15 (12), 9464–9472.

Ghosh, S., Patil, S., Chopade, N.B., Luikham, S., Kitture, R., Gurav, D.D., Patil, A.B., Phadatare, S.D., Sontakke, V., Kale, S., Shinde, V., Bellare, J., Chopade, B.A., 2016a. *Gnidia glauca* leaf and stem extract mediated synthesis of gold nanocatalysts with free radical scavenging potential. J. Nanomed. Nanotechnol. 7, 358.

Ghosh, S., Harke, A.N., Chacko, M.J., Gurav, S.P., Joshi, K.A., Dhepe, A., Dewle, A., Tomar, G.B., Kitture, R., Parihar, V.S., Banerjee, K., Kamble, N., Bellare, J., Chopade, B.A., 2016b. *Gloriosa superba* mediated synthesis of silver and gold nanoparticles for anticancer applications. J. Nanomed. Nanotechnol. 7, 4.

Ghosh, S., Chacko, M.J., Harke, A.N., Gurav, S.P., Joshi, K.A., Dhepe, A., Kulkarni, A.S., Shinde, V.S., Parihar, V.S., Asok, A., Banerjee, K., Kamble, N., Bellare, J., Chopade, B.A., 2016c. *Barleria prionitis* leaf mediated synthesis of silver and gold nanocatalysts. J. Nanomed. Nanotechnol. 7, 4.

Ghosh, S., Gurav, S.P., Harke, A.N., Chacko, M.J., Joshi, K.A., Dhepe, A., Charolkar, C., Shinde, V.S., Kitture, R., Parihar, V.S., Banerjee, K., Kamble, N., Bellare, J., Chopade, B.A., 2016d. *Dioscorea oppositifolia* mediated synthesis of gold and silver nanoparticles with catalytic activity. J. Nanomed. Nanotechnol. 7, 5.

Hamburg, M.A., 2012. Science and regulation. FDA's approach to regulation of products of nanotechnology. Science 336 (6079), 299–300.

Huang, Z., Rajasekaran, P., Ozcan, A., Santra, S., 2018. Antimicrobial magnesium hydroxide nanoparticles as an alternative to cu biocide for crop protection. J. Agric. Food Chem. 66 (33), 8679–8686.

Jamdade, D.A., Rajpali, D., Joshi, K.A., Kitture, R., Kulkarni, A.S., Shinde, V.S., Bellare, J., Babiya, K.R., Ghosh, S., 2019. *Gnidia glauca* and *Plumbago zeylanica* mediated synthesis of novel copper nanoparticles as promising antidiabetic agents. Adv. Pharm. Sci. 2019, 9080279.

Jampílek, J., Králová, K., 2017. Nanopesticides: preparation, targeting, and controlled release. In: Grumezescu, A.M. (Ed.), New Pesticides and Soil Sensors. Academic Press, Elsevier, USA, pp. 81–127.

Jo, Y.K., Kim, B.H., Jung, G., 2009. Antifungal activity of silver ions and nanoparticles on phytopathogenic fungi. Plant Dis. 93 (10), 1037–1043.

Karmakar, S., Ghosh, S., Kumbhakar, P., 2020. Enhanced sunlight driven photocatalytic and antibacterial activity of flower-like ZnO@MoS_2 nanocomposite. J. Nanopart. Res. 22, 11.

Kitture, R., Chordiya, K., Gaware, S., Ghosh, S., More, P.A., Kulkarni, P., Chopade, B.A., Kale, S.N., 2015. ZnO nanoparticles-red sandalwood conjugate: a promising anti-diabetic agent. J. Nanosci. Nanotechnol. 15, 4046–4051.

Lade, B.D., Gogle, D.P., Lade, D.B., Moon, G.M., Nandeshwar, S.B., Kumbhare, S.D., 2019. Nanobiopesticide formulations: application strategies today and future perspectives. In: Koul, O. (Ed.), Nano-Biopesticides Today and Future Perspectives. Academic Press, pp. 179–206.

Li, Y., Jin, Q., Yang, D., Cui, J., 2018. Molybdenum sulfide induce growth enhancement effect of rice (*Oryza sativa* L.) through regulating the synthesis of chlorophyll and the expression of aquaporin gene. J. Agric. Food Chem. 66 (16), 4013–4021.

Liu, R., Zhang, H., Lal, R., 2016. Effects of stabilized nanoparticles of copper, zinc, manganese, and iron oxides in low concentrations on lettuce (*Lactuca sativa*) seed germination: nanotoxicants or nanonutrients? Water Air Soil Pollut. 227, 42.

Mittal, D., Kaur, G., Singh, P., Yadav, K., Ali, S.A., 2020. Nanoparticle-based sustainable agriculture and food science: recent advances and future outlook. Front. Nanotechnol. 2, 579954.

Mukhopadhyay, S.S., 2014. Nanotechnology in agriculture: prospects and constraints. Nanotechnol. Sci. Appl. 7, 63–71.

Nel, A., Xia, T., Mädler, L., Li, N., 2006. Toxic potential of materials at the nano level. Science 311 (5761), 622–627.

O'Brien, J.A., Benková, E., 2013. Cytokinin cross-talking during biotic and abiotic stress responses. Front. Plant Sci. 4, 451.

Ocsoy, I., Paret, M.L., Ocsoy, M.A., Kunwar, S., Chen, T., You, M., Tan, W., 2013. Nanotechnology in plant disease management: DNA-directed silver nanoparticles on graphene oxide as an antibacterial against *Xanthomonas perforans*. ACS Nano 7 (10), 8972–8980.

Oliveira, H.C., Stolf-Moreira, R., Martinez, C.B.R., Grillo, R., de Jesus, M.B., Fraceto, L.F., 2015. Nanoencapsulation enhances the post-emergence herbicidal activity of atrazine against mustard plants. PLoS One 10 (7), e0132971.

Parisi, C., Vigani, M., Rodrı'guez-Cerezo, E., 2015. Agricultural nanotechnologies: what are the current possibilities? Nano Today 10 (2), 124–127.

Raj, S.N., Anooj, E.S., Rajendran, K., Vallinayagam, S., 2021. A comprehensive review on regulatory invention of nano pesticides in agricultural nano formulation and food system. J. Mol. Struct. 1239, 130517.

Ranpariya, B., Salunke, G., Karmakar, S., Babiya, K., Sutar, S., Kadoo, N., Kumbhakar, P., Ghosh, S., 2021. Antimicrobial synergy of silver-platinum nanohybrids with antibiotics. Front. Microbiol. 11, 610968.

Robkhob, P., Ghosh, S., Bellare, J., Jamdade, D., Tang, I.M., Thongmee, S., 2020. Effect of silver doping on antidiabetic and antioxidant potential of ZnO nanorods. J. Trace Elem. Med. Biol. 58, 126448.

Rokade, S.S., Joshi, K.A., Mahajan, K., Tomar, G., Dubal, D.S., Parihar, V.S., Kitture, R., Bellare, J., Ghosh, S., 2017. Novel anticancer platinum and palladium nanoparticles from *Barleria prionitis*. Glob. J. Nanomed. 2 (5), 555600.

Rokade, S., Joshi, K., Mahajan, K., Patil, S., Tomar, G., Dubal, D., Parihar, V.S., Kitture, R., Bellare, J. R., Ghosh, S., 2018. *Gloriosa superba* mediated synthesis of platinum and palladium nanoparticles for induction of apoptosis in breast cancer. Bioinorg. Chem. Appl. 2018, 4924186.

Salunke, G.R., Ghosh, S., Santosh, R.J., Khade, S., Vashisth, P., Kale, T., Chopade, S., Pruthi, V., Kundu, G., Bellare, J.R., Chopade, B.A., 2014. Rapid efficient synthesis and characterization of AgNPs, AuNPs and AgAuNPs from a medicinal plant, *Plumbago zeylanica* and their application in biofilm control. Int. J. Nanomedicine 9, 2635–2653.

Sant, D.G., Gujarathi, T.R., Harne, S.R., Ghosh, S., Kitture, R., Kale, S., Chopade, B.A., Pardesi, K.R., 2013. *Adiantum philippense* L. frond assisted rapid green synthesis of gold and silver nanoparticles. J. Nanoparticles 2013, 182320.

SCENIHR, 2007. Opinion on the Appropriateness of the Risk Assessment Methodology in Accordance With the Technical Guidance Documents for New and Existing Substances for Assessing the Risks of Nanomaterials. Scientific Committee on Emerging and Newly Identified Health Risks, Brussels. https://ec.europa.eu/health/ph_risk/committees/04_scenihr/docs/scenihr_o_010.pdf. Accessed on 13/08/2021.

Servin, A.D., White, J.C., 2016. Nanotechnology in agriculture: next steps for understanding engineered nanoparticle exposure and risk. NanoImpact 1, 9–12.

Shakil, N.A., Singh, M.K., Pandey, A., Kumar, J., Pankaj, Parmar, V.S., Singh, M.K., Pandey, R.P., Watterson, A.C., 2010. Development of poly (ethylene glycol) based amphiphilic copolymers for controlled release delivery of carbofuran. J. Macromol. Sci., Pure Appl. Chem. 47, 241–247.

Shende, S., Joshi, K.A., Kulkarni, A.S., Shinde, V.S., Parihar, V.S., Kitture, R., Banerjee, K., Kamble, N., Bellare, J., Ghosh, S., 2017. *Litchi chinensis* peel: a novel source for synthesis of gold and silver nanocatalysts. Glob. J. Nanomed. 3 (1), 555603.

Shende, S., Joshi, K.A., Kulkarni, A.S., Charolkar, C., Shinde, V.S., Parihar, V.S., Kitture, R., Banerjee, K., Kamble, N., Bellare, J., Ghosh, S., 2018. *Platanus orientalis* leaf mediated rapid synthesis of catalytic gold and silver nanoparticles. J. Nanomed. Nanotechnol. 9, 2.

Shinde, S.S., Joshi, K.A., Patil, S., Singh, S., Kitture, R., Bellare, J., Ghosh, S., 2018. Green synthesis of silver nanoparticles using *Gnidia glauca* and computational evaluation of synergistic potential with antimicrobial drugs. World J. Pharm. Res. 7 (4), 156–171.

Stadler, T., Buteler, M., Valdez, S.R., Gitto, J.G., 2018. Particulate nanoinsecticides: a new concept in insect pest management. In: Begoum, G. (Ed.), Insecticides: Agriculture and Toxicology. IntechOpen, London, UK, pp. 83–105.

Stirling, D.A., 2018. The Nanotechnology Revolution: A Global Bibliographic Perspective. Jenny Stanford Publishing, CRC Press, pp. 253–405.

Transparency Market Research, 2017. Nanotechnology in Agriculture Market—Global Industry Analysis, Size, Share, Growth, Trends, and Forecast 2017–2025. https://www.transparencymarketresearch.com/nanotechnology-agriculture-market.html. Accessed on 13/08/2021.

Tyler, L.C., 2012. FDA issues draft guidance on use of nanotechnology in food and food packaging. Nanotechnol. Law Bus. 9 (2), 149–155.

Vankova, R., Landa, P., Podlipna, R., Dobrev, P.I., Prerostova, S., Langhansova, L., Gaudinova, A., Motkova, K., Knirsch, V., Vanek, T., 2017. ZnO nanoparticle effects on hormonal pools in *Arabidopsis thaliana*. Sci. Total Environ. 593–594, 535–542.

Wani, A.H., Shah, M.A., 2012. A unique and profound effect of MgO and ZnO nanoparticles on some plant pathogenic fungi. J. Appl. Pharm. Sci. 2 (3), 40–44.

Yan, A., Chen, Z., 2019. Impacts of silver nanoparticles on plants: a focus on the phytotoxicity and underlying mechanism. Int. J. Mol. Sci. 20 (5), 1003.

Yu, Z., Sun, X., Song, H., Wang, W., Ye, Z., Shi, L., Ding, K., 2015. Glutathione-responsive carboxymethyl chitosan nanoparticles for controlled release of herbicides. Mater. Sci. Appl. 6, 591–604.

Zhao, L., Sun, Y., Hernandez-Viezcas, J.A., Servin, A.D., Hong, J., Niu, G., Peralta-Videa, J.R., Duarte-Gardea, M., Gardea-Torresdey, J.L., 2013. Influence of CeO_2 and ZnO nanoparticles on cucumber physiological markers and bioaccumulation of Ce and Zn: a life cycle study. J. Agric. Food Chem. 61, 11945–11951.

Index

Note: Page numbers followed by *f* indicate figures and *t* indicate tables.

P

Q

R

Kolibri 360 GmbH
Ferdinand-Jühlke-Straße 7
99095 Erfurt, Deutschland
produktsicherheit@kolibri360.de